Plant Propagation

Plant Propagation
principles and practices

third edition

Hudson T. Hartmann
University of California, Davis

Dale E. Kester
University of California, Davis

Prentice-Hall, Inc., Englewood Cliffs, New Jersey

Library of Congress Cataloging in Publication Data

HARTMANN, HUDSON THOMAS
    Plant propagation.

    Includes bibliographies.
    1. Plant propagation. I. Kester, Dale E., joint
author. II. Title.
SB119.H3 1975   631.5'3   74–9832
ISBN 0–13–680991–X

Printed in the United States of America

10   9

Prentice-Hall International, Inc., *London*
Prentice-Hall of Australia, Pty. Ltd., *Sydney*
Prentice-Hall of Canada, Ltd., *Toronto*
Prentice-Hall of India Private Limited, *New Delhi*
Prentice-Hall of Japan, Inc., *Tokyo*

Contents

reader has a knowledge of the general principles of botany and horticulture. On the other hand, the chapters describing the techniques and equipment used in the different methods of propagation have been written in terms understandable to those with little botanical background.

The term, "cultivar" (**cultiv**ated **var**iety)—now widely used throughout the world—is used in this book to replace the less precise term, "variety," which does not distinguish between the botanical varieties and the commonly cultivated varieties of horticultural interest.

A rather extensive representative bibliography of the research literature is included for each subject considered. In addition, suggested supplementary readings are listed for most of the subjects considered. These are generally specialized works which deal with the topic more extensively than is possible in this book. These references should be valuable for those wishing to study the subject in greater detail.

In preparing the third edition of this book, we have depended upon the help of authorities in the various fields of propagation and related subjects. They gave their time most generously in reading sections of the manuscript and in offering suggestions. We wish to acknowledge the assistance of Dr. C. J. Alley, Dr. R. S. Bringhurst, Dr. R. Carlson, Dr. A. deHertogh, Dr. T. Furuta, Mr. R. Garner, Dr. W. Hackett, Dr. B. Haissig, Dr. B. Howard, Dr. L. Lider, Dr. R. O. Magie, Dr. G. Nyland, Mr. K. Opitz, Dr. P. Read, Dr. R. Sachs, and Dr. H. B. Tukey, Jr.

We especially wish to thank Dr. W. P. Tufts, former chairman of the Department of Pomology, University of California, for his interest and encouragement during the initial writing of this book. Marilyn Hartman prepared some of the drawings in the third edition.

Hudson T. Hartmann
Dale E. Kester

Introduction 1

The propagation of plants is a fundamental occupation of mankind. **Civil**ization may have started when ancient man learned to plant and grow **kinds** of plants which fulfilled nutritional needs for himself and his animals. As civilization advanced he added to the variety of plants, cultivating not only additional food crops but also those which provided fibers, medicine, recreational opportunities, and beauty (*8*). From the great diversity and variation in plant life he has been able to select kinds of plants particularly useful to his welfare.

Much progress in plant improvement was made long before the modern period of plant breeding (*1*). Our cultivated plants originated mainly by three general methods. First, some kinds of plants were selected directly from wild species but, under the selective hand of man, evolved into types that differ radically from their wild relatives. Examples of this group include lima bean, tomato, barley, and rice. Second, other kinds of plants arose as hybrids between species, accompanied by changes in chromosome number. These plants are completely unique to cultivation and have no single wild relative. Examples of this group include maize, wheat, tobacco, pear, strawberry, and prune. Third, another group of plants occur naturally as rare monstrosities. Although unadapted to a native environment, they may be useful to man. Among these are heading cabbage, broccoli, and Brussels sprouts.

This progress in plant improvement would have been of little significance, however, without methods whereby improved forms could be maintained in cultivation. Consequently there has been a process of invention and discovery of techniques for *plant propagation*. Most cultivated plants either will be lost or will revert to less desirable forms unless they are propagated under controlled conditions that preserve the unique characteristics that make them useful. Through history, as new kinds of plants became available,

1

the development of knowledge and techniques to preserve them had to be learned; conversely, as new advances in propagation techniques developed, the number of plants that became available for cultivation increased.

Listed below are the general methods of propagating plants. Many, if not most, of them antedate recorded history. It is probably no mere coincidence that some of the oldest fruit crops—grape, olive, mulberry, quince, and fig—are also the easiest to propagate by means of the simple techniques using hardwood cuttings. To grow most other tree fruits, budding and grafting had to be learned. Invention of glass houses in the nineteenth century made the rooting of leafy cuttings possible. More recently, the discovery of root-inducing chemicals and mist propagation has revolutionized many nursery procedures. Likewise, production of seed crops has been revolutionized by the discovery of genetic principles leading to the production of hybrid seed. It is likely that a new step forward has occurred with the development of the micro-propagation techniques described in Chapter 16.

Outline of Methods of Propagating Plants with Typical Examples

I Sexual
  A Propagation by seed—annuals, biennials, and many perennial plants.
II Asexual (vegetative)
  A Propagation by apomictic embryos—citrus
  B Propagation by runners—strawberry
  C Propagation by suckers—red raspberry, blackberry
  D Layering
    (1) Tip—trailing blackberry, black raspberry
    (2) Simple—honeysuckle, spirea, filbert
    (3) Trench—apple, pear, cherry
    (4) Mound or stool—gooseberry, apple
    (5) Air (pot or Chinese)—India rubber plant, lychee
    (6) Compound or serpentine—grape, honeysuckle
  E Separation
    (1) Bulbs—hyacinth, lily, narcissus, tulip
    (2) Corms—gladiolus, crocus
  F Division
    (1) Rhizomes—canna, iris
    (2) Offsets—houseleek, pineapple, date
    (3) Tubers—Irish potato
    (4) Tuberous roots—sweet potato, dahlia
    (5) Crowns—everbearing strawberry, phlox
  G Propagation by cuttings
    (1) Root cuttings—red raspberry, horseradish
    (2) Stem cuttings
      (a) Hardwood—fig, grape, gooseberry, quince, rose, forsythia
      (b) Semi-hardwood—lemon, olive, camellia, holly
      (c) Softwood—lilac, forsythia, weigela
      (d) Herbaceous—geranium, coleus, chrysanthemum
    (3) Leaf cuttings—*Begonia rex, Bryophyllum, Sansevieria,* African violet
    (4) Leaf-bud cuttings—blackberry, hydrangea

H Grafting
  (1) Root grafting
    (a) Whip or tongue graft—apple and pear
  (2) Crown grafting
    (a) Whip or tongue graft—Persian walnut
    (b) Cleft graft—camellia
    (c) Side graft—narrow-leaved evergreens
  (3) Top grafting
    (a) Cleft graft—various fruit trees
    (b) Saw-kerf or notch graft—various fruit trees
    (c) Bark graft—various fruit trees
    (d) Side graft—various fruit trees
    (e) Whip or tongue graft—various fruit trees
  (4) Approach grafting—mango

I Budding
  (1) T-budding—stone and pome fruit trees, rose
  (2) Patch budding—walnut and pecan
  (3) Ring budding—walnut and pecan
  (4) I-budding—walnut and pecan
  (5) Chip budding—grape, mango

J Micro-propagation
  (1) "Meristem" culture—orchid, carnation
  (2) Tissue culture—tobacco
  (3) Embryoids—tobacco
  (4) "Embryo" culture—orchid

LIFE CYCLES IN PLANTS

Plant propagation involves the control of two basically different types of developmental life cycles—*sexual* and *asexual*. Preservation of the unique characteristics of a plant or group of plants depends upon the transmission from one generation to the next of a particular combination of genes present on the chromosomes in the cells. The sum total of these genes make up the *genotype* of the plant. The genotype, in combination with the environment, produces a plant of a given outward appearance (the *phenotype*). Therefore, the function of any plant propagation technique is *to preserve a particular genotype or combination of genotypes* that will reproduce the particular kind of plant being propagated.

The **sexual cycle** utilizes seed propagation by which new individual offspring plants are created whose characteristics reflect the genetic contributions of the two parents. Reproduction by seed can be expected to result in a certain amount of variation among plants. Consequently in seed propagation, the propagator must deal with the problem of controlling the genetic variation within populations of plants. Such procedures are described in Chapter 4.

The **asexual cycle,** on the other hand, utilizes various vegetative methods of propagation, as listed in the previous outline. By use of these techniques the unique characteristics of any individual plant are preserved in the offspring plants and, in addition, the genotype of the source plant can be preserved intact.

The **apomictic cycle** (see page 81) is an exception. The embryo originates directly from cells of the mother plant by a vegetative, or asexual, process and is not created from the union of the male and female sex cells (gametes). The apomictic process is described in Chapter 3.

Phases of the sexual cycle   Growth and development of the seedling [1] plant occurs in three phases leading to the production of flowers on the new plant with the formation of sex cells to create the next generation, as shown in Figure 1–1.

The embryo phase begins with the union of male and female gametes in the flower (Figure 1–4) to form a single-celled zygote. The subsequent

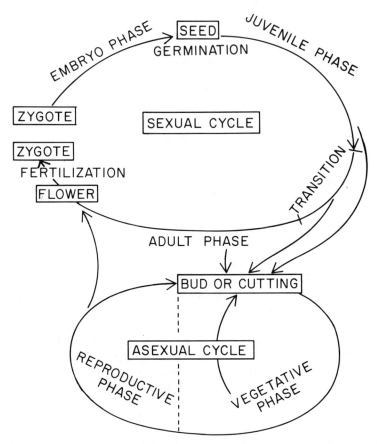

Figure 1–1   Reproduction in higher plants by means of sexual and asexual cycles.

[1] In this context, the term *seedling* refers to any plant originating from seed, as opposed to the vegetatively propagated plant. This term would apply throughout the life of the plant.

growth and development of the resulting embryo within the fruit and seed is described in Chapter 3.

The juvenile phase begins with the germination of the seed and involves growth of the embryo into a *juvenile* plant (13). Vegetative growth predominates as the seedling plant enlarges in size through elongation in stem and root and increases in cross-sectional area. This phase may be accompanied by the development of unique morphological characteristics of the plant—leaf shape, growth habit, and often thorniness (in woody plants) (see Chapter 8). Physiologically, the juvenile plant may possess enhanced vegetative regeneration abilities, a property of great significance in some propagation techniques (see Chapter 9). In general, juvenile plants do not respond to flower-inducing stimuli. In the *adult* phase the plant enters a stage in which reproduction by seed predominates. The plant reaches its ultimate size and develops flowers in response to signals from the environment or from its internal physiological or hormonal state.

A *transitional* phase marks the separation between the juvenile phase, as the plant gradually loses the characteristics of juvenility and acquires the characteristics of the adult phase. This shift may be signaled by changes in morphological appearance,—leaf shape, growth habit, reduction in thorniness, loss in ability to regenerate vegetatively, and increase in the ability to respond to flower induction stimuli. In some plants, the change occurs abruptly and quickly; in others, it may take place slowly over a period of many years. In some plants both phases may be found in the same plant at one time. Although the time required to reach the flowering phase is genetically controlled it can also be manipulated to some extent by the environment and by specific horticultural practices. *Precocity* is the tendency for some plants to become reproductive sooner than others of the same kind.

Phases of the asexual cycle   An asexual cycle can be initiated by removing a section of a plant (bud, scion, cutting, or other vegetative structure) and from it regenerating a new plant. Any part of the seedling plant at any phase of the sexual cycle—juvenile, transitional, or adult—can be chosen as the starting plant, as shown in Figure 1–1. Agricultural and horticultural plants repeatedly propagated by vegetative means are the result of continuous recycling through the asexual cycle. Some kinds of vegetatively propagated plants, particularly those selected by man because of their vegetative characteristics, may continue to show juvenile characteristics. Other kinds of plants chosen because of their flowering and fruiting characteristics may no longer retain any juvenile or transitional characteristics and remain biologically "adult" indefinitely.

Consequently, it is desirable to refer to the two developmental phases of the asexual cycle as the *vegetative* and *flowering* phases. The vegetative phase involves the growth of the plant by elongation of the roots and stems, increase in plant volume, and expansion of leaves. In general, such plants that are in the adult phase can respond readily to appropriate flower-inducing stimuli. In the flowering phase, elongation of stems ceases and some of the growing points differentiate into flower buds that eventually produce flowers, fruits, and seeds.

Types of life cycles   Life cycles of plants can be classified as annual, biennial, or perennial depending upon the period of time that the plants require to grow through the sexual cycle from zygote to seed production. Variations in these cycles are adaptations to natural seasonal or climatic cycles.

Annual plants go through the entire sequence, from germination of seeds to flowering and production and dissemination of seeds, in one growing season and then die. *Summer annuals* include species that require warm temperatures. These are usually grown in regions having a warm-summer, cold-winter cycle. Seeds germinate in early spring, the plants flower and produce seeds in the long days and high temperatures of summer. The seeds overwinter in the ground. *Winter annuals* include species that tolerate low temperatures. These can be grown in regions having mild, moist winters. Seeds germinate in the fall or winter, the plants flower and produce seeds in late winter or spring, then die during the summer.

Biennial plants have a two-year life cycle and require chilling. Seeds of these plants germinate during summer or fall and remain in a vegetative or juvenile state during the same season. The plants become dormant in winter; the chilling they receive stimulates the transition to the reproductive stage. Flowers and seeds are produced in the second season; then the plant dies.

Perennial plants live for more than two years and have an annual regeneration of the vegetative-reproductive cycle. Consecutive growth and dormancy cycles are related either to warm-cold or to wet-dry climatic changes. *Herbaceous perennials* are those in which the shoots die during the winter or in dry periods. The plants survive such dormant periods as specialized underground structures, such as bulbs, rhizomes, or crowns (see Chapters 14 and 15). *Woody perennial* plants continue to increase in size each year by growth of the shoot and root tips or by lateral cambium growth, or both.

CELLULAR BASIS FOR PROPAGATION
Mitosis and Asexual Reproduction
Asexual propagation is possible because each cell of the plant contains all the genes necessary for growth and development and, during the cell division (mitosis) that occurs during growth and regeneration, the genes are replicated in the daughter cells. Regeneration of a new organism by asexual methods occurs readily in higher plants but not in higher animals. In some forms of lower animal life, however, such as the flatworm, *Planaria,* in the phylum Platyhelminthes, asexual multiplication can take place. A flatworm cut in half will develop into two worms, each half regenerating the missing part.

The details of mitosis are shown in Figure 1–2, its principal feature being that individual chromosomes split longitudinally, the two identical parts going to two daughter cells. As a result, the complete chromosome system of an individual cell is duplicated in each of its two daughter cells (with certain exceptions). The chromosomes produced will be the same as in the cell from which they came. Consequently the characteristics of the new plant that grows will be the same as that from which it originated.

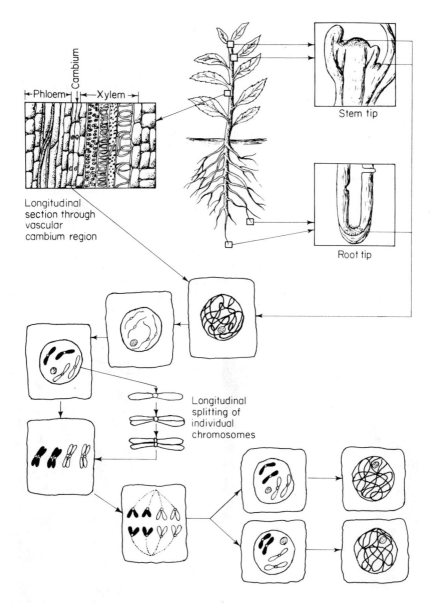

Figure 1-2   Diagrammatical representation of the process by which growth and asexual reproduction take place in a dicotyledonous plant. Mitosis occurs in three principal growing regions of the plant: the stem tip, the root tip of primary and secondary roots, and the cambium. A meristematic cell is shown dividing to produce two daughter cells whose chromosomes will (usually) be identical with those of the original cell.

Mitosis occurs in specific growing points or areas of the plant to produce growth. (See Figure 1–2.) These are the *shoot apex,* the *root apex,* the *cambium,* and the *intercalary zones* (internode bases of monocotyledonous plants). Mitosis also occurs when *callus* forms on a wounded plant part and when new growing points are initiated on root and stem pieces. Callus parenchyma consists of new cells proliferating from cut tissues in response to wounding. When new growing points are initiated on a vegetative structure, such as root, stem, or leaf, they are referred to as *adventitious roots* or *adventitious shoots.* (See Figure 1–3.)

Adventitious roots are those that arise from aerial plant parts, from underground stems, or from relatively old roots. All roots other than those arising from the embryo axis and all their branches formed in normal sequence can be considered adventitious roots. Adventitious shoots are those appearing on roots or internodally on stems after the terminal and lateral growing points are produced. New shoots (*watersprouts*) sometimes arise from *latent growing points* or buds that are not adventitious but originated along with the branch on which they occur. These are common on old branches of woody plants and can be stimulated into active growth if the part terminal to it is removed.

Mitosis is the basic process of normal vegetative growth, regeneration, and wound healing which makes possible such vegetative propagation techniques as cuttage, graftage, layerage, separation, and division. These methods of propagation are important because they permit large-scale multiplication of an *individual plant* into as many separate plants as the amount of parent

Figure 1–3   Regeneration in asexual propagation. *Left:* adventitious shoots growing from a root cutting. *Center:* adventitious roots developing from the base of a stem cutting. *Right:* callus tissue produced to give healing of a graft union.

material will permit. Each separate plant produced by such means is (in most cases) genetically identical with the plant from which it came. The primary reason for using these vegetative propagation techniques is to reproduce exactly the genetic characteristics of any individual plant, although there may be additional advantages from the standpoint of culture.

Meiosis and Sexual Reproduction

Sexual reproduction involves the union of male and female sex cells, the formation of seeds, and the creation of individuals with new genotypes (Figures 1–4 and 1–5). The cell division (*meiosis*) which produces the sex cells involves reduction division of the chromosomes, in which their number is reduced by half. The original chromosome number is restored during fertilization, resulting in new individuals containing chromosomes from both the male and the female parents. Offspring may resemble either, neither, or both of the parents, depending upon their genetic similarities. Among the progeny from a particular combination of parents, considerable variation may occur.

The outward appearance (phenotype) of a plant and the way characteristics are inherited from generation to generation are controlled through the action of genes present on the chromosomes. Some traits are controlled by a single gene as shown for pea height in Figure 1–6. Figure 1–7 shows the more complex case where two independent genes affect the appearance of peach fruit. The analysis of inheritance for traits controlled by many genes, which is the usual case, is more complex and requires statistical analysis to show how closely the offspring resembles the parents. The kind of genes present and how much the environment influences their effect must be taken into account (*1, 2*).

The two terms *homozygous* and *heterozygous* are useful to describe the genotype of a particular plant. If a high proportion of the genes present on one chromosome are the same as those on the opposite member of the chromosome pair (*homologous chromosomes*), the plant is homozygous and will "breed true" if self-pollinated or if the other parent is genetically similar. This means that the particular traits or characteristics that the plant possesses will be transmitted to its offspring, and the offspring will resemble the parent. On the other hand, if a sufficient number of genes on one chomosome differ from those on the other member of the chromosome pair, then the plant is said to be heterozygous. In such a case, important phenotypic traits of the parent may not be transmitted to its offspring, and the seedlings may differ in appearance not only from the parent but also from each other. The amount of seedling variation may be very great in some kinds of plants.

To minimize variation and to be sure that seedling offspring will possess the particular characteristics for which they are to be grown, certain procedures must be followed during seed production. These are discussed in Chapter 4.

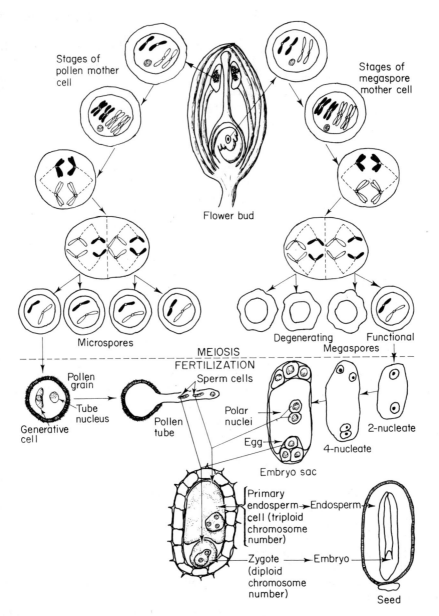

Figure 1-4 Diagrammatical representation of the sexual cycle in angiosperms. Meiosis occurs in the flower bud in the anther (male) and the pistil (female) during the bud stage. During this process the pollen mother cells and the megaspore mother cells, both diploid, undergo a reduction division in which homologous chromosomes segregate to different cells. This is followed immediately by a mitotic division which produces four daughter cells, each with half the chromosomes of the mother cells. In fertilization a male gamete unites with the egg to produce a zygote, in which the diploid chromosome number is restored. A second male gamete unites with the polar nuclei to produce the endosperm.

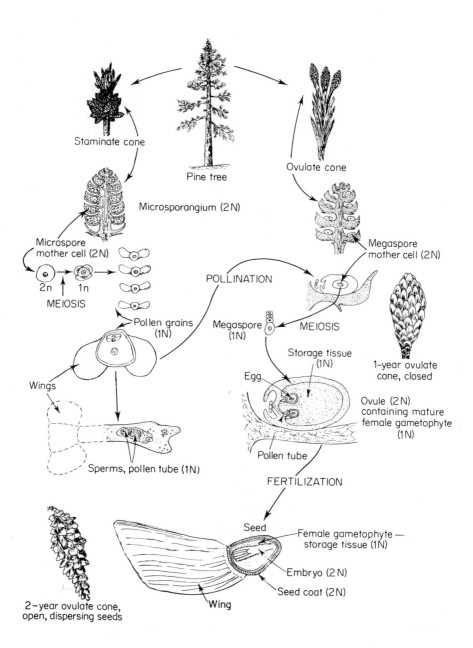

Staminate cone

Pine tree

Ovulate cone

Microsporangium (2N)

Microspore mother cell (2N)

2n   1n

MEIOSIS

Pollen grains (1N)

Megaspore mother cell (2N)

POLLINATION

Megaspore (1N)

MEIOSIS

1-year ovulate cone, closed

Storage tissue (1N)

Egg

Wings

Ovule (2N) containing mature female gametophyte (1N)

Pollen tube

Sperms, pollen tube (1N)

FERTILIZATION

Seed

Female gametophyte — storage tissue (1N)

Embryo (2N)

Seed coat (2N)

2-year ovulate cone, open, dispersing seeds

Wing

Figure 1-5   Diagrammatical representation of the sexual cycle in a gymnosperm (pine) showing meiosis and fertilization.

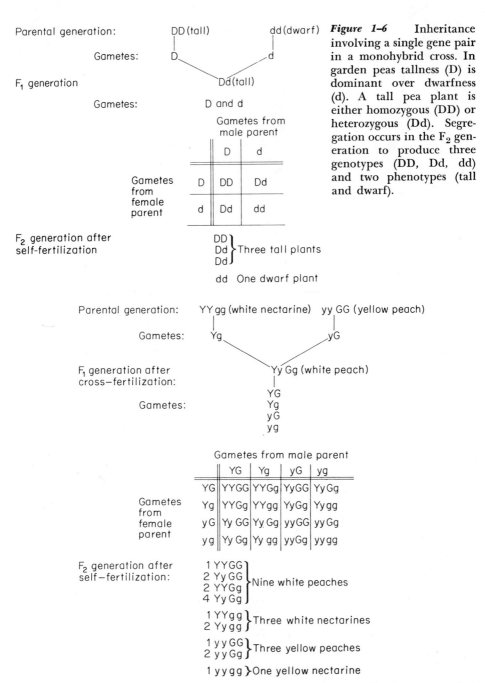

Figure 1-6 Inheritance involving a single gene pair in a monohybrid cross. In garden peas tallness (D) is dominant over dwarfness (d). A tall pea plant is either homozygous (DD) or heterozygous (Dd). Segregation occurs in the F_2 generation to produce three genotypes (DD, Dd, dd) and two phenotypes (tall and dwarf).

Figure 1-7 Inheritance in a dihybrid cross involving peach (*Prunus persica*). Fuzzy skin (G) of a peach is dominant over the glabrous (i.e., smooth) skin of the nectarine (g). White flesh color (Y) is dominant over yellow flesh color (y). In the example shown, the phenotype of the F_1 generation is different from either parent. Segregation in the F_2 generation produces nine genotypes and four phenotypes.

PLANT NOMENCLATURE

Since plant propagation involves the preservation of genotypes that are important to man, it is essential to have some means of labeling them. The problem of supplying plants with names and then correctly identifying them by these names is a continuing one. However, a system of nomenclature has gradually evolved through the efforts of botanists and horticulturists that provides the basis for uniform worldwide plant identification. The system is embodied in the *International Code of Botanical Nomenclature (6)* and the *International Code of Nomenclature for Cultivated Plants (3).*

Botanical Classification

Classification of plants in nature is the function of taxonomists. The system of classification is based upon increasing specialization and complexity in structure and organization resulting from evolutionary processes. For example, the plant **kingdom** is divided into **divisions**—first, with one-celled plants—i.e., Schizophyta (bacteria and blue-green algae)—through somewhat more complex plants, such as the Bryophyta (mosses, liverworts), to specialized higher plants, Pterophyta.

This text is principally concerned with the Pterophyta, which includes three **classes:** Filicinae (ferns), Gymnospermae (gymnosperms, as *Ginkgo* and conifers) and Angiospermae (flowering plants). Classification within these classes is based principally on flower structure. The gymnosperms produce seeds which are not enclosed. The angiosperms include plants in which the seed is produced within an enclosed structure—the ovary. The angiosperms are divided into two **subclasses**—the monocotyledons (e.g., grasses, palms, orchids) and the dicotyledons (e.g., beans, roses, peaches). These subclasses are divided into **orders,** the orders into **families,** families into **genera,** and genera into **species.**

The species is the fundamental unit customarily used by taxonomists to designate groups of plants that can be recognized as distinct kinds. In nature, individuals within one species normally interbreed freely but do not interbreed with a separate species because they are separated either by distance or by some other physiological, morphological, or genetic barrier that prevents the interchange of genes between the two species *(11).* Consequently it is possible to reproduce a species by seed and to maintain it through propagation. However, if one reproduces a species from individual plants or from different parts of the natural range of that species, he may find that there is a great deal of natural variability in appearance and adaptability among individuals of that species *(4, 5, 10, 12).* Consequently, to get a total picture of species variability, one should examine individuals from all parts of the range rather than relying on a few.

A distinct morphological subgrouping within a species (usually resulting from geographical separation) may be recognized taxonomically as a **botanical variety.** The Code also makes provision for certain other naturally occurring subdivisions within the species such as **subspecies, subvariety, form,** and **individual.**

Natural variation among native plants of a species can also be described with the two terms, *cline* and *ecotype (5, 10)*. A cline refers to the continuous differences in genetically controlled physiological and morphological characteristics that occur within a species in different parts of its range. The differences are related to continuous variations in the environment and result from the evolution of populations of plants adapted to these variations. Where the differences are distinct and discontinuous, the term ecotype is used.

Classification of Cultivated Plants

The classification and naming of those special kinds of plants cultivated by man has a different basis from that for plants growing in nature. The reason for separating one kind from another in cultivation is not because of naturally occurring variation but because each kind has some practical significance for man. The group of plants representing each kind has invariably arisen as a minor variant within a species (or sometimes as a hybrid between species) and very often is derived from a single individual plant which is reproduced asexually. Such a group of plants representing a single propagatable type is commonly referred to as a *variety* (English), or *variété* (French), *variedad* (Spanish), *sorte* (German), *sort* (Scandinavian), *ras* (Russian), and *razza* (Italian). Such groups of plants are identified by a name usually bestowed by its originator.

Naming of cultivated plants can be handled best if it conforms to a widely agreed-upon system of nomenclature. The International Code of Botanical Nomenclature governs the use of botanical Latin names for both cultivated and wild species. The International Code of Nomenclature for Cultivated Plants extends these rules to cover the special categories and names of cultivated plants. It recognizes as a special taxonomic category the **cultivar,** that designates a group of cultivated plants which are significant in agriculture, forestry, or horticulture, which are clearly distinguished by any characteristic and which, when reproduced, retain such distinguishing characteristics *(3, 7, 9)*. The word *cultivar* is a contraction of the phrase "cultivated variety" and should be distinguished from the analogous, naturally occurring category, "botanical variety." The term *variety* is synonymous with *cultivar.*

The complete scientific name of any plant developed or maintained under cultivation includes the (a) genus, (b) species, and (c) cultivar name, the first two in the usual Latin form and the latter in words of a common language. The cultivar name can be set off by the abbreviation cv. or by single quotation marks and can be attached to the common name.

> *Syringa vulgaris* cv. Mont Blanc
> *Syringa vulgaris* 'Mont Blanc'
> Lilac 'Mont Blanc'

A cultivar should have a proper name that can be recognized by propagators everywhere and that will not confuse its identification with other cultivars. Multiple names for the same cultivar, or one name for several similar cultivars, arising either by accident or from deliberate changes in name, can only lead to confusion and misrepresentation. The most important principle in naming cultivars is that normally the earliest name applied should have priority. A second principle is that once a name is correctly applied it should be changed only for exceptional reasons. There are additional rules which can be of assistance in choosing a plant name; for these the Code of Nomenclature for Cultivated Plants should be consulted (*3*).

Categories of Cultivars

Cultivars can be differentiated as those which are *sexually reproduced* and those which are *asexually reproduced*. Within these two groups, cultivars may differ as to origin, genetic nature, or conditions of reproduction, depending upon the category.

Asexually reproduced cultivars:

(a) A *clone* is a group of plants originating from a single individual and reproduced by vegetative means, such as by cuttings, layers, or grafts (see Chapter 8). Examples of clones are 'Elberta' peach, 'King Alfred' daffodil, and 'Bliss Triumph' potato.

(b) *Apomictic* cultivars (or *apomicts*), are biologically unique kinds of plants that reproduce by seed but are asexual because of complete or partial apomixis (see page 61).

Sexually reproduced cultivars: These are propagated by seed, and specific production programs geared to the genetic characteristics of individual cultivars may be necessary to maintain their genetic identity. Chapter 4 describes such programs.

(a) A *line* cultivar is a group of self-fertilizing plants that maintains its genetic identity from generation to generation naturally. Examples are 'Rosy Morn' petunia, 'Marglobe' tomato, and 'Marquis' wheat.

(b) An *inbred line* is a group of naturally cross-fertilizing plants maintained as self-fertilizing lines through artificial restraints on cross-pollination. These are generally used to produce hybrid cultivars.

(c) *Hybrid cultivars* are groups of plants grown from seed produced by cross-pollinating two or more parental breeding stocks which are maintained either as inbred lines or as clones. Examples of hybrid cultivars are 'Granex' onion, derived from crossing two onion inbred lines, and 'U.S. 13' corn, produced by consecutive crossing involving four inbred lines.

(d) A cultivar may consist of a *seedling mixture* of cross-fertilized individuals which, as a group, may be more or less variable genetically but which possess one or more common phenotypic characteristics. For example, *Phlox drummondi* 'Sternenzauber' is a mixture of different color forms, but all have the same star-like corolla shape.

(e) *Synthetic cultivars* are a special category of seedling mixture in which separately developed seedling lines are combined (see page 69). An example is 'Ranger' alfalfa, a cultivar derived from intercrossing among five seed-propagated lines which were previously developed and maintained in isolation. (*2*).

REFERENCES

1 Allard, R. W., *Principles of Plant Breeding.* New York: John Wiley & Sons, Inc., 1960.

2 Briggs, F. N., and P. F. Knowles, *Introduction to Plant Breeding.* New York: Reinhold, 1967.

3 Gilmour, J. S. L., et al., *International Code of Nomenclature for Cultivated Plants.* Regnum Vegetabile, Vol. 64. 1969.

4 Harlin, J. R., Distribution and utilization of natural variability in cultivated plants, *Genetics in Plant Breeding,* Brookhaven Symposia in Biol., 9:191–206. 1956.

5 Langlet, O., Ecological variability and taxonomy of forest trees, in T. T. Kozlowski (ed.), *Tree Growth.* New York: The Ronald Press Company, 1962, pp. 357–69.

6 Lanjouw, J. (ed.), *International Code of Botanical Nomenclature,* Regnum Vegetabile, 46:402. 1966.

7 Lawrence, G. H. M., The term and category of cultivar, *Baileya,* 3:177–82. 1955.

8 Sauer, C. O., *Agricultural Origins and Dispersals.* Cambridge, Mass.: The Massachusetts Institute of Technology Press, 1969.

9 Stearn, W. T., International code of nomenclature for cultivated plants, *Rpt. 13th Int. Hort. Cong. 1952,* pp. 42–68. 1953.

10 Stebbins, G. L., *Variation and Evolution in Plants.* New York: Columbia University Press, 1950.

11 Stebbins, G. L., *Processes of Organic Evolution,* 2nd ed. Englewood Cliffs, N.J.: Prentice-Hall, Inc., 1971.

12 Vavilov, N. I., Wild progenitors of the fruit trees of Turkestan and the Caucasus and the problem of the origin of fruit trees, *Rpt. and Proc. IX Inter. Hort. Cong.,* London, pp. 271–86. 1930.

13 Zimmerman, R. H., Juvenility and flowering in woody plants: a review. *HortScience,* 7(5):447–55. 1972.

SUPPLEMENTARY READING

Frankel, O. H., and E. Bennett, *Genetic Resources in Plants. Their Exploration and Conservation.* Blackwell Scientific Publication. 1970.

Heiser, C. B., Jr., *Seed to Civilization.* San Francisco: W. H. Freeman and Company, 1973.

Hodgson, R. E., "Germ Plasm Resources," *American Association for Advancement of Science Publication No. 66.* Washington, D.C., 1961.

Schery, R. W., *Plants for Man,* 2nd ed. Englewood Cliffs, N.J.: Prentice-Hall, Inc., 1972.

Schwanitz, F., *The Origin of Cultivated Plants* (English translation from German edition of 1957). Cambridge, Mass.: Harvard University Press, 1966.

Weir, T. E., C. R. Stocking, and M. G. Barbour, *Botany: An Introduction to Plant Biology,* 5th ed. New York: John Wiley & Sons, Inc., 1974.

Propagating Structures, Media, Fertilizers, Soil Mixtures, and Containers

2

Facilities required for propagating many plant species by seed, cuttings, or grafts include two basic units. One is a structure with temperature control and ample light, such as a greenhouse or hotbed, where seeds can be germinated or cuttings rooted. The second unit is a structure into which the young, tender plants can be moved for hardening, preparatory to transplanting out-of-doors. Cold frames or lathhouses are useful for this purpose. Either greenhouses or cold frames may, at certain times of the year and for certain species, serve for both purposes.

PROPAGATING STRUCTURES

Greenhouses

There are a number of types of greenhouses. The simplest is a shed-roof lean-to construction, utilizing one side, preferably the east, of another building as one wall. Small, inexpensive greenhouses can also be constructed from a number of standard 3-ft by 6-ft hotbed sashes fastened to a 2-by-4 wood framework.

Commercial greenhouses are usually independent structures of even-span, gable-roof construction, proportioned so that the space is well utilized for convenient walkways and propagating benches (16). In large operations, a number of single greenhouse units are often attached side by side, eliminating the cost of glassing-in the adjoining walls. Arrangements of benches in greenhouses vary considerably. Some well-run propagation installations do not have permanently attached benches, their placement varying according

17

to the type of equipment, such as lift trucks, electric carts, etc., used to bring flats and plants in and out of the propagation houses.

In greenhouse construction, a wood or metal framework is built, to which are fastened wood or metal sash bars to support panes of glass embedded in putty. All-metal prefabricated aluminum greenhouses are also widely used. In Europe, and to some extent in the United States, a translucent type of glass, which tends to give a uniform, diffuse light, is used for greenhouse construction.

A means of providing air movement and air exchange is necessary in all greenhouses to aid in controlling temperature and humidity. A ridge ventilator is almost always used, and in some cases, side ventilators are also used. Automatic, thermostatically operated controls are available to open and close the ventilators as the temperature fluctuates during the day and night. Forced-air ventilation is usually used in large installations.

Traditionally greenhouses have been heated by steam or hot water from a central boiler through runs of pipes (some finned to increase radiation surface) suitably located in the greenhouse. However, unit heaters, with fans for improved air circulation, are also used. If oil or gas heaters are used they must be vented to the outside as the combustion products are toxic to plants. In large greenhouses heated air is sometimes blown into large (12 to 24 in.) 4 mil polyethylene tubes hung overhead and running the length of the greenhouse. Small—2 to 3 in.—holes spaced throughout the length of these tubes allow the hot air to escape, thus giving uniform heating all through the house (see Figure 2–1). These same tubes can be used for ventilation in summer, eliminating the need for mechanical side and top vents.

Greenhouses can be mechanically cooled in the summer at low cost by the use of large evaporative cooling units, as shown in Figure 2–1. The "pad and fan" system, in which a wet pad of some material, such as aspen-wood excelsior, comprises a large portion of one side (or end) of a greenhouse and large exhaust fans the other, has proved to be one of the best methods of cooling large greenhouses (*19*).

As a general practice, greenhouses are painted (or sprayed) on the outside at the onset of warm weather in the spring with a thin layer of whitewash (see p. 420) or a white cold-water paint. This reflects much of the radiant energy from the sun, thus preventing the buildup of excessively high temperatures within the greenhouse. Too heavy a coating of whitewash, however, can reduce the light intensity to undesirably low levels.

It is best, if possible, to have the greenhouse heating, self-opening ventilators, and evaporative cooling systems controlled by thermostats. As a general rule, a minimum night temperature of 60° F (15.5° C) is suggested, for which the heating thermostat should be set. Ventilator thermostats should be set to open at about 72° F (22° C) and the evaporator cooler thermostats set to start operating the blowers at 75° F (24° C).

Glass-covered greenhouses are expensive, but for a permanent installation are likely to be more satisfactory than the lower-cost plastic-covered houses, as described in the following section.

Figure 2–1 Completely automated heating and cooling systems installed in fiberglass covered greenhouse. *Upper left:* hot air from hot-water heaters (at top) is blown into polyethylene distribution tubes. *Upper right:* distribution tubes, which have outlet holes spaced to disperse heated air uniformly, extend the length of the house. *Lower left and right:* the opposite end wall of greenhouse has an insert of a wettable pad *(right)* through which air is pulled by exhaust fans for cooling. Automatic closure panels *(left)* shut off outside air movement into the house through this pad when heating is required. All components of both heating and cooling systems are thermostatically controlled.

Plastic-Covered Greenhouses (17, 47, 52, 54, 58)

Lightweight frames covered with various types of plastic film are popular for small home-garden structures as well as for large commercial installations. Several kinds of plastic materials are available, some quite light and inexpensive, permitting construction of greenhouses at much lower cost than when glass is used as the covering material. Plastic houses are usually of temporary construction, except when the more permanent, high-cost coverings, such as fiberglass panels, are used (*47, 52*).

Plastic-covered greenhouses tend to be much tighter than glass-covered ones, with a consequent buildup of high humidity and, especially in winter, an undesirable water drip on the plants. This trouble can be overcome, however, by maintaining adequate ventilation (*15*).

Polyethylene

This is the least expensive covering material, but has the shortest life. It breaks down in summer and must be replaced once a year or oftener; this is generally done in the fall for use during the winter. Ultraviolet-ray-resisting polyethylene lasts longer but costs somewhat more. A thickness of 4 to 6 mils (1 mil = 0.001 in.) is recommended. For better insulation and lowered winter heating costs, a double layer is used with a 1-in. air gap between, kept separated by air pressure from a small blower; 2-to 4-mil material is used for the inside layer (*53*). This double, inflated layer extends the life of the polyethylene by reducing wind flapping and tearing. It also reduces moisture condensation and dripping within the house. A single-layer polyethylene-covered greenhouse will lose more heat at night or in winter than a glass-covered house since polyethylene allows passage of heat energy from the soil and plants inside the greenhouse much more readily than glass does. Glass stops most infrared radiation, whereas polyethylene is transparent to it. Polyethylene is available in widths up to 40 ft. Many installations, especially in windy areas, use a supporting material, usually welded wire mesh, for the polyethylene film. Occasionally other supporting materials, such as Saran cloth, are used. Chicken wire tends to tear the plastic.

A tough, white, opaque film consisting of a mixture of polyethylene and vinyl plastic is available. This film blocks the sun's radiation better and stays more flexible under low winter temperatures than does polyethylene. Because temperature fluctuates less under such film than under clear plastic, it is good for winter storage of container-grown plants.

A considerable savings in time and labor in the annual task of securing the polyethylene to the frame has resulted from the availability of two pieces of extruded aluminum, made specifically for this purpose, between which the polyethylene sheeting is clamped.

Fiberglass

Rigid panels, corrugated or flat, of fiberglass sheets embedded in plastic are widely used for greenhouse construction. Fiberglass is strong, long-lasting, lightweight, and easily applied, coming in a variety of widths, lengths, and thicknesses. Only the clear material—especially made for greenhouses and in a thickness of 0.038 in. or more and weighing 4 to 5 oz. per sq. ft.—should be used. Light transmission of this material tends to decrease over the years; this can be a serious problem. Fiberglass is the most expensive of the plastic materials described here.

Hotbeds

The hotbed is often used for the same purpose as a greenhouse. Seedlings can be started and leafy cuttings rooted in such structures early in the season. Heat is provided below the propagating medium by electric heating cables, hot water, steam pipes, or hot air flues. As in the greenhouse, close attention must be paid to shading and ventilation, as well as temperature and humidity control.

The hotbed may consist of a large wood box or frame with a sloping tight-fitting lid made of window sash or, preferably, regular hotbed sash. It should be placed in a sunny but protected and well-drained location. The size of the frame usually conforms to the size of the sash available. A standard size is 3 ft. by 6 ft. If one of the plastic films is used as the covering, any convenient dimensions can be used. The frame can be easily built of 1-in. or 2-in. lumber nailed to 4 by 4 corner posts set in the ground. Decay-resistant wood such as redwood, cypress, or cedar should be used, and preferably treated with a wood preservative, such as copper naphthenate. This compound retards decay for many years and does not give off fumes toxic to plants. Creosote should not be used on wood structures in which plants will be grown, since the fumes released, particularly on hot days, are toxic to plant tissue. Publications are available giving in detail the construction of such equipment (59). Hotbeds can be used throughout the year, except in areas with severe winters, where their use may be restricted to spring, summer, and fall.

Lead- or plastic-covered electric soil-heating cables are quite satisfactory for providing bottom heat in hotbeds. Automatic temperature control can be obtained with inexpensive thermostats. For a hotbed 6 ft by 6 ft, about 60 ft of heating cable is required. The details of a typical installation are shown in Figure 2–2. To insure safety, the wiring of these units should be done by a qualified person. Low-voltage soil-heating systems are sometimes used, especially in Europe. A transformer reduces the regular line voltage to about 30 v, lessening the danger of electrical shock. The heating element consists of low-cost No. 8 or No. 10 bare, galvanized wire (49).

The hotbed is filled with 4 to 6 in. of the rooting or seed-germinating medium over the heating cables. Alternatively, flats containing the medium can be used; these are placed directly on a thin layer of sand covering wire netting, placed over the heating cables for protection from tools.

Cold Frames (59)

Cold-frame construction (Figure 2–3) is almost identical with that of hotbeds, except that no provision is made for supplying bottom heat. The standard glass 3-ft by 6-ft hotbed sash is often used as a covering for the frame, although lightweight, less expensive coverings can be made, utilizing frames covered with one of the plastic materials. The covered frames should fit tightly in order to retain heat and to obtain high-humidity conditions. Cold frames should be placed in locations protected from winds, with the sash covering sloping down from north to south.

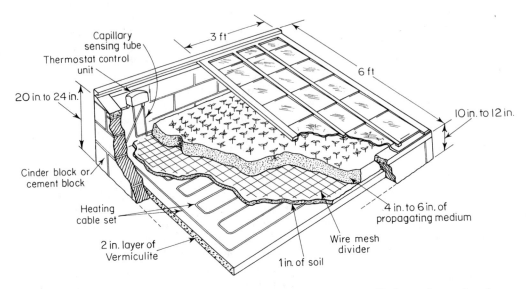

Figure 2–2   Construction of a hotbed showing the installation of an electric heating cable and thermostat. Courtesy General Electric Co.

A primary use of these structures is in conditioning or hardening rooted cuttings or young seedlings preceding field, nursery-row, or container planting. They may also be used for starting new plants in late spring, summer, or fall when no artificial supply of heat is necessary. In cold frames only the heat of the sun, retained by the transparent covering, is utilized.

Close attention to ventilation, shading, watering, and winter protection is necessary for success with cold frames. When young, tender plants are first placed in a cold frame, the coverings are generally kept closed tightly to maintain a high humidity, but as the plants become adjusted, the sash frames can be gradually raised to permit more ventilation and dryer conditions. Frequent sprinkling of the plants in a cold frame is essential in maintaining humid conditions. During sunny weather temperatures can build up to excessively high levels in closed frames unless some ventilation and shading are provided. Spaced lath, muslin-covered frames, or reed mats are useful to lay over the sash to provide protection from the sun.

In areas where extremely low temperatures occur, plants being overwintered in cold frames may require additional protective coverings.

Lathhouses
These structures (Figure 2–4) are very useful in providing protection for container-grown nursery stock, especially in areas of high summer temperatures and high light intensity. Although protection is particularly important just after transplanting, well-established plants also can require lathhouse protection. In holding shade plants for any length of time, a lathhouse is almost a necessity. For tender species the lathhouse is used as an intermediate

step between the cold frame and field planting. At times a lathhouse, in which watering needs are relatively low, is used simply to hold plants for sale. Removable lath is often convenient, especially in the northern regions, to compensate for changing light conditions.

Lathhouse construction varies widely. Aluminum prefabricated lathhouses are available but may be more costly than wood structures. More commonly, wood or pipe supports are used, set in concrete with the necessary supporting cross-members. Shade is provided by thin wood strips about 2 in. wide, placed to give $\frac{1}{3}$ to $\frac{2}{3}$ cover, depending on the need. The sides, as well as the top, are usually covered. Rolls of snow fencing attached to a supporting framework can be utilized for inexpensive construction.

Figure 2-3　A series of cold frames on the south side of a greenhouse. Hotbed construction is similar except that provision is made for bottom heat.

A woven plastic material—Saran fabric—is also used widely in covering structures to provide shade. This material is available in different densities, thus allowing various intensities of light on the plants. It is lightweight and can be attached to heavy wire fastened to supporting posts. A similar material, polypropylene fabric, is also available for this purpose; it is stronger and lighter in weight than Saran.

In the cooler latitudes, "shading with water" is sometimes done, using sprinklers to cool the plants on the few hot days which occur, rather than building expensive lathhouses, which give too much shade most of the time under these conditions.

Figure 2-4　Two lathhouse constructions. *Left:* wooden slats supported by a metal framework. *Right:* large area covered by Saran cloth supported by wires stretched between metal poles.

Miscellaneous Propagating Structures

Fluorescent Light Boxes

Young plants of many species grow satisfactorily under the artificial light from fluorescent lamps or other light sources, such as high-pressure metal halide lamps. These units are used, too, for starting young seedlings and rooting cuttings of many plant species. By enclosing the lamps in boxes, it is possible to maintain a high humidity (57). The "cool-white" fluorescent tubes are generally preferable to other types (36). In some basement rooms which have a fairly high humidity, a closed case may be unnecessary. The lamp fixture can be suspended over a table on which the flats of cuttings are placed. With such equipment it is often helpful to provide bottom heat, either from thermostatically controlled soil-heating cables or by the use of a heater or a lamp in the air space below the rooting medium. Although adequate growth of many plant species may be obtained under fluorescent lamps, in general it is unlikely that the results will equal or surpass those obtained under good greenhouse conditions.

For the home gardener's use as an indoor facility for propagating and growing plants, Cornell University has developed an automated, controlled environment plant grower using fluorescent lamps as a light source. Detailed directions for the construction of this unit are available (38).

Figure 2–5   A simple but effective enclosed propagating frame in the greenhouse can be constructed just by draping a sheet of polyethylene film over a supporting wire netting. It is necessary to use white plastic or to provide shading over such structures to prevent excessive heat build-up.

Propagating Cases

Even in a greenhouse, humidity is often not sufficiently high to permit satisfactory rooting of leafy cuttings. Enclosed frames or cases covered with glass or one of the plastic materials may be necessary for successful rooting (see Figure 2–5). There are many variations of such devices, called Wardian cases in earlier days. Such cases are also useful for completed grafts of small potted nursery stock, enabling the retention of high-humidity conditions during the callusing process.

Bell jars are large inverted glass jars to set over a container of cuttings to be rooted. Humidity can be kept high in such devices, but attention is necessary to provide shading from the sun and to allow ventilation as soon as rooting starts. As shown in Figure 2–6, polyethylene plastic bags can be inverted and tied over such containers to provide an inexpensive protective

cover to prevent water loss when rooting a few cuttings.

Polyethylene permits the passage of
oxygen and carbon dioxide, necessary
for the growth processes of the plant
material.

In using all such structures, care is
necessary to avoid the buildup of disease organisms. The warm, humid
conditions, combined with lack of air
movement and the relatively low light
intensity, provide excellent conditions for the growth of various fungi
and bacteria. Cleanliness of all materials placed in such units is important, but in addition spraying with
fungicides is sometimes necessary.

MEDIA FOR PROPAGATING
NURSERY PLANTS

Several materials and mixtures of different materials are available for germinating seeds and rooting cuttings.
For good results the following characteristics are required *(51)*:

(a) The medium must be sufficiently
firm and dense to hold the cuttings
or seeds in place during rooting or
germination. Its volume must be fairly

Figure 2–6   A small unit for rooting
a few cuttings. A polyethylene bag has
been placed over a framework made
from two wire coat hangers. The bag is
folded under the pot to give a tight
seal. This unit should be set in a fairly
light place but never in the direct sunlight, which would cause overheating
within the bag.

constant when either wet or dry. That is, excessive shrinkage after drying is
undesirable. (b) It must be sufficiently retentive of moisture that watering
does not have to be too frequent. (c) It must be sufficiently porous that excess
water drains away, permitting adequate aeration. (d) It must be free from
weed seeds, nematodes, and various noxious organisms. (e) It must not have
a high salinity level. (f) It should be capable of being sterilized with steam
without deleterious effects, and (g) for seed germination there should be
adequate nutrient availability.

Soil

A soil is composed of materials in the solid, liquid, and gaseous states. For
satisfactory plant growth these materials must exist in the proper proportions.

The solid portion of a soil is comprised of both inorganic and organic
forms. The inorganic part consists of the residue from parent rock after de-

composition due to the chemical and physical process of weathering. Such inorganic components vary in size from gravel down to extremely minute colloidal particles of clay, the texture of the soil being determined by the relative proportions of particles of different size. The coarser-sized particles serve mainly as a supporting framework for the remainder of the soil, whereas the colloidal clay fractions of the soil serve as storehouses for nutrients which may be absorbed by plants. The organic portion of the soil consists of both living and dead organisms. Insects, worms, fungi, bacteria, and plant roots generally constitute the living organic matter, whereas the remains of such animal and plant life in various stages of decay make up the dead organic material. The residue from such decay (termed *humus*) is largely colloidal and assists in holding the water and plant nutrients.

The liquid part of the soil, the soil solution, is made up of water containing dissolved minerals in various quantities, as well as oxygen and carbon dioxide. Mineral elements, water, and possibly some carbon dioxide enter the plant from the soil solution.

The gaseous portion of the soil is important to good plant growth. In poorly drained, waterlogged soils, water replaces the air in the soil, thus depriving plant roots as well as certain desirable aerobic microorganisms of the oxygen necessary for their existence.

The texture of a soil will depend upon the relative proportions of *sand* (2 to 0.05 mm particle diameter), *silt* (0.05 to 0.002 mm particle diameter), and *clay* (less than 0.002 mm particle diameter). The principal texture classes are sand, loamy sand, sandy loam, silt loam, clay loam, and clay. A typical sandy loam might consist of 75 percent sand, 14 percent silt, and 11 percent clay; whereas a clay loam might have 34 percent sand, 39 percent silt, and 27 percent clay.

In contrast to *soil texture,* which refers to the individual soil particles, *soil structure* refers to the arrangement of those particles in the entire soil mass. These individual soil grains are held together in aggregates of various sizes and shapes. Maintenance of a favorable granular and crumb soil structure is very important. For example, working heavy clay soils when they are too wet can so change the soil structure that the heavy clods formed may remain for years.

Sand

Sand consists of small rock grains, from about 0.05 to about 2.0 mm in diameter, formed as the result of the weathering of various rocks, its mineral composition depending upon the type of rock. Quartz sand is generally used for propagation purposes, consisting chiefly of a silica complex. The type used in plastering is the grade ordinarily the most satisfactory for rooting cuttings. Sand is the heaviest of all rooting media used, a cubic foot of dry sand weighing about 100 lb. It should preferably be fumigated or heat-treated before use, as it may contain weed seeds and various harmful fungi. Sand contains virtually no mineral nutrients and has no buffering capacity. It is used mostly in combination with organic materials.

Peat

Peat consists of the remains of aquatic, marsh, bog, or swamp vegetation which has been preserved under water in a partially decomposed state. Composition of different peat deposits varies widely, depending upon the vegetation from which it originated, state of decomposition, mineral content, and degree of acidity (*39, 48*).

There are three types of peat: moss peat, reed sedge, and peat humus.

Moss peat (usually referred to in the market as "peat moss") is the least decomposed of the three types and is derived from sphagnum, hypnum, or others mosses. It will vary in color from light tan to dark brown. It has a high moisture-holding capacity (10 times its dry weight), is high in acidity (pH of 3.8 to 4.5) and contains a small amount of nitrogen (about 1.0 percent) but little or no phosphorus or potassium. This type of peat comes from Canada or Europe but some is being produced in certain northern states.

Reed sedge peat consists of the remains of grasses, reeds, sedges, and other swamp plants. This type of peat varies considerably in composition and color, ranging from reddish-brown to almost black. The pH ranges from about 4.5 to 7.0.

Peat humus is in such an advanced state of decomposition that the original plant remains cannot be identified and can originate from either hypnum moss or reed sedge peat. It is dark brown to black in color with a low moisture-holding capacity but with 2.0 to 3.5 percent nitrogen.

When peat moss is to be used in mixtures, it should be broken apart and moistened before adding to the mixture. Continued addition of coarse organic materials such as peat moss or sphagnum moss to greenhouse soil mixtures can cause a decrease in wettability. Water will not penetrate easily, and many of the soil particles will remain dry even after watering. No good method for preventing this nonwettability is known, although the repeated use of commercial wetting agents may improve water penetration (*40*).

Sphagnum Moss

Commercial sphagnum moss is the dehydrated young residue or living portions of acid-bog plants in the genus *Sphagnum*, such as *S. papillosum, S. capillaceum,* and *S. palustre.* It is relatively sterile, light in weight, and has a very high water-holding capacity, being able to absorb 10 to 20 times its weight of water. The stem and leaf tissues of sphagnum moss consist largely of groups of water-holding cells. This material is generally shredded, either by hand or mechanically, before it is used as a propagating medium. It contains such small amounts of minerals that plants grown in it for any length of time require added nutrients. Sphagnum moss has a pH of about 3.5. It contains a specific fungistatic substance, or substances, which accounts for its ability to inhibit damping-off of seedlings germinated in it (*18, 22*).

Vermiculite

This is a micaceous mineral which expands markedly when heated. Extensive deposits in the United States are found in Montana and in North Carolina. Chemically, it is a hydrated magnesium-aluminum-iron silicate. When expanded, it is very light in weight (6 to 10 lb per cu ft), neutral in reaction with good buffering properties, and insoluble in water; it is able to absorb large quantities of water—3 to 4 gal. per cu ft. Vermiculite has a relatively high cation exchange capacity and thus can hold nutrients in reserve and later release them. It contains enough magnesium and potassium to supply most plants. In the crude vermiculite ore, the particles consist of a great many very thin, separate layers which have microscopic quantities of water trapped between them. When run through furnaces at temperatures near 2000° F, the water turns to steam, popping the layers apart, forming small, porous, sponge-like kernels. Heating to this temperature gives complete sterilization. Horticultural vermiculite is graded to four sizes: No. 1 has particles from 5 to 8 mm in diameter; No. 2, the regular horticultural grade, from 2 to 3 mm; No. 3, from 1 to 2 mm; and No. 4, which is most useful as a seed-germinating medium, from 0.75 to 1 mm. Expanded vermiculite should not be pressed or compacted when wet, as this will destroy its desirable porous structure.

Perlite

This gray-white silicaceous material is of volcanic origin, mined from lava flows. The crude ore is crushed and screened, then heated in furnaces to about 1400° F, at which temperature the small amount of moisture in the particles changes to steam, expanding the particles to small, sponge-like kernels which are very light, weighing only 5 to 8 lb per cu ft. The high processing temperature gives a sterile product. A particle size of $\frac{1}{16}$ to $\frac{1}{8}$ in. in diameter is usually used in horticultural applications. Perlite will hold three to four times its weight of water. It is essentially neutral with a pH of 6.0 to 8.0 but with no buffering capacity; unlike vermiculite, it has no cation exchange capacity and contains no mineral nutrients. It is most useful in increasing aeration in a mixture.

Compost

In the home garden a compost mixture may be useful as a moisture-holding humus material having limited value as a plant nutrient. It may be mixed with soil to add organic matter. Grass clippings, leaves, and garden refuse are accumulated and allowed to decompose, preferably in a 4 x 6 ft bin with slatted sides to give good aeration. Moisture should be added from time to time during the dry seasons; decomposition will be hastened if some nitrogenous fertilizer is sprinkled through each batch of newly added material. The mass of material should be stirred every 5 to 10 days to ensure even decomposition. Several bins are preferable—one for newly started material, one for material undergoing decomposition, and one for completely decomposed compost, ready to use. Compost such as this may contain weed seeds and nematodes, as well as noxious insects and diseases, so preferably it should be sterilized before use.

Shredded Bark, Sawdust, and Wood Shavings

These materials, consisting of redwood, cedar, fir, or pine, can be used in soil mixes, serving much the same purposes as peat moss except that their rate of decomposition is slower. One widely used material is a nitrified redwood sawdust. Nitrogen is added in an amount sufficient to take care of the decomposition requirements of the sawdust, plus an additional amount for use by the plants. Rate of decomposition varies with the wood species. Due to its low cost it is widely used as a soil amendment, although some types, especially when fresh, may contain materials toxic to plants.

SOIL MIXTURES FOR CONTAINER
GROWING (24, 44)

In propagation procedures, young seedlings or rooted cuttings are sometimes planted directly in the field, but frequently they are started in a soil mix in some type of container. For various reasons, loam soils alone are unsatisfactory for this purpose. They are often heavy and poorly aerated, or tend to become sticky after watering. Upon drying they may shrink, forming a hard and cracked surface. Such soils draw away from the sides of the container during drying, and subsequent added water then runs down the inner sides of the container and out the drainage holes rather than rewetting the soil mass.

To provide potting mixtures of better textures, sand and some organic matter, such as peat moss or sawdust or shredded bark, are usually added. In preparing these mixtures, the soil should be screened to make it uniform and to eliminate large particles. If the materials are very dry, they should be moistened slightly. This applies particularly to peat, which if once mixed when dry, will absorb moisture very slowly. The soil should not be wet and sticky, however. In mixing, the various ingredients may be placed in a pile arranged in layers and turned with a shovel. A power-driven cement mixer, soil shredder or skip-loader is used in large-scale operations. Preparation of the soil mixture should preferably take place at least a day prior to use. During the ensuing 24 hours the moisture will tend to become equalized throughout the mixture. The soil mixture should be just slightly moist at the time of use so that it does not crumble; on the other hand, it should not be sufficiently wet to form a ball when squeezed in the hand. Some dry fertilizers, as lime and superphosphate, may be added during the mixing.

Traditional potting mixtures that have been used in the past are:

1 For potting rooted cuttings and young seedlings:
   1 or 2 parts sand
   1 part loam soil
   1 part peat moss (or shredded bark or leaf mold)
2 For general container-grown nursery stock:
   1 part sand
   2 parts loam soil
   1 part peat moss (or shredded bark or leaf mold)

However, the physical and chemical properties of mixes such as these vary so much from batch to batch that many problems in management and cultural practices arise.

After starting new plants by rooting cuttings or germinating seeds, many commercial producers of nursery stock grow many of these plants on to a salable size in containers, using a modified sandy soil mix. These mixes vary widely throughout the industry, but generally include a fine sand mixed in varying proportions with such materials as peat moss, redwood sawdust, or shredded fir bark. Such mixes require added fertilizer supplements and continued feeding of the plants until the plants become established in their permanent locations.

Numerous formulations of artificial "soils" have been developed and used by commercial nurseries. For example, one successfully used mix for small seedlings, rooted cuttings, and bedding plants consists of equal parts of shredded fir bark, nursery grade peat moss, perlite, and sand. To this is added gypsum, superphosphate, dolomite lime, and potash. Nitrogen is added subsequently in the irrigation water.

The U.C. Soil Mixes (43)

These series of mixtures were originally developed by plant pathologists and others at the University of California, Los Angeles, to provide a growing medium that could be readily prepared in large quantities for the commercial nurseryman and would be an integral part of a pathogen-free propagating and cultural program. Since the U.C. mixes are based upon materials that are uniform, generally available, and require no previous preparation, they can easily be duplicated. The basic components are (a) an inert type of fine sand, (b) finely shredded peat moss—mixed with each other in varying proportions, and (c) fertilizer mixtures as described below.

The sand consists of round, wind-blown particles, uniform in size and relatively small (0.5 to 0.05 mm in diameter), thus having a rather high moisture-holding capacity. Such sand does not tend to compact even though the particles are small, owing to their round shape and uniformity. The absence of colloidal clay particles in sand further tends to prevent compaction or shrinkage. Any ocean beach sand of high salt content would, of course, be unsuitable for soil mixtures, although satisfactory sands may sometimes be found only a short distance from beaches.

The chief purpose of the peat moss in this soil mixture is to increase its moisture and nutrient-holding capacities. In a mixture of equal parts of sand and peat moss, the maximum moisture-holding capacity is about 48 percent.

Basic fertilizer additives recommended (43) for a U.C. Mix of 50 percent fine sand and 50 percent peat moss are as follows:

1   *If the mix is to be stored for an indefinite period before using.* This furnishes a moderate supply of available nitrogen, but the plants will soon require supplemental feeding. To each cubic yard of the mix, add

4 oz potassium nitrate

4 oz potassium sulfate

2½ lb single superphosphate

7½ lb dolomite lime

2½ lb calcium carbonate lime

2 *If the mix is to be planted within one week of preparation.* This furnishes available nitrogen as well as a moderate nitrogen reserve. To each cubic yard of the mix, add

2½ lb hoof and horn or blood meal

  (13 percent nitrogen)

4 oz potassium nitrate

4 oz potassium sulfate

2½ lb single superphosphate

7½ lb dolomite lime

2½ lb calcium carbonate lime

The organic nitrogen is omitted if the mix is to be stored for a time before using, since such organic forms will break down during storage, releasing a high content of water soluble nitrogen, which may cause plant injury. Other forms of organic nitrogen, such as cottonseed meal (7 percent nitrogen) or fish tankage (6 to 10 percent nitrogen), may be substituted for the hoof and horn or blood meal, provided that a comparable amount of nitrogen is supplied.

In using the U.C. Mix, the fine sand, shredded peat moss, and fertilizer must be mixed together thoroughly. The peat moss should be moistened before mixing. If the mixing is done well, the peat moss will not separate and float to the top when the mixture is saturated with water. Different combinations of sand and peat are used; 75 percent sand and 25 percent peat moss is suitable for bedding plants and nursery container-grown stocks; a 50 percent sand and 50 percent peat mixture is satisfactory for potted plants.

This mixture, including the fertilizer, can be safely sterilized by steam or chemicals without the subsequent harmful effects to the plants that often occur when other soils or mixes are sterilized.

The John Innes Soil Mixes

Two basic soil mixtures, one for seeds and one for potting, were developed many years ago at the John Innes Horticultural Institution in England (2).

The John Innes Seed Compost	The John Innes Potting Compost
2 parts loam (by volume)	7 parts loam (by volume)
1 part peat moss (by volume)	3 parts peat moss (by volume)
1 part clean sand (by volume)	2 parts clean sand (by volume)
To each cubic yard of this mixture is added	To each cubic yard of this mixture is added
2 lb superphosphate and	5 lb John Innes Base,* and
1 lb ground limestone	1 lb ground limestone

* The John Innes Base consists of
    2 parts by weight—hoof and horn meal, ⅛ in. grist (13 percent nitrogen)
    2 parts by weight—superphosphate of lime (18 percent phosphoric acid)
    1 part by weight—potassium sulfate (48 percent potash)

The loam used in the John Innes mixtures should be taken from well-drained, slightly acid, medium clay pasture turf and cut into 9 by 12 in. sections 4 to 5 in. thick. These are stacked in late spring in layers about 9 in. thick, each layer interspersed with a 2-in. layer of strawy manure and, if desired, layers of limestone, potassium sulfate, and superphosphate. The layers are well-moistened as stacking proceeds until a mound 6 ft high and 6 to 8 ft across is formed. The stack is covered to keep off rain. Decomposition continues for about 6 months. When it is ready for use, the stack is cut through from top to bottom, to minimize variation. Before using the loam, one should put it through a 3/8-in. sieve or mechanical shredder. The most desirable sand for this mix is one in which 60 to 70 percent of the particles are between 1/8 and 1/16 in. in size. Before the compost mixture is made up, the basic loam alone should be steam sterilized. While it has long been considered that soil from under grass is a desirable constituent of the John Innes mixtures, studies (3) have shown that soil taken from bare, fallow land performs just as well.

The Cornell "Peat-Lite" Mixes (11, 55)

These are used primarily for seed germination and for container-growing of spring bedding plants and annuals, using components that are lightweight, uniform, and readily available, and have chemical and physical characteristics suitable for the growth of plants. Excellent results have been obtained with these mixes, and commercial ready-mixed preparations are available. Mixes of this type have an important advantage of not requiring any decontamination before use.

Peat-Lite Mix A (to make 1 cu yd)
Shredded German or Canadian sphagnum peat moss—11 bu (88 gal.)
Horticultural grade vermiculite (No. 2 or 4)—11 bu (88 gal.)
Ground limestone (preferably dolomitic)—5 lb
Superphosphate (20%), preferably powdered—1 lb
5-10-5 fertilizer (nitrogen, phosphorus, potassium)—2 to 12 lb †
Peat-Lite Mix B (same as A, except that horticultural perlite is substituted for the vermiculite.
Peat-Lite Mix C (for germinating seeds)
Shredded German or Canadian sphagnum peat moss—1 bu (8 gal.)
Horticultural grade vermiculite No. 4 (fine)—1 bu (8 gal.)
Ammonium nitrate—1½ oz (4 level tbsp)
Superphosphate (20%), powdered—1½ oz (2 level tbsp)
Ground limestone, dolomitic—7½ oz (10 level tbsp)

The materials should be mixed thoroughly, with special attention given to wetting the peat moss during mixing. Adding a non-ionic wetting agent,

---

† If 12 lb of the 5-10-5 fertilizer are used, this will supply nutrients to the plants for 5 or 6 weeks during the growing season. With less than 12 lb of the fertilizer added, a program of liquid feeding should be used.

such as Aqua-Gro (1 oz per 6 gal. of water) to the initial wetting usually will aid in wetting the peat moss.

PRE-PLANTING SOIL TREATMENTS

Soils may contain weed seeds, nematodes, and various fungi and bacteria harmful to plant tissue. The so-called "damping-off" commonly encountered in seed-beds is caused by soil fungi, such as species of *Pythium* and *Rhizoctonia*. To avoid loss from these pests, it is desirable to treat the soil or soil mixture before it is used for growing plants. Along with clean soil, the use of noninfected plants; treatment of seeds with fungicides (see p. 37); disinfection of flats, greenhouse benches, soil bins, and tools; and general cleanliness are also necessary to avoid recontamination. It is useless to put clean soil in contaminated containers or to use it for diseased plants. Tools can be sterilized by soaking in some disinfectant such as Clorox (diluted 1 to 10 with water), 2 percent formaldehyde, rubbing alcohol, or even boiling water. Flats and benches can be treated with live steam, or drenched with boiling water, 2 percent formaldehyde, or copper naphthenate.

Soil can be heated, or fumigated with chemicals, to eliminate weeds, nematodes, and disease organisms. However, heating soil mixes high in manure, leaf mold, or compost will hasten decomposition of the organic matter, especially if it is already partially rotted. This leads to the formation of compounds toxic to plant growth, necessitating leaching with water or a 3- to 6-week delay in planting. However, undecomposed materials like the brown types of peat moss are relatively unaffected.

Certain of the complex chemical compounds in the soil are also broken down by excessive heat—above 185° F (85° C)—yielding increased amounts of soluble salts of nitrogen, manganese, phosphorus, potassium, and others. Some, particularly nitrogen in the form of ammonia, may be present during the first few weeks after steaming in such quantities as to be toxic to the plants. Later the ammonia is converted to nitrate nitrogen, which reaches a peak in about 6 weeks. The presence of superphosphate in the soil mix ties up the excessive manganese released during heating, preventing injury from manganese toxicity.

Heat Treatment (4, 5, 6, 7, 27)

Although the term "soil sterilization" is established by common usage, a more accurate word is "pasteurization," since the recommended heating processes do not kill all organisms.

Steam is the best and most common heat source for soil treatment. The moist heat is advantageous; it can be injected directly into the soil in covered bins or benches from perforated pipes placed 6 to 8 in. below the surface. In heating the soil, which should be moist but not wet, a temperature of 180° F (82° C) for 30 minutes has been a standard recommendation, since this will kill most harmful bacteria and fungi as well as nematodes, insects, and most weed seeds, as indicated in Figure 2–7. However, a lower

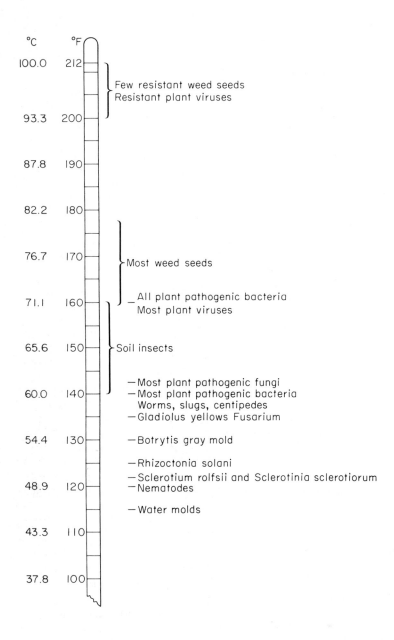

Figure 2-7   Soil temperatures required to kill weed seeds, insects, and various plant pathogens. Temperatures given are for 30 min under moist conditions. From University of California Division of Agricultural Sciences, *Manual 23 (4)*.

temperature, such as 140° F (60° C) for 30 minutes is more desirable since, while killing pathogens, it will leave many antagonistic beneficial organisms which, if present, will prevent explosive growth of harmful organisms if recontamination occurs. Using the lower temperature also tends to avoid toxicity problems, such as the release of excess ammonia and nitrite, as well as manganese injury (*10, 63*), which is often encountered at higher steaming temperatures.

Air mixed with steam (aerated steam) in the ratio of 4.1 to 1 by volume will give a temperature of 140° F (*8, 46*). This temperature will not kill many weed seeds but, if held for 30 minutes, and if the soil is moist, it will kill most pathogenic bacteria and fungi. Designing equipment to mix air and steam in the proper proportions to give this temperature presents some difficulties but successful equipment has been developed (*1, 13*).

Fumigation with Chemicals (45, 60)

Chemical fumigation will kill organisms in the propagating mixes without disrupting their physical and chemical characteristics to the extent to which this occurs with heat treatments. Ammonia production may increase following chemical fumigation, however, owing to the removal of organisms antagonistic to the ammonifying bacteria. The mixes should be moist (between 40 percent and 80 percent of field capacity) and at temperatures of 65° to 75° F (18° to 24° C) for satisfactory results. After chemical fumigation a waiting period for dissipation of the fumes of 2 days to 2 weeks, depending upon the material, is required before use.

Formaldehyde

This is a good fungicide with strong penetrating powers. It will kill some weed seeds, but is not reliable for killing nematodes or insects. A mixture of 1 gal. of commercial formalin (40 percent strength) with 50 gal. of water is applied to the soil at the rate of 2 to 4 qt per sq ft (or 1 gal. per bu of soil). The treated area should be covered immediately with an airtight material and left for 24 hours or more. Following this treatment, about 2 weeks should be allowed for drying and airing, but the soil should not be planted until all odor of formaldehyde has disappeared.

For small-scale treatments, commercial formalin can be applied at a rate of 2½ tbsp per bu of a light soil mixture or 1 tbsp per standard size flat. Dilute with five to six parts of water, apply to soil and mix thoroughly. Let stand 24 hours, plant seeds, and water thoroughly.

Chloropicrin (Tear Gas)

This is a liquid ordinarily applied with an injector, which should put 2 to 4 ml into holes 3 to 6 in. deep, spaced 9 to 12 in. apart. It may also be applied at the rate of 5 ml per cu ft of soil. Chloropicrin changes to a gas which penetrates through the soil. The gas should be confined by sprinkling the soil surface with water and then covering it with an airtight material, which is then left for 3 days. Seven to 10 days is required for thorough aeration of the soil before it can be planted. Chloropicrin is effective against nematodes, insects, some weed seeds, *Verticillium,* and most other resistant fungi. Chloropicrin fumes are very toxic to living plant tissue.

Chloropicrin and methyl bromide are hazardous materials to use, especially in confined areas. They should be applied only by persons trained in their use and who will take the necessary precautions as stated in the instructions on the containers or in the accompanying literature.

Methyl Bromide
This odorless material is very volatile and very toxic to humans. It should be used mixed with other materials and applied only by those trained in its use. Most nematodes, insects, weed seeds, and some fungi are killed by methyl bromide, but it will not kill *Verticillium.* It is often used by injecting the material at 1 to 4 lb per 100 sq ft from pressurized containers into an open vessel under a plastic cover placed over the soil to be treated. The cover is sealed around the edges with soil, and should be kept in place for 48 hours. Penetration is very good, and the sterilization effect will extend to a depth of 12 in. For treating bulk soil, methyl bromide at 10 ml per cu ft or 4 lb per 100 cu ft can be used.

Methyl Bromide–Chloropicrin Mixtures
Proprietary materials are available containing both methyl bromide and chloropicrin. Such combinations are more effective than either material alone in controlling weeds, insects, nematodes, and soil-borne disease organisms. Aeration for 10 to 14 days is required following applications of methyl bromide-chloropicrin mixtures.

Vapam (Sodium N-methyl Dithiocarbamate Dihydrate)
This is a water-soluble soil fumigant which will kill weeds, germinating weed seeds, most soil fungi and, under the proper conditions, nematodes. It undergoes rapid decomposition to produce a very penetrating gas. Vapam is applied by sprinkling on the soil surface, through irrigation systems, or with standard injection equipment. For seed-bed fumigation, 1 qt of the liquid formulation of Vapam in 2 to 3 gal. of water is used, sprinkled uniformly over 100 sq ft of area. After application, the Vapam is sealed with additional water or with a roller. Two weeks after application the soil can be planted. Although Vapam has a relatively low toxicity to man, care should be taken to avoid inhaling fumes or splashing the solution on the skin.

Fungicidal Soil Drenches

Certain chemical formulations are available which can be applied to soil in which young plants are growing or are to be grown and will inhibit growth of many soil-borne fungi. These materials may be applied to the soil or to the plants. It is very important in using such chemicals to read and follow the manufacturer's directions carefully, and to try them on a few plants first before going to large-scale applications. Examples of these materials are:

Dexon (*p*-dimethylaminobenzene diazo sodium sulfonate)   This will control some of the water molds, *Phytophthora* and *Pythium,* which often invade sterilized soils.

Terraclor (pentachloronitrobenzene)   This is useful for inhibiting growth of certain soil-borne fungi, such as *Rhizoctonia*. It is almost insoluble in water and has a long residual action, remaining effective for 6 to 12 months. It is best applied to the soil before planting.

Benomyl   This systemic fungicide will inhibit growth of a number of soil pathogens as *Rhizoctonia, Fusarium,* and *Verticillium*. A 50 percent wettable powder is available for use on ornamentals.

SUPPLEMENTARY FERTILIZERS

Even with a good soil mixture, complete with added mineral nutrient components, continued growth of plants in containers necessitates the addition at intervals of supplementary minerals, especially nitrogen.

A satisfactory feeding program for growing container plants is to combine a slowly available dry fertilizer in the original soil mix with a liquid fertilizer applied at frequent intervals during the growing season or with controlled-release fertilizers added as top dressings at intervals, as needed. Of the three major elements—nitrogen, phosphorus, and potassium—nitrogen has the most control on the amount of vegetative growth.

Some organic fertilizers which supply nitrogen are blood meal or hoof and horn meal applied at about 1 heaping tsp. for a plant growing in a 1-gal. container. Twice this amount of cottonseed meal should be used.

To supply nitrogen, phosphorus, and potassium in dry form, 2 heaping tsp. of the following mixture is recommended *(43)* for plants growing in 1-gal. containers:

4 lb hoof and horn or blood meal
4 lb single superphosphate
1 lb potassium sulfate

As a supplement to this, a dilute inorganic nutrient feeding at weekly intervals throughout the growing season is desirable. A simple solution may be prepared by dissolving 1 tsp. of potassium nitrate and one of ammonium nitrate in 1 gal. of water. Adding 2 tsp of a mixed fertilizer, such as 10–6–4, to 1 gal. of water will also make a satisfactory solution.

Unless labeled "biuret-free," urea-containing formulations should not be used in fertilizing container-grown plants. Biuret is quite toxic, especially to conifers, ericaceous plants, pineapple, and citrus (*43, 61*); typical symptoms are leaf burning, chlorosis, and stunted growth.

Complete nutrient solution for fertilizing container-grown plants:

water	100 gal.
ammonium nitrate (or urea)	6 oz
monoammonium phosphate	6 oz
potassium nitrate	6 oz

Nutrient solution for addition of nitrogen. Use any *one* of the following in 100 gal. of water:

urea	1 lb
ammonium nitrate	1 lb
calcium nitrate	2 lb
ammonium sulfate	2 lb

For large-scale operations, it is more feasible to prepare a liquid concentrate and inject it into the regular watering or irrigating system by the use of a proportioner (*12*). For this, a nutrient concentrate formula such as the following could be used, although since soluble forms of phosphorus are expensive, it may be best to add this as superphosphate in the soil mix and use only potassium nitrate and ammonium nitrate in the liquid concentrate. Indicator dyes are sometimes added to the concentrate to serve as a visual guide to the fertilizer flow.

water	10 gal
ammonium nitrate	15 lb
monoammonium phosphate	4 lb
potassium chloride	6 lb

The above chemicals should be thoroughly dissolved. *This concentrate should then be diluted—1 part to 200 parts water—before being applied to the plants.*

Controlled-Availability Fertilizers (*14, 32, 41, 42, 62*)

These materials provide nutrients to the plants gradually over a long period of time and reduce the possibility of injury from excessive applications. They are expensive, however, and are used chiefly on high-value container-grown plants.

In one of the most promising methods of preparation, granules of soluble fertilizers are coated with some type of membrane which regulates the rate at which soluble fertilizers pass through it and become available to the plant. In other types, granular fertilizer pellets are coated with a resin. In a moist soil mix, water penetrates the capsule coating and dissolves the nutrients inside, causing the capsule to swell. The nutrient solution inside diffuses through the resin into the growing medium where it becomes available to the plants.

Another method of obtaining slow-release nutrients is by using slightly soluble materials, as a combination of granulated magnesium-ammonium-phosphate plus magnesium-potassium-phosphate. These materials mixed with the soil will supply nutrients slowly for up to two years. Similarly, potassium glass frit is a relatively soft glass which will slowly supply potassium at adequate levels for up to 18 months. Certain synthetic organic compounds, such as urea-formaldehyde (UF), are available commercially and will supply nitrogen slowly over a long period of time, owing to their slow mineralization (32). Tests (9) have shown that 5 to 10 gm of UF applied to the soil surface of 1-gal. containers is satisfactory for small transplanted ornamental liners. At 10 to 20 gm, UF was suitable for year-old established plants in gallon containers, furnishing nitrogen for about 2 months.

Phosphorus for container-grown plants can be supplied by adding soluble phosphate salts to a fertilizer concentrate injected into the irrigation water, or, at less cost, by prior incorporation of slowly available forms into the growing mixture. Tests with rooted cuttings of several ornamental plants grown in containers of perlite and sphagnum moss showed good results with phosphate in solution applied as ammonium phosphate (50 or 100 ppm) plus superphosphate at $2\frac{1}{2}$ lb per cu yd previously incorporated into the growing mixture (23).

SALINITY IN SOIL MIXES (21, 31, 56)

Excessive salts in the soil mixes or irrigation water (over 2 milli-mho/cm)* can reduce plant growth, burn the foliage, or even kill the plants. The required fertilization programs will also contribute to high salt accumulations. Overfertilization will cause rapid and severe salinity symptoms, starting with foliage wilting and tip and marginal leaf burning. These symptoms may be accompanied by a white surface accumulation of salts on the soil. To prevent salt buildup in the soil, the containers or benches should be leached with water periodically. If the irrigation water contains 250 ppm of salts, the leaching should be done every 12 weeks; for 500 ppm, every 6 weeks; and for 1000 ppm every 3 weeks. In addition, fertilizers should not be used which tend to contribute to excess salinity; for example, use potassium nitrate rather than potassium chloride.

WATER QUALITY

Water quality is an important factor in rooting cuttings, germinating seeds and growing-on the young plants (20). For good results the available water supply must not contain total soluble salts in excess of 1400 ppm (approx. 2 milli-mhos/cm).—ocean water averages about 35,000 ppm. The salts are com-

---

* Salinity levels in water extracts of the soil (saturation-extract method) can be measured by electrical conductivity using a "Solubridge." Such readings are expressed as milli-mhos per centimeter ($EC_e \times 10^3$). Readings of less than 2 (1400 ppm) would indicate no salinity problem; readings much over 4 indicate a level at which most plants are likely to be affected. At readings over 8, only salt-tolerant plants will grow.

binations of such cations as sodium, calcium, and magnesium with such anions as sulfate, chloride, and bicarbonate. Water containing a high proportion of sodium to calcium and magnesium can adversely affect the physical properties and water-absorption rates of soils and should not be used for irrigation purposes.

Although not itself detrimental to plant tissue, so-called "hard" water contains relatively high amounts of calcium and magnesium (as bicarbonates and sulfates) and can be a problem in mist-propagating units or in evaporative water cooling systems as deposits build up wherever evaporation occurs. When water is much over 6 grains per gallon (100 ppm) in hardness it is often run through a "water softener" for household use. Some types of equipment are based upon the replacement of the calcium and magnesium in the water by sodium ions. Such "soft," high-sodium water is toxic to plant tissue and, as mentioned above, causes breakdown in soil structure. A better, but more costly, method of improving water quality is the "de-ionization" process. Here calcium, magnesium, and sodium are removed by substituting hydrogen ions for them. Water passes over an absorptive medium charged with H ions, which absorbs calcium and other ions in exchange for hydrogen. For further de-ionization the water is passed through a second filter charged with hydroxyl (OH) ions, which replace carbonates, sulfates, and chlorides. Such de-ionized water may then have the proper nutritive ions added back in suitable amounts.

Boron salts are not removed by de-ionization units and if present in water in excess of 1 ppm they can cause plant injury. There is no satisfactory method for removing excess boron from water. The best solution is to acquire another water source.

Municipal treatment of water supplies with chlorine (0.1 to 0.6 ppm) or with sodium fluoride (1 to 2 ppm) is not sufficiently high to cause plant injury.

SOIL *pH*

Soil reaction (or pH) is a measure of the concentration of hydrogen ions in the soil. Although not directly influencing plant growth, it has a number of indirect effects, such as the availability of various nutrients and the activity of beneficial microbial activity. A pH range of 5.5 to 7.0 is best for growth of most plants (7.0 is neutral—below this is acid and above is alkaline). To lower pH of an alkaline soil, use ammonium sulfate fertilizer; to raise the pH of acid soils use calcium nitrate.

CARBON DIOXIDE (CO_2) FERTILIZATION IN THE GREENHOUSE (33, 34, 35, 65)

Carbon dioxide is one of the required ingredients for the basic photosynthetic process which accounts for the dry-weight materials produced by the plant:

$$6CO_2 + 12H_2O \xrightarrow[\text{green plant cell}]{\text{light energy}} C_6H_{12}O_6 + 6O_2 + 6H_2O$$

Carbon dioxide exists normally in the atmosphere at about 300 ppm (0.03 percent). Sometimes the concentration in winter in closed greenhouses may drop to 200 ppm, or lower, during the sunlight hours, owing to its use by the plants. Under conditions whereby CO_2 is limiting the photosynthetic rate (at adequate light intensities and relatively high temperatures—about 85° F [29.5° C]) an increase in the CO_2 concentration, to between 1000 and 2400 ppm, can be expected to result in an increase in photosynthesis, as much as 200 percent over the rate found at 300 ppm. To take full advantage of this potential increase in dry-weight production, plant spacing must be made adequate to prevent shading of overcrowded leaves. When supplementary CO_2 is used during periods of sunny weather, the growing temperature in the greenhouse should be kept relatively high. Added CO_2 would be of little benefit whenever the light intensity drops to very low levels. Adding CO_2 at night is of no value. Good circulation of air in the greenhouse will prevent undesirable lowering of the CO_2 level just at the leaf surface. A tightly closed greenhouse is necessary to be able to increase the ambient CO_2.

There are several possible sources of supplementary carbon dioxide. Gaseous CO_2 can be introduced through plastic-tube distribution systems from dry ice or from tanks of gaseous or liquid carbon dioxide. A much less expensive and apparently satisfactory system uses specially constructed gas-combustion units utilizing air from the outside atmosphere and burning relatively pure forms of natural gas, fuel oil, or propane to release CO_2 and water vapor as combustion products. Unless pure fuels are used, however, toxic products, as sulfur dioxide or ethylene, may be released.

During the winter, when greenhouses remain closed much of the time, especially with tight, plastic-covered houses, and with now-available sources of CO_2 free from contaminants, it is possible that fertilization of nursery plants in the greenhouse with CO_2 can develop into a useful practice to promote rapid growth of young rooted cuttings and seedlings. The economics of adding supplementary CO_2 must be carefully considered; the increased returns from the production of plants must offset the cost of supplying the CO_2 and of the control and distribution system. A point is reached with warm weather at which the amount of outside ventilation required to control the temperature will make it uneconomical to add CO_2.

Some plants, such as carnations, chrysanthemums, and snapdragons, have responded especially well to increased CO_2 levels; other plants have not.

In addition to increasing the CO_2 level, controlling temperature and light at optimum values can greatly accelerate plant growth (see Figure 2–8). Young plants held under such controlled conditions, as shown in Figure 2–9, make dramatic increases in growth and maturity (*35*).

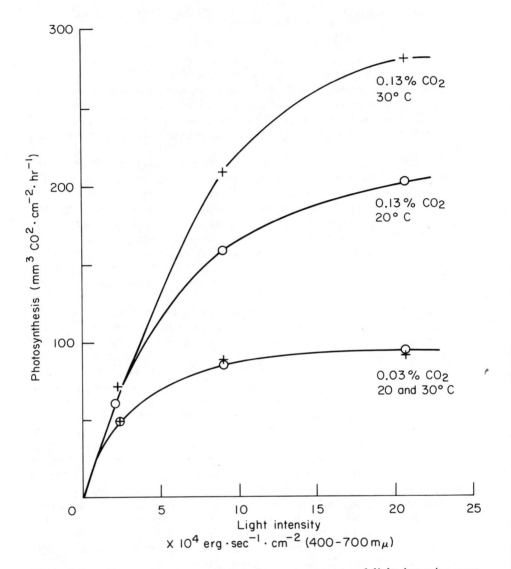

Figure 2-8   For maximum growth benefits, temperature and light intensity must be considered when increasing CO_2 levels. For highest photosynthetic rates, both light intensity and temperature should be high, as well as CO_2 *(upper curve)*. Increasing CO_2—but not temperature—*(middle curve)* does not give maximum benefits. At limiting CO_2 levels *(lower curve)*, neither increased temperature nor light intensity promotes photosynthesis. In fact, in this situation, increased temperature can be harmful due to higher respiration rates. Redrawn from Gaastra *(26)*.

Figure 2–9   Comparative growth of 'Pink Cascade' petunia plants under green-house and growth chamber conditions. Both plants are 35 days old. Plant on left was grown in the greenhouse under natural days at ambient CO_2. Plant on right was grown from seed in the growth chamber for 18 days under elevated tempera-tures (85/75°F day/night), high-intensity light (2000 ft-c of cool white fluorescent and supplemental incandescent light), and CO_2-enriched atmosphere (2000 ppm), and then kept in the greenhouse for 17 days. Courtesy Donald T. Krizek *(35)*.

CONTAINERS FOR GROWING YOUNG PLANTS
Flats
These are essentially shallow wood, plastic, or metal trays, with drainage holes in the bottom. They are useful for germinating seeds or rooting cuttings, since they permit young plants to be moved around easily when this is required. Durable wood, as cypress, cedar, or redwood, should be used for flats. Galvanized-iron and plastic flats are available in various sizes. Both types will nest, thus requiring relatively little storage space. In the trend toward mechanization, large "flats," several feet in dimensions, are built as pallets which can be moved about by forklift trucks after cuttings have rooted.

Clay Pots
The familiar red clay flower pots, long used for growing young plants, are porous, lose moisture readily, and are heavy. They are easily broken, and their round shape is not economical of space. After continued use, toxic salt accumulations build up, requiring soaking in water before reuse. How-ever, they can be steam sterilized, are reusable, and are a popular container.

Plastic Pots

Plastic pots, round and square, have numerous advantages; they are re-usable, lightweight, use little storage space since they will "nest," and are nonporous. Some types are fragile, however, and require careful handling, although other types, made from polyethylene, are flexible and quite sturdy. Square pots are also made up into "packs" of 8 or 12 for easier handling. Plastic pots (and flats) cannot be steam sterilized, but some of the more common plant pathogens can be eliminated by a hot water dip (158° F; 70° C) for 3 minutes without damage to the plastic.

Fiber Pots

Small pots 2 to 4 in. in size, round or square, are available pressed into shape from peat plus wood fiber, with fertilizer added. They are dry and will keep indefinitely. Such pots are biodegradable and are set right in the soil with the plant inside. Peat pots find their best use where plants are to be held for a relatively short time and then put in a larger container or in the field. Peat pots with plants growing in them will eventually deteriorate, owing to the constant moisture, and may fall apart when moved. On the other hand, unless the pots are kept moist, roots will fail to penetrate the walls of the pot and will grow into an undesirable spiral pattern. Units of six or twelve square peat pots fastened together are available. When large numbers of plants are involved, time and labor are saved in handling by the use of these units.

Larger-size pots made of molded asphalt-impregnated fiber are available for growing nursery stock; these can be used for direct planting as they also deteriorate in the soil.

Peat or Fiber Blocks

Blocks of solid material, sometimes with a prepunched hole, have become popular as a germinating medium for seeds or as a rooting medium for cuttings, especially for such plants as chrysanthemums and poinsettias. Fertilizers are usually incorporated into the material. One type (Figure 2–10) is made of highly compressed peat and, when water is added, swells to its usable size and is soft enough for the cutting or seed to be inserted. Such blocks essentially become a part of the plant unit and are set in the soil along with the plant. These blocks replace not only the pot but the propagating mix also.

Paraffined Paper and Styrofoam Cups

When these are punched with drainage holes, they make satisfactory temporary containers for growing and transferring young plants. They are cheap and lightweight, and require little storage space.

Asphalt-Coated Felt Paper Containers

These are available commercially in a number of sizes, and are suitable temporary plant containers. They are inexpensive, sturdy, lightweight,

Figure 2–10   Use of solid block rooting medium. *Left:* compressed sphagnum peat discs, encased in a plastic netting and containing some added mineral nutrients. *Right:* adding water causes peat to swell to size shown. Chrysanthemum cuttings inserted into full size pellets rooted rapidly and are ready to be planted in soil medium.

and easily nested. Containers can be constructed from asphalt-treated building paper, shaping the correct size paper pieces into square, open-bottom cups by placing them in removable "egg-crate" cells in a flat.

Metal Containers

Hundreds of thousands of nursery plants are grown and marketed each year in 1-gal. and—to a lesser extent—in 3-gal. and 5-gal. cans. Generally, these are used containers salvaged from canneries, large restaurants, and bakeries. Such cans are also available especially prepared for growing plants. One type is crimped and tapered for nesting, and has drainage holes punched. They are enameled to retard rusting. Machine planters have been developed, utilizing such containers, in which rooted cuttings or seedlings can be planted as rapidly as 10,000 or more a day. Plants are easily removed from tapered containers by inverting and tapping. Untapered metal containers must be cut down each side with can shears or tin snips to permit removal of the plant.

In areas having high summer temperatures, use of light-colored (white or aluminum) containers may improve root growth by avoiding heat damage to the roots, often encountered with dark-colored containers which will absorb considerable heat if exposed to the sun. In addition, soil temperatures in metal cans tend to be higher than that in plastic containers.

Polyethylene Bags

Small polyethylene bags with holes punched in the bottom for drainage may be filled with a porous rooting medium, such as perlite plus vermiculite, and placed in the propagating bench with the bag tops open. A single cutting of an easily rooted species such as chrysanthemum is placed in each bag. These may then be placed under mist (see p. 297) and when rooted, hardened off, then transferred in the bag for planting.

HANDLING CONTAINER-GROWN PLANTS

Watering of container-grown nursery stock is a major expense. Hand watering of individual cans with a large-volume, low-pressure applicator on a hose several times a week is expensive, and is used only for small-scale operations. In large operations, overhead or travelling sprinklers are sometimes used, although the runoff may be tremendous.

Fertilizer solutions are usually injected into the irrigation system (Figure 2–11). After the container stock leaves the nursery the retailer still has to maintain the stock by adequate irrigation and fertilization until the plants are set in the ground by the purchaser (25).

In areas with severe winters attention must be given to the problems of winter injury (37). The amount of injury will vary with the species, the less hardy ones generally requiring some type of mulch around the containers to protect the roots, especially from rapid temperature fluctuations. The chances of cold injury can be lessened by having the plants well established in the containers before the onset of winter. In addition, setting the plants close together in large groups tends to prevent the damaging rapid fluctuation of temperature. In some cases, laying the plants on their sides horizontally and letting them be covered with a snow mulch has given good protection. However, some form of winter protection should be used, such as one or more of the following: (a) Place a mulch covering, such as ground corncobs, straw, or hay, over the tops of the containers. (b) Place the containers inside a protective structure, such as a cold frame, hotbed, or greenhouse. (c) Plunge the cans into the soil; this will give good root protection but is expensive to do on a large scale (50). However, the most dependable winter protection is to construct, as shown in Figure 2–12, a temporary frame over the plants and cover it with polyethylene sheeting, 2 to 4 mils in thickness. Often two layers of polyethylene, separated by a 1 or 2 in. space, are used.

In most woody plant species the roots do not develop as much winter hardiness as the tops, hence low-temperature damage of container stock

Figure 2–11   Automatic watering system used for container plants in the greenhouse. A small "spaghetti" tube, inserted into a plastic water line, goes to each individual plant. Turning on one main valve can water hundreds of plants simultaneously. Rabbits will gnaw through small plastic tubing so out-of-door installations may have to be fenced or the tubes kept high, out of their runways. "Spaghetti"-size tubing should only be used in the greenhouse and larger tube systems used out-of-doors.

Figure 2–12 *Above:* winter protection of broad- and narrow-leaved evergreen nursery stock in severe winter areas. Plants are placed close together in beds and covered in late fall with 2 mil polyethylene on pipe framing. *Right:* a basin made from polyethylene is filled with water which releases heat upon freezing and absorbs heat upon thawing, acting as a buffer against sudden temperature changes. (Courtesy The Conard-Pyle Co., West Grove, Pa.)

occurs primarily to the root system. Winter-hardiness of roots varies with the species. Table 2–1 gives the minimum safe temperature and the killing temperatures for roots of some of the common ornamental species (*30*).

As shown in Figure 2–13, plants kept in containers too long will form an undesirable constricted root system from which they may never recover

Figure 2–13 One disadvantage of growing trees in containers is the possibility of producing poorly shaped root systems. Here a defective, twisted root system resulted from holding the young nursery tree too long in a container before transplanting. Such spiraling roots retain this shape after planting and are unable to firmly anchor the tree in the ground.

Table 2–1   The minimum safe root temperatures and killing temperatures for some common ornamental species over-wintered as container-grown stock. From data of Havis *(30)*.

	Safe Temperature		Killing Temperature	
Magnolia soulangeana	26°F	−3.5°C	23°F	−5.0°C
Magnolia stellata	26	−3.5	23	−5.0
Cornus florida	24	−4.5	20	−6.5
Daphne cneorum	24	−4.5	20	−6.5
Ilex crenata 'Convexa'	24	−4.5	20	−6.5
Ilex crenata 'Hetzi'	24	−4.5	20	−6.5
Ilex opaca	24	−4.5	20	−6.5
Ilex crenata 'Stokesi'	24	−4.5	20	−6.5
Cotoneaster horizontalis	22	−5.5	18	−7.8
Pyracantha coccinea	22	−5.5	18	−7.8
Cryptomeria japonica	20	−6.5	16	−9.0
Cotoneaster adpressa praecox	20	−6.5	16	−9.0
Viburnum carlesi	20	−6.5	15	−9.5
Cytisus praecox	20	−6.5	15	−9.5
Acer palmatum 'Atropurpureum'	17	−8.5	14	−10.0
Taxus media 'Nigra'	15	−9.5	10	−12.3
Azalea 'Gibraltar'	15	−9.5	10	−12.3
Azalea 'Hinodegiri'	15	−9.5	10	−12.3
Kalmia latifolia	15	−9.5	*	*
Leucothoe catesbaei	15	−9.5	*	*
Mahonia aquifolium	15	−9.5	*	*
Pieris floribunda	15	−9.5	*	*
Pieris japonica	15	−9.5	*	*
Rhododendron carolinianum	15	−9.5	*	*
Rhododendron catawbiense	15	−9.5	*	*
Thuja occidentalis	15	−9.5	*	*
Euonymus fortunei 'Colorata'	10	−12.2	5	−15.0
Juniperus horizontalis 'Plumosa'	10	−12.2	0	−17.8
Juniperus horizontalis 'Douglasi'	10	−12.2	0	−17.8
Rhododendron 'PJM'	10	−12.2	0	−17.8

* Not determined

when planted in their permanent location *(28)*. The plants should be shifted to larger containers before such "root spiralling" occurs. Judicious root pruning, early transplanting, and careful potting during the early transplanting stages can do much to develop a good root system by the time the young plant is ready for setting in its permanent location *(29)*.

Often when a container plant, grown in an almost synthetic, lightweight "soil" is transplanted into the home garden—usually into a much heavier loam or clay soil—a problem is encountered in maintaining sufficient moisture to the plant. If the root ball is covered with the heavier native soil, and water is then applied to the plant, possibly in a shallow basin at ground level, the water tends to stay in the heavier soil with its smaller, more absorptive, pores, while the coarser-textured mix containing the roots remains completely dry. To avoid this, the soil mix in the root ball should remain exposed at the top so that applied water must pass through it and so wet the roots.

REFERENCES

1 Aldrich, R. A., and P. E. Nelson, Equipment for aerated steam treatment of small quantities of soil and soil mixes, *Plant Dis. Rpt.,* 53(10):784–788. 1969.

2 Alvey, N. G., Soil for John Innes composts, *Jour. Hort. Sci.,* 36:228–40. 1961.

3 Anonymous, Seed and potting composts, *Minist. Agr., Fish., and Food (London) Advisory Leaf. 471,* 1964.

4 Baker, K. F., and C. N. Roistacher, Heat treatment of soil. Section 8 in *Calif. Agr. Exp. Sta. Man. 23,* 1957.

5 ———, Principles of heat treatment of soil. Section 9 in *Calif. Agr. Exp. Sta. Man. 23,* 1957.

6 ———, Equipment for heat treatment of soil. Section 10 in *Calif. Agr. Exp. Sta. Man. 23,* 1957.

7 ———, Principles of heat treatment of soil and planting material, *Jour. Australian Inst. Agr. Sci.,* 28(2):118–26. 1962.

8 ———, Treatment of soil by aerated steam, *Proc. Turf, Lands., and Nurs. Conf.,* pp. 18–1 to 18–3. Univ. of Calif. (Davis). 1966.

9 Benjamin, L. P., L. C. Chadwick, and K. W. Reisch, The effectiveness of urea-formaldehyde as a source of nitrogen for container-grown woody ornamental plants, *Proc. Amer. Soc. Hort. Sci.,* 84:636–47. 1964.

10 Birch, P. D. W., and D. J. Eagle, Toxicity to seedlings of nitrite in sterilized composts, *Jour. Hort. Sci.,* 44:321–30. 1969.

11 Boodley, J. W., and R. Sheldrake, Jr., Cornell "Peat-Lite" mixes for container growing, Dept. Flor. and Orn. Hort., Cornell Univ. Mimeo Rpt. 1964.

12 Boodley, J. W., C. F. Gortzig, R. W. Langhans, and J. W. Layer, Fertilizer proportioners for floriculture and nursery crop production management. *Cornell Ext. Bul.* 1175. 1966.

13 Brazelton, R. W., Sterilizing soil mixes with aerated steam, *Agric. Eng.,* 49(7): 400–401. 1968.

14 Cochrane, R. D., and O. A. Matkin, Nutrient supply characteristics of slow release fertilizers, *Plant Prop. (Int. Plant Prop. Soc.),* 12(4): 2–10, 1967.

15 Cotter, D. J., and J. N. Walker, Climate-humidity relationships in plastic greenhouses, *Proc. Amer. Soc. Hort. Sci.,* 89:584–93. 1966.

16 Courter, J. W., Home greenhouses for year-round gardening pleasure, *Ill. Agr. Ext. Ser. Cir. 879,* 1964.

17 ———, Plastic greenhouses, *Ill. Agr. Ext. Ser. Cir. 905,* 1965.

18 Creech, J. L., R. F. Dowdle, and W. O. Hawley, Sphagnum moss for plant propagation, *USDA Farmers' Bul. 2085,* 1955.

19 DeWerth, A. F., and R. C. Jaska, Greenhouse cooling, *Texas Agr. Exp. Sta. M. P. 163* (rev.), 1958.

20 Fireman, M., and H. E. Hayward, Irrigation water and saline and alkali soils, *USDA Yearbook of Agriculture—Water,* pp. 321–27. 1955.

21 ———, and R. L. Branson, Salinity in greenhouse soils, *Calif. Agr. Ext. Ser. OSA 68* (rev.), 1963.

22 Fleming, G., and C. E. Hess, The isolation of a damping-off inhibitor from sphagnum moss, *Proc. Inter. Plant Prop. Soc.,* 14:153–54. 1965.

23 Flint, H. L., Effects of different soil levels and methods of application of phosphorus on growth of selected woody ornamental species in containers, *Proc. Amer. Soc. Hort. Sci.,* 81:552–55. 1962.

24 Furuta, T., Soil mixtures, Section 6 in *Nursery Management Handbook, Univ. Calif. Agr. Ext. Serv.* 1970.

25 ———, Fertilizer and irrigation for plants on retail display, *Univ. Calif. Agr. Ext Serv., AXT–361,* 1971.

26 Gaastra, P., Climatic control of photosynthesis and respiration, in L. T. Evans (ed.), *Environmental Control of Plant Growth.* New York: Academic Press, 1963.

27 Griffin, R., R. Maire, and W. Humphrey, Sterilizing nursery soils with steam, *Calif. Agr. Ext. Serv. AXT–177,* 1965.

28 Harris, R. W., D. Long, and W. B. Davis, Root problems in nursery liner production, *Calif. Agr. Ext. AXT–244,* 1967.

29 Harris, R. W., W. B. Davis, N. W. Stice, and D. Long, Effects of root pruning and time of transplanting in nursery liner production, *Calif. Agr.,* 25(12):8–10. 1971.

30 Havis, J. T. Tolerance of plant roots in winter storage, *Amer. Nurs.* 139(1):10. 1974.

31 Kelley, J. D., Effects of over-fertilization on container-grown plants, *Proc. Plant Prop. Soc.,* 10:58–63. 1960.

32 ———, Response of container-grown woody ornamentals to fertilization with urea-formaldehyde and potassium frit, *Proc. Amer. Soc. Hort. Sci.,* 81:544–51. 1962.

33 Kohl, H. C., Carbon dioxide fertilization, *Proc. Inter. Plant Prop. Soc.,* 15:300–306. 1966.

34 Krizek, D. T., Controlled atmospheres for plant growth, *Trans. Amer. Soc. Agr. Eng.,* 13(3):237–68 1970.

35 ———, W. A. Bailey, H. H. Klueter, and H. M. Cathey, Controlled environments for seedling production, *Proc. Inter. Plant Prop. Soc.,* 18:273–80. 1968.

36 LaCroix, L. J., D. T. Canvin, and J. Walker, An evaluation of three fluorescent lamps as sources of light for plant growth, *Proc. Amer. Soc. Hort. Sci.,* 89:714–22. 1966.

37 Lanphear, F. O., Overwintering container-grown plants, *Amer. Nurs.,* 133(2): 40, 42, 44, 46. 1971.

38 Lechner, A., D. Sprague, E. Schaufler, B. Shalucha, R. Langhans, and P. Hammer, The Cornell automated plant grower, *Cornell Agr. Ext. Serv. Bul. 40.* 1972.

39 Lucas, R. E., P. E. Riecke, and R. S. Farnham, Peats for soil improvement and soil mixes, *Mich. Coop. Ext. Ser. Bull. No. E–516,* 1971.

40 Lunt, O. R., R. H. Sciaroni, and W. Enomoto, Organic matter and wettability for greenhouse soils. *Calif. Agr.,* 17(4):6. 1963.

41 Lunt, O. R., A. M. Kofranek, and S. B. Clark, Nutrient availability in soil. Availability of minerals from magnesium-ammonium-phosphates, *Agr. and Food Chem.,* 12:497–504. 1964.

42 Lunt, O. R., Controlled availability fertilizers, *Farm Tech.,* 21(4):11, 26. 1965.

43 Matkin, O. A., and P. A. Chandler, The U. C. type soil mixes. Section 5 in *Calif. Agr. Exp. Sta. Man. 23,* 1957.

44 Matkin, O. A., Soil mixes today, *Proc. Inter. Plant Prop. Soc.*, 21:162–63. 1971.

45 Munnecke, D. E., Chemical treatment of nursery soils. Section 11 in *Calif. Agr. Exp. Sta. Man. 23*, 1957.

46 Olsen, C. M., Aerated steam treatment of soils—its principles and application, *Proc. Inter. Plant Prop. Soc.*, 14:305–8. 1964.

47 Parsons, R. A., Small plastic greenhouses, *Univ. Calif. Agr. Ext. Serv. AXT–328*, 1971.

48 Patek, J. M., Peat moss, *Amer. Hort. Mag.*, 44:132–41. 1965.

49 Peterson, H., Low voltage heating, *N. Y. State Flower Growers' Bul. 115*, 1955.

50 Reisch, K. W., and L. C. Chadwick, Suggestions for over-wintering container-grown nursery stock, *Ohio Farm Home Res.*, 46(5):67, 79. 1961.

51 Richards, S. J., J. E. Warneke, and F. K. Aljibury, Physical properties of soil mixes used by nurseries, *Calif. Agr.*, 18(5):12–13. 1964.

52 Ritchie, R. M., Jr., A. A. Banadyga, and B. L. James, Plastic greenhouses, *N. Car. State Ext. Cir. 434*, 1964.

53 Sheldrake, R., Jr., and R. W. Langhans, Heating requirement of plastic greenhouses, *Proc. Amer. Soc. Hort. Sci.*, 80:666–69. 1962.

54 Sheldrake, R., Jr., The Cornell "Twenty-one" plastic greenhouse, *Veg. Crops Bul. 3*, Cornell Univ., Ithaca, N. Y. 1964.

55 Sheldrake, R., Jr., and J. W. Boodley, Commercial production of vegetable and flower plants, *Cornell Ext. Bul. 1056*. 1965.

56 Schoonover, W. R., and R. H. Sciaroni, The salinity problem in nurseries. Section 4 in *Calif. Agr. Exp. Sta. Man. 23*, 1957.

57 Stoutemeyer, V. T., and A. W. Close, Rooting cuttings and germinating seeds under fluorescent and cold cathode lighting, *Proc. Amer. Soc. Hort. Sci.*, 48:309–25. 1946.

58 Trickett, E. S., and J. D. S. Goulden, The radiation transmission and heat conserving properties of some plastic films, *Jour. Agr. Eng. Res.*, 3(4):281–87. 1958.

59 U.S.D.A., Hotbed and propagating frames, *USDA Misc. Pub. 986*, 1965.

60 Vaartaja, O., Chemical treatment of seedbeds to control nursery diseases, *Bot. Rev.*, 30:1–91. 1964.

61 Webster, G. C., R. A. Berner, and A. N. Gansa, The effect of biuret on protein synthesis in plants, *Plant Phys.*, 32:60–61. 1957.

62 White, D. P., and B. G. Ellis, Nature and action of slow release fertilizers as nutrient sources for forest tree seedlings, *Mich. Quart. Bul.*, 47(4):606–14. 1965.

63 White, J. W., Interaction of nitrogenous fertilizers and steam on soil chemicals and carnation growth, *Jour. Amer. Soc. Hort. Sci.*, 96(2):134–37. 1971.

64 Wiebe, J., Phytotoxicity as a result of heat treatment of soil, *Proc. Amer. Soc. Hort. Sci.*, 72:331–38. 1958.

65 Wittwer, S. H., and W. Robb, Carbon dioxide enrichment of greenhouse atmospheres for food crop production, *Econ. Bot.*, 18:34–56. 1964.

SUPPLEMENTARY READING

Anonymous, *Gardening in Containers*. Menlo Park, Ca.: Lane Magazine and Book Company. 1967.

Baker, K. F. (ed.), "The U.C. System for Producing Healthy Container-Grown Plants," *California Agricultural Experiment Station Manual 23,* 1957.

Ball, V. (ed.), *The Ball Red Book,* 12th ed. West Chicago, Ill.: Geo. J. Ball, Inc., 1972.

Bewley, W. F., *Commercial Glasshouse Crops,* 2nd ed. London: Country Life, 1963.

Brooklyn Botanic Garden, Greenhouse Handbook for the Amateur, *Plants & Gardens* 19(2):1–97. 1970.

Courter, J. W., "Plastic Greenhouses," *Illinois Agricultural Extension Service Circular 905,* 1965.

Furuta, T., *Nursery Management Handbook,* Berkeley: University of California Agricultural Extension Service. 1970.

Krüssman, G., *Die Baumschule (The Nursery)* (in German), 3rd ed. Berlin: P. Parey, 1964.

Laurie, A., D. C. Kiplinger, and K. S. Nelson, *Commercial Flower Forcing,* 7th ed. New York: McGraw-Hill Book Company, 1968.

McGuire, J. J., "Growing Ornamental Plants in Containers: A Handbook for the Nurseryman," *University of Rhode Island Cooperative Extension Service Bulletin 197,* 1972.

Nelson, K. S., *Flower and Plant Production in the Greenhouse*. Danville, Ill.: Interstate Printers & Publishers, Inc., 1966.

Patterson, J. M., *Container Growing,* Chicago: American Nurseryman, 1969.

Renard, W., Chairman, "Symposium on Glasshouse Construction and Automation," *International Society for Horticulture Science, Technical Communication No. 2,* 1965.

Toovey, F. W., et al., "Commercial Glasshouses," *Ministry of Agriculture, Fisheries, and Food (London) Bulletin No. 115,* 1964.

The Development of Fruits, Seeds, and Spores

3

PRODUCTION OF THE FLOWER

The life cycle of a seed-bearing plant is divided into two broad phases, vegetative and reproductive (flowering), whether it starts in a sexual or an asexual cycle. The plant is first vegetative, during which phase the predominant processes are elongation of the stem and roots and increase in plant cross-sectional area. When first grown from seed, while the plant is in the juvenile phase, it appears that it must grow to a given size or age to attain the ability to flower. Once past the juvenile stage, the vegetative plant may respond to flowering stimuli at any age.

The initiation of flowering marks the end of the vegetative phase. The flowering phase begins with *flower induction,* an internal physiological change that occurs prior to any morphological change. Some of the growing points of the stem then proceed to develop into flowers rather than remaining vegetative growing points. Very often flower induction results from environmental signals that are received by the plant which induce prerequisite changes. Such signals may be a certain amount of chilling (*vernalization*) or longer or shorter day lengths. In many instances woody plants that have passed the juvenile period may be induced to flower by girdling, ringing, heading-back shoots, root pruning, or grafting onto dwarfing or restrictive rootstocks. Removing strips of bark, inverting, and regrafting them may stimulate flowering, as may tying slender branches into knots.

Following induction, the growing points develop flower parts and eventually a flower is produced (see Chapter 1). The sequence during flower differentiation and development varies with different kinds of plants and with particular environmental cycles. For example, in biennials and in some woody perennials the development of flowers takes place in the elongating stem in the growth period following vernalization. Among other plants, particularly many woody plants, differentiation begins in the warm period of one growing season, but the buds must then be subjected to a particular regime of chilling temperatures in order for normal progression of development to occur. Many spring flowering plants of the temperate zone behave in this manner.

PRODUCTION OF THE EMBRYO

The embryo is the end result of the sexual cycle, shown diagrammatically in Figure 1–4 (see p. 10). These events occur in the flower, the basic structure of which is shown diagrammatically in Figure 3–1. During flowering, pollen is transferred from the anther to the stigma (*pollination*); it germinates and grows down the style until it reaches the embryo sac. (Figure 3–2.) Male gametes from the pollen tube unite with female gametes in the embryo sac (*fertilization*) to produce the embryo and endosperm.

To produce a viable seed, both pollination and fertilization must take place. In some cases, however, the fruit may mature and contain only shriveled and empty seed coats with no embryo or with one that is thin and shrunken. Such "seedlessness" may result from several causes: (a) *parthenocarpy* (the development of the fruit without pollination or fertilization); (b) *embryo abortion* (the death of the embryo during its development); or (c) the inability of the embryo to accumulate the required food reserves. If embryo abortion occurs early, it is most likely that the fruit will soon drop or will not grow to its normal size. (*20*).

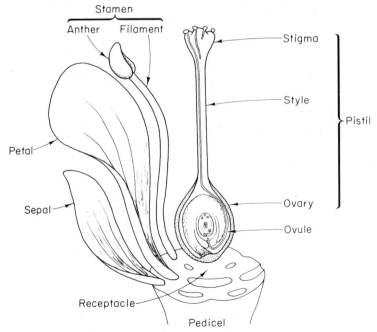

Figure 3–1   Diagram of basic flower structure of an angiospermous plant.

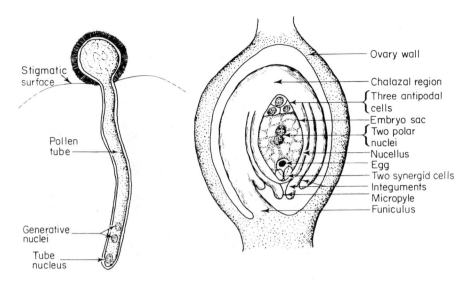

Figure 3–2   *Left:* pollen grain germinating on stigma. *Right:* an ovary showing a single ovule with a mature embryo sac and its eight nuclei. An angiospermous plant has double fertilization. One of the generative nuclei of the pollen grain combines with the egg nucleus to form the zygote, which develops into the embryo; the other combines with the two polar nuclei to form the triploid endosperm.

FRUIT AND SEED DEVELOPMENT
Morphological Development

The relationship between the structure of the flower and the structure of the fruit and seed in angiosperms is as follows:

Ovary ··> fruit (sometimes composed of more than
                                                          one ovary plus additional tissues)
  Ovule ··> seed (sometimes coalesces with fruit)
    integuments ·····························> testa (seed coats)
    nucellus ································> perisperm (usually absent or reduced;
    2 polar nuclei +   sometimes storage tissue)
      sperm nucleus ····················> endosperm (triploid–3N)
    egg nucleus + sperm
      nucleus ·······>   zygote ·······> embryo (diploid–2N)

Figures 3–3 and 3–4 show the growth relationship among the different parts of a seed and fruit during development in an angiosperm plant. The embryo develops from the zygote initially as a microscopically small mass of cells embedded in the endosperm, which is in turn embedded in the nucellus. The embryo undergoes continuous morphological and physiological development. Three broad stages of development can be observed in

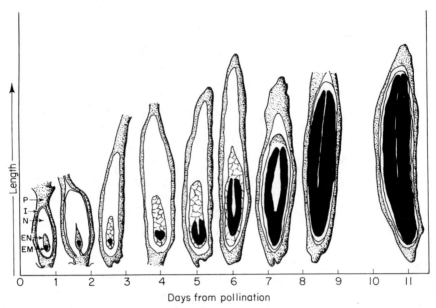

Figure 3–3 Growth of the fruit (an indehiscent achene) and seed of lettuce. P—pericarp, I—integuments, N—nucellus, EN—endosperm, EM—embryo. Redrawn from Jones *(15).*

embryos of most plants *(32)*. In the early *proembryo* stage, the embryo is very small (microscopic) and globular in shape; it becomes *heart*-shaped as cotyledons begin to develop. This period is indicated by the 1– to 4–day lettuce embryo in Figure 3–3 or the 10– to 12–day *Datura* embryo in Figure 3–4. In the second stage the cotyledons continue to enlarge and the embryo becomes *torpedo*-shaped with its length increasing to nearly fill the seed. Five- to eight-day lettuce embryos (Figure 3–3) and 15– to 20–day *Datura* embryos (Figure 3–4) illustrate this stage. In the third stage, the embryos are full-sized morphologically and are increasing in weight.

The endosperm functions in a nutritional capacity for the embryo, although no vascular connections exist between the two structures *(4, 22, 32)*. Growth of the embryo is preceded by growth of the endosperm, which digests the inner nucellar tissue as it grows. The endosperm is, in turn, digested by the developing embryo, although in some species a large part of the endosperm remains in the mature seed.

Failure of the endosperm to develop properly results in retardation or arrest of embryo development, and embryo abortion can result. This phenomenon commonly occurs when two genetically different individuals are hybridized, either from two different species *(9, 25, 32)* or from two individuals of different polyploid constitution *(10)* . It therefore can be a barrier to hybridization. If such embryos are excised from the developing fruit before they abort, they can sometimes be saved by being grown in aseptic culture *(19, 26)* (see Chapter 16).

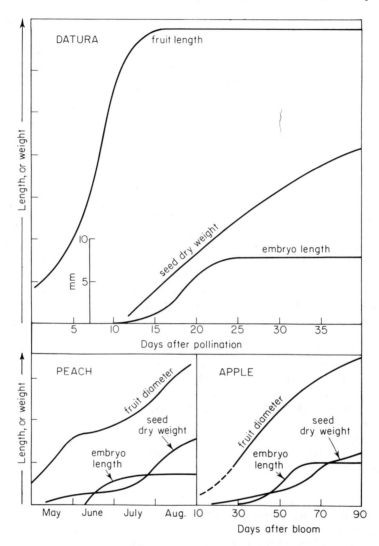

Figure 3–4 Comparative development of three different types of fruits: *Datura* (dry capsule), peach (fleshy drupe), and apple (fleshy pome). Note that changes in seed dry weight and embryo length are essentially the same for all species shown. *Datura* data from Rietsema et al. *(23);* apple data in part from Luckwill *(16).*

Seed development in the gymnosperms is somewhat different. The ovules are exposed within the ovular cone and are not enclosed in an ovary. The ovular cone develops into the mature cone. Gymnosperms do not undergo double fertilization. The embryo is nourished by the haploid female gametophyte and is not dependent upon a triploid endosperm *(27).* (See p. 11.)

Various external agents can prevent embryo development, even though the fruit itself continues to develop. Numerous kinds of insects (*3*) attack the developing seed and fruit, particularly in forest trees. The developing seed of carrot and other Umbelliferous plants is attacked by *Lygus* bugs which can penetrate the fruit and feed on the embryo. Adverse weather, such as frosts during early fruit development, sometimes kills the embryo, but the fruit itself continues to develop (*12, 28*). Growth tensions, causing the hardening endocarp to split or crack, can result in abortion of stone fruit seeds (*11*). This condition is associated with excessive vigor, reduced crop density, root restriction, or tree injuries resulting in girdling.

Accumulation of Food Reserves in the Seed

The accumulation of storage materials in the seed can be measured by changes in dry weight of the seed, although in the earlier part of the fruit development period, increase in weight may occur because of increase in size. Later, when the seed has attained its full size, the increase in dry weight is a measure of this accumulative process (*32*). These reserve materials originate as carbohydrates produced by photosynthesis in the leaves and translocated to the fruits and seeds, where they are converted to complex storage products—carbohydrates, fats, and proteins (*31*). This process largely takes place in the final developmental periods of fruit growth.

This accumulative process must take place properly if high-quality seeds are to be produced. Such seeds should be plump and heavy for their size. Since the initial growth of the seedling depends upon these reserve materials, heavier seeds should result in better germination and produce more vigorous seedlings. If conditions interfere with this storing process so that few reserve materials accumulate, the seeds will be thin, shrunken, and light in weight. The more severe this condition, the less the seeds can survive storage periods, the poorer the germination, and the weaker are the seedlings that are produced.

Ripening and Dissemination of Seeds

Specific physical and chemical changes take place during ripening which lead to the senescence of the fruit and the dissemination of the seed. One of the most obvious changes is the drying of the fruit tissues. In certain fruits, this leads to dehiscence and the discharge of the seeds from the fruit. Changes may take place in the color of the fruit and the seed coats, and softening of the fruit may occur.

Seed dispersal is accomplished by many agents. Fish, birds, rodents, and bats consume and carry seeds in their digestive tract. Fruits with spines or hooks become attached to the fur of animals and are often moved considerable distances. Wind dispersal of seed is facilitated in many plant groups by "wings" on dry fruits; tumbleweeds can move long distances by rolling in the wind. Seeds carried by moving water—streams or irrigation canals—can be taken great distances and often become a source of weeds in cultivated fields. Some plants themselves have mechanisms for short-distance dispersal, such as explosive liberation of spores from fern sporangia. The activities of man in purposeful shipment of seed lots all over the world are, of course, of considerable effectiveness in seed dispersal (*30*).

THE MATURE SEED

Botanically, in the angiosperms the seed is a matured ovule enclosed within the ovary, or fruit. Seeds and fruits of different species vary greatly in appearance; size; shape; location and structure of the embryo; and presence of storage tissues (Figure 3–5). These points are useful in identification (*13, 18*).

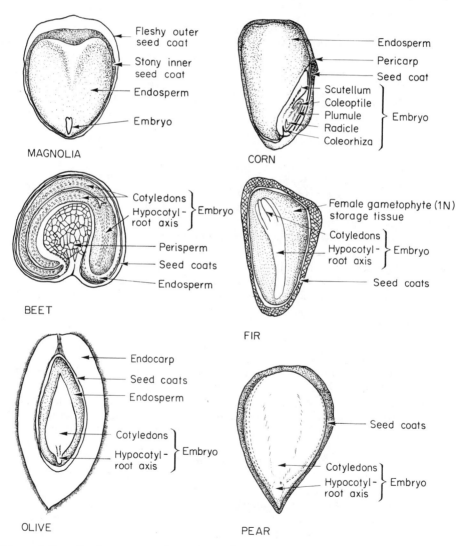

Figure 3–5   Seed structure of representative species.

From the standpoint of seed handling, it is not always possible to separate the fruit and seed, since they are sometimes joined in a single unit. In such cases, the fruit itself is treated as the "seed," as in wheat or corn.

Parts of the Seed

The seed has three basic parts: (a) embryo, (b) food storage tissues, and (c) seed coverings.

Embryo

The embryo is a new plant resulting from the union of a male and female gamete during fertilization. Its basic structure consists of an *axis* with growing points at each end, one for the shoot and one for the root, and one or more seed leaves (*cotyledons*) attached to the embryo axis. Plants are classified by the number of cotyledons. Monocotyledonous plants (such as the grasses or onion) have a single cotyledon, dicotyledonous plants (such as the bean or peach) have two, and gymnosperms (such as the pine or *Ginkgo*) may have as many as fifteen.

Storage Tissues

The storage tissues of the seed may be the cotyledons, the endosperm, the perisperm, or, in gymnosperms, the haploid female gametophyte. Seeds in which the endosperm is large and contains most of the stored food are referred to as *albuminous* seeds; seeds in which the endosperm is lacking or reduced to a thin layer surrounding the embryo are referred to as *exalbuminous* seeds. In the latter case, the stored food is usually within the cotyledons, the endosperm being digested by the embryo during development. The *perisperm,* originating from the nucellus, occurs in only a few families of plants, such as the Chenopodiaceae and the Caryophyllaceae. Usually it is digested by the developing endosperm during seed formation.

Seed Coverings

The seed coverings may consist of the seed coats, the remains of the nucellus and endosperm, and sometimes parts of the fruit. The seed coats, or *testa,* usually one or two (rarely three) in number, are derived from the integuments of the ovule. During development the seed coats become modified, and at maturity they present a characteristic appearance. Usually the outer seed coat becomes dry, somewhat hardened and thickened, and brownish or otherwise colored. On the other hand, the inner seed coat will usually be thin, transparent, and membranous. Remnants of the endosperm and nucellus are found within the inner seed coat, sometimes making a distinct, continuous layer around the embryo.

In some plants, parts of the fruit remain attached to the seed so that the fruit and seed are commonly handled together as the "seed." In certain kinds of fruits, e.g., achenes, caryopsis, samaras, and schizocarps, the fruit and seed layers are contiguous. In others, such as the acorn, the fruit and seed coverings separate but the fruit covering is indehiscent. In still others, such as the "pit" of stone fruits or the shell of walnuts, the covering is a hardened portion of the pericarp, but it is dehiscent and can usually be removed without much difficulty.

The seed coverings provide mechanical protection for the embryo, making it possible to handle seeds without injury, and thus permitting transportation for long distances and storage for long periods of time. The seed coverings can also play an important role in influencing germination, as discussed in Chapter 6.

APOMIXIS

In some species embryos are produced not as a result of meiosis and fertilization but by certain asexual processes as described below. The occurrence of an asexual reproductive process in place of the normal sexual reproductive processes of reduction division and fertilization is known as *apomixis*. Seedling plants produced in this manner are known as *apomicts*. Plants that produce only apomictic embryos are known as *obligate apomicts;* those that produce both apomictic and sexual embryos are *facultative apomicts*.

Types of Apomixis (1, 2, 14, 17, 21)

Recurrent Apomixis
An embryo sac (female gametophyte) develops from the egg mother cell (or from some adjoining cell, the egg mother cell disintegrating), but complete meiosis does not occur. Consequently, the egg has the normal diploid number of chromosomes, the same as the mother plant. The embryo subsequently develops directly from the egg nucleus without fertilization.

This series of events is known to occur in some species of *Crepis, Taraxacum* (dandelion), *Poa* (bluegrass), and *Allium* (onion) without the stimulus of pollination; in others. e.g., species of *Parthenium* (guayule), *Rubus* (raspberry), *Malus* (apple), some *Poa* species, and *Rudbeckia,* pollination appears to be necessary, either to stimulate embryo development or to produce a viable endosperm.

Adventitious Embryony
This is also known as *nucellar embryony* or *nucellar budding*. In this type of apomixis the embryos rise from a cell or group of cells either in the nucellus (usually) or in the integuments, as shown in Figure 3-6. It differs from recurrent apomixis in that such embryos develop outside of the embryo sac and in addition to the regular embryo. In some plants, *Citrus* for instance, fertilization takes place normally, and a sexual plus several apomictic embryos may develop. In others, e.g., some species of *Opuntia,* embryos develop spontaneously, no pollination or fertilization apparently needed.

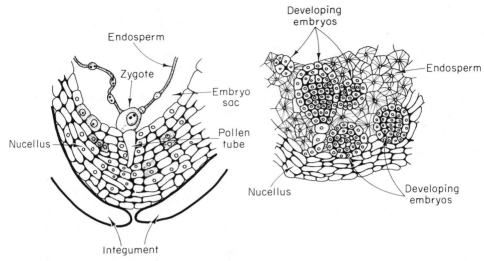

Figure 3–6   The development of nucellar embryos in Citrus. *Left:* stage of development just after fertilization showing zygote and remains of pollen tube. Note individual active cells (those shaded) of the nucellus which are in the initial stages of nucellar embryony. *Right:* a later stage showing developing nucellar embryos. The large one may be a sexual embryo. Redrawn from Gustafsson *(14)*.

Nonrecurrent Apomixis
In this case an embryo arises directly from the egg nucleus without fertilization. Since the egg is haploid, the resulting embryo will also be haploid. This case is rare and primarily of genetic interest. It does not consistently occur in any particular kind of plant as do recurrent apomixis and adventitious embryony.

Vegetative Apomixis
In some cases vegetative buds or *bulbils* are produced in the inflorescence in place of flowers. This occurs in *Poa bulbosa* and some *Allium, Agave,* and grass species.

Polyembryony
The phenomenon in which two or more embryos are present within a single seed is called *polyembryony (2, 17)*. It may result from one of several causes. Nucellar embryony, as described for *Citrus,* is one cause (see Figure 3–7). The occasional development of more than one nucleus within the embryo sac (in addition to the egg nucleus) is another. Cleavage of the proembryo during the early stages of development is a common occurrence in conifers and leads to multiple embryos.

Significance of Apomixis

Apomixis occurs in nature in many different plant families *(14, 21)*. Usually polyploids or complex hybrids are involved. Apomixis occurs in some kinds of plant cultivars. The significance of apomixis is twofold. It provides a means of assuring uniformity in seed propagation since any apomictic cultivar is actually a type of clone. Many *Citrus* species and cultivars produce apomictic seedlings which are used as rootstocks. These invariably are uniform and vigorous in contrast to the occasional weaker, variable hybrid seedlings that may also occur. *Malus toringoides* and *M. sikkimensis* are apple species which are apomictic if self-pollinated, although they pro-

Figure 3–7 Polyembryony in citrus seeds as indicated by two seedlings from each seed. One or both could be a nucellar seedling.

duce large proportions of hybrid seeds if cross-pollinated *(8)*. Some grass species are facultative apomicts. For instance, Kennedy bluegrass (*Poa pratensis*) plants consist of both apomicts and sexually reproduced individuals; in one study *(5)* 85 percent of such plants were found to be apomictically reproduced. Certain pasture grasses, such as 'King Ranch' bluestem (*Andropogon*), 'Argentine' Bahiagrass (*Paspalum notatum*), and 'Tucson' side oats grama (*Bouteloua curtipendula*), are examples of apomictic cultivars *(6)*.

A second significance of apomixis is that many virus diseases are not transmitted by seed. Consequently, growing apomictic seedlings provides a means of rejuvenating an old clone that has become infected with virus diseases *(7)*. This procedure has been developed most extensively in improving citrus cultivars (see p. 540). These nucellar seedlings are in a juvenile state and are extremely vigorous and thorny; such plants are not suitable horticulturally until they attain the adult state (see p. 184).

SPORE DEVELOPMENT

Reproduction by spores occurs in ferns *(24)*. Two separate stages or generations are involved as shown in Figure 3–8. One is the asexual, or *sporophyte* generation, in which the plant has conspicuous roots, stems, and leaves (fronds). The other is the sexual, or *gametophyte,* generation, in which the plant is small; inconspicuous; without roots, stems, or leaves; and is called a *prothallium.* Spores are produced on the underside of fronds in clusters of *sporangia,* or spore cases, that appear as brownish "dots." Groups of sporangia are called *sori* and sometimes have a cover known as the *indusium.* Within each sporangium 16 *spore mother cells* form, each of which undergoes meiosis to produce four *spores.* Each spore is haploid, with half the normal chromosome number of the species.

Spores are discharged and under favorable temperature and moisture conditions "germinate" to produce the *prothallus,* a flat green plate of cells

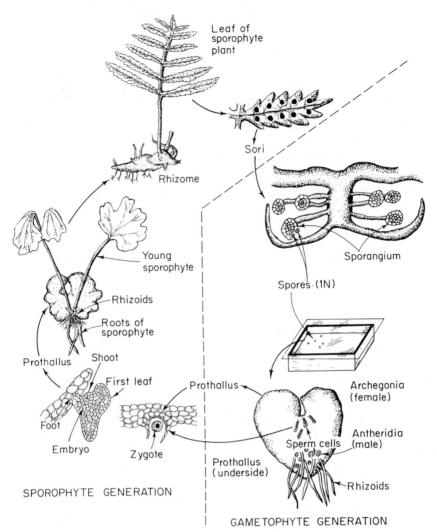

Figure 3–8 Development of spores in the reproduction cycle of a fern. See text for explanation.

with small rootlike structures (*rhizoids*). This grows to about ¼ in. in diameter in about 3 months. Male (*antheridia*) and female (*archegonia*) structures are produced on the underside of the same prothallus (except in two economically unimportant families). Sperm cells are discharged and, in the presence of water, are attracted into the archegonium to fuse with the egg and subsequently produce a zygote.

The *zygote* develops into an embryo, which grows to produce the fern plant (sporophyte), which has the diploid chromosome number for the species. In its initial development, the embryo develops a *foot*, through which it absorbs moisture and nutrients from the prothallus. A *root* is produced which grows downward into the soil. Also produced is the *primary leaf*, which acts as a temporary photosynthetic organ, and the *stem*, which develops into the rhizome from which fronds and permanent roots arise.

Propagation procedures for ferns are given in Chapter 19.

REFERENCES

1   Battaglia, E., Apomixis, in P. Maheshwari (ed.), *Recent Advances in the Embryology of Angiosperms.* Delhi, India: Univ. of Delhi, Int. Soc. of Plant Morph., 1963, pp. 221–64.

2   Bhatnager, S. P., and B. M. Johri, Development of angiosperm seeds, in T. T. Kozlowski (ed.), *Seed Biology,* Vol. I. New York: Academic Press, 1972, pp. 77–149.

3   Bohart, G. E., and T. W. Koerber, Insects and seed production, in T. T. Kozlowski (ed.), *Seed Biology,* Vol. III. New York: Academic Press, 1972, pp. 1–54.

4   Brink, R. A., and D. C. Cooper, The endosperm in seed development, *Bot. Rev.,* 13:423–541. 1947.

5   Brittingham, W. H., Type of seed formation as indicated by the nature and extent of variation in Kentucky bluegrass, and its practical application, *Jour. Agr. Res.,* 67:255–64. 1943.

6   Burton, G. W., and G. F. Sprague, Use of hybrid vigor in plant improvement, in R. E. Hodgson (ed.), *Germ Plasm Resources.* Amer. Assoc. Adv. Sci. Publ. No. 66. Washington, D.C., 1961, pp. 191–203.

7   Cameron, J. W., R. K. Soost, and H. B. Frost, The horticultural significance of nucellar embryony in citrus, in J. Wallace (ed.), *Citrus Virus Diseases.* Berkeley: Univ. of Calif., Division of Agr. Sci., 1959, pp. 191–96.

8   Campbell, A. J., and D. Wilson, Apomictic seedling rootstocks for apples: progress report, III. *Ann. Rept. Long Ashton Hort. Res. Sta. (1961):* 68–70. 1962.

9   Cooper, D. C., and R. A. Brink, Somatoplastic sterility as a cause of seed failure after interspecific hybridization, *Genetics,* 25:593–617. 1940.

10   ———, Seed collapse following matings between diploid and tetraploid races of *Lycopersicon pimpinellifolium, Genetics,* 30:375–401. 1945.

11   Davis, L. D., Size of aborted embryos in the Phillips Cling peach. *Proc. Amer. Soc. Hort. Sci.,* 37:198–202. 1939.

12   Ehrenberg, C., A. Gustafsson, C. P. Forshell, and M. Simak, Seed quality and the principles of forest genetics. *Hereditas,* 41:291–366. 1955.

13   Gunn, C. R., Seed collecting and identification, in T. T. Kozlowski (ed.), *Seed Biology,* Vol. III. New York: Academic Press, 1972, pp. 55–144.

14   Gustafsson, A., Apomixis in higher plants, Parts I–III. *Lunds Univ. Arsskrift,* N.F. Avid. 2 Bd 42, Nr. 3:42(2); 43(2); 43(12). 1946–1947.

15   Jones, H. A., Pollination and life history studies of the lettuce *(Lactuca sativa* L.), *Hilgardia,* 2:425–79. 1927.

16   Luckwill, L. C., The hormone content of the seed in relation to endosperm development and fruit drop in the apple, *Jour. Hort. Sci.,* 24:32–44. 1948.

17   Maheshwari, P., and R. C. Sachar, Polyembryony, in P. Maheshwari (ed.), *Recent Advances in the Embryology of Angiosperms.* Delhi, India: Univ. of Delhi, Int. Soc. of Plant Morph., 1963, pp. 265–96.

18   Martin, A. C., The comparative internal morphology of seeds, *Amer. Midland Nat.,* 36:512–660. 1946.

19   McLean, S. W., Interspecific crosses involving *Datura ceratocaula* obtained by embryo dissection, *Amer. Jour. Bot.,* 33:630–38. 1946.

20   Nitsch, J. P., Perennation through seeds and other structures, in F. C. Steward

(ed.), *Plant Physiology*, Vol. VIA. New York: Academic Press, 1971.

21   Nygren, A. Apomixis in the angiosperms II, *Bot. Rev.*, 20:577–649. 1954.

22   Rangaswamy, N. S., Control of fertilization and embryo development, in P. Maheshwari (ed.), *Recent Advances in the Embryology of Angiosperms.* Delhi, India: Univ. of Delhi, Int. Soc. of Plant Morph., 1963, pp. 327–49.

23   Rietsema, J., S. Satina, and A. F. Blakeslee, Studies on ovule and embryo growth in *Datura.* I. Growth Analysis, *Amer. Jour. Bot.*, 42:449–54. 1955.

24   Roberts, D. J., Modern propagation of ferns, *Proc. Int. Plant Prop. Soc.*, 15:317–22. 1965.

25   Sanders, M. E., Embryo development in four *Datura* species following self and hybrid pollinations, *Amer. Jour. Bot.*, 35:525–32. 1948.

26   Sanders, M. E., and P. R. Burkholder, Influence of amino acids on growth of *Datura* embryos in culture, *Proc. Natl. Acad. Sci.* (U.S.), 34:516–26. 1948.

27   Singh, H., and B. M. Johri, Development of gymnosperm seeds, in T. T. Kozlowski (ed.), *Seed Biology,* Vol. I:22–77. 1972.

28   Shepherd, P. H., The kernel told the tale. *Amer. Fruit Grower,* 75:37. 1955.

29   Stoutemyer, V. T., Juvenility and flowering potential in woody plants, *Amer. Hort. Mag.,* 43:161–67. 1964.

30   Van der Pijl, L., *Principles of Dispersal in Higher Plants,* Berlin: Springer-Verlag, 1969.

31   Varner, J. E., Seed development and germination, in J. Bonner and J. E. Varner (eds.), *Plant Biochemistry.* New York: Academic Press, Inc., 1965, Chap. 29.

32   Wardlaw, C. W., *Physiology of embryonic development in cormophytes,* in *Encyclopedia of Plant Physiology,* Vol. XV, 1965, pp. 844–965.

SUPPLEMENTARY READING

Free, J. B., *Insect Pollination of Crops.* New York: Academic Press, 1970.

Fruit Committee of Horticultural Education Association, *The Pollination of Fruit Crops;* reprint from *Scientific Horticulture,* XIV and XV, pp. 1–68. Canterbury, England: Gibbs and Sons, 1959, 1960.

Heslop-Harrison, J., Sexuality of angiosperms, Chap. 9 in F. C. Steward (ed.), *Plant Physiology,* Vol. VIC. New York: Academic Press, 1972, pp. 134–289.

Kozlowski, T. T. (ed.), *Seed Biology,* Vol. I: *Importance, Development and Germination.* New York: Academic Press, 1972.

Martin, A. C., and W. D. Barkley, *Seed Identification Manual.* Berkeley: Univ. of Calif. Press, 1961.

Musil, A. F., *Identification of Crop and Weed Seeds.* U.S. Department of Agriculture Handbook 219. Washington, D.C.: U.S. Government Printing Office, 1963.

Nitsch, J. P., Perennation through seeds and other structures, Chap. 4 in F. C. Steward (ed.), *Plant Physiology,* Vol. VIA. New York: Academic Press, 1971.

U.S. Department of Agriculture, *Seeds. Yearbook of Agriculture.* Washington, D.C.: U.S. Government Printing Office, 1961.

Production of Genetically Pure Seed

4

Higher plants reproduce naturally by seeds. One characteristic of seed reproduction is the variation which can exist within groups of seedlings. In nature variability is important, since it provides the genetic material that allows the continued adaptation of a particular species to the environment; those individuals within each generation that are best adapted to that environment tend to survive and produce the next generation. Propagation of cultivated plants, however, requires that variability during seed reproduction be controlled or the value of a cultivar may be lost. Characteristics important for selection may not be consistently perpetuated into the next generation unless the principles and procedures as outlined in this chapter are followed.

USES OF SEEDLINGS IN PROPAGATION

(a) The growing of seedlings is used in propagating more species and cultivars than any other propagation method. Annuals and biennials must be grown from seed, and account for essentially all of the grains, most of the vegetables, and many garden and florist's plants (see Chapter 19). For the production of ornamental bedding plants there is a very large propagation industry, which relies largely on seeds. Herbaceous and woody perennials may also be grown commercially from seed, even though vegetative methods might also be available. To reproduce species and botanical varieties, seeds are used. To reproduce cultivars, seeds are used for those in which variability can be controlled, or where either alternate vegetative methods are not feasible or mass production by such a method would be uneconomical and impractical. For example, trees and shrubs grown for reforestation, woodlots, wildlife cover, and roadsides are primarily seedlings, as are most forage and turf and pasture grasses. Some trees and shrubs used for landscape purposes, and some fruit and nut crops, are grown as seedlings in some parts of the world. However, many if not most cultivars in the fruit, nut, and ornamental plant category are clones that are vegetatively propagated.

(b) Seedlings are extensively used in nurseries to provide rootstocks upon which to graft or bud selected clones of fruit and ornamental plants (see p. 182).

(c) In plant breeding, growing seedlings is the most important means of developing new cultivars. This is due to the variability which results from segregation attending sexual reproduction.

POLLINATION REQUIREMENTS OF PLANTS

Pollination is the transfer of pollen from an anther to the stigma of a flower. A pollen tube grows down the style into the ovule, where fertilization takes place. Seed development may result from (a) *self-pollination,* in which the pollen may come from the same flower, from different flowers on the same plant, or from different plants of the same clone; (b) *cross-pollination,* in which pollen comes from a different plant or from a different clone; or (c) *apomixis,* in which an asexual process is substituted for a sexual one (see p. 61). In addition, in some plants, either self- or cross-pollination can occur, the amount of either varying with the kind of plant and environmental factors. In some cases, seed production is partially sexual and partially apomictic. It is important to know which process predominates in any plant being grown for the production of seed (*18*).

Self-Pollinated Plants

Cultivars of self-pollinated crop plants can generally be maintained by seed propagation with little difficulty even when grown in close proximity to other closely related cultivars. Cross-pollination is usually less than 4 percent (*18*).

The ability of self-pollinated cultivars to be maintained readily by seed propagation is due to the fact that they are largely homozygous. With self-pollination, offspring of homozygous plants are also homozygous and inherit the characteristics of the parent. On the other hand, heterozygous plants, if self-pollinated, segregate (for instance, like the tall and dwarf offspring in Figure 1–6). However, with continued self-pollination the proportion of homozygous plants increases and that of the heterozygous decreases by a factor of one-half each generation. After six to ten generations, the group of descendants from the original parent segregate into more or less true-breeding *lines*. Most cultivars of self-pollinated plants represent such lines which were started from relatively few plants selected to a particular standard and subjected to consecutive generations of self-pollination. Individuals not conforming to this standard are systematically eliminated. The descendants of a single more or less homozygous plant would be a *pure line*. Sometimes two or more morphologically similar lines are combined to produce a *multiline*.

Cross-Pollinated Plants

Most species and cultivars are cross-pollinated, although in many cases either self- or cross-pollination is possible. Some important economic plants in this category whose varieties are propagated by seed are listed on p. 71. Others which are also cross-pollinated include: sugar beet, cucurbits, cabbage and related plants (*Brassica*), most root vegetables, many flowering plants used as ornamentals, most shrubs, and fruit, nut, forest, and shade trees.

Seed propagation of cross-pollinated cultivars poses more production problems since such plants tend to be heterozygous to some degree and groups of seedlings are often variable. However the genetic identity of cross-pollinated cultivars of herbaceous species can be maintained to almost the same degree as self-pollinated species if the cultivars are selected to a standard and maintained under conditions preventing cross-pollination from unwanted sources. Such cultivars are not necessarily homozygous, but variation is controlled.

In most cross-pollinated plants, enforced self-pollination and its attending inbreeding usually result in the production of inferior plants and a decline in vigor in succeeding generations, even though the group may eventually be more uniform. Enforced self-pollination is used by plant breeders to produce an *inbred line*. To produce phenotypic uniformity, and yet to assure sufficient heterozygosity to maintain vigor, several special categories of cultivars have been utilized by plant breeders (*6*). *Hybrid cultivars* are the first generation progeny between two genetically different plants or inbred lines. Such plants are heterozygous, normally uniform within the group, and may exhibit hybrid vigor. Hybrid cultivars, however, cannot be used as seed sources in the next generation (F_2) since that generation would be extremely variable. The parental stocks are maintained and the cross must be continuously repeated. *Synthetic cultivars* are made up of seedling plants arising from a combination of a number of genetically distinct but phenotypically similar lines or clones, which have been allowed to cross-pollinate at random. The first generation of seeds is called *Syn-1*, the second generation, the *Syn-2*, and so on. The original stocks are maintained and usually these cultivars are allowed a limited number of seed generations from the original *Syn-0*.

Some plants have mechanisms that completely prevent self-pollination. This situation occurs in *dioecious* plants, such as holly, date, asparagus, spinach, in which pistillate (female) and staminate (male) flowers are produced on different plants. A similar situation occurs with *dichogamy*, where the stamens do not shed pollen at a time when the pistil is receptive. This occurs in some walnut and pecan cultivars. With *incompatibility*, the pollen tube is unable to grow properly in the style of a flower on the same plant (or clone), although the pollen is normal and will grow properly in the style of another plant. Some combinations of individual plants, likewise are cross-incompatible and cannot pollinate each other. This phenomenon is based on genetic factors and occurs in such diverse species as the sweet cherry, almond, *Hemerocallis*, lily, petunia, and cabbage.

Self-sterility of either pollen or pistil can result from specific genes, and produces a dioecious individual. Such genes may be incorporated into specific seed stocks or individuals to facilitate cross-fertilization (*6*).

Cross-pollination is accomplished by various agencies. (a) *Wind pollination* is the rule with many plants with inconspicuous flowers. Examples are grasses, conifers, and catkin-bearing trees such as the walnut, oak, alder, and cottonwood. The pollen produced from such plants is generally light and dry, and in some cases, is carried long distances in wind currents. (b) *Insect pollination* is the rule with plants with brightly colored, fragrant, and otherwise conspicuous flowers which attract insects. The honey bee is one of the most important pollinating insects, although wild bees, butterflies, moths, and flies also obtain pollen and nectar from the flower. Generally, pollen is heavy and sticky, adhering to the insect. (c) *Controlled hand-pollination* by man is an important practice in breeding programs and is also used in producing seed of a few kinds of plants, for instance, tuberous begonia and certain kinds of petunia. (d) *Other pollinating agents,* such as birds, bats, snails, and water, are effective with certain kinds of plants.

Lines of cross-pollinated cultivars may be maintained genetically true to type under proper conditions. Plants of the cultivar should be relatively homozygous, and cross-pollination should take place from similar individuals to produce offspring similar to both parents.

A different procedure is necessary to produce seed of *hybrid cultivars.* Here cross-pollination takes place between genetically distinct parental lines, selected to produce a given type of seedling offspring.

If heterozygous plants are used as a seed source, the seedlings from them cannot be depended upon to reproduce the characteristics of the parents. Many herbaceous and woody perennials are clones wich exhibit such behavior. These include most fruit and nut cultivars, such as peach, apple, and pear; many herbaceous ornamentals, such as tulip, dahlia, and chrysanthemum; and numerous woody ornamentals, such as rose and camellia. Seed propagation of such cultivars is not used because they do not "breed true"— that is, the essential characteristics of the cultivar are not transmitted to its seedling offspring and the value of the cultivar is lost in the succeeding generation. Consequently they are propagated vegetatively.

SEED PRODUCTION OF HERBACEOUS PLANTS
Control of the seed source and production procedure is necessary to maintain the genetic identity of a seed-propagated cultivar. Contamination may result from chance-pollination with plants of a different genotype, or from a mixture with other crop seed of lesser value or with weed seed. Foreign seed similar to the pure crop seed cannot be easily removed with seed-cleaning equipment.

Methods of Maintaining Genetic Identity
Isolation
Isolation is used (a) to prevent contamination by cross-pollination with a different but related cultivar, and (b) to prevent mechanical mixing of the seed during harvest. It is primarily achieved through distance, but it also

can be attained by enclosing plants or groups of plants in cages, enclosing individual flowers, or removing male flower parts and then employing artificial pollination.

Not as much isolation is needed in the production of self-pollinated plants as in the production of cross-pollinated plants. As a general rule, two or more closely related cultivars of self-pollinated plants can be planted close to each other without one contaminating the other. The principal reason for separation in this case is to prevent mechanical mixing of seed during harvest. The minimum distance usually specified between plots of self-pollinated seed-producing plants is 10 ft (see Table 4–1).

Seed of different cultivars must be kept separate during harvest. Careful cleaning of the harvesting equipment when a change is made from one cultivar to another is required. Likewise, sacks and other containers used to hold the seed must be carefully cleaned to remove any seed which may have remained from previous lots.

Isolation is essential in the production of plants cross-pollinated by wind or insects. The minimum distance between cultivars depends upon a num-

Table 4–1   Minimum isolation requirements for seed production of certain species of field and vegetable crops *(3)*.

Type of Pollination	Species	Seed Class Founda-tion	Regis-tered	Certi-fied
Self-pollinated	barley, oats, wheat, rice, peanut, soybean, lespedeza, field pea, garden bean, cowpea, flax grasses (self-pollinated and apomictic species)	Fields should be separated by a definite boundary adequate to prevent mechanical mixture		
		60 ft	30 ft	15 ft
Self-pollinated but to a	cotton (upland type)	100 ft from cultivars which differ markedly.		
lesser degree	cotton (Egyptian type)	1320 ft	1320 ft	660 ft
than those	pepper	200 ft	100 ft	30 ft
in the above	tomato	200 ft	100 ft	30 ft
list	tobacco	150 ft or by four border rows of each cultivar. Isolation between cultivars of different types should be 1320 ft.		
Cross-pollinated by insects	alfalfa, birdsfoot trefoil, red clover, white clover, sweet clover	600 or 900 ft*	300 or 450 ft*	165 ft
	millet	1320 ft	1320 ft	660 ft
	onion	5280 ft	2640 ft	1320 ft
	watermelon	2640 ft	2640 ft	1320 ft
Cross-pollinated by wind	hybrid field corn	660 ft (may be reduced if field is surrounded by specified numbers of border rows and the cultivars nearby are of same color and texture).		
	grasses	900 ft	300 ft	165 ft

* First number if plot is less than 5 acres; second number if plot is more than 5 acres.

ber of factors, such as the degree of natural cross-pollination, the pollination agency, direction of prevailing winds, and the number of insects present. The required distances specified by seed-regulating agencies varies with the class of seed produced. That is, greater distances should be used with seeds in which a higher degree of genetic purity is desired (see Table 4–1). The minimum distance for insect-pollinated plants is $\frac{1}{4}$ mi. to 1 mi.

The distance that should separate cultivars of wind-pollinated plants varies with the kind of plants. The distance usually specified for corn is $\frac{1}{8}$ mi. but this distance may be reduced by planting border rows of the pollinator cultivar. With beets, $\frac{1}{2}$ to 1 mi. is recommended and, to produce stock seed, 2 mi. is preferable (*24, 36*). Seed-producing fields of two different cultivars should not be in the line of prevailing winds.

Cross-pollination takes place between certain cultivars and not between others. In general, any cultivar can contaminate any other of the same species; it may or may not contaminate cultivars of a different species but in the same genus; and rarely will it contaminate cultivars belonging to another genus. Since the horticultural classification may not indicate taxonomic relationships, the seed producer should be familiar with botanical relationships among the cultivars he grows.

Roguing

Off-type plants, plants of other cultivars, and weeds in the seed production field should be eliminated. Although a low percentage of such plants may not seriously affect the performance of any one lot of seed, their continued presence will lead to deterioration of the cultivar over a period of time. The removal of such plants is referred to as "roguing."

Off-type plants may arise because some recessive genes are present in a heterozygous condition even in highly homozygous cultivars. Recessive genes may arise by mutation, a process continually occurring at a low rate. The effect of a mutant recessive gene controlling a given plant characteristic may not be immediately observed in the plant in which it arises. The plant becomes heterozygous for that gene, and in a later generation the gene segregates and the character appears in the offspring. Some cultivars have mutable genes that continuously produce specific off-type individuals (*35*). Off-type individual plants should be rogued out of the seed production fields before pollination occurs. Regular supervision of the fields by trained personnel is required.

Volunteer plants arising from accidentally planted seed or from seed produced by earlier crops is another source of contamination. Fields for producing seed of a particular cultivar should not have grown a potentially contaminating cultivar for a number of preceding years.

Testing

Cultivars being grown for seed production should be periodically tested for genetic identity to make sure that they are being maintained in their true form. Seed companies maintain test gardens for this purpose under the supervision of trained personnel.

Genetic Shifts

Sometimes a seed-propagated cultivar that is produced initially in one location may undergo a *genetic shift* if it is grown for several consecutive generations in a different location. Such a change takes place because those seedling plants better adapted to the new environment survive in greater amounts than others. Physiological changes affecting yielding ability, disease resistance, and environmental adaptation are difficult to detect without testing. To minimize the opportunity for such shifts to occur in seeds of forage crops, only one generation of seed production is allowed in regions of mild winters if the crop seed is to be planted in regions with cold winters *(20)*.

Seed Certification

Genetic purity in commercial seed production may be regulated through a system of *seed certification (3, 11)*. Such programs exist in most states of the U.S. through the cooperative efforts of public research, extension, and regulatory agencies in agriculture and a state seed-certifying agency often known as a Crop Improvement Association. The agency is usually designated by law to certify seeds. Its members include growers as well as other interested individuals involved in the production of "certified seed." These individual state organizations are coordinated through the Association of Official Seed Certifying Agencies (AOSCA) *(3)* in the United States and Canada. International certification is regulated through the Organization for Economic Cooperation and Development (OECD).

The principal objective of seed certification is to protect the genetic qualities of a cultivar. To accomplish this purpose the seed-certifying agency determines the eligibility of particular cultivars; sets up production standards for isolation, presence of off-type plants, and quality of harvested seed; and makes regular inspections of the production fields to see that the standards are maintained. Supervision of seed processing is also involved. Recommended minimum standards exist for many crops (see Table 4–1). Since specific requirements for certification may vary from state to state, the local, state, and national regulations should be examined. These usually specify the cultivars which can be certified, the amount of isolation required of any particular crop, required inspections, and standards for the seed quality after harvest.

Genetic purity is maintained by utilizing certain classes of seed which designate the generations allowed away from the original source. Their purposes are to allow an increase in seed supplies under conditions to preserve genetic identity and purity.

Breeder's seed   Breeder's seed is that which originates with the sponsoring plant breeder or institution and provides the initial source of all the certified classes.

Foundation seed   This is the progeny of breeder's seed and is so handled as to maintain the highest standard of genetic identity and purity. It is the source of all other certified seed classes but can also be used to produce additional foundation seed plants. *Select seed* is a comparable seed class used in Canada.

Registered seed   This is the progeny of foundation seed (or sometimes breeder's seed or other registered seed) that is produced under specified standards approved and certified by the certifying agency and designed to maintain satisfactory genetic identity and purity.

Certified seed   This is the progeny of registered seed (or sometimes breeder's, foundation, or other certified seed) and is that which is produced in the largest volume and sold to growers. It is produced under specified standards designed to maintain a satisfactory level of genetic identity and purity and is approved and certified by the certifying agency.

Bags of the different seed classes are identified by different color tags attached so that they are difficult to remove without defacing the bags.

Bags of certified seed have a blue tag attached, distributed by the seed-certifying agency as evidence of the genetic identity and purity of the seed contained therein. Bags of registered seed are labeled with a purple tag or a blue tag marked with the word "registered." Likewise, foundation seed is labeled with a white tag or a regular certified seed tag with the word "foundation." The International OECD scheme includes *basic* (equivalent to either foundation or registered) seed. It also allows for *certified first generation* (blue tag) and *second generation* (red tag) seed.

Vegetable and Flower Seed Production (24)

In the U.S. vegetable and flower seed is largely produced by commercial seed companies who either grow the seed themselves or contract with private growers to produce seed for them. The company preserves genetic purity by specifying the particular cultivar an individual contractor will grow, by supervising the fields during production, and by maintaining test gardens.

Although such seed is not necessarily produced as certified seed, similar procedures are used to maintain genetic purity. Essentially the same classes of seed are used, although they may be designated by different names. The primary seed source of the cultivar, the foundation seed, is usually maintained by the seed company. Trained technicians have the responsibility for maintaining the foundation seed at the highest level of purity. Seed provided by the company to the contractor-grower for growing his crop is known as *stock seed*. The seed actually harvested from the fields of the contractor-grower will be placed on the market and sold commercially by the seed company.

The Production of Seed for Hybrid Cultivars

Hybrid cultivars have become an increasingly important category of cultivated plants. These are the F_1 progeny produced by the repetitive crossing of two or more parental lines that are maintained either (a) by seed, such as inbred lines, or (b) asexually, such as by clones. The group of plants produced are highly heterozygous but are very uniform in appearance and may show hybrid vigor and other desirable characteristics providing the cultivar has been properly selected. In the next generation (F_2) and succeeding generations the cultivar deteriorates rapidly. Plants will not maintain the vigor and the group will be more or less variable. To produce commercial hybrid seed, the parental lines must be grown side by side so that cross-pollination takes place between them. The seed produced (the F_1 progeny of the cross) is the seed used to grow commercial crops. This cross must be repeated every time the seeds are produced.

Hybridization may be made between two inbred lines (*single-cross*), two single-crosses (*double-cross*), an inbred line and an open pollinated cultivar (*top-cross*), or between a single-cross and an inbred line (*three-way cross*). These combinations are mostly used in corn production.

Controlled Hand-Pollination

Controlled hand pollination is a basic technique used in plant breeding to produce seeds from selected parents. It is also used to produce commercial hybrid seed for some annual flower crops, such as petunia, pansy, and snapdragon, where the high cost of seed production is offset by the high value of the seed (*21*). The specific technique used depends upon flower structure and the pollination characteristics of the individual species, but it follows certain principles (*39, 43*).

Accidental contamination by unwanted pollen can be prevented by washing the hands and instruments with alcohol before handling pollen or flowers. Pollen should be collected from flower buds immediately preceding bloom and before the anthers open (dehisce). This eliminates the chance of contamination by unwanted pollen. Anthers are extracted by pulling them from the flowers with tweezers, by squeezing the flower between the fingers, or by rubbing the buds across a wire screen. Catkins (walnut, birch, aspen) or staminate cones (pine) will open on drying and will shed large quantities of pure, dry pollen if placed on a sheet of paper in a warm room. Anthers are dried on a sheet of paper until their opening can be observed under a magnifying glass. Pollen and anthers can be screened through a fine sieve to remove extraneous material; they are then stored in a glass vial or bottle.

Most kinds of pollen will remain viable for only a few days or weeks at warm temperatures, but many kinds can be preserved for several months to several years if stored at low temperatures and relatively low humidity. Effec-midity and a temperature of 32° to 50° F (0° to 10° C). Moisture content of tive storage conditions are a combination of 10 to 50 percent relative hu-the pollen can be controlled by storing over a desiccant, such as calcium chloride or sulfuric acid. Some pollen—that of grasses, for instance—is best stored at 90 to 100 percent R.H. Pollen can be effectively stored at about 0° F (−18° C), as in a home freezer.

The stigma at the tip of the pistil must be exposed to apply the pollen. If the plant is self-fertile, the stamens must be removed (emasculated), Petals and stamens can be removed at the bud stage either with tweezers or by cutting the base of the flower between the fingernails. If the flower is to be self-fertilized or is a self-sterile cultivar, the flower need not be emasculated but only enclosed to keep out pollinating insects or wind-blown pollen.

Pollen is applied to the sticky, receptive stigma of insect-pollinated flowers with a fine-hair brush, a glass rod, a pencil eraser, or the tip of the finger. Removal of all the flower parts with the exception of the pistil reduces or eliminates the likelihood of further chance pollination by insects. Complete protection can be obtained by covering the individual flower, the pistil, or the branch with cellophane, plastic, paper, or a cloth bag.

For wind-pollinated plants, e.g., grasses, pines, and walnut, female flowers (or cones) must be carefully sealed throughout the flowering period. They can be covered with cellophane, paper, or a tightly woven cloth bag. Pollen can be transferred by inverting a paper bag containing pollen over the exposed flower; or dry pollen can be blown into the bag with an atomizer or inserted with a hypodermic needle.

The flower, or the branch on which it is borne, should be carefully labeled to maintain a record of the pollen used. The usual method is to identify the cross as follows: *"seed parent × pollen parent."*

Controlled Pollination Systems

Mass production of hybrid seed requires some system of sterility that prevents self-pollination and enforces cross-pollination. Plant breeders have utilized various naturally occurring self-sterility systems for particular plants and have discovered others as mutant genes which are then incorporated into seed production lines. If parental lines can be economically propagated vegetatively, then self-sterility can be maintained intact with no difficulty. Seed-propagated lines, however, are more difficult to manage and knowledge of inheritance in the self-sterility system is required. Methods must be used to insure that only the desired self-sterile genotype is produced when needed, or procedures must be used to identify the unwanted pollen-producing individual plants early enough in their development so that they can be rogued out of the seed-producing fields.

(a) Removal of male flowers of monoecious species   Plants of monoecious species such as corn (*Zea mays*), were first adapted to hybrid seed production because the male flowers (tassels) were separate from the female flowers (silks) and could be readily removed by hand. Parental inbred lines are planted in separate rows, one male row to 3 seed-producing rows or some similar arrangement (42). Tassels were pulled from the plant by hand before pollen was shed (see Figure 4–1). This procedure has largely been replaced by planting male-sterile lines.

(b) Male sterility   Genetic factors for male sterility in which pollen production is inhibited have been discovered in many crop plants (6). These factors have been incorporated by plant breeders into hybrid cultivars of a number of crop species. Methods to maintain the male-sterile parental line must also be worked out, based on inheritance of the gene (12, 19). The

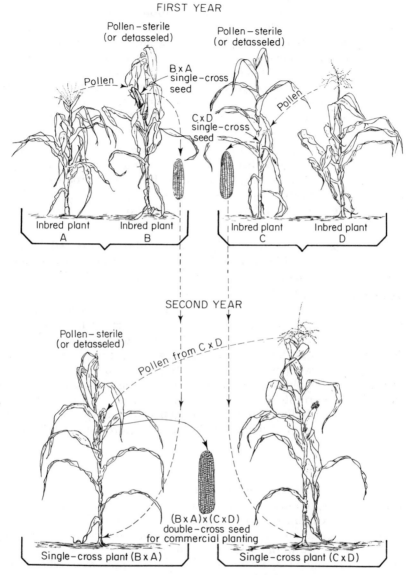

Figure 4–1 The principles of hybrid seed production are illustrated by corn *(Zea mays)*. Four inbred lines are involved in this particular hybrid. Parental lines are maintained in isolation. To produce hybrid seed, parental lines are grown side by side. The seed-producing line either is male-sterile or the tassel is removed by hand to insure that pollen comes from the other line. In this example, the seed produced from the original crosses (F_1 offspring) are combined in a second generation to produce a so-called *double-cross* hybrid. The seed produced is used to grow the commercial crop. Redrawn from *USDA Yearbook of Agriculture, 1937.* Washington, D.C.: U.S. Govt. Printing Office.

most useful self-sterility system involves a combination of sterility factors of the cytoplasm and of nuclear genes. In such a system the male-sterile parental line is relatively easy to maintain on a large scale. A male-sterile individual has a genetic makeup of *Smsms* (*S* means sterile cytoplasm; *msms* means male-sterile genes). When crossed with an *Nmsms* (*N* means normal cytoplasm which is not inherited through pollen), all offspring are *Smsms*. This system was first used in producing onion hybrids (*28, 29*),but is also used for field corn, sugar beets, sorghum, and pearl millet (*19*) with prospects for use with wheat and barley (*6*).

(c) Self-incompatibility   Self-incompatibility has been used in the production of some cabbage hybrids (*34*). Some grass hybrids of Pensacola Bahiagrass (*Paspalum notatum*), for instance, have been produced by interplanting self-incompatible (but cross-compatible) parental clones (*7*). The plants spread vegetatively to occupy the field and all seeds are harvested together as hybrids.

(d) Dioecious plants   Dioecious plants can also be utilized effectively. 'Mesa' buffalograss (*Buchloe dactyloides*) is produced by interplanting vegetatively propagated male and female parents in isolated fields (*7*).

(e) Gynoecious hybrids   Cucumbers are normally monoecious but on a given vine "femaleness," i.e., pistillate flowers, tends to predominate at the first nodes, with a shift toward maleness in other parts of the plant. Inbred lines have been produced that are called "gynoecious," that is, the plants produce predominantly female flowers (*37*). where a completely gynoecious cultivar exists, a certain percentage (about 10 percent) of seed from a monoecious cultivar is mixed in to produce some pollinizing plants (*10*). In another instance, a hermaphrodite line has been developed which when crossed with a gynoecious line produced 100 percent gynoecious offspring (*38*). Otherwise, chemical induction of staminate flowers is required, as described in the next section.

Chemical Control of Pollination Systems

Certain growth regulators can shift sex expression in certain monoecious plants, as various cucurbits (*47*). Seed-production lines of gynoecious cucumbers produce staminate flowers when sprayed with gibberellin at 1000 ppm 3 times per week beginning with the expansion of the first true leaf. By treating one row out of 3, F_1 hybrid seed can be produced—where the induced male flowers provide the pollen.

Other potentially useful chemicals are selective gametocides, as have been utilized in cotton with certain commercial compounds (*13*).

PLANT VARIETY PROTECTION ACT

In the United States, breeders of a new seed-reproduced plant variety (cultivar) may retain exclusive propagation rights through the protection given by the Plant Variety Protection Act, which became effective in 1970 (*40*).

The breeder applies to the U.S. Department of Agriculture for a Plant Variety Protection Certificate. For one to be granted, the cultivar must be "novel." It must differ from all known cultivars by one or more morphological, physiological, or other characteristics. It must be uniform; any variation must be describable, predictable, and acceptable. It must be stable; i.e., essential characteristics must remain unchanged during propagation. A certificate is good for 17 years. The applicant may designate that the cultivar be certified and that reproduction continue only for a given number of seed generations from the Breeder or Foundation stock. If designated that the cultivar be certified, it becomes unlawful under the Federal Seed Act to market seed by cultivar name unless it is certified.

SEED SOURCES FOR WOODY PERENNIALS

Most trees and shrubs are heterozygous and cross-pollinated and have considerable potentiality for genetic variability. Consequently, seedling production may result in variabilities among plants and the inability to transmit specific characteristics from seed-source tree to offspring. Control of both of these problems and improvement in tree crops may be achieved by following particular seed-selection practices.

Seedling grown plants are used for many purposes, such as for landscaping, to produce rootstocks of fruit trees and ornamentals, for Christmas tree growing, and for replanting natural areas, as in reforestation. The amount of allowable variability may differ with intended use. Where the characteristics of the individual plant are important, as in landscaping, careful seed selection is essential or vegetative methods are necessary. Likewise uniformity among rootstocks is important for fruit-tree growing. Some seedling variation may be beneficial in reforestation where initial seed planting density is high. Close spacing promotes good stem form; competition among plants gradually eliminates the weakest trees. However, forestry practices do involve careful seed selection to reproduce trees of desired timber quality.

Principles of Selection
Seed Origin
Seed origin refers to the particular climatic and geographical locality where the seed is obtained. The original geographic source of seed, referred to as *provenance,* is particularly important in reforestation and is second in importance only to choice of species (*16, 22, 30, 46*). The significance of seed origin was first recognized during the nineteenth century when extensive plantings of forest trees from unselected seed were made in Europe. Observers noticed that trees grown from imported seeds were generally inferior to trees grown from seeds of local origin. In one instance, thousands of acres of Scotch pine (*Pinus sylvestris*) had to be eradicated because of their inferior characteristics. After that, many European countries enacted laws covering the importation and use of forest tree seeds, in part requiring the labeling of seed as to origin.

Origin is important when plants of single species grow over a wide range of ecologically different areas in nature. Variation in morphology, physiology, adaptation to climate and soil, and resistance to diseases and insects may exist that represent specific races, or *ecotypes,* that are characteristic of a particular locality. Variation may also exist in a continuous gradation, or *cline* (*30*) (see p. 14). A locality may be defined by its latitude, longitude, and elevation (*2, 17*). Seeds of a given species collected from one locality may produce plants completely unadapted to a different locality. For example, seed collected from trees in warm climates or at low altitudes is likely to produce seedlings that will be injured when grown in colder regions even if the species is the same and the plants are similar in appearance, Although the reverse situation—collecting seed from colder areas for growth in warmer regions—would be more satisfactory, it might result in a net reduction in growth due to the inability of the trees to fully utilize the growing season.

Distinct ecotypes have been identified in many native forest tree species, including Douglas fir, ponderosa pine, lodgepole pine, red pine, eastern white pine, slash pine, loblolly pine, shortleaf pine, and white spruce (*4*). Some of these ecotypes have been shown to be superior seed sources and, in fact, preferable to local sources—for example, the East Baltic race of Scotch pine, the Hartz Mountain source of Norway spruce, the Sudeten (Germany) strain of European larch, the Burmese race of teak, Douglas fir from the Palmer area in Oregon, ponderosa pine from the Lolo Mountains in Montana, and white spruce from the Pembroke, Ontario (Canada) area.

An important principle in seed selection has been to use local seed where possible unless studies have demonstrated that another source is better. In 1939 the U.S. Department of Agriculture adopted the following policy regarding the use of seed stocks and individual clones obtained by them for forest, shelter-belt, and erosion-control plantings (*46*). It serves as a guide for the selection of woody plant seed from natural sources.

1 Only seed of known locality of origin or nursery stock grown from such seed is to be used.

2 Evidence for the place and year of origin is to be required of the vendor when seeds are bought.

3 An accurate record of the following data is to be required of all shipments: (a) lot number; (b) year of seed crop; (c) species; (d) seed origin as to state, county, locality, and range of elevation; and (e) proof of origin.

4 Whenever available, local seed from natural stands is to be used unless it is demonstrated that another source can give desirable plants. "Local seed" means seed from an area subject to similar climatic influences, and this may usually be considered to mean within 100 mi. of the planting and within 1000 ft of its elevation.

5 When local seed is not available, seed should be used from a region having as nearly as possible the same climatic characteristics, such as length of growing season, mean temperature of growing season, frequency of summer droughts, and latitude.

Seed-collection zones designating particular areas that have a specific climatic and geographical basis have been established in a number of areas in the U.S. For example, California is divided into six physiographic and climatic regions, 32 subregions, and 85 seed-collection zones *(41)*. Similar zones are established in Washington and Oregon and in the central states area of the United States *(31)* .

Choice of seed source is highly important in growing conifers and various hardwood species of trees in the nursery—for instance, for landscape purposes or for Christmas trees *(15, 25)*. These differences can be demonstrated in nursery tests *(5)*. For example (Figure 4–2), Douglas fir has at least three recognized races, *viridis, caesia,* and *glauca,* with various geographical strains within them *(22)*. The *viridis* strain from the U.S. west coast was not winter-hardy in New York tests but it is well-suited to Western Europe *(22)*. Strains collected further inland were winter-hardy and vigorous; those from Montana and Wyoming were very slow growing. Trees of the *glauca* (blue) strain from the Rocky Mountain region were winter-hardy but varied in growth rate and appearance. Similarly, tree differences due to source occurred in Scotch pine, mugho pine, Norway spruce, and others.

Seed-Tree Selection
Although local seed sources of forest species can produce the best adapted plants for a given site, selection of *individual seed trees* may be necessary to improve quality in other characteristics *(32)*. In forest trees there is a great deal of evidence that the phenotype of the individual seed tree is a good index of the phenotype of its offspring *(31)*. Parent tree selection for many

Figure 4–2 Two-year Douglas fir *(Pseudotsuga menziesii)* seedlings from different seed sources growing side by side in nursery plots. *Left:* two fast-growing green strains from Washington and British Columbia. *Center:* four slow-growing strains from Montana and Wyoming, mostly gray-green in color. *Right:* fast-growing strain from Arizona, deep blue in color. Seedlings from different sources vary in height from 2 to 4 in. up to 10 to 16 in. Courtesy C. E. Heit.

characteristics, such as stem form, branching habit, growth rate, resistance to diseases and insects, presence of surface defects, and certain other lumber qualities, can be effective.

Selected trees in native stands showing a superior phenotype are referred to by foresters as "plus" trees. Such individual trees may be used in natural reseeding or as seed sources for other areas.

The initial step in selecting a seed source is to evaluate the phenotypic characteristics of the seed trees. Although plants possessing the desired characteristics are most likely to pass them on to their offspring, this assumption cannot be safely made without progeny testing and consideration of pollen sources.

Consequently, a second step in selecting a seed source is to evaluate the characteristics of surrounding trees that could cross-pollinate the seed trees. Most woody tree species are cross-pollinated; in general, better quality seeds will likely result (*14*).

The most desirable procedure is to collect seeds from groups of the same kind of plants with desired characteristics growing in a *pure stand,* where cross-pollination can come from a number of plants of the same type. Seedling plants grown from such sources, although not necessarily homozygous, would be most likely to reproduce the characteristics observed among the parent seed trees.

Collecting seed from an isolated plant, particularly if it is growing near others of related species, as it might in an arboretum (*48, 50*), is usually undesirable.

Cross-pollination with an undesirable pollen source can nullify the potential value of a given seed source and produce excessive and unpredictable variability. For example, before about 1915, "Old French" pear seedlings were grown in the U.S. from seed obtained in Europe from *Pyrus communis* trees. Much variability resulted from hybridization with nearby *Pyrus nivalis* trees. Similarly, "Oriental pear" rootstock used in California about the same time originated from seeds obtained from Asia with little knowledge about the characteristics of either the seed tree or the pollen source. Plants grown from these sources have turned out to be a heterogeneous group with many species and hybrids (*8, 23*).

The differences among flame eucalyptus (*Eucalyptus ficifolia*) trees arising from separate collections of a particular source in Australia have been explained by the fact that seeds collected from the center of the block were likely to reproduce the true species and produce upright, uniform groups of trees, while those from the edge of the block were likely to be hybrids with surrounding *E. calophylla* trees, and consequently produced many dwarfed and off-type plants (*44*).

Control of both seed and pollination source can be achieved by collecting from seed orchards with known pollination arrangements. Seedling rootstocks of fruit trees are usually produced from seed collected from known clones, whether as by-products of fruit production or from specially selected rootstock-producing clones. For example, *Pyrus communis* seedlings, e.g., the so-called "French pear," seeds are obtained from trees of 'Winter Nelis' grown for cross-pollination in 'Bartlett' ('Williams Bon Cretien') pear orchards. All seedlings would be 'Winter Nelis' × 'Bartlett' hybrids. Likewise, 'Lovell' peach seeds or 'Royal' apricot seeds collected from fruit-drying yards come from known clones that most likely are grown in a solid block of trees where self-pollination occurred.

Progeny Testing

The true genetic value of a seed source can only be established through a *progeny test* which will identify a superior genotype (*5*). A representative sample of seeds is planted, and the resulting progeny are grown under test conditions that will identify essential characteristics or demonstrate superiority to other sources, or both. Foresters refer to seed-source trees with a superior genotype, as demonstrated by a progeny test, as "elite" trees.

Progeny testing has been used to identify clones that will transmit high yielding characteristics—as in tung (*Aleurites*) (*33*)—or important rootstock qualities, such as nematode resistance (*9*). Nursery tests can identify and characterize specific seed sources for landscape and Christmas tree uses (*25*). Progeny testing is an essential phase of identifying seed sources of superior "elite" forest trees (*5*).

Nursery Selection

Desired individuals in variable seedling populations may be identified through selection in the nursery. Variability involves identifiable characters of vigor or appearance, or both. For example, Paradox hybrid walnut seedlings can be identified in nursery plantings of *Juglans hindsii* seedlings (see Figure 4–3). "Blue" seedlings of the Colorado spruce (*Picea pungens*) appear among others having the usual green form. Differences in fall coloring of *Liquidamber* and *Pistacia* trees among seedlings necessitate fall selection of individual trees for landscaping. Variation among seedlings grown for rootstocks may be reduced by grading to a size and eliminating the weak, small seedlings, a usual practice in U.S. nurseries (see p. 315).

Figure 4–3 Paradox hybrid (*Juglans hindsii* x *J. regia*) seedlings *(far left)* sometimes appear among Northern California black walnut *(J. hindsii)* seedlings *(center)* being grown for rootstocks. Identification of the hybrid seedlings in the nursery row is by their lighter bark color and larger leaves. Leaves of Persian walnut *(top right)*, Paradox hybrid *(center right)*, and Northern California black walnut *(below right)*.

Procedures in Selection of Woody Plant Seed
Seed-Collection Areas

A *seed-collection* area is one where the trees have promise as a potential seed source but have not been progeny-tested (*31*). This should be a pure stand of uniform trees large enough to insure adequate cross-pollination without inbreeding and separated from other plants having undesirable characteristics. Such areas may not be managed or treated specifically for seed production.

Seed-Production Areas (*31*)

This refers to an area enclosing a group of trees that has been set aside specifically as a seed source. Use of such an area for a seed source applies primarily to forest trees, but the procedure could be utilized in any woody plant seed enterprise. Seed trees within the area are evaluated for the characteristics desired. Off-type trees or those that do not meet minimum standards are removed. Other trees or shrub species that would interfere with the operations may be removed. It may be desirable to eliminate additional trees to provide adequate space for tree development and seed production. An isolation zone at least 400 ft wide from which off-type trees are removed should be established around the area. Trees may or may not have been progeny tested prior to establishment of the area.

Seed Orchards (*1, 26, 31*)

Orchards or plantations may be established specifically for seed production, propagated from seed trees of acceptable origin and quality, and preferably from progeny-tested seed trees. Seed orchards are used directly or indirectly in the production of fruit-tree rootstock seeds (*9*) and in the production of forest trees (*26*). Seed orchards are useful for maintaining rootstock seed trees involved in virus-control programs (see p. 197).

There are three general types of seed orchards (*1, 31*): (a) seedling trees produced from selected parents through natural or controlled pollination; (b) clonal seed orchards in which selected clones are propagated by grafting, budding, or rooting cuttings; and (c) seedling-clonal seed orchards in which certain clones are grafted into branches of some of the trees. The choice of which type should be used will depend upon the species that is being grown.

Details of setting up a seed orchard vary with the species. A site should be selected for maximum efficient production. For most native species enough different tree selections should be included in a suitable arrangement to insure cross-pollination and to decrease effects of possible inbreeding, which may result in offspring less vigorous than the offspring of cross-pollination. If the species is dioecious, pollen-producing trees must be included. Fruit-tree rootstock clones that are self-pollinated can be planted as solid blocks (*9*). An isolation zone at least 400 ft wide should be established around the orchard. The size of this zone may be reduced if a buffer area of the same kind of tree is established around the orchard.

Hybrid Seed Production

Production of F_1 trees may result from crosses between species or within species (*16, 21, 26, 49*). Hybridization between the parental plants to produce the desired F_1 offspring may be achieved by various means.

(a)   Hybrids may be produced by hand pollination (see p. 75). This procedure is used mostly to produce F_1 hybrids for testing purposes. Mass production of seeds by this method would normally be too expensive, although hybrids of *Pinus rigida* × *P. taeda* have been produced this way in Korea.

(b)   Hybrids may be produced in seed orchards by interplanting the parental trees (*1*). Hybrids of *Larix decidua* and *L. leptolepsis* have been produced in seed orchards in Europe. In Idaho, a western white pine (*Pinus monticola*) seed orchard was established in the late 1950s containing 13 parental clones selected by progeny testing (*49*). Their F_1 offspring showed an average of 30 percent resistance to blister rust. The F_2 generation showed an even higher average level of resistance in tests and will therefore be utilized as a seed source.

(c)   Production of Paradox hybrid walnut seed is obtained by collecting seed from individual Northern California black walnut (*Juglans hindsii*) seed trees where natural cross-pollination is known to take place with pollen coming from nearby English walnut (*Juglans regia*) trees (*45*). Seed-source trees are identified by progeny testing. Individual Paradox seedlings are identified in the nursery by vigor and leaf characteristics, as shown in Figure 4–3, and other trees are rogued out.

(d)   Seed from F_1 seed trees gives F_2, or second-generation hybrids. Collection of seed from many such sources, particularly if interspecific hybrids are involved, produces a group of seedlings that are highly variable in regard to vigor and other characteristics. As a group these may fail to reproduce the desirable attributes of the F_1 hybrid generation, although some individuals may be highly desirable. Such variability is undesirable if the hybrid plants are to be used for rootstocks or for landscaping where uniformity and predictability of the final form is important. On the other hand, in forest trees, some variability in vigor commonly associated with second-generation hybrids may have advantages (*16*). In tree plants with relatively high initial seedling density, competition among plants will favor the most vigorous and tend to eliminate the weaker, less desirable plants. Consequently, planting seeds from F_1 hybrids (the F_2) may be an economical way to obtain a natural stand of vigorous hybrids. F_2 hybrids of selected lines within species may be quite uniform and desirable as in the blister-rust-resistant, white pine hybrids described above (*49*). The value of any F_1 hybrid tree as a seed source should be established by progeny testing, as should that of any other seed source.

Tree-Certification Programs (11)

Certification of tree seeds is available, similar to that for crop seed (see p. 73). This program had a start in 1959 when the Georgia Crop Improvement Association adopted standards for certifying forest tree seeds. Recommended minimum standards are given by the Association of Official Seed Certifying Agencies (*3*). A number of states have such programs, as do a number of European countries. A program to control seed shipped in international commerce that includes much the same provisions has been initiated by the Office for Economic Cooperation and Development (*27*).

These programs seek to provide for improved, progeny-tested sources of forest-tree seed that are maintained and produced under standards of a regulatory body. Categories of seed developed for this program include the following:

Source-Identified Tree Seed

Seed is collected from natural stands where the geographic origin (source and elevation) is known and specified or from seed orchards or plantations of known provenance, specified by seed-certifying agencies. These seeds carry a yellow tag.

Selected Tree Seed

Seed is collected from trees which have been rigidly selected for promising phenotypic characteristics but which have not been progeny tested. The source and elevation must be stated. These seeds are given a green label.

Certified Tree Seed

Seed shall be collected from trees of proven genetic superiority, as defined by a certifying agency, and produced under conditions that assure genetic identity. These could come from trees in a seed orchard, or from superior ("plus") trees in natural stands with controlled pollination.

REFERENCES

1 Anonymous, Agricultural and horticultural seeds. *FAO Agricultural Studies No. 55*. Rome: Food and Agriculture Organization of the United Nations, 1961.

2 ———, *Forest Tree Seed Directory*. Rome: Food and Agriculture Organization of the United Nations, 1975.

3 Assoc. Off. Seed Cert. Agencies, *AOSCA Certification Handbook*. Publ. No. 23. June 1971.

4 Barber, J., R. Z. Callahan, P. C. Wakely, and P. O. Rudolf, The seed we use: Part I, *Jour. For.*, 61:181–84. 1963.

5 Barker, S. C., Progeny testing forest trees for seed certification programs, *Ann. Rpt. Int. Crop Imp. Assoc.*, 46:83–87. 1964.

6 Briggs, F. N., and P. Knowles, *Introduction to Plant Breeding*. New York: Reinhold, 1967.

7 Burton, G. W., and G. F. Sprague, Use of hybrid vigor in plant improvement, in R. E. Hodgson (ed.), *Germ Plasm Resources*. Washington, D.C.: Amer. Assoc. Adv. Sci., 1961, pp. 191–204.

8 Catlin, P. B., and E. A. Olsson, Identification of some *Pyrus* species after paper chromatography of leaf and bark extracts, *Proc. Amer. Soc. Hort. Sci.*, 88:127–44. 1966.

9 Cochran, L. C., W. C. Cooper, and E. C. Blodgett, Seed for rootstocks of fruit and nut trees, in *Seeds. Yearbook of Agriculture*. Washington, D.C.: U.S. Govt. Printing Office, 1961, pp. 233–39.

10 Conner, L. J., and E. C. Martin, Staminate: pistillate flower ratio best suited to the production of gynoecious hybrid cucumbers for machine harvest. *Hort-Science* 6(4):337–39. 1971.

11 Cowan, J. R., Seed certification, in T. T. Kozlowski (ed.), *Seed Biology*, Vol. III. New York: Academic Press, 1972, pp. 371–97.

12 Duvick, D. N., Influence of morphology and sterility on breeding methodology, in K. J. Frey, *Plant Breeding*, Ames, Iowa: Iowa State University Press, pp. 85–138. 1966.

13 Eaton, F. M., Selective gametocide opens way to hybrid cotton, *Science*, 126:74–75. 1957.

14 Ehrenberg, C., A. Gustafsson, C. P. Forshell, and M. Simak, Seed quality and the principles of forest genetics, *Heredity*, 41:291–366. 1955.

15 Flint, H., Importance of seed source to propagation, *Proc. Int. Plant Prop. Soc.* 20:171–78. 1970.

16 Fowells, H. A., Making better forest trees available, in *Seeds Yearbook of Agriculture*. Washington, D.C.: U.S. Govt. Printing Office, 1961, pp. 378–82.

17 ———, *Silvics of Forest Trees of the United States*. U.S. Dept. of Agric. Handbook No. 271. Washington, D.C.: U.S. Govt. Printing Office, 1965. pp. 1–762.

18 Fryxell, P. A., Mode of reproduction of higher plants, *Bot. Rev.*, 23:135–233. 1957.

19 Gabelman, W. H., Male sterility in vegetable breeding, in *Genetics in Plant Breeding*, Brookhaven Symposia in Biology No. 9, pp. 113–22. 1956.

20 Garrison, C. S., and R. J. Bula, Growing seeds of forages outside their regions of use, in *Seed. Yearbook of Agriculture*. Washington, D.C.: U.S. Govt. Printing Office, 1961, pp. 401–6.

21 Goldschmidt, G. A., Current developments in the breeding of F_1 hybrid annuals, *HortScience* 3(4):269–71. 1968.

22 Haddock, P. G., The importance of provenance in forestry, *Proc. Int. Plant Prop. Soc.*, 17:91–98. 1968.

23 Hartman, H., Historical facts pertaining to root and trunkstocks for pear trees. *Oregon State Univ. Agr. Expt. Sta. Misc. Paper 109*, 1–38. 1961.

24 Hawthorn, L. R., and L. H. Pollard, *Vegetable and Flower Seed Production*. New York: Blakiston Co., 1954.

25 Heit, C. E., The importance of quality, germinative characteristics and source for successful seed propagation and plant production, *Proc. Int. Plant Prop. Soc.*, 14:74–85. 1964.

26 Hoekstra, P. A., E. P. Merkel, and H. R. Powers, Jr., Production of seeds of forest trees, in *Seeds. Yearbook of Agriculture*, Washington, D.C.: U.S. Govt. Printing Office, 1961, pp. 227–32.

27   Horne, F. R., Forest tree seeds in Europe and the OECD proposals, *Ann. Rpt. Int. Crop Imp. Assoc.,* 46:90–94. 1964.

28   Jones, H. A., and A. E. Clarke, Inheritance of male sterility in the onion and the production of hybrid seed, *Proc. Amer. Soc. Hort. Sci.,* 43:189–94. 1943.

29   ———, The story of the hybrid onion, *USDA Yearbook of Agriculture, 1943– 1947,* Washington, D.C.: U.S. Govt. Printing Office, pp. 320–26. 1947.

30   Langlet, O., Ecological variability and taxonomy of forest trees, in T. T. Kozlowski (ed.), *Tree Growth.* New York: The Ronald Press Company, 1962, pp. 357–69.

31   Linstrom, G. A., Interim forest tree improvement guides for the central states. *U.S. Forest Service Research Paper CS-12,* pp. 1–63. 1965.

32   Mergen, F., Selection of superior forest trees, in T. T. Kozlowski (ed.), *Tree Growth.* New York: The Ronald Press Company, 1962, pp. 327–44.

33   Merrill, S., Jr., et al., Relative growth and yield of budded and seedling tung trees for the first seven years in the orchard, *Proc. Amer. Soc. Hort. Sci.,* 63:119–27. 1954.

34   Odland, M. L., and C. L. Noll, The utilization of cross-incompatibility and self-incompatibility in the production of F_1 hybrid cabbage, *Proc. Amer. Soc. Hort. Sci.,* 55:391–402. 1950.

35   Pearson, O. H., Unstable gene systems in vegetable crops and implications for selection, *HortScience* 3(4):271–74. 1968.

36   Pendleton, R. A., H. R. Fennell, and F. C. Reimer, Sugar beet seed production in Oregon, *Ore. Agr. Exp. Sta. Bul. 437.* 1950.

37   Peterson, C. E., A gynoecious inbred line of cucumber, *Mich. Agri. Exp. Sta. Quart. Bul.,* 43:40–42. 1960.

38   Pike, L. M., and M. A. Mulkey, Use of hermaphrodite cucumber lines in development of gynoecious hybrids, *HortScience* 6(4):339–40. 1971.

39   Rangaswamy, N. S., Control of fertilization and embryo development, in P. Mahashwari (ed.), *Recent Advances in the Embryology of Angiosperms.* Delhi, India: Univ. of Delhi, Int. Soc. Plant Morph., pp. 327–53. 1963.

40   Rollins, S. F., The Plant Variety Protection Act. *Seed World,* 109(9): 8–9. 1971.

41   Schubert, G. H., and R. S. Adams, *Reforestation Practices for Conifers in California.* Sacramento: California State Division of Forestry, 1971.

42   Sprague, G. F., Production of hybrid corn, *Iowa Agr. Exp. Sta. Bul. P48,* pp. 556–82. 1950.

43   Stanley, R. G., Physiology and uses of tree pollen, *Agr. Sci. Rev.,* 3:9–17. 1965.

44   Stoutemyer, V. T., Seed propagation as a nursery technique, *Proc. Plant Prop. Soc.,* 10:251–55. 1960.

45   Stuke, W., Seed and seed handling techniques in production of walnut seedlings, *Proc. Plant Prop. Soc.,* 10:274–77. 1960.

46   U.S. Dept. Agr., *Woody Plant Seed Manual.* Misc. Publ. 654. Washington, D.C.: U.S. Govt. Printing Office. 1948.

47   Weaver, R. J., *Plant Growth Substances in Agriculture.* San Francisco: W. H. Freeman, 1972.

48 Westwood, M. N., Arboretums—a note of caution on their use in agriculture, *HortScience,* 1:85–86. 1966.

49 Wright, J. W., New forest tree varieties. *Agr. Sci. Rev.,* 1:27–37. 1963.

50 Wyman, D. Seeds of woody plants, *Arnoldia,* 13:41–60. 1953.

SUPPLEMENTARY READING

Briggs, F. N., and P. Knowles, *Introduction to Plant Breeding.* San Francisco: W. H. Freeman, 1967.

Seeds. Yearbook of Agriculture. Washington, D.C.: U.S. Govt. Printing Office, 1961.

U.S. Dept. of Agriculture, Forest Service. *Seeds of Woody Plants in the United States.* C. S. Schopmeyer, ed. Agr. Handbook No. 450. 1974.

Techniques of Seed Production and Handling 5

The production of high-quality seed is of prime importance to the propagator, whether he collects or produces the seed himself, or obtains the seed from others. In the production of any crop, the cost of the seed is usually minor compared to other production costs. Yet no single factor is as important in determining the success of the operation.

SOURCES OF SEED
Commercial Seed Production
The commercial production of seeds is a large, specialized industry, the success of whose efforts is a vital concern of the plant propagator. The commercial seed industry produces cereal and forage crop seed, vegetable seed, and annual, biennial, and perennial flower seeds for both commercial growers and home gardeners.

Commercial seed is raised primarily in areas where the environmental conditions are suited for such production, for unfavorable climatic conditions before and during harvest might impair viability (2). For instance, much vegetable and flower seed is produced in somewhat limited areas that are especially suitable but whose climate may not necessarily correspond with that of the locality in which the plants are later grown. Even within a species, some cultivars may differ from others in their requirements for seed production. Low humidity and lack of summer rainfall are desirable conditions for crops which must be dried for harvesting. Too low a humidity, on the other hand, may be undesirable for some plants, since it can cause premature shattering of the seed pods and cracking of seeds during harvesting. For this reason, much flower seed production in the U.S. is located in the coastal areas of the West where the moist air from the nearby ocean and frequent night and morning fogs tend to prevent the pods from dehiscing during harvest.

Conditions of low atmospheric moisture make it relatively easy to control fungus and bacterial diseases. For instance, two seedborne diseases, anthracnose and bacterial blight, are serious problems in bean seed production in all but the drier parts of the country. The mountain states and the central valley of California are particularly desirable for bean seed production because of low humidity. Furrow irrigation may be more desirable than sprinklers for seed production because it allows better disease control.

Adequate isolation of cross-pollinated plants is also important and may influence the choice of seed-production areas. For example, little sweet corn seed is produced in the Midwestern states of the U.S., owing to the large amount of field corn grown there which would contaminate the seed.

Seed Collecting

Tree and shrub seed may be collected from plants not grown specifically for their seeds. Seeds of native species of plants for forest planting and similar purposes may be obtained from natural stands in the forests and other wild areas. Such seed may be collected from trees felled in logging operations, from standing trees, or from squirrel caches. Seed-production areas (see p. 84) have been designated in certain regions. Other sources may include parks, roadways, streets, or woodlots. Seed selection for genetic adaptability to a growing area is essential, as described in Chapter 4.

Seed Orchards

Seed orchards or plantations (see p. 84) can maintain a particular seed source under careful genetic control. They can also maintain a seed source of a particular item that might otherwise be difficult to locate, produce higher seed yields than would otherwise be found in nature, and prevent virus contamination (see p. 197). This procedure is used in producing fruit-tree rootstocks. It also has importance in forestry, where scions of genetically superior trees are grafted into orchards for seed production.

Fruit-Processing Industries

Seeds of fruit plants, which are used for producing rootstocks, may be obtained as by-products of fruit-processing industries. For instance, pear seed may be obtained from canneries, apple seed from canneries or cider presses, and some cultivars of apricot and peach seed from fruit-drying yards. An example of this is the 'Lovell' peach, which has been an important drying peach in California and also an important rootstock for a number of stone fruits. Peach pits are collected in large numbers at drying yards, cleaned, and sold to nurserymen. This source can be undesirable because it is difficult to know the genetic identity or the virus status of the seed source.

HARVESTING AND PROCESSING SEEDS
Maturity and Ripening

A seed becomes mature when it has reached a stage at which it can be removed from the plant without impairing the seed's germination. This usually means that it has reached a stage on the plant when no further increase in dry weight will occur (2). If the fruit ripens or the seed is harvested when the embryo is insufficiently developed, the seed is apt to be thin, light in weight, shriveled, poor in quality and short-lived. If harvesting is delayed too long, the fruit may dehisce or "shatter," drop to the ground, or be eaten or carried off by birds or animals.

Fruit ripening may follow three patterns (*24*):

Type 1 includes dry fruits that dehisce, "shatter" regularly, or open at maturity and distribute single, dry seeds. Plants of this group include many that are uncultivated or, if cultivated, are not grown for their seeds or fruits. This includes follicles, pods, capsules, siliques, cones of conifers, some grasses (caryopses) and Compositae plants (achenes) that have not been subject to agricultural selection. To avoid seed losses in general, such plants must be harvested before the fruits are fully ripe and cured or dried before the seeds are extracted. The collector or harvester must know rather closely the criteria that indicate the optimum yield and quality for that particular kind of seed. A problem with Type 1 seeds is that those on a single plant may not develop uniformly and at harvest a portion of the seeds will be immature. Much of this poor seed can be eliminated by screening or blowing to remove the lightweight seed.

Early seed harvest may be desirable for seeds of some species of woody plants that produce dormant seeds (see p. 115) which germinate the first spring after harvest. If seeds become dry and the seed coats harden, they may not germinate until the second spring (*33*).

Type 2 includes fruits that do *not* dehisce or shatter and *do not* disseminate their seeds immediately upon maturity. This pattern is found with many crop plants (corn, beans, wheat) which are grown for their seeds and have undergone considerable selection for agricultural purposes. In many cases, the nearly dry seeds are harvested directly from the mature, standing plant. Other plants having a tendency for the stems to "lodge," or bend over, are cut and piled for drying and curing prior to harvesting.

Type 3 includes seeds produced in fleshy fruits. In nature, many of these fruits are eaten by birds and the seeds disseminated through their digestive tracts. Such fruits must become sufficiently soft to facilitate removal of the fleshy portion. Overly ripe fruit that fall to the ground may produce injury to the seed, through either heating or the effects of microorganisms. Fruits allowed to dry around the seed may contribute to a very hard covering that increases dormancy problems.

Harvesting Procedures for Seeds
of Herbaceous Plants

Plants with dry seeds and fruit that dehisce or shatter readily (Type 1) are cut (often by hand) and placed on a canvas or tray to dry for one to three weeks. This procedure is used for many flower crops, such as delphinium, pansy, and petunia, a number of vegetables, as onion, cabbage and other cruciferous plants, and okra. Where only a few plants from a backyard garden are involved, they may be cut and dried by hanging them upside down in a paper bag.

In threshing, the plants are beaten, flailed, or rolled to loosen the seed from the plant. Special seed harvesting machines that are modified for individual crops are used to minimize mechanical damage. The essential parts are a revolving cylinder, which acts as a beater to loosen the seeds, and devices to separate the good seeds from the rest of the plant along with chaff, dirt, and other debris. Threshing may also be accompanied by pulling large rollers across the plants. For

Figure 5–1   Threshing box useful for cleaning small lots of dry seed.

small lots of seeds, beating, flailing, or screening by hand may be utilized. Figure 5–1 shows a small threshing box that may be convenient to use.

Further cleaning may be required following threshing to remove dirt, chaff, extraneous plant parts, and weed and other crop seeds of different shapes and sizes. Small lots of seed can be cleaned by screening or by pouring back and forth from one container to another, letting the wind blow away the lighter materials. Commercial seed cleaning and processing utilizes various kinds of specialized equipment, such as screens of different sizes, air blasts, and gravity separators (*31*). These take advantage of differences in such physical characteristics between the seeds and the materials to be removed as size, thickness, weight, friction, and color.

Seeds of field-grown crops that produce indehiscent fruits, such as cereals, grasses, and corn (Type 2), can be harvested by cutting and threshing the standing plant in a single operation by a combine. These machines are developed for particular crops or modified in order to minimize damage. However, some crops may first be cut, windrowed, or placed in piles or "shocks," and allowed to dry before harvesting. Low humidity is important during this period to facilitate drying and curing. Rain at harvest may cause damage that can result in seeds which later show low vigor. Seeds of this group may be somewhat difficult to remove and the force required to dislodge them may result in mechanical damage. Such injury can reduce viability and result in abnormal seedlings (see Figure 5–2). Some of these injuries are internal and not noticeable at the time, but they result in low viability after storage (*1, 15, 24*). Damage is a potential factor in any operation involving beating or flailing of the seeds, and is most likely to occur if the machinery used is not properly adjusted. Usually less injury occurs if the seeds are somewhat moist at harvest (around 12 to 15 percent).

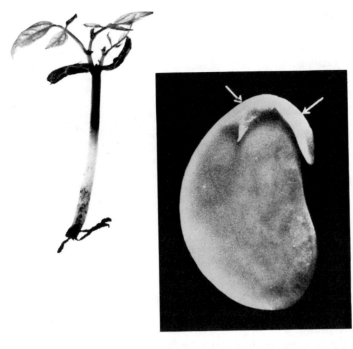

Figure 5–2   Seed injury during harvest can affect viability and produce seedling abnormalities. A break in the radicle below the cotyledons *(right arrow)* can prevent germination. Baldhead in lima beans results from a break in the stem of the embryo immediately above the cotyledons *(left arrow).* The slower development of the shoots *(left)* arising from the cotyledon axils results in delayed maturity and lower yields. From R. W. Allard, "Production of Dry, Edible Lima Beans in California," *California Agriculture Experimental Station Circular 423.*

Most seeds must be dried following harvest. If left in bulk for even a few hours, seeds that have more than 20 percent moisture will heat; this impairs viability. Drying may occur either naturally in open air if the humidity is low, or artificially with heat or other devices (*13*). Drying temperatures should not exceed 110° F (43° C); if the seeds are quite wet, 90° F (32° C) is better. Too-rapid drying can cause shrinkage and cracking and can sometimes produce hard seed coats. The minimum safe moisture content for most seeds is in the range of 8 to 15 percent.

Herbaceous plants with fleshy fruits include tomato, pepper, eggplants, and the various kinds of cucurbits. Fruits are harvested ripe, or in some cases overripe (e.g., cucumber and eggplant). For small lots of seeds, the fruits may be cut open and the seeds within scooped out, cleaned, and dried. Commercial harvesting of these crops utilizes machines which macerate the fruit. The pulp and seeds are then separated by fermentation, mechanical means, or washing through screens.

Fermentation is used in extracting tomato seed. The macerated fruits are placed in large barrels or vats and allowed to ferment for about 4 days at about 70° F (21° C), with occasional stirring. If the process is continued too long, sprouting may result. Higher temperatures during fermentation can shorten the required time. As the pulp releases the seeds, the heavy, sound seeds sink to the bottom of the vat, and the pulp remains at the surface. Following extraction, the seeds are washed and dried either in the sun or in dehydrators. Additional cleaning is sometimes necessary to remove dried pieces of pulp and other materials. Extraction by fermentation is particularly desirable for tomato seed, because it controls bacterial canker.

Special machines have been developed to extract and clean the seeds from the pulp of cucumber and other vine crops. Following separation, the seeds are washed and dried as with the fermentation process.

Harvesting Procedures for Tree and Shrub Seed

Both dry and fleshy fruits can be collected from standing trees by shaking them onto canvas, by knocking with poles, by using cone hooks attached to long poles (as for conifers), or by hand picking. Collection may be made from trees felled in logging operations. In some cases squirrel caches yield high-quality seeds. Seeds of some street trees—elm and hackberry, for instance —can be swept up from the street. Seeds on low trees and shrubs can be harvested by hand picking, clipping seed stalks, or knocking. Cones from tall conifer trees can be removed by mechanical tree shakers.

Seed viablity of tree and shrub species varies considerably from year to year, from locality to locality, and from plant to plant. Before seeds are collected from any particular source, it is desirable to cut open a number of fruits and examine the seed contents to determine the percentage with sound, well-matured embryos. Such an examination is known as a *cutting test*. Although not necessarily a reliable viability test, it avoids taking seed from a source which produces only empty, unsound seed. X-ray examination can also accomplish this purpose (see p. 154).

Seeds are extracted from dry, dehiscent fruits as pods and capsules of such plants as certain woody legumes (honey locust), *Caragana, Ceanothus,* poplar, and willow. The fruits of these plants are dried by being spread in shallow layers on canvas, on cloth, on the floor or shelves of open sheds, or in screen-bottom trays. Air-drying takes one to three weeks if the relative humidity is low.

Extraction may be accomplished by beating the pods with a flail, treading them under foot, or rubbing them through screens. For larger operations, commercial threshers are more suitable. A macerator, developed by the USDA Forest Service for this purpose, has also been described (*30*). Made of metal, it is sufficiently watertight that running water can be used in it when macerating fleshy fruits. Fruits and seeds pass through the hopper and are macerated by means of a revolving cylinder like that of a threshing machine. Such machines will extract and clean 500 lb of seed per hour. Following extraction, the seeds may require additional cleaning to remove extraneous materials, using conventional screening or fanning mills.

Extraction of conifer seeds requires special procedures. Cones of some species will open if they are dried in the open air for 2 to 12 weeks. Others must be force-dried at higher temperatures in special heating kilns. Under such conditions, the cones will open within several hours or at most 2 days. The temperature of artificial drying should be 115° to 140° F (46° to 60° C) depending upon the species, although a few require even higher temperatures. Jack pine (*Pinus banksiana*) and red pine (*P. resinosa*), for example, need temperatures of 170° F (77° C) for 5 to 6 hours. Caution must be used with high temperatures; overexposure will damage seeds. After the cones have been dried, the scales open, exposing the seeds. The cones must then be shaken by tumbling or raking to remove the seeds. A revolving wire tumbler or a metal drum is used where large numbers of seeds are to be extracted. The seeds should be removed immediately upon drying, or the cones may close.

Some seeds have appendages or coverings which are removed. Carrot seeds have spines and tomato seeds have hairs. Conifer seeds have wings, which are removed except in species whose seed coats are easily injured, such as incense cedar (*Calocedrus*). Fir (*Abies*) seed is easily injured, but the wings can be removed if the operation is done gently. Redwood (*Sequoia* and *Sequoiadendron*) seed have wings which are an inseparable part of the seed. For small lots of seed, dewinging can be done by rubbing the seeds between moistened hands or trampling or beating the seeds packed loosely in sacks. For larger lots of seeds, special dewinging machines are used. The seeds are cleaned after extraction to remove the wings and other light chaff. As a final step, separation of heavy, filled seed from light seed is accomplished by gravity or pneumatic separators.

Fleshy fruits include berries (grape), drupes (peach, plum), pomes (apple, pear), aggregate fruits (raspberry, strawberry), and multiple fruits (mulberry). The flesh must be removed promptly to prevent spoilage and injury to the seeds (*14, 27*). Cleaning by hand, treading in tubs, and rubbing through screens are suitable methods for small lots of seed. Relatively large fruits can be conveniently cleaned by placing the fruits in a wire basket and washing them with water from a high-pressure spray machine. For larger lots of seed, a macerator is convenient to use. This (see Figure 5–3) is constructed with a watertight feeder; water is passed through it along with the fleshy fruits, and the pulverized mass is diverted into a tank where the pulp and seeds can be separated by flotation.

Figure 5–3   Power-operated seed cleaner for fleshy fruits. The whole fruits are placed in top of cleaner with side door closed. A stream of water from a hose washes over the fruit. A rapidly rotating plate with low vertical flanges is raised slightly from the bottom of the cleaner and removes the flesh from the fruit. The flesh washes out through the bottom of the cleaner leaving inside the cleaned seeds, which can then be removed through the side door.

Flotation involves placing the seeds and pulp in water so that the heavy, sound seeds will sink to the bottom and the lighter pulp, empty seeds, and other extraneous materials will float off the top. It can also be used for removing the poor seeds and other materials from dry fruits, such as acorn fruits infested with weevils.

Fruit-tree seeds are sometimes collected from the wastes of drying yards, canneries, or juice presses. These seeds should be separated from the pulp, washed as quickly as is convenient, and not allowed to ferment or heat in the piles. Such seeds can be handled by flotation or washing with high-pressure spray machines.

The small berries of some species, such as *Cotoneaster, Juniperus,* and *Viburnum,* are somewhat difficult to process because of their size and the difficulty in separating the seeds from the pulp. One way to handle such seeds is to crush the berries with a rolling pin, soak them in water for several days, and then remove the pulp by flotation. A better method for small-seeded fleshy fruits is to use an electric mixer or blender (*28*). To avoid injuring the seeds, the metal blade of the latter machine can be replaced with a piece of rubber, $1\frac{1}{2}$ in. square, cut from a tire casing. It is fastened at right angles to the revolving axis of the machine (*35*). A mixture of fruits and water is placed in the mixer and stirred for about 2 minutes. When the pulp has separated from the seed, the pulp is removed by flotation. The above procedure is satisfactory for fruits of such species as (*28*):

Amelanchier (serviceberry)	*Ligustrum* (privet)
Berberis (barberry)	*Lonicera* (honeysuckle)
Celtis (hackberry)	*Magnolia*
Cornus (dogwood)	*Nyssa* (tupelo)
Crataegus (hawthorn)	*Parthenocissus* (Virginia creeper)
Diospyros (persimmon)	*Rhamnus* (buckthorn)
Elaeagnus angustifolia (Russian olive)	*Rosa* (rose)
Elaeagnus commutata (silverberry)	*Sassafras*
Fragaria (strawberry)	*Shepherdia* (buffalo berry)
Gaylussacia (huckleberry)	*Sorbus* (mountain ash)
Ilex (holly)	*Symphoricarpos* (snowberry)
Juniperus (juniper)	*Taxus* (yew)

SEED STORAGE

Seeds are usually stored for varying lengths of time after harvest. Viability at the end of any storage period is the result of (a) the initial viability at harvest, as determined by factors of production and methods of handling, and (b) the rate at which deterioration takes place. This rate of physiological change, or aging, is associated with (a) the kind of seed and (b) the environmental conditions of storage, primarily temperature and humidity.

Seeds of certain species are short-lived if they are not allowed to germinate immediately in their natural habitat (*13*). Their period of viability may be as short as a few days, months, or at most a year. This group particularly includes certain spring-ripening seeds of temperate zone trees as poplar (*Populus*), some maple (*Acer*) species, willow (*Salix*), and elm (*Ulmus*). Their seeds drop to the ground and normally germinate immediately. Seeds of many tropical plants grown under high temperature and humidity conditions are short-lived. The group includes such plants as sugar cane, rubber, jackfruit, macadamia, avocado, loquat, citrus, many palms, litchi, mango, tea, choyote, cocoa, coffee, tung, kola, and others. Another group with short-lived seeds includes many aquatic plants of the temperate zone, such as wild rice (*Zizania aquatica*), pondweeds, arrowheads, rushes, and others. Many tree nut and similar species produce seeds with large fleshy cotyledons and are relatively short-lived, particularly if allowed to dry out—hickories and pecan (*Carya*), birch (*Betula*), hornbeam (*Carpinus*), hazel and filbert (*Corylus*), chestnut (*Castanea*), beech (*Fagus*), oak (*Quercus*), walnut (*Juglans*), and buckeye (*Aesculus*). Seed longevity of many of these species can be increased significantly with proper handling and storage, as will be described later.

The following are some woody plant species whose seeds are short-lived (*35*):

Acer (some species) (maple)	*Mahonia* (Oregon grape)
Alnus (alder)	*Myrica* (bayberry)
Amelanchier (serviceberry)	*Nandina*
Ampelopsis	*Nyssa*
Aralia	*Ostrya* (hophornbeam)
Asimina (paw paw)	*Populus* (poplar)
Cedrus (cedar)	*Potentilla* (cinquefoil)
Cercidiphyllum (Katsura tree)	*Rhus* (sumac)
Chamaecyparis lawsoniana (false cypress)	*Salix* (willow)
Clerodendron (glory bower)	*Sassafras*
Cryptomeria	*Shepherdia* (buffalo berry)
Davidia (dove tree)	*Sophora*
Diospyros (persimmon)	*Spiraea*
Franklinia	*Staphylea* (bladder-nut)
Halesia (silver bell)	*Stewartia*
Lindera (spice bush)	*Styrax*
Liriodendron	*Taxus* (yew)
Liquidambar (sweetgum)	*Ulmus* (elm)
Lycium (wolfberry)	*Zelkova*
Magnolia	

Seeds which may considered medium-lived are those which remain viable for periods of 2 or 3 up to perhaps 15 years, providing the seeds are stored at low humidity and, preferably, at low temperatures. Seeds of most conifers and commercially grown vegetables, flowers, and grains fall in this group.

Seeds which are long-lived, even at warm temperatures, generally have hard seed coats that are impermeable to water. If the hard seed coat remains undamaged, such seeds should remain viable for at least 15 to 20 years. The maximum life can be as long as 75 to 100 years and perhaps more. Records exist of seeds being kept in museum cupboards for 150 to 200 years, some still retaining viability.

Indian lotus (*Nelumbo nucifera*) seeds which had been buried in a Manchurian peat bog for an estimated 1000 years germinated perfectly when the impermeable seed coats were cracked (*23, 34*). Some weed seeds retain viability for many years (50 to 70 years or more) while buried in the soil, even though they have imbibed moisture (*25*). Longevity seems related to dormancy induced in the seeds by the environmental conditions deep in the soil (see page 116).

Factors Affecting Seed Viability During Storage

The storage conditions that maintain seed viability are those which slow respiration and other metabolic processes without injuring the embryo. The most important conditions for achieving this are reduced moisture content of the seed, reduced storage temperature, and modification of the storage atmosphere. Of these, the temperature-moisture relationships have the most practical significance.

Moisture Content

Many kinds of the short-lived seeds listed previously lose viability if the moisture content becomes low. For instance, in silver maple (*Acer saccharinum*) seeds, the moisture content was 58 percent in the spring when the seeds matured. Viability was lost when the moisture content dropped below 30 to 34 percent (*18*). Citrus seeds can withstand only slight drying (*5, 9*). The same is true with seeds of some water plants, such as wild rice, which can be stored directly in water at low temperatures (*21*). The large fleshy seeds of oaks (*Quercus*), hickories (*Carya*), and walnut (*Juglans*) lose viability if allowed to dry following ripening. They are normally stored moist for no more than 1 year (*30*).

On the other hand, medium- to long-lived seeds of most other species must be dry to survive long periods of storage. A 4 to 6 percent moisture content is favorable for prolonged storage (*10*), although a somewhat higher moisture level is allowable if the temperature is reduced (*29*) (see Table 5–1). However, if the moisture content of the seed is low (1 to 2 percent), loss in viability and reduced germination rate can occur in some kinds of seeds (*13*). Some seeds can be stored at these low moisture levels but must be rehydrated with water vapor before planting (*22*). Moisture in seeds is in equilibrium with the relative humidity of the storage atmosphere. This moisture percentage varies with the kind of seed, which is a function of the kind of storage reserves within the seed (*3*). Longevity of seed is maximum if stored in a relative humidity range of 20 to 25 percent (*13*).

Fluctuations in seed moisture during storage will reduce seed longevity (*4*). Consequently the ability to successfully store seeds exposed to the open atmosphere varies greatly in different climatic areas. Dry climates are conducive to increased longevity; in areas with high relative humidity, seed life is shorter. In open storage in tropical areas, seed viability is particularly difficult to maintain.

Table 5–1   Moisture-temperature relationships
of vegetable seeds in storage for a one-year period.*

Kind of Seed	Estimated Maximum Safe Seed Moisture (Percent) for Average Storage Temperature Indicated		
	40° to 50° F (4.5° to 10° C)	70° F (21° C)	80° F (26.5° C)
Bean, kidney	15	11	8
Bean, lima	15	11	8
Beet	14	11	9
Cabbage	9	7	5
Carrot	13	9	7
Celery	13	9	7
Corn, sweet	14	10	8
Cucumber	11	9	8
Lettuce	10	7	5
Okra	14	12	10
Onion	11	8	6
Pea, garden	15	13	9
Peanut (shelled)	6	5	3
Pepper	10	9	7
Spinach	13	11	9
Tomato	13	11	9
Turnip	10	8	6
Watermelon	10	8	7

* Adapted from Table 2 in E. H. Toole (29).

Various storage problems arise with increasing seed moisture (13). At 8 or 9 percent or more insects are active and reproduce; above 12–14 percent (65 percent R.H. or more) fungi are active; above 18–20 percent heating may occur; and above 40–60 percent germination occurs.

Storage in sealed, moisture-resistant containers is advantageous for long storage, but the moisture content of the seed must be low at the time of sealing. In fact, a seed moisture content in a sealed container of 10 to 12 percent (in contrast to 4 to 6 percent) is worse than storage in an unsealed container (10).

Temperature
Reduced temperature invariably lengthens the storage life of seeds and, in general, can offset the adverse effect of a high moisture content. Harrington has given two "rules of thumb" (13): (a) for seeds not adversely affected by low moisture conditions, each 1 percent decrease in seed moisture, between 5 and 14 percent, doubles the life of the seed; and (b) each decrease of 9° F (5° C), between 32° and 112° F (0° and 44.5° C) in storage temperature, also doubles seed storage life. On the other hand, seeds stored at low temperature but at a high relative humidity may lose viability rapidly when moved to a higher temperature (6).

Table 5-2 Approximate moisture content * of seeds in equilibrium with air at various relative humidities.†

Seeds	Relative Humidity (%)				
	15	30	45	60	75
Cereals (starchy)					
Rye	7.0	8.5	10.5	12.0	15.0
Rice (milled)	6.5	9.0	10.5	12.5	14.5
Sorghum	6.5	8.5	10.5	12.0	15.0
Corn (maize)	6.5	8.5	10.5	12.5	14.5
Wheat	6.5	8.5	10.0	11.5	14.5
Barley	6.0	8.5	10.0	12.0	14.5
Oats	5.5	8.0	9.5	12.0	14.0
Vegetables (starchy)					
Spinach	7.0	8.0	9.5	11.0	13.0
Pea	5.0	7.0	8.5	11.0	14.0
Bean, snap	5.0	6.5	8.5	11.0	14.0
Oil seeds					
Soybean	—	6.5	7.5	9.5	13.0
Flaxseed	4.5	5.5	6.5	8.0	10.0
Vegetables (oily)					
Tomato	6.0	7.0	8.0	9.0	11.0
Carrot	5.0	6.0	7.0	9.0	11.5
Cucumber	6.0	7.0	7.5	8.0	9.5
Lettuce	4.0	5.0	6.0	7.0	9.0
Cabbage	3.5	4.5	6.0	7.0	9.0

* 77°F (25°C) moisture content wet basis, in percent.
† From Harrington (13).

Subfreezing temperatures, at least down to 0° F (−18° C), will increase storage life of most kinds of seeds but moisture content should be in equilibrium with 70% R.H. or lower, or the free water in the seeds may freeze and cause injury (13). Such storage is particularly useful for conifer seeds (26, 30). Refrigerated storage should be combined with dehumidification or sealing dried seeds in moisture-proof containers.

Storage Atmosphere

Modification of the storage atmosphere has been attempted in order to increase longevity of various short-lived seeds. Procedures for changing the atmosphere are to create a vacuum, increase the carbon dioxide level, or replace oxygen with nitrogen or other gases. In a number of studies, no consistent benefits were demonstrated when such procedures were compared with other good storage methods (8). However, benefit has been noted for very short-lived seeds of some tropical plants. Viability of rubber (*Hevea brasiliensis*) seeds was prolonged by sealed storage in 40 to 45 percent carbon dioxide (19). Likewise, sugar cane seed viability was lengthened by sealing the air-dry seeds in cans with 9 grams of calcium chloride for each liter of space after displacing the air with carbon dioxide and storing at near-freezing temperature (32).

Types of Seed Storage
Open Storage, Without Moisture or Temperature Control
Many kinds of seeds that are used in large commercial volumes are stored in bins or in sacks or other containers. Under these conditions, seed longevity depends on the relative humidity and temperature of the storage atmosphere, although it also depends upon the kind of seed and their condition at the beginning of storage. Retention of viability consequently varies with the climatic factors of the area in which storage occurs. Poorest conditions are found in warm, humid climates; best storage conditions occur in dry, cold regions. Fumigation or insecticidal treatments may be necessary to control insect infestations.

Uncontrolled storage can be used for many kinds of commercial seeds for at least a year—i.e., to hold seeds from one season to the next. Seeds of many species will retain viability for longer periods except under the most adverse conditions. Most agricultural seeds, such as cereals, grasses, or forage crops retain viability for 4 to 5 years or more.

Approximate storage periods for vegetable seeds in open storage in the eastern United States are as follows (*20*):

1 year—sweet corn, onion, parsley, parsnip
2 years—beet, pepper
3 years—asparagus, bean, celery, carrot, lettuce, pea, spinach, tomato
4 years—cabbage, cauliflower, eggplant, okra, pumpkin, radish, squash
5 years—cucumber, endive, muskmelon, watermelon

The following information concerning storage of various flower seeds is based upon individual samples stored over a period of 10 years in a relatively dry climate (*11*). It gives an approximate guide for handling such seeds and shows the range of longevity represented by such seeds. These seeds retained 50 percent of their original viability at the times indicated.

1 year or less—*Delphinium* (perennial), *Iberis umbellata* (candytuft), *Kochia*
2 years—*Callistephus chinensis* (aster), *Helichrysum monstrosus* (straw flower), *Kochia trichophylla* (summer cypress)
3 years—*Callistephus chinensis* (aster), *Centaurea candidissima* (dusty miller), *Delphinium chinensis*, *Phlox drummondi*, *Verbena*
4 years—*Delphinium* (annual), *Iberis*
5 to 10 years—

Althea rosea	*C. segetum*
Alyssum maritimum	*Cosmos*
Arctotis grandis (African daisy)	*Dianthus*
Calendula	*Eschscholtzia californica* (California
Calliopsis	poppy)
Centaurea	*Gilia capitata*
Chrysanthemum leucanthemum	*Lathyrus odoratus* (sweet pea)
(Shasta daisy)	*Mathiola* (stock)

Nigella damascena (love-in-a-mist)	*Tagetes* (marigold)
Papaver (Shirley poppy)	*Tropaeolum majus* (nasturtium)
Petunia	*Verbena*
Salpiglossis	*Viola* (pansy)
Scabiosa grandiflora	*Zinnia*
Schizanthus (butterfly flower)	

Seeds which have a water-impervious seed coat will retain viability in open storage for many years (i.e., 10 to 20 years or more) once they have been dried. Open storage is adequate. Some of the woody plants whose seeds are handled in this manner are:

Acacia sp.	*Eucalyptus sp.*
Albizia sp. (albizzia)	*Koelreuteria paniculata* (golden rain
Amorpha fruticosa (indigo bush)	tree)
Caragana arborescens	*Rhus ovata* (sumac)
(Siberian pea shrub)	*Robina pseudoacacia* (locust)
Elaeagnus sp. (Russian olive)	*Tilia* (linden)

Warm Storage with Humidity Control

Improved seed storage can be achieved by drying the seeds, then storing them in humidity-controlled rooms. Toole (*29*) recommends for vegetable seeds: (a) seeds exposed to 80° F (27° C) for more than a few days should be in air with relative humidity no higher than 45 percent; (b) seeds exposed to 70° F (21° C) should be at no higher humidity than 60 percent; (c) very short-lived seed (onion, peanut), old seed, or those contaminated with fungi should be kept at an even lower humidity.

Dry seeds may be stored in sealed moisture-resistant containers. Various containers are used which vary in durability and strength, cost, protective capacity against rodents and insects, and ability to retain or transmit moisture. Different materials vary in moisture transmitting qualities (*12*). Those completely resistant to moisture transmission include tin cans (if properly sealed), aluminum cans, hermetically sealed glass jars, and aluminum pouches. Those almost as good (80 to 90 percent effective) are polyethylene (3-mil or thicker) and various types of aluminum-paper laminated bags. Somewhat less desirable, in regard to moisture transmission, are asphalt and polyethylene laminated paper bags and friction-top tin cans. Paper and cloth bags give no protection against moisture change.

Seeds kept in sealed containers for long periods should have a very low moisture content. Listed below are the maximum seed moisture contents under sealed storage for a number of kinds of vegetable seed:

5.0 percent: tomato, pepper, cabbage, cauliflower	6.5 percent: parsley
	7.0 percent: carrot, peas
5.5 percent: celery, lettuce	7.5 percent: beet
6.0 percent: cucumber, watermelon, cantaloupe, onion, eggplant	8.0 percent: spinach, sweet corn, beans, lawn grasses

Cold Storage With or Without Humidity Control

Almost without exception, seed longevity in the species listed in the two previous categories would be enhanced by reducing the storage temperature to 50° F (10° C) or less. Toole (*29*) recommends that for vegetable seed stored at 40° to 50° F (4.5° to 10° C), the relative humidity be no higher than 70 percent, preferably no higher than 50 percent. When removed from storage at a relative humidity of more than 50 percent, seeds should be dried to a safe moisture content, unless they are planted immediately.

Low storage temperatures, down to freezing or lower, might be desirable if the need justifies the added cost. Below-freezing temperatures can be used for very long storage. In such low-temperature storage, R.H. should be 70 percent or lower (*13*). Sealed containers should be used. Seeds may gain moisture and form ice crystals that damage seeds. Such seeds may deteriorate quickly unless brought out of storage.

The most effective storage procedure is to dry seeds to a low moisture content (3 to 8 percent), place them in sealed containers, and store them at a very low temperature. Such a procedure is best for seeds which deteriorate rapidly or when the maximum possible storage life is needed for a particular reason, for instance, to maintain seeds of breeding stocks.

Cold storage of tree and shrub seed used in nursery production is generally advisable if the seeds are to be held for longer than 1 year (*16, 17, 26, 30*), except for hard-coated seeds listed previously. Seed storage is useful in forestry because of the uncertainty of good seed-crop years. Seeds of the following species are best stored under cold and dry conditions (*35*):

Abies sp. (fir)	*Platanus sp.* (sycamore)
Acer (some species) (maple)	*Populus sp.* (poplar)
Arbutus sp. (madrone)	*Prunus sp.* (stone fruits)
Berberis sp. (barberry)	*Pseudotsuga menziesii* (Douglas fir)
Ceanothus sp. (ceanothus)	*Ptelea trifoliata* (hoptree)
Celtis sp. (hackberry)	*Rhus sp.* (sumac)
Cercis sp. (redbud)	*Rubus sp.* (blackberry)
Cupressus sp. (cypress)	*Sambucus sp.* (elder)
Fraxinus sp. (ash)	*Sassafras albidum* (sassafras)
Gaylussacia (huckleberry)	*Sequoia sempervirens* (redwood)
Gleditsia dioca (Kentucky coffee tree)	*Sequoiadendron sp.* (sequoia)
Hamamelis sp. (witch hazel)	*Sorbus sp.* (mountain ash)
Juniperus sp. (juniper)	*Symphoricarpos sp.* (snowberry)
Larix sp. (larch)	*Thuja sp.* (arborvitae)
Liquidambar sp. (sweetgum)	*Tsuga sp.* (hemlock)
Maclura pomifera (osage orange)	*Vitex sp.* (chaste tree)
Malus sp. (apple)	*Vitis sp.* (grape)
Picea sp. (spruce)	*Zanthoxylum* (prickly ash)
Pinus sp. (pine)	

Cold Moist Storage

The storage temperature should be 32° to 50° F (0° to 10° C). Seeds should not be dried but should either be stored in a container that will maintain their high moisture content, or be mixed with moisture-retaining material. The relative humidity in storage should be 80 to 90 percent. The procedure is similar to moist-chilling (stratification) (see page 157). Acorns and large nuts can be dipped in paraffin or sprayed with latex paint before storage to preserve their moisture content (*17*).

Examples of species whose seeds require this storage treatment are:

Acer saccharinum (maple)
Aesculus sp. (buckeye)
Carpinus caroliniana (American hop hornbeam)
Carya sp. (hickory)
Castanea sp. (chestnut)
Corylus sp. (filbert)
Citrus sp. (citrus)

Eriobotrya japonica (loquat)
Fagus sp. (beech)
Juglans sp. (walnut)
Litchi
Nyssa silvatica (tupelo)
Persea (avocado)
Quercus sp. (oak)

REFERENCES

1 Anonymous, A study of mechanical injury to seed beans, *Asgrow Monograph No. 1*. New Haven: Associated Seed Growers, Inc., 1949.

2 Austin, R. B., Effects of environment before harvesting on viability, in E. H. Roberts (ed.), *Viability of Seeds*. Syracuse, N. Y.: Syracuse Univ. Press, 1972.

3 Barton, L. V., Relation of certain air temperatures and humidities to viability of seeds, *Contrib. Boyce Thomp. Inst.*, 12:85–102. 1941.

4 ———, Effect of moisture fluctuations on the viability of seeds in storage, *Contrib. Boyce Thomp. Inst.*, 13:35–45. 1943.

5 ———, The storage of some citrus seeds, *Contrib. Boyce Thomp. Inst.*, 13:47–55. 1943.

6 ———, Seed storage and viability, *Contrib. Boyce Thomp. Inst.*, 17:87–103. 1953.

7 ———, Storage and packeting of seeds of Douglas fir and Western hemlock, *Contrib. Boyce Thomp. Inst.*, 18:25–37. 1954.

8 Bass, L. N., D. C. Clark, and E. James, Vacuum and inert-gas storage of lettuce seed, *Proc. Assoc. Off. Seed Anal.*, 52:116–22. 1963.

9 Childs, J. F. L., and G. Hrnciar, A method of maintaining viability of citrus seeds in storage, *Proc. Fla. State Hort. Soc.*, 64:69. 1948.

10 Crocker, W., and L. V. Barton, *Physiology of Seeds*. Waltham, Mass.: Chronica Botanica, 1953.

11 Goss, W. L., Germination of flower seeds stored for ten years in the California state seed laboratory, *Calif. Dept. Agr. Bul.*, 26:326–33. 1937.

12   Harrington, J. F., The value of moisture-resistant containers in vegetable seed packaging, *Calif. Agr. Exp. Sta. Bul. 792*, pp. 1–23. 1963.

13   ———, Seed storage and longevity, in T. T. Kozlowski (ed.), *Seed Biology*. New York: Academic Press, 1972, pp. 145–245.

14   Haut, I. C., and F. E. Gardner, The influence of pulp disintegration upon viability of peach seeds, *Proc. Amer. Soc. Hort. Sci.*, 32:323–27. 1934.

15   Hawthorn, L. R., and L. H. Pollard, *Vegetable and Flower Seed Production*. New York: Blakiston Co., 1954.

16   Heit, C. E., Propagation from seed, Part 10: Storage methods for conifer seed, *Amer. Nurs.*, 126(20):14–15. 1967.

17   ———, Propagation from seed, Part 11: Storage of deciduous tree and shrub seed, *Amer. Nurs.*, 126(21):12–13, 86–94. 1967.

18   Jones, H. A., Physiological study of maple seeds, *Bot. Gaz.*, 69:127–52. 1920.

19   Kidd, F., The controlling influence of carbon dioxide in the maturation, dormancy, and germination of seeds, Part II, *Proc. Royal Soc. Lond., Ser. B.*, 87:609–25. 1914.

20   MacGillivray, J. H., *Vegetable Production*. New York: Blakiston Co., 1953.

21   Muenscher, W. C., Storage and germination of seeds of aquatic plants. *New York (Cornell Univ.) Agr. Exp. Sta. Bul.*, 652:1–17. 1936.

22   Nutile, G. E., Effect of dessication on viability of seeds, *Crop Science*, 4:325–28. 1964.

23   Ohga, I., The germination of century old and recently harvested Indian lotus with special reference to the effect of oxygen supply, *Amer. Jour. Bot.*, 13: 754–59. 1926.

24   Pollock, B. M., and E. E. Roos, Seed and seedling vigor, in T. T. Kozlowski (ed.), *Seed Biology*, Vol. I. New York: Academic Press, 1972.

25   Roberts, E. H., Dormancy: A factor affecting seed survival in the soil, in E. H. Roberts (ed.), *Viability of Seeds*. London: Chapman and Hall, Ltd., pp. 320–59. 1972.

26   Schubert, G. H., and R. S. Adams, *Reforestation Practices for Conifers in California*. State of California Division of Forestry, 1971.

27   Scott, D. H., J. G. Waugh, and F. P. Cullinan, An injurious effect of peach juice on germination of the seed, *Proc. Amer. Soc. Hort. Sci.*, 40:283–85. 1942.

28   Smith, B. C., Cleaning and processing seeds, *Amer. Nurs.*, 92(11):13–14, 33–35. 1950.

29   Toole, E. H., Storage of vegetable seeds, *USDA Leaflet No. 220* (rev.). 1958.

30   USDA, C. S. Schopmeyer (ed.), *Seeds of Woody Plants in the United States*. Agriculture Handbook No. 450. 1974.

31   Vaughn, C. E., B. R. Gregg, and J. C. Delauche, *Seed Processing and Handling*, State College, Miss.: Miss. State Univ. Seed Technology Library, 1968.

32   Verret, J. A., Sugar cane seedlings, *Assoc. Hawaii Sugar Technol. Rpts.*, 7:15–23. 1928.

33 Wells, J. S., *Plant Propagation Practices.* New York: Macmillan, 1955.

34 Wester, H. V., Further evidence on age of ancient viable lotus seeds from Tulantien deposit, Manchuria. *HortScience,* 8:371–77. 1973.

35 Wyman, D., Seeds of woody plants, *Arnoldia,* 13:41–60. 1953.

SUPPLEMENTARY READING

Barton, L. V., *Bibliography of Seeds.* New York: Columbia Univ. Press, 1967.

Hawthorn, L. R., and L. H. Pollard, *Vegetable and Flower Seed Production.* New York: Blakiston Co., 1954.

James, E., *An Annotated Bibliography on Seed Storage and Deterioration.* U.S. Dept. of Agr., Agr. Res. Serv. No. 34–15–1, 1961, and No. 34–15–2, 1963.

Kozlowski, T. T. (ed.), *Seed Biology,* Vols. 1, II, III. New York: Academic Press., 1972.

Roberts, E. H., (ed.), *Viability of Seeds.* London: Chapman and Hall, Ltd., 1972.

Seed World, published monthly. 434 S. Wabash Ave., Chicago, Ill. 60605.

U.S. Department of Agriculture, *Seeds. The Yearbook of Agriculture.* Washington, D.C.: U.S. Govt. Printing Office, 1961.

U.S. Department of Agriculture, Forest Service, C. S. Schopmeyer (ed.), *Seeds of Woody Plants in the United States.* Agri. Handbook No. 450. 1974.

Principles of Propagation by Seeds 6

A seed consists of an embryo and its stored food supply, surrounded by protective seed coverings. At the time the seed separates from the parent plant, metabolism is at a low level and there is no apparent growth activity within the seed. During seed germination, cell metabolism increases, the embryo resumes active growth, seed coverings rupture, and the seedling plant emerges. At fertilization genetic information was passed to the new embryo through DNA contained in the chromosomes. During development of the seedling, genes either become active or remain inactive according to a coded, predetermined schedule. Activated genes will engage in protein synthesis until they are again deactivated at some further point in development. Consequently, the specific enzymes and structural proteins available at a given time are the basis for differential growth and development.

Seeds of most plants are unable to germinate while they are enclosed within the fruit attached to the parent plant, or for a period of time after fruit ripening and seed dispersal. Seeds whose germination is prevented by their own internal mechanisms are said to be *dormant.* If the seed is capable of immediate germination when subjected to proper environmental conditions, the seed is said to be *quiescent,* or nondormant. The distinction between dormant and quiescent seeds is that, in the first, germination control is due to mechanisms internal to the seed, and in the second to environmental factors external to the seeds.

Consequently, three conditions must be fulfilled before germination will begin. *First,* the seed must be viable; that is, the embryo must be alive and capable of germination. *Second,* internal conditions of the seed must be favorable for germination; that is, physical or chemical barriers to germination must have disappeared. *Third,* the seed must be subjected to appropriate environmental conditions. The essential requirements are available water, proper temperature, a supply of oxygen, and sometimes light. However, internal conditions within seeds may change with time. As a consequence, environmental requirements may also change because they can be affected by the internal state of the seed.

THE GERMINATION PROCESS (17, 92)

The **first stage** of germination—awakening or activation—may be complete within a matter of minutes or hours.

Water is absorbed by the dry seed and the moisture content increases rapidly, then stabilizes. Initial water absorption involves the imbibition of water by colloids of the dry seed, which softens the seed coverings and causes hydration of the protoplasm. As a result the seed swells and the seed coats may break; since water absorption is largely a physical process, it can take place even in nonviable seeds.

Components (i.e., various DNA and RNA molecules) of the protein-synthesizing system of cells become active. These were formed during seed development and became inactive as the seed matured. After uptake of water, however, this system is reactivated to permit protein synthesis to continue. The enzymes produced by protein synthesis control metabolic activities of the cell (19). Some enzymes were produced during seed development and must be reactivated. Others are synthesized after germination begins.

Energy becomes available for protein synthesis from high-energy bonds in adenosine triphosphate (ATP) located in the mitochondria. Some of these systems were formed during seed development, preserved in the dormant seed, and reactivated during cell hydration.

The **second stage** of germination involves digestion and translocation. Water uptake and respiration now continue at a steady state. The existing cell systems have been activated and the protein-synthesizing system is functioning to produce various new enzymes, structural materials, regulator compounds, nucleic acids, etc., to carry on the cell functions and synthesize new materials. Enzymes appear and begin to digest reserve substances (fats, protein, carbohydrates) in the storage tissues (cotyledons, endosperm, perisperm, or megagametophyte) to simpler chemical compounds. These compounds are then translocated to the growing points of the embryonic axis to be used for growth and the production of new plant parts.

The metabolic patterns in different plant species depend largely upon the type of chemical reserves in the seed. Fats and oils—the major food constituents in the seeds of most higher plants—are converted enzymatically to fatty acids and eventually to sugar. Storage proteins, present in most seeds, are a source of nitrogen essential to the growing seedling. Starch, present in many seeds as an energy source, is converted to sugar. The sequence of the metabolic patterns that occur during germination involves the activation of specific enzymes at the proper time and regulation of their activity. Control may be exercised by various biochemical processes within cells and may depend on the presence of specific chemicals.

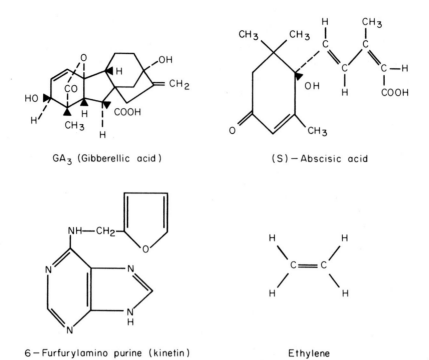

GA₃ (Gibberellic acid)

(S) – Abscisic acid

6 – Furfurylamino purine (kinetin)

Ethylene

Figure 6–1 Chemical structure of various plant regulators involved in germination or germination control. Gibberellic acid, abscisic acid, kinetin, and ethylene.

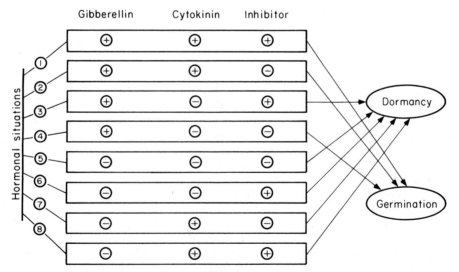

Figure 6–2 According to this model, germination occurs only in the presence of gibberellin. If an inhibitor is present it offsets the effects of gibberellin and germination does not take place (No. 3). But if cytokinin is then added, it blocks the effects of the inhibitor and permits germination to take place (No. 1). From Khan (54). Copyright 1971 by the American Association for the Advancement of Science.

Four classes of plant hormones have been implicated in control of seed germination *(36)* (see Figure 6–1). The *gibberellins* (GA) have been most directly implicated as having a role in germination. For example, control of the food-mobilizing system of stage two has been shown to occur in experiments with barley seeds. When the quiescent dry seed imbibes water, gibberellin appears in the embryo and is translocated to the *aleurone* (a layer 3 to 4 cells thick surrounding the endosperm), where it causes new production of α-amylase. This enzyme moves to the endosperm, where it converts starch to sugar, which is in turn translocated to the growing points of the embryo to provide energy for growth. In barley seeds GA also promotes the induction or stimulation of other specific enzymes.

Inhibitors are substances that can prevent germination. A significant compound of this class is the naturally occurring *abscisic acid* (ABA), which can block the stimulation of germination by gibberellin *(54)*. There is evidence from studies with barley that ABA interferes with GA enhancement of α-amylase production by inhibiting RNA synthesis *(20)*.

Cytokinins are another class of natural endogenous hormones that appear to control seed germination, probably at the level of the DNA→RNA transcription system *(54)*. In some plants these compounds can overcome the action of ABA in inhibiting gibberellin action *(54)*.

Ethylene has been implicated in seed germination and may have a control function in seeds of some plants *(31, 53, 85)*.

Many plant physiologists believe that germination is regulated by balances among various endogenous promoting and inhibiting substances, with gibberellin the major promoter and ABA the inhibitor *(2, 103)*. (See Figure 6–2.) Cytokinin has been described as having a "permissive" role in that it overcomes inhibition and enables gibberellin to function *(54)*. In some cases kinetin appears to act synergistically with ethylene *(85)* to promote germination.

The **third stage** of germination consists of *cell division* in the separate growing points of the embryo axis followed by the expansion of the seedling structures. (*Cell elongation* and *emergence of the radicle* are early indicators of germination and may mark the end of the first stage. The initiation of cell division in the growing points appears to be independent of the initiation of cell elongation *(10, 39)*, and may not be directly involved in initial radicle emergence.) Once growth begins in the embryo axis, fresh weight and dry weight of the seedling increase but weight of storage tissue decreases. Respiration, as measured by oxygen uptake, increases steadily with advance in growth. Storage tissues of the seed eventually cease their metabolic activities except in plants where the cotyledons become active in photosynthesis.

As germination proceeds, the structure of the seedling soon becomes evident. The embryo consists of an axis bearing one or more seed leaves, or *cotyledons*. The growing point of the root, the *radicle,* emerges from the base of the embryo axis. The growing point of the shoot, the *plumule,* is at the upper end of the embryo axis, above the cotyledons. The seedling stem is divided into the section below the cotyledons—the *hypocotyl*—and the section above the cotyledons—the *epicotyl.*

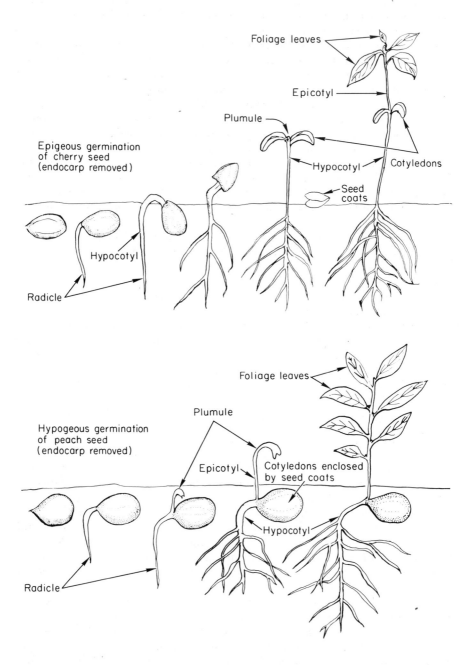

Figure 6–3   Seed germination in dicotyledonous plants. *Top:* epigeous germination of cherry. The cotyledons are above ground. *Below:* hypogeous germination of peach. The cotyledons remain below ground.

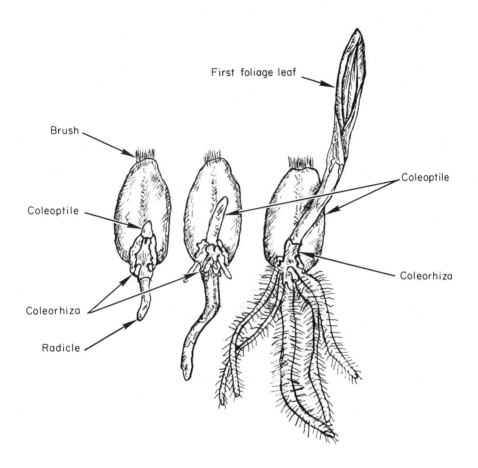

First foliage leaf

Brush

Coleoptile

Coleoptile

Coleorhiza

Coleorhiza

Radicle

Coleorhiza

Figure 6–4   Seed germination in a monocotyledonous plant—wheat. Redrawn from Weier, Stocking, and Barbour, *Botany—An Introduction to Plant Biology.* 4th ed. John Wiley & Sons, Inc., 1970.

The initial growth of the seedling follows one of two patterns. In one type—*epigeous* germination—the hypocotyl elongates and raises the cotyledons above the ground. In the other type—*hypogeous* germination—the lengthening of the hypocotyl does not raise the cotyledons above the ground, and only the epicotyl emerges. The germination pattern differs between dicotyledonous plants (Figure 6–3) and monocotyledonous plants (Figure 6–4).

QUALITY OF SEEDS
A supply of viable seed is essential in successful seed propagation. However, the difference between a live and a dead seed may not be distinct, but may be characterized by gradual decline in vigor and the appearance of

necrosis or injuries in localized areas of the seed *(47, 78)*. *Viability* is represented by the *germination percentage,* which expresses the number of seedlings which can be produced by a given number of seeds. Additional characteristics of viability are that germination should be prompt, the growth of the seedling vigorous, and its appearance normal. Vigor of seed and seedling are, therefore, important attributes of quality but may be somewhat difficult to measure. A low germination percentage and a low germination rate are often associated. A reduction in both may result from genetic differences among cultivars, incomplete seed development on the plant, injuries during harvest, improper processing and storage, disease, and aging. Loss in viability is usually preceded by a period of declining vigor *(97)*.

Seeds with low vigor may not be able to withstand unfavorable conditions in the seed bed, may succumb to attacks by disease organisms, or may lack the strength to emerge if planted too deeply or if the soil surface is crusted. Field survival of low-vigor seeds is apt to be less than a laboratory germination percentage would indicate.

MEASUREMENT OF SEED QUALITY

If one measures the time sequence of germination of a given lot of seeds, or the emergence of seedlings from a seed bed, one usually finds a pattern that is like the germination curve shown in Figure 6–5. There is an initial delay in the start of germination, then a rapid increase in the number of seeds that germinate, followed by a decrease in the rate of appearance. When viability is less than 100 per cent, the exact end point may be difficult to ascertain.

Germination is measured on two parameters—the *germination percentage* and the *germination rate*. Vigor may be indicated by these measurements, but seedling growth rate and morphological appearance must also be con-

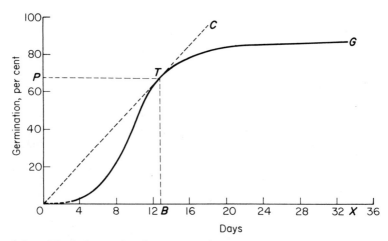

Figure 6–5   Typical germination curve of a sample of slow germinating seeds. After an initial delay, the number of seeds germinating increases, then decreases. Such a curve can be used to measure germination. See text for explanation. Redrawn from Czabator *(24)*.

sidered. Sometimes abnormally growing seedlings appear and reduce seed quality *(47)*.

In cases of slow germination, statements of germination percentage should involve a time element, indicating the number of seedlings produced within a specified length of time. Germination rate can be measured by several methods. One determines the number of days required to produce a given germination percentage. Another method calculates the average number of days required for radicle or plumule emergence as follows:

$$\text{Mean days} = \frac{N_1 T_1 + N_2 T_2 + \ldots N_x T_x}{\text{total number or seeds germinating}}$$

N values are the numbers of seeds germinating within consecutive intervals of time; T values indicate the times between the beginning of the test and the end of the particular interval of measurement. Kotowski *(58)* has used the reciprocal of this formula multiplied by 100 to determine a *coefficient of velocity*. Gordon *(37)* has suggested the term *germination resistance* as the time (hours or days) to average germination, based on seeds that germinate.

Czabator *(24)* has suggested another measurement for seeds of woody perennials where germination may be slow: the *germination value (GV)*. It includes both the germination rate and percentage. To calculate GV, a germination curve, as shown in Figure 6–5, must be obtained by periodic counts of radicle or plumule emergence. The important values on the curve are T—the point at which the germination rate begins to slow down—and G—the final germination percentage. These points divide the curve into two parts—a rapid phase and a slow phase. Peak value (PV) is the germination percentage at T, divided by the days to reach that point. Mean daily germination (MDG) is the final germination percentage divided by number of days in the test. For example:

$$GV = PV \times MDG$$
$$GV = \frac{68}{13} \times \frac{85}{34}$$
$$= 5.2 \times 2.5$$
$$= 13.0$$

REGULATION OF GERMINATION: DORMANCY

Ripening of seeds includes the development of internal mechanisms that control the onset of germination so as to coincide with periods during the year having environmental conditions most likely to favor survival of the seedlings. In seeds of most plants, reduction in moisture to a level below that required for germination is one method of control, but most freshly harvested seeds have additional regulating mechanisms that prevent germination even if environmental conditions are seemingly favorable.

Under some conditions, the seeds of certain plants may germinate while still attached to the mother plant. The phenomenon is called *vivipary*. Interesting adaptations of this phenomenon are found in mangrove trees, swamp-growing species in which embryos are produced that germinate on the tree to produce seedlings with a long javelin-shaped root. These eventually fall and become embedded in the mud below (*89*). Premature sprouting may occur in some grain crops during periods of wet weather during harvest (*109*). However, the tendency toward vivipary is inherited and is automatically selected against as a defective character.

The term *dormancy* has broad applications in plant physiology to refer to lack of growth in any part due either to internally or externally induced factors (*102*). Seed technologists, on the other hand, define a *dormant seed* in a somewhat narrower sense as being due to conditions within the seed (other than nonviability) (*100, 101*). In this sense, a *dormant* seed is one that fails to germinate even though it has absorbed water and is exposed to favorable temperature and oxygen levels. If the seed can germinate immediately upon the absorption of water without a barrier to germination, the embryo is said to be *quiescent,* or nondormant (*15*). The internal conditions of the dormant seeds may be quite different from those of the quiescent seeds.

In the "Rules for Testing Seeds," the Association of Official Seed Analysts further distinguishes between "hard seeds" and "dormant seeds" (*4*). *Hard seeds* include those which cannot absorb moisture because of their impermeable seed coat. Dormant seeds include those that fail to germinate because restrictive influences within the seed block some physiological reaction in the embryo even though moisture is absorbed.

When conditions that prevent germination exist within the seed at the time it matures on the plant, the state of the seed has been called *primary dormancy* (*21*). The physiological changes that occur within the seed so that germination can take place are referred to in horticultural literature as *after-ripening*. Dormancy may be transitory, lasting only for a few days, disappearing with normal handling in dry storage; on the other hand, dormancy removal may require complex and time-consuming treatments. Once a seed has passed through an after-ripening period it may again become dormant if the imbibed seed is subjected to particular unfavorable conditions. This is referred to as *secondary dormancy* (*21*).

Ecologically, germination-controlling mechanisms appear to have arisen as adaptations for natural survival. The particular requirements for germination are related to the environmental conditions where the plant species has evolved (*57, 93, 113*).

Such mechanisms are particularly important in plants growing in desert or cold regions where environmental conditions may not be favorable for germination immediately following dissemination of the seed. In the temperate zone, spring-ripening seed on such trees as elm, maple, or willow germinates immediately, and the seedling establishes itself the same season. If fall-ripening seeds of trees growing in the same region germinated immediately, the seedlings would probably be killed during the winter. Seeds of many plants of the latter type require winter chilling for germination to take place, and consequently do not germinate until the following spring.

Seeds of some species can remain buried in the ground for many years and remain dormant even though imbibed with moisture. Germination occurs when the ground is disturbed and the seeds are exposed to a new set of environmental conditions, such as light, modified gaseous atmosphere, or fluctuating temperatures (*79*).

Propagators of cultivated plants have long recognized these germination-delaying phenomena and have learned to cope with them through the adoption of appropriate pregermination and handling procedures arrived at through trial and error.

The domestication of many seed-propagated cultivars has undoubtedly included selection for easier propagation.

Consequently, in propagation most difficulty in germination regulation is found in seeds of trees and shrubs of native plants, or those recently introduced into cultivation. Particularly difficult is getting quick and reliable results in testing seed for viability, especially when this is attempted shortly after the seed is harvested.

Categories of Seed Dormancy

Seed dormancy results from a number of different physiological causes. Nikolaeva (72) has provided a useful classification of dormancy systems according to their underlying physiological causes. In using this classification one must recognize that intergrading categories exist, so that it may sometimes be difficult to place a given plant species within a specific category. Likewise, more that one underlying mechanism may exist, although only one may be in control at a given time.

Group I Seeds where regulation occurs in the nonliving external seed coverings but the embryo itself is quiescent.

A *Hard seed coverings impermeable to moisture* (**seed coat dormancy**). Seeds fail to absorb water until the covering is modified by natural or artificial methods. "Hard seeds" are common in certain plant groups.

B *Hard seed coverings resistant to embryo expansion.* Probably few seeds fail to germinate solely because of this mechanism, but it may be a factor in delaying germination in seeds with hard shells (e.g., walnuts), "pits" (stone fruits and olive), or hardened pericarps *(Crataegus).*

C *Seed coverings containing chemical inhibitors.* Specific chemical substances that prevent germination are produced by many plants. These are commonly found in the pericarp, such as in juices of fleshy fruits or in the dry coverings that are retained in seeds of some plants. Inhibitors also may occur in the seed coats, endosperm, or the embryo, and may be involved in some other dormancy categories as well. Many tropical plants (72) and certain desert plants appear to produce specific inhibitors. These are reduced or eliminated by leaching with water or adsorption by soil.

Group II *Seeds with morphologically undeveloped (rudimentary) embryos.* Embryo size varies from very small to those completely filling the seed coats. The proportion to the storage tissue (endosperm, perisperm) also varies. Embryos that are very small at the time of fruit ripening must increase in size before germination occurs. This situation is common among tropical plant species, e.g., palms (72) and orchids, but somewhat less common in plants found in temperate zones (60).

Group III *Seeds with internal (endogenous) dormancy.* Germination is regulated by the inner tissues of the seed, that is, the embryo, the enclosing endosperm and inner integumental layer, or both. Seed coverings appear to play a role in all the subclasses of this group; differences among the subclasses result from variations in depth of dormancy within the embryo.

A   *Physiologically shallow dormancy.* This type is present in most freshly harvested seed and disappears with dry storage over a period of days or months. Regulation appears to come from the physiological activity of the inner seed coat or endosperm layers, with the embryo itself being relatively quiescent. Such seeds are likely to be light- and temperature-sensitive and responsive to mechanical abrasions, and to various chemicals, such as potassium nitrate, gibberellic acid, and kinetin. This kind of dormancy is common among herbaceous plants, both cultivated and native, and is probably present in most freshly harvested seeds.

B   *Physiologically intermediate dormancy.* Moist-chilling stimulates germination but may not be essential to overcome dormancy. This type of dormancy is found in seeds of various conifers and other woody plants. Regulation by the seed coverings appears to be more significant than the conditions within the embryo.

C   *Physiologically deep dormancy.* This type of dormancy disappears with prolonged moist-chilling. Regulation is predominantly within the embryo, although the seed coverings appear also to be involved. This class is common in seeds of trees and shrubs and those of some herbaceous plants of the temperate zone and colder climates, where seeds overwinter in the ground and germinate in the spring. Within this group there are variations in embryo size relative to endosperm ranging from (a) small, through (b) intermediate, to (c) full-sized.

Two other subgroups are known: (a) seeds that require a warm period prior to the cold, moist period for root and hypocotyl growth (various lily *(Lilium)* species, *Viburnum,* peony), and (b) seeds requiring a cold period followed by a warm period for the root to grow, then followed by a second cold period to stimulate the shoot (various native perennials of the temperate zone).

Group IV   Combined, or *double dormancy.* Both seed coat (external) dormancy and embryo (internal) dormancy occurs, and the required treatments must be given in sequence. This class includes seeds of various woody tree and shrub species; these are some of the most difficult for propagators to handle because of the long period before germination will occur, up to two years in some cases.

ENVIRONMENTAL FACTORS AFFECTING GERMINATION
Water Availability

A water-absorption curve for seeds has three parts: (a) an initial rapid uptake which is mostly imbibitional, (b) a slow period, and (c) a second increase as the radicle emerges and the seedling develops. Because of their colloidal nature, dry seeds have great absorptive power for water, both in storage (see Chapter 5), and in the germination medium, depending upon the nature of the seed, the permeability of the seed covering, the availability of water in the surrounding medium, and the temperature (86, 87). Higher temperatures increase water uptake. Once the seed germinates, the water supply to the seedling depends upon the ability of the radicle to grow into the germination medium and for the new roots to absorb water.

The moisture supplied to the germinating seed may affect both the germination percentage and the germination rate (28, 41, 48). Germination percentage tends to be equal over most of the range of available soil moisture from field capacity (FC) to permanent wilting percentage (PWP). Differences among species become evident as the moisture content of the soil approaches dryness (*PWP*). Some seeds germinate only above *PWP;* others can germinate below *PWP.* For instance, vegetable plants can be grouped according to the moisture requirements for seed germination (*43*) as follows:

Group 1 Seeds that will germinate in soils with moisture from permanent wilting percentage (or a little above) to moisture content above field capacity.

cabbage	sweet corn	muskmelon	pepper
turnip	squash	cucumber	onion
radish	watermelon	tomato	carrot

Group 2 Seeds that will germinate in soils from intermediate moisture content to above field capacity.

snap bean	pea	endive
lima bean	lettuce	beet

Group 3 Seeds that will germinate only in soils near field capacity.
celery

Group 4 Seeds that germinate well at lower moisture contents but show reduced germination near field capacity.

spinach   New Zealand spinach

Rate of seedling emergence from a seed bed is influenced particularly by the available moisture supply. From a point approximately halfway through the range from *FC* to *PWP* there is a decline in rate, as illustrated in Figure 6–6 (*5, 28, 41*).

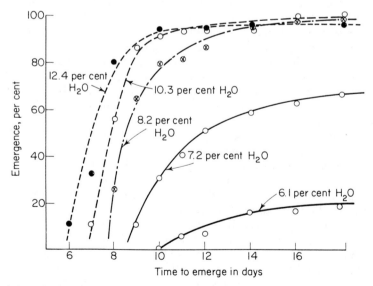

Figure 6–6 The effect of different amounts of available soil moisture on the germination (emergence) of 'Sweet Spanish' onion seed in Pachappa fine sandy loam. From Ayers (*5*).

Availability of water to the germinating seed may be limiting under conditions where excess or free water is not present. Two properties of the germination medium, referred to as the "matrix potential" and the "osmotic potential" (57, 76), are important.

The *matrix potential* is the ability of the water to move by capillarity from the pores of the soil to the seed. Rate of movement depends upon the pore structure of the germination medium and the closeness and distribution of the soil-seed contact. As moisture is removed from the soil by the seed, the area nearest the seed becomes dry and must be replenished by water in pores further away.

Osmotic potential depends upon the presence of solutes (salts) in the soil solution. Excess soluble salts in the germination medium may inhibit germination and reduce seedling stands (5). Such salts may originate in the soil and other materials used in the germination medium, the irrigation water, or in excessive fertilization. Since the effects of salinity become more acute when the moisture supply is low and the concentration of salts thereby increased, it is particularly important to maintain a high moisture supply in the seed bed where the possibility of high salinity exists. Surface evaporation from subirrigated beds can result in the accumulation of salts at the soil surface even under conditions in which salinity would not be expected. Planting seeds several inches below the crown of a sloping seed-bed can minimize this hazard (*11*).

Maintaining an adequate continuous moisture supply can be difficult because germination takes place in the upper surface of the germination medium, which is subject to fluctuations in temperature and moisture supply. The problem is greater with the necessarily shallow planting of small seeds, or where the germination rate is slow.

Methods of maintaining a uniform moisture supply to the seed are: (1) frequent or continuous watering, such as by a mist system, (2) use of a rooting medium of proper density, properly compacted about the seed, (3) deep seed planting (where moisture fluctuation is less), and (4) the surface application of mulch. On the other hand, excess watering accompanied by poor drainage can be deleterious, because it reduces aeration in the germination medium and favors "damping-off."

Seeds are sometimes soaked before planting to initiate the germination process and to shorten the time required for seedlings to emerge from the soil. Such a treatment may be advantageous with seeds which are normally slow to germinate or are hard and dry, or when certain dormancy conditions exist (see page 115). However, seeds which have imbibed water are easily injured and may be more difficult to plant.

Prolonged soaking can result in injury to the seed and can reduce germination (*6*). These harmful results have been attributed principally to the effects of microorganisms and to a reduced oxygen supply, although there seem to be other effects that are not well understood (7). If soaking is to be prolonged, the water should be changed at least every 24 hours.

Temperature

Temperature is perhaps the single most important environmental factor that regulates germination and controls subsequent seedling growth. Seeds of different species not only have upper (maximum) and lower (minimum) temperature limits for germination but may also respond to specific seasonal (summer-winter) cycles or diurnal (day-night) fluctuations. Such temperature requirements are not necessarily constant but may change with time (*102*) or interact with some other environmental factor, such as light.

In nature, temperature requirements largely determine the time of year that germination takes place and are a major factor in the distribution of species. For example, the Mediterranean climate of southern Europe is characterized by a hot, dry summer with a cool and wet fall, winter, and spring. Seeds of many species growing in this region ripen in the spring, remain dormant all summer, even if an occasional rain occurs, but then germinate in the cooler temperatures of fall when adequate rains occur (*93*). Many other similar relationships between temperature requirements and time of germination, some of them very intricate relationships, have been recorded (*57, 88, 93, 110*).

Temperature limits In cultivated plants, the temperature limits for seed germination are characteristic of the species but may vary with the particular cultivar. In general, temperature requirements reflect the environmental requirements of the original wild species from which they descended. However, generations of selection may have modified these characteristics to adapt them to conditions under cultivation and, in some cases, tended to eliminate some of the more precise requirements of related species in nature. Nevertheless, temperature requirements are very important and must be taken into account both in seed propagation and in seed-testing laboratories.

It is difficult to determine these temperature requirements precisely. For one thing, temperature affects both germination percentage and germination rate (*58*). Germination rate usually increases directly with temperature; that is, rate is very low at low temperature but increases continuously as temperature rises, like a chemical rate-reaction curve (*57*). Above an optimum level, where the rate is most rapid, a decline occurs as the temperatures approach a lethal limit and seed is injured (*42*). Germination percentage, on the other hand, may remain relatively constant, at least over the middle part of the temperature range, if sufficient time is allowed for germination to occur.

Three temperatures (minimum, optimum, and maximum) have usually been designated for seeds (*29*). Often these are determined at constant temperatures by the use of a seed germinator and may give misleading information, since in field propagation or in nature the seed is subjected to varying diurnal fluctuations which can produce results different from those obtained at constant temperature.

Nevertheless, seeds of different species, whether cultivated or native, can be usefully categorized in broad temperature-requirement groups, with specificity indicated for individual species or cultivar as needed. Seeds of many cultivated plants, after the usual seed handling operations, including a period of dry storage, will germinate over a wide temperature range from about 40° F (4.5° C) (or sometimes near freezing) to a maximum of the lethal limit—from 86° F (30° C) to about 104° F (40° C) (*42*). Exposure to higher temperatures for short periods may be possible for disease control (see page 200).

One group of seeds can be classified as *cool-season* because of their ability to germinate at relatively low temperatures. This category includes the large group of vegetable plants that originated in temperate climates. Seeds of some cool-season plants require low temperatures and fail to germinate at temperatures higher than about 77° F (25° C) or more (*12, 42, 46*). High temperature sensitivity, or *thermodormancy,* is found to occur in seeds of a number of important crops, such as lettuce, celery, endive, as well as in many flower species and a number of woody perennial plants. As described earlier in this section, this phenomenon is apparently a natural adaptation which, in nature, prevents germination during hot, dry summers immediately after seed maturity (*93*). Thermodormancy of dry seeds is commonly associated with light-sensitivity and other dormancy phenomena and tends to disappear with dry storage. It is particularly troublesome in seed-testing laboratories but can also interfere with establishment of good stands of particular crops, such as lettuce, when seeds are planted during warm periods. Procedures to overcome this type of dormancy are described in Chapter 7.

Seeds of another broad group of species, primarily from subtropical or tropical regions, are classed as *warm-season,* with a minimum germination requirement of about 50° F (10° C), such as asparagus, sweet corn, and tomato; or 60° F (15° C), such as beans, eggplant, pepper, and cucurbits. Seeds of some of these species, such as lima bean, cotton, soybean, and sorghum, are susceptible to "chilling injury." Exposure of the seed to temperatures of 50° to 60° F (10° to 15° C) during initial imbibition, as could occur if they are planted in a cold soil, can injure the embryo axis and result in abnormal seedlings (*78*). Chilling injury is more severe if the seed is very dry at the start of imbibition, or if the oxygen supply is limited.

Optimum temperatures are those most favorable for seed germination as well as seedling growth. This should be the range at which the largest percentage of seedlings are produced at the highest rate. On this basis, the optimum for quiescent seeds of many plants is between 80° and 95° F (26.5° and 35° C). The optimum may shift after germination begins since seedling growth may have different requirements from seed germination. The usual practice is to shift the seedlings to a somewhat lower temperature regime in order to prepare plants for transplanting and to reduce disease problems in the seed bed.

If seeds are germinated and the seedlings grown at high temperatures, it is important that other environmental conditions be favorable. Plants should have increased light, preferably long photoperiods, have adequate fertilization, and be under sterile conditions to eliminate disease pathogens. Increased CO_2 is also a useful component of this system (see page 24).

Alternating temperatures   Fluctuating day-night temperatures sometimes give better results than constant temperatures for both seed germination and seedling growth—a fact that has been known for a long time. Fluctuating temperatures are often standard practice in seed-testing laboratories, even for seeds not requiring it. The alternation should be of an 18° F (10° C) difference (*101*). This requirement is particularly important with dormant, freshly harvested seeds. Seeds of a few species will not germinate at all at constant temperatures. It has been suggested that one of the reasons imbibed seeds deep in the soil do not germinate is that soil temperature fluctuations disappear with soil depth (*79*).

Exchange of Gases Between Embryo and Atmosphere

Gases in the germination medium that can affect seed germination are oxygen (O_2), carbon dioxide, and possibly ethylene. Oxygen is essential for the respiratory processes in the germinating seed, and O_2 uptake can be measured shortly after imbibition begins. Rate of oxygen uptake is an indicator of germination progress and has been suggested as a measure of seed vigor (*78*). In general, O_2 uptake is proportional to the amount of metabolic activity taking place.

Carbon dioxide is a product of respiration and under conditions of poor aeration can accumulate. At lower depths in the soil increased CO_2 may inhibit germination to some extent but probably has a minor role in maintaining dormancy of such seeds (*79*). On the other hand, CO_2 levels higher (0.5 to 5.0 percent) than atmospheric (0.03 percent) have stimulated germination of subterranean clover (*Trifolium subterranean*) and some other legumes (*57*), and such higher levels interact with ethylene to overcome thermodormancy in lettuce seeds (*73*). It has been suggested that the production of these gases by clusters of seeds can stimulate germination by overcoming severe crusting of the soil (*31*).

Ethylene alone has been found to be given off by seeds of subterranean clover and may function in an independent role in stimulating germination of seed clusters under conditions of soil crusting (*31*). Ethylene may overcome dormancy in Virginia type peanut seeds, which are produced underground (*53*). Ethylene may also be used to overcome high temperature dormancy in lettuce, particularly if used in combination with kinetin (*85*), carbon dioxide (*70, 71*) or light (*73*) (see Figure 6–7).

Aeration in the germination medium   The amount of oxygen in the germination medium is affected by its low solubility in water and its slow diffusability into the medium. Thus, gaseous exchange between the germination medium and the atmosphere, where the O_2 concentration is 20 percent, can be reduced significantly by soil depth and, in particular, by a hard crust on the surface which can limit oxygen diffusion (*41*).

Oxygen supply is most decisively limited where there is excessive water in the medium. Poorly drained seed beds, particularly after heavy rains or irrigation, can result in the pore spaces of the soil being so filled with water that little oxygen is available to the seeds.

Seeds of different species vary in their ability to germinate at low oxygen pressure, as occurs under water (*68, 69*). Seeds of some water plants germinate readily under water, with germination inhibited in air. Rice seeds can germinate in a shallow layer of water. At low oxygen levels, rice seedlings, however, develop differently than those of other monocots. Shoot development is stimulated and the plumule grows to extend up through the water into the air; root growth is suppressed and poor anchorage results unless the water layer is drained away (*16*).

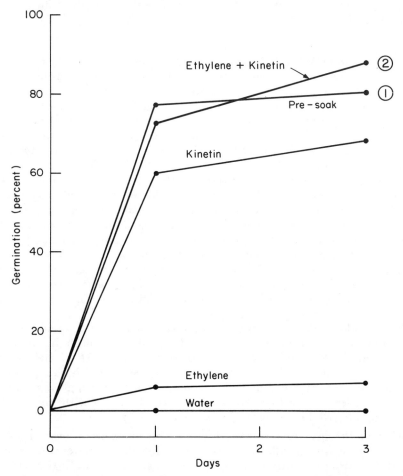

Figure 6–7   Germination of lettuce seed at high temperature (95°F; 35°C). No germination occurred in water, but 80% germination took place if seeds were presoaked in water for 24 hours at 77°F (25°C) (No. 1). Presoaking could be partially replaced by ethylene or by kinetin treatments and completely by a combination of both (No. 2). Treatments were 3 minutes in 100 mg/l kinetin, followed by germination on filter paper soaked with 100 mg/l ethephon, an ethylene-releasing chemical. From data of Sharples (*85*).

Aeration restriction by seed coats   In most seeds there is probably some physical restriction in the movement of gases (particularly oxygen) to the imbibed embryo, due either to the inner membranous seed coat or to the enclosing nucellus or endosperm (*15*). This situation has been used to explain the germination control by dormant embryos of some seeds (*18, 59, 96, 104*). For example, freshly-harvested dormant seeds which respond to light and are temperature sensitive will germinate if the embryo is excised, the seed coat altered, or the seed subjected to an oxygen level higher than that found in the atmosphere. The effect of after-ripening in dry storage has been attributed to an increase in permeability of the seed coverings. These permeability effects are operable only during the initial stages of germination because once germination occurs and the seed coat is ruptured the gas exchange capacity is completely altered.

Dormancy in cocklebur (*Xanthium*) seeds was long considered to be due to lack of oxygen because of inner seed coat impermeability (*95*). The two seeds that are present in a single bur differ in their level of dormancy (see Figure 6–8), as shown by a difference in response to an oxygen supply increased over that of the atmosphere. Later, the situation was shown to be more complex. Two water-soluble chemical inhibitors were found in the more dormant seeds (*108*). If these are removed by leaching, or the seed is subjected to high oxygen pressure, then germination will occur. The two seeds also differ in the seed coat strength and different germinating forces required to rupture them (*30*). Furthermore, treatment with kinetin will stimulate germination while abscisic acid will inhibit it (*56*).

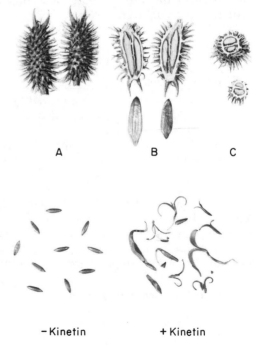

Figure 6–8   Germination of cocklebur *(Xanthium).* Two burs are shown at A, each of which contains two seeds (B and C); the smaller one is dormant but the other is not. Early experiments showed that low permeability to gases was a dormancy inducing factor. Later experiments *(108)* showed that the smaller, dormant, seed contains two water-insoluble inhibitors that prevent germination. These inhibitions can be overcome by kinetin. From Khan, et al. *(56)*.

Light

It has been known since the mid-nineteenth century that light can stimulate or inhibit germination of seeds of some plants (*23*). Plants whose seeds have an absolute requirement for light, and lose viability in a few weeks without it, include mistletoe (*Viscum album*), strangling fig (*Ficus aurea*), and *Areuthobium oxycedri,* all epiphytic * plants. Seeds of many species and cultivars with the type IIIA and IIIB dormancy (see p. 118) require light for germination for a period of time following harvest. The light requirement tends to disappear with dry storage and can often be overcome by chilling, alternating temperatures, or chemical treatments, including potassium nitrate (*4*), kinetin, gibberellic acid, or thiourea (*56*). Light stimulation will not occur if the germination temperatures are too high—about 75° F (24° C) or more (*99*). Light-requiring seeds include those of celery, lettuce, tobacco, most grasses, many conifers, many herbaceous flower plants, and many native "weed" species (*4*).

Seeds of another group are inhibited by light; these include some species of *Phacelia, Nigelia, Allium, Amaranthus,* and *Phlox.* Seeds of still other species respond to length of day (photoperiod). Eastern hemlock (*Tsuga canadensis*) (*88*) and birch (*Betula*) (*107*), for instance, have "long-day" seeds, but germination is also promoted by low temperatures. Photoperiodic control can partially replace a seed-chilling requirement, and vice versa. Seed germination of some other species, such as *Nemophila, Nigella, Veronica persica,* and *Eschscholzia californica* is inhibited by long days (*66*).

Germination control is exerted through a photochemically reversible reaction involving the response of a pigment (phytochrome) to light of particular wavelengths (*13, 98*):

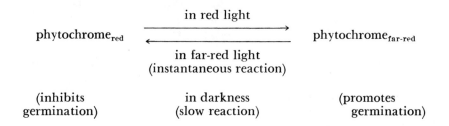

Exposure of the imbibed seed to red light (6400 Å to 6700 Å) causes the phytochrome$_r$ in it to change to phytochrome$_{fr}$, whereas exposure to far-red causes the phytochrome$_{fr}$ to change to phytochrome$_r$. The first step of this reaction can occur at moisture levels lower than that essential for germination, but the second step, where the photoreaction is linked to metabolic

---

* Nonparasitic plant growing on another.

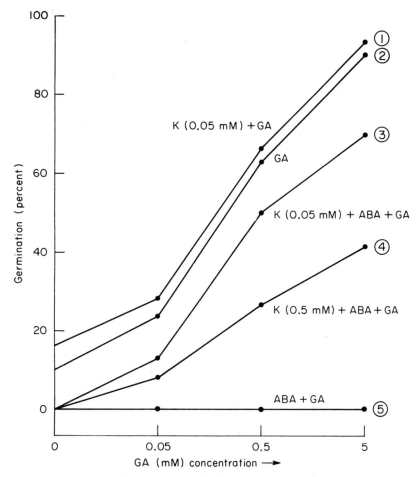

Figure 6–9   Germination of 'Grand Rapids' lettuce seed does not occur in darkness, as under the conditions of the experiment illustrated here, but is promoted
by increasing concentrations of gibberellic acid (GA) (No. 2). Abscisic acid (ABA)
completely inhibits gibberellin-promoted germination (No. 5). Kinetin (K) partially
overcomes the effect of ABA (Nos. 3 and 4). Kinetin does not, however, further
stimulate gibberellin-promoted germination in the absence of ABA (No. 1). Redrawn from Khan, et al. *(56)*.

reactions controlling germination, requires full seed imbibition. Protein or
enzyme induction involving gibberellin production is evidently involved
(94, 103).

A light requirement can be replaced by gibberellin (see Figure 6–9) to
produce germination in the dark which can, however, be inhibited by
abscisic acid (ABA). In turn this inhibition can be offset by kinetin *(56)*. In

the light, ABA-inhibited germination can be stimulated by kinetin (see Figure 6–10). These interactions can be predicted from Figure 6–2.

Exposure to far-red light (7200 Å to 7500 Å) produces a change to the alternate form, phytochrome$_r$, which inhibits germination. In the absence of any light, in the presence of oxygen, and at favorably low temperatures, a slow change to P_r takes place. In natural sunlight, red wavelengths dominate over far-red at a ratio of 2:1 so that the phytochrome tends to remain in the active P_{fr} form. Under a foliage canopy, far-red is dominant and the red/far-red ratio may be as low as 0.12:1.00 to 0.70:1.00, which mimicks darkness, thus inhibiting seed germination (79). Light sensitivity can be induced in nonsensitive seeds by exposing the imbibed seeds to conditions inhibiting germination, such as high temperature, high osmotic pressure, or germination-inhibiting gases in the soil (112).

Light-inhibited seeds are less common, but evidently their response is to the same phytochrome system. Blue light may also be involved, but the reactions are not known (67). Light control appears to involve the seed coats or the endosperm layer, since the light requirements disappear if the embryos are excised (33).

Light sensitivity is now recognized as a major ecological factor in adaptation of plant species. Light-sensitive seeds are often small and germination is favored by placement near the soil surface where the seedlings can emerge quickly and begin photosynthesis. Such seeds would not have the capacity to emerge from greater depths. Red light penetrates less deeply into the soil than far-red so that the red/far-red ratio becomes lower with depth until eventually darkness is complete (94). Imbibed light-sensitive seeds buried in the soil will remain dormant until such time as the soil is cultivated or disturbed so as to expose them to light. Similarly, seedling survival is not favored if the seed germinates in close proximity to other plants where there would be intense competition for light, nutrients, and water by the already established plant population.

In contrast, plants producing light-inhibited seeds tend to be found in dry desert environments where seedling survival would be enhanced if the seeds germinated at somewhat greater soil depths where there is less heat and more moisture.

For cultivated plants light stimulation has considerable significance in obtaining prompt seed germination in seed-testing laboratories (see Chapter 7).

Light and Seedling Growth

Light affects the growth processes in the seedling and has an important role in its emergence through the soil. In the darkness of the seed bed, the hypocotyl or epicotyl is elongated and etiolated and the leaves are unexpanded. There is a hook toward the end of the plumule which pushes through the ground. When the seedling shoot emerges into the light, shoot elongation decreases, the plumular hook straightens out, leaves expand, and normal growth takes place.

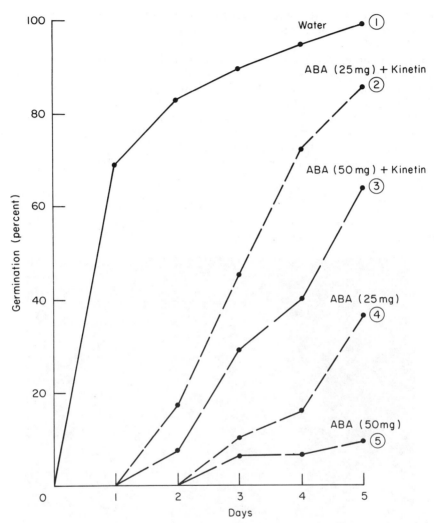

Figure 6–10 Germination of 'Grand Rapids' lettuce seed will occur in continuous light as in water control (No. 1). Abscisic acid (ABA) inhibits this light-promoted germination in proportion to concentration (Nos. 4 and 5). Germination is partially restored by treatment with kinetin (K) (Nos. 2 and 3). Redrawn from Khan, et al. *(56)*.

In the early germination stages, the seedling utilizes the reserve supply of the seed. Later growth depends upon the production of carbohydrates and other materials resulting from photosynthesis. Light of a relatively high intensity is necessary to produce sturdy, vigorous plants. Low light intensity results in etiolation and reduced photosynthesis.

An extremely high light intensity, on the other hand, can result in high temperatures which may produce heat injury to the seedling. Consequently, excessively high as well as excessively low light intensities need to be avoided. Partial shading is necessary for many plants during their early seedling growth out-of-doors.

Supplementary artificial light to maintain high light intensity and to increase the photoperiod can increase seedling growth particularly during winter when the day length is short and light intensity is low. Artificial light can be used as the sole illumination source *(91, 92)*. Light intensities should be 2000 to 2500 ft-candles or more, and the photoperiod should usually be at least 16 hours long. To obtain maximum benefits from high light intensity, temperature and humidity should be relatively high, mineral fertilizers should be added, and the ambient air should be enriched with CO_2 (see Figure 6–11).

A certain amount of blue light is necessary to produce stocky seedlings; red light alone produces tall, "leggy" plants. Satisfactory artificial light sources are fluorescent lamps of the "daylight" type (3500° K), or "white" (4500° K) types, or a mixture of the two, or the "cool-white" alone.

Figure 6–11   Effect of environmental factors during first three months on growth of paper birch *(Betula papyrifera)* seedlings. *Left:* seedlings grown in greenhouse under natural short-day winter conditions from November to February (Beltsville, Md.). Day/night temperatures, 76°/65°F (24°/18°C). CO_2 level, 350 ppm. (natural level, 300 ppm). *Center:* seedlings grown in greenhouse but with supplementary fluorescent light (cool white at 200 f.c.) to provide 16 hr day length. Temperatures same as described for first group. CO_2 level, 400 ppm. *Right:* seedlings grown in controlled environment propagation unit with 16 hr day length provided by 2000 f.c. from cool white fluorescent tubes plus supplementary incandescent light. Day/night temperatures, 79°/67°F (26°/19°C). CO_2 level, 400 ppm. Courtesy Donald Krizek.

DORMANCY FACTORS AFFECTING
SEED GERMINATION
Hard Seed Covers

Impermeability to water   Impermeability of the seed coats to water is a major factor in maintaining seed dormancy in species of certain families, including Leguminoseae, Malvaceae, Cannaceae, Geraniaceae, Chenopodiaceae, Convolvulaceae, and Solanaceae. The embryo is quiescent (nondormant) but is sealed inside a water impermeable covering that can preserve the seed at low moisture contents for many years even at warm temperatures. Among cultivated crops hardseededness is chiefly found in the herbaceous legumes, including clover, alfalfa, etc., as well as many woody legumes (*Robinia, Acacia,* etc.). Hardseededness depends on the genetic nature of the species and cultivar, environmental conditions during seed maturation, and environmental conditions during seed storage. Drying at high temperatures during ripening will increase hardseededness. Harvesting slightly immature seeds that are then prevented from drying can reduce or overcome hardseededness.

Impermeability of the seed coat is due to a layer of palisade-like macrosclereid cells, especially thick-walled on their outer surfaces, and having a layer of waxy, cuticular substances external to this (see Figure 6–12). Disintegration of the caps of such cells or mechanical stress separating the cells may allow water to enter and produce germination (*14*). The *hilum* (point of attachment between seed and funiculus) appears to act as a one-way valve, opening to allow water to escape in a dry atmosphere but closing in a moist atmosphere to prevent water uptake (*49*). Studies have indicated that a small opening near the hilum (*strophiole*) is sealed with a cork-like plug which can be dislodged with vigorous shaking or impaction (*40*). Seeds of other species, such as olive, cotoneaster, Russian olive (*Eleagnus*), *Symphoricarpos*, and *Rosa*, are enclosed in hardened portions of the pericarp.

In nature, seed coverings are softened by various agents of the environment, including mechanical abrasion, alternate freezing and thawing, attack by soil microorganisms, passage through the digestive tracts of birds or mammals, or fire. Softening of seed coverings by fungi and bacteria is most effective at temperatures above 50° F (10° C) and in a nonsterile medium. Chapter 7 lists various methods by which the seed coats can be artificially modified. (See also Figure 6–12.)

Mechanically resistant seed coverings   In general, once water is absorbed by the seed, if the embryo is not dormant the expanding force of germination will rupture the seed coats and break apart any outer covering. The shells of walnuts, pits of stone fruits, and other hard fruit coverings may have some delaying effect on germination but it has been hard to demonstrate that such mechanical effects are a primary cause of dormancy, once other factors have been overcome. Any effects of mechanically resistant coats can be overcome by the same treatments as used for impermeable seed coats.

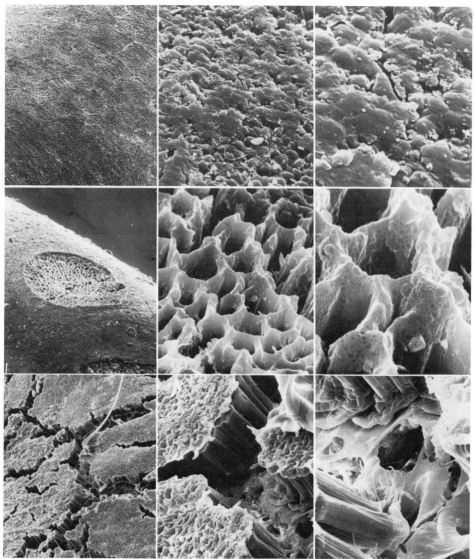

Figure 6–12   Scanning electron micrographs of 'Penngift' crownvetch seed showing scarification effects on seed coats. *Top row:* unscarified seed. Left to right: 170×, 1700×, and 4250× magnification. Note intact platy surface covering. *Center row:* after 30 min. acid scarification. Left to right: 170×, 1700×, and 4250× magnification. Caps of macrosclereid cells have been destroyed, exposing the lumens of the cells. *Bottom row:* after seed immersion in boiling water for 30 sec. with agitation. Left to right: 170×, 850×, 1700× magnification. Note columnar structure of exposed macrosclereid cells. Courtesy R. E. Brant, G. W. McKee, and R. W. Cleveland *(14).*

Chemical Inhibitors

Chemicals that act as seed *germination inhibitors* have been extracted from many plant parts and identified (*32*). During fruit and seed development, such chemicals are produced and some accumulate in the fruit, seed coverings, and embryo. Two general classes of substances are produced. One class includes by-products of metabolic processes whose presence may be incidental to a role in regulating germination (*2*). A second class includes naturally occurring plant hormones that control not only seed germination (see page 111) but plant growth and development in general. However, in particular environmental niches in nature it is tempting to believe that different species might evolve with genetic systems that would select specific endogenous chemicals that become important to regulate germination and insure survival in particular adverse environments (*2*).

Most fleshy fruits, or juices from them, strongly inhibit seed germination. This occurs, for instance, in citrus, cucurbits, stone fruits, apples, pears, grapes, and tomatoes. Likewise, many dry fruits and fruit coverings, such as the hulls of guayule, *Pennisetum ciliare,* and wheat (*Triticum*), and the capsules of mustard (*Brassica*), can produce inhibition. Undoubtedly these substances play an important biological role in preventing premature germination while the seeds are present on the parent plant. Where the fruit covering remains with the seed, the inhibiting effect can persist into the germination period. For instance, beet "seeds" contain substances giving off ammonia, which interferes with germination in seed testing; but if such seeds are planted in soil, leaching or absorption of the ammonia by soil particles overcomes this factor (*101*). Dormancy in *Iris* seeds is due to a water- and ether-soluble germination inhibitor in the endosperm which can be leached from the seeds with water or avoided by embryo excision (*3*). Germination inhibitors are reported to be widespread in seeds of tropical species (*72*).

Specific seed-germination inhibitors play a role in the ecology of certain desert plants (*57, 110, 111*). Inhibitors are leached out by heavy soaking rains (which would insure survival of the seedlings). A light rain shower is insufficient. Such substances have been referred to as "chemical rain gauges."

Inhibitors appear to be a part of the germination-controlling system involved in the dormancy of freshly harvested seeds of Class III (see p. 117), where the control site either is in the seed coverings—and is influenced by light and temperature—or is within the embryo. Abscisic acid has been found to be closely associated with germination inhibition in such cases, although the mechanism by which this control is exerted is not certain. Treatment with cytokinin appears to overcome the effects of such inhibitors, at least in some plants (*56*), as shown in Figures 6–9 and 6–10.

Presence of Rudimentary Embryos

In some species the embryos are not completely developed morphologically at the time of seed ripening and normally undergo further growth within the seed after its removal from the plant. Orchid seeds, for instance, are minute in size, consisting of an embryo and seed coat with little reserve food. Germination in nature requires a symbiotic relationship with certain fungi to supply nutrients. Commercially, orchid seeds are propagated by aseptic culture methods with nutrients supplied (see page 517). In most seeds of this group, however, the rudimentary or partly developed embryos are embedded in adequate reserve food materials in the endosperm. These occur in such families as Palmaceae, Magnoliaceae, Araliaceae, Paeoniaceae, etc. (72). In temperate zone species, seeds with rudimentary embryos often have other forms of dormancy as well, including hard seed coats and dormant embryos requiring chilling. Holly (*Ilex opaca*) has small, largely undifferentiated embryos that for development require a period of warm, moist conditions following ripening; but other germination-delaying factors, as a hard, tough seed covering and a subsequent requirement of the embryo for chilling, are involved (38). Similarly, warm temperatures, followed by chilling, are required by seeds of *Fraxinus* species where the embryo enlarges twice its original length following harvest (103). Seeds with rudimentary embryos are common in tropical species (72). For instance, African oil palm (*Elaeis guineensis*) seed germinates over a period of several years during which the embryo grows to full size. Maintaining high temperatures (100°–104° F; 38°–40° C) will reduce this period to a few months, and excising the embryo can give very rapid development. Other examples include *Actinidia*, whose seed requires two months warmth, and *Anona squamosa* seed, which requires 3 months.

Presence of Physiologically Active Seed Layers

Many, if not most, freshly harvested seeds have physiologically active seed coverings, made up of the inner seed coat (integument) and the endosperm. Germination control by these two layers tends to disappear with time, particularly if the seeds are in dry storage. This phenomenon is the basis for Class III seeds (see page 117). Seeds with this characteristic are sensitive to various environmental influences described in the preceding section, such as temperature, light, and concentration of gases, as well as the presence of various chemicals. Excision of the embryo avoids these germination blocks, invariably producing immediate germination.

The mechanism(s) by which such control over germination is exerted is not clear, but systems that have been experimentally shown to be operative in controlling germination in such seeds include gas permeability effects as well as the presence of endogenous inhibitor-promoter systems that respond to environmental signals (see Figures 6–9 and 6–10). These germination-control influences have been described in various parts of this chapter.

Presence of Dormant Embryos That Respond to Chilling

Plant propagators have known since early times that moist seeds of many fall-ripening tree and shrub species of the temperate zone must be chilled as they overwinter in the ground before they will germinate in the spring. This requirement led to the horticultural practice termed *stratification,* in which seeds are placed between layers of moist sand or soil in boxes (or in the ground) and exposed to chilling temperatures either out-of-doors or in refrigerators. A more accurate term used by some propagators for this procedure is *moist-chilling.*

Conditions required to overcome this kind of embryo dormancy include: (a) imbibition of moisture by the seed; (b) exposure to chilling, but not necessarily freezing, temperatures; (c) aeration; and (d) a certain amount of time.

A dry seed with a dormant embryo initially absorbs moisture by imbibition, although the rate may be somewhat slow. During the ensuing chilling period, the moisture content stays relatively constant or may increase gradually. Near the end of the dormant period the embryo absorbs water rapidly as germination begins. Drying during chilling is reported either to cause injury *(44)*, stop after-ripening changes, and cause secondary dormancy *(26, 105)*, or to preserve an attained state of after-ripening if dried at low temperature *(25, 51)*.

Temperatures just above freezing (35° to 45° F; 2° to 7° C) are generally most effective, and higher and lower temperatures produce changes at a slower rate. The minimum effective temperature is reported to be about 23° F (−5° C) *(26)*, and the maximum 62° F (17° C) *(1)*. At temperatures higher than this, the seeds not only fail to germinate but revert to secondary dormancy *(1, 83, 84, 105)*. The processes involved in after-ripening and those leading to induction of secondary dormancy are said to be in equilibrium at 62° F (17° C) in apples *(1)*. This *compensation temperature* has been determined in dormant seeds of other plants *(90)*, but the exact temperature may vary with different species and different stages of after-ripening *(84)*. Figure 6–13 shows the response of partially after-ripened apple seeds to various germination temperatures. At low temperatures the seeds germinate slowly but the percentage is high. At higher temperatures, germination rates are faster, but the germination percentage decreases in proportion to the increase in temperature *(26, 83)*. Nongerminating seeds revert to secondary dormancy. The temperature conditions most effective for germination may be similar to those in the natural environment, in which soil temperatures gradually increase with the advance of spring. Secondary dormancy resulting in inhibition of germination due to late planting or an early hot period can be advantageous to seed survival in nature.

Good aeration during seed stratification is necessary to maintain afterripening progress. Reduced oxygen will produce secondary dormancy, but there appears to be an interaction among embryo oxygen requirements, seed-coat permeability, and temperature *(18, 102)*. The moist seed coverings of dormant inbibed seeds of some species, such as apple, apparently restrict oxygen uptake due to low oxygen solubility in water. In addition, oxygen fixation by phenolic substances in the seed coats may occur. At chilling temperatures, the embryo's oxygen requirements are low and not restrictive. As the temperature increases, the oxygen requirement of the embryo increases, the solubility of oxygen in water is less, and the amount fixed by the phenols increases. As a result secondary dormancy develops at high

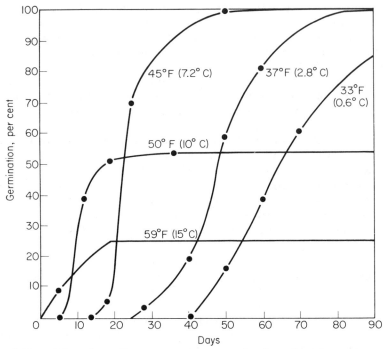

Figure 6-13   The effect of temperature on germination of apple seeds previously stratified for 65 days at 37°F (3°C). The abscissa represents the days following stratification. Data from DeHaas and Schander (26).

temperatures and prevents germination. Since seed coverings have been shown to contain other kinds of inhibitors (63), the precise role of the seed coats in controlling germination in these kinds of dormant seeds is not certain.

Whatever the cause, removal of the seed coverings invariably produces an improved germination response but not necessarily normal growth of the excised embryo. This procedure is the basis of a rapid viability test for dormant seeds (see p. 151) and is also sometimes used to circumvent dormancy. The degree of response indicates the depth of dormancy (22, 72). Embryos of some species will germinate in an essentially normal fashion while those of other species show some degree of abnormality. Typical responses include swelling, enlargement, and greening of the cotyledons; thick, short radicles with no epicotyl development; or development of normal root systems but dwarfed abnormal epicotyls. The latter have been called *physiological dwarfs* (see Figure 6-14) and can remain in such a condition for many years at warm temperatures (8). Normal shoot growth will resume if the dwarfed plant is given chilling or if lateral shoots develop from below the apex.

Pollock (75) demonstrated that peach seedlings were dwarfed when the apical meristem of the excised embryo was exposed to germination temperatures between 73° and 81° F (23° and 27° C) but were normal when exposed

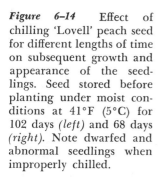

Figure 6–14 Effect of chilling 'Lovell' peach seed for different lengths of time on subsequent growth and appearance of the seedlings. Seed stored before planting under moist conditions at 41°F (5°C) for 102 days *(left)* and 68 days *(right)*. Note dwarfed and abnormal seedlings when improperly chilled.

to lower temperatures. Dwarfing has been offset by exposure of the seedlings to long photoperiods or continuous light (*34, 61, 106*), provided that this action is taken before the apical meristem becomes fully dormant. Repeated application of gibberellic acid has also overcome dwarfing (*8, 9, 34*).

The time required to after-ripen seeds of most woody perennial species is from one to three months, although for certain species five to six months are necessary. The time requirement is a genetic characteristic that reflects not only the genotype of the seed-source tree (*113*) but also that of the pollen source (*52*). Moreover, germination time can be influenced by the environmental exposure of the mother plant during seed development (*103*). Consequently, variation in the time requirement can occur not only among species but among different seed sources of the same species, depending upon the year the seed was produced and its location. Variation occurs among individual seeds of a single seed lot and probably accounts for the temperature response shown in Figure 6–13.

The biological mechanism whereby embryo dormancy is overcome by chilling is not completely understood, although much information has been obtained about changes taking place in response to chilling (*45, 72, 74, 101*). Reported changes include increased water-absorbing power, increased enzyme activity, increased acidity and changes in complex storage materials (*22*). These represent the increasing ability of the seed to germinate and not causal mechanisms that control release from dormancy. Chilling causes an increase in the ability of the embryo to utilize phosphorus in high-energy nucleotides and nucleic acids (*77*) and to synthesize RNA (*50, 55*).

Many studies support the concept that embryo dormancy of this kind is controlled through levels of growth promoting and growth inhibiting substances that exist at a given time in different parts of the seed and change in response to environment. Species may differ somewhat in the patterns of change of these endogenous materials and variability may exist in control mechanisms.

For example, high levels of growth-inhibiting substances, identified usually as abscisic acid (ABA), have been found in dormant seeds of walnut (*64*), peach (*27, 63*), plum (*62*), and hazel (*114*). These often occur in highest concentrations in the seed coverings, particularly in the freshly harvested seed. They invariably decrease during chilling, as shown in Figure 6–15 in plum (or sometimes they can be leached out with water). However, their disappearance may not necessarily coincide with the beginning of germination. Application of ABA to chilled seed ready to germinate invariably prevents germination.

Likewise, growth promoting substances, identified usually as gibberellin, are at a low level in dormant seeds of a number of species, such as plum (*62*) and peach (*65*). Gibberellin levels may increase during chilling (*65*), as shown in Figure 6–15 for plum seed (*62*). In hazel (*Corylus avellana* L.) separation between growth inhibiting and growth promoting systems has been shown to occur. The embryo of the freshly harvested seed is not dormant and has a significant quantity of gibberellin (*81*). The seed, however, is dormant because of the ABA in the seed coverings (*114*). Following harvest, the embryo becomes dormant with dry storage; the gibberellin level becomes very

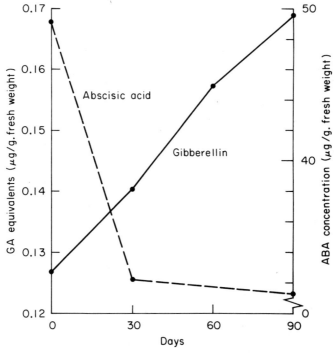

Figure 6–15   Changes in the level of abscisic acid (ABA) and gibberellin-like substances (GA) found in plum seeds during moist-chilling (stratification) at 45°F (7.2°C). Redrawn from Lin and Boe (*62*).

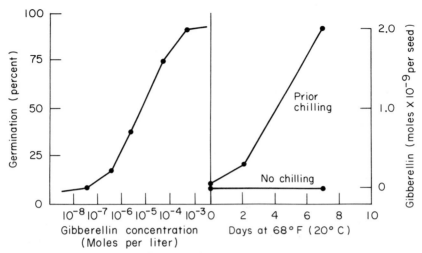

Figure 6-16   Embryos in hazel *(Corylus avellana* L.) seeds become dormant in dry storage following harvesting and then require moist chilling (stratification) for germination. Application of gibberellic acid to the embryo can replace the chilling requirement *(left).* After chilling, gibberellin does not increase until *after* the seeds are placed at warm temperature to allow germination to begin *(right).* Reproduced by permission from A. W. Galston and P. S. Davies, *Control Mechanisms in Plant Development.* Englewood Cliffs, N.J.: Prentice-Hall, Inc., 1970.

low and does not increase until after a moist-chilling period that produces germination *(80),* as shown in Figure 6–16, *right.* Gibberellic acid applied to the dormant seed will replace the chilling requirement, as shown in Figure 6–16, *left (13).* ABA will overcome the effect of GA and prevent germination *(82).*

REFERENCES

1 Abbott, D. L., Temperature and the dormancy of apple seeds, *Rpt. 14th Int. Hort. Cong.* Vol. 1: 746–53. 1955.

2 Amen, R. D., A model of seed dormancy, *Bot. Rev.,* 34:1–31. 1948.

3 Arditti, J., and P. R. Pray, Dormancy factors in iris (Iridaceae) seeds. *Amer. Jour. Bot.,* 56(3):254–59. 1969.

4 Assoc. Off. Seed Anal., Rules for seed testing, *Proc. Assoc. Off. Seed Anal.,* 54(2):1–112. 1965

5 Ayers, A. D., Seed germination as affected by soil moisture and salinity, *Agron. Jour.,* 44:82–84. 1952.

6 Barton, L. V., Relation of different gases to the soaking injury of seeds, *Contrib. Boyce Thomp. Inst.,* 16(2):55–71. 1950.

7 ————, Relation of different gases to the soaking injury of seeds, II, *Contrib. Boyce Thomp. Inst.,* 17(1):7–34. 1952.

8 ————, Growth response of physiologic dwarfs of *Malus arnoldiana* Sarg. to gibberellic acid, *Contrib. Boyce Thomp. Inst.,* 18:311–17. 1956.

9 ————, and C. Chandler, Physiological and morphological effects of gibberellic acid on epicotyl dormancy of tree peony, *Contrib. Boyce Thomp. Inst.,* 19:201–14. 1957.

10 Berlyn, G. P., Seed germination and morphogenesis, in T. T. Kozlowski (ed.), *Seed Biology,* Vol. III. New York: Academic Press, 1972.

11 Berstein, L., A. J. MacKenzie, and B. A. Krantz, The interaction of salinity and planting practice on the germination of irrigated row crops, *Proc. Soil Science Soc. Amer.,* 19:240–43, 1955.

12 Borthwick, H. A., S. B. Hendricks, E. H. Toole, and V. K. Toole; Action of light on lettuce seed germination, *Bot. Gaz.* 115:205–225. 1954.

13 Bradbeer, J. W. and N. J. Pinfield, Studies in seed dormancy. III. The effects of gibberellin on dormant seeds of *Corylus avellana* L., *New Phytol.* 66:515–23. 1967.

14 Brant, R. E., G. W. McKee, and R. W. Cleveland, Effect of chemical and physical treatment on hard seed of Penngift crown vetch. *Crop Sci.,* 11:1–6. 1971.

15 Brown, R., Germination, in F. C. Steward, (ed.), *Plant Physiology,* Vol. VIC:3–48. New York: Academic Press, 1972.

16 Chapman, A. L., and M. L. Peterson, The seedling establishment of rice under water in relation to temperature and dissolved oxygen. *Crop. Sci.,* 2:391–95. 1962.

17 Ching, Te May, Metabolism of germinating seeds, in T. T. Kozlowski, (ed.), *Seed Biology,* Vol. II. New York: Academic Press. 1972.

18 Comé, D., and T. Tissaoui, Interrelated effects of imbibition, temperature, and oxygen on seed germination, in W. Heydecker (ed.), *Seed Ecology.* University Park, Pa.: Pennsylvania State Univ. Press, 1973.

19 Conn, E. C., and P. K. Stumpf, *Outlines of Biochemistry.* 3rd ed. Chap. 19: Biosynthesis of proteins. New York: John Wiley & Sons, Inc., 1972.

20 Chrispeals, M. J., and J. E. Varner. Hormonal control of enzyme synthesis: On the mode of action of gibberellic acid and abscisin in aleurone layers of barley. *Plant Phys.,* 42:1008–16. 1967.

21 Crocker, W., Mechanics of dormancy in seeds, *Amer. Jour. Bot.,* 3:99–120. 1916.

22 ————, *Growth of Plants.* New York: Reinhold Publishing Corp., 1948.

23 ————, Effect of the visible spectrum upon the germination of seeds and fruits, in *Biological Effects of Radiation.* New York: McGraw-Hill Book Company, 1930, pp. 791–828.

24 Czabator, F., Germination value: an index combining speed and completeness of pine seed germination, *For. Sci.,* 8:386–96. 1962.

25 DeHaas, P. G., Neue Beitrage zur Samlingsanzucht bei Kernobst, *Rpt. 14th Int. Hort. Cong.,* Vol. I: 1185–95. 1955.

26 DeHaas, P. G., and H. Schander, Keimungsphysiologische Studien an Kernobst. I. Kamen and Keimung, *Z. f. Pflanz.,* 31(4):457–512. 1952.

27 Diaz, D. H., and G. C. Martin, Peach seed dormancy in relation to endogenous inhibitors and applied growth substances. *J. Amer. Soc. Hort. Sci.,* 97(5): 651–54. 1972.

28 Doneen, L. D., and J. H. MacGillivray, Germination (emergence) of vegetable seed as affected by different soil conditions, *Plant Phys.,* 18:524–29. 1943.

29 Edwards, T. J., Temperature relations of seed germination, *Quart. Rev. Biol.,* 7:428–43. 1932.

30 Esashi, Y. and A. C. Leopold, Physical forces in dormancy and germination of Xanthium seeds. *Plant Phys.,* 43:871–76. 1968.

31 ——, Dormancy regulation in subterranean clover seeds by ethylene. *Plant Phys.,* 44:1470–72. 1969.

32 Evenari, M., Germination inhibitors, *Bot. Rev.,* 15:153–94, 1949.

33 ——, and G. Newman. The germination of lettuce seed. II. The influence of fruit coats, seed coat and endosperm upon germination, *Bul. Res. Council, Israel,* 2:75–78. 1952.

34 Flemion, F., and E. Waterbury, Further studies with dwarf seedlings of non-after-ripened peach seeds, *Contrib. Boyce Thompson Inst.,* 13:415–422. 1945.

35 Frankland, B., and P. F. Wareing, Hormonal regulation of seed dormancy in hazel *(Corylus avellana* L.) and beech *(Fagus sylvatica), J. Exp. Bot.* 17:596–611. 1971.

36 Galston, A. W., and P. J. Davies, Hormonal regulation in higher plants, *Science,* 163:1288–97. 1969.

37 Gordon, A. G., The rate of germination, in W. Heydecker (ed.), *Seed Ecology.* University Park, Pa.: Pennsylvania State Univ. Press, 1973, pp. 391–410.

38 Giersbach, J., and W. Crocker, Germination of *Ilex* seeds, *Amer. Jour. Bot.,* 16:854–55. 1929.

39 Haber, A. H., and H. J. Luippold, Separation of mechanisms initiating cell division and cell expansion in lettuce seed germination, *Plant Phys.,* 35:168–73. 1960.

40 Hamly, D. H., Softening the seeds of *Melilotus alba, Bot. Gaz.,* 93:345–75. 1932.

41 Hanks, R. S., and F. C. Thorp, Seedling emergence of wheat as related to soil moisture content, bulk density, oxygen diffusion rate and crust strength, *Proc. Soil Sci. Soc. Amer.,* 20:307–10. 1956.

42 Harrington, J. F., The effect of temperature on the germination of several kinds of vegetable seeds, *XVIth Int. Hort. Cong.,* Vol. II:435–41. 1962.

43 Harrington, J. F., and P. A. Minges, *Vegetable Seed Germination,* Calif. Agr. Ext. Serv. Mimeo. Leafl., 1954.

44 Haut, I. C., The influence of drying on after-ripening and germination of fruit tree seeds, *Proc. Amer. Soc. Hort. Sci.,* 29:371–74. 1932.

45 ——, Physiological studies on after-ripening and germination of fruit tree seeds, *Md. Agr. Exp. Sta. Bul. 420,* 1938.

46 Heit, C. E., Germination of sensitive flower seed kinds and varieties with suggested methods for testing in the laboratory, *Proc. Assoc. Off. Seed Anal.,* 40:107–17. 1950.

47 Heydecker, W., Vigour, in E. H. Roberts (ed.), *Viability of Seeds.* Syracuse, N. Y.: Syracuse Univ. Press, 1972.

48 Hunter, J. B., and A. E. Erickson, Relation of seed germination to moisture tension, *Agron. Jour.,* 44:107–9. 1952.

49 Hyde, E. O. C., the function of some Papilionaceae in relation to the ripening of the seed and permeability of the testa. *Ann. Bot.,* 18:241–56. 1956.

50 Jarvis, B. C., B. Frankland, and J. H. Cherry. Increased nucleic acid synthesis in relation to the breaking of dormancy of hazel seed by gibberellic acid, *Planta,* 83:257–66. 1968.

51 Kamininski, W., and R. Rom, Secondary dormancy in stratified peach embryos, *HortScience,* 8(5):401. 1973.

52 Kester, D. E., Pollen effects on chilling requirements of almond and almond hybrid seeds, *Jour. Amer. Soc. Hort. Sci.,* 94:318–21. 1969.

53 Ketrick, D. L., and P. W. Morgan, Physiology of oil seeds. I: Regulation of dormancy in Virginia-type peanut seeds, *Plant Phys.,* 45:268–73. 1970.

54 Khan, A. A., Cytokinins: permissive role in seed germination, *Science,* 171:853–59. 1971.

55 ———, C. E. Heit, and P. C. Lippold. Increase in nucleic acid synthesizing capacity during cold treatment of dormant pear embryos, *Biochem. Biophys. Res. Commun.,* 33:391–96. 1968.

56 Khan, A. A., C. E. Heit, E. C. Waters, C. C. Anojulu, and L. Anderson, Discovery of a new role for cytokinins in seed dormancy and germination, in *Search,* New York Agr. Exp. St. (Geneva), 1(9):1–12. 1971.

57 Koller, D., Environmental control of seed germination, in T. T. Kozlowski (ed.), *Seed Biology,* Vol. II. New York: Academic Press. 1972.

58 Kotowski, F., Temperature relations to germination of vegetable seeds, *Proc. Amer. Soc. Hort. Sci.,* 23:176–84. 1926.

59 Kozlowski, T. T., and A. C. Gentile, Influence of the seed coat on germination, water absorption and oxygen uptake of eastern white pine seed, *For. Sci.,* 5:389–95. 1959.

60 Kozlowski, T. T., *Growth and development of trees.* Vol. I. New York: Academic Press, 1971.

61 Lammerts, W. E., Effect of photoperiod and temperatures on growth of embryo-cultured peach seedlings, *Amer. Jour. Bot.,* 30:707–11. 1943.

62 Lin, C. F., and A. A. Boe, Effects of some endogenous and exogenous growth regulators on plum seed dormancy, *Jour. Amer. Soc. Hort. Sci.,* 97:41–44. 1972.

63 Lipe, W., and J. C. Crane, Dormancy regulation in peach seeds, *Science,* 153:541–42. 1966.

64 Martin, G. C., H. Forde, and M. Mason, Changes in endogenous growth substances in the embryo of *Juglans regia* during stratification, *Jour. Amer. Soc. Hort. Sci.,* 94:13–17. 1969.

65 Mathur, D. D., G. A. Couvillon, H. M. Vines, and C. H. Hendershott. Stratification effects of endogenous gibberellic acid (GA) in peach seeds, *HortScience,* 6:538–39. 1971.

66   Mayer, A. M., and A. Poljakoff-Mayber, *The Germination of Seeds*. New York: Macmillan, 1963.

67   Mohr, H., Primary effects of light on growth, *Ann. Rev. Plant Phys.*, 13:465–88. 1962.

68   Morinaga, T., Germination of seeds under water, *Amer. Jour. Bot.*, 13:126–31. 1926.

69   ———, The favorable effect of reduced oxygen supply upon the germination of certain seeds, *Amer. Jour. Bot.*, 13:150–65. 1926.

70   Negm, F. B., O. E. Smith, and J. Kumamoto, Interaction of carbon dioxide and ethylene in overcoming thermodormancy of lettuce seeds, *Plant Phys.*, 49:869–72. 1972.

71   ———, The role of phytochrome in an interaction with ethylene and carbon dioxide in overcoming lettuce seed thermodormancy, *Plant Phys.*, 51:1089–94. 1973.

72   Nikolaeva, M. G., *Physiology of deep dormancy in seeds*, Acad. of Sciences of the USSR, V. L., Komarov Bot. Inst., Leningrad 1967 (translated from Russian, Jerusalem, 1969).

73   Olatoye, S. T., and M. A. Hall, Interaction of ethylene and light on dormant weed seeds, in W. Heydecker (ed.), *Seed Ecology*. University Park, Pa.: Pennsylvania State Univ. Press, 1973.

74   Olney, H. O., and B. M. Pollock, Studies of rest period. II. Nitrogen and phosphorus changes in embryonic organs of after-ripening cherry seed, *Plant Phys.*, 35:970–75. 1960.

75   Pollock, B. M., Temperature control of physiological dwarfing in peach seedlings, *Plant Phys.*, 37:190–97. 1962.

76   ———, Effects of environment after germination, in E. H. Roberts (ed.), *Viability of Seeds*. Syracuse, N. Y.: Syracuse Univ. Press, 1972.

77   ———, and H. O. Olney, Studies of the rest period. I. Growth, translocation, and respiratory changes in the embryonic organs of the after-ripening cherry seed, *Plant Phys.*, 34:131–42. 1959.

78   Pollock, B. M., and E. E. Roos, Seed and seedling vigor, in T. T. Kozlowski (ed.), *Seed Biology*, Vol. III. New York; Academic Press. 1972.

79   Roberts, E. H., Dormancy: a factor affecting seed survival in the soil, in E. H. Roberts (ed.), *Viability of Seeds*, Syracuse, N. Y.: Syracuse Univ. Press, 1972.

80   Ross, J. D., and J. W. Bradbeer, Concentrations of gibberellin in chilled hazel seeds, *Nature* (Lond.) 220:85–86. 1968.

81   ———, Studies in seed dormancy. V. The concentrations of endogenous gibberellins in seeds of *Corylus avellana* L., *Planta* (Berl.) 100:288–302. 1971.

82   ———, Studies in seed dormancy. VI. The effects of growth retardants on the gibberellin content and germination of chilled seeds of *Corylus avellana* L., *Planta* (Berl.) 100:303–8. 1971.

83   Schander, H. Keimungsphysiologische Studien an Kernobst. III. Sortenvergleichende Untersuchungen über die Temperature-ansprüche stratifizierten Saatgutes von Kernobst und über die Reversibilitat der Stratifikationsvorgange, *Z. f. Pflanz*, 35:89–97. 1955.

84   Semeniuk, P., and R. N. Stewart, Temperature reversal of after-ripening of rose seeds, *Proc. Amer. Soc. Hort. Sci.*, 80:615–21. 1962.

85   Sharples, G. C., Stimulation of lettuce seed germination at high temperatures by ethephon and kinetin, *Jour. Amer. Soc. Hort. Sci.*, 98(2):207–9. 1973.

86   Shull, C. A., Measurement of the surface forces in soils, *Bot. Gaz.*, 62:1–29. 1916.

87   ———, Temperature and rate of moisture uptake in seeds, *Bot. Gaz.*, 69:361–90. 1920.

88   Stearns, F., and J. Olson, Interactions of photoperiod and temperature affecting seed germination in *Tsuga canadensis*, *Amer. Jour. Bot.*, 45:53–58. 1958.

89   Stephens, W., The mangrove, *Oceans Mag.*, 2(5):51–55. 1969.

90   Stewart, R. N., and P. Semeniuk, The effect of the interaction of temperature with after-ripening requirement and compensating temperature on germination of seed of five species of *Rosa*, *Amer. Jour. Bot.*, 52:755–60. 1965.

91   Stoutemyer, V. T., and A. W. Close, Rooting cuttings and germinating seeds under fluorescent and cold cathode lighting, *Proc. Amer. Soc. Hort. Sci.*, 48:309–25. 1946.

92   ———, Propagation by seedage and grafting under fluorescent lamps, *Proc. Amer. Soc. Hort. Sci.*, 62:459–65. 1953.

93   Thompson, P. A., Geographical adaptation of seeds, in W. Heydecker, (ed.), *Seed Ecology*. University Park, Pa.: Pennsylvania State University Press, 1973.

94   Thomas, H., Control mechanisms in the resting seed, in E. H. Roberts (ed.), *Viability of Seeds*. Syracuse, N. Y.: Syracuse Univ. Press, 1972.

95   Thornton, N. C., Factors influencing germination and development of dormancy in cocklebur seeds, *Contrib. Boyce Thompson Inst.*, 7:477–96. 1935.

96   ———, Importance of oxygen supply on secondary dormancy and its relation to the inhibiting mechanism regulating dormancy, *Contrib. Boyce Thomp. Inst.* 13:497–500. 1945.

97   Toole, E. H., V. K. Toole, and E. A. Gorman, Vegetable seed storage as affected by temperature and relative humidity. *USDA Tech. Bul.* 972. 1948.

98   ———, Photocontrol of *Lepidium* seed germination, *Plant Phys.*, 30:15–21, 1955.

99   Toole, E. H., V. K. Toole, H. A. Borthwick, and S. B. Hendricks, Interaction of temperature and light in germination of seeds, *Plant Phys.*, 30:473–78. 1955.

100   U.S. Dept. Agriculture, Forest Svce, C. S. Schopmeyer (ed.), *Seeds of Woody Plants in the United States*. Agri. Handbook No. 450. 1974.

101   ———, *Manual for testing agricultural and vegetable seeds*, USDA Agr. Handbook 30, Washington, D.C.: U.S. Govt. Printing Office, 1952.

102   Vegis, A., Dormancy in higher plants, *Ann. Rev. Plant Phys.*, 15:185–224. 1964.

103   Villiers, T. A., Seed dormancy, in T. T. Kozlowski (ed.), *Seed Biology*, Vol. II: pp. 220–82. New York: Academic Press. 1972.

104   Visser, T., The role of seed coats and temperature in after-ripening, germination and respiration of apple seeds, *Proc. Koninkl. Nederl. Akad. van Wetens*, Series C, 59:211–22. 1956.

105 ———, Some observations on respiration and secondary dormancy in apple seeds, *Proc. Koninkl. Akad. van Wetens,* Series C, 59:314–24. 1956.

106 ———, The growth of apple seedlings as affected by after-ripening, seed maturity and light, *Proc. Koninkl. Nederl. Akad. van Wetens,* Series C, 59:325–34. 1956.

107 Wareing, P. F., The germination of seeds, in *Vistas in Botany,* III. New York: Macmillan, 1963, pp. 195–227.

108 ———, and H. A. Foda, Growth inhibitors and dormancy in *Xanthium* seed, *Physiol. Plant.,* 10(2):266–80. 1957.

109 Wellington, P. S., and V. W. Durham, Varietal differences in the tendency of wheat to sprout in the ear. *Empire Jour. Exp. Agr.,* 26:47–54. 1958.

110 Went, F. W., Ecology of desert plants. II. The effect of rain and temperature on germination and growth, *Ecology,* 30:1–13. 1949.

111 ———, and M. Westergaard, Ecology of desert plants, III, Development of plants in the Death Valley National Monument, California, *Ecology.* 30:26–38. 1949.

112 Wesson, G., and P. F. Wareing, The induction of light sensitivity in weed seeds by burial, *Jour. Exp. Bot.,* 20(63):414–25. 1969.

113 Westwood, M. N., and H. O. Bjornstad, Chilling requirement of dormant seeds of fourteen pear species as related to their climatic adaptation, *Proc. Amer. Soc. Hort. Sci.,* 92:141–49. 1948

114 Williams, P. M., J. D. Ross, and J. W. Bradbeer, Studies in seed dormancy, VII. The abscisic acid content of the seeds and fruits of *Corylus avellana* L. *Planta* (Berl.) 110:303–10. 1973.

SUPPLEMENTARY READING

Amen, R. D., "A Model of Seed Dormancy," *Botanical Review,* Vol. 34 (1968), pp. 1–31.

Barton. L. V., *Bibliography of Seeds.* New York: Columbia University Press, 1967.

Crocker, W., and L. V. Barton, *Physiology of Seeds.* Waltham, Mass.: Chronica Botanica, 1953.

Galston, A. W., and P. J. Davies, *Control Mechanisms in Plant Development,* Englewood Cliffs, N. J.: Prentice-Hall, 1970.

Heydecker, W., (ed.), *Seed Ecology.* London, Butterworth & Co., Ltd. 1973.

Koller, D., A. M. Mayer, A. Poljakoff-Mayber, and S. Klein, "Seed Germination," *Annual Review of Plant Physiology,* Vol. 13 (1962), pp. 437–64.

Kozlowski, T. T. (ed.), *Seed Biology,* Vols. I, II, III. New York: Academic Press. 1972.

Mayer, A. M., and A. Poljakoff-Mayber, *The Germination of Seeds,* 2nd ed. Oxford: Pergamon Press, 1975.

Roberts, E. H. (ed.), *Viability of Seeds.* Syracuse, N. Y.: Syracuse Univ. Press, 1972.

Stokes, P., "Temperature and Seed Dormancy," in *Handbuch der Pflanzenphysiologie,* Vol. 15 (Part 2): pp. 746–803, Berlin, Heidelberg, New York: Springer-Verlag. 1965.

Techniques of Propagation by Seeds 7

Seed propagation involves careful management of germination conditions and facilities and a knowledge of the requirements of individual kinds of seeds. Its success depends upon the degree to which the following conditions are fulfilled: (a) The seed must maintain the particular cultivar or species which the propagator wishes to grow. This can be accomplished by obtaining seed from a reliable dealer, buying certified seed, or—if producing one's own—following the principles of seed selection described in Chapter 4. (b) The seed must be viable and capable of germination. It also should germinate rapidly and vigorously to withstand possible adverse conditions in the seedbed. Viability can be determined by seed tests, but seed vigor is difficult to predict. (c) Any dormancy condition of the seed which would inhibit germination must be overcome by applying any necessary pregermination treatment. The propagator should know the requirements of the seed with which he is concerned. A germination test will be helpful in indicating the necessity for any pregermination treatment. In the absence of specific knowledge, the propagator should try to duplicate the natural environmental conditions associated with good seed germination of this particular kind of plant. (d) Assuming that the seed is capable of prompt germination, propagation success depends upon supplying the proper environment—moisture, temperature, oxygen, and light or darkness—to the seed and the resulting seedling until it is established in its permanent location. A proper environment also includes control of diseases and insects.

SEED TESTING
Good-quality seed has the following characteristics: It is genetically true to species or cultivar; capable of high germination; free from disease and insects; and free from mixture with other crop seeds, weed seeds, and inert and extraneous material. The germination capacity and purity of the seed can

be determined by conducting a seed test on a small representative sample drawn from the seed lot in question (*2, 29, 46*).

In the U.S., state laws regulate the shipment and sale of agricultural and vegetable seeds within that state. Seeds entering interstate commerce or those sent from abroad are subject to the Federal Seed Act adopted in 1939 (*1*). Such regulations require labeling by the shipper of commercially produced seeds as to: name and cultivar; origin; germination percentage; and the percentage of pure seed, other crop seed, weed seed, and inert material. The regulations may set minimum standards of quality, germination percentage, and freedom from weed seeds. Shipment and sale of tree seed is regulated by law in some states (*39*) and in most European countries.

Seed testing provides information to meet legal standards, determines seed quality, and establishes the rate of sowing for a given stand of seedlings. It is desirable to retest seeds that have been in storage for a prolonged period.

Procedures for testing agriculture and vegetable seed in reference to the Federal Seed Act are given by the U.S. Department of Agriculture (*1*). The Association of Official Seed Analysts also publishes procedures for testing these seeds in addition to procedures for testing seeds of many flower, tree, and shrub species (*2*). International rules for testing seeds of many tree, shrub, agricultural, and vegetable species are published by the International Seed Testing Association (*26*). The Western Forest Tree Seed Council also has published testing procedures for tree seed (*50*).

Sampling

The first step in carrying out a seed test is to obtain a uniform sample that represents the entire lot under consideration. Equal portions are taken from evenly distributed parts of the seed lot, such as a sample from each of several sacks in lots of less than five sacks or from every fifth sack with larger lots. The seed samples are thoroughly mixed and then redivided into smaller lots to produce the *working sample,* i.e., the sample upon which the test is actually to be run. The amount of seed required for the working sample varies with the kind of seed and is specified in the Rules for Seed Testing.

Testing Purity

Purity is the percentage by weight of the "pure seeds" present in the sample. "Pure seed" refers to the principally named kind, cultivar, or type of seed present in the seed lot. After the working sample has been weighed, it is divided visually into (a) the pure seed of the kind under consideration; (b) other crop seed; (c) weed seed; and (d) inert material, including seed-like structures, empty or broken seeds, chaff, soil, stones, and other debris (see Figure 7–1). In some cases, it is possible to check the genuineness of the seed or trueness to cultivar or species by visual inspection. Often, however, identification cannot be made except by growing the seeds and observing the plants.

At the time of making the purity test, the number of pure seeds per pound can be calculated. These data are necessary as a guide to seeding rates.

Figure 7-1   Purity of seeds is determined by visual examination of individual seeds in a weighed sample taken from the larger lot in question. Impurities may include other crop seed, weed seed, and inert, extraneous material. Courtesy E. L. Erickson Products, Brookings, S.D.

Testing Viability

The *germination percentage* is the relative number of normal seedlings produced by the pure seed (the kind under consideration). To produce a good test, it is desirable to use at least 400 seeds picked at random and divided into lots of one hundred. If any two of these lots differ by more than 10 percent, a retest should be carried out. Otherwise, the average of the four tests becomes the germination percentage (see Figure 7–2).

Figure 7-2   Germination testing of seeds. *Top:* 100 seeds from the sample to be tested are placed on a moistened blotter. In this case placing the seeds evenly and quickly is made possible by an automatic vacuum counter. *Below:* after one or more weeks in a germinator the number of germinated seeds is counted. Note that this test consists of four lots of 100 seeds each. Courtesy E. L. Erickson Products, Brookings, S.D.

Germination Test

In a germination test the seeds are placed under optimum environmental conditions of light and temperature to induce germination (Figure 7–3). The conditions required to meet legal standards are specified in the rules for testing the different kinds of seeds.

Various techniques are used for germinating seed. In seed-testing laboratories the seeds are commonly placed on germination trays (not galvanized steel, which contains toxic zinc salts), either between two thicknesses of blotters or on top of blotters, and then placed in germinators where the temperature, moisture, and light are controlled. To discourage the growth of microorganisms, all materials and equipment should be kept scrupulously clean, sterilized when possible, and the water amount carefully regulated. No water film should form around the seeds; neither should the germinating medium be so wet that a film of water appears when that medium is pressed with a finger.

Plastic boxes, paraffined cardboard boxes, and covered glass Petri dishes are useful germinators. Media in them can be specially made paper blotters (two layers), absorbent cotton, specially made paper toweling (five layers), filter paper (five layers) and sand, vermiculite, or soil (5/8 in.).

Another method for testing seed is the *rolled towel test.* Paper toweling, 11 by 14 in. or a similar size, is moistened. Seeds are spaced along one side so that the edge of the towel can cover them. More rows of seeds are spaced along the towel as it is rolled up with the seeds inside. Rolls should not be tight, about five layers being desirable. They are placed either horizontally or vertically in germination trays. Germination blotters or flannel cloth could be substituted for the paper toweling.

Tree seed can be tested by germination in sterile sand in flats placed in a greenhouse. Another procedure is to place the seeds on peat moss in the bottom of a glass baking dish covered with a sheet of glass. The seeds are planted uncovered in shallow grooves with enough water added so that a shallow layer remains on the bottom after the peat has absorbed all it can hold. This method is particularly desirable for seeds with a long chilling (stratification) requirement (*45*). Seeds of many tree species, particularly

Figure 7–3   Commercial seed germinator with light and temperature control for testing viability. Courtesy C. E. Heit.

Table 7–1   Tree and shrub seeds (with examples) grouped according to germinative characteristics when tested in the laboratory under different environmental conditions or when given other special treatment.*

Group 1. Seeds that germinate within a wide range of temperature and without light exposure.

Beefwood *(Casuarina glauca)*
Italian cypress *(Cupressus sempervirens)*
many species of *Eucalyptus*
Honeylocust *(Gleditia triacanthos)*†

some spruce species *(Picea abies, P. asperata, P. polita)*
Chinese and Siberian elm *(Ulmus pavifolia, U. pumila)*

Group 2. Seeds that have specific temperature requirements but do not require light.

68° to 86° F *(20° to 30° C) diurnally alternating:*
Catalpa
Ailanthus
Red pine *(Pinus resinosa)*
    or 77° F (25° C) constant
50° to 86° F (10° to 30° C)
    diurnally alternating:
Curleaf mahogany *(Cercocarpus ledifolius)*
Cliffrose *(Cowania stansburiana)*
Antelope bitterbush *(Purshia tridentata)*

68° F *(20° C) constant:*
several pine species *(Pinus cembroides, P. halepensis, P. pinea)*
Lilac *(Syringa vulgaris)*
Arborvitae *(Thuja orientalis)*

Group 3. Seeds that germinate in 7 to 12 days within a wide temperature range if exposed to artificial light.

several spruce species *(Picea engelmannii, P. mariana, P. omerika)*
several pine species *(Pinus banksiana, P. nigra, P. mugo var. mughus, P. rigida, P. sylvestris, P. ponderosa scopulorum)*

---

conifers, can be tested by the same procedures used for crop seed, with particular attention paid to light and temperature (*2, 24, 50*) listed in Table 7–1.

A test usually runs from ten days to four weeks but may continue as long as three months for some slow-germinating seeds. A normal seedling generally should have a well-developed root and shoot, although the criteria for a "normal seedling" vary with different kinds of seeds. In addition, in the seed lot there may be abnormal seedlings, hard seeds, dormant seeds, and dead or decayed seeds. Abnormal seedlings can be caused by declining vitality due to age or to poor storage conditions; by insect, disease, or mechanical injury; by overdoses of fungicides; by frost damage; by mineral deficiencies (manganese and boron in peas and beans); or by toxic materials sometimes present in metal germination trays, substrata, or tap water. Dormant seeds can usually be distinguished from nonviable seeds; the dormant seeds are firm, swollen, free from molds, or may show erratic sprouting. Any ungerminated seed should be examined to determine the possible reason.

Dormancy of freshly harvested seeds produces difficulties in direct testing and can prolong the testing period, impose specific environmental require-

Table 7–1   Continued

Group 4.   Seeds that germinate in 14 to 28 days if exposed to artificial light and to warm alternating temperatures of 68° to 86° F (20° to 30° C). Seeds may also respond to moist-chilling.

Birch *(Betula)*
Elm *(Ulmus americana)*
Larch *(Larix sibirica)*
Mulberry *(Morus alba, M. nigra)*
Liquidambar styraciflua
some spruce series *(Picea glauca,*
   P. orientalis, P. rubens, P.
   sitchensis)

some pine species *(Pinus densiflora,*
   P. echinata, P. elliotti, P. taeda,
   P. thunbergii, P. virginiana)
Rhododendron
Sequoia and *Sequoiadendron*
Thuja plicata, T. occidentalis

Group 5.   Seeds that require 3 to 4 weeks moist-chilling at 37° F (3° C) before germination at alternating 68° to 86° F (20° to 30°C) (except as noted) in light for 2 to 4 weeks.

Fir species *(Abies balsamea, A.*
   fraseri, A. grandis, A. homo-
   lepsis, A. procera)
Cedrus species (68° F) (20° C)
some pine species *(Pinus flexilis, P.*
   glabra, P. leucodermis, P. stro-
   bus)

some sources of Douglas fir *(Pseudo-*
   tsuga menziesii)
Rosa multiflora (50° to 86° F) (10°
   to 30° C)
Eastern hemlock *(Tsuga canadensis)*
   (59° F) (15° C)
Sumac *(Rhus aromatica)* †

Group 6.   Seeds that require 2 to 6 months moist-chilling as a minimum requirement prior to germination. Some also have other dormancy problems. Embryo excision or a tetrazolium test may be useful in determining germinative capacity.

Sensitive to germination tempera-
   ture (use 68° F) (20° C):
Maple *(Acer* sp.)
Apple *(Malus* sp.)
Pear *(Pyrus* sp.)
Peach, cherry, etc. *(Prunus* sp.)
Yew *(Taxus* sp.)

Not sensitive to germination tem-
   perature:
some pine species *(Pinus cembra,*
   P. lambertiana, P. monticola,
   P. peuce)

---

* From C. E. Heit (24).
† Must be treated also for hard seed coats.

ments, and sometimes interfere with the reliability of the test. Conditions effective in overcoming dormancy which are recommended by the Rules for Seed Testing for particular kinds of seeds include: prechilling, germination at low temperatures (59° F, or 15° C), germination at high (over 86° F, or 30° C) or alternating temperatures, subjection to light, moistening with 0.1 or 0.2 percent potassium nitrate, modification of seed coats, leaching with water, or pre-drying.

Excised-Embryo Test

The excised-embryo test is used to test the germination of seeds of woody shrubs and trees whose embryos require long periods of after-ripening before germination will take place *(13, 19)*. In this test the embryo is excised from the rest of the seed and germinated alone. A viable embryo will either germinate or show some indication of activity whereas a nonviable embryo becomes discolored and deteriorates (see Figure 7–4).

The excision must be done carefully to avoid injury to the embryo. Any hard, stony seed coverings, such as the endocarp of stone fruit seeds, must first be removed. The seeds are then soaked for one to four days, the water being changed once or twice a day. Storing seeds in moist peat for three days to two weeks at cool temperatures is also satisfactory in preparing seeds for excision.

The seed coats are cut with a sharp scalpel or knife, and the embryo carefully removed. If a large endosperm is present, the seed coats may be slit and the seeds covered with water, and after about a half hour the embryo will float out or can easily be removed.

Procedures for germinating excised embryos are similar to those for germinating intact seeds. Petri dishes filled with a good moist substratum, such as blotting or filter paper are used. The embryos are placed on the filter paper so that they do not touch. The dishes are kept in the light at a temperature of 64° to 74° F (18° to 22° C). At higher temperatures, growth of molds may become excessive and interfere with the test. The time required for the test varies from 3 days to 3 weeks rather than several months, as is sometimes required for the direct germination test.

Nonviable embryos become soft, brown, and moldy within two to ten days; viable embryos remain firm and show some indication of viability, depending upon the species. Types of response that occur include spreading of the cotyledons, development of chlorophyll, and growth of the radicle and plumule. The rapidity and degree of development may give an indication of the vigor of the seed.

Figure 7–4   The excised embryo method of testing seed germination. *Top:* germination of apple seeds showing the range of vigor from strong to weak to dead. Actual germination percentages for the four lots were (left to right): 100, 70, 44, and 0. *Below:* vigor and germination of four peach stocks (left to right): strong seed, vigorous growth (80 percent viability); good seed, fair vigor (52 percent viability); old seed, weak (18 percent viability); and dead seed. Courtesy C. E. Heit.

Tetrazolium Test

The tetrazolium test is a biochemical method in which viability is determined by the red color appearing when the seeds are soaked in a 2,3,5-triphenyltetrazolium chloride (TTC) solution. In cells of living tissue, it is changed to an insoluble red compound (chemically known as formazan); non-living tissue remains uncolored (Figure 7–5). This test was developed in Germany by Lakon (*33*) who referred to it as a topographical test, since loss in embryo viability begins to appear at the extremity of the radicle, epicotyl, and cotyledon tips. The reaction takes place equally well in dormant and nondormant seed. Results can be obtained within 24 hours, sometimes in two or three hours. TTC is soluble in water, making a colorless solution. Although the solution deteriorates with exposure to light, it will remain in good condition for several months if protected. It should be discarded if it becomes yellowish. A 0.1 to 1.0 percent solution is commonly used. The pH should be 6 or 7.

The test distinguishes between living and dead tissues within a single seed and can indicate weakness before germination is actually impaired. Necrotic areas may be attacked by pathogenic organisms, and seeds with such dead tissues may decay during stratification or give reduced germination under unfavorable soil conditions. In the hands of a skilled technologist this test can be used for seed-quality evaluation and as a tool in seed research (*37, 38*).

On the other hand, this test may not adequately measure certain types of injury which could lead to seedling abnormality—for instance, an overdose of chemicals, seedborne diseases, frost or heat injury. It requires standardization of procedures and skill in evaluating results.

In using this test, certain procedures must be followed, although details vary with different kinds of seeds (*26, 38*).

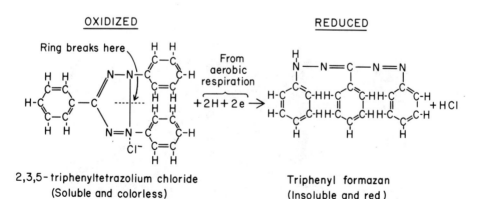

OXIDIZED   REDUCED

2,3,5-triphenyltetrazolium chloride   Triphenyl formazan
(Soluble and colorless)   (Insoluble and red)

Figure 7–5 Chemical reaction in the tetrazolium test for seed viability. Living and respiring tissues cause the reaction to take place with the formation of the red-colored triphenyl formazan. In dead, nonrespiring tissues the reaction does not occur and no color change takes place.

(a)   Any hard covering such as an endocarp, wing, or scale must be removed. Tips of dry seeds of some plants, such as *Cedrus,* can be clipped.

(b)   Seeds should be soaked in the dark; they must be moistened to activate enzymes and to facilitate cutting or removal of seed coverings. Seeds with fragile coverings, such as snap beans or citrus, must be softened slowly on a moist medium to avoid fracturing.

(c)   Most seeds require preparation for TTC absorption. Some embryos with large cotyledons, such as *Prunus,* apple, and pear, are excised completely. Seeds may be cut longitudinally to expose the embryo (corn and large seeded grasses, larch, some conifers); or transversely ¼ to ⅓ at the end away from the radicle (small seeded grasses, juniper, *Carpinus, Cotoneaster, Crataegus,* rose, *Sorbus, Taxus*). Seed coats can be removed, leaving the large endosperm intact (some pines, *Tilia*). Some seeds (legumes, timothy) require no alteration.

(d)   Seeds are soaked in TTC solution for 2 to 24 hours. Cut seeds require a shorter time; those with exposed embryos somewhat longer; intact seeds 24 hours or more.

(e)   Interpretation of results depends upon the kind of seed and its morphological structure. Completely colored embryos indicate good seed. Conifers must have both megagametophyte and embryo stained. In grass and grain seeds only the embryo colors. Seeds with declining viability may have uncolored spots, or be unstained at the radicle tip and the extremities of the cotyledons. Nonviability depends upon the amount and location of necrotic areas and correct interpretation depends upon standards worked out for specific seeds *(37).*

(f)   If the test continues too long, even tissues of known dead seeds become red due to respiration activities of infecting fungi and bacteria. Likewise, the solution can become red because of contamination.

X-ray Analysis

X-ray photographs of seeds *(30, 41, 42)* can be used to detect mechanical disturbances in seeds, insect infestation and empty seeds, numbers of embryos per seed (polyembryony), development of the embryo and endosperm, and cracks and fissures due to weathering damage. (Figure 7–6). Contrast

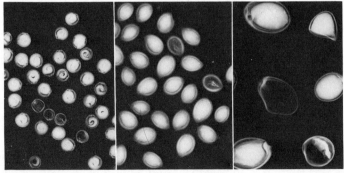

Figure 7–6   X-ray radiographs of seeds showing presence and condition of structures within seed coats. X-ray photos taken of dry seeds exposed for 5 min. at 20 KVP (kilo volt peak). *Left: Koelreuteria paniculata* (Goldenrain tree). *Center: Prunus cerasifera* (Myrobalan plum). *Right: Camellia japonica* (camellia).

agents, such as solutions of salts or heavy metals may be used. These will enter damaged seeds and penetrate dead cells, but not living cells because of their semipermeability. X-rays penetrate through the living cells more easily and produce an exposure on the film. Fairly consistent and reliable correlations with germination percentages can be obtained with fresh conifer seeds, but consistent results may not occur with stored seed, even with contrast agents *(10)*. Seeds may be soaked in water 16 hours, transferred to a concentrated solution (20 to 30 percent) of barium chloride for 1 to 2 hours, washed to remove all excess material, and dried.

PRECONDITIONING SEEDS TO STIMULATE GERMINATION

Mechanical Scarification

Mechanical scarification is done to modify hard or impervious seed coats. Scarification is any process of breaking, scratching, or mechanically altering the seed covering to make it permeable to water and gases. Although some scarification probably occurs during harvesting, extraction, and cleaning, germination of most hard-coated seeds is improved by additional artificial treatment.

Rubbing the seeds on sandpaper, cutting with a file, or cracking the seed covering with a hammer or between the jaws of a vise are simple methods useful for small amounts of relatively large seed. For large-scale operations, special mechanical scarifiers are used. Small-seeded legumes, such as alfalfa and clover, are often treated in this manner to increase germination. Tree seeds may be tumbled in drums lined with sandpaper or in concrete mixers, combined with coarse sand or gravel *(45)* (Figure 7–7). The sand or gravel should be of a different size than the seed to facilitate separation.

Scarification should not proceed to the point at which the seeds are injured. To determine the optimum time, a test lot can be germinated, the seeds may be soaked to observe swelling, or the seed coats may be examined with a hand lens. The seed coats generally should be dull but not so deeply pitted or cracked as to expose the inner parts of the seed.

Figure 7–7   Disk scarifier used to modify hard seed coats. Construction based on one described by U.S. Dept. Agr. for use with tree seeds. Scarifier consists of five disks covered with abrasive paper mounted on a shaft and enclosed within a metal cylinder also lined with abrasive paper. The bottom third of the cylinder is filled with seeds and the disks are rotated at a speed of 500 to 900 rpm. Duration of treatment must be established for the particular seed lot and kind of seed.

Mechanical scarification is simple and effective with many species if suitable equipment is available. The seeds are dry after treatment and may be stored or planted immediately by mechanical seeders, although scarified seed is more susceptible to injury from pathogenic organisms and will not store as well as comparable nonscarified seed.

Soaking Seeds in Water

Soaking seeds is done to modify hard seed coats, remove inhibitors, soften seeds, and reduce the time of germination. This treatment will overcome seed coat dormancy and stimulate germination in some cases. Some impermeable seed coats can be softened by placing the seeds in four to five times their volume of hot water (170° to 212° F, 77° to 100° C). The heat is immediately removed, and the seeds allowed to soak in the gradually cooling water for 12 to 24 hours. Following this the unswollen seeds can be separated from the swollen ones by suitable screens and either re-treated or subjected to some other method of treatment. The seeds should usually be planted immediately after the hot-water treatment. However, honey locust seeds have been carefully dried and stored for later planting without impairing the germination percentage, although the rate of germination was reduced. (See Figure 6–12.)

Seeds have been boiled in water for a period of several minutes in some cases, but such a procedure is hazardous. Exposure to such a high temperature is likely to injure the seeds.

Inhibitors present in some seeds can be leached out by washing or soaking in water. For instance, the procedure for laboratory germination of beet seeds calls for pre-soaking the seeds for 2 hours in 250 ml of water for each 100 seeds, followed by washing with water and blotting dry. This procedure is unnecessary in field planting, because the inhibitors are adsorbed by soil particles.

Soaking seeds prior to germination may shorten the time for emergence if the seeds are normally slow to germinate. Cold soaking the seeds of some conifers, such as Coulter pine, Monterey pine, and Douglas fir, for 24 hours just above freezing is sometimes used in conditioning for germination.

Acid Scarification

This procedure is useful in modifying hard or impermeable seed coverings. Soaking seeds in concentrated sulfuric acid is effective in doing this. Sulfuric acid must be used with care, because it is strongly corrosive and reacts violently with water, causing high temperatures and splattering. Protective clothing should be worn, and the operator should be aware of the dangers to skin and eyes.

Dry seeds are placed in glass or earthenware containers and covered with concentrated sulfuric acid (specific gravity 1.84) in a ratio of about one part seed to two parts acid. Separatory funnels are useful containers for small lots of seed, enabling easy removal of the acid. The mixture should be stirred cautiously at intervals during the treatment to produce uniform results and to prevent the accumulation of the dark, resinous material from the seed coats which is sometimes present. Since stirring tends to raise the temperature, vigorous agitation of the mixture should be avoided, or injury to the seeds and splattering of the acid may result.

The length of treatment should be carefully standardized *(23)*. This will depend upon temperature, the kind of seed, and sometimes the particular lot of seed. Large lots of seed should be thoroughly mixed prior to treatment to insure uniformity. The time of treatment may vary from as little as ten minutes for some species to as much as 6 hours or more for other species. If a large lot of seed is to be treated, the optimum period of treatment should be determined by preliminary tests. With thick-coated seeds that require long periods, the progress of the acid treatment may be followed by drawing out samples at intervals and checking the thickness of the seed coat. When it becomes paper thin, the treatment should be terminated immediately.

At the end of the treatment period the acid is poured off, and the seeds are washed to remove the acid. Copious amounts of water should be added immediately to dilute the acid as quickly as possible, reduce the temperature, and avoid splattering. Washing for ten minutes with running water should be sufficient. The seeds can either be planted immediately when wet, or dried and stored for later planting. (Waste acid should be poured out on unused soil, never down a sink drain.) (See Figure 6–12.)

Moist-Chilling (Stratification)

The primary purpose of this treatment is to provide the exposure to low temperatures that is often required to bring about prompt and uniform seed germination. This treatment is necessary for germination of seeds of many woody tree and shrub species. It permits physiological changes within the embryo to occur (after-ripening; see p. 134). This process requires that exposure to low temperatures (32° to 50° F; 0° to 10° C), moisture and air be continued for a certain length of time. Although stratification has some benefit in softening the seed coats, it is usually more effective to subject the seeds to a moist, warm treatment prior to chilling if hard seed coats are present.

Dry seeds should be soaked in water for 12 to 24 hours, drained, mixed with a moisture-retaining medium, then stored for the required period of time. The usual storage temperature is 35° to 45° F (2° to 7° C). Higher temperatures may be satisfactory but may result in premature sprouting. Lower temperatures increase the length of storage required but will prevent early sprouting. Below-freezing temperatures are less effective. Almost any medium which holds moisture, provides aeration, and contains no toxic substances is suitable. These include well-washed sand, peat moss, chopped sphagnum moss, vermiculite, and well-weathered sawdust. (Fresh sawdust may contain toxic substances.) A good material is a mixture of one part sand and one part peat moss, moistened and allowed to stand 24 hours before use. Any medium used should be moist but not so wet that water can be squeezed out.

Seeds are mixed with one to three times their volume of the medium or may be stratified in layers ½ to 3 in. thick, alternating with equally thick

Figure 7–8 Seeds prepared for stratification in polyethylene bags filled with moist vermiculite. It is important that the bags be made of polyethylene and not Saran. Polyethylene is more than 100 times permeable to oxygen—which the seeds need—than Saran of equal thickness.

layers of the medium. Suitable containers are boxes, cans, glass jars (with perforated lids), or other containers which provide aeration, prevent drying, and protect against rodents. Polyethylene bags are excellent containers (Figure 7–8). A fungicide may be added as a seed protectant (see p. 164).

Seeds may be stratified in refrigerators or out-of-doors over winter in covered boxes or in shallow pits in the ground. They should be protected against freezing, drying, and rodents (*43*).

The time required for completion of after-ripening depends on the kind of seed, and sometimes upon the individual lot of seed as well. For most seeds, between 1 and 4 months are required for low-temperature stratification. During this time the seeds should be examined periodically; if they are dry, the medium should be remoistened.

At the end of the after-ripening period some of the seeds may begin to germinate in storage. For planting, the seeds are removed from the containers and separated from the medium, care being taken to prevent injury to the moist seeds. A good method is to use a screen which will allow the medium to pass through while retaining the seeds. The seeds should be planted without drying.

Seeds should be germinated at relatively cool temperatures or they may revert to secondary dormancy. Flats of seeds exposed to the sun in a greenhouse, for instance, may have too high a temperature.

Combinations of Two or More Pregermination Treatments

Two or more treatments are combined either in order to overcome the effects of an impervious seed covering plus a dormant embryo (*double dormancy*) or to promote germination of seeds with complex embryo dormancy.

The combination of mechanical scarification, acid scarification, or hot water soaking followed by moist-chilling is effective for seeds which have both a hard, impermeable seed covering and a dormant embryo. Any of the three seed-coat modifying treatments could be used. An even more effective treatment in some cases is to interpose several weeks of moist-warm conditions between the seed coat treatment and the moist-chilling period. Softening the seed coats must precede chilling so that water can penetrate and be absorbed by the embryo. With fall-ripening seeds, this combination of treatments usually results in prompt germination the following spring.

A period of moist-warm conditions of perhaps several months, followed by a period at low temperatures (usually several months), is an effective treatment for seeds of a number of plants. The warm period, 68° F (night) to 86° F (day) (20° to 30° C), results in decomposition of the seed coverings due to activity of microorganisms. Sterile media, such as vermiculite or perlite, lacks organisms which are helpful in breaking down seed coats. At effective after-ripening temperatures, i.e., 32° to 50° F (0° to 10° C), micro-organism activity is low, and little decomposition occurs. A constant intermediate temperature near 50° F may be effective in some cases, since it is within the lower temperature range of microorganism activity and also within the upper temperature range of after-ripening, although the required time period would usually be greater than if the two pre-treatments were used consecutively.

The procedure for preparing seeds for warm stratification does not differ essentially from that for moist-chilling. Seeds are planted in flats and kept at the desired temperature for the required time. In germinating large numbers of seeds, the most practical method is to plant out-of-doors as described in the following section.

Timing the Planting

Seed is planted out-of-doors directly in the seedbed, cold frame, or nursery row at a time of the year when the natural environment provides the necessary conditions for after-ripening. However, when seeds remain for a long period in an outdoor seedbed, they must be protected from drying, adverse weather conditions, rodents, birds, diseases, and competition from weeds, although herbicides can be used to control the latter.

Several different categories of seeds can be handled in this way with good germination in the spring following planting. Seeds which require a cold treatment may be fall-planted, whereupon they after-ripen in the soil during winter. They will germinate promptly in the spring when the soil begins to warm up, but while the soil temperature is still low enough to inhibit damping-off organisms and to avoid high-temperature inhibition. Seeds which require high temperatures followed by chilling can be planted in the summer sufficiently early to fulfill their warm-temperature requirements, while the subsequent winter period in the soil satisfies the chilling requirement.

Germination can be facilitated for some seeds (e.g., juniper and some *Magnolia* species) if the fruit is harvested when ripe and the seeds planted immediately without drying. Once the seeds become dry, the seed coats harden and germination may be delayed, perhaps until the second spring after they ripen. Seeds which ripen early in the growing season and lose viability rapidly should be collected and planted in spring or summer as soon as they mature. Where effective treatments are not known, the propagator should attempt to reproduce the natural seeding habits of the plant and provide the germinating conditions of its natural environment.

Dry Storage

Freshly harvested seeds of many annual and perennial herbaceous plants fail to germinate until after a period of normal dry storage. Such post-harvest dormancy may last from a few days to several months, depending upon the species of plant. Since dry storage is the usual method of handling and storing most cereal, vegetable, and flower seeds, this dormant period has usually passed by the time the seeds reach the hands of the propagator.

Where seeds are to be germinated soon after harvesting, as in seed-testing laboratories, this form of dormancy may cause difficulty. Drying for three days at 104° F (40° C) or five days at 99° F (37° C) has stimulated germination (*62*).

Temperature Control During Germination

Germination of freshly harvested dormant seeds of certain species can occur only below particular temperatures, although this requirement tends to disappear after dry storage. Seeds of some vegetable and flower species have this requirement. Seeds sensitive to high temperatures can be germinated by maintaining constant reduced temperatures of approximately 55° to 65° F (13° to 18° C), by planting seeds at favorable times of the year (during spring or fall), or by germinating the seeds in a cool climate area. If such seeds are subjected, while moist, to somewhat lower temperatures (50° F, or 10° C) for 3 to 5 days, they can often be germinated subsequently at the higher temperatures. Since aeration is necessary, the seeds should not be submerged in water during this chilling period.

Daily alternation of temperatures is effective in stimulating germination of freshly harvested seeds of many species. The usual combinations of temperature are 59° to 86° F (15° to 30° C), or 68° to 86° F (20° to 30° C), with seeds being held at the lower temperatures for 16 hours and at the higher temperature for 8 hours. Such temperature fluctuations are found normally out-of-doors at certain seasons of the year. Controlled fluctuating temperatures are of great importance in seed-testing laboratories.

A high constant temperature will overcome dormancy in freshly harvested seeds of some species (*46*). The seed-testing rules for dormant citron melon (*Citrullus vulgaris* var. *citroides*), for instance, specify a temperature of 86° F (30° C). Likewise, alyce clover (*Alysicarpus*) is germinated at 95° F (35° C), although germination occurs at lower temperatures as the seed ages.

Chemical Stimulants

Gibberellins (see Figure 6–1)   This group of plant hormones has significant activity in seed physiology. Gibberellic acid (GA_3) will promote germination in some kinds of dormant seeds, increase the germination rate, stimulate seedling growth, and overcome dwarfing of dormant epicotyls (*49*). The latter effect may be transitory and produce abnormal seedling growth.

The response to this treatment may vary, depending upon the kind of seed. Seeds are treated with GA by soaking 24 hrs. in water solution at concentrations from 100 to 10,000 ppm. Removal of restrictive seed coverings may be necessary to allow penetration. Large-scale use should be preceded by preliminary trials. Gibberellic acid is produced commercially by fungus cultures and is available as the potassium salt (*49*).

Cytokinins (see Figure 6–1)   These natural growth hormones appear to be active in stimulating germination of some kinds of seed (see page 111). A commercial preparation, *kinetin* (6-furfurylamino purine) is available. Dissolve first in a small amount of HCl, then dilute with water. Other available synthetic cytokinins are BA (6-benzylamino purine) and PBA (6-benzylamino)-9-(2-tetrahydropyranyl)-9H-purine); these are more active for higher plants than is kinetin (*49*). These materials may stimulate germination and overcome high temperature dormancy of certain seeds, such as lettuce (see page 122). Seeds are generally soaked in 100 ppm kinetin solutions for three minutes. Large-scale treatments should be preceded by trials at varying concentrations. Cytokinins are sometimes effective in promoting germination when in combination with gibberellic acid and with ethylene-producing compounds.

Ethylene (see Figure 6–1)   Ethylene occurs naturally in plants and is known to have growth-regulating properties. Ethylene applied to seeds has stimulated germination of some seeds experimentally (see page 123). With the availability of ethylene-generating chemicals, such as ethephon (*49*), the practical commercial use of ethylene to stimulate seed germination is possible.

Potassium nitrate   Many freshly harvested dormant seeds germinate better after soaking in a potassium nitrate solution. The technique is used largely in seed-testing laboratories. Seeds are placed in germination trays or Petri dishes and the substratum moistened with 0.2 percent potassium nitrate. For Kentucky bluegrass (*Poa pratensis*) or Canada bluegrass (*P. compressa*) a 0.1 percent solution should be used. If they are rewatered, tap or distilled water is used rather than additional nitrate solution.

Thiourea   This chemical—$CS(NH_2)_2$—has been used to stimulate germination of some dormant seeds, particularly those that do not germinate in darkness or at high temperatures, or that require a moist-chilling treatment. Water solutions at 0.5 to 3 percent are used. Since thiourea is somewhat inhibitory to growth, it is desirable to soak the seeds no longer than 24 hours and then rinse in water.

Sodium hypochlorite   This material is used to stimulate germination of rice seed, apparently overcoming a water-soluble inhibitor in the hull (*35*). A proportion of 1 gal. of commercial concentrate to 100 gal. of water is used.

Light Requirements

Exposure to light can stimulate germination of many kinds of seeds, depending upon age, previous handling, and accompanying temperature. Light sensitivity is strongest just following harvesting and tends to disappear with dry storage. Consequently light influence is quite important in seed testing. The Rules for Seed Testing (2) specify light exposure for seeds of 42 percent of the 493 species cited. These include most grasses and conifers, many flowers, and a number of vegetables, including lettuce and endive.

Light should be provided by cool-white fluorescent lamps at an intensity of at least 75 to 125 ft-c (800 to 1345 lux) for at least eight hours daily. Seeds must have imbibed moisture at the time of light exposure and should be placed on top of the medium.

DISEASE CONTROL DURING SEED GERMINATION

The control of disease during seed germination is one of the most important tasks of the propagator. The most universally destructive pathogens are those resulting in "damping-off," which may cause serious losses of seeds, seedlings, and young plants. Most of the following discussion deals with this problem. In addition, there are a number of fungus, virus, and bacterial diseases that are seedborne and may infect certain plants (4, 46). In such cases, specific methods of control are required during propagation.

Damping-Off

Damping-off is caused by certain fungi, primarily *Pythium ultimum* and *Rhizoctonia solani,* although other fungi—for example, *Botrytis cinerea* and *Phytophthora sp.*—may also be involved. Mycelia from these organisms occur in soil, in infected plant tissues, or on seeds, from which they contaminate clean soil and infect clean plants. *Pythium* and *Phytophthora* produce spores that are moved about in water.

Damping-off occurs at various stages during seed germination and subsequent seedling growth. (a) The seed may decay or the seedling may rot before emergence from the soil (*pre-emergence damping-off*). (b) The seedling may develop a stem rot near the surface of the medium and fall over (*post-emergence damping-off*). (c) The seedling may remain alive and standing but the stem will become girdled and the plant stunted, eventually dying (*wire-stem*). (d) Rootlets of larger plants may be attacked; the plants will become stunted and eventually die (*root rot*).

Symptoms resembling damping-off are also produced by certain unfavorable environmental conditions in the seedbed. Drying, high soil temperatures, or high concentrations of salts in the upper layers of the germination medium can cause injuries to the tender stems of the seedlings near the ground level. The collapsed stem tissues have the appearance of being "burned off." These symptoms may be confused with those caused by pathogens.

The control of damping-off involves two separate procedures: (a) the complete elimination of the organisms during propagation and (b) the control of plant growth and environmental conditions, which will minimize the effects of damping-off or give temporary control until the seedling has passed its initial vulnerable stages of growth. Where possible, it is best to eliminate the organisms completely. Otherwise, there may continue to be a source of infection which can cause difficulty later if conditions become favorable to it.

Control of the organisms can be obtained by properly treating the germination medium, treating the seed, and following good sanitation practices. Soil treatments are considered in Chapter 2, seed treatment later in this section. Seeds should always be treated if a treated germination medium is used; otherwise, recontamination will result, and losses may be more severe than if an untreated medium were used in the first place. On the other hand, planting treated seed in untreated soil can be effective, since many of the materials used protect the seed for a time after planting. Sphagnum moss inhibits damping-off and is an effective germination medium (25).

Good sanitation in the general area where seeds are being germinated is important. Any decayed or diseased plants or other organic matter should be disposed of promptly. Walks, benches, and walls should be kept clean and sprayed or washed with a disinfectant.

Much benefit can be derived by using good seed and controlling the environment during germination. Many of the practices developed by seed propagators over the years have permitted seedling production despite the presence of disease organisms in the medium. The difficulty of control depends upon the quantity of the organisms present. If there is an initially high disease potential, the control procedures described below may not be effective.

High-vitality seed that will germinate rapidly is less likely to be attacked by damping-off fungi than weak, slow-germinating seeds. A slow rate of germination could be due to age, conditions of development, or other factors affecting seed vitality. Some kinds of seeds are more apt to be attacked than others either because of naturally slow germination or because of inherent susceptibility.

The environmental conditions prevailing during the germination period will affect the growth rate of both the fungi and the seedling. For instance, the optimum temperatures for the growth of *Pythium ultimum* and *Rhizoctonia solani* are between approximately 68° and 86° F (20° and 30° C), with a decrease in activity at higher or lower temperatures. Seeds which have a high minimum temperature for germination (warm-season plants) are particularly susceptible to damping-off, because at lower or intermediate temperatures (less than 75° F, or 23° C) their growth rate is low at a time when the activity of the fungi is high. At higher temperatures, not only do the seeds germinate faster, but also the activity of the fungi is less. Field planting of such seeds should be delayed until the soil is warm. On the other hand, seeds of cool-season plants germinate (although slowly) at temperatures of less than 55° F (13° C), but since there is little or no activity of the fungi, they can escape the effects of damping-off. As the temperature increases, their susceptibility increases, because the activity of the fungi is relatively greater than that of the seedling.

The moisture content of the germination medium is of great importance in determining the incidence of damping-off. Conditions usually associated with damping-off include over-watering, poor drainage, lack of ventilation, and high relative humidity. Experiments have shown that the damping-off fungi can grow in concentrations of soil solutes high enough to seriously inhibit the growth of seedlings. Where salts accumulate in the germination medium, damping-off can thus be particularly serious.

If damping-off begins after seedlings are growing, it may sometimes be controlled by treating that area of the medium with a fungicide. The ability to control attacks depends on their severity and on the modifying environmental conditions. (See Chapter 2.)

Seed Treatments

Seed treatment to control diseases is of three types: disinfestation, disinfection, and seed protection (5, 16, 34, 47).

Disinfestants eliminate organisms present on the surface of the seed. Materials which have this sole action are useful if the seeds or embryos are to be grown in aseptic culture (see p. 516) or in some type of sterile medium. Materials which have been used for this purpose include calcium hypochlorite, Merthiolate (described on p. 516), bromide water, and mercuric chloride. Likewise, materials listed below as disinfectants or protectants are likely to be disinfestants.

Calcium hypochlorite (53) In this treatment, 10 gm of calcium hypochlorite are placed in 140 ml of water and shaken for 10 minutes or allowed to stand for an hour. The filtrate containing about 2 percent hypochlorite is usually used, although this is sometimes diluted by half. The time of contact which will disinfest and still avoid injury varies with different kinds of seeds, usually being between 5 and 30 minutes. Adjusting the pH to 8 to 10 produces the most consistent results. Commercial bleaching preparations, such as Clorox (containing 5.25 percent sodium hypochlorite), have also been used, diluted with water 1:9.

Disinfectants eliminate organisms within the seed itself. Treatments of this type include hot water, formaldehyde, and aerated steam.

Hot water treatment (3, 4) Dry seeds are immersed in hot water (120° to 135° F, or 49° to 57° C) for 15 to 30 minutes, depending upon the species. After treatment, the seeds are cooled and dried rapidly by spreading out in a thin layer. Temperature and timing should be regulated precisely, or the treatment may injure the seeds. Old, weak seeds should not be treated. Also, a seed protectant should be used. This treatment is effective for specific seedborne diseases of vegetables and cereals, e.g., *Alternaria* blight in broccoli and onion, loose smut of wheat and barley. Aerated steam is an alternate method that is less apt to injure seeds, is easier to handle, and is less expensive (4).

Protectants are materials applied to the seed which protect it from soil fungi. These materials are also applied as a soil drench either before or after seed planting. Numerous materials for seed treatment are available commercially. Among the older materials are certain zinc and copper fungicides. Copper oxide (red or yellow) has been used as a seed protectant for vegetable seeds. Machine planting is more difficult with such treated seeds, because the adhering dust slows down the flow of seed. Other fungicides in use include chloranil, thiram, ferbam, benomyl, captan, and zinc trichlorophenate. All are available under various trade names.

The combination of an insecticide and a fungicide has been found beneficial in field plantings of certain crops to control attacks by wire-worms and the seed corn maggot. Insecticides used for this purpose include lindane, aldrin, dieldrin, heptachlor, and benzenehexachloride. These materials should be combined with a fungicide (captan, thiram, chloranil), since they tend to increase the susceptibility of the seed to fungus attack. Various proprietary products involving such combinations are available.

All state and federal regulations governing the use of such materials on crop plants should be determined and followed carefully.

In using the various trade preparations, the directions of the manufacturer should be followed carefully, since if improperly used, these materials can cause injury or death to the seedlings of some plants. Since some tend to be hazardous to the person using them, they should be handled carefully so as to avoid contact with the skin or breathing of the dust. In many cases, special machines must be used for treating the seeds.

There are several ways to apply pest controlling materials. (a) For small lots of seed, the dry or dust method is probably the most convenient. For treating large amounts of seed, the dust is applied in a rotating barrel, cement mixer, or other mechanical mixer. (b) In the liquid fixation method, $\frac{1}{4}$ pt to $\frac{1}{2}$ gal. of water is sprayed on each 100 lb of seed by means of a sprayer after the dust has been applied. (c) In the "quick-wet" method, a concentrated solution of a volatile chemical is added to the seed and thoroughly mixed with it. (d) The slurry method involves the treating of the seed with a thick suspension of chemicals in a special slurry treater. (e) Soluble fungicides may be sprayed directly on the seed in a continuous or batch spray treater. (f) Seed pelleting is used particularly for some conifers (*16*). An adhesive is first applied to the seed. The fungicide (and sometimes insecticide) is then added. Bird and rodent repellents may also be applied. Pelleted seed is particularly useful in precision machine planting.

INDOOR SEEDLING PRODUCTION
Indoor seedling production has many important uses. It is required for all greenhouse and indoor plant growing. It is used to produce plants for later transplanting to the field or garden and is an important technique in flower and vegetable production. The very large bedding plant industry is based on greenhouse seedling production. Indoor production has become more attractive as a propagation procedure with the availability of costly F_1 hybrid flower and vegetable seeds, artificial soil mixes (see page 32) and low-cost plastic containers. Similar methods may be used for producing container-grown stock of woody plants for landscaping, reforestation, or orchards, particularly for hard-to-transplant nursery species.

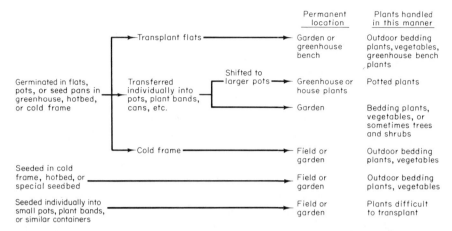

Figure 7–9   Methods of handling seedlings in relation to indoor culture.

Indoor seedling production provides better germination control than direct seeding out-of-doors and makes it possible to program precise production for particular seasons and uses. On the other hand, expensive facilities, such as greenhouses, covered frames, hotbeds, coldframes, etc., are required (see Chapter 2). Seedlings to be transplanted bare-root, as are many vegetables, undergo a certain "transplanting shock" which checks their growth and, if too severe, can prevent this technique from being used. Transplanting from containers also has disadvantages. Seedlings of many woody plants have tap roots and do not form good root systems unless properly handled. Figure 7–9 shows various uses of this procedure in producing plants.

Producing Seedlings of Herbaceous Plants

Seedlings may be started in various containers, ranging from a small flat or box in a sunny window to special propagating structures, such as a greenhouse, cold frame, or hotbed, as described in Chapter 2. Any container that will hold the germination medium and provide drainage can be used. Many types are available, from wood flats to various plastic containers and special units largely composed of peat moss (see Chapter 2). Polyethylene bags can be effectively used to produce seedlings. All containers should be sterile when used.

The germination medium should be sterile, and should be porous to hold moisture and provide good aeration. Although soil mixes, such as equal parts loamy soil, sand, and peat, have been used for many years, special soil mixes made from various artificial or inert media have come into wide use (see Chapter 2). Where the latter are used, it is important that nutrients be added to the mix and supplementary fertilization be started relatively soon, care being taken not to build up harmful concentrations of salts.

Seed flats of soil should be filled completely, worked carefully to fill corners, and the excess soil should be removed with a straight board across the top. Soil is then tamped with a block to provide a uniformly firm seed bed to a level about ½ inch below the top of the flat. The flat should be watered from above or presoaked, drained, covered, such as with polyethylene, to hold moisture until planting. Medium to large seeds are sown two to three times their minimum diameter. Very fine seeds may be dusted on the surface (see Figure 7–10).

Maintaining proper germination temperature in the medium is very important. For most seeds it should be relatively high for uniform, prompt germination, preferably 70°–75° F (21°–24° C) (6), although for some kinds of seed 65° F (18° C) is better, and for others, particularly warm-season and tropical plants, 80° F (26.5° C) is preferred.

Maintaining proper moisture conditions during the germination period is essential so that the medium neither dries out at any time nor remains too wet so that damping-off becomes a problem. Cover the seed flat to prevent water loss, or else water frequently. A method widely used commercially is a fine intermittent mist spray (see page 297) set for about eight seconds of water every ten minutes during the day.

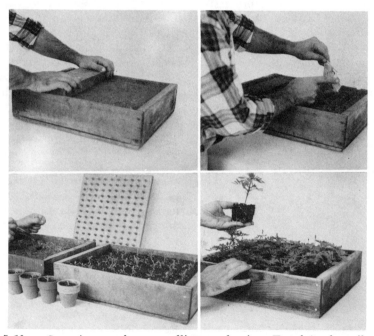

Figure 7–10   Steps in greenhouse seedling production. *Top left:* the soil medium is firmed to eliminate air pockets. *Top right:* the seed is planted into shallow furrows (on surface for very fine seed). *Below left:* when seedlings have produced the first true leaves, they are pricked off into a transplant flat at wider spacing or into individual containers. The peg board in the background is convenient for producing uniform spacing. *Below right:* after hardening, the seedlings are transplanted into the garden or other permanent location. A cube of soil from the flat is removed with the plant to avoid disturbing the root system.

Once seeds have germinated, the principal objectives of production are to prevent damping-off and to develop stocky, vigorous plants capable of being transplanted with little check in growth. The usual procedure in production is to immediately shift flats of germinated seed to somewhat lower temperatures and to expose them to good light. High temperatures and low light tend to produce spindly, elongated plants that will not survive transplanting (see Figure 7–11).

Once root systems have developed sufficiently to grow into the medium, irrigation schedules can be reduced to keep the surface of the medium somewhat dry without allowing the medium underneath to become dry. This helps prevent damping-off and produces sturdy seedlings.

Before closely planted seedlings in the container become crowned they must be transplanted into a transplant flat or a check in growth will occur. Transplanting normally is done no later than after two to four true leaves appear.

Success in transplanting depends to a large extent upon the previous handling. When the permanent location is in the greenhouse, environmental conditions are controlled and are not different from those to which the seedling has been exposed during development. When movement is from a greenhouse or hotbed to the open field, the operation is somewhat more critical and requires that the plant be "hardened" prior to its shift to the open field. "Hardening" involves a checking of growth resulting in the accumulation of carbohydrates which makes the plant better able to withstand adverse conditions. This process can be brought on by withholding the supply of moisture, reducing the temperature, and gradually shifting from the greenhouse or hotbed to the environment of the permanent location. A cold frame or lathhouse is useful for this purpose. Flats of seedlings can be moved into them and allowed to remain for seven to ten days prior to placing in the field.

Figure 7–11   Tomato plants ready for transplanting. The two plants on the left are spindly and undesirable for transplanting. The three on the right are healthy, well-formed plants suitable for transplanting. The latter show the effects of sufficient space and desirable growing conditions. Tomato plants are transplanted bare-root. From *Calif. Agr. Ext. Cir. 167.*

Before being moved into the field, the plants should be watered thoroughly. Planting is done in the field by hand or, in some cases, by transplanting machines. During transplanting, it is desirable to retain as much soil about the roots as practical to avoid disturbing the root system. Afterward, the plants should be thoroughly watered and, if practical, temporary shade should be provided. For the first few days, until the plants have become established, they should be watched for wilting, and watered as necessary.

The use of "booster" or "starter" solutions containing nitrogen, phosphorus, and potassium shortly before or after transplanting is sometimes beneficial in establishing plants in the field. Such solutions should be used with caution, since if the soil is low in moisture at transplanting time, high concentrations of fertilizer will develop and injure the plants. (See Chap. 2.)

Dramatically increased growth of high-quality plants can be produced under controlled environmental conditions as compared to conventional methods of seedling handling. This is shown in Figures 2–8 and 6–11. Such ideal environmental conditions are initiated at the time of seeding and have been called the "headstart" program (*32*). From time of germination seedlings are given high temperatures (about 86° F; 30° C), high light intensity (2000 ft c. minimum from artificial lights) photo-period with a day length of at least 16 hours, an increase in ambient carbon dioxide concentration (2000 ppm) as described on page 40, and a relative humidity of 60 percent or more. Under these conditions it is essential to eliminate the pathogens and to have proper nutritional conditions in the medium, since if any one factor becomes limiting, the large increases in production will not occur.

Container-Grown Seedling Plants

Production of seedling trees and shrubs in containers for later transplanting follows principles and procedures similar to those for other plants. Seeds are usually planted in a germination medium in a flat. Seeds may have been pretreated before planting, depending upon their germination requirements, or they may be planted and then the flats stored at the appropriate temperature to after-ripen seeds (when needed) until germination starts. It is important that the flats be protected or covered so the germination medium does not dry out if the time period is very long.

Considerable variation in handling procedures can follow. A usual procedure is to transplant the seedlings to a small liner pot, such as one made with pressed peat, and then later transplant them to a one-gallon container. A major problem with propagating seedling plants of many species is that a long tap root with little branching is produced. The root may grow in circles or become kinked if the root system is restricted by the small size of the container (see Figure 2–13). Once established, the root system may be permanently impaired, although the constriction that is produced may not be apparent until some time later when the seedling is planted outdoors.

Root pruning is desirable and should be done early. Copper or plastic screen-bottomed flats (*14*) stimulate formation of branch roots. Root pruning should be done at the first transplanting, which should be soon after the roots reach the bottom of the flat (*17, 18*). Further root pruning should take place when the liner flat is transplanted, which should be done when the roots protrude through the pot about an inch. These roots are removed. Any delay in transplanting at either stage, or omission of pruning, will increase the incidence of poor root systems.

Planting into flats or liner pots is desirable because it makes possible the production of a better root system and also makes possible the selection of more vigorous plants from a variable group of seedlings. These procedures are particularly useful for plants with long, unbranched root systems, such as pistacia (*27*). In this case it would be desirable to use a longer container, such as one made with a roll of felt roofing paper stapled into a cylinder six inches in diameter and 16 inches long. Various other kinds of containers have been used. For instance, a small tube 3 inches long and about ½ inch in diameter made with styrene-latex and slit along one side has been used in propagating some forest tree species (*36*). Organic soils are useful as a germinating medium, and placing a copper screen in the bottom of the tube is particularly effective in stimulating root development. Seeds are planted in the tube and then transplanted directly to the field at two or three months.

DIRECT SEEDING INTO A PERMANENT LOCATION
Direct Seeding in the Field or Garden
Direct seeding is the primary method of commercially propagating annual, biennial, and sometimes perennial field crops and vegetables. It is more difficult to control germination and obtain a uniform stand of the desired density by direct seeding than it is by transplanting. However, if well handled, direct seeding gives continuous and rapid development of the seedling with none of the checks in growth associated with transplanting. Where carefully controlled densities are required, thinning of seedling plants is a major cost later in the operation. Amateur vegetable and flower gardeners may find direct seeding preferable to transplanting since it is simple, less expensive, and does not make demands on time and facilities.

Seed tapes are available for many flower and vegetable cultivars. Seeds are attached to a plastic tape spaced at the proper distance. The tapes are buried in a shallow furrow. The tapes disintegrate from moisture as the seeds germinate.

Good seedbed preparation is essential for successful field and garden sowing. A good seedbed should have a loose physical texture that produces close contact between seed and soil so that moisture can be supplied continuously to the seed. Such a soil should provide good aeration, but not too much or it dries too rapidly. Adequate moisture should be available to carry the seeds through the germinating and early seedling stage. At the same time there should be good drainage so that water does not stand and impede the oxygen supply to the seed. A medium texture, not too sandy and not too fine, is best. An excellent seedbed is one in which three-fourths of the soil particles (*aggregates*) range from 1 to 12 mm in diameter (*28*). Organic matter in the soil will improve texture, and the soil may be conditioned by incorporating a greenmanure crop or animal manure (allowing time for decomposition) or incorporating peat moss, or other organic materials. These should be mixed uniformly with the soil so that no stratified layers exist that will impede movement of water.

In an ideal seedbed, the soil is moist but not wet, uniformly pulverized to 6 to 10 inches in depth, and without air spaces which would increase evaporation and water loss. Field preparation of seedbeds is by various agricultural implements, such as, plows, discs, listers, etc. On a smaller scale, a rototiller type of implement is useful, or spading and forking by hand is used. For best operation, moisture content should be such that the soil will crumble when squeezed. Working the soil when too wet can lead to compaction and later difficulty in water penetration and aeration. As a rule, if the soil forms a tight ball when squeezed with the hand, it is too wet.

Problems in seedbed preparation occur where the surface soil has a large proportion of fine aggregates less than 0.5 mm diameter which will form surface crusts and clods when they become wet and then dry. When cultivated very dry, the fine aggregates flow among the larger aggregates and limit air and water movement and seedling emergence. When wetted by rainfall or sprinkling the fine aggregates are distributed to form a layer that seals like cement. When broken up, large clods result. A method known as *aggresizing* has been described to combat this problem (*28*). First, the field is cultivated in conventional manner when dry, beds are then sprinkled for four to eight hours (up to 1 inch irrigation) and left to dry overnight. When surface clods dry and show white the bed is cultivated with a rototiller. This breaks up the soil in uniformly sized aggregates. If properly done, three fourths of the aggregates will be between 1 and 12 mm in diameter, (ideal for most seeds) in the top 2 to 3 inches.

Surface mulches of various kinds may also be used to maintain surface moisture and prevent crusting—paper of special manufacture, clear or black polyethylene or polyvinyl plastic film, perlite, vermiculite treated with polyvinyl acetate are used (*15, 44*). Seedbed preparation may include fumigation and other soil treatments to control harmful insects, disease organisms, and weed seeds (see Chapter 2).

In areas of low rainfall, excess soluble salts may accumulate to toxic levels due to exaporation. Overhead sprinkling and planting seed below the crest of sloping seed beds may eliminate or reduce this problem.

Time for planting is determined by the need to provide proper germination temperature and to meet production schedules. Temperature requirements vary with the particular kind of seed. Low soil temperatures can result in slow and uneven germination, disease problems, injury to the seedling resulting in abnormalities of growth. High soil temperatures can result in excessive drying, injury or death to the seedling, or induction of thermo-dormancy (see p. 122).

Depth of planting is a critical factor that will determine the rate of emergence and perhaps stand density. If too shallow, the seed may be in the upper surface that dries out rapidly; if too deep, emergence is delayed. A seedling may not be able to grow sufficiently until the stored food of the embryo is exhausted. Depth varies with the kind and size of seed, and to some extent the condition of the seedbed and the environment at the time of planting. Exposure to light may be a factor with some seeds. A rule of thumb is to plant approximately three to four times the diameter of the seed, being sure the seeds are in the moist layers of the soil.

Rate of sowing to establish the required stand density is another critical factor in direct sowing. If plant density is too low, yields will be reduced; if too high, size and quality of the finished plant may be reduced due to competition among plants for available space, sunlight, water, and nutrients.

Rate of sowing can be determined by the following formula:

$$\begin{array}{l}\text{Pounds (ounces) of seed}\\ \text{required per unit area}\end{array} = \frac{\text{Number of plants per unit area desired}}{\begin{array}{c}\text{Number of seeds per pound (ounces)}\\ \times \text{ percentage germination}\\ \times \text{ percentage purity}\end{array}}$$

This rate is a minimum and should be adjusted to account for expected losses in the seedbed, determined by previous experience at that site.

Vegetable crops usually must be thinned to a given plant density. This expensive process could be eliminated or reduced by precision planting using various devices that place the seed at exact locations. Success of the operation requires seed of high viability and good vigor. Seeds may be "pelletted" with adhesive materials, such as clay, possibly including nutrients and other chemicals, which makes them uniformly round to be used in precision planters. Seed emergence and final stands are sometimes reduced, however, apparently due to the restriction in available oxygen (*11*). Placing vermiculite over the seeds in the planting row has been effective in precision planting (*51*); activated carbon added to the vermiculite protects the seeds against pre-emergence weed sprays (*52*).

Direct Seeding of Trees and Shrubs

Trees and shrubs may be directly seeded into permanent locations for reforestation and planting natural areas (*8*), for planting in permanent places in the landscape (*7*), and occasionally for planting in orchards.

Direct seeding of forest trees is an important reforestation technique because its costs and labor requirements are lower than those for transplanting seedlings, provided soil and site conditions favor the operation (*8*). The major difficulty is the very heavy losses of seeds and young plants that result from predation by insects, birds, and animals and from drying, hot weather and disease (*31, 40*). A proper seedbed is essential and an open mineral soil with competing vegetation removed is best. The latter may be done by burning, disking or furrowing. Seeds may be broadcast by hand or by special planters, or drilled with special seeders. Seeds should be coated with a bird and rodent repellent before sowing. An available mixture for this purpose is a combination of thiram and endrin.

In certain situations, trees and shrubs can effectively be directly seeded into a natural setting for landscape setting purposes (*7*). Procedures are to dig a 4 to 5 inch hole, carefully pulverizing soil if compacted, or, if on a rocky slope, place seeds in a crevice or pocket of soil. If on a slope, make a slight backslope to avoid erosion and to prevent the seed hole from being covered with loose soil from above. Place 1 gram of slow-release fertilizer in the bottom of the hole. Replace soil, leaving a slight depression for the seed. Plant 2 to 20 seeds, depending upon size, and cover with pulverized soil to $\frac{1}{8}$ to $\frac{1}{2}$ inches deep depending upon size. Contact herbicides or mulches around the seed site are essential to remove weed competition. Mulches can be coarse organic materials, (wood chips, coarse bark) or sheet materials (uncoated and polyethylene-coated mulching paper, black polyethylene film, asphalt roofing paper, or kraft building paper). With the coarse organic mulches use a tin can (size 2½), milk carton, or asphalt or kraft paper collar around the seed site. Cut a hole in the center of the sheet mulches to allow the seedlings to emerge. Small wire cages may be needed to cover the site for rodent and bird protection.

GROWING SEEDLINGS IN FIELD NURSERIES

Propagating trees and shrubs by seed is an important nursery operation either to produce plants to be used directly in landscape or reforestation, or to produce rootstocks upon which to bud or graft selected clones or cultivars of fruit, nut, or shade trees. (See Figure 7–12).

Time of Planting

Seeds may be planted in the fall or in the spring depending upon their germination requirements and upon the management practices of the nursery. Most tree species can be placed into three general categories (*21*) in regard to their seed-germination requirements.

Group I includes species (e.g., apple, pear, *Prunus,* and yew) whose seeds require moist-chilling (stratification) but after the completion of after-ripening, are sensitive to high germination temperatures (see Table 7–1). Germination temperatures of 50° to 62° F (10° to 17° C) would be optimum. High soil temperature results in damping-off, seedling injury, and a tendency to

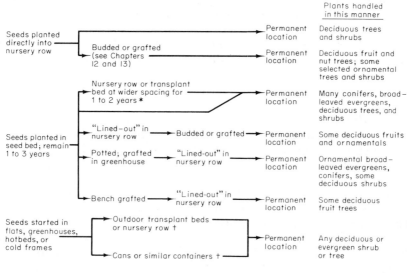

* May be reset into new transplant beds after 2 years.

† May be used to produce stocks for grafting.

Figure 7–12   General schedules for the propagation of seedling and grafted nursery stock.

revert to secondary dormancy. Such seeds could be effectively planted in the fall and sprouting would not take place until early in the spring. If they are kept in stratification storage overwinter, planting should take place as early in the spring as practical, but after frost danger is over.

Group II includes most fir, pine, and spruce species and many deciduous hardwood species whose seeds require varying amounts of moist-chilling (unless exposed to light; see Table 7–1) but that, after this is over, do not germinate in the spring until after the soil becomes warm. Optimum germination temperatures are 68° to 86° F (20° to 30° C). Seeds of these species do not germinate in the cool fall season and not until late in the spring. Such seeds can be handled without difficulty either by fall planting or by moist winter storage and spring planting.

Group III includes those species (e.g., black locust, Colorado and Norway spruce, Chinese and Siberian elm, European larch, bristlecone pine, and Douglas fir) whose seeds are not temperature sensitive and are able to germinate over a range of 60° to 90° F (15° to 32° C). Some of these species require moist-chilling or have hard coats requiring special treatment. If fall-planted, seeds of this group will usually germinate prematurely and the seedling plants can be injured by winter cold. They are best planted in spring, actual time to plant not being as critical as for the other two groups.

If winters are not too severe, fall planting has important advantages compared to spring planting. It eliminates need for storage facilities and chilling during winter to overcome dormancy. Planting time and labor requirements are not critical, whereas with spring planting delays in the spring due to bad weather may occur. With fall planting, as soon as the environmental conditions are favorable in the spring, germination takes place and is not delayed by late planting, resulting in better stands and a longer growing season (*20, 22*). On the other hand, losses sometimes result from unfavorable winter conditions, freezing, drying, disease, or rodents. Also, herbicidal weed control is necessary.

Seedbed Culture

Seedlings of many species are grown in special out-of-door seedbeds during the first one or two years in the nursery. A common size is 3½ to 4 ft. in width with a walkway between seedbeds, the length varying with the size of the operation (see Figure 7–13). In some cases, sideboards are placed alongside the seedbed after sowing to maintain the shape of the bed and to provide support for glass frames or lath shade.

Seed may be either broadcast over the surface of the bed or drilled into closely spaced rows. Regulating the density of planting is essential in planting a seedbed. For economy, seeds should be planted as closely together as feasible. On the other hand, overcrowding leads to greater difficulty with damping-off and reduces vigor and size of the seedlings (*22*), resulting in thin, spindly trees and small root systems. Seedlings with these characteristics do not transplant well (*20*).

The optimum density depends on the species and on the purposes of propagation. In other words, if a high percentage of the seedlings are to

Figure 7-13 Growing seedlings in seedbeds. Olive *(Olea europaea)* seedlings in Italy planted thickly in un-shaded beds.

reach a desired size for field planting or for use as rootstocks for grafting, lower densities might be desired. If the seedlings are to be transplanted into other beds for additional growth, higher densities (with smaller seedlings) might be more practical. Once the actual density is determined, the necessary rate of sowing can be calculated from data obtained from a germination test and from the experiences of the operator at that particular nursery. The following formula is used *(45)*:

$$W = \frac{A \times S}{D \times P \times G \times L}$$

where W = Weight of seeds in pound to sow per given area.

A = area of seedbed in square feet.

S = number of living seedlings desired per square foot.

D = average number of seeds per pound at the moisture content at which the seeds are sown.

P = average purity percentage of seeds (expressed as a decimal).

G = average effective germination percentage in the laboratory (expressed as a decimal).

L = average percentage of germinating seeds that will be living trees at the end of the season (expressed as a decimal). This is a correction factor which takes into account the expected losses which experience at that nursery indicates will occur with that species.

Seeds of a particular lot should be thoroughly mixed before planting to be sure that the density in the seedbed will be uniform. Treatment with a fungicide for control of damping-off is desirable. Small conifer seeds may be pelleted for protection against disease, insects, birds, and rodents (see p. 165). Seeds are planted either by hand or by machine. Depth of planting varies with the size of the seed. In general, a depth of about two to four times the thickness of the seed is a fairly safe estimate, but this also varies with the kind of seed. During early stages of development, seedlings generally must be protected against drying, heat, or cold. In the case of tender plants, glass frames can be placed over the beds, although for most plants a lath shade is sufficient. With some plants, shade should be provided all through the first season; with others, shade is necessary only during the first part of the season. Sprinkling with water to reduce ground temperature during the hot part of the day is useful (*9*).

A mulch applied to the seedbed after planting helps to protect against drying, crusting, and cold, and discourages weed growth. It is particularly desirable for fall-sown seeds, for seeds which are to remain for long periods in the seedbed before germination, or for seeds sown in cold areas. Materials which have been used include sawdust, burlap, pine litter, straw, hay, ground corncobs, wood shavings, sandpaper, canvas, shredded pine cones, sand, manure, snow, and old rooting media of sand and peat. Polyethylene film as a seedbed cover has also been used. The seedbed may be sown in the fall, covered lightly with sawdust, well watered, and then covered with a sheet of polyethylene film, on top of which burlap is placed. The following spring the film is removed as soon as germination starts.

During the first year in the nursery, it is important to keep the seedlings growing continuously without any check in development. A continuous moisture supply, cultivation to control weeds, and proper disease and insect control contribute to successful seedling growth. Fertilization with nitrogen is usually necessary, particularly when a mulch has been applied. Decomposition of organic material can produce nitrogen deficiency.

Weed control can be facilitated by careful seedbed preparation, cultivation, and chemical sprays. Three types of chemical controls are available. *Pre-planting* fumigation with methyl bromide, chloropicrin, D-D, Vapam, steam, or the like (see p. 33) is effective and also kills disease organisms and nematodes. *Pre-emergence* herbicides are applied to the soil before crop seeds emerge. *Post-emergence* herbicides are applied after the crop or nursery seedlings have emerged and are growing. A wide range of selective and non-selective commercial products are available. However, such materials should be used with caution since improper use can cause injury to the young plants. Not only should directions of the manufacturer be followed, but preliminary trials should be made before large-scale use (*12, 48*).

The seedlings may remain in the seedbed for one to three years, depending upon the kind of plant. Many plants are dug at the end of the first year and placed in *transplant beds* or "lined-out" in the nursery bed for further development. Deciduous plants are planted 4 to 6 in. apart in nursery rows 18 in. to 4 ft apart. Nursery plants in rows are often undercut at the end of one year to stimulate subsequent production of a more fibrous root system. Conifers usually go into transplant beds similar to the seedbed but at wider spacing. Spacing may be several inches apart in rows 6 to 8 in. apart. Other cases involve field planting directly from the nursery. Variations in the general procedure described may be followed, as indicated in Figure 7–13.

Seedlings produced in a nursery are often designated by numbers to indicate the length of time in a seedbed and the length of time in a transplant bed. For instance, a designation of 1-2 means a seedling grown one year in a seedbed and two years in a transplant bed or field. Similarly, a designation of 2-0 means a seedling produced in two years in a seedbed and none in a transplant bed.

Nursery Row Culture

Seeds of many tree species are planted directly in nursery rows rather than in special seedbeds. This procedure is used to produce rootstocks for fruit and nut cultivars as well as many shade trees. Where plants are to be budded or grafted in place, the width between rows is about 4 ft and the seeds are planted 3 to 4 in. apart in the row. (See Figure 7–14) If the seeds show poor germinability, the seeds must be planted closer together to get the desired stand of seedlings. Large seed (walnut) is usually planted 4 to 6 in. deep, medium-sized seed (apricot, almond, peach, and pecan) about 3 in., and small (myrobalan plum) about 1½ in. This may vary with soil type. Plants to be grown as seedlings without budding could be spaced at closer intervals and in rows closer together. If germination is low, and a poor stand results, the surviving trees may, owing to the wide spacing, grow too large to be suitable for budding.

Usually, no shading is required by tree seedlings grown in the nursery row, although other cultural operations are similar to seedbed culture. Plants generally remain in place in the nursery row until they are to be transplanted to their permanent location. (See Figure 10–29.)

A variation of this procedure is sometimes followed with species whose seeds germinate well but require time to grow to budding size. Seeds are either planted close together to get a stand of 10 to 15 seedlings per ft, or planted in a narrow band about 4 in. wide in the nursery row. The seedlings remain relatively small during the first season, after which they are lifted and lined-out in a nursery row. Cherry, apple, and pear seedlings are sometimes produced in this manner.

Figure 7–14   Direct seeding in nursery row, illustrated by peach seedlings being grown for rootstocks. These seedling plants will be budded to peach cultivars.

REFERENCES

1   Anonymous, Rules and regulations under the Federal Seed Act, *U.S. Dept. Agr., Agr., Mark. Serv.* 156:1–77. 1963.

2   Assoc. Off. Seed Anal., Rules for seed testing, *Proc. Assoc. Off. Seed Anal.,* 60:1–116. 1970.

3   Baker, K. F., Thermotherapy of planting material, *Phytopathology,* 52:1244–55. 1962.

4   ———, Seed pathology, in T. T. Koslowski (ed.), *Seed Biology.* Vol. II. New York: Academic Press, 1972.

5   ———, and P. A. Chandler, Development and maintenance of healthy planting stock, in K. F. Baker (ed.), *The U.C. System for Producing Healthy Container-Grown Plants,* Calif. Agr. Exp. Sta. Man. 23: pp. 217–36. 1957.

6   Ball, V. (ed.), *The Ball Red Book,* 12th ed. Chicago Ill.: Geo. J. Ball, Inc. 1972.

7   Chan, F. J., R. W. Harris, and A. T. Leiser, Direct seeding of woody plants in the landscape, Univ. of Calif. Ext. Svce. AXT–n27. 1971.

8   Deer, H. J., and W. F. Mann, Jr., *Direct Seeding Pines in the South,* U.S. Agr. Handbook No. 391. Washington, D.C.: U.S. Govt. Printing Office. 1971.

9   Eden, C. J., Conifer seed—from cone to seed bed, *Proc. Plant Prop. Soc.,* 12:208–14. 1962.

10   ———, Use of X-ray technique for determining sound seed, *U.S. Forest Service, Tree Planters' Notes,* 72:25–27. 1965.

11   Ellis, J. E., Prevention of stand losses in tomato due to soil crust formation, *Proc. Amer. Soc. Hort. Sci.,* 87:433–37. 1965.

12   Elmore, C., Management of undesirable plants in ornamental containers and ground covers, *Proc. Inter. Plant Prop. Soc.,* 21:184–90. 1971.

13   Flemion, F., A rapid method for determining the viability of dormant seeds, *Contrib. Boyce Thomp. Inst.,* 9:339–51. 1938.

14   Frolich, E. F., The use of screen bottom flats for seedling production, *Proc. Inter. Plant Prop. Soc.,* 21:79–80. 1971.

15   Gowans, K. D., D. Ririe, and J. Vomocil, Soil crust prevention aids lettuce seed emergence, *Calif. Agr.,* 19(1):6–7. 1965.

16   Hansen, E. W., E. D. Hansing, and W. T. Schroeder, Seed treatments for control of disease, in *Seeds. Yearbook of Agriculture.* Washington, D.C.: U.S. Govt. Printing Office, 1961, pp. 272–80.

17   Harris, R. W., W. B. Davis, N. W. Stice, and D. Long, Root pruning improves nursery tree quality, *Jour. Amer. Soc. Hort. Sci.,* 96:105–9. 1971.

18   ———, Influence of transplanting time in nursery production, *Jour. Amer. Soc. Hort. Sci.,* 96:109–10. 1971.

19   Heit, C. E., The excised embryo method for testing germination quality of dormant seed, *Proc. Assoc. Off. Seed Anal.,* 45:108–17. 1955.

20   ———, The importance of quality, germinative characteristics and source for successful seed propagation and plant production, *Proc. Int. Plant Prop. Soc.* 14:74–85. 1964.

21   ———, Propagation from seed, Part 3. Ten ways laboratory seed tests can help growers, *Amer. Nurs.,* 124(12):10–11, 40–49. 1966.

22 ———, Propagation from seed, Part 5. Control of seedling density. *Amer. Nurs.,* 125(8):14–15, 56–59. 1967.

23 ———, Propagation from seed, Part 6. Hardseededness, a critical factor, *Amer. Nurs.,* 125(10):10–12, 88–96. 1967.

24 ———, Thirty five years testing of tree and shrub seed, *New York Agr. Exp. Sta. (Geneva) Mimeographed leaflet* (undated).

25 Hope, C., V. T. Stoutemyer, and A. W. Close, The control of damping-off by the use of sphagnum for seed germination, *Proc. Amer. Soc. Hort. Sci.,* 39:397–406. 1941.

26 International Seed Testing Association, International rules for seed testing, *Proc. Int. Seed Test. Assoc.,* 31:1–152. 1966.

27 Joley, L. E., and K. W. Opitz, Further experiments with propagation of *Pistacia., Proc. Int. Plant Prop. Soc.,* 21:67–76. 1971.

28 Hoyle, B. J., H. Yamada, and T. D. Hoyle, Aggresizing—to eliminate objectionable soil clods, *Calif. Agri.,* 26(11):3–5. 1972.

29 Justice, O. L., Essentials of seed testing, in T. T. Kozlowski (ed.), *Seed Biology,* Vol. III. New York: Academic Press, 1972.

30 Kamra, S. K., The use of x-rays in seed testing. *Proc. Int. Seed Testing Assoc.,* 29:71–79. 1964.

31 Kozlowski, T. T., *Growth and Development of Trees.* Vol. I. New York: Academic Press, 1971.

32 Krizek, D. T., W. A. Bailey, and H. H. Klueter, A "head start" program for bedding plants through controlled environments, *Proc. 3rd Natl. Bedding Plant Conference.* East Lansing, Mich.: Michigan State Univ., October 1970.

33 Lakon, G., The topographical tetrazolium method for determining the germinating capacity of seeds, *Plant Phys.,* 24:389–94. 1949.

34 Leukel, R. W., Treating seeds to prevent diseases, in *Plant Diseases. Yearbook of Agriculture.* Washington, D.C.: U.S. Govt. Printing Office, 1953, pp. 134–45.

35 Mikkelson, D. S., and M. N. Sinah, Germination inhibition in *Oryza sativa* and control by preplanting soaking treatments, *Crop Science,* 1:332–35. 1961.

36 Maclean, N. M., Propagation of trees by tube technique, *Proc. Int. Plant Prop. Soc.,* 18:303–9. 1968.

37 MacKay, D. B., The measurement of viability, in E. H. Roberts (ed.), *Viability of Seeds.* Syracuse, N. Y.: Syracuse Univ. Press, 1972.

38 Moore, R. P., Tetrazolium staining for assessing seed quality, in W. Heydecker (ed.), *Seed Ecology.* London: Butterworth and Co., 1973, pp. 347–66.

39 Rudolf, P. O., State tree seed legislation, *U.S. Forest Service, Tree Planters' Notes* 72:1–2. 1965.

40 Schubert, G. H., and R. S. Adams, *Reforestation Practices for Conifers in California.* Sacramento: Division of Forestry, 1971.

41 Simak, M., The x-ray contrast method for seed testing, *Medd. f. Stat. skogsforssr. Inst.,* 47(4):1–22. 1957.

42 Stark, R. W., and R. S. Adams, X-ray inspection technique aids forest tree seed production, *Calif. Agr.,* 17(7):6–7. 1963.

43 Stuke, W., Seed and seed handling techniques in production of walnut seedlings, *Proc. Plant Prop. Soc.*, 10:274–77. 1960.

44 Takatori, F. H., L. F. Lippert, and F. L. Whiting. The effect of petroleum mulch and polyethylene films on soil temperature and plant growth, *Proc. Amer. Soc. Hort. Sci.*, 85:532–40. 1964.

45 U.S. Dept. of Agriculture, Forest Service. *Seeds of Woody Plants in the United States.* C. S. Schopmeyer (ed.). Agr. Handbook No. 450. 1974.

46 ———, *Manual for Testing Agriculture and Vegetable Seeds,* USDA Agr. Handbook No. 30. Washington, D.C.: U.S. Govt. Printing Office, 1952.

47 Vaartaja, O., Chemical treatment of seed beds to control nursery diseases. *Bot. Rev.*, 30:1–91. 1964.

48 Wakeley, P. C., Planting the southern pines, *USDA Monograph No. 18.* Washington, D.C.: U.S. Govt. Printing Office, 1954.

49 Weaver, R. J., *Plant Growth Substances in Agriculture.* San Francisco: W. H. Freeman, 1972.

50 Western Forest Tree Seed Council, *Sampling and Service Testing Western Conifer Seeds.* Portland, Ore.: Western Forestry and Conservation Association, 1966, 36 pp.

51 Wilcox, G. E., and P. E. Johnson, An integrated tomato seedling system, *HortScience,* 6:214–16. 1971.

52 Williams, R. D., and R. R. Romanowski, Vermiculite and activated carbon adsorbents protect direct-seeded tomatoes from partially selective herbicides, *Jour. Amer. Soc. Hort. Sci.,* 97:245–49. 1972.

53 Wilson, J. K., Calcium hypochlorite as a seed sterilizer, *Amer. Jour. Bot.,* 2:420–27. 1915.

SUPPLEMENTARY READING

Association of Official Seed Analysts, Proceedings.

Ball, V. (ed.), *The Ball Red Book,* 12th ed. Chicago: Geo. J. Ball, Inc. (1972).

International Plant Propagators Society, Proceedings of Annual Meetings.

International Seed Testing Association, Proceedings.

Kozlowski, T. T. (ed.), *Seed Biology,* Vols. I, II, III. New York: Academic Press, (1972).

Roberts, E. H. (ed.), *Viability of Seeds.* Syracuse: Syracuse Univ. Press, 1972.

Stoeckeler, J. H., and P. E. Slabaugh, Conifer Nursery Practice in the Prairie States, *U.S. Dept. of Agriculture, Agricultural Handbook No. 279.* Washington, D.C.: U.S. Govt. Printing Office (1965).

U.S. Dept. of Agriculture, *Manual for Testing Agricultural and Vegetable Seeds.* U.S. Dept. of Agriculture, Agricultural Handbook No. 30. Washington, D.C.: U.S. Govt. Printing Office (1952).

———, *Seeds. Yearbook of Agriculture.* Washington, D.C.: U.S. Govt. Printing Office (1961).

———, Forest Service, *Seeds of Woody Plants of the United States.* C. S. Schopmeyer (ed.). Agricultural Handbook No. 450 (1974).

General Aspects
of
Asexual Propagation

8

NATURE AND IMPORTANCE OF ASEXUAL PROPAGATION

Asexual propagation involves reproduction from vegetative parts of plants and is possible because the vegetative organs of many plants have the capacity for regeneration. Stem cuttings have the ability to form adventitious roots. Root cuttings can regenerate a new shoot system. Leaves can regenerate new roots and new shoots. A stem and a root (or two stems) can be grafted together to form a continuous vascular connection and a new plant when properly combined as a graft.

New plants can start from a single cell. Cells of tobacco pith (*90*) and carrot root (*79*) in aseptic cultures have regenerated entire plants identical with the one from which the original cells were taken. Any living cell of the plant appears to have all the genetic information needed to regenerate the complete organism.

The term "ortet" is sometimes used, especially in forestry, in reference to the plant from which the vegetative propagating unit, termed the "ramet," is taken.

REASONS FOR USING ASEXUAL PROPAGATION

Asexual propagation maintains clones. Such propagation involves mitotic cell division (see p. 6) in which there is replication (usually) of the complete chromosome system and associated cytoplasm from the parent cell to the two daughter cells. Consequently plants propagated vegetatively reproduce, by means of DNA replication, all the genetic information of the parent plant. This is why the unique characteristics of any single plant are perpetuated in the propagation of a clone. The asexual process is particularly important in horticulture because the genetic makeup (genotype) of most fruit and ornamental cultivars is highly heterozygous, and the unique characteristics of such plants are immediately lost if they are propagated by seed.

Asexual propagation is necessary to grow cultivars that produce no viable seeds, such as some bananas, figs, oranges, and grapes.

Propagation of some species may be easier, more rapid, and more economical by vegetative methods than by seed. For example, cotoneaster seed has complex dormancy conditions but leafy stem cuttings root rapidly and in high percentages. Seedlings of some species grow more slowly than rooted cuttings.

Some plants grown from seed have a long juvenile stage (see p. 184). During this time the plant not only fails to produce flowers and fruit, but may exhibit other undesirable morphological features (e.g., thorniness) that are not present when propagation is by the use of material which has reached the adult stage. On the other hand, it may be desirable to maintain the juvenile stage indefinitely so as to facilitate propagation of difficult-to-root cuttings (see p. 239).

THE CLONE

Most fruit and many ornamental cultivars consist of groups of plants propagated vegetatively, starting with one individual plant, usually grown from seed, or a part of a plant, e.g., a *bud sport* (see p. 192). Such a group of plants taken collectively has been given the name *clone* (75, 78, 81, 95). A *clone* can be defined as "genetically uniform material derived from a single individual and propagated exclusively by vegetative means, such as cuttings, divisions, or grafts" (see p. 15). A discovery with any individual member of the clone, e.g., a method of propagation, a certain cultural practice, a method of disease control, or a cross-pollination requirement, applies in the same manner to all other members of that clone.

When a member plant of a certain clone requires cross-pollination to set fruit, it must be pollinated by a plant of a different clone, rather than by a different plant of the same clone, which in reality is just another part of the same plant.

Many clones of horticultural interest have been discovered and perpetuated by man (58). The 'Bartlett' ('Williams Bon Chrétien') pear clone originated from a seedling in England about 1770 and has been maintained asexually ever since. The 'Delicious' apple clone originated about 1870 as a chance seedling in Mr. Jesse Hiatt's apple orchard in Peru, Iowa. Buds taken from this tree and budded to rootstocks produced new trees having tops of the same genetic constitution as the original seedling tree. By continuing this process of asexual propagation, thousands upon thousands of trees have since been grown whose budded tops, collectively, make up the 'Delicious' apple clone.

Clones exist in nature, reproducing naturally by such structures as bulbs, rhizomes, runners, stolons, and tip layers (see Chapters 14 and 15). Apomixis (see p. 61) in some species of Rosaceae, Gramineae, and Compositeae makes possible the maintenance of some clones naturally by seeds. A clone can perpetuate itself successfully in nature, sometimes better than can seed-propagated plants, as long as the environment remains reasonably constant. If the environment changes drastically, however, a clonally reproduced species will be at a disadvantage because it has no opportunity to evolve forms better adapted to the new environment. Similarly, cultivated clones have a disadvantage in that an adverse situation, such as disease or insect attack, can affect all members of the clone equally and may even destroy it. If a major food crop in a country is based upon one or two cultivars, the results of such attacks could be disastrous. A diversity of quite genetically different cultivars would be a safer situation.

The concept of the clone, however, does not mean that all individual members are necessarily identical in all characteristics. The actual appearance and behavior of a plant, i.e., its *phenotype,* results from the interaction of its genes (*genotype*) with the *environment* in which the plant is growing. Consequently, within a given clone the appearance of the plants, or of fruits and flowers on different plants, will vary due to the effects of climate, soil, disease, and the like. In many years 'Bartlett' pear trees grown in California tend to produce round, apple-shaped fruit, whereas 'Bartlett' pears grown in Washington and Oregon are relatively long and narrow; this difference is believed to be due to different climatic factors (*88*). Some plants produce leaves of different appearance when growing in shade than when growing in the sun. Certain water plants produce leaves quite different in appearance on the part of the plant submerged than on the part in the air. Within a single orchard, fruit trees of the same clone often differ noticeably because of differences in soil, water availability, rootstock, or competition from surrounding plants (*35, 71*). Although the environment can modify the growth and appearance of individual members of a clone, such changes are not permanent, since the genotype of the plants is unaffected by the environmental modifications. Further in this chapter, however, several causes of more or less permanent changes within clones are discussed.

The life of a clone is theoretically unlimited. An early belief that has long prevailed is that a clone deteriorates with age and can only be rejuvenated by seed propagation (*52, 60*). Evidence now shows that modification, sometimes leading to deterioration, can indeed occur in particular clones, and must be guarded against in propagation (see p. 197). The most significant factor seems to be virus infection (*87*). In fact, virtually any clone that is grown for any length of time is likely to become infected, its survival being determined by its ability to tolerate the virus. Genetic changes (mutations) may occur in the clone which, although not always degenerative in a strict sense, can produce off-type individuals that will reduce the value of the clone (*73*). A continually unfavorable environment may lead to progressive deterioration of a clone. For example, lack of sufficient winter chilling for certain strawberry cultivars (*10*), or an insufficiently long post-bloom vegetative period for some bulbous plants (see Chapter 15) may lead to gradual decline in vigor and productivity. If a clone is maintained in a proper environment, and procedures are carried out whereby viruses, other pathogens, and off-type mutants are eliminated, a clone could be perpetuated indefinitely. Such procedures, described later in this chapter, are an important aspect of plant propagation.

Another belief, proposed by some people, is that permanent genetic changes in the clone can be induced by the environment or by the influence of one genetically different plant on another if the two are grafted together (57). Although some aspects of the latter theory have not perhaps been examined critically, the prevailing belief among most scientists is that such permanent genetic changes in clonal material do not occur.

CHANGES IN CLONES ASSOCIATED WITH AGE

Most clones originate from vegetative material taken from some part of a seedling plant. In Chapter 1 the sexual cycle in plants was described as involving a morphological and physiological change from a juvenile to an adult phase (see Figure 1–1). Propagation of the clone then involves the asexual cycle and particular phase of the plant may be reproduced. A basic understanding of the mechanism involved in the juvenile-to-adult phase changes is still needed. Such changes apparently take place in the shoot apical meristem as it undergoes cell division. The juvenile-to-adult changes are referred to as *ontogenetic,* being more or less permanent although the basic genetic information is not altered. New seedlings of the next sexual cycle again revert to the beginning of the juvenile phase (*11*).

Nucellar seedlings provide evidence that no permanent genetic change is involved since they are apomictic (see page 61) and, therefore, asexual, but such seedlings also go through the same juvenile-to-adult phase change (*33*).

Figure 8–1   The juvenile phase in a seedling plant often is indicated by leaf characteristics, as illustrated in the two kinds of plants shown here. *Left: Acacia melanoxylon.* In this seedling plant the bipinnate leaves at the base are characteristic of the juvenile condition; the "leaves," which are actually expanded petioles (phyllodia), at the top of the seedling are characteristic of the adult form. Note the gradual change in the leaves from base to top. Right: *Eucalyptus.* Juvenile leaves at the base are large, broad, and have no petiole (sessile); mature leaves on the upper part of the shoots are elongated and have a distinct petiole.

The juvenile-to-adult change is not the same as the vegetative-to-reproductive change of the asexual cycle, although the two kinds of changes may be correlated *(70, 93)*. Likewise, the development of senescence is different than either of these phenomena *(70)*.

The juvenile and adult phases may differ distinctly in appearance or there may be a transitional development from one phase to the other. Leaf shape is a common means of identifying phase changes *(1)*, as illustrated in *Acacia* and *Eucalyptus* in Figure 8–1. The juvenile phase of many plants— for example, citrus *(12)*, pear *(91)*, apple *(8, 66, 91)*, and honey locust *(14)*— is characterized by nonflowering, excessive vigor, and thorniness. Plants of these species in the adult phase flower and fruit, vigor decreases, and thorns disappear. The juvenile phase of English ivy *(Hedera helix)* (Figure 8–2), which is a classic example of this phenomenon, is a trailing vine with alternate palmate leaves. The mature, flowering form is an erect or semi-erect shrub with entire, ovate leaves, produced oppositely around the stem. In some conifers, the juvenile forms produce needle-shaped leaves, but the adult produces scale-like leaves.

Cuttings taken from juvenile parts of many woody perennial plants can produce adventitious roots and shoots easily, whereas cuttings taken from the adult form of the same plant are much less, if at all, capable of forming adventitious roots or shoots (see page 239).

The ontogenetic change from juvenile to adult as the seedling grows older occurs in the somatic (vegetative) cells and results in differences in the apical meristem in separate parts of the plant at various stages of development *(22, 33, 42, 93)*. Prevailing evidence suggests that a plant must attain a certain size before the adult phase appears *(91, 93, 100)*. Consequently the growing points produced in different parts of an individual plant where such changes are taking place may differ considerably in their ontogenetic age *(66)*. This explains why one portion of an individual plant may be in the juvenile phase and another portion in the adult phase.

The phenomenon that different growing points at different parts of a plant may perpetuate particular phases (juvenile or adult) if used in propagation has been called *topophysis (60)*. For example, buds taken from the lower, juvenile portion of seedling trees of pear, apple, citrus or

Figure 8–2   Adult *(left)* and juvenile *(right)* forms of a variegated (chimeral) ivy *(Hedera helix)*. Note flowering in adult and vine-like growth of juvenile with adventitious roots along stem.

honey locust, if used in budding, will produce nursery trees that are vigorous, thorny, and slow to flower; buds from the upper portion of the same tree can be used to produce nursery trees that are less vigorous, smooth-barked, and thornless, and that flower quickly. This difference is illustrated in Figure 8–3. By proper selection of propagation material from a seedling plant, it is possible to reproduce leaf-persistent (juvenile) or deciduous (adult) beech trees, or to reproduce fungus (*Keithia thusima*) -resistant (juvenile) or fungus-susceptible (adult) forms of *Thuja plicata* (*22*).

The direction of growth produced by a cutting of some species may differ, depending upon whether the cutting was taken from an upright or a lateral branch. For example, cuttings of *Araucaria*, or of coffee, taken from lateral shoots will continue to grow in a horizontal direction, whereas cuttings made from upright shoots of the same plant develop into upright plants.

Figure 8–3   Two nursery trees of 'Kara' mandarin propagated from single buds (arrows) taken from the same tree, a nucellar seedling. Thorny tree on left grew from a bud taken from the thorny, juvenile portion of the tree; the thornless tree on right grew from a bud taken from the thornless, adult part of the tree. Courtesy H. B. Frost (*33*). From H. J. Webber and L. D. Batchelor, *The Citrus Industry*, Vol I. Berkeley: Univ. of California Press.

Rapid juvenile-to-adult phase changes are important in the initial establishment of a clone from a seedling plant, or in obtaining early flowering in breeding programs. Various procedures can speed this transition (*83*). For example, birch seedlings grown continuously under long days in the greenhouse during the winter will pass through the juvenile into the adult flowering stage much faster than seedlings under the normal short-day winter conditions (*69*). Crabapple (*Malus hupehensis*) seedlings grown continuously in the greenhouse until the plants reach a certain height (2.5 to 3.0m) change from the juvenile to the adult form and start to flower much earlier (about 13 months from seed germination) than field-grown plants which require up to four growing seasons before flowering. Onset of flowering is also hastened by treatments to induce rapid seedling growth, such as an ample mineral nutrient supply and supplementary lights to provide long-day conditions. Even greater growth stimulation is obtained with added CO_2 (2000 ppm) along with increased growing temperatures (75°–86° F; 24°–30° C) (*100, 101*).

Adult material for propagation may, most likely, be obtained from the upper, ontogenetically older part of the plant. Most well-established clones have attained the adult phase since considerable vegetative propagation has already taken place. Consequently plants propagated from material taken from different parts of grafted or own-rooted plants, young or old, do not exhibit juvenile and adult differences. However, sometimes juvenile characteristics can persist for several vegetative generations (*12*).

It may be desirable to maintain the juvenile phase of the clone (*83*). This can be done to some extent by proper selection of cutting material. The basal parts of many woody plants will remain juvenile even though the upper parts are in the adult form. Likewise, adventitious buds from roots or from heavily cut back stems, or from "sphaeroblasts" (wart-like growths containing meristematic and conductive tissue) (see p. 240) also tend to produce juvenile growth (*84*).

GENETIC VARIATION IN ASEXUALLY PROPAGATED PLANTS

Mutations

Gene and chromosome changes can occur within somatic (vegetative) cells and, if followed by mitotic division, may lead to a permanent change in the clone if subsequent dauhter cells occupy a substantial portion of a growing point.

There are many types of mutational and chromosomal changes that can occur. These may result from chemical alterations of the chromosome material at specific locations on the chromosomes (*point mutations*), from gross structural changes of the chromosomes (*deletions, duplications, inversions*), from addition or subtraction of one or several chromosomes of a set (*aneuploidy*), or from multiplication of the entire set of chromosomes (*polyploidy*). The cytoplasm also contains replicable units, such as *plastids,* that are independently involved in determining plant characteristics (*49*). Permanent changes can occur in them and lead to changes within the clone (*50, 80*).

Any single mutation of chromosomal change is itself a relatively rare event. However, since vegetative growth of a clone involves billions of cell divisions, the chances that some type of spontaneous change will occur within a clone are reasonably good if it is maintained for any length of time. The fact that the genotypes of most important clones used in horticulture are highly heterozygous and often complex hybrids seems to increase the chance that these changes can occur and, if they do, that they will be detected. In a heterozygous plant the mutation of a dominant gene to the recessive state will lead to immediate expression of the recessive condition (see Chapter 1). This change will only be detected, however, if the mutated cell in the growing point leads to the development of a shoot. If the change is in external characteristics, such as fruit color, size, or growth habit, its detection may be assured. Buds or scions taken from this altered branch would form a new clone which could be propagated asexually to give any number of plants of the new form.

Mutations and chromosomal changes can be induced artificially. Such techniques include treating the growing points with mutagenic agents, such as colchicine *(25)*, or bombarding seeds, or scions for grafting, with radio-active emissions such as X-rays or gamma rays. *(92)*. These procedures are used to develop new characteristics in crop plants.

Chimeras

Mutations that occur within a plant often affect only a segment of the meristem and give rise to sectors or layers of the mutated tissue. A plant propagated from such a meristem is thus composed of two or more geneti-cally distinct tissues growing adjacent to each other. Plants of this type are termed *chimeras*. They may be horticultural curiosities or (sometimes) eco-nomically important plants *(6)*. Plants with variegated foliage, as found in *Citrus, Vitis, Pelargonium, Chrysanthemum, Hydrangea, Dahlia, Coleus, Euonymous, Bouvardia, Sansevieria*, and others, are examples of chimeras. In these plants the plastids in part of the leaf tissue lack the capacity to produce chlorophyll, whereas other leaf cells are normal. The resulting pat-tern shows distinct green and white (or yellow) areas in the leaf. (Some kinds of variegation are not chimeras but are due to virus diseases, an im-portant distinction *(94)*. In this case, the leaf pattern is white mottled or mosaic.)

Fruit chimeras like the orange in Figure 8–4 are frequently observed. Apple fruits that have mixtures of sweet and sour flesh and peach fruits that have fuzzy and smooth (nectarine) surfaces have been discovered *(26)*. Red forms of certain apple *(21)* and potato cultivars are chimeras in which the red-skin-color mutation is present in the outer layer or epidermis, the inner tissue being the original unmutated form. Likewise, thornless black-berry cultivars exist in which the thornlessness is restricted to the outer stem layer. Chromosomal chimeras *(cytochimeras)* may occur in which cells in part of the tissue of the stem are diploid, i.e., they have the normal chromo-some number for the species, and cells in other parts of the stem are poly-ploid, i.e., they have some larger number of chromosomes.

A chimera may originate in several ways *(86)*: (a) A spontaneous mutation may take place in a cell of the plant within one of the layers of the growing point. This change does not necessarily affect the entire growing point but only those parts of the stem resulting from further division of the mutated cell. (How this occurs is described on page 190.) (b) Attempts to produce mutations artificially by colchicine or other agents invariably affect only

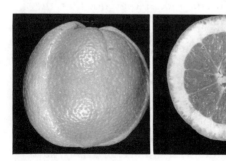

Figure 8–4   Chimera in 'Washington Navel' orange. Left portion of the fruit has developed from the mutation, producing a thick rather than a thin rind and a yellow in-stead of an orange skin color. This is an undesirable mutant.

certain cells in the growing points and result in chimeras that are basically the same as those occurring naturally. (c) Chimeras may result from inheritance. Variegated plants which have a tendency to produce both normal and defective plastids are often observed in seedling populations. For example, certain *Pelargonium* cultivars showing variegation have been developed in this way. The variegation pattern is unstable in the young seedling with normal green, albino, and variegated shoots developing on different parts of the plant. (d) Chimeras may be produced artificially by grafting. (This procedure is described on p. 193.)

There are several kinds of chimeras. Success in reproducing them, either sexually or asexually, depends upon their structure and stability. All are relatively unstable and may revert to plants showing one or the other type of tissue of which they are formed. The degree of stability depends upon their structure, which can be described as follows:

First, the cells in the shoot tips of most flowering plants are arranged in separate, more or less distinct, concentric layers over a central body (*23, 24, 26*) as illustrated in Figure 8–5. Usually there are three layers but there may be more. Each layer produces a more or less specific part of the stem. For convenience, these layers are numbered. The outer layer (L-I) usually produces the epidermis and remains one or, at times, several layers thick. The next layer (L-II) produces the outer cortex and some portion of the vascular cylinder. The reproductive cells in the anthers and ovules are produced by this layer. The third layer (L-III) usually gives rise to the inner cortex, vascular cylinder, and pith. As seen in Figure 8–5 the parts of the stem arising from the L-II and L-III layers are not always the same. A mutation that occurs in a cell in one of these layers will, in general, affect only that part of the stem arising from that layer. Thus, if a tetraploid cell arises in the L-II layer of a diploid plant, the resulting plant is described as a 2–4–2 cytochimera.

Second, the development of chimeras involves the direction of cell division. A cell in the growing point divides either at right angles to the outer surface (*anticlinally*) or parallel (*periclinally*) to the outer surface. Cells in L-I usually divide anticlinally and generally a distinct, more or less stable, outer cell layer results. Cells in L-II divide most often anticlinally near the apex but divide randomly farther away from the apex. Cells in L-III divide in both directions.

Occasionally a cell produced in one of the three layers becomes displaced into an adjoining layer. If this shift occurs relatively close to the growing tip so that the displaced cell, as it divides, produces a significant part of the new stem or leaf, a change in chimeral structure will result. For example, a shift could occur from a 2–4–2 cytochimera to one of 2–4–4 (or from 4–2–2 to 4–4–2), although the change might not be readily observed externally. A change from 2–4–2 to 4–4–2 would probably be more readily detected (since it would effect a change in the epidermis), but this change occurs less commonly (*26*).

Third, the location of lateral buds directly affects the development of the chimera. If a mutation occurs in a cell on one side of the growing point, only a small part of that stem may be affected. A lateral shoot that develops

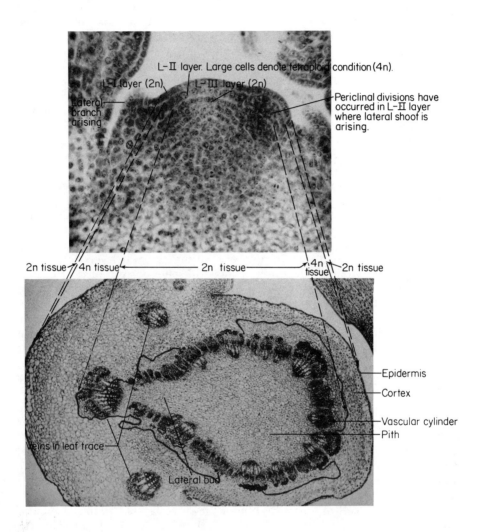

Figure 8–5   Photomicrographs of a 2–4–2 chimera in peach showing an enlarged, longitudinal section of the shoot tip (top) and a cross section of the stem under the tip (below). *Top:* three cell layers in the tip are identified as L–I, L–II, and L–III (see text). In this example, the L–I is diploid (2n), the L–II is tetraploid (4n), and the L–III is diploid (2n). Note that the L–I layer is a single row of cells, but that the L–II layer has increased in thickness on each side because of periclinal divisions at the point of leaf initiation. *Below:* the epidermis of the stem is 2n because it was derived from the L–I layer. The cortex and part of the vascular system are 4n, being derived from the L–II layer. The remainder of the vascular system and the pith are 2n, being derived from the L–III layer. From Derman *(26).*

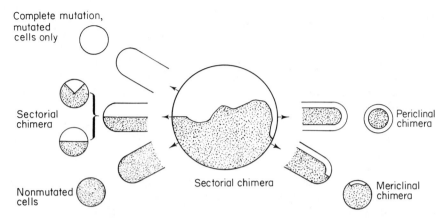

Complete mutation,
mutated
cells only

Sectorial
chimera

Nonmutated
cells

Sectorial chimera

Periclinal
chimera

Mericlinal
chimera

Figure 8–6   Types of chimeras. Buds arising at different positions on a sectorial chimera may produce shoots consisting entirely of mutated cells or entirely of non-mutated cells; or the shoots may be a sectorial, mericlinal, or periclinal chimeral, depending upon their location. Adapted from W. N. Jones, *Plant Chimeras and Graft Hybrids.* Courtesy Methuen & Co., London, Publishers.

from the mutated part of the stem may be completely changed; a lateral shoot developing from the opposite side of the original stem may be un-affected. Figure 8–6 shows how the development of a chimera is affected by the location of the lateral shoots in relation to the chimeral structure of the original plant.

Sectorial Chimera
In this type, the growing point of the shoot is composed of two genetically different tissues situated side by side occupying distinct sectors of the stem. Leaves and lateral buds arising from such a shoot may be composed of two tissues combined in various ways, depending upon their location. This type is uncommon in shoots but usual in roots. It must occur early in the develop-ment of the embryo before distinct layers develop in the growing point. It is unstable, and with continued growth, different parts of the plant develop either as normal, as mutated, or as a periclinal chimera (Figure 8–6).

Periclinal Chimera
In this type, tissues of one genetic composition occur as a relatively thin "skin," one or several cell layers in thickness, over a genetically different "core." This is the most common and relatively stable chimeral type. The genus *Rubus,* for example, has numerous thornless blackberry forms in which the epidermal layer lacks the gene for thorniness *(17).* Such plants will usually retain this characteristic if propagated by stem cuttings or by tip layering. In propagating such thornless chimeras by stem cuttings, it is significant that the adventitious roots arise endogenously beneath the mu-tated tissue, with the result that the root system of such thornless plants is

composed entirely of nonmutated cells. If such roots are subsequently used as a source of root cuttings, the resulting plants are thorny. Likewise, seedlings developing from seeds taken from the thornless plant are thorny because the gametes are produced from cells of the tissue which originated from the L-II (nonmutated) layer. Some thornless forms, however, do involve the L-II layer in the mutation, and these transmit the altered characteristics to the seedling offspring (*18*).

Some potato cultivars are periclinal chimeras. Their nature can be demonstrated experimentally by removing the buds ("eyes") from tubers, and forcing adventitious buds to arise from internal tissues. Removing the buds from 'Noroton Beauty,' which has mottled tubers, gives 'Triumph,' which has red tubers (*2*). 'Golden Wonder,' which has tubers with a thick brown russet skin, is also a chimera, having inner tissues characteristic of 'Langworthy,' whose tubers have a thin, white, smooth skin (*15*).

Propagation of a periclinal chimera by some types of leaf cuttings, as in variegated *Sansevieria,* may also result in reversion to the nonmutated form, since adventitious shoots and roots, in this case, arise from the inner, nonmutated tissues.

Mericlinal Chimera

This type is similar to the periclinal chimera except that the outer layer of a different tissue does not extend completely around the shoot, occupying only a segment of the circumference, as shown in Figure 8–6. This type of chimera is probably one of the most frequent to occur naturally, since a mutation in a single cell of a growing point would be apt to produce this type. It is not permanent, however, and with continued asexual propagation either changes to a periclinal chimera or reverts to a non-chimeral shoot. A lateral bud arising from the chimeral portion would become a periclinal chimera, whereas lateral buds from the normal part produce normal shoots. Only in shoots from the terminal growing point and from lateral shoots at the edge of the mutated portion would the mericlinal chimera continue.

Bud Sports

A branch which shows changes from the remainder of the plant in one or more inheritable characters that can be perpetuated by asexual means is termed a *bud sport.* Bud sports can originate by any of the somatic mutations or chromosomal changes mentioned earlier. Many of these sports are chimeras and are discovered when a lateral shoot develops from a mutated area. In addition, a bud sport may result when there is a rearrangement of tissues of a chimera such that a shoot may arise from a deeper layer, either from the displacement of epidermal cells with cells of the L-II layer, or by formation of adventitious buds.

Somatic variants such as these can be starting points of new clones and, if introduced as cultivars, are given names. Numerous bud sports have been discovered, and a great many more have undoubtedly gone undetected (*73*).

Some have become important new clones in both fruit and ornamental species. In Florida, pink-fleshed grapefruits were found on a single branch of a tree in a grove of thousands of trees producing white-fleshed grapefruits: an obvious mutation. The seedless 'Washington Navel' orange probably arose as bud mutation of the Brazilian orange, 'Laranja Selecta' (*58*). The first citrus fruit to be patented (see p. 204) originated in 1929 as a mutation: the red-fleshed 'Ruby' grapefruit. Mutations in apples are common, often resulting in desirable changes in fruit color, such as the deep red-colored 'Starking' and the 'Richared' apples, bud sports of the original 'Delicious' apple. The "spur-type" apples, such as the 'Starkrimson,' a sport of 'Delicious,' and the 'Stark-spur,' a sport of 'Golden Delicious,' are valuable cultivars giving natural dwarf trees, about two-thirds the usual size with short internodes and a compact growth habit. These "spur-type" variants appear either on the original apple clones or on one of their highly colored mutants (*73, 89*).

Polyploidy may give rise to "giant" sports in certain plants. Individual tetraploid grapevines occur in vineyards, growers referring to such vines with prodigious growth habits and poor fruitfulness as 'bulls," "males," or "giants" (*65*). Chromosome counts have shown them to be often only partial mutations (periclinal chimeras, see p. 191), consisting of a single outer layer of diploid cells surrounding the mutated tetraploid inner tissues (*25, 28*). Such giant sport canes in the grape are observed to arise at pruning wounds or near areas where a bud or shoot has been injured or killed. These mutant shoots seem to arise from deeply embedded dormant bud initials (*64*).

Certain giant apple sports (*19*) are diploid-tetraploid periclinal chimeras, with a $2n$ layer (34 chromosomes per cell), one, two, or three cell layers in thickness, over a $4n$ tetraploid (68 chromosomes per cell) interior.

Many (possibly most) mutations produce sports inferior to the parent plant in one or more characteristics (*24, 29, 74*). If detected by the propagator, such sports are discarded. Sometimes, as in citrus, these may be propagated unknowingly and cause serious economic loss after the trees come into production (*77*).

Graft Chimeras

While most chimeras occur naturally, they can also be established artificially by grafting (*7, 53*). In a young, grafted plant, if the scion is cut back severely, almost to the stock, an adventitious bud may sometimes arise from the callus at the junction of the stock and scion. The resulting shoot is a chimera in which the cells of the two graft components remain genetically distinct regardless of how intermingled they become.

Various graft chimeras have been reported in the horticultural literature.

Many years ago Winkler (*98, 99*) artificially produced numerous graft chimeras of tomato (*Lycopersicon esculentum*) on black nightshade (*Solanum nigrum*) and black nightshade on tomato, both of which are easily grafted and produce callus readily (see Figure 8–7). Winkler gave such mixed shoots the name "chimera" after the mythological monster that was part lion and part dragon.

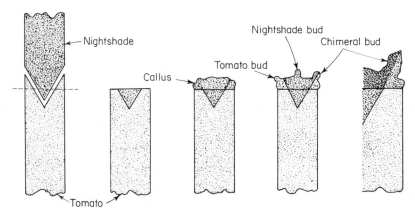

Figure 8–7 Stages in Winkler's method of producing a graft-chimera shoot between the nightshade and tomato. Adapted from W. N. Jones, *Plant Chimeras and Graft Hybrids.* Courtesy Methuen & Co., London, Publishers.

Natural graft chimeras have been known to exist and have been perpetuated vegetatively. The "bizzaria" orange, a fruit half-orange and half-citron was discovered in 1644 in Florence, Italy. It apparently originated from an adventitious bud arising from callus at the junction of a sour orange (*Citrus aurantium*) graft on a citron (*Citrus medica*) rootstock (*32, 85*). Similarly, medlar (*Mespilus germanica*) grafted on hawthorn (*Crataegus monogyna*) has led to several chimeras known as hawmedlars (*7, 41*).

Existence of these chimeras have given rise to proposals of the "graft hybrid" concept in which new plants result from the fusion of nuclei of vegetative cells. Much early literature was written on this concept (*16, 98, 99*). However, the evidence tended to support the development of morphological chimeras as being the best explanation.

True somatic hybridization, which would be comparable to a "graft hybrid" was achieved in 1972 with tobacco (*Nicotiana*) species in aseptic plant cell culture (*13*). The protoplasts of somatic cells from two different species were isolated, fused, and induced to regenerate plants which flower and produce fertile seed. These so-called "parasexual hybrids" were similar in all respects to those obtained by means of the usual hybridization methods.

Genetic Disorders with Viruslike Effects

A group of plant disorders occurs that are usually classed as *genetic disorders* because they are characteristically found only in certain clones, are noninfectious and some, at least, are hereditary (*72, 97*). Important examples, named after their symptoms, are *noninfectious bud-failure in almond, cherry crinkle, strawberry yellows, wood pocket of lemon,* and *rusty blotch of plum.* They also occur in apples and pears and certain other crops (*67, 94*). These off-type variants may appear sporadically in clones—in different parts of a single plant, or in different plants in a uniformly propagated group—and will show different degrees of severity. These disorders affect plants in different ways, but usually some degree of degeneration and nonproductivity results. One disorder—*noninfectious bud-failure in almond*—has been found to develop progressively in response to the tree's continued exposure to high growing temperatures (*51*). Characteristically, these disorders have been associated with particular sources of propagating material. Such disorders have been controlled to some extent by careful selection of propagating material, but this may not be a completely reliable method, since the potential for the disorder may be present without symptoms appearing. If possible, susceptible clones should be avoided entirely. If this is not practical, careful selection among sources of propagating material may reduce the incidence of the disorder. Information concerning the behavior of progeny plants developed from various sources of propagating material should be obtained.

Effects of Viruses and Mycoplasmas in Clones

Any clone that has been propagated for a long period of time probably has become infected with one or more viruses or by mycoplasma-like organisms * (*27, 43, 48*). These may influence growth, appearance, and production, not only of the infected plant but also of plants propagated from it. These organisms may remain with the clone, as would a genetic change but they are also infectious. Nursery propagation practices using budwood or seedling stocks infected with these organisms can be held partly responsible for the widespread distribution of viruses and mycoplasmas (*38, 46, 55*).

How severely a virus affects the clone depends upon the characteristics of the virus, the tolerance of the particular clone to the virus, and sometimes the accompanying environmental conditions. For example, *Prunus* ring spot virus (PRSV) produces a "shock" symptom that can sharply reduce bud survival during budding in the nursery (*38*). However, recovery then occurs, and subsequent symptoms are much less noticeable. The severity of PRSV "shock" differs with species, plum and apricot being less affected than almond, peach, or cherry. There are a number of recognized virus diseases of strawberry, and there is a great difference in the tolerance of particular cultivars to them.

Rather commonly, plant viruses produce distinct symptoms in the plant at low temperatures, but these are not detectable when it is grown at higher temperatures. This is a reason, for instance, for the production of potato propagation stock in cooler climates where possible diseased plants, particularly those affected by mosaic virus, can be easily detected.

Multiple infections of several virus components may be involved in many (if not most) virus diseases. For instance, three latent viruses—A, B, and C—

---

* Mycoplasmas are extremely small parasitic organisms that will live in plant or animal cells and may be found in the phloem tissues of plants. They have no constant shape, consisting only of a membrane enclosing the living protoplasm. These organisms multiply rapidly, utilizing the plant's food for their own growth and they also disrupt translocation of food materials in the phloem, causing the plant to wilt and start yellowing. In some diseases due to these organisms, hormonal disturbances in affected plants result in witches'-brooms and other malformations, such as conversion of flower parts to leafy structures.

if present individually in strawberry may cause slight or no symptoms in a given clone, but will produce a severe reaction when present in combinations of two or three. This situation indicates a danger of distributing clones carrying a single "nonpathogenic" virus, since later infections by additional viruses can have serious consequences.

A number of plant diseases formerly thought to be due to viruses are now known to be caused by mycoplasma-like bodies. These include *aster yellows* in strawberries and certain vegetables and ornamentals, *blueberry stunt disease, pear decline,* and *X-disease* in stone fruits. In addition, rickettsia-like organisms * have been found to be associated with certain plant diseases, such as Pierce's disease of grapes (*39*) and Phony disease of peach (*63*). Since multiplication of mycoplasmas and rickettsiae can be inhibited by tetracycline and other types of antibiotics, chemical control of diseases from these causes is more likely than similar control of virus-caused diseases.

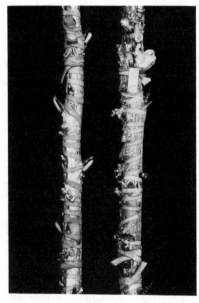

Identification of the presence of a virus can be made by two methods: (a) If the virus produces recognizable plant symptoms, it can be identified by *visual inspection.* Careful selection of propagation material, omitting obviously infected sources, can go far toward eliminating certain viruses. (b) On the other hand, some clones are so tolerant of certain viruses that no symptoms are apparent. Identification of a virus latent in such carriers requires *indexing* (*62*). A widely used method of indexing is to graft or bud propagating material from a suspected carrier to a healthy, susceptible host (*an indicator plant*) that subsequently produces identifiable symptoms. If the tissue is infected, the virus is transmitted across the graft union to the healthy plant and produces symptoms in it (see Figure 8–8). In indexing procedures, for correct determinations, the temperature level must be taken into consideration, since virus-induced symptoms may be pronounced at one temperature and absent at another (*31*). Another method is to inoculate

Figure 8–8   Test for indexing stone fruits in the genus *Prunus* for the presence of virus diseases. Several buds from plants to be tested are T-budded into shoots of indicator trees (*Prunus serrulata* 'Shirofugen'). *Left:* "clean" buds—no gumming around inserted buds. *Right:* buds from virus-infected plant, as shown by symptoms of profuse gumming around inserted buds. Photo courtesy George Nyland.

* Rickettsiae are considered to be highly modified bacteria, parasitic principally in arthropods, but man and other mammals are known hosts as well as certain homopterous insects (*20*).

Figure 8–9 Grafting procedure used in excised leaf method of transferring strawberry viruses. *Left:* terminal leaf removed, petiole split, and scion (excised leaf) inserted. *Center:* union partly wrapped with latex tape. *Right:* enlarged view showing healed union 40 days after grafting. Courtesy R. S. Bringhurst.

an herbaceous host plant, such as cucumber, tobacco, or bean, with juice from the infected plant and to observe symptoms in the host.

Viruses and mycoplasmas are infectious and can spread from diseased to healthy plants. One of the most important methods of spread is by insect vectors, notably aphids, and leaf hoppers. Natural root grafting between nearby trees also can result in virus transmission (*40*). There is evidence too that pollen transfer from plant to plant is a path of natural spread (*37*).

Grafting techniques for detecting viruses in strawberries (9, 44, 59) Some of the cultivated strawberry cultivars are relatively tolerant of viruses and even though infected may show few or on symptoms. On the other hand, the wood strawberry, *Fragaria vesca,* is very sensitive to viruses and shows symptoms in several weeks after inoculation.

In the "excised leaf" method of grafting (*9*), mature, well-developed leaves should be used. Only the terminal leaflet of the "scion" leaf is used, and two-thirds of this is removed. The petiole is trimmed with a razor blade into a tapered wedge 8 to 10 mm long. This "scion" leaf must be kept constantly moist until inserted.

The "stock" leaf is prepared by removing the terminal leaflet, and the petiole is split equally between the remaining two lateral leaflets. The scion petiole is then inserted in this split, and the union is tightly wrapped from the base upward. A self-adhesive latex bandage of crepe rubber works very well for wrapping. (See Figure 8–9.)

PRODUCTION AND MAINTENANCE OF PATHOGEN-FREE, TRUE-TO-TYPE CLONES

As emphasized earlier in this chapter, a clone may be modified during continued propagation because of systemic infection by various pathogens such as fungi, bacteria, nematodes, viruses, or mycoplasma-like organisms (*27, 43*) or because of genetic modification. Historically, many propagation and production practices had their basis in minimizing the effects of disease. Clones (sometimes highly desirable ones) subject to deterioration due to any of these causes have tended to disappear. Clonally propagated cultivars that have survived are those that grow and produce in a given environment despite the fact they may be carrying latent viruses or are exposed to various agents of disease. However, when a clone satisfactory in one area is transferred to a different environment, it may become exposed to new disease agents for which it has no resistance. Or it may contribute agents of disease which it is carrying to other species in the new environment. Bringing one diseased plant into proximity to other clean stock can result in infection of all. Quarantine procedures are designed to minimize disease spread that is due to movement of planting and propagating stock carrying harmful pathogens.

The concept of pathogen-free propagation stock, although not new, has been put into practice only in recent years (*4, 5, 96*). In *disease-free* stock the disease is controlled, although the causal organism may still be present. *Pathogen-free* stock, however, involves the elimination of the organism itself. Research activities in this area and the control programs that resulted have been a major advance in propagation. Food crops for which control programs have been in operation in varying degrees include potato, strawberry, citrus, grape, and deciduous tree fruits and nuts. Ornamental crops with systems of disease control include carnation, chrysanthemum, lily, gladiolus, rose, flowering *Prunus* species, and orchid. In propagating and maintaining these plants, such procedures have become standard operations. In fact, continued high-level economical production of most of these crops would be unlikely without them. In other crop plants propagated as clones for which precise systems of control are not available, the general principles and procedures listed here still apply. Considerable emphasis should be placed by propagators on the proper selection of propagation sources to avoid diseases.

Any program to maintain pathogen-free, true-to-type clones for propagation involves the following three phases: (a) initial selection of a source of propagation stock that is true-to-type and free of serious pathogens; (b) maintenance of such stock in a block with adequate safeguards against reinfection or genetic change; and (c) a system of propagation and distribution whereby such stock is disseminated without reinfection before it reaches the grower.

To carry out programs of the kind described here requires technically trained individuals and an opportunity to recover added costs if they occur. Improved production may, however, absorb these costs. Programs of this nature have been carried out in several ways. In certain crop industries, such as carnations and chrysanthemums for greenhouse culture, pathogen-free propagating stock is produced by *specialist-growers* who supply rooted or unrooted cuttings to commercial growers. In other crops, as potato, strawberry, grapes, and tree fruits and nuts, *certification programs* are available in some states of the U.S. and in some European countries, developed and administered through cooperative efforts of commercial nurserymen and research and regulatory governmental agencies; these are similar to programs of seed certification (see Chapter 4). Disease control through use of pathogen-free stock may be carried out by an *individual commercial nursery*. These operations are most significant for those kinds of plants for which established control programs are not available or where a nursery is sufficiently large to have its own technically trained personnel. However, some degree of propagation control as outlined in this section is essential in any propagation procedure.

Absolute freedom from all pathogens is not necessarily achieved and may not be a practical ideal *(82)*. Procedures have been directed toward eliminating particularly serious pathogens that adversely affect the plant. In certification programs, propagating stock is usually not certified as "disease-free" but is *certified to have been produced under conditions designed to eliminate certain harmful pathogens.*

Initial Selection

Initial selection of a suitable propagation source has three possible steps. The first is the selection of a uniform source plant (or plants) that is correctly identified and genetically true-to-type. A full-grown bearing plant, preferably one with a history of superior performance, is usually the best starting place. Individual plants should be carefully examined for genetic disorders, bud sports, and symptoms of virus or other pathogenic diseases. Furthermore, besides yield performance of original plants, information about how plants propagated from a given source have performed would be of value. The source plant and the propagating material taken from it should be labeled with a correct name. *Propagation stock may become mixed, or labels changed, and the mistake not discovered until after thousands of new plants have been propagated from it.*

The second step is to index the plant source for viruses or other pathogens, using a recommended minimum indicator host range and prescribed procedures *(67, 94)* (see Figures 8–8, 8–9, and 8–10). If the tests are negative, that is, if no evidence of infection is found by such tests, the plant is suitable as an initial propagation source for future "clean" stock. This procedure, however, does not prove that the plant is "pathogen-free," since the tests might not be sufficiently sensitive to demonstrate the presence of all pathogens.

If a "clean" plant is not found, a third step to eliminate the pathogen from some part of the plant is needed. As little as a single small part, such as a cutting, a bulb scale, or a single growing point can be the starting point of good propagation stock. There are a number of techniques that have been used to obtain "clean" stock, not all of which are equally effective for all plants or with all pathogens *(5, 46)*.

a Selection of uninfected parts. Some parts of a plant may be infected and others not. Soilborne organisms, such as *Phytophthora*, can be avoided by taking only tip cuttings from tall growing stock plants that are off the ground.

b Shoot apex culture. The terminal growing point of a plant is often free of virus, even if the rest of the plant is infected. Excision and culture of this small segment can produce a "virus-free" source; this was first shown to be true for dahlia *(47)* and subsequently has been used in aseptic culture methods with a number of herbaceous plants, such as carnation, chrysanthemum, hops, garlic, rhubarb, strawberry, and orchid (see p. 500), and in woody plants, such as gooseberry and lemon. In a similar technique *(61)* using the virus-free characteristics of shoot tips, where plant regeneration from shoot-tip culture may be difficult to accomplish, such apices

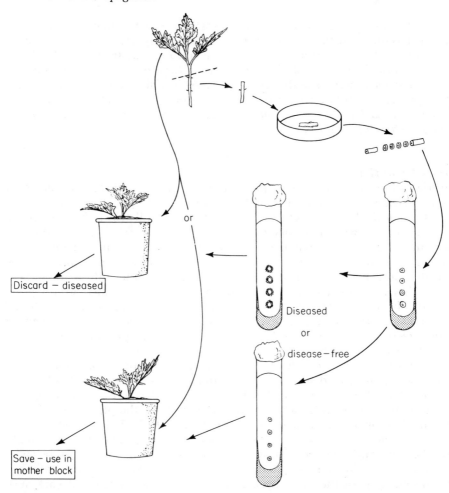

Figure 8–10   Testing propagating sources for disease organisms (fungi, bacteria) as illustrated for *Verticillium* wilt in chrysanthemum cuttings. From Baker and Chandler *(5)*.

can be excised and grafted *in vitro* onto disease-free rootstock seedlings, thus obtaining virus-free clonal plants. (Shoot apex culture is a laboratory method requiring special equipment and training in its techniques.)

c **Heat treatments of short duration.**   This procedure has been used widely in freeing many kinds of plants and seeds from fungi, bacteria, and nematodes *(3, 54)*. The plant is subjected to temperatures high enough and of long enough duration to destroy the pathogen but not so high as to kill the plant. Only one plant need survive to provide a beginning of pathogen-free material. (Less drastic treatments are needed where all the planting stock is to be treated, as, for instance, in commercial planting of gladiolus corms.) Treatment may vary with different plants from 110° to 135° F (43.5° to 57° C) for one-half to four hours. Methods include hot-water soaking and exposure to hot air or to aerated steam. For example, the causal agent for Pierce's disease in grapevines can be eliminated by immersion of the propagating wood in hot water for three hours at 113° F (45° C), thus preventing the possibility of moving this disease to new areas by diseased wood *(39)*. Hardening-off vegetative material or reducing the moisture content in seeds is generally desirable prior to such heat treatment.

d Heat treatments of low intensity and long exposure. This procedure will free many kinds of plants from virus diseases *(3)*. Plants are grown in containers until they are well established and have ample carbohydrate reserves; then they are held in a chamber (Figure 8–11) at 98° to 100° F (37° to 38° C) for two to four weeks or longer. Buds may be taken from the treated plants and inserted into "virus-free" rootstocks, or cuttings may be taken and rooted. Another procedure is to insert the bud from the infected plant into a rootstock prior to treatment.

e Combinations of heat treatment and shoot apex culture. In the strawberry certain viruses are heat tolerant and are not eliminated by the usual heat treatments, while shoot apex culture alone will not eliminate viruses from all plants. A combination of the two methods has been successful in eliminating the viruses. Infected plants are heat-treated at 104° to 106° F (40° to 41° C) for four to six weeks then apical shoot tips about 0.33 mm long are aseptically removed from the heat-treated plants

Figure 8–11 Chamber for heat-treating plants to eliminate viruses. This unit, 12 × 20 ft, which will hold about 100 plants in three-gallon containers, is covered with fiberglass. Warm air from a residence-type gas heater is distributed by means of a continuously operating fan through 10 in.-diameter ducts on each side and one end of the chamber through 2 in. openings in the ducts, spaced 11 in. apart. Containers are set on slatted benches and the house is provided with automatic watering facilities. An evaporative-type air cooler on a separate thermostat forces air through a plastic tube near the roof when cooling is needed. Temperatures can be maintained at the required 100 to 103°F (37.5 to 39.5°C) with this equipment. Photo courtesy Charles Howard, Howard Rose Company.

and placed on filter-paper bridges in test tubes containing a nutrient medium. Subsequent growth from these shoot tips results in nuclear stock mother plants which become the source for "clean" cultivars *(76)*.

f   Chemical treatment of propagation stock. This can sometimes be used to eradicate externally carried pathogens. For example, white calla lily can be freed of the *Phytophthora* root fungus by soaking the rhizomes for an hour in a weak formaldehyde solution.

g   Growing seedlings (apomictic and nonapomictic). Since most viruses are not transmitted through seeds (although there are exceptions), seedling growing may be used to produce new virus-free clones to replace older cultivars that have deteriorated from virus infections. Growing apomictic seedlings, in species where they occur, provides a means of both preserving an existing clone and eliminating viruses from the clone. This procedure has been utilized in citrus. Nucellar seedlings are the basis of new "virus-free" strains of old cultivars that had become severely affected with serious viruses *(12)*. In citrus species that do not naturally produce nucellar embryos, aseptic culture of nucellar tissue has induced nucellar embryos to form *(68)*.

Maintenance of Propagation Stock

Once a source of pathogen-free, true-to-type material has been obtained, it should be multiplied and maintained under conditions that prevent recontamination and allow detection of any significant change from the original source. A planting in which stock is maintained permanently under rigid control to be used as the primary source for all subsequent propagations is a *foundation block*. Its main function is to serve as a primary repository for "clean," true-to-type stock and not as a direct source of propagating material itself. For example, the U.S. Government and some states maintain such a repository for cultivars of fruit and nut crops *(30, 56)*. A planting that maintains pathogen-free stock as a source for commercial propagation is referred to as a *mother block*. It could originate either from a foundation block or directly from the originally developed "clean" stock.

Preservation of pathogen-free propagation stock has three aspects: isolation, sanitation, and periodic inspection and testing. Isolation is necessary to separate the plants from agents of recontamination. A minimum requirement is to separate the propagation source block from the propagation area. It can best be achieved by growing plants in containers and keeping them in an insect-proof screenhouse or a special greenhouse with restricted entry. Tree crops should be at least one-half mile from potential sources of virus diseases. Strawberry nurseries in California are grown in isolated mountain valleys far from production areas.

Sanitation is essential to eliminate disease agents in or on unclean equipment, tools, soil mixes, and so on (see Chapter 2).

Source plants should be tested periodically to detect evidence of recontamination and to see that the plants conform to their original standards. Visual inspections in conjunction with indexing and culturing methods discussed earlier are used. Specific requirements must be established for individual crops.

Distribution Systems of Propagation Materials

The aim of distribution systems that utilize the control programs discussed in this section is to provide nursery plants that meet some minimum standard of cleanliness and genetic purity. However, the development of pathogen-free material and its maintenance in foundation and mother blocks and subsequent propagation will increase cost. Such added expense can be justified in that the value of the propagated material is thereby increased, since increased production from the nursery material may result.

Figure 8–12 shows a general control system incorporating the three phases outlined in this chapter as meeting the technical requirements for pathogen-free material. Variations in procedure are possible to meet the needs of the individual propagator and the particular crop he is producing. Foundation blocks and mother blocks must be maintained under particularly rigid standards of isolation and inspection.

Procedure 1 should provide maximum protection from recontamination *if it is accompanied by appropriate inspections, indexing, and adherence to isolation and sanitation requirements.* It is used in the production of grapes and of fruit and nut crops in many countries of the world (*56*). Procedure 2 might be utilized by a commercial nursery to develop a source of propagation stock which would then be maintained in a mother block under its own control. Procedure 3 could be used in particular crop plants, such as carnations (*45*), where new "clean" stock can be periodically obtained.

Certification of planting stock has been utilized in production of some clonal material, with objectives similar to that described for certification of seed-propagated material (see p. 73). Such programs have been used in potato growing and in production of some grape and some tree fruit and nut crops. Commercial planting stock can be referred to as *certified stock* if it is grown under supervision of a legally designated agency with prescribed regulations designed to maintain minimum standards of cleanliness and clonal identity.

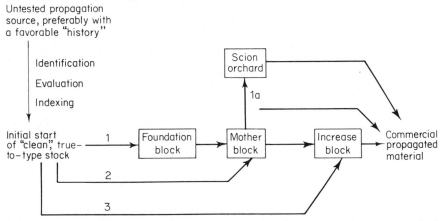

Figure 8–12   Generalized control system for developing, maintaining, and distributing pathogen-free, true-to-type plant materials. Three modifications in procedure are shown.

THE PLANT PATENT LAW

An amendment to the United States Patent Law was enacted in 1930 which enabled the originators of new plant forms in the U.S. to obtain a patent on them. This amendment added impetus to the development and introduction of new and improved plants by establishing the possibility of monetary rewards to private plant breeders and alert horticulturists for valuable plant introductions. It has stimulated growers of fruit and ornamentals to be on the lookout for improved plant forms and has handsomely paid numerous fortunate or observant persons for their finds.

Essentially, what can be patented as stated in the statute is "any distinct and new variety of plant, including cultivated sports, mutants, hybrids, and newly found seedlings, other than a tuber-propagated plant or a plant found in an uncultivated state." For a new plant to be patentable, it must be one that has been asexually reproduced and can be so propagated commercially, as by cuttings, layering, budding, or grafting. Protection is also given to some seed propagated cultivars (see below).

Characteristics which would cause a plant to be "distinct and new" and thus patentable include such things as growth habit; immunity from disease; resistance to cold, drought, heat, wind, or soil conditions; the color of the flower, leaf, fruit, or stem; flavor; productivity, including everbearing qualities in the case of fruits; storage qualities; form; and ease of reproduction.

The applicant for a plant patent must be the person who invented or discovered and subsequently reproduced the new cultivar of plant for which the patent is sought. If a person who was not the inventor applied for a patent, the patent, if it were obtained, would be void; in addition, such a person would be subject to criminal penalties for committing perjury. A plant found growing wild in nature is not considered patentable.

A plant patent issued to an individual is a grant consisting of the right to exclude others from propagation of the plant or selling or using the plant so reproduced. Essentially, it is a grant by the United States Government, acting through the Patent Office, to the inventor (or his heirs or assigns) of certain exclusive rights to his invention for a term of 17 years throughout the United States and its territories and possessions. A U.S. plant patent affords no protection in other countries. The mere fact that a patent has been issued on a new plant does not imply any endorsement by the government of high quality or merit. The only implication of a patent on a plant is that it is "distinct and new."

The U.S. Plant Variety Protection Act, which became effective in 1970, extends plant patent protection to certain sexually propagated cultivars which can be maintained as "lines" (see p. 68), such as those of cotton, alfalfa, soybeans, marigolds, bluegrass, and others.

REFERENCES

1 Ashby, E., Leaf shape, *Sci. Amer.*, 181:22–24. 1949.

2 Asseyera, T., Bud mutations in the potato and their chimerical nature, *Jour. Gen.*, 19:1–26. 1927.

3 Baker, K. F., Thermotherapy of planting material, *Phytopathology*, 52:1244–55. 1962.

4 Baker, K. F., Disease-free propagation in relation to standardization of nursery stock, *Proc. Inter. Plant Prop. Soc.*, 21:191–98. 1972.

5 Baker, K. F., and P. A. Chandler, Development and maintenance of healthy planting stock, in K. F. Baker (ed.), *The U.C. System for Producing Healthy Container-Grown Plants.* Calif. Agr. Exp. Sta. Man. 23: pp. 217–36. 1957.

6 Bateson, W. Root cuttings, chimeras and "sports," *Jour. Gen.*, 6:75–80. 1916.

7 Baur, E., Pfropfbastarde Periclinal Chimaeren und Hyperchimaeren, *Ber. Deuts. Bot. Ges.*, 27:603–5. 1909.

8 Blair, D. S., M. MacArthur, and S. H. Nelson, Observations in the growth phases of fruit trees, *Proc. Amer. Soc. Hort. Sci.*, 67:75–79. 1956.

9 Bringhurst, R. S., and V. Voth, Strawberry virus transmission by grafting excised leaves, *Plant Disease Rpt.*, 40(7):596–600. 1956.

10 Bringhurst, R. S., V. Voth, and D. van Hook, Relationship of root starch content and chilling history to performance of California strawberries, *Proc. Amer. Soc. Hort. Sci.*, 75:373–81. 1960.

11 Brink, R. A., Phase change in higher plants and somatic cell heredity, *Quart. Rev. Biol.*, 37(1):1–22. 1962.

12 Cameron, J. W., R. K. Soost, and H. B. Frost, The horticultural significance of nucellar embryony in citrus, in J. M. Wallace (ed.), *Citrus Virus Diseases,* Univ. of Calif.: Div. of Agr. Sci., pp. 191–96. 1957.

13 Carlson, P. S., H. H. Smith, and R. D. Dearing, Parasexual interspecific plant hybridization, *Proc. Nat. Acad. Sci.*, 69(8):2292–94. 1972.

14 Chase, S. B., Propagation of thornless honey locust, *Jour. For.*, 45:715–22. 1947.

15 Crane, M. B., Note on a periclinal chimera in the potato, *Jour. Gen.*, 32:73–77. 1936.

16 Daniel, L., Les hybrides de greffe, Int. Tuinbouw. Cong., Amsterdam, pp. 175–93. 1923.

17 Darrow, G. M., Notes on thornless blackberries, *Jour. Hered.*, 19:139–42. 1928.

18 ———, A productive thornless sport of the Evergreen blackberry, *Jour. Hered.*, 22:404–6. 1931.

19 Darrow, G. M., R. A. Gibson, W. E. Toenjes, and H. Dermen, The nature of giant apple sports, *Jour. Hered.*, 39:45–51. 1948.

20 Davis, R. E., and R. F. Whitcomb, Mycoplasmas, rickettsiae, and chlamydiae: possible relation to yellows diseases and other disorders of plants and insects, *Ann. Rev. Phytopath.*, 9:119–54. 1971.

21 Dayton, D. F., Genetic heterogeneity in the histogenic layers of apple, *Jour. Amer. Soc. Hort. Sci.*, 94(6):592–95. 1969.

22 de Muckadell, M. S., Juvenile stages in woody plants. *Physiol. Plant.*, 7:782–96. 1954.

23 Dermen, H., Periclinal cytochimeras and histogenesis in cranberry, *Amer. Jour. Bot.*, 34:32–43. 1947.

24 ———, Periclinal cytochimeras and origin of tissues in stem and leaf of peach, *Amer. Jour. Bot.*, 40:154–68. 1953.

25 ———, Colchiploidy in grapes, *Jour. Hered.*, 45:159–72. 1954.

26 ———, Nature of plant sports, *Amer. Hort. Mag.*, 39:123–73. 1960.

27 Doi, Y., M. Teranaka, K. Yora, and H. Asuyama, Mycoplasma or PLT group-like microorganisms found in the phloem elements of plants infected with mulberry dwarf, potato witches' broom, aster yellows, or paulownia witches' broom, *Ann. Phytopath. Soc. Jap.*, 33:259–66. 1967.

28 Einset, J., and C. Pratt, "Giant" sports of grapes, *Proc. Amer. Soc. Hort. Sci.*, 63:251–56. 1954.

29 ———, Spontaneous and induced apple sports with misshapen fruit, *Proc. Amer. Soc. Hort. Sci.*, 73:1–8. 1959.

30 Fridlund, P. R., IR–2, a germ plasm bank of virus-free fruit tree clones, *HortScience*, 3(4):227–29. 1968.

31 ———, Temperature effects on virus disease symptoms in some *Prunus, Malus,* and *Pyrus* cultivars, *Wash. Agr. Exp. Sta. Bul. 726,* 1970.

32 Frost, H. B., Polyembryony, heterozygosis, and chimeras in *Citrus, Hilgardia,* 1:365–402. 1926.

33 ———, Nucellar embryony and juvenile characters in clonal varieties of citrus, *Jour. Hered.*, 29:423–32. 1938.

34 Gardner, V. R., Studies in the nature of the pomological variety, *Mich. Agr. Exp. Sta. Tech. Bul. 161,* 1938.

35 ———, Studies in the nature of the clonal variety. III. Permanence of strain and other differences in the Montmorency cherry. *Mich. Agr. Exp. Sta. Tech. Bul.*, 186:1–20. 1943.

36 ———, A study of the sweet and sour apple chimera and its clonal significance, *Jour. Agr. Res.*, 68:383–94. 1944.

37 George, J. A., and T. R. Davidson, Pollen transmission of necrotic ring spot and sour cherry yellows viruses from tree to tree, *Canad. Jour. Plant Sci.*, 43:276–88. 1963.

38 Gilmer, R. M., K. D. Brase, and K. G. Parker, Control of virus diseases of stone fruit nursery trees in New York, *New York Agr. Exp. Sta. Bul. 779,* 1957.

39 Goheen, A. C., G. Nyland, and S. K. Lowe, Association of a rickettsia-like organism with Pierce's disease of grapevines and alfalfa dwarf and heat therapy of the disease in grapevines, *Phytopathology*, 63:341–45. 1973.

40 Guengerich, H. W., and D. F. Millikan, Root grafting, a potential source of error in apple indexing, *Plant Dis. Rep.*, 49:39–41. 1965.

41 Haberlandt, G., Das Wesen der *Crataegomespili* Sitzber, *Preuss. Akad. Wiss.*, 20:374–94. 1930.

42 Haberman, H. M., Grafting as an experimental approach to the problem of physiological aging in *Helianthus annuus* L., *XVI Int. Hort. Cong.* Vol. IV, pp. 243–51. 1962.

43 Hampton, R. O., Mycoplasmas as plant pathogens: perspectives and principles, *An. Rev. Plant Physiol.*, 23:389–418. 1972.

44 Harris, R. V., Grafting as a method for investigating a possible virus disease of the strawberry, *Jour. Pom. and Hort. Sci.,* 10:35–41. 1932.

45 Holley, W. D., and R. Baker, *Carnation production,* Dubuque, Iowa: Brown, 1963.

46 Hollings, M., Disease control through virus-free stock, *Ann. Rev. Phytopath.,* 3:367–96. 1965.

47 Holmes, F. O., Elimination of spotted wilt from a stock of dahlia, *Phytopathology,* 38:314. 1948.

48 Hooper, G. R., and M. L. Lacy, Mycoplasma: new causes for old diseases in Michigan, *Mich. Agr. Ext. Bul. E-644,* 1971.

49 Jinks, J. L., *Extrachromosomal inheritance,* Englewood Cliffs, N. J.: Prentice-Hall, Inc., 1964.

50 Jones, C. M., and W. H. Gabelman, Cytoplasmic inheritance in *Epilobium, Jour. Hered.,* 56(3):139–42. 1963.

51 Kester, D. E., and R. Hellali, Variation in distribution of non-infectious bud failure (BF) in almond as a function of temperature and growth, *HortScience,* 7(2):322(abst.). 1972.

52 Knight, Thomas Andrew, Observations on the grafting of trees, *Phil. Trans. Roy. Soc., London,* 85:290. 1795.

53 Krenke, N. P., *Wundkompensation Transplantation und Chimären bei Pflanzen* (translated from Russian). Berlin: Springer, 1933.

54 Kunkel, L. O., Heat treatments for the cure of yellows and other virus diseases of peach, *Phytopathology,* 26:809–30. 1936.

55 Lazar, A. C., and P. R. Fridlund, The incidence of latent viruses in Washington peach orchards, *Plant Dis. Rpt.,* 51(2):1063–65. 1967.

56 Mather, S. M., Nursery stock registration and certification in California, *Calif. Dept. of Agr. Quart. Bul.,* L(3):173–84. 1961.

57 Michurin, I. V., *Summary of sixty year's labor* (translated by A. J. Bruman and C. Frogen, USDA, 1936), from *Itogi Shestidesiatiletnikh Rabot* (in Russian).

58 Miller, E. V., The natural origins of some popular varieties of fruit, *Econ. Bot.,* 8:337–48. 1954.

59 Miller, P. W., Technique for indexing strawberries for viruses by grafting to *Fragaria vesca, Plant Disease Rpt. 36:* 94–96. 1952.

60 Molisch, H., *The longevity of plants* (1928). Lancaster, Pa.: E. Fulling (English translation), 1938.

61 Murashige, T., W. P. Bitters, T. S. Rangan, E. M. Nauer, C. N. Roistacher, and P. B. Holliday, A technique of shoot apex grafting and its utilization towards recovering virus-free citrus clones, *HortScience,* 7(2):118–19. 1972.

62 Nyland, G. A., and J. A. Milbrath, Obtaining virus-free stock by index techniques, *Phytopathology,* 52:1235–39. 1962.

63 Nyland, G., A. C. Goheen, S. K. Lowe, and H. C. Kirkpatrick, The ultrastructure of a rickettsialike organism from a peach tree affected with phony disease, *Phytopathology,* 63(10):1275–78. 1973.

64   Olmo, H. P., Breeding tetraploid grapes, *Proc. Amer. Soc. Hort. Sci.,* 59:285–90. 1952.

65   ——, Bud mutation in the vinifera grape. II. Sultanina gigas. *Proc. Amer. Soc. Hort. Sci.,* 33:437–39. 1936.

66   Passecker, F., Zur Frage der jugend Formen der Apfel, *Zuchter,* 19:311. 1949.

67   Posnette, A. F., and R. Cropley, Genetical disorders of apple with virus-like effects, in *Virus diseases of apple and pear,* Tech. Comm. Bur. Hort. and Plant. Crops, E. Malling, 30: pp. 79–86. 1963.

68   Rangan, T. S., T. Murashige, and W. P. Bitters, *In vitro* initiation of nucellar embryos in monoembryonic *Citrus, HortScience,* 3:226–27. 1968.

69   Robinson, L. W., and P. F. Wareing, Experiments on the juvenile-adult phase change in some woody species, *New Phytol.,* 68:67–78. 1969.

70   Sax, K., Aspects of aging in plants, *Ann. Rev. Plant Phys.,* 13:489–506. 1962.

71   ——, and J. W. Gowen, Permanence of tree performance in a clonal variety and a critique of the theory of bud mutation, *Genetics,* 8:179–211. 1923.

72   Schneider, H., J. W. Cameron, R. K. Soost, and E. C. Calavan, Classifying certain diseases as inherited, *Proc. 2nd Conf. Int. Org. Citrus Virol.* Gainesville, Fla.: University of Florida Press, pp. 15–21, 1961.

73   Shamel, A. D., and C. S. Pomeroy, Bud mutations in horticultural crops, *Jour. Hered.,* 27:487–94. 1936.

74   Shamel, A. D., C. S. Pomeroy, and R. E. Caryl, Bud selection in the Washington Navel orange; progeny tests of limb variations, *USDA Tech. Bul. 123,* 1929.

75   Shull, C. H., "Phenotype" and "clone," *Science,* N. S. 35:182–83. 1912.

76   Smith, S. H., R. E. Hilton, and N. W. Frazier, Meristem culture for elimination of strawberry viruses, *Calif. Agric.,* 24(8):8–10, 1970.

77   Soost, R. K., J. W. Cameron, W. P. Bitters and R. G. Platt, Citrus bud variation, *Calif. Citrograph,* 46:176, 188–93. 1961.

78   Stern, W. T., The use of the term "clone," *Jour. Roy. Hort. Soc.,* 74:41–47. 1943.

79   Steward, F. C., L. M. Blakely, A. E. Kent, and M. A. Mapes, Growth and organization in free cell colonies, in *Brookhaven Symp. in Biol.* No. 16, pp. 73–88. 1963.

80   Stewart, R. N., The origin and transmission of a series of plastogene mutants in *Dianthus* and *Euphorbia, Genetics,* 52:925–47. 1965.

81   Stout, A. B., The nomenclature of cultivated plants, *Amer. Jour. Bot.,* 27:339–47. 1940.

82   Stout, G. L., Maintenance of "pathogen-free" planting stock, *Phytopathology,* 52:1255–58. 1962.

83   Stoutemyer, V. T., The control of growth phases and its relation to plant propagation, *Proc. Plant Prop. Soc.,* 12:260–64. 1962.

84   ——, Regeneration in various types of apple wood, *Iowa Agr. Exp. Sta. Res. Bul.,* 220:308–52. 1937.

85   Tanaka, T. Bizzarria—a clear case of periclinial chimera, *Jour. Gen.,* 18:77–85. 1927.

86 Tilney-Bassett, R. A., The structure of periclinal chimeras, *Heredity*, 18:265–85. 1963.

87 Tincker, M. A. H., Propagation, degeneration, and vigor of growth, *Jour. Roy. Hort., Soc.*, 70:333–37. 1945.

88 Tufts, W. P., and C. J. Hansen, Variation in shape of Bartlett pears, *Proc. Amer. Soc. Hort. Sci.*, 28:627–33. 1931.

89 Upshall, W. H., *North American Apples: Varieties, Rootstocks, Outlook*. E. Lansing, Mich.: Michigan State Univ. Press, 1970.

90 Vasil, V., and A. C. Hildebrandt, Differentiation of tobacco plants from single, isolated cells in microculture, *Science*, 150:889–92. 1965.

91 Visser, T., Juvenile phase and growth of apple and pear seedlings. Meded. 224. *Inst. voor de Vereh, vontuinb. Wageningen* (Nederland), 1966.

92 Wallace, A. T., Mutagenic agents, *Agr. Sci. Rev.*, 2:1–8. 1964.

93 Wareing, P. F., Problems of juvenility and flowering in trees, *J. Linn. Soc. Bot.*, 56:282–89. 1960.

94 U.S. Dept. Agr., *Virus and other disorders with virus-like symptoms of stone fruits in North America*, USDA Agr. Handbook No. 10. Washington, D.C.: U.S. Govt. Printing Office, 1951.

95 Webber, H. J., New horticultural and agricultural terms, *Science*, N.S., 18:501–2. 1903.

96 Wilhelm, S., Symposium on pathogen-free stock. *Phytopathology*, 52:1234–35. 1962.

97 Wilson, E. E., and R. D. Schein, The nature and development of noninfectious bud-failure in almond, *Hilgardia*, 24:519–42. 1956.

98 Winkler, H., Uber Pfropfbastarde und pflanzliche Chimaren, *Ber. Deuts. Bot. Ges.*, 25:568–76. 1907.

99 ———, Uber die Nachkommenschaft der *Solanum* Pfropfbastarde und die Chromosomenzahlen ihrer Keimzellen, *Zeits. f. Bot.*, 2:1–38. 1910.

100 Zimmerman, R. H., Flowering in crabapple seedlings; methods of shortening the juvenile phase, *Jour. Amer. Soc. Hort. Sci.*, 96(4):404–11. 1971.

101 ———, Shortening the juvenile phase in crabapple seedlings, *Proc. Inter. Plant Prop. Soc.*, 21:434–36. 1972.

SUPPLEMENTARY READING

Baker, K. F. (ed.), "The U.C. System for Producing Healthy Container-Grown Plants," *California Agricultural Experiment Station Manual 23*. (1957).

Corbett, M. K., and H. D. Sisler (eds.), *Plant Virology*. Gainesville, Fla.: University of Florida Press, 1964.

Dermen, H., "Colchiploidy and Cytochimeras in the Study of Ontogenic Problems," *Proceedings of the XVII International Horticultural Congress*, Vol. III (1966), pp. 3–14.

Doorenbos, J., "Juvenile and Adult Phases in Woody Plants," in *Handbuch der*

Pflanzenphysiologie, Vol. 15 (Part I), pp. 1222–35. Berlin, Heidelberg, New York: Springer-Verlag. 1965.

Frazier, N. W. (ed.), *Virus Diseases of Small Fruits and Grapevines,* Berkeley: Division of Agricultural Sciences, University of California, 1970.

Hinds, H. V. (ed.), Special Issue on Vegetative Propagation, *New Zealand Journal of Forestry Science,* Vol. 4, No. 2 (1974), 120–458.

Kenneth, N., and J. Katan (eds.), Symposium on Production of Healthy Plants by Therapeutic and other Methods and their Maintenance and Use, *Proceedings of the XVIII International Horticultural Congress,* Vol. III (1970).

Kneen, O. H., "Patent Plants Enrich Our World," *National Geographic Magazine,* Vol. 93 (1948), pp. 357–78.

Marston, M. E., "The History of Vegetative Propagation," *Report 14th International Horticultural Congress,* Vol. II (1955), pp. 1157–64.

Neilson-Jones, W., *Plant Chimeras,* 2nd ed. London: Methuen & Company, 1969.

Nyland, G., and A. C. Goheen, Heat Therapy of Virus Diseases of Perennial Plants, *Annual Review of Phytopathology,* Vol. 7 (1969), pp. 331–54.

Weiss, F. E., The Problem of Graft Hybrids and Chimaeras, *Biological Review,* Vol. 5 (1930), pp. 231–71.

Welsh, M. F., Control of Fruit Plant Virus Diseases, *Proceedings of the XVII International Horticultural Congress,* Vol. III (1966), pp. 95–102.

Zimmerman, R. H., "Juvenility and Flowering in Woody Plants: A Review," *HortScience,* Vol. 7, No. 5 (1972), 447–55.

Anatomical and Physiological Basis of Propagation by Cuttings

9

In propagation by *stem* and *leaf-bud cuttings,* it is only necessary that a new root system be formed, since a potential shoot system—a bud—is already present. *Root cuttings* must initiate a new shoot system—from an adventitious bud—as well as an extension of the existing root piece. In *leaf cuttings,* both a new root and a new shoot system must be regenerated. Many cells, even in mature plant parts, have the capability of returning to the meristematic condition (dedifferentiation) and producing a new root or shoot system, or both. This fact makes propagation by cuttings possible. In fact, a single, living, vegetative cell contains all the information necessary to regenerate a complete new plant similar to the plant from which it came (*213*).

Cutting propagation is ordinarily used for dicotyledonous plants, but cuttings of some monocots, such as asparagus, can be rooted under the proper conditions (*3, 66*). Adventitious roots commonly occur naturally in many monocotyledonous plants—for example, "brace" roots in corn, which arise at the intercalary region at the base of the internodes. Adventitious roots, too, comprise the root system of the bulbous types of plants discussed in Chapter 15.

The various types of cuttings and methods of preparing and caring for them are discussed in Chapter 10.

THE ANATOMICAL DEVELOPMENT OF ROOTS AND SHOOTS IN CUTTINGS

Stem Cuttings

A knowledge of the internal structure of the stem is necessary in order to understand the origin of adventitious roots. In Figure 9–1 cross sections of an herbaceous stem and a woody stem are shown, with the principal tissues indicated.

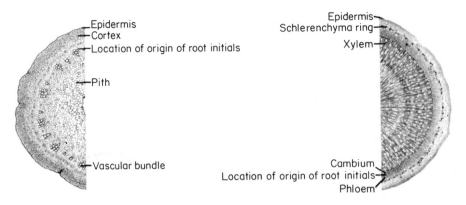

Figure 9–1   Stem cross sections showing the usual location of origin of adventitious roots. *Left:* young, herbaceous, dicotyledonous plant. *Right:* young, woody plant.

The process of development of adventitious roots in stem cuttings can be divided into three stages: (1) cellular dedifferentiation followed by the initiation of groups of meristematic cells (the *root initials*); (2) the differentiation of these cell groups into recognizable *root primordia;* and (3) the growth and emergence of the *new roots,* including rupturing of other stem tissues, and formation of vascular connections with the conducting tissues of the cutting.

Initiation of Root Primordia
The precise location inside the stem where adventitious roots originate has intrigued plant anatomists for centuries. Probably the first definite study of this was made by a French dendrologist, Duhamel du Monceau in 1758 *(44)*; a great many studies have subsequently been made, covering a wide range of plant species *(60)*. In most plants, the initiation of adventitious roots takes place after the cutting is made. These are sometimes called "induced" or "wound" roots, since they occur following some type of wounding, such as cutting off a stem piece, or girdling the stem—as in air layering. The origin of adventitious roots in stem cuttings is in certain groups of cells which become meristematic. Tissues involved at the site of origin vary widely, depending upon the kind of plant.

In *herbaceous plants* adventitious root origin is usually just outside and between the vascular bundles *(162)*. These small groups of cells, the *root initials,* continue dividing, forming groups of many small cells which develop into *root primordia.* Cell division continues, and soon each group of cells takes on the appearance of a root tip. A vascular system develops in the new root primordium and becomes connected to the adjacent vascular bundle *(48)*. The root tip grows outward, through the cortex, emerging from the epidermis of the stem (see Figure 9–2). For example, adventitious roots in chrysanthemum cuttings are first observed in the interfascicular region, as shown in Figure 9–3. Root initials in carnation cuttings arise in a layer of

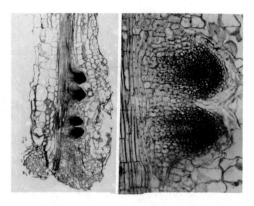

Figure 9-2   Development of adventitious roots in stem segments of tobacco. Root primordia originated in cambial region. *Left:* group of four primordia. Pith is at left, cortex at right. Longitudinal section ✕ 18. *Right:* enlarged view of two primordia. Xylem is at left, phloem at right. Longitudinal section ✕ 94. Courtesy Clarence Sterling.

parenchymatous cells inside a fiber sheath; the developing root tips, upon reaching this band of fiber cells, do not push through it, but turn downward, emerging from the base of the cutting *(181)*. The origin of adventitious roots in pumpkin and tomato is in phloem parenchyma *(160)*.

In *woody perennial plants,* where one or more layers of secondary xylem and phloem are present, adventitious roots in stem cuttings usually originate

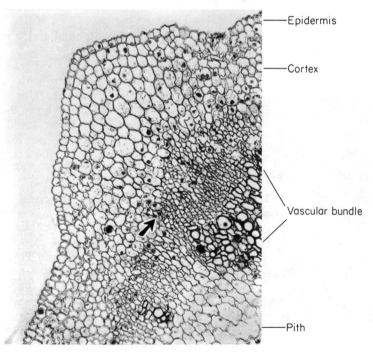

Figure 9-3   Photomicrograph of a portion of chrysanthemum stem. Arrow points to a root initial in a very early stage showing its close relationship to a vascular bundle. (Enlarged 200 times.) Courtesy B. B. Stangler.

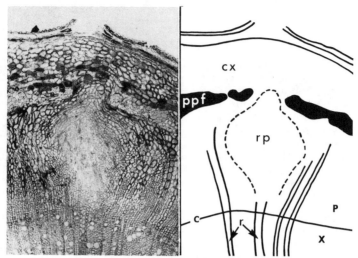

Figure 9–4   Tissues involved in adventitious root formation in 'Brompton' plum hardwood cuttings. cx = cortex; ppf = primary phloem fibers; rp = root primordium; p = phloem; r = rays; c = cambium; x = xylem. Courtesy A. Beryl Beakbane, East Malling Research Station *(10)*.

in the young, secondary phloem (Figure 9–4), although they may also arise from other tissues as vascular rays, cambium, or pith *(31, 36, 121, 144, 216)*. Generally, the origin and development of adventitious roots takes place next to and outward from the central core of vascular tissue. Upon emergence from the stem (Figure 9–5), the adventitious roots have developed a root cap and the usual root tissues, as well as a complete vascular connection with the originating stem *(48)*

Usually, adventitious roots on stems arise *endogenously;* that is, they originate within the stem tissue and grow outward, but in *Tamarix* stems such roots were observed to arise in the lenticels, with subsequent connection of procambial strands to those of the parent stem *(59)*.

The time at which root initials develop after cuttings are placed in the propagating bed varies widely. In one study *(181)* they were first observed microscopically after three days in the chrysanthemum, five days in the carnation, and seven days in the rose.

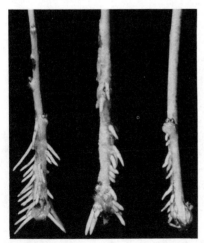

Figure 9–5   Emergence of adventitious roots in plum stem cuttings. Observe the tendency of the roots to form in longitudinal rows, which appear directly below buds.

Visible roots emerged from the cuttings after ten days for the chrysanthe-mum, but three weeks were required for the carnation and rose.

Preformed (Latent) Root Initials

In some plants adventitious root initials form during early stages of intact stem development and are already present at the time the cuttings are made (*23, 130, 207*). These are termed *preformed,* or *latent, root initials* and generally lie dormant until the stems are made into cuttings and placed under environmental conditions favorable for further development and emergence of the primordia as adventitious roots. However, in some species they develop into aerial roots on the intact plant and become quite prom-inent. Such preformed root initials occur in a number of easily rooted genera, such as willow (*Salix*), hydrangea (*Hydrangea*), poplar (*Populus*), jasmine (*Jasminum*), currant (*Ribes*), citron (*Citrus medica*), and others (*60*). The position of origin of these preformed root initials is essentially the same as that of other adventitious roots (*22, 129*). In old trees of some apple and quince cultivars, these preformed latent roots cause swellings, called "burr knots" (*195*). Species with preformed root initials generally root rapidly and easily, but cuttings of many species without such root initials root just as easily.

In willow, latent root primordia remain dormant, embedded in the inner bark for years if the stems remain intact on the tree (*22*). Their location can be observed by peeling off the bark and noting the protuberances on the wood, with corresponding indentations on the inside of the removed bark.

When experimental work is being done with cuttings of species having preformed root initials, it must be remembered that the differential treat-ments may only be affecting the development of root primordia rather than their actual initiation.

Callus

After stem cuttings have been made and placed under environmental con-ditions favorable for rooting, *callus* will usually develop at the basal end of the cutting. Callus is an irregular mass of parenchyma cells in various stages of lignification. This callus growth arises from young cells in the region of the vascular cambium, although various cells of the cortex and pith may also contribute to its formation. Frequently the first roots appear through the callus, leading to the belief that callus formation is essential for rooting. In most cases, the formation of callus and the formation of roots are independent of each other; that they often occur simultaneously is due to their dependence upon similar internal and environmental con-ditions.

In some species, however, e.g., *Pinus radiata* (*21*), *Sedum* (*236*), and *Hedera helix* (*61*) (adult phase) (see **Figure** 9–6), adventitious roots have been found to originate in the callus tissue itself which has formed at the

basal end of the cutting; hence in such cases callus formation is a precursor of adventitious root initiation.

There is evidence that the pH * of the rooting medium can influence the type of callus produced, which in turn can affect emergence of newly formed adventitious roots. In studies (*32, 33*) using stem cuttings of balsam poplar at pH 6.0 the callus cells were large and somewhat soft, and the cuttings rooted readily. With increasing alkalinity, the callus masses became smaller until at pH 11.0, the callus cells were small and compactly arranged with a firm calcareous structure. Such cuttings were rootless, although upon sectioning, well-formed root initials were found beneath the callus.

Figure 9–6 Adventitious root (arrow) forming in callus tissue at base of an *Hedera helix* (adult phase) cutting, as seen in longitudinal section. P = phloem; X^1 = primary xylem; X^2 = secondary xylem; V = vessels. Courtesy R. M. Girouard (*61*).

Leaf Cuttings

Many plant species, including both monocots and dicots, can be started by leaf cuttings (*74*). Although the origin of new shoots and new roots in leaf cuttings is quite varied, they can be grouped generally as developing from primary or secondary meristems, the latter type being the most common.

Primary meristems are groups of cells that are direct descendants of embryonic cells which have never ceased to be involved in meristematic activity.

Secondary meristems are groups of cells which have differentiated and functioned in some mature tissue system and then again resumed meristematic activity.

Leaf Cuttings with Primary Meristems

In Figure 10–16 detached leaves of *Bryophyllum* are shown with small plants arising from the notches around the leaf margin. These small plants originate from so-called foliar "embryos," which are formed in the early stages of leaf development from small groups of cells at the edges of the leaf. As the leaf expands, a foliar embryo develops until it consists of two rudimentary leaves with a stem tip between them, two root primordia, and a "foot" which extends toward a vein (*86, 235*). As the leaf matures,

---

* pH is a measure of relative acidity or alkalinity. A pH of 7.0 is neutral; decreasing values below this indicate increasing acidity; increasing values above 7.0 indicate increasing alkalinity.

cell division in the foliar embryo ceases, and it remains dormant. If the leaf is detached and placed in close contact with a moist rooting medium, the young plants rapidly break through the leaf epidermis and become visible in a few days. Roots extend downward, and after several weeks many new independent plants form while the original leaf dies. The new plants thus develop from latent primary meristems—from cells that have not assumed mature characteristics. Such production of new plants from leaf cuttings by the renewed activity of primary meristems is found, too, in other species, as the piggy-back plant (*Tolmiea*) and walking fern (*Camptosorus*).

Leaf Cuttings with Secondary Meristems

In leaf cuttings of such plants as *Begonia rex, Sedum,* African violet (*Saintpaulia*), snake plant (*Sansevieria*), *Crassula,* and lily, new plants may develop from secondary meristems arising from mature cells at the base of the leaf blade or petiole.

In *Lilium longiflorum* and *L. candidum,* the bud primordium originates in parenchyma cells in the upper side of the bulb scale, whereas the root primordium arises from parenchyma cells just below the bud primordium. Although the original scale serves as a source of food for the developing plant, the vascular system of the young bulblet is independent of that of the parent scale, which eventually shrivels and disappears (*218*).

In African violet (*Saintpaulia*), new roots and shoots arise by the formation of meristematic cells from mature cells in the leaves. The roots are produced endogenously from thin-walled cells lying between the vascular bundles. The new shoots arise from cells of the epidermis and the cortex immediately below the epidermis. The roots emerge, form branch roots, and continue to grow for several weeks before the shoots appear. Although the original leaf supplies nutrient materials to the young plant, it does not become a part of the new plant (*149*).

In several species, for example, sweet potato, *Peperomia,* and *Sedum,* new roots and new shoots on leaf cuttings arise in callus tissue which develops over the cut surface through activity of secondary meristems. The petiole of leaf cuttings of *Sedum* forms a considerable pad of callus within a few days after the cuttings are made. Root primordia are organized within the callus tissue, and shortly thereafter four or five roots develop from the parent leaf. Following this, stem primordia arise on a lateral surface of the callus pad and develop into new shoots (*236*).

Adventitious roots form on leaves much more readily than do adventitious buds. In some plants, such as the India rubber fig (*Ficus elastica*) and jade plant (*Crassula argentea*) (See Figure 10–17), the cutting must include a portion of the old stem containing an axillary bud because although adventitious roots may develop at the base of the leaf, an adventitious shoot is not likely to form. In fact, rooted leaves of some species will survive for years without producing an adventitious shoot. Treatments with a cytokinin, (see p. 220) as benzyladenine, may however initiate buds and shoots (*14*).

Root Cuttings

In root cuttings production of adventitious shoots and, in many cases, adventitious roots must take place. In many plants, adventitious buds form readily on intact roots, especially if they are wounded. In young roots, such buds may arise in the pericycle near the vascular cambium (230). In old roots buds may arise exogenously in a callus-like growth from the phellogen; or they may appear in a callus-like proliferation from ray tissue (48). Bud primordia may also develop from wound callus tissue which proliferates from the cut ends or injured surfaces of the roots (162).

Bud initiation on root pieces is stimulated by applications of cytokinins (37). Kinetin treatments of root segments of *Convolvulus* (bindweed) at 0.1 ppm in a sterile nutrient medium induced initiation of buds, especially in light (16). On the other hand, auxins, such as indoleacetic acid, tend to inhibit shoot formation on root cuttings.

Regeneration of new root meristems is often more difficult than the production of adventitious buds. New roots may not be adventitious but develop from latent root initials contained in old branch roots which may be present on the root piece (177). Generally, such branch roots arise from mature cells of the pericycle or endodermis, or both, adjacent to the central vascular cylinder (48). Adventitious root initials in roots have been observed to arise in the region of the vascular cambium. Studies on the origin of branch roots from root segments grown in nutrient media under aseptic conditions have shown (12) that such root formation requires the presence of auxin, minerals, and sucrose—as a carbon source.

Regeneration of new plants from root cuttings takes place in different ways, depending upon the species. The most common type is for the root cutting to produce first an adventitious shoot, roots appearing later, often from the base of the new shoot rather than from the original root piece itself. Sometimes these adventitious shoots can be removed, then rooted as stem cuttings. In other plants, a well-developed root system has formed by the time the first shoots appear. Root cuttings of some species form a strong adventitious shoot, but no new roots develop, and the cutting eventually dies. In certain species, root cuttings will produce a strong new root system, but no adventitious shoot arises, so the cutting finally dies (107).

Root cuttings taken from very young seedling trees are much more successful than those taken from older trees. Failures in the latter case are apparently due to the inability of the root pieces to regenerate a new root system. This situation is probably related to the phenomenon of juvenility (see p. 184), which is also involved in root formation in stem cuttings.

One of the chief advantages claimed for asexual propagation is the faithful reproduction of all characteristics of the parent plant. However, with root cuttings this does not always hold true. In plants which are periclinal chimeras (see p. 191), where the cells of the outer layers are of a different genetic makeup from those of the inner tissues, the production of a new plant by root cuttings taken from such a plant results in a plant differing from the parent. This is well illustrated in the thornless boysenberry, in which stem or leaf-bud cuttings produce plants which retain the thornless condition, but root cuttings develop into thorny plants.

A list of plants commonly propagated by root cuttings is given on p. 288.

PHYSIOLOGICAL BASIS OF ROOT INITIATION IN CUTTINGS

Plant Growth Substances

Certain concentrations in the plant of the various naturally occurring materials having hormonal properties are more favorable than others for adventitious root initiation. Much study has been given to determining these relationships. In distinguishing between *plant hormones* * and *growth regulators* † it can be said that all hormones regulate growth but not all growth regulators are hormones. Various classes of growth regulators, such as the auxins, cytokinins, gibberellins, inhibitors (e.g., abscisic acid), and ethylene influence root initiation. Of these, auxins have had the greatest effect on root formation on cuttings. In addition to these groups there are, no doubt, other naturally occurring materials which have a part in adventitious root initiation.

Auxins

In the mid-1930s and later, studies of the physiology of auxin action showed that auxin was involved in such varied plant activities as stem growth, root formation, lateral bud inhibition, abscission of leaves and fruits, activation of cambial cells, and others.

Indole-3-acetic acid (IAA) was identified in 1934 as a naturally occurring compound having considerable auxin activity (*118, 198*) and was soon found to promote adventitious root formation (*201, 203, 227*). This action of IAA was originally shown by a biological test using etiolated pea epicotyls under a set of standard conditions (*226, 228, 229*).

Synthetic indoleacetic acid (See Figure 9–18) was subsequently tested for its activity in promoting roots on stem segments and in 1935 several investigators (*28, 124, 201*) demonstrated the practical use of this material in stimulating root formation on cuttings. About the same time it was shown (*197, 241*) that two similar materials, indolebutyric acid (IBA) and naphthaleneacetic acid (NAA) (see Figure 9–18)—although not naturally occurring—were even more effective than the naturally occurring indoleacetic acid for this purpose. It is now well accepted and has been subsequently confirmed many times (*31*) that auxin, natural or artificially applied, is a requirement for initiation of adventitious roots on stems (*57, 150*) and, indeed, it has been shown that the division of the first root initial cells are dependent upon either applied or endogenous auxin (*72*).

Formation of preformed root initials (see p. 215) in stems apparently is dependent upon the native auxins in the plant plus an auxin synergist; together these lead to synthesis of ribonucleic acid (RNA) which is involved in initiation of the root primordia (*71*).

---

* *Plant hormones* are organic compounds other than nutrients produced by plants which, in low concentrations, regulate plant physiological processes. They usually move within the plant from a site of production to a site of action.

† *Plant growth regulators* are either synthetic compounds or plant hormones that modify plant physiological processes. They regulate growth by mimicking hormones, by influencing hormone synthesis, destruction, or translocation, or (possibly) by modifying hormonal action sites.

Cytokinins

Cytokinins are plant growth hormones involved in cell growth and differentiation. Various natural and synthetic materials (See Figure 6–1) such as zeatin, kinetin, and 6-benzyl adenine have cytokinin activity. Generally, applied synthetic cytokinins have not stimulated or prevented root initiation. However, cytokinin at relatively low concentrations, when applied to decapitated pea cuttings at an early developmental stage, promoted root initiation while higher concentrations inhibited initiation. At a later stage in root initiation such inhibition did not occur. The influence of cytokinins in root initiation may thus depend upon the particular stage of initiation and the concentration *(47)*. Cytokinins relate to auxins in controlling organ differentiation *(70, 178, 231)* as shown in Figure 9–7 for studies with tobacco stem segments.

Cytokinins strongly promote bud initiation. For example, in root cuttings of *Isatis tinctoria* (a biennial herb) grown in a sterile medium, after several subcultures, shoot formation on the root pieces did not occur unless the medium was supplied with kinetin *(37)*.

Leaf cuttings provide good test material for studying auxin-cytokinin relationships, since such cuttings must initiate both roots and shoots. In a study *(85)* with *Begonia* leaf cuttings under aseptic conditions, cytokinin at relatively high concentrations (about 13 ppm) promoted bud formation and inhibited root formation. Auxins, at high concentrations, gave the opposite effect. There were interacting relationships, however, between auxins and cytokinins. At low concentrations (about 2 ppm), IAA promoted bud formation, enhancing the cytokinin influence. Also, at low concentrations (about 0.8 ppm), kinetin stimulated the effect of IAA on root promotion.

Temperature was an influencing factor in these relationships. High temperature, 81° F (27° C), itself inhibited bud formation and opposed the

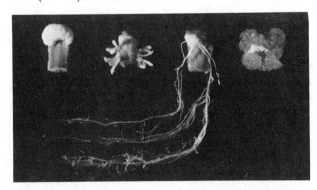

Figure 9–7   Effects of adenine sulfate (a cytokinin) and indoleacetic acid (auxin) on growth and organ formation in tobacco stem segments. *Far left:* control. *Center left:* adenine sulfate, 40 mg per liter. Bud formation with decrease in root formation. *Center right:* indoleacetic acid, 0.02 mg per liter. Root formation with prevention of bud formation. *Far right:* adenine sulfate, 40 mg per liter plus indoleacetic acid, 0.02 mg per liter. Growth stimulation but without organ formation. Courtesy Folke Skoog.

stimulatory effects of cytokinin on this process as well as the suppressing effect of cytokinin on root formation. On the other hand, the auxin effects were stimulated under long days at this temperature as compared to lower temperatures [60° F (15° C)].

Under low light intensities, however, neither auxin level nor regeneration ability were affected by temperaure. It would seem that the considerable seasonal changes in the regenerative ability of *Begonia* leaf cuttings are due to a complex interaction of temperature, photoperiod, and light intensity, controlling the levels of endogenous auxins and other growth regulators (*88*).

Buds are initiated in the leaf notches of *Bryophyllum* at an early stage in leaf formation (see p. 286). Shoots arising from the leaves originate from these buds. Cytokinin applications have a stimulatory effect on bud development while auxin applications inhibit bud development but stimulate root formation (*86*).

Gibberellins

The gibberellins (Figure 6–1) are a group of closely related, naturally occurring compounds, first isolated in Japan in 1939, and known principally for their effects in promoting stem elongation. At relatively high concentrations (up to 10^{-3}M) they have consistently inhibited adventitious root formation. There is evidence that this inhibition is a direct local effect which prevents the early cell divisions involved in transformation of mature stem tissues to a meristematic condition (*20*). Gibberellins have a function in regulating nucleic acid and protein synthesis and may be suppressing root initiation by interfering with these processes (*117*). At lower concentrations, however, (10^{-11} to 10^{-7}M) gibberellin has promoted root initiation in pea cuttings (*45, 46*).

In *Begonia* leaf cuttings, gibberellic acid was noted (*89*) to inhibit both adventitious bud and root formation, probably by blocking the organized cell divisions which initiate formation of bud and root primordia. There is evidence (*72*) in willow stem cuttings that applied gibberellin blocks auxin activity in root primordium development subsequent to the earlier initiation phase.

Lowering the natural levels of gibberellin in the tissues should stimulate adventitious root formation in cuttings. In fact, promotion of rooting has been done experimentally by various chemical substances that interfere with gibberellin activity such as Alar (SADH) (*165, 234*), abscisic acid (*9, 24*), gonadotropins (*133*), and EL 531 [α cyclopropyl α (4 methoxyphenyl-5-pyrimidine methanol], a gibberellin antagonist (*116*).

Other Naturally Occurring Materials

Many compounds, some identified and some not, are being found in extracts of various plant tissues which affect adventitious root production. Some of these possibly are the same as the rooting co-factors discussed in the next section. Although there can be extracted from plants many substances that influence root initiation, cellular compartmentalization may normally exclude such extracted substances from physiological processes. Thus, searches for "mysterious substances" that influence root initiation require cautious interpretation.

Ethylene (C_2H_4) is produced by plants and has a number of hormonal effects although it does not exactly fit the definition of a hormone (*1*). In 1933 Zimmerman and Hitchcock (*239*) showed that applied ethylene at about 10 ppm can cause root formation on stem and leaf tissue as well as the development of preexisting latent roots on stems. These and other workers (*241*) also showed about the same time that auxin applications can regulate ethylene production and suggested that auxin-induced ethylene may account for the ability of auxin to cause root initiation. Kawase (*111, 113*), by centrifuging *Salix* cuttings in water, or by just soaking in hot or cold water, stimulated ethylene production in the tissues as well as root development, suggesting a possible causal relationship between ethylene production and subsequent root development. However, Mullins (*147*), in studies of root initiation in mung bean cuttings, found that ethylene, from 0 to 1000 ppm, decreased root initiation. He concluded that ethylene inhibits formation of adventitious roots in cuttings but promotes root emergence in stems having latent root primordia.

In contrast to these findings Krishnamoorthy (*122*) found that applications of ethephon, an ethylene-generating compound, to mung bean cuttings did stimulate root formation. The relationships between auxin, ethylene, and adventitious root formation apparently are complex, involving more than a simple alteration in ethylene concentration and will require more study to resolve these contradictions.

Heliangine, a sesquiterpenic lactone isolated from leaves of *Helianthus tuberosus,* was found by Japanese workers (*176*) to promote root formation in *Phaseolus* and *Azukia* cuttings.

Portulal, a root-promoting substance isolated from leaves of *Portulaca grandiflora* by researchers, also in Japan (*143*), has been identified as a bicyclic diterpene active in promoting root initiation in *Azukia, Vigna,* and *Phaseolus.*

Water-soluble substances obtained from such woody plants as *Cotoneaster, Euonymus, Symplocos,* and *Salix* by centrifugal diffusion or by water extraction from ground, freeze-dried cuttings were found to promote rooting when tested in the mung bean bioassay (*114, 115*). These substances were composed of four root-promoting fractions which could be separated by paper chromotography. Root-promoting substances have also been found (*5*) in tomato seedings, appearing to originate first in the cotyledons and later in the young, plumular leaves. By applying triiodobenzoic acid—known to block polar transport of auxin (*232*)—to the tomato stem below the cotyledon, and observing a reduction in root initiation, it was assumed that these rooting substances were auxin-like in nature.

It is likely that such water-soluble, root-promoting substances commonly occur in tissues of various plant species.

The Effects of Leaves and Buds:
Rooting Co-factors (Auxin synergists)

Duhamel du Monceau (*44*) explained in 1758 the formation of adventitious roots on stems on the basis of the downward movement of sap. In extending this concept, Sachs, the German plant physiologist, postulated (*169*) in 1882 the existence of a specific root-forming substance manufactured in the leaves which moves downward to the base of the stem, where it promotes root formation. It was shown (*129, 131*) by van der Lek in 1925 that strongly sprouting buds promote the development of roots just below the buds in cuttings of such plants as willow, poplar, currant, and grape. It was assumed that hormone-like substances were formed in the developing buds and transported through the phloem to the base of the cutting, where they stimulated root formation.

The existence of a specific root-forming factor was first determined by Went (*226*) in 1929 when he found that if leaf extracts from the *Acalypha* plant were applied back to *Acalypha* or to *Carica* tissue they would induce root formation. Bouillenne and Went (*17*) in 1933 found substances in cotyledons, leaves, and buds which stimulated rooting of cuttings; they termed this material "rhizocaline."

In Went's pea test (*228*) for root-forming activity of various substances, it is significant that the presence of at least one bud on the pea cutting was essential for root production. A budless cutting would not form roots even when treated with an auxin-rich preparation. This indicates again that a factor other than auxin, presumably produced by the bud, was needed for root formation. In 1938, Went postulated the presence of specific factors other than auxin which he believed were manufactured in the leaves and were necessary for root formation.

That the amount of some naturally occurring root-forming substance(s) other than auxin, as yet unidentified but essential for root initiation, may be abundant in some plants and slight or even lacking in others, is shown by other experiments. In Cooper's studies (*29*) in 1938 apple and lemon cuttings were treated with auxin, and after analysis for auxin in the two groups of cuttings, it was found that there was little difference in the amount recovered, yet none of the apple cuttings rooted, whereas the lemon cuttings did. It was assumed that the apple cuttings were lacking in certain unidentified natural substances necessary for root formation, while the lemon cuttings had this substance(s) in abundance.

From their studies on the rooting of coniferous evergreen cuttings, Thimann and Delisle (*200*) agreed that some unknown factor(s), other than auxin, is involved in root initiation. They believed that this factor might exist in larger quantities in young plants, such as one-year-old seedlings, thus accounting for the comparative ease of rooting cuttings taken from young plants—the "juvenility effect" (see p. 184).

Removal of the buds from cuttings in certain plants will stop root formation almost completely, especially in species without preformed root initials (*129, 226*). In some plants, if a ring of bark is removed down to the wood just below a bud, root formation is reduced, indicating that some influence

Figure 9–8 Effect of leaves on the rooting of 'Lisbon' lemon cuttings. Both groups were rooted under intermittent mist and were treated with indolebutyric acid at 4000 ppm by the concentrated-solution-dip method.

travels through the phloem from the bud to the base of the cutting, where it is active in promoting root initiation. It has been shown (*50, 131*) that if hardwood cuttings are taken in midwinter when the buds are in the rest period,* they had no stimulating effect on rooting; if the cuttings are made in early fall or spring, however, when the buds are active and not in the "rest" influence, they showed a strong root-promoting effect. It has been shown, too, with cuttings of apple and plum rootstocks that the capacity of the shoots to regenerate roots increased during the winter, reaching a high point just before bud-break in the spring; this is believed to be associated with a decreasing level of bud dormancy following winter chilling (*104*).

It has long been known, and there is considerable supporting experimental evidence (*29, 164, 226*), that the presence of leaves on cuttings exerts a strong stimulating influence on root initiation. (See Figures 9–8 and 9–9.)

Carbohydrates translocated from the leaves undoubtedly contribute to root formation. However, the strong root-promoting effects of leaves and buds are due, probably, to other, more direct, factors. Leaves and buds are known to be powerful auxin producers, and the effects are observed directly below them, showing that polar apex-to-base transport is involved.

Cuttings of certain clones are easily rooted but cuttings of other, closely related, clones root with considerable difficulty. It would seem that grafting a leafy portion of the easily rooted clone onto a basal stem portion of the difficult-to-root clone and then preparing the combination as a cutting, would cause it to root readily. The rooting factors provided by the leaves or buds of the easily rooted clone could perhaps stimulate rooting of the difficult-to-root basal part. Such experiments have been tried but with conflicting results. Van Overbeek and others (*214, 215*) rooted the difficult-to-root 'Purity' (white) hibiscus by previously grafting onto it sections of the easily

* The "rest period" is a physiological condition of the buds of many woody perennial species beginning shortly after the buds are formed. While in this condition, they will not expand into flowers or leafy shoots even under suitable growing conditions. But after exposure to sufficient cold, the "rest" influence is broken, and the buds will develop normally with the advent of favorable temperatures.

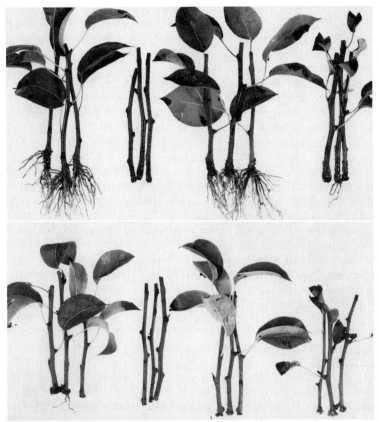

Figure 9–9   Effect of leaves, buds, and applied auxin on adventitious root formation in leafy 'Old Home' pear cuttings. *Top:* cuttings treated with auxin (indolebutyric acid at 4000 ppm for 5 sec.). *Bottom:* untreated cuttings. Left to right: with leaves; leaves removed; buds removed; one-fourth natural leaf area. Courtesy W. Chantarotwong.

rooted 'Brilliant' (red) hibiscus, then applying auxins. Auxin-treated 'Purity' cuttings, ungrafted, did not root, the leaves soon dropping. Auxin-treated ungrafted 'Brilliant' cuttings rooted readily and retained their leaves.

Later experiments with the same hibiscus cultivars by Ryan and others (*167*) gave different results. With cuttings held for six weeks, they obtained 100 percent rooting with 'Brilliant,' 90 per cent with 'Purity,' and 84 percent with cuttings made from 'Brilliant' on 'Purity' grafts. It is possible that in van Overbeek's studies, difficulty in rooting 'Purity' cuttings was due primarily to the early loss of leaves under the conditions of their experiments. Leaves of the 'Purity' cultivar may stimulate rootings just as well as those of the 'Brilliant' if they can be retained long enough.

Bouillenne and Bouillenne-Walrand (*17*) in 1955 proposed that "rhizocaline" be considered as a complex of three components: (1) a specific factor, translocated from the leaves, and characterized chemically as an ortho-dihydroxy phenol; and (2) a nonspecific factor (auxin), which is translocated and is found in biologically low concentrations; and (3) a specific enzyme located in cells of certain tissues (pericycle, phloem, cambium), which is probably of the polyphenol-oxidase type.

They further proposed that the ortho-dihydroxy phenol reacts with auxin wherever the enzyme is present, giving rise to the complex "rhizocaline," which may be considered one step in a chain of reactions leading to root initiation (Figure 9–10). Libbert (*134*) in 1956 also developed evidence showing that auxin, forming a complex with a mobile factor "X," would result in root initiation, but he rejected the concept that a non-mobile enzyme was involved. He asserted that auxin itself acts to cause dedifferentiation of cells, determining the site of root formation.

Hess (*91, 92, 93, 95*) in 1962 isolated various rooting co-factors from cuttings, using chromatography together with mung bean (*Phaseolus aureus*) bioassay techniques. He worked with cuttings of the easily rooted juvenile form and the difficult-to-root mature form of *Hedera helix* (English ivy). He also used easy and difficult-to-root cultivars of chrysanthemum and of the red-flowered and white-flowered forms of *Hibiscus rosa-sinensis*. These co-factors are naturally occurring substances which appear to act synergistically with indoleacetic acid in promoting rooting. The easily rooted forms of plants he has worked with have a larger content of such co-factors than the difficult ones.

One of these so-factors (No. 4) represents a group of active substances, tentatively characterized as oxygenated terpenoids. Another (No. 3) was identified in 1965 as chlorogenic acid (*94*). Further work in 1972 showed that there were three lipid-like root-promoting compounds in juvenile ivy tissue which contain functional alcohol and nitrile groups. All three were

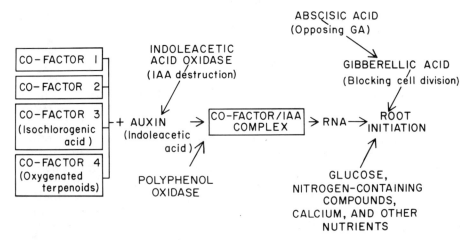

Figure 9–10   Hypothetical relationships of various components leading to adventitious root initiation. In addition, specific root-inhibiting factors may be present which interfere with root development.

colorless and unstable, breaking down to orange-yellow compounds and losing their root-promoting activity (*97*).

In testing the biological activity of compounds structurally related to co-factor 4, Hess (*92*) found that the phenolic compound catechol reacts synergistically with indoleacetic acid in root production in the mung bean bioassay. Since, as he points out, catechol is readily oxidized to a quinone, and since the mung bean itself is a good source of phenolase, it may be that oxidation of an ortho-dihydroxy phenol is one of the first steps leading to root initiation, as suggested earlier by Bouillenne and Bouillenne-Walrand (*17*). In addition, various other compounds have been found to react synergistically with auxin in promoting rooting (*67*).

One of the postulated rooting co-factors could possibly be abscisic acid which can promote root initiation (*24*), perhaps by antagonizing gibberellic acid which, in certain concentrations, is known to inhibit root formation.

Rooting co-factors were found in hardwood cuttings of 'Crab C' and 'E.M. 26' apple rootstocks by the mung bean biossay (*25*). Increased co-factor activity was found when cuttings of 'E.M. 26' were subjected to elevated temperatures (65° F; 18.5° C) during the winter storage period, a practice known to increase rooting. Hardwood stem tissue of the easily rooted 'MM 106' was found (*4*) to show strong root promoting factors in the mung bean bioassay, whereas in the difficult-to-root 'EM 2' such factors not only were present in lower amounts but rooting inhibitors appeared.

Fadl and Hartmann (*49, 50*) in 1967 isolated an endogenous root-promoting factor from basal sections of hardwood cuttings of an easily rooted pear cultivar ('Old Home'). This highly active root-promoting material appeared only in cuttings having buds and which had been treated with indolebutyric acid. It was found in largest amounts about ten days after the cuttings had been made, treated with IBA, and placed in the rooting medium. Extracts from basal segments of similar cuttings of a difficult-to-root cultivar ('Bartlett'), treated with IBA, did not show this rooting factor. It did not appear either in 'Old Home' cuttings which had the buds removed or at a time of year when the buds were in a deep "rest" condition. Tests using UV spectrum analysis and infrared spectroscopy indicated that this rooting factor is a complex structure of high molecular weight and possibly is a condensation product between the applied auxin and a phenolic substance produced by the buds.

The action of these phenolic compounds in root promotion could be, at least partly, in protecting the root-inducing, naturally occurring auxin—indoleacetic acid—from destruction by the enzyme, indoleacetic acid oxidase (*41*). This was presumed to be a possibility in rooting (*69*) juvenile *Hedera helix*, where the phenol, catechol, showed remarkable synergism with IAA in promoting rooting. However, when naphthaleneacetic acid (NAA), which is not affected by IAA oxidase, was used as the auxin, NAA alone was as effective as NAA plus catechol. This implies that catechol was protecting IAA from destruction by the enzyme.

The role played by the phenolic rooting cofactors in adventitious root initiation is controversial, however, with some studies (*136*) failing to show any correlation between rooting cofactors and rooting response.

Endogenous Rooting Inhibitors

Cuttings of certain difficult-to-root plants may fail to root because of naturally occurring rooting inhibitors. This was found many years ago *(180)* to be the case with grapes, in which chromatographic studies suggested the presence of two inhibitors associated with rooting response. Leaching the cuttings with water enhanced the quantity and quality of roots. An inhibitor was released into the water during leaching which had a detrimental effect on rooting cuttings of the easily rooted *Vitis vinifera*. Shy-rooting cuttings of *V. berlandieri* seemed to possess a high inhibitor content.

'Bartlett' pear hardwood cuttings are difficult to root under treatments which give good rooting of 'Old Home' pear hardwood cuttings *(49)*. Extracts taken from cuttings of both varieties 20 days after being treated with IBA and placed in a rooting medium showed distinctly different amounts of inhibitors and promoters, as shown in Figure 9–11.

Other evidence implicating rooting inhibitors is given in studies *(35, 154)* in Australia with the difficult-rooting adult tissues of *Eucalyptus grandis* which contained compounds that blocked adventitious root formation. In these studies three such inhibitors were found and considerable informa-

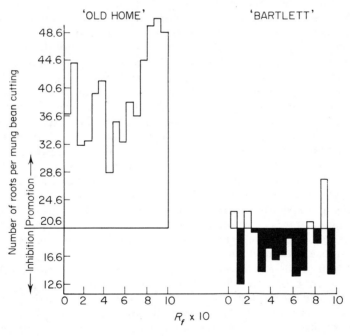

Figure 9–11 A comparison between promoters and inhibitors—as determined by the mung bean bioassay—in paper strips prepared from chromatograms loaded with extracts from bases of 'Old Home' pear cuttings *(left)* and 'Bartlett' pear cuttings *(right)* 20 days after the cuttings had been treated with IBA (indolebutyric acid) and placed in the rooting medium .The easily rooted 'Old Home' cuttings had large amounts of promoters at all R_f levels. Bases of the difficult-to-root 'Bartlett,' on the other hand, not only lacked promoters but also showed strong inhibition at most R_f levels *(49, 50)*.

tion was developed concerning their structure and physical properties. They were determined to be naturally occurring derivatives of the 2, 3-dioxabicycle [4,4,0] decane system. These inhibitors were not present in the easily rooted juvenile tissue of *E. grandis.*

From studies *(11)* with cuttings taken from easily rooted and difficult-to-root dahlia cultivars it was determined that in plants whose cuttings were difficult to root, inhibitors formed in the roots and moved upwards, accumulating in the shoots, subsequently interfering with root formation. In cultivars whose cuttings rooted easily, inhibitor levels were low.

Abscisic acid, known to be an inhibitor in plants, was observed to interfere with root formation in *Begonia* leaf cuttings at 20 ppm, although it increasingly stimulated adventitious bud formation from $1\frac{1}{4}$ to 20 ppm. *(87).*

Plants could be divided into three groups in regard to their relation to materials involved in adventitious root initiation. (a) Those in which the tissues provide all the various native substances, including auxin, essential for root initiation. When cuttings are made and placed under proper environmental conditions, rapid root formation occurs. (b) Those in which the naturally occurring co-factors are present in ample amounts but auxin is limiting. With the application of auxin, rooting is greatly increased. (c) Those that lack the activity of one or more of the internal co-factors although natural auxin may or may not be present in abundance. The external application of auxin gives little or no response, owing to lack of the effects of one or more of the naturally occurring materials essential for root formation. Concerning the last group, Haissig *(73)* postulates that lack of root initiation in response to applied auxin (or even to native auxin) may be due to one or more of the following:

1   a lack of necessary enzymes to synthesize the root-inducing auxin-phenol conjugates,
2   lack of enzyme activators,
3   presence of enzyme inhibitors,
4   lack of substrate phenolics, or
5   physical separation of enzyme reactants due to cellular compartmentalization.

Biochemical Changes Associated with the Development of Newly-formed Adventitious Roots

Once adventitious roots are initiated in cuttings considerable metabolic activity takes place as new root tissues are developed and the roots grow through and out of the surrounding stem tissue to become external functioning roots. It is well known that applications of the chemical, cycloheximide, to plant tissue will interfere with processes leading to protein synthesis; likewise, applications of actinomycin D will inhibit RNA (ribonucleic acid) synthesis *(116)*. Protein synthesis and RNA production were both shown indirectly to be involved in adventitious root development from studies *(108)* with etiolated stem segments of *Salix tetrasperma* which revealed that treatments with either cycloheximide (blocking protein synthesis) or actinomycin D (blocking RNA synthesis) reduced adventitious root production. However, if the stem segments were also treated with an auxin (indolebutyric acid) plus glucose, higher concentrations of the inhibiting chemicals were required to obtain blockage of root production as compared to water controls. IBA alone, without glucose, was comparatively ineffective in overcoming the repressive effects of cycloheximide and actinomycin D. These results imply the involvement of protein synthesis in root production, mediated through mRNA production. Auxin, or an auxin-co-factor complex, may be acting by overcoming the repression of genes, resulting in synthesis of new enzymes (proteins) required for increased root production. The fact that auxin action requires the presence of nutritional factors (glucose) is due to the requirement of a carbon source for the biosynthesis of nucleic acids and proteins.

Some significant studies (*145, 146*) have been made of the biochemical changes taking place during the *development* of preformed root initials in hydrangea into emerging roots. These studies followed, in particular, the changing patterns of DNA and enzyme levels as the roots developed.

Root initials were found to originate in the phloem ray parenchyma, with roots emerging 10 to 12 days after the cuttings were made. The total protein content of the root initials increased over 100 percent in the first four days after the cuttings were made but there was no pronounced increase in the DNA content of the cells until the sixth day. Apparently, then, considerable protein synthesis preceded large-scale replication of nuclear DNA.

Pronounced increases in enzyme activity took place during adventitious root formation in hydrangea (*145*); the enzymes peroxidase, cytochrome oxidase, succinic dehydrogenase, and starch-hydrolyzing enzymes were observed in the phloem and xylem ray cells of the vascular bundles. Then, during root development, enzyme activity shifted from the vascular tissues to the periphery of the bundles. Such increases in enzyme activity occurred two to three days after the cuttings were made.

Starch was found to disappear from the endodermis, phloem and xylem rays, and pith in the region of the developing root primordia, apparently being utilized as a carbohydrate food source. Starch evidently plays an important nutritional role in adventitious root development.

Thus, as preformed root initials in hydrangea develop into emerging adventitous roots, it appears that the following events occur: Synthesis of both DNA and proteins takes place before cell division in the root initials appears. The subsequent development of enzyme activity may result from prior activities of certain fractions of the early synthesized proteins. It is likely that some of the proteins play specific roles in the activation of the root primordia. As an example, peroxidase may be responsible for the destruction of certain inhibitors, which may be blocking metabolic processes leading to adventitious root formation. Increases in the activity of other enzymes may only be associated with resumption of rapid general cellular activity. Succinic dehydrogenase and cytochrome oxidase are involved in cellular respiration, while amylase releases sugars from starch as a general substrate for various continuous synthetic processes.

In the development of adventitious roots on IBA-treated plum cuttings, it was determined (*18*), using radioactive CO_2 applied to leaves, that as soon as callus and roots started forming, pronounced sugar increases—and starch losses—occurred at the base of the cuttings, the ^{14}C appearing in sucrose, glucose, fructose, and sorbitol. The developing callus and roots apparently act as a "sink" for the movement of soluble carbohydrates from the top of the cutting.

In other studies (*2*), using cuttings prepared from bean seedlings, it was noted that added IAA promoted sugar accumulation at the base of the cuttings, increasing the downward movement of ^{14}C-labelled assimilates from the leaves.

ANATOMICAL RELATIONSHIPS TO ROOTING

Although the ease or difficulty with which cuttings develop adventitious roots is most likely due to biochemical factors, not to be overlooked is the relation of the anatomical structure of the stem to rooting. For example, in some plants preformed root initials are present in the stem, and in others root production follows certain patterns corresponding to the anatomical structure of the stem. This is illustrated in the grape, in which adventitious roots on stem cuttings often appear in longitudinal rows corresponding to the primary rays in which they originate, running the entire length of the internode.

Continuous sclerenchyma rings (Figure 9–12) between the phloem and cortex, exterior to the point of origin of adventitious roots, may constitute an anatomical barrier to rooting. In a study of olive stem cuttings (*26, 27*), such a ring was associated with types of cuttings difficult to root, while easily rooted types were characterized by discontinuity of this sclerenchyma ring. Leafy cuttings of difficult-to-root types having a continuous sclerenchyma ring showed, when placed under mist (see p. 297) for rooting, active prolifer-

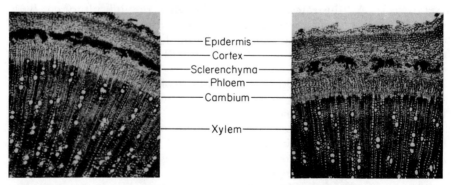

Epidermis
Cortex
Sclerenchyma
Phloem
Cambium

Xylem

Figure 9–12   Transverse sections of olive stem cuttings, $\times$ 30. *Left:* the continuous sclerenchyma ring is believed by some to constitute a barrier to emergence of adventitious roots. *Right:* the sclerenchyma ring is broken into groups of fibers separated by parenchyma cells, which would permit emergence of roots without difficulty. Courtesy Ciampi and Gellini (*26*).

ation of parenchyma ray cells, resulting in breaking of the continuity of the sclerenchyma ring, making rooting possible in anatomically unsuitable stems. Instances of such associations between shy-rooting cultivars and the presence of a heavily lignified continuous sclerenchyma ring in cuttings of other species have also been noted (*10*).

While a sheath of lignified tissue in stems may in some cases act as a mechanical barrier to root emergence, there are so many exceptions that this certainly cannot be a primary cause of rooting difficulty. These exceptions were pointed out by Sachs et al. (*170*) who noted that mist propagation and auxin treatments cause considerable cell expansion and proliferation in the cortex, phloem, and cambium, resulting in breaks in continuous sclerenchyma rings, yet in shy-rooting cultivars of several fruit species there was still no formation of root initials.

In studies with difficult-to-root cuttings of the mature stage of *Hedera helix*, English ivy, Girouard observed intense groups of discontinuous schlerenchyma fibers in the cortex, but adventitious roots had no difficulty growing through them and again in this case, some other factor must be the explanation for the low rooting performance (*62*).

Easily rooted carnation cultivars have a band of sclerenchyma present in the stems, yet the developing root primordia emerge from the cuttings by growing downward and out through the base (*181*). In other plants in which an impenetrable ring of sclerenchyma could block root emergence this same possibility is open. Rooting is more likely to be related to the actual formation of root initials than to the mechanical restriction of a sclerenchyma ring barring root emergence.

Certain types of stem structure or tissue relationships within the stem seem to be more favorable to the initiation of root primordia than others. This is shown by studies (*141*) with the easily rooted citron (*Citrus medica*), which produces roots profusely from preformed root initials along the entire stem after a short time in the rooting medium, and the sour orange (*C. aurantium*), which forms only a few roots at the base of the cutting after several weeks. As illustrated in Figure 9–13, when the difficult-to-root bark piece of the sour orange was grafted on the citron stem—where it would presumably be able to obtain any translocatable essential rooting factors from the leaves of the easily rooted citron—it still failed to form roots more readily than it did normally.

Similar studies were conducted (*174*) with shy-rooting 'Santa Maria' pear and easily rooted 'LePage E' quince cuttings. 'Santa Maria' shoots grafted at their bases with a ring of bark from 'LePage E' and subsequently used as cuttings rooted readily but only from the region of the grafted quince bark. 'LePage E' quince shoots ring-grafted with bark from 'Santa Maria' pear produced roots, but only from the quince tissues, above the inserted bark piece, when made into cuttings.

Adventitious root formation may be limited by certain inherent nontranslocatable factors already in the tissues. It is likely, however, that interactions between certain fixed or nonmobile factors located within the cells—perhaps certain enzymes—and easily conducted nutrients and endogenous rooting factors take place to establish conditions favoring root initiation.

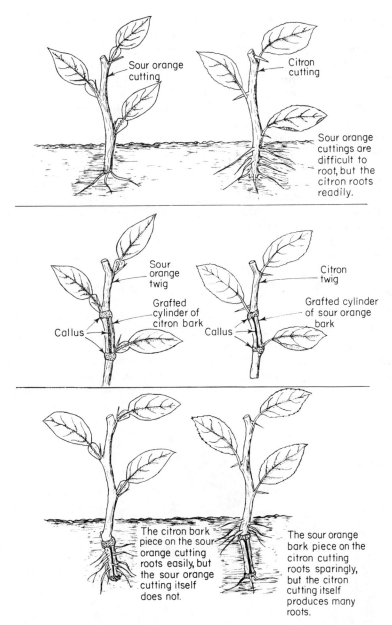

Figure 9–13   Anatomical conditions occurring in the tissues of some plants apparently are more conducive to adventitious root formation than are those found in other species. In this case, the citron roots easily but the sour orange with difficulty. Reciprocal transfers of grafted bark pieces showed that this same situation prevailed even when the bark piece was grown on the other species. Adapted from Mes *(141)*. Courtesy Brooklyn Botanical Garden.

POLARITY

The polarity inherent in shoots and roots is dramatically shown in rooting cuttings (Figures 9–14 and 9–15). Stem cuttings form shoots at the distal end (nearest the shoot tip), and roots at the proximal end (nearest the crown of the plant). Root cuttings form roots at the distal end and shoots at the proximal end. Changing the position of the cuttings with respect to gravity does not alter this tendency (*13*) (Figure 9–15).

In Vöchting's early studies (*217*) on the polarity of regeneration in plants, he pointed out that stem tissue is strongly polarized. The theory was then advanced that this property could be attributed to the individual cellular components, since no matter how small the piece, regeneration was consistently polar. When a root piece is cut into two segments, the two surfaces touching before the cut would be similar in all respects, yet upon regeneration of roots and shoots, one surface of the cut produces a shoot and the other produces roots. Vöchting also concluded that the intensity of the polarity effect varied considerably among the different plant organs. Stems showed strong regeneration polarity, with somewhat weaker polarity by roots and much weaker regeneration polarity by leaves. It is commonly observed in leaf cuttings that roots and shoots arise at the same position, usually the base of the cutting, showing that little, if any, polarity influence is present (see Figure 10–15).

When tissue segments are cut, the physiological unity is disturbed. This must cause a redistribution of some substance, probably auxin, thus accounting for the different responses observed at previously adjacent surfaces. The correlation of polarity of root differentiation with auxin movement has been noted in several instances (*178, 197, 219*). It is also known that the polarity in auxin transport varies in intensity among different tissues, being particularly

Figure 9–14   Results of planting a cutting of red currant *(Ribes sativum)* upside down (reversed polarity). *Left:* several months after starting. *Right:* one year later. The shoot from the center bud with new roots at its base has become the main plant and is growing with correct polarity. The shoot from the top bud, while still alive, has failed to develop normally.

Figure 9–15   Polarity of root regeneration in grape hardwood cuttings. Cuttings at left were placed for rooting in an inverted position, but roots still developed from the morphologically basal (proximal) end. Cuttings on right were placed for rooting in the normal, upright orientation with roots forming at the basal end.

weak in leaf petioles. The polar movement of auxins is an active transport process and apparently is a secretion activity, with its basis found within the structural features of the individual cells (*132*).

FACTORS AFFECTING THE REGENERATION OF PLANTS FROM CUTTINGS

Great differences exist among species and among cultivars in the rooting ability of cuttings taken from them. It is difficult to predict whether or not cuttings of a certain clone will root easily. Although botanical relationships give a general indication, empirical trials with each clone are necessary. This has already been done with most plants of economic importance. Stem cuttings of some cultivars root so readily that the most simple facilities and care give high rooting percentages. On the other hand, cuttings of many cultivars or species have yet to be rooted. Cuttings of other "difficult" cultivars can be rooted only if various influencing factors are taken into consideration and maintained at the optimum condition. The environmental factors to be discussed in this section are of great importance to this group, and the attention given to them makes the difference between success or failure in obtaining satisfactory rooting.

Selection of Cutting Material
Physiological Condition of the Stock Plant

There is considerable evidence that the nutrition of the stock plant exerts a strong influence on the development of roots and shoots from cuttings taken from such plants (*157, 161, 172*). Kraus and Kraybill (*120*) observed long ago in making tomato cuttings that those with yellowish stems, high in carbohydrates but low in nitrogen, produced many roots but only feeble shoots, whereas those with greenish stems, containing ample carbohydrates but higher in nitrogen, produced fewer roots but stronger shoots. Green, succulent stems, very low in carbohydrates but high in nitrogen, all decayed without producing either roots or shoots.

Many internal factors, such as auxin levels, rooting co-factors, and carbohydrate storage, can, of course, influence root initiation of cuttings. In a study (185) where all these factors were determined in easy and difficult-to-root cultivars of chrysanthemum, the only correlation that could be obtained was in the carbohydrate storage in the stems, the easily rooted cultivars having the highest storage levels. Softwood cuttings of the hop plant, when held under low light intensities—at about the compensation point—responded to sugar pre-treatments by large increases in rooting, illustrating the need for ample carbohydrates for root production (105). In other studies, either glucose (57, 150) or sucrose (63) in the medium was essential for root formation on stem segments placed for rooting under aseptic conditions.

Quite often the most suitable cutting material, as far as the carbohydrate content is concerned, can be determined by stem firmness. Those undesirably low in carbohydrates are soft and flexible, while those higher in carbohydrates are firm and stiff, and break with a snap rather than bending. This desirable condition may be confused, however, with firmness due to the maturing of the tissues, caused by thickening and lignification of the cell walls. A method of determining cutting material having a desirable high starch content is the iodine test. Freshly cut ends of a bundle of cuttings are immersed for one minute in a 0.2 percent solution of iodine in potassium iodide. Cuttings having the highest starch content stain the darkest color. A rough grouping of cuttings high, medium, and low in starch can then be made. In tests with grapes (231), 63 percent of the high-starch cuttings rooted, 35 percent of the medium-starch cuttings rooted, and 17 percent of the low-starch cuttings rooted.

Evidence concerning the effect of nitrogen levels in the stock plants on rooting behavior of cuttings taken from them is conflicting. Pearse, in rooting studies (157, 159) with grapes showed that when stock plants were grown under phosphorus, potassium, magnesium, or calcium deficiency, root formation in cuttings taken from them was poorer than from full-nutrient plants; but with reduced nitrogen in the stock plants, root formation by the cuttings was increased. However, extreme nitrogen deficiency of the stock plants lowered rather than increased rooting. This was substantiated by tests (82) with geranium in which stock plants were grown at three levels of nitrogen, phosphorus, and potassium. The nitrogen nutrition of the stock plants had a greater effect on rooting response of the cuttings than did phosphorus or potassium, with the low and medium levels of nitrogen resulting in higher percentages of rooted cuttings than the high level.

However, for root initiation to take place nitrogen is necessary for nucleic acid and protein synthesis, so there would be a cut-off level of nitrogen availability, below which root initiation would be impaired; in such cases added nitrogen would promote rooting.

Grape cuttings taken from vines fertilized with zinc rooted in higher percentages and were of better quality than cuttings taken from untreated vines (172). This could be due to an increase in native auxin production resulting from the increased level of tryptophan (an auxin precursor) found in the treated plants. (Zinc is required for tryptophan production.) In fact, applications of synthetic tryptophan have increased rooting of vine cuttings. Beneficial effects of zinc applications to stock plants have also been noted in South Africa in the propagation of 'Marianna' plum by hardwood cuttings.

It is not clear why high levels of nitrogen in cuttings are not conducive to good rooting, but it is probable that tissues with high nitrogen content are rank-growing, soft, and succulent, with low carbohydrate storage. Such rapidly growing shoots may also be low in other components necessary for rooting.

The low-nitrogen–high-carbohydrate balance in stock plants, which in many cases seems to favor rooting, can be achieved in several ways:

(a) Reduce the nitrogen supply to the stock plants, thus reducing shoot growth and allowing for carbohydrate accumulation. This can be done by withholding nitrogenous fertilizers and allowing the stock plants to grow in full sunlight. Any type of root restriction of the stock plants, such as occurs when they are grown in containers or close together in hedge rows, tends to reduce excessive vegetative growth and permits the accumulation of carbohydrates.

(b) Select portions of the plant for cutting material which are in the desirable nutritive stage. For example, take lateral shoots in which rapid growth has decreased and carbohydrates have accumulated, rather than succulent terminal shoots.

(c) Select regions of the shoot which are known to have a high carbohydrate content. In a chemical analysis (208) of rose shoots of the type used for making cuttings, the nitrogen content increased uniformly from the base of the shoot to the tip. Conversely, there was a gradient of decreasing starch from base to tip. Basal portions of such a shoot would therefore have the low-nitrogen–high-carbohydrate balance favorable for good rooting. It cannot be said, however, that a high carbohydrate content in cuttings is invariably associated with ease of rooting. There may be other, stronger influencing factors present.

For plants difficult to root, there are several treatments that can be used to alter the physiological and/or nutritional condition of the stock plant or portions of the plant. These treatments often result in increased rooting of cuttings taken from such plants, and include such practices as etiolation * of shoots, or wiring or girdling the shoots some time before they are made into cuttings.

Etiolation
The practice of etiolation has long been known to be remarkably effective in increasing adventitious root formation in stem tissue (53); this was shown

---

* Etiolation is the development of plants or plant parts in the absence of light. This results in such characteristics as small, unexpanded leaves, elongated shoots, and lack of chlorophyll with the development of a yellowish or whitish color.

experimentally by Sachs in 1864 (*168*). The procedures of trench and mound layering (see p. 464), in which shoot bases are kept in the dark, covered by soil, is based upon etiolation of stem tissue. Even rooting of stem cuttings where the bases are in the dark, within the rooting medium, is no doubt promoted by stimulation from etiolation effects.

In studies (*90*) with *Hibiscus* and Red Kidney bean cuttings etiolated stems had a reduction in starch content, mechanical strengthening tissues, cell-wall thickness, cell-wall deposits, and vascular tissues, in comparison with nonetiolated stems. Etiolated stems had more parenchyma cells and more tissue in a less differentiated condition. Etiolated bean tissue was found to react strongly to indolebutyric acid in rooting, implying the existence of some other materials in the tissues, acting synergistically as a rooting co-factor. Cuttings under etiolation were found to have a higher level of en-dogenous auxin (IAA) at the etiolation site during the period of root ini-tiation. (*112*).

Adult ivy (*Hedera helix*) stem tissue is known to be difficult to root. In a study (*69*) in rooting adult shoot apices under aseptic conditions, excellent rooting was obtained under low light intensities or darkness, especially with IAA plus catechol treatments. These same treatments, but under high light intensity, gave little rooting. Other studies (*150*) showed root formation in stem segments of two rhododendron cultivars maintained under aseptic conditions to be completely inhibited by continuous light.

A significant study (*123*) with bean hypocotyls implicates the destruction of root-promoting factors by light. DNP (2,4-dinitrophenol) (known to be an uncoupler of oxidative phosphorylation), applied to the tissues, which were then maintained in darkness by covering with opaque rubber tubing, caused root primordia to form along the stem, increasing in amount with concentration. No roots formed on treated but uncovered areas or on those treated but covered by transparent tubing. Adding an auxin (IAA) greatly stimulated the DNP effect. There are theories (*49, 93*) (see p. 226) that phe-nolic compounds, interacting with auxin, promote adventitious root forma-tion and, too, certain auxin effects have been shown (*204*) to be enhanced by phenolic compounds. In the presence of ascorbate and chlorophyll a (natu-rally occurring materials)—and in the presence of light—DNP is reduced to 2-amino, 4-nitrophenol (*140*), a compound which is inactive in causing roots to form on bean hypocotyls (*123*) either in the light or dark. Thus, the promotive influence of the absence of light on tissues during root initiation (the etiolation effect) may possibly be explained by photoinactivation of an essential component in a complex required for root initials to form.

In stimulating root formation by etiolation (*56*), shoots which later are to be made into cuttings may be allowed to develop during their initial stages in complete darkness, thus assuming the elongated, whitish charac-teristics of etiolation. The basal portion of the shoots, where the root initials develop, can be retained under this etiolated condition by being covered with black adhesive tape; the terminal leafy portion of the shoots is allowed to develop further in the light, thus assuming a normal type of growth. After the shoots attain sufficient length, they can be cut off, either immediately or at a later time, and placed in a rooting bed.

The preformed root primordia found in the stems of Lombardy poplar (*Populus nigra,* var. *italica*) develop and emerge if the cuttings are placed in darkness, but they fail to emerge if the cuttings are exposed to light each day. Red light (about 6800 Angstrom units) is more inhibitory than blue, green, or far-red light (*175*).

Girdling. Since a high carbohydrate level in shoots is conducive to root formation, treatments that block the downward movement of carbohydrates —and other root-promoting factors—such as girdling or constriction of the phloem with wire, should increase root initiation. There are reports of success with such treatments. For example, rooting of citrus and hibiscus cuttings was stimulated by girdling or binding the base of the shoots with wire several weeks before taking the cuttings (*109, 184*). Girdling was found (*183*) to cause substantial increase in a rooting co-factor above the girdle in an easily rooted hibiscus clone.

Juvenility Factor or Phase Change (Age of the Stock Plant)

In plants difficult to root, the age of the stock plants can be a very important factor. (See p. 184.) Almost always, either stem or root cuttings taken from young seedling plants (in the juvenile growth phase) will root much more readily than those taken from older plants (in the adult growth phase) (*55, 99, 173*). (See Figure 1–1). Experiments with apple, pear, Eucalyptus, Douglas fir, and other species have shown that the ability of cuttings to form adventitious roots decreases with increasing age of the plants from seed. In one study (*55*) cuttings made from one-year-old seedlings rooted readily, but if the seedlings were only two years old, rooting decreased markedly.

In a study (*200*) of rooting cuttings of certain coniferous and deciduous species known to root only with extreme difficulty, it was concluded that the most important single factor affecting root initiation was the age of the tree from which the cuttings were taken. Some old English garden books list great numbers of plants that were rooted from cuttings—plants now considered very difficult, if not impossible, to propagate by cuttings. It is likely that in earlier days, such species were imported from distant countries as seeds; cuttings were subsequently made from young seedlings, the juvenility factor accounting for their ease in rooting.

In species of some trees, such as oak, spruce, and beech, leaf retention late into the fall occurs on the basal parts of the tree and indicates the part still in the juvenile stage. Cuttings should be taken from this type of wood (*68*).

Any treatment which maintains the juvenile growth phase would be of value in preventing the decline in rooting potential as the stock plant ages. The hedging or shearing treatments given *Pinus radiata* trees (Figure 9–16) was quite effective in maintaining the rooting potential of cuttings taken from them as the trees aged, compared to nonhedged trees (*135*). Such benefits in maintaining rooting potential by hedging may possibly be explained by the prevention of the normal phase change from the juvenile to the adult form. In England it has been a practice for many years to use as stock plants for hardwood cuttings of fruit-tree rootstocks, special plants

Figure 9–16   Stock plants for *Pinus radiata* cuttings maintained in a hedge form by shearing (compare with unsheared trees in background). Such "hedged" trees can yield high numbers of cuttings which maintain the high rooting percentages, root quality, and growth potential normally associated with the younger stock trees *(135)*. Photo courtesy W. J. Libby.

which are maintained as hedges rather than allowed to grow to a tree form *(58)*.

Juvenility in relation to rooting may possibly be explained by the increasing production of rooting inhibitors as the plant grows older. Stem cuttings taken from young seedlings of a number of eucalyptus species root easily, but as the stock plants become older rooting decreases dramatically; studies *(154)* in Australia showed that there was a direct and quantitative association between such decreased rooting and the production of a rooting inhibitor in the tissues at the base of the cuttings. In easily rooted young seedling stems this inhibitor was absent, as it was absent in adult stem tissue of the easily rooted *Eucalyptus deglupta*.

Reduced rooting potential as plants age may possibly be a result of lowering phenolic levels. Phenols are postulated (see p. 226) as acting as auxin co-factors or synergists in root initiation. In certain plants, lower phenolic levels were noted in mature forms than in the juvenile forms *(62)*.

In rooting cuttings of difficult species it would be useful to be able to induce rejuvenation of the easily rooted juvenile stage from plants in the adult form. This has been done in several instances by the following methods:

Juvenile forms of apple can be obtained from mature trees by causing adventitious shoots to develop from root pieces, which are then made into softwood stem cuttings and rooted *(186)*.

In some plants juvenile wood from mature plants can be obtained by forcing juvenile growth from sphaeroblasts (wart-like protuberances containing meristematic and conductive tissues sometimes found on trunks or branches).

These are induced to develop by disbudding and heavily cutting back stock plants. These juvenile shoots then may be easily rooted under the usual conditions *(224)*. By using the mound layering (stooling) method (see

p. 464) on these rooted sphaeroblast cuttings, rooted shoots are produced which continue to possess the juvenile characteristics. In the stooling method of propagation the juvenile phase may be continually maintained.

Grafting adult forms of ivy onto juvenile forms have induced a change of the adult to the juvenile stage, provided the plants are held at fairly high temperatures *(42, 192)*; such transmission of the juvenile rooting ability from seedlings to adult forms by grafting has also been accomplished in rubber trees *(Hevea brasiliensis) (148)*.

Gibberellin sprays on ivy *(Hedera)* plants in the adult growth phase caused substantial stimulation of growth and reversion of some of the branches to the juvenile stage *(166)*, as well as improved rooting of cuttings taken from the sprayed plants *(192)*.

It is important to remember that the juvenile condition is found in such tissues as those originating from young seedlings, those arising from adventitious (not latent) buds on stems, or those caused to revert to juvenility either by gibberellin treatments or by grafting to juvenile wood. Tissues from young plants which have been propagated from material taken from plants in the adult phase are not in a truly juvenile state.

Type of Wood Selected for Cuttings

There are many choices of the type of material to use, ranging (in woody perennials) from the very succulent terminal shoots of current growth to large hardwood cuttings several years old. Here, as with most of the other factors affecting rooting of cuttings, it is impossible to state any one type of cutting material that would be best for all plants. What may be ideal for one plant would be a failure for another. What has been found to hold true for certain species, however, often may be extended to other related species.

(a) Differences between individual seedling plants   In rooting cuttings taken from individual plants of a species which ordinarily is propagated by seed, experience has shown that wide differences may exist among individuals in the ease with which cuttings taken from them form roots. Just as seedlings differ in many respects, it is not surprising that this difference in root-forming ability should also exist. Differences in the rooting ability of clones are, of course, recognized. Likewise, such differences should be anticipated when woody plants usually propagated by seed—as most forest tree species—are propagated by cuttings. In rooting cuttings taken from old seedling trees of Norway spruce *(Picea abies)*, white pine *(Pinus strobus)*, and red maple *(Acer rubrum)* marked differences occurred in the rooting capacity of shoots taken from individual trees *(39, 179)*.

(b) Differences between lateral and terminal shoots   Experiments in rooting plum stem cuttings compared the rooting of different types of softwood cuttings taken in the spring. The results showed a marked superiority in rooting of lateral shoots; the terminals had only 10 percent rooting; laterals in active growth, 19 percent; and laterals which had ceased active growth, 35 percent.

Similarly, lateral branches of white pine and Norway spruce, both with and without auxin treatments, gave consistently higher percentages of rooted cuttings than did terminal shoots *(39, 51)*. In the rhododendron, too, thin cuttings made from lateral shoots consistently give higher rooting percentages than those taken from vigorous, strong terminal shoots.

Plants which exhibit marked horizontal branching will produce very diverse plants from rooted cuttings, depending upon whether vertical or horizontal cutting material is used (see p. 185). For example, to obtain the desirable upright type of plant of *Taxus cuspidata* 'capitata,' cuttings should be taken only from erect terminals or upright-growing laterals, rather than from horizontal side shoots. Another example is Norway spruce cuttings. New growth from lateral shoots after rooting is at an angle, while that from terminal shoots is erect. This phenomenon is also noticeable in rooting coffee cuttings (*Coffea arabica*). If cuttings are made from upright-growing shoots, they produce the desired vertical plants, but if taken from drooping lateral branches, the cuttings, after rooting, produce only branches spreading along the ground.

(c) Differences between different parts of the shoot   In some woody plants, hardwood cuttings are made by sectioning shoots several feet long and obtaining four to eight cuttings from a single shoot. Marked differences in the chemical composition of such shoots are known to exist from base to tip. Variations in root production on cuttings taken from different portions of the shoot are often observed, with the highest rooting, in many cases, found in cuttings taken from the basal portions of the shoot. In propagation of the olive (*Olea europaea*) from one-year-old leafy stem cuttings, woody basal portions of the shoot root more readily than soft terminal sections *(139)*. In the same manner, cuttings prepared from shoots of three cultivars of the highbush blueberry (*Vaccinium corymbosum*) were significantly more successful if taken from the basal portions of the shoot rather than from terminal portions *(152)*. The number of preformed root initials in woody stems has been determined (in some plants at least) as distinctly decreasing from the base to the tip of the shoot *(130)*. Consequently the rooting capacity of basal portions of such shoots would be considerably higher than that of the apical parts.

In studies with a different type of wood, however, cherries (*Prunus cerasus, P. avium, P. mahaleb*) rooted under mist by softwood cuttings prepared from succulent new growth gave the following percentages of rooted cuttings: 'Stockton Morello,' basal—30 percent, tip—77 percent; 'Bing,' basal —0 percent, tip—100 percent; and 'Montmorency,' basal—10 percent, tip—90 percent *(76)*.

It may well be that in woody stems, a year or more in age, where carbohydrates have accumulated at the base of the shoots and, perhaps, where some root initials have formed in the basal portions, possibly under the influence of root-promoting substances from buds and leaves, the best cutting material is the basal portions of such shoots. An entirely different physiological situation exists in the succulent shoots of deciduous plants which are used for softwood cuttings. Here carbohydrate storage and preformed root initials are not present. The better rooting of shoot tips may be explained by the possibility of higher concentrations of endogenous root-promoting substances arising in the terminal bud. There is also less differentiation in the terminal cuttings, with more cells capable of becoming meristematic.

This factor is of less importance, however, in easily rooted species where entirely satisfactory rooting is obtained regardless of the position of the cutting on the shoot.

(d) Flowering or vegetative wood   In most plants, cuttings could be made from shoots that are in either a flowering or a vegetative condition. Again, with easily rooted species it makes little difference which is used, but in difficult-to-root species this can be an important factor. For example, in the blueberry (*Vaccinium atrococcum*), hardwood cuttings from shoots bearing flower buds did not root nearly so well as those bearing only leaf buds *(151)*. When vegetative wood was used, 39 percent of the cuttings rooted, but when cuttings contained one or more flower buds, not one rooted. Removing the flower buds prior to rooting did not increase the rooting percentage, indicating that it is not the actual presence of flower buds that inhibits rooting, but rather some previous physiological or anatomical condition associated with the presence of flower buds.

In the rhododendron, early removal of potential flower buds increased rooting, presumably due to the elimination of a flower-promoting stimulus antagonistic to rooting; later flower bud removals still enhanced rooting, possibly by eliminating competition for materials necessary for rooting *(110)*.

Some antagonism apparently exists between vegetative regeneration and flowering. The basis for this is probably found in auxin relationships, since it is known that high auxin levels, which are favorable for adventitious root formation in stem cuttings, tend to inhibit flower initiation *(205)*.

It has been consistently noted for both root and stem cuttings of many species that better regeneration takes place when cuttings are taken either before or after, rather than during, the flowering period *(43)*. In some instances cuttings taken any time while the stock plants were in the vegetative state rooted very well, but as soon as the stock plants started initiating flower buds, the cuttings failed to root.

Considerable information has been obtained concerning the effect of flowering on adventitious root formation in studies *(171)* in which cuttings of the tomato (a day-neutral plant), *Perilla crispa* (a short-day plant), and red clover (a long-day plant) were used. In the tomato, in which it is impossible to prevent flower formation by either long- or short-day treatments, the onset of flowering had no effect on the rooting of cuttings. In *Perilla,* however, not only flower initiation but also flower development inhibited rooting. In red clover, too, flower initiation inhibited root formation. It was concluded that adventitious root formation and flowering are antagonistic systems in both short-day and long-day plants, owing to the distribution of auxin during flower initiation and development. This may possibly hold true for day-neutral plants with a terminal flower bud but not for day-neutral plants with a terminal vegetative bud.

In order to have the highest regenerative capacity, stock plants should be showing active vegetative growth (and not be entering the flowering stage). In England *(58)* the usual sources for hardwood cuttings of fruit tree rootstocks are hedges which are severely pruned in winter in collecting the cutting material, thus maintaining a state of vigorous growth for the following year.

(e) "Heel" vs. "non-heel" cuttings   In preparing cuttings, it is often recommended that a "heel" (a small slice of older wood) be retained at the base of the cutting in order to obtain maximum rooting. For hardwood cuttings of some plants, this may be true. In quince (*Cydonia oblonga*), considerably better rooting was obtained with the heel type of cutting, probably owing in this case to the presence of preformed root initials in the older wood. Narrow-leaved evergreen cuttings often, but not always, root more readily if a heel of old wood is retained at the base of the cuttings.

Presence of Virus Diseases
Since procedures have been developed for eliminating viruses from clones by heat treatment (see p. 201), it has been possible to show the depressing effects of viruses on adventitious root initiation. This was found to be true for both apple (*106*) and gooseberry (*211*) cuttings, where those from virus-infected clones rooted considerably more poorly than those from "clean" stock. The presence of viruses not only reduces rooting percentages but numbers of roots forming on the cuttings. Poor results often obtained in rooting cuttings could thus be traced to the use of virus-infected cutting material and may account for the variable results often obtained in different rooting tests with the same cultivar.

Time of Year in Which the Cuttings Are Taken
This can have, in some instances, dramatic influences on the results obtained in rooting cuttings, and may provide the key to highly successful rooting. It is possible, of course, to make cuttings at any time during the year. In propagating deciduous species, hardwood cuttings could be taken during the dormant season, or leafy softwood or semi-hardwood cuttings could be prepared during the growing season, using succulent or partially matured wood. The narrow- and broad-leaved evergreen species have one or more flushes of growth during the year, and cuttings could be obtained at various times in relation to these flushes of growth.

Certain species, such as the privets, can be rooted readily if cuttings are taken almost any time during the year; on the other hand, for example, excellent rooting of leafy olive cuttings under mist can be obtained during late spring and summer, whereas rooting drops almost to zero with similar cuttings taken in midwinter (*79*). Softwood cuttings of deciduous woody species taken during spring or summer usually tend to root more readily than hardwood cuttings procured in the winter. For plants difficult to root, it is thus often necessary to resort to the use of softwood cuttings. In rooting rhododendron stem segments under controlled aseptic conditions, rooting occurred only with segments cut out of young, soft shoots (*150*). In tests (*76*) with cherries, not one hardwood cutting taken in winter was induced to root, whereas softwood cuttings made in the spring gave satisfactory rooting with most cultivars. This situation is also well illustrated in the lilacs (*Syringa sp.*). About the only way cuttings of these plants can be rooted is to make softwood cuttings during a short period in the spring when the shoots are several inches long and in active growth. The Chinese fringe tree (*Chionanthus retusus*) is notoriously difficult to root, but by taking cuttings during a short period in mid-spring, high rooting percentages can be obtained (*187*).

The effect of timing is also strikingly shown by difficult-to-root deciduous azalea cuttings. These root readily if the cuttings are taken from succulent growth in early spring; by late spring, however, the rooting percentages rapidly decline (*119*). For any given plant, empirical tests are required to determine the optimum time of taking cuttings, which is more related to the physiological condition of the plant than to any given calendar date.

It is possible for hardwood cuttings of deciduous species to be taken at any time from just before leaf fall until the buds start development in the spring; for easily rooted species, it makes little difference when the cuttings are made during the dormant season. Rapidly developing buds sometimes tend to promote root formation, whereas buds in the "rest period" may inhibit root development (*131*). Often the effects of timing are merely a reflection of the response of the cuttings to the existing environmental conditions at the different times of the year. When hardwood cuttings of deciduous species are taken and planted in the nursery in early spring after the rest period of the buds has been broken by the winter chilling, the results are quite often a complete failure, since the buds quickly open with the onset of warm days. The newly developing leaf area will start transpiration and remove the moisture from the cuttings before they have the opportunity of forming roots; thus they soon die. If cuttings can be taken and planted in the fall while the buds are still in the rest period, roots may form and be well established by the time the buds open in the spring. A preplanting moist, warm (60° to 70° F; 15° to 21° C) storage period is often helpful in starting adventitious root initiation (*77, 78, 80, 81, 104*). (See p. 277.)

For softwood cuttings of deciduous species, best results are generally secured if the cuttings are taken in the spring just as soon as the leaves are fully expanded and the shoots have attained some degree of maturity. Good results are sometimes obtained when softwood cuttings are taken from potted plants that have started growth early in the spring after being moved into a greenhouse.

Broad-leaved evergreens usually root most readily if the cuttings are taken after a flush of growth has been completed and the wood is partially matured. This occurs, depending upon the species, from spring to late fall.

In rooting cuttings of narrow-leaved evergreens, best results may be expected if the cuttings are taken during the period from late fall to late winter. For some species (Figure 9–17), day length—changing during the seasons—apparently is not a factor accounting for this striking seasonal variability in rooting; juniper cuttings held under long and short days and under

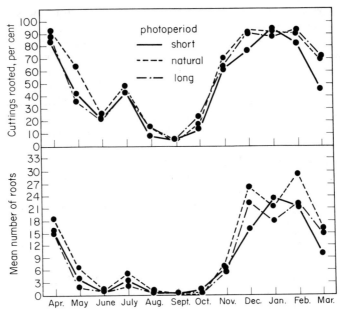

Figure 9–17   Rooting of 'Andorra' juniper cuttings taken at monthly intervals throughout the year at Lafayette, Indiana. Note that the photoperiod under which the cuttings were rooted had very little effect. Courtesy Lanphear and Meahl *(126)*.

natural day lengths rooted with equal ease or difficulty. Rooting was lowest during the season of active vegetative growth and highest during the dormant period. Furthermore, the low temperatures occurring at the time when such coniferous evergreens root best apparently is not a requirement; *Juniperus* stock plants held in a warm greenhouse from early fall to midwinter produced cuttings which rooted better than those taken from out-of-door stock plants *(127)*.

The season of the year in which root cuttings are taken is very important with some species; in others it makes no difference. For example, in the red raspberry (*Rubus idaeus*), regeneration from roots taken from autumn to spring was almost 100 percent successful. Cuttings taken during the summer months failed to survive. On the other hand, root cuttings of horseradish or dandelion were completely successful regardless of the time of year in which they were taken *(107)*.

Treatment of Cuttings
Growth Regulators and Other Materials
Before the use of synthetic root-promoting growth regulators (auxins) in rooting stem cuttings, many other chemicals were tried with varying degrees of success. In some of the earlier studies *(36)* of this type, treatments of tomato and privet cuttings with sugar as well as with compounds of manganese, iron, and phosphorus were tried. Improved rooting often resulted,

especially with potassium permanganate. The effect of such compounds, however, was so slight and variable that they are no longer used.

It was shown in 1933 by Zimmerman (*239*), before auxins were discovered, that certain unsaturated gases, such as ethylene, carbon monoxide and acetylene, stimulate initiation of adventitious roots as well as development of latent, preexisting root initials. Cuttings of many herbaceous plants respond to these gases with increased rooting.

The discovery in 1934 and 1935 that auxins, such as indoleacetic acid, were of real value in stimulating the production of adventitious roots in stem and leaf cuttings was a major milestone in propagation history (see p. 219). The response, however, is not universal; cuttings of some species which are difficult to root will still root poorly after treatment with auxin. It is believed that certain naturally occurring materials (rooting co-factors; see p. 223) are limiting in such cases.

A practice of some European gardeners in early days was to embed grain seeds into the split ends of cuttings to promote rooting. This seemingly odd procedure had a sound physiological basis, for it is now known that germinating seeds are good producers of auxin, which, of course, would materially aid in root formation in the cuttings.

Some of the phenoxy compounds having auxin activity promote root formation even at very low concentrations (*100, 102, 191*). The weed-killer, *2,4-dichlorophenoxyacetic acid* (2,4-D), is quite potent in inducing rooting of certain species, but it has the disadvantage of tending to inhibit shoot development.

Mixtures of root-promoting substances are sometimes more effective than either component alone. For example, equal parts of indolebutyric acid and naphthaleneacetic acid, when used on a number of widely diverse species, were found to induce a higher percentage of cuttings to root and more roots per cutting than either material alone (*101*).

Adding a small percentage of certain of the phenoxy compounds to either indolebutyric acid or naphthaleneacetic acid has caused excellent rooting in some species (*102*) and has produced root systems qualitatively better than those obtained with phenoxy compounds alone.

The use of the salts of some of the growth regulators rather than the acid may be desirable in some instances, owing to their comparable activity and greater solubility in water (*240*). The acid form of most of these compounds is relatively insoluble in water, but can be used by dissolving in a few drops of alcohol or ammonium hydroxide before adding to the water.

For general use in rooting stem cuttings of the majority of plant species, naphthaleneacetic acid (NAA), and indolebutyric acid (IBA), particularly the latter, are recommended (Figure 9–18). To determine the best material and optimum concentration for rooting any particular species under a given set of conditions, empirical trials are necessary (see p. 291).

Application of synthetic auxins to stem cuttings at high concentrations can inhibit bud development, sometimes to the point at which no shoot growth will take place even though root formation has been adequate. Also, applications of rooting substances to root cuttings may inhibit the development of shoots from such root pieces.

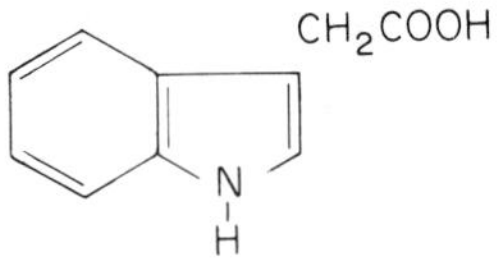

CH_2COOH

3—Indoleacetic acid
beta—Indoleacetic acid
(IAA)

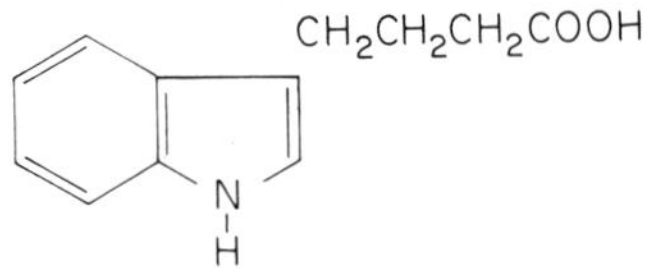

CH_2CH_2CH_2COOH

3—Indolebutyric acid
gamma—(Indole—3)—butyric acid
(IBA)

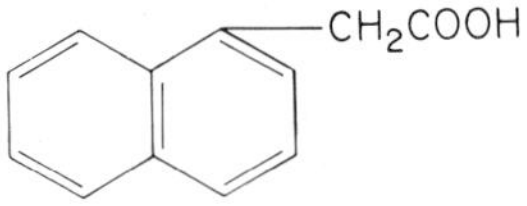

CH_2COOH

1—Naphthaleneacetic acid
alpha—naphthaleneacetic acid
(NAA)

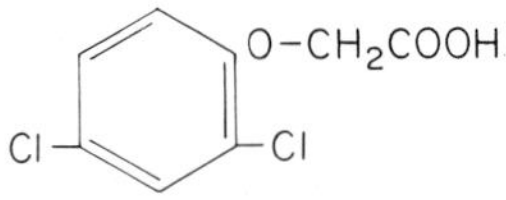

O—CH_2COOH

Cl   Cl

2,4—Dichlorophenoxyacetic acid
(2,4—D)

Figure 9–18   Structural formulas of some growth regulators (auxins) active in promoting adventitious root development in cuttings.

There is often a question of how long the various root-promoting preparations will keep without losing their strength. Bacterial destruction of indoleacetic acid occurs readily in unsterilized solutions. A 9-ppm concentration was found to disappear in 24 hours and a 100-ppm concentration in 14 days. In sterile solutions, this material remains active for several months. A widely distributed species of *Acetobacter* destroys IAA, but the same organism has no effect on indolebutyric acid. Uncontaminated solutions of naphthaleneacetic acid and 2,4-dichlorophenoxyacetic acid maintained their strength for as long as a year. Indoleacetic acid is sensitive to light, strong sunlight destroying a 10-ppm concentration in about 15 minutes. IBA is much more light-stable than IAA, a 20-hour exposure to strong sunlight causing only a slight change in concentration. Both NAA and 2,4-D seem to be entirely light-stable. With their resistance to bacterial decomposition and light destruction, the latter two compounds are more likely to maintain their effectiveness over a long period of time than the indole compounds. When dilute indoleacetic acid solutions are prepared, they should be used immediately, owing to their rapid deterioration (*142*). In the plant, too, the enzyme system, indoleacetic acid oxidase, will break down IAA but it has no effect on indolebutyric acid.

The natural auxin flow in stem tissue is in a basipetal direction (apex to base). In early work, synthetic auxin applications were made to cuttings at the upper end to conform to the natural downward flow. As a practical matter, however, it was soon found that basal applications gave better results. Sufficient movement apparently resulted to carry the applied auxin into the parts of the cutting where it stimulated root production. In tests using radioactive indoleacetic acid (4000 ppm for 5 seconds) for rooting leafy plum cuttings, it was noted (*193*) that IAA was absorbed and distributed throughout leafy cuttings in 24 hours, whether application was at the apex or base. However, with basal application, most of the radioactivity remained in the basal portion of the cuttings, this being the case from 24 hours after treatment, until rooting occurred. Leafless cuttings absorbed the same amount of IAA as leafy cuttings, indicating that transpiration "pull" was not the chief cause of absorption and translocation. Twenty-eight days after the cuttings were made and placed under mist there was still considerable radioactivity throughout the cuttings and in the adventitious roots which subsequently developed. Studies (*210*) in the U.S.S.R. with radioactive NAA showed that radioactivity was found in all stem tissues of black currant cuttings directly after treatment and persisted there up to complete rooting. In cherry cuttings, which root with more difficulty, absorption and redistribution of NAA were less.

In respiration studies of tissues at the basal ends of IBA-treated and of control cuttings it was found that by the time roots had formed on treated cuttings, their respiration rate was four times as great as in untreated cuttings. In addition, IBA-treated cuttings had a considerably higher level of amino acids at their bases 48 hours after treatment than untreated cuttings. This pattern continued, with nitrogenous substances accumulating in the basal part of treated cuttings, apparently mobilized in the upper part and translocated as asparagine (*194*).

Tests have been conducted by research workers all over the world, using most of the species of plants whose propagation is of interest, to determine the value of the various synthetic root-promoting substances in inducing roots on cuttings. Several compilations (*6, 158, 199*) listing the results of this vast amount of work are available.

(*For application methods of root-promoting growth regulators, see p. 291.*)

Treatments with vitamins   It is known (*15*) that vitamin B_1 (thiamin chloride) is necessary for growth in sterile media of detached roots of many species. In intact plants, however, this material is produced in the leaves, and transported to the roots, where it enters into growth processes. Although it is necessary for root growth, apparently in most plants an ample supply is already stored in the cutting or can be manufactured by the leaves. In one study (*9*) where the vitamins—thiamine, pyroxidine, riboflavin, nicotinic acid, and ascorbic acid—were used either alone or in conjunction with the auxins, indoleacetic or indolebutyric acids, no benefit in rooting was obtained from the added vitamins. However, used with the auxin, naphthalene-acetic acid, nicotinic acid (or amide) at ½, 1, or 2 ppm significantly promoted rooting of *Justicia* cuttings.

Treatments with mineral nutrients   The rooting of cuttings has been distinctly promoted by the addition of *nitrogenous* compounds in a number of different plants. The addition of several nitrogen compounds, both organic and inorganic, was found (*40*) to have a beneficial effect on the rooting response of rhododendron cuttings. Also, in experiments (*202*) in rooting bean leaf cuttings, treatments with two organic forms of nitrogen—asparagine and adenine—were found to be very effective. Leafless hibiscus cuttings, when treated wih arginine or ammonium sulfate in combination with sucrose, were stimulated markedly in initiating roots, provided that they were also treated with auxin (*215*). The fact that concentrations as low as 0.05 ppm are sometimes quite effective supports the hypothesis that the role of these nitrogenous materials may be in hormonal interactions.

Boron stimulates root production in cuttings, at least for some plants, owing, generally, to its promotion of root growth rather than to an effect on root initiation. Experiments (*83*) in rooting cuttings prepared from bean hypocotyls showed that when cuttings were placed in nutrient solutions completely lacking in boron, visible roots failed to appear, although in complete nutrient solutions or in solutions lacking in certain other trace elements, adequate rooting took place. The use of boron—in combination with indolebutyric acid—increased the rooting percentage, the number and length of roots, and the speed of rooting of English holly cuttings taken in the fall. Apparently a synergistic reaction with IBA occurred, since boron alone had no effect (*221*). While the mechanism of the action of boron in stimulating rooting is not precisely known, there is evidence (*222*) to support the belief that boron is acting in oxidative processes, possibly increasing mobilization of oxygen-rich citric and iso-citric acids into the rooting tissues.

Treatments with fungicides   During the rooting period cuttings are subject to attacks by various fungi. Treatments with fungicides should give some protection, and should result in both better survival and improved root quality. This has proved to be the case in a number of instances. There is a question whether improvement is due to protection from fungus attacks or whether there was a direct stimulus from the fungicide in root initiation or growth, or both. However, studies (*75*) in which *Prunus* hardwood cuttings were treated under sterile conditions, where no fungus was involved, with (*1*) a root-promoter, indolebutyric acid alone, (*2*) captan—a fungicide— alone, and (*3*) a combination of both, indicated, as shown in Figure 9–19, that the captan does not show any root-initiating or stimulating properties. It probably acts by protecting the newly formed roots from fungal attacks and giving increased survival. Other reports (*212, 225*) indicate considerable improvement in cutting survival and in the quality of the rooted cuttings resulting from the use of captan (N-trichloromethyl mercapto-4-cyclohexene-1,2-dicarboximide). This may be used as a powder dip following an IBA treatment, or the IBA in talc may be mixed with the captan powder. Captan is especially suitable for treating cuttings, since it does not decompose easily and has a long residual action.

Benomyl [1-(butylcarbamoyl)-2-benzimidazole carbamic acid, methyl ester] is a very effective systemic fungicide, controlling many fungi on a wide range of host plants and its use as a pre-planting soak for cuttings has been shown to promote cutting survival (*52*). (See p. 295.) This was also true in tests (*196*) with cuttings of *Pinus strobus,* where treatments with a mixture of benomyl (5 percent) and captan (25 percent) in talc gave maximum rooting. In this

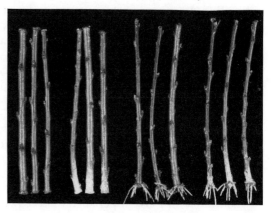

Figure 9–19 Typical peach hardwood cuttings placed for rooting in sterile peat moss. Treatments, left to right: control; captan alone, 25 percent; indolebutyric acid alone, 4000 ppm, 5 sec dip; IBA, 4000 ppm, 5 sec dip plus 25 percent captan. In this case the captan was not responsible for stimulation of root initiation.

trial adding IBA did not increase rooting. Bases of untreated cuttings had decayed in 60 days, while benomyl-treated cuttings were still green after 160 days. It was noted (*103*), too, that untreated rhododendron cuttings became heavily infected by the fungus, *Cylindrocladium,* and did not root, while treatments with benomyl inhibited development of the fungus and greatly improved rooting.

Wounding

Basal wounding, as described on p. 290, is beneficial in rooting cuttings of certain species, as rhododendrons and junipers, especially cuttings with older wood at the base. Following wounding, callus production and root development frequently are heavier along the margins of the wound. Evidently, wounded tissues are stimulated into cell division and production of root primordia. This is due, perhaps, to a natural accumulation of auxins and carbohydrates in the wounded area and to an increase in the respiration rate. In addition, injured tissues from wounding would be stimulated to produce ethylene, which is known to promote adventitious root formation (*122, 239*).

Wounded cuttings probably absorb more water from the medium than unwounded, and wounding would also permit greater absorption of applied growth regulators by the tissues at the base of the cuttings. In stem tissue of some species there is a sclerenchymatic ring of tough fiber cells in the cortex external to the point of origin of adventitious roots (see Figure 9–12). There is some evidence (*10, 26*) that roots have difficulty penetrating this band of cells. A shallow wound would cut through these cells and perhaps permit outward penetration of the developing roots more readily.

Environmental Conditions During Rooting
Water Relations

Although the presence of leaves on cuttings is a strong stimulus to root initiation, loss of water from the leaves may reduce the water content of the cuttings to such a low level as to cause them to die before root formation can take place. In cuttings, the natural water supply to the leaf from the roots has been cut off, yet the leaf still carries on transpiration. In cuttings of species which root rapidly, quick root formation soon permits water uptake to compensate for that removed by the leaves, but in more slowly rooting species, water loss from the leaves must be reduced to a very low rate to keep the cuttings alive until roots form. To reduce transpiration of the leaves on cuttings to a minimum, the vapor pressure of the water in the atmosphere surrounding the leaves should be maintained nearly equal to the water vapor pressure in the intercellular spaces within the leaf.

Traditionally it has been a standard practice in propagating frames and greenhouses to sprinkle the cuttings frequently, as well as the walls and floor, so as to maintain a high humidity. Automatically operated devices which disperse a fog-like mist are sometimes used in greenhouses. These methods of humidification give a beneficial effect primarily in increasing the amount of water vapor in the air.

Mist propagation   A major advance in propagation of plants, starting about 1940, occurred with the development of techniques for rooting leafy cuttings under *mist* (*54, 153, 163, 190*). Such sprays maintain a film of water on the leaves, which not only results in a high relative humidity surrounding the leaf, but also lowers the air and leaf temperature—all factors tending to lower the transpiration rate. In tests in which leaf temperatures were recorded by thermocouples, leaves under mist were found to be 10° to 15° F (5.5° to 8.5° C) cooler than leaves not under mist (*125*). In other comparisons, the air temperature in a mist bed in the greenhouse was very uniform, averaging about 70° F (21° C), whereas in an adjacent polyethylene-covered closed propagating frame, wide temperature fluctuations occurred, reaching almost 90° F (32.2° C) during the hottest part of the day. Cooling from water sprays is so effective that propagating beds can be placed in full sun without an appreciable temperature increase of the leaves. The high light intensity obtained by cuttings in the sun increases the photosynthetic activity of the leaves in comparison with that occurring in shaded propagating beds.

A distinction should be made between *humidification* and *mist*. Under systems which only increase the relative humidity, the water vapor pressure in the area around the leaves is increased. Under mist systems this also occurs, but the leaf itself is covered with a film of water, which has the additional benefit of reducing leaf temperature, due to cooling by the water and to evaporation of water from the leaf surfaces, which absorbs heat; this cooling reduces the internal water vapor pressure in the leaf and consequently the transpiration rate.

Under mist, conditions are ideal for rooting leafy cuttings. Transpiration is reduced to a low level, but the light intensity can be high, thus promoting full photosynthetic activity; the temperature of the entire cutting is relatively low, thereby reducing the respiration rate. On the other hand, in the closed propagating bed, temperatures tend to build up, and some ventilation and shading are necessary. Otherwise the cuttings would "burn up." Under these conditions of higher temperature, the transpiration rate increases. The low light intensity resulting from shading reduces photosynthesis, whereas

Table 9–1 Some comparisons between *Cornus florida* (dogwood) cuttings rooted under mist and under glass.*

	Mist	Glass frame
Leaf temperature	75° F (24° C)	86° F (30° C)
Water vapor pressure within leaf (mm mercury)	23.76	31.82
Water vapor pressure of surrounding air (mm mercury)	17.50	17.50
Vapor pressure differential	6.26	14.32
Light intensity (foot-candles)	7000	240
Increase in photosynthetic rate under mist (mg $CO_2/hr/cm^2$)	6.93	—
Increase in respiration rate over the photosynthetic rate (mg $CO_2/hr/cm^2$)	—	5.26
Increase in carbohydrates (mg per cutting)	138.0	17.2
Percentage of cuttings rooted	96	22

* From data of Hess and Snyder *(96)*.

the higher temperature increases respiration throughout the cutting. It can be using up its reserve food faster than it is being manufactured, which, of course, will soon result in its death. Cuttings under mist, conversely, can be synthesizing food in excess of that used in respiration, such nutrients being very important in promoting the initiation and development of new roots *(96)*. (See Table 9–1.) The leaves of certain species, however, will not tolerate the continual wetting which occurs under mist and bud deterioration can occur with such high moisture levels.

Although mist nozzles should maintain a film of water on the leaves at all times during the daylight hours, the application of additional water as mist is of no advantage; in fact, it is likely to be harmful. The large amounts of water used in mist systems operating continuously will reduce the temperature of the tissues of the cuttings as well as that of the rooting medium to that of the water, which may be too low for optimum rooting. Under the commonly used *intermittent mist* systems, in which water is applied at frequent, short intervals, relatively little water is used, and the temperature of the rooting medium is not adversely low. Temperatures in the rooting area of cuttings under intermittent mist are higher than those under constant mist, thus being more favorable for rooting *(96)*.

While the primary benefit in rooting under mist is in controlling water loss, a secondary benefit could be in sufficiently changing the physiological conditions of the tissues to promote an increase in naturally occurring rooting factors. It has been shown *(128)* that *Euonymus* plants grown under mist accumulated natural root-inducing phenolic and flavenoid compounds so that cuttings taken from them rooted much more readily than those from nonmisted plants.

Studies *(138, 209)* with radioactive isotopes have shown that subjecting leaves to natural or artificial rains, or just soaking in water, will remove nutrients, both organic and inorganic, although considerable variability exists among plant species in regard to their susceptibility to losses of min-

Figure 9–20   Propagation under mist enables the use of large cuttings with greater leaf area. This often causes stronger rooting, which results in a larger rooted plant. These cuttings of the olive were treated with indolebutyric acid at 4000 ppm by the concentrated-solution-dip method.

erals by leaching (*7, 65*). Nutrients added to the mist can replenish, to some extent, those lost by leaching (*233*). Root initiation itself is not increased under nutrient mist but root quality is better and subsequent growth of the rooted cuttings is enhanced (see p. 303).

Development of disease organisms under mist might be expected, but such has not been the case; in fact, the reverse has been true. For example, in greenhouse roses, no mildew was found to develop on leaves under mist, but mildew did develop on those not under mist. This is explained by the failure of mildew spores to germinate in water (*125*). Water sprays inhibit the development of powdery mildew (*Sphaerotheca pannosa*), and it may be that other disease organisms are held in check in the same manner (*237*).

With mist, softwood cuttings may be taken early in the season, at a stage most favorable for rooting. Under conventional equipment, such immature cuttings would be difficult to maintain without wilting. As illustrated in Figure 9–20, the reduction in transpiration under mist permits the rooting of large cuttings with a considerable leaf area.

(For details in setting up mist propagating equipment, see p. 297.)

Temperature
Daytime air temperatures of 70° to 80° F (21° to 27° C) with night temperatures about 60° F (15° C) are satisfactory for rooting cuttings of most species, although some root better at lower temperatures. Excessively high air temperatures tend to promote bud development in advance of root development and increase water loss from the leaves. It is important to have root development ahead of shoot development. In cutting beds, some type of thermostatically controlled heat applied below the cuttings, as described on p. 22, is beneficial in maintaining the temperature at the base of the cutting higher than that of the buds.

Light
In all types of plant growth light is, of course, of major importance, since it is the source of energy in photosynthesis. In rooting leafy cuttings, the products of photosynthesis are important for root initiation and growth.

Light intensity and duration must be great enough so that carbohydrates will accumulate in excess of those used in respiration. In leafless hardwood cuttings, dependence is upon stored carbohydrates.

It is well known (as noted on p. 237) that the absence of light on stem tissue (etiolation) in the region where roots are expected to form is conducive to root initiation.

White fluorescent lamps, providing intensities of 150 to 200 foot-candles, as a light source have given good rooting of cuttings (*190*). Although this is relatively low (full sunlight is about 10,000 footcandles), it seems to be sufficient for satisfactory root formation in some species.

Radiation in the orange-red end of the spectrum seems to favor rooting of cuttings more than that in the blue region (*188*), but in one test (*189*), when stock plants were exposed for six weeks to light sources of different quality before taking the cuttings, those from plants exposed to blue light rooted most readily.

There is some evidence that the *photoperiod* under which the stock plant is grown may exert an influence on the rooting of cuttings taken from it. This may be related to carbohydrate accumulation, the best rooting being obtained under photoperiods promoting carbohydrate increase, although in some cases (*182*) stock plants held under short photoperiods have produced the best rooting cuttings. In some species, the photoperiod under which the cutting is actually rooted may affect root initiation, long days or continuous illumination generally being more effective than short days (*188*), although in other species photoperiod has no influence (*182*).

This situation can become quite complex, however, since photoperiod can involve shoot development as well as root initiation. For example, in propagation by leaf cuttings there must be development of both adventitious roots and shoots. Using *Begonia* leaf cuttings (*84*), where the light intensity was adjusted so that the total light energy was about the same under both long days and short days, it was found that short days and relatively low temperatures promoted adventitious bud formation on the leaf pieces, whereas short days suppressed adventitious root formation. Roots formed best under long days with relatively high temperatures.

In rooting cuttings of 'Andorra' juniper, pronounced variations in rooting occurred during the year, but the same variations took place whether the cuttings were maintained under long days, short days, or the natural day length (*126*); see Figure 9–16. A number of tests have been made of the effect of photoperiod on root formation in cuttings, but the results are conflicting; it is difficult to make any generalization (*8, 126, 220*).

In some plants photoperiod will control growth after the cuttings have been rooted. Certain plants cease active shoot growth in response to natural changes in day length. This is the case with spring cuttings of deciduous azaleas and dwarf rhododendrons which had rooted and were potted in late summer or early fall. Considerably improved growth of such plants was obtained during the winter in the greenhouse if they were placed under continuous supplementary light, in comparison with similar plants subjected only to the normal short winter days. The latter plants, without added day length, remained in a dormant state until the following spring (*34, 64, 223*).

Rooting Medium

The rooting medium has three functions: (a) to hold the cutting in place during the rooting period, (b) to provide moisture for the cutting, and (c) to permit penetration of air to the base of the cutting.

An ideal rooting medium provides sufficient porosity to allow good aeration, has a high water-holding capacity, and yet is well drained. For tender softwood and semi-hardwood cuttings, it should be free from harmful fungi and bacteria.

The rooting medium can affect the type of root system arising from cuttings. Cuttings of some species, when rooted in sand, produce long, unbranched, coarse, and brittle roots, but when rooted in a mixture, such as sand and peat moss, or perlite and peat moss, develop roots that are well branched, slender, and flexible, a type much more suited for digging and repotting.

Experiments (137) to determine which of the characteristic differences between peat moss and sand were responsible for the different types of root system produced indicated that it was the difference in the moisture content. Determinations of the air and moisture content of peat moss and sand when each was at a point considered optimum for rooting showed that, on a volume basis, peat moss contained over twice as much air and three times as much moisture as sand.

The pH level of the rooting medium can be an important consideration in the production of adventitious roots. Studies with *Thuja occidentalis* cuttings rooted in perlite saturated with solutions maintained at various pH values gave best rooting at pH 7. Increasing acidity of the medium markedly inhibited rooting but high alkaline levels did not significantly reduce rooting (20).

The level of exchangeable calcium in peat moss, used as a rooting medium for chrysanthemum cuttings, had an influence on the number of roots per cutting and on root length. Increasing the exchangeable calcium (complementary ion—hydrogen) from 0 to 100 percent caused root numbers per cutting to decrease linearly from 15 to 4, respectively. Root length per cutting was at a maximum with 37.5 percent calcium, decreasing at higher or lower calcium levels. In a rooting medium for chrysanthemum cuttings, between 37.5 and 75 percent calcium saturation should give the best results (155).

In an extension of such studies to several kinds of woody plants, it was noted (156) that cuttings of different species did not respond the same to different levels of exchangeable calcium in the sphagnum peat rooting medium, so generalizations would be difficult. Some plants, as *Euonymus* and *Pyracantha,* were strongly affected by the calcium level, while *Osmanthus* and *Rhododendron* were not. Available oxygen in the rooting medium is essential for root production, although the requirement varies with different species. Willow cuttings form roots readily in water with an oxygen content as low as 1 ppm, but English ivy requires about 10 ppm for adequate root growth (238). Rooting of carnation and chrysanthemum cuttings increased markedly as the water in which they were rooted was aerated with increasing

Figure 9–21   Necessity of oxygen for adventitious root development in willow cuttings. Started October 24 and photographed November 16. Water depths in tubes are 2, 8, 15 in. *Left:* water not aerated. *Right:* water aerated with oxygen from a commercial gas cylinder. Roots on cuttings in tubes at the left developed only at the water surface, where oxygen is available. Cuttings in oxygenated water at the right produced roots throughout their length. Courtesy P. W. Zimmerman.

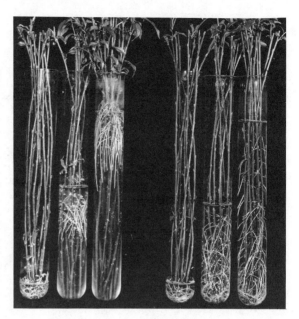

amounts of oxygen, from 0 to 21 percent (*206*). When roots are produced only near the surface of the rooting medium, it is likely that the oxygen supply in the medium is inadequate. (See Figure 9–21.)

The different kinds and mixtures of rooting media commonly used are discussed on p. 288.

REFERENCES

1   Abeles, F. B., *Ethylene in Plant Biology.* New York: Academic Press, 1973.

2   Altman, A., The role of auxin in root initiation in cuttings, *Proc. Inter. Plant Prop. Soc.,* 22:280–94. 1972.

3   Andreassen, D. C., and J. H. Ellison, Root initiation of stem tip cuttings from mature asparagus plants, *Proc. Amer. Soc. Hort. Sci.,* 90:158–62. 1967.

4   Ashiru, G. A., and R. F. Carlson, Some endogenous rooting factors associated with rooting of East Malling II and Malling-Merton 106 apple clones, *Proc. Amer. Soc. Hort. Sci.,* 92:106–12. 1968.

5   Aung, L. H., The nature of root-promoting substances in *Lycopersicon esculentum* seedlings, *Physiol. Plant.,* 26:306–9. 1972.

6   Avery, G. S., Jr., et al., *Hormones and Horticulture.* New York: McGraw-Hill Book Company, 1947.

7   Bahn, K. C., A. Wallace, and O. R. Lunt, Some mineral losses from leaves by leaching, *Proc. Amer. Soc. Hort. Sci.,* 73:289–93. 1959.

8   Baker, R. L., and C. B. Link, The influence of photoperiod on the rooting of cuttings of some woody ornamental plants, *Proc. Amer. Soc. Hort. Sci.,* 82:596–601. 1963.

9   Basu, R. N., B. N. Roy, and T. K. Bose, Interaction of abscisic acid and auxins in rooting of cuttings, *Plant & Cell Physiol.*, 11:681–84. 1970.

10   Beakbane, A. B., Relationships between structure and adventitious rooting, *Proc. Inter. Plant Prop. Soc.*, 19:192–201. 1969.

11   Biran, I., and A. H. Halevy, Endogenous levels of growth regulators and their relationship to the rooting of dahlia cuttings, *Physiol. Plant.*, 28:436–42, 1973.

12   Blakely, L. M., S. J. Rodaway, L. B. Hollen, and S. G. Croker, Control and kinetics of branch root formation in cultured root segments of *Haplopappus ravenii*, *Plant Physiol.*, 50:35–49. 1972.

13   Bloch, R., Polarity in plants, *Bot. Rev.*, 9:261–310. 1943.

14   Boe, A. A., R. B. Steward, and T. J. Banko, Effects of growth regulators on root and shoot development of Sedum leaf cuttings, *HortScience*, 74(4):404–5. 1972.

15   Bonner, J., Vitamin B_1, a growth factor for higher plants, *Science*, 85:183–84. 1937.

16   Bonnett, H. T., Jr., and J. G. Torrey, Chemical control of organ formation in root segments of *Convolvulus* cultured in vitro, *Plant Phys.*, 40:1228–36. 1965.

17   Bouillenne, R., and M. Bouillenne-Walrand, Auxines et bouturage, *Rpt. 14th Int. Hort. Cong.* Vol. I, pp. 231–38. 1955.

18   Breen, P. J., and T. Muroaka, Effect of indolebutyric acid on distribution of ^{14}C photosynthate in softwood cuttings of Marianna 2624 plum, *Jour. Amer. Soc. Hort. Sci.*, 98(5):436–39. 1973.

19   Brian, P. W., H. G. Hemming, and D. Lowe, Inhibition of rooting of cuttings by gibberellic acid, *Ann. Bot. n.s.*, 24:407–9. 1960.

20   Bruckel, D. W., and E. P. Johnson, Effects of pH on rootability of *Thuja occidentalis*, *Plant Prop.* (Inter. Plant Prop. Soc.), 15(4):10–12. 1969.

21   Cameron, R. J., and G. V. Thomson, The vegetative propagation of *Pinus radiata:* root initiation in cuttings. *Bot. Gaz.*, 130(4):242–51. 1969.

22   Carlson, M. C., Nodal adventitious roots in willow stems of different ages, *Amer. Jour. Bot.*, 37:555–61. 1950.

23   Carpenter, J. B., Occurrence and inheritance of preformed root primordia in stems of citron *(Citrus medica* L.), *Proc. Amer. Soc. Hort. Sci.*, 77:211–18. 1961.

24   Chin, T. Y., M. M. Meyer, Jr., and L. Beevers, Abscisic acid stimulated rooting of stem cuttings, *Planta*, 88:192–96. 1969.

25   Challenger, S., H. J. Lacey, and B. H. Howard, The demonstration of root promoting substances in apple and plum rootstocks, *Ann. Rpt. E. Malling Res. Sta. for 1964*, pp. 124–28. 1965.

26   Ciampi, C., and R. Gellini, Studio anatomico sui rapporti tra struttura e capacita di radicazione in talee di olivo (Anatomical study on the relationship between structure and rooting capacity in olive cuttings), *Nuovo Giorn. Bot. Ital.*, 65:417–24. 1958.

27   ———, Insorgenza e sviluppo delle radici avventizie in *Olea europaea* L.: importanza della struttura anatomica agli effetti dello sviluppo delle radichette (Formation and development of adventitious roots in *Olea europaea* L.: signifi-

cance of the anatomical structure for the development of radicles, *Nuovo Giorn. Bot. Ital.,* 70:62–74. 1963.

28 Cooper, W. C., Hormones in relation to root formation on stem cuttings, *Plant Phys.,* 10:789–94. 1935.

29 ———, Hormones and root formation, *Bot. Gaz.,* 99:599–614. 1938.

30 ———, The concentrated-solution-dip method of treating cuttings with growth substances, *Proc. Amer. Soc. Hort. Sci.,* 44:533–41. 1944.

31 Corbett, L. C., The development of roots from cuttings, *W. Va. Agr. Exp. Sta. Ann. Rpt.,* 9(1895–96):196–99. 1897.

32 Cormack, R. G. H., The effect of calcium ions and pH on the development of callus tissue on stem cuttings of balsam poplar, *Canad. Jour. Bot.,* 43:75–83. 1965.

33 ———, and P. L. Lemay, A further study of callus tissue development on stem cuttings of balsam poplar, *Canad. Jour. Bot.,* 44:47–50. 1966.

34 Crossley, J. H., Light and temperature trials with seedlings and cuttings of *Rhododendron molle, Proc. Int. Plant Prop. Soc.,* 15:327–34. 1965.

35 Crow, W. D., W. Nicholls, and M. Sterns, Root inhibitors in *Eucalyptus grandis:* naturally occurring derivatives of the 2,3-dioxabicyclo (4,4,0) decane system. Tetrahedron Letters 18. London: Pergamon Press, 1971, pp. 1353–56.

36 Curtis, O. F., Stimulation of root growth in cuttings by treatment with chemical compounds, *Cornell Univ. Agr. Exp. Sta. Mem. 14,* 1918.

37 Danckwardt-Lillieström, C., Kinetin-induced shoot formation from isolated roots of *Isatis tinctoria, Physiol. Plant.,* 10:794–97. 1957.

38 Delisle, A. L., Histological and anatomical changes induced by indoleacetic acid in rooting cuttings of *Pinus strobus* L., *Va. Jour. Sci.,* 3:118–24. 1942.

39 Deuber, C. G., Vegetative propagation of conifers, *Trans. Conn. Acad. Arts and Sci.,* 34:1–83. 1940.

40 Doak, B. W., The effect of various nitrogenous compounds on the rooting of rhododendron cuttings treated with naphthaleneacetic acid, *New Zealand Jour. Sci. and Tech.,* 21:336A–43A. 1940.

41 Donoho, C. W., A. E. Mitchell, and H. N. Sell, Enzymatic destruction of C^{14} labelled indoleacetic acid and naphthaleneacetic acid by developing apple and peach seeds, *Proc. Amer. Soc. Hort. Sci.,* 80:43–49. 1962.

42 Doorenbos, J., Rejuvenation of *Hedera helix* in graft combinations, *Proc. Kon. Ned. Akad. Wet.,* Series C., 57:99–102. 1954.

43 Dore, J., Seasonal variation in the regeneration of root cuttings, *Nature,* 172:1189. 1953.

44 Duhamel du Monceau, H. L., *La physique des arbres,* Vols. I and II. Paris: Guerin and Delatour, 1758.

45 Ericksen, E. N., Promotion of root initiation by gibberellin, *Kgl. Vet. -og Landbohøjsk. Arsskr.* (Copenhagen). 1971. pp. 50–59. 1970.

46 Ericksen, E. N., Root initiation in pea cuttings: interactions between tryptophane and gibberellin, *Kgl. Vet. -og Landbohøjsk Arsskr.* (Copenhagen). 1972. pp. 115–26. 1971.

47 Ericksen, E. N., Root formation in pea cuttings III. The influence of cyto-kinin at different developmental stages, *Physiol. Plant.* 30(2):163–67. 1974.

48 Esau, K., *Plant Anatomy*, 2nd ed. New York: John Wiley & Sons, Inc., 1965, pp. 513–14.

49 Fadl, M. S., and H. T. Hartmann, Isolation, purification, and characterization of an endogenous root-promoting factor obtained from the basal sections of pear hardwood cuttings, *Plant Physiol.*, 42:541–49, 1967.

50 ——, Relationship between seasonal changes in endogenous promoters and inhibitors in pear buds and cutting bases and the rooting of pear hardwood cuttings, *Proc. Amer. Soc. Hort. Sci.*, 91:96–112. 1967.

51 Farrar, J. H., and N. H. Grace, Vegetative propagation of conifers. XI. Effects of type of cutting on the rooting of Norway spruce cuttings. *Can. Jour. Res.*, Sect. C, 20:116–21. 1942.

52 Fiorino, P., J. N. Cummins, and J. Gilpatrick, Increased production of rooted *Prunus besseyi* Bailey softwood cuttings with pre-planting soak in benomyl, *Proc. Inter. Plant Prop. Soc.*, 19:320–36. 1969.

53 Frolich, E. F., Etiolation and the rooting of cuttings, *Proc. Int. Plant Prop. Soc.*, 11:277–83. 1961.

54 Gardner, E. J., Propagation under mist, *Amer. Nurs.*, 73(9):5–7. 1941.

55 Gardner, F. E., The relationship between tree age and the rooting of cuttings, *Proc. Amer. Soc. Hort. Sci.*, 26:101–4. 1929.

56 ——, Etiolation as a method of rooting apple variety stem cuttings, *Proc. Amer. Soc. Hort. Sci.*, 34:323–29. 1937.

57 Gautheret, R. J., Investigations on the root formation in the tissues of *Helianthus tuberosus* cultured in vitro, *Amer. Jour. Bot.*, 56(7):702–17. 1969.

58 Garner, R. J., and E. S. J. Hatcher, Regeneration in relation to vegetative growth and flowering, *Proc. 16th Int. Hort. Cong.*, pp. 105–11, 1962.

59 Ginzburg, C., Organization of the adventitious root apex in *Tamarix aphylla*, *Amer. Jour. Bot.*, 54:4–8. 1967.

60 Girouard, R. M., Anatomy of adventitious root formation in stem cuttings, *Proc. Inter. Plant Prop.* Soc., 17:289–302. 1967.

61 ——, Initiation and development of adventitious roots in stem cuttings of *Hedera helix*, *Canad. Jour. Bot.*, 45:1883–86. 1967.

62 ——, Physiological and biochemical studies of adventitious root formation. Extractible rooting co-factors from *Hedera helix*, *Canad. Jour. Bot.*, 47(5):687–99. 1969.

63 Greenwood, M. S., and G. P. Berlyn, Sucrose-indole-3-acetic acid interactions on root regeneration by *Pinus lambertiana* embryo cuttings, *Amer. Jour. Bot.*, 60(1):42–47. 1973.

64 Goddard, W., Forestalling dormancy and inducing continuous growth of *Azalea molle* with supplementary light for winter propagation, *Proc. Int. Plant Prop. Soc.*, 13:276–78. 1963.

65 Good, G. L., and H. B. Tukey, Jr., Leaching of nutrients from cuttings under mist, *Proc. Int. Plant Prop. Soc.*, 14:138–42. 1964.

66 Gorter, C. J., Vegetative propagation of *Asparagus officinalis* by cuttings, *Jour. Hort. Sci.,* 40:177–79. 1965.

67 ———, Auxin-synergists in the rooting of cuttings, *Physiol. Plant.* 22:497–502, 1969.

68 Grace, N. H., Rooting of cuttings taken from the upper and lower regions of a Norway spruce tree, *Canad. Jour. Res.,* 17(C): 172–80. 1939.

69 Hackett, W. P., The influence of auxin, catechol, and methanolic tissue extracts on root initiation in aseptically cultured shoot apices of the juvenile and adult forms of *Hedera helix, Jour. Amer. Soc. Hort. Sci.,* 95(4):398–402. 1970.

70 Haissig, B. E., Organ formation *in vitro* as applicable to forest tree propagation, *Bot., Rev.,* 31:607–26. 1965.

71 ———, Influence of indole-3-acetic acid on incorporation of ^{14}C-uridine by adventitious root primordia in brittle willow, *Bot. Gaz.,* 132(4):263–67. 1971.

72 ———, Meristematic activity during adventitious root primordium development. Influences of endogenous auxin and applied gibberellic acid, *Plant Physiol.,* 49:886–92. 1972.

73 ———, Influence of hormones and auxin synergists on adventitious root initiation, in *Proc. I.U.F.R.O. Working Party on Reprod. Processes,* Rotorua, New Zealand. 1973.

74 Hagemann, A., Untersuchungen an Blattstecklingen, *Gartenbauwiss.,* 6:69–202. 1932.

75 Hansen, C. J., and H. T. Hartmann, The use of indolebutyric acid and captan in the propagation of clonal peach and peach-almond hybrid rootstocks by hardwood cuttings, *Proc. Amer. Soc. Hort. Sci.,* 92:135–40. 1968.

76 Hartmann, H. T., and R. M. Brooks, Propagation of Stockton Morello cherry rootstock by softwood cuttings under mist sprays, *Proc. Amer. Soc. Hort. Sci.,* 71:127–34. 1958.

77 Hartmann, H. T., and C. J. Hansen, Effect of season of collecting, indolebutyric acid, and pre-planting storage treatments on rooting of Marianna plum, peach, and quince hardwood cuttings, *Proc. Amer. Soc. Hort. Sci.,* 71:57–66. 1958.

78 ———, Rooting pear and plum rootstocks, *Calif. Agr.,* 12(10): 4, 14, 15. 1958.

79 Hartmann, H. T., and F. Loreti, Seasonal variation in the rooting of olive cuttings, *Proc. Amer. Soc. Hort. Sci.,* 87:194–98. 1965.

80 Hartmann, H. T., W. H. Griggs, and C J. Hansen, Propagation of own-rooted Old Home and Bartlett pears to produce trees resistant to pear decline, *Proc. Amer. Soc. Hort. Sci.,* 82:92–102. 1963.

81 Hatcher, E. S. J., and R. J. Garner, Aspects of rootstock propagation. IV. The winter storage of hardwood cuttings, *Ann. Rpt. E. Malling Res. Sta. for 1956,* pp. 101–8. 1957.

82 Haun, J. R., and P. W. Cornell, Rooting response of geranium *(Pelargonium hortorum,* Bailey, var. Ricard) cuttings as influenced by nitrogen phosphorus, and potassium nutrition of the stock plant, *Proc. Amer. Soc. Hort. Sci.,* 58:317–23. 1951.

83 Hemberg, T., Rooting experiments with hypocotyls of *Phaseolus vulgaris* L., *Physiol. Plant.,* 11:1–9. 1951.

84   Heide, O. M., Photoperiodic effects on the regeneration ability of *Begonia* leaf cuttings, *Physiol. Plant.,* 18:185–90. 1965.

85   ———, Interaction of temperature, auxin, and kinins in the regeneration ability of *Begonia* leaf cuttings, *Physiol. Plant.,* 18:891–920. 1965.

86   ———, Effects of 6-benzylamino-purine and 1-naphthaleneacetic acid on the epiphyllous bud formation in *Bryophyllum, Planta,* 67:281–96. 1965.

87   ———, Stimulation of adventitious bud formation in *Begonia* leaves by abscisic acid, *Nature,* 219(5157):960–61. 1968.

88   ———, Auxin level and regeneration of *Begonia* leaves, *Planta,* 81:153–59. 1968.

89   ———, Non-reversibility of gibberellin-induced inhibition of regeneration in *Begonia* leaves. *Physiol. Plant.,* 22:671–79. 1969.

90   Herman, D. E., and C. E. Hess, The effect of etiolation upon the rooting of cuttings. *Proc. Int. Plant Prop. Soc.,* 13:42–62. 1963.

91   Hess, C. E., A physiological analysis of root initiation in easy and difficult-to-root cuttings, *Proc. 16th Int. Hort. Cong.,* pp. 375–81. 1962.

92   ———, Characterization of the rooting co-factors extracted from *Hedera helix* L. and *Hibiscus rosa-sinensis* L., *Proc. 16th Int. Hort. Cong.,* pp. 382–88. 1962.

93   ———, Naturally-occurring substances which stimulate root initiation, *Col. Int. du Centre Nat. Recherche Sci.,* No. 123, pp. 517–27. Paris, 1963.

94   ———, Phenolic compounds as stimulators of root initiation, *Plant Physiol.,* (suppl.) 40 XLV. 1965.

95   ———, Internal and external factors regulating root initiation. Root Growth (Proc. 15th Easter Sch. Agr. Sci., U. of Nott.). London: Buttersworth, 1968.

96   ———, and W. E. Snyder, A physiological comparison of the use of mist with other propagation procedures used in rooting cuttings, *Rpt. 14th Int. Hort. Cong.,* pp. 1133–39. 1955.

97   Heuser, C. W., and C. E. Hess, Isolation of three lipid root-initiating substances from juvenile *Hedera helix* shoot tissue, *Jour. Amer. Soc. Hort. Sci.,* 97(5): 571–74. 1972.

98   Hiller, Charlotte, A study of the origin and development of callus and root primordia of *Taxus cuspidata* with reference to the effects of growth regulator, Master's thesis, Cornell Univ., Ithaca, N.Y., 1951.

99   Hitchcock, A. E., and P. W. Zimmerman, Relation of rooting response to age of tissue at the base of green wood cuttings, *Contrib. Boyce Thomp. Inst.,* 4:85–98. 1932.

100   ———, A sensitive test for root formation, *Amer. Jour. Bot.,* 24:735–36. 1937.

101   ———, Effects obtained with mixtures of root-inducing and other substances, *Contrib. Boyce Thomp. Inst.,* 11:143–60. 1940.

102   ———, Root inducing activity of phenoxy compounds in relation to their structure, *Contrib. Boyce Thomp. Inst.,* 12:497–507. 1942.

103   Hoitink, H. A. J., and A. F. Schmitthenner, Disease control in rhododendron cuttings with benomyl or thiabendazole mixtures, *Plant Dis. Rpt.,* 54:427–30. 1970.

104 Howard, B. H., Increase during winter in capacity for root regeneration in detached shoots of fruit tree rootstocks, *Nature,* 208:912–13. 1965.

105 ———, and J. Y. Sykes, Regeneration of the hop plant *(Humulus lupulus* L.) from softwood cuttings, II. Modification of the carbohydrate resources within the cutting, *Jour. Hort. Sci.,* 41:155–63. 1966.

106 ———, Depressing effects of virus infection on adventitious root production in apple hardwood cuttings, *Jour. Hort. Sci.,* 47:255–58. 1972.

107 Hudson, J. P., The regeneration of plants from roots, *Proc. 14th Int. Hort. Cong.,* Vol. II, pp. 1165–72. 1955.

108 Jain, M. K., and K. K. Nanda, Effect of temperature and some antimetabolites on the interaction effects of auxin and nutrition in rooting etiolated stem segments of *Salix tetrasperma, Physiol. Plant.,* 27:169–72. 1972.

109 Jauhari, O. S., and S. F. Rahman, Further investigations on rooting in cuttings of sweet lime *(Citrus limettoides)* Tanaka, *Sci. and Cult.,* 24:432–34. 1959.

110 Johnson, C. R., The nature of flower bud influence on root regeneration in the *Rhododendron* shoot. Ph.D. Dissertation. Ore. State Univ., Corvallis, Ore., 1970.

111 Kawase, M., Centrifugation, rhizocaline, and rooting in *Salix alba, Physiol. Plant.,* 17:855–65. 1964.

112 ———, Etiolation and rooting in cuttings, *Physiol. Plant.,* 18:1066–76. 1965.

113 ———, Causes of centrifugal root promotion, *Physiol. Plant.,* 25: 64–70. 1971.

114 ———, Root promoting substances in *Salix alba, Physiol. Plant.,* 23:159–70. 1970.

115 ———, Diffusible rooting substances in woody ornamentals, *Jour. Amer. Soc. Hort. Sci.,* 96(1):116–20. 1971.

116 Kefford, N. P., Effect of a hormone antagonist on the rooting of shoot cuttings, *Plant Physiol.,* 51:214–16. 1973.

117 Key, J. L., Hormones and nucleic acid metabolism, *Ann. Rev. Plant Physiol.,* 20:449–74. 1969.

118 Kögl, F., A. J. Haagen-Smit, and H. Erxleben, Uber ein neues Auxin ("Heteroauxin") aus Harn. XI, *Mitteilung. Z. physiol. Chem.,* 228:90–103. 1934.

119 Kraus, E. J., Rooting azalea cuttings, *Nat. Hort. Mag.,* 32:163–64. 1953.

120 ———, and H. R. Kraybill, Vegetation and reproduction with special reference to the tomato, *Ore. Agr. Exp. Sta. Bul. 149,* 1918.

121 Kraus, E. J., N. A. Brown, and K. C. Hamner, Histological reactions of bean plants to indoleacetic acid, *Bot. Gaz.,* 98:370–420. 1936.

122 Krishnamoorthy, H. N., Promotion of rooting in mung bean hypocotyl cuttings with Ethrel, an ethylene releasing compound, *Plant & Cell Physiol.,* 11:979–82. 1970.

123 Krul, W. R., Increased root initiation in Pinto bean hypocotyls with 2,4-dinitrophenol, *Plant Physiol.,* 43(3):439–41. 1968.

124 Laibach, F., and O. Fischnich, Künstliche Wurzelneubildung mittels Wuchsstoffpaste, *Ber. Deuts. bot. Ges.,* 53:528–39. 1935.

125　Langhans, R. W., Mist for growing plants, *Farm Res.* (Cornell Univ.), 21(3): 1955.

126　Lanphear, F. O., and R. P. Meahl, The effect of various photoperiods on rooting and subsequent growth of selected woody ornamental plants, *Proc. Amer. Soc. Hort. Sci.,* 77:620–34. 1961.

127　Lanphear, F. O., and R. P. Meahl, Influence of the stock plant environment on the rooting of *Juniperus horizontalis* 'Plumosa,' *Proc. Amer. Soc. Hort. Sci.,* 89:666–71. 1966.

128　Lee, C. I., and H. B. Tukey, Jr., Induction of root promoting substances in *Euonymus alatus* 'Compactus' by intermittent mist, *Jour. Amer. Soc. Hort. Sci.,* 96(6):731–36. 1971.

129　Lek, H. A. A., van der, Root development in woody cuttings, *Meded. Landbouwhoogesch. Wageningen,* 38(1). 1925.

130　———, Anatomical structure of woody plants in relation to vegetative propagation, *Proc. IX. Int. Hort. Cong.,* pp. 66–76. 1930.

131　———, Over den invloed der knoppen op de wortelvorming der stekken, *Meded. Landbouwhoogesch. Wageningen,* 38(2):1–95. 1934.

132　Leopold, A. C., The polarity of auxin transport, in Meristems and Differentiation, *Brookhaven Symposia in Biology, Rpt. No. 16,* pp. 218–34. Upton, N.Y.: Brookhaven Nat'l Lab., 1964.

133　Lesham, Y., and B. Lunenfield, Gonadotropin in promotion of adventitious root production on cuttings of *Begonia semperflorens* and *Vitis vinifera, Plant Physiol.,* 43:313–17. 1968.

134　Libbert, E., Untersuchungen über die Physiologie der Adventivewurzelbildung, I. Die Wirkungsweise einiger Komponenten des "Rhizokalinkomplexes", *Flora,* 144:121–50. 1956.

135　Libby, W. J., A. G. Brown, and J. M. Fielding, Effects of hedging Radiata pine on production, rooting, and early growth of cuttings, *New Zealand Jour. For. Sci.,* 2(2):263–83. 1972.

136　Lipecki, J. and F. G. Dennis, Growth inhibitors and rooting cofactors in relation to rooting response of softwood apple cuttings, *HortScience* 7(2):136–38. 1972.

137　Long, J. C., The influence of rooting media on the character of roots produced by cuttings, *Proc. Amer. Soc. Hort. Sci.,* 29:352–55. 1933.

138　Long, W. G., D. V. Sweet, and H. B. Tukey, The loss of nutrients from plant foliage by leaching as indicated by radioisotopes, *Science,* 123:1039–40. 1956.

139　Loreti, F., and H. T. Hartmann, Propagation of olive trees by rooting leafy cuttings under mist, *Proc. Amer. Soc. Hort. Sci.,* 85:257–64. 1964.

140　Massini, P., and G. Voorn, The effect of ferrodoxin and ferrous ion on the chlorophyll sensitized photoreduction of dinitrophenol, *Photochem.–Photobiol.,* 6:851–56. 1967.

141　Mes, M. G., Cuttings difficult to root, *Plants and Gardens,* 7(2):95–97. 1951.

142　———, Plant hormones, *Prog. Rpt. Plant Phys. Res. Inst.,* 1950–51, Univ. of Pretoria, 1951.

143 Mitsuhashi, M. H., Shiboaka, and M. Shimokoriyama, Portulal: a root promoting substance, *Plant Physiol.* (suppl.), p. 26. 1969.

144 Mittempergher, L. Indagini sull' origine delle radici avventizie in talee legnose di pero, *Riv. Ortoflorofrutticoltura Ital.,* 48:39–44. 1964.

145 Molnar, J. M., and L. J. LaCroix, Studies of the rooting of cuttings of *Hydrangea macrophylla:* enzyme changes, *Canad. Jour. Bot.,* 50(2):315–22. 1972.

146 ———, Studies of the rooting of cuttings of *Hydrangea macrophylla:* DNA and protein changes, *Canad. Jour. Bot.,* 50(3):387–92. 1972.

147 Mullins, M. G., Auxin and ethylene in adventitious root formation in *Phaseolus aureus* (Roxb.), in D. J. Carr (ed.), *Plant Growth Substances—1970.* Berlin: Springer-Verlag, 1972.

148 Muzik, T. J., and H. J. Cruzado, Transmission of juvenile rooting ability from seedlings to adults of *Hevea brasiliensis, Nature,* 181:1288. 1958.

149 Naylor, E. E., and B. Johnson, A histological study of vegetative reproduction in *Saintpaulia ionantha, Amer. Jour. Bot.,* 24:673–78. 1937.

150 Olieman, van der Meer, R. L. M. Pierik, and S. Roest, Effects of sugar, auxin, and light on adventitious root formation in isolated stem explants of *Phaseolus* and *Rhododendron, Med. Facult. Landbouw, Weten. Gent.,* 36(1):511–18. 1971.

151 O'Rourke, F. L., The influence of blossom buds on rooting of hardwood cuttings of blueberry, *Proc. Amer. Soc. Hort. Sci.,* 40:332–34. 1940.

152 ———, Wood type and original position on shoot with reference to rooting in hardwood cuttings of blueberry, *Proc. Amer. Soc. Hort. Sci.,* 45:195–97. 1944.

153 ———, Mist humidification and the rooting of cuttings, *Mich. Agr. Exp. Sta. Quart. Bul. 32,* pp. 245–49. 1949.

154 Paton, D. M., R. R. Willing, W. Nichols, and L. D. Pryor, Rooting of stem cuttings of eucalyptus: a rooting inhibitor in adult tissue, *Austral. Jour. Bot.,* 18:175–83. 1970.

155 Paul, J. L., and L. V. Smith, Rooting of chrysanthemum cuttings in peat as influenced by calcium, *Proc. Amer. Soc. Hort. Sci.,* 89:626–30. 1966.

156 Paul, J. L., and A. T. Leiser, Influence of calcium saturation of sphagnum peat on the rooting of five woody species, *Hort. Res.,* 8(1):41–50. 1968.

157 Pearse, H. L., The effect of nutrition and phytohormones on the rooting of vine cuttings, *Ann. Bot.,* n.s., 7:123–32. 1943.

158 ———, Growth substances and their practical importance in horticulture, *Commonwealth Bur. Hort. and Plant. Crops, Tech. Comm. No. 20,* 1948.

159 ———, Rooting of vine and plum cuttings as affected by nutrition of the parent plant and treatment with phytohormones, *Sci. Bul. 249, Dept. of Agr. Union of S. Afr.,* 1946.

160 Petri, P. S., S. Mazzi, and P. Strigoli, Considerazione sulla formazione delle radici avventizie con particolare riguardo a: *Cucurbita pepo, Nerium oleander, Menyanthes trifoliatae, Solanum lycopersicum, Nuovo Giorn. Bot. Ital.,* 67:131–75. 1960.

161 Preston, W. H., J. B. Shanks, and P. W. Cornell, Influence of mineral nutrition on production, rooting and survival of cuttings of azaleas, *Proc. Amer. Soc. Hort. Sci.*, 61:499–507. 1953.

162 Priestley, J. H., and C. F. Swingle, Vegetative propagation from the standpoint of plant anatomy, *USDA Tech. Bul. 151*, 1929.

163 Raines, M. A., Some uses of a spray chamber in experimentation with plants, *Amer. Jour. Bot.*, Suppl. to Vol. 27, No. 10, p. 185. 1940.

164 Rappaport, J., The influence of leaves and growth substances on the rooting response of cuttings, *Natuurw. Tijdschr.*, 21:356–59. 1940.

165 Read, P. E., and V. C. Hoysler, Stimulation and retardation of adventitious root formation by application of B-Nine and Cycocel, *Jour. Amer. Soc. Hort. Sci.*, 94:314–16. 1969.

166 Robbins, W. J., Further observations on juvenile and adult *Hedera*, *Amer. Jour. Bot.*, 47:485–91. 1960.

167 Ryan, G. F., E. F. Frolich, and T. P. Kinsella, Some factors influencing rooting of grafted cuttings, *Proc. Amer. Soc. Hort. Sci.*, 72:454–61. 1958.

168 Sachs, J., Ueber die Neubildung von Adventivwurzelin durch Dunkelheit, *Verhandlingen des naturhistorischen Vereines der preussischen Rheinlande und Westphalens*, pp. 110–11. Abs. in *Bull. Soc. Bot. de France*, 12, Part 2, p. 221. 1865.

169 ———, Stoff und Form der Pflanzenorgane. I and II. *Arb. bot. Inst. Würzburg*, 2:452–88 and 4:689–718. 1880 and 1882.

170 Sachs, R. M., F. Loreti, and J. DeBie, Plant rooting studies indicate sclerenchyma tissue is not a restricting factor, *Calif. Agr.*, 18(9):4–5. 1964.

171 Selim, H. H. A., The effect of flowering on adventitious root formation, *Meded. Landbouwhoogesch. Wageningen*, 56(6):1–38. 1956.

172 Samish, R. M., and P. Spiegel, The influence of the nutrition of the mother vine on the rooting of cuttings, *Ktavim*, 8:93–100. 1957.

173 Sax, K., Aspects of aging in plants, *An. Rev. Plant Phys.*, 13:489–506. 1962.

174 Scaramuzzi, F., Alcune osservazioni sul potere rizogeno delle talee (Some observations on the rooting ability of cuttings), *Riv. Ortoflorofruitticoltura Ital.*, 44:187–93. 1960.

175 Shapiro, S., The role of light in the growth of root primordia in the stem of Lombardy poplar, in K. V. Thimann (ed.), *The Physiology of Forest Trees*. New York: The Ronald Press Company, 1958.

176 Shiboak, H. M. Mitsuhashi, and M. Shimokoriyama, Promotion of adventitious root formation by heliangine and its removal by cysteine, *Plant & Cell Physiol.*, 8:161–70. 1967.

177 Siegler, E. A., and J. J. Bowman, Anatomical studies of root and shoot primordia in 1-year apple roots, *Jour. Agr. Res.*, 58:795–803. 1939.

178 Skoog, F., and C. Tsui, Chemical control of growth and bud formation in tobacco stem and callus, *Amer. Jour. Bot.*, 35:782–87. 1948.

179 Snow, A. G., Jr., Clonal variation in rooting response of red maple cuttings, *USDA Northeastern Forest Exp. Sta. Tech. Note 29*, 1939.

180 Spiegel, P., Auxins and inhibitors in canes of *Vitis*, *Bull. Res. Counc., Israel*, 4:176–83. 1954.

181   Stangler, B. B., An anatomical study of the origin and development of adventitious roots in stem cuttings of *Chrysanthemum morifolium* Bailey, *Dianthus caryophyllus* L., and *Rosa dilecta* Rehd., Ph.D. dissertation, Cornell Univ., Ithaca, N. Y., 1949.

182   Steponkus, P. L., and L. Hogan, Some effects of photoperiod on the rooting of *Abelia grandiflora* Rehd. 'Prostrata' cuttings, *Proc. Amer. Soc. Hort. Sci.,* 91: 706–15. 1967.

183   Stoltz, L. P., and C. E. Hess, The effect of girdling upon root initiation; auxin and rooting co-factors, *Proc. Amer. Soc. Hort. Sci.,* 89:744–51. 1966.

184   ———, The effect of girdling upon root initiation: carbohydrates and amino acids, *Proc. Amer. Soc. Hort. Sci.,* 89:734–43. 1966.

185   ———, Factors influencing root initiation in an easy- and difficult-to-root chrysanthemum, *Proc. Amer. Soc. Hort. Sci.,* 92:622–26. 1968.

186   Stoutemyer, V. T., Regeneration in various types of apple wood, *Iowa Agr. Exp. Sta. Res. Bul.,* 220:309–52. 1937.

187   ———, The propagation of *Chionanthus retusus* by cuttings, *Nat. Hort. Mag.,* 21(4):175–78. 1942.

188   Stoutemyer, V. T., and A. W. Close, Rooting cuttings and germinating seeds under fluorescent and cold cathode light, *Proc. Amer. Soc. Hort. Sci.,* 48:309–25. 1946.

189   ———, Changes of rooting response in cuttings following exposure of the stock plants to light of different qualities, *Proc. Amer. Soc. Hort. Sci.,* 49:392–94. 1947.

190   Stoutemyer, V. T., and F. L. O'Rourke, Spray humidification and the rooting of greenwood cuttings, *Amer. Nurs.,* 77(1):5–6, 24–25. 1943.

191   ———, Rooting of cuttings from plants sprayed with growth-regulating substances, *Proc. Amer. Soc. Hort. Sci.,* 46:407–11. 1945.

192   Stoutemyer, V. T., O. K. Britt, and J. R. Goodin, The influence of chemical treatments, understocks, and environment on growth phase changes and propagation of *Hedera canariensis, Proc. Amer. Soc. Hort. Sci.,* 77:552–57. 1961.

193   Strydom, D. K., and H. T. Hartmann, Absorption, distribution, and destruction of indoleacetic acid in plum stem cuttings, *Plant Phys.,* 35:435–42. 1960.

194   ———, Effect of indolebutyric acid on respiration and nitrogen metabolism in Marianna 2624 plum softwood stem cuttings, *Proc. Amer. Soc. Hort. Sci.,* 76: 124–33. 1960.

195   Swingle, C. F., Burr knot formation in relation to the vascular system of the apple stem, *Jour. Agr. Res.,* 34:533–44. 1927.

196   Thielges, B. A., and H. A. J. Hoitink, Fungicides and rooting of eastern white pine cuttings, *For. Sci.,* 18(1):54–55. 1972.

197   Thimann, K. V., On an analysis of activity of two growth-promoting substances on plant tissues, *Proc. Kon. Ned. Akad. Wet.,* 38:896–912. 1935.

198   ———, On the plant growth hormone produced by *Rhizopus suinus, Jour. Biol. Chem.,* 109:279–91. 1935.

199   ———, and J. Behnke-Rogers, *The Use of Auxins in the Rooting of Woody*

Cuttings, Harvard Forest, M. M. Cabot Foundation, Petersham, Mass. 1950.

200 Thimann, K. V., and A. L. Delisle, The vegetative propagation of difficult plants, *Jour. Arnold Arb.,* 20:116–36. 1939.

201 Thimann, K. V., and J. B. Koepfli, Identity of the growth-promoting and root-forming substances of plants, *Nature,* 135:101–2. 1935.

202 Thimann, K. V., and E. F. Poutasse, Factors affecting root formation of *Phaseolus vulgaris, Plant Phys.,* 16:585–98. 1941.

203 Thimann, K. V., and F. W. Went, On the chemical nature of the root-forming hormone, *Proc. Kon. Ned. Akad. Wet.,* 37:456–59. 1934.

204 Thomaszewski, M., and K. V. Thimann, Interactions of phenolic acids, metallic ions, and chelating agents on auxin-induced growth, *Plant Physiol.,* 41: 1443–54. 1966.

205 Thurlow, J., and J. Bonner, Inhibition of photoperiodic induction in *Xanthium, Amer. Jour. Bot.,* 34:603–4. 1947.

206 Tinga, J. H., The effect of five levels of oxygen on the rooting of carnation cuttings in tap water culture, M.S. Thesis, Cornell Univ., Ithaca, N. Y., 1952.

207 Trécul, A. Recherches sur l' origine des racines, *Ann. Sci. Nat. Bot. Ser.,* 3:340–50. 1846.

208 Tukey, H. B., and E. L. Green, Gradient composition of rose shoots from tip to base, *Plant Phys.,* 9:157–63. 1934.

209 Tukey, H. B., Jr., H. B. Tukey, and S. H. Wittwer, Loss of nutrients by foliar leaching as determined by radioisotopes. *Proc. Amer. Soc. Hort. Sci.,* 71:496–506. 1958.

210 Turetskaya, R. K., Radioactive carbon investigations of the uptake and distribution of growth regulators in plant cuttings (translated title), *Fiziologiya Rasteny,* 4:42–51. 1957.

211 Vander der Meer, F. A., Nerfvergelingsmozaick bij kruisbessen, *Fruitteelt.,* 55:245–46. 1965.

212 Van Doesburg, J., Use of fungicides with vegetative propagation, *Proc. 16th Int. Hort. Cong.,* Vol. IV, pp. 365–72. 1962.

213 Vasil, V., and A. C. Hildebrandt, Differentiation of tobacco plants from single, isolated cells in microcultures, *Science,* 150:889–92. 1965.

214 Van Overbeek, J., and L. E. Gregory, A physiological separation of two factors necessary for the formation of roots on cuttings, *Amer. Jour. Bot.,* 32:336–41. 1945.

215 Van Overbeek, J., S. A. Gordon, and L. E. Gregory, An analysis of the function of the leaf in the process of root formation in cuttings, *Amer. Jour. Bot.,* 33:100–107. 1946.

216 Van Tieghem, P., and H. Douliot, Recherches comparatives sur l'origine des membres endogènes dans les plantes vasculaires, *Ann. Sci. Nat. Bot.,* VII, 8:1–160. 1888.

217 Vöchting, H., *Uber Organbildung im Pflanzenreich.* Bonn: Verlag Max Cohen & Son, 1878, pp. 1–258.

218 Walker, R. I., Regeneration in the scale leaf of *Lilium candidum* and *L. longiflorum, Amer. Jour. Bot.,* 27:114–17. 1940.

219 Warmke, H. E., and G. L. Warmke, The role of auxin in the differentiation of root and shoot primordia from root cuttings of *Taraxacum* and *Cichorium*, *Amer. Jour. Bot.*, 37:272–80. 1950.

220 Waxman, S., and J. P. Nitsch, Influence of light on plant growth, *Amer. Nurs.*, 104(10):11–12. 1956.

221 Weiser, C. J., and L. T. Blaney, The effects of boron on the rooting of English holly cuttings, *Proc. Amer. Soc. Hort. Sci.*, 75:704–10. 1960.

222 ———, The nature of boron stimulation to root initiation and development in beans, *Proc. Amer. Soc. Hort. Sci.*, 90:191–99. 1967.

223 Weiser, C. J., Rooting and night-lighting trials with deciduous azaleas and dwarf rhododendrons, *Amer. Hort. Mag.*, 42:95–100. 1963.

224 Wellensiek, S. J., Rejuvenation of woody plants by formation of sphaeroblasts, *Proc. Kon. Ned. Akad. Wet.*, 55:567–73. 1952.

225 Wells, J. S., The use of captan in rooting rhododendrons, *Proc. Int. Plant Prop. Soc.*, 13:132–35. 1963.

226 Went, F. W., On a substance causing root formation, *Proc. Kon. Ned. Akad. Wet.*, 32:35–39. 1929.

227 ———, A test method for rhizocaline, the root-forming substance, *Proc. Kon. Ned. Akad. Wet.*, 37:445–55. 1934.

228 ———, On the pea test method for auxin, the plant growth hormone, *Proc. Kon. Ned. Akad. Wet.*, 37:547–55. 1934.

229 ———, Hormones involved in root formation, *Proc. 6th Int. Bot. Cong.*, 2:267–69. 1935.

230 Wilkinson, R. E., Adventitious shoots on saltcedar roots, *Bot. Gaz.*, 127:103–4. 1966.

231 Winkler, A. J., Some factors influencing the rooting of vine cuttings *Hilgardia*, 2:329–49. 1927.

232 Winter, A., 2,3,5-triiodobenzoic acid and the transport of 3-indoleacetic acid, in F. Wightman, and G. Setterfield (eds.), *Biochemistry and Physiology of Plant Growth Substances*. Ottawa, Canada: Runge Press, 1968, pp. 1063–76.

233 Wott, J. A., and H. B. Tukey, Jr., Influence of nutrient mist on the propagation of cuttings, *Proc. Amer. Soc. Hort. Sci.*, 90:454–61. 1967.

234 Wylie, A. W., K. Ryugo, and R. M. Sachs, Effects of growth retardants on biosynthesis of gibberellin precursors in root tips of peas, *Pisum sativum* L., *Jour. Amer. Soc. Hort. Sci.*, 95(5):627–30. 1970.

235 Yarborough, J. A., Anatomical and developmental studies of the foliar embryos of *Bryophyllum calycinum*, *Amer. Jour. Bot.*, 19:443–53. 1932.

236 ———, Regeneration in the foliage leaf of *Sedum*, *Amer. Jour. Bot.*, 23:303–7. 1936.

237 Yarwood, C. E., Control of powdery mildews with a water spray, *Phytopath.*, 29:288–90. 1939.

238 Zimmerman, P. W., Oxygen requirements for root growth of cuttings in water, *Amer. Jour. Bot.*, 17:842–61. 1930.

239 ———, Initiation and stimulation of adventitious roots caused by unsaturated hydrocarbon gases, *Contrib. Boyce Thomp. Inst.*, 5:351–69. 1933.

240 ———, Comparative effectiveness of acids, esters, and salts as growth substances and methods of evaluating them, *Contrib. Boyce Thomp. Inst.,* 8:337–50. 1937.

241 ———, and F. Wilcoxon, Several chemical growth substances which cause initiation of roots and other responses in plants, *Contrib. Boyce Thomp. Inst.,* 7:209–29. 1935.

SUPPLEMENTARY READING

Argles, G. K., "Root Formation by Stem Cuttings," *Nurseryman and Garden Centre,* Vol. 148, Nos. 18, 19; Vol. 149, No. 3. 1969.

Dore, J., "Physiology of Regeneration in Cormophytes," *Handbuch der Pflanzenphysiologie,* Vol. 15 (Part 2): pp. 1–91. Berlin: Springer-Verlag. 1965.

Fernqvist, I., "Studies on Factors in Adventitious Root Formation," *Lantbrukshogskolans Annaler* (Annals of Agricultural College of Sweden, Uppsala), Vol. 32 (1966), pp. 109–244.

Galston, A. W., and P. J. Davies, "Hormonal Regulation in Higher Plants," *Science,* Vol. 163 (1969): pp. 1288–97.

Haissig, B. E., "Origins of Adventitious Roots," Vol. 4, No. 2 ; "Influences of Auxins and Auxin Synergists on Adventitious Root Primordium Initiation and Development," Vol. 4, No. 2 ; "Consideration of Metabolism During Adventitious Root Primordium Initiation and Development," Vol. 4, No. 2 . *New Zealand Journal of Forestry Science.* 1974.

International Plant Propagators' Society, *Proceedings of Annual Meetings.*

Klein, R. M., and D. T. Klein, *Research Methods in Plant Science.* New York: The Natural History Press, 1970.

Komissarov, D. A., *Biological Basis for the Propagation of Woody Plants by Cuttings* (translated from Russian). Springfield, Va.: U.S. Dept. Commerce, Clearinghouse Fed. Sci. 1964.

Weaver, R. J., *Plant Growth Substances in Agriculture.* San Francisco: W. H. Freeman, 1972. Chapter 5: Rooting and Propagation.

Techniques
of Propagation
by Cuttings

10

In propagation by cuttings, a portion of a stem, root, or leaf is cut from the parent plant, after which this plant part is placed under certain favorable environmental conditions and induced to form roots and shoots, thus producing a new independent plant which, in most cases, is identical with the parent plant.

THE IMPORTANCE AND ADVANTAGES
OF PROPAGATION BY CUTTINGS

This is the most important method of propagating ornamental shrubs—deciduous species as well as the broad- and narrow-leaved types of evergreens. Cuttings are also used widely in commercial greenhouse propagation of many florists' crops and are commonly used in propagating several fruit species.

For species that can be easily propagated by cuttings, this method has numerous advantages. Many new plants can be started in a limited space from a few stock plants. It is inexpensive, rapid, and simple, and does not require the special techniques necessary in grafting or budding. There is no problem of compatibility with rootstocks or of poor graft unions. Greater uniformity is obtained by absence of the variation which sometimes appears owing to the variable seedling rootstocks of grafted plants. The parent plant is usually reproduced exactly, with no genetic change.

It is not always desirable, however, to produce plants on their own roots by cuttings even if it is possible to do so. It is often advantageous or necessary to use a rootstock resistant to some adverse soil condition or soilborne organism, or to utilize available dwarfing or invigorating rootstocks.

TYPES OF CUTTINGS

Cuttings are made from the vegetative portions of the plant, such as stems, modified stems (rhizomes, tubers, corms, and bulbs), leaves, or roots. Cuttings can be classified according to the part of the plant from which they are obtained:

> Stem cuttings
>> Hardwood
>>> Deciduous
>>> Narrow-leaved evergreen
>> Semi-hardwood
>> Softwood
>> Herbaceous
> Leaf cuttings
> Leaf-bud cuttings
> Root cuttings

Many plants can be propagated by several of these different types of cuttings with satisfactory results. The type used would depend upon the individual circumstances, the least expensive and easiest usually being selected.

If the plant being propagated roots easily by hardwood stem cuttings in an outdoor nursery, this method is ordinarily used, because of its simplicity and low cost. Root cuttings of some species are also satisfactory, but cutting material may be difficult to obtain in large quantities. For species more difficult to propagate, it is necessary to resort to the more expensive and elaborate facilities required for rooting the leafy types of cuttings.

In selecting cutting material it is important to use stock plants that are free from diseases, moderately vigorous, and of known identity. Stock plants that are injured by frost or drought, that have been defoliated by insects, that have been stunted by excessive fruiting, or that have made rank, overly vigorous growth should be avoided.

A commendable practice for the propagator is the establishment of stock blocks as a source of propagating material, where uniform, true-to-type, pathogen-free mother plants can be maintained and held under the proper nutritive condition for the best rooting of cuttings taken from them (see p. 235).

Stem Cuttings

This is the most important type of cutting. They can be divided into four groups, according to the nature of the wood used in making the cuttings: *hardwood, semi-hardwood, softwood,* and *herbaceous.* In propagation by stem cuttings, segments of shoots containing lateral or terminal buds are obtained with the expectation that under the proper conditions adventitious roots (see p. 212) will develop and thus produce independent plants.

The type of wood, the stage of growth used in making the cuttings, the time of year in which the cuttings are taken, and several other factors can be very important in securing satisfactory rooting of some plants. Information concerning these factors is given in Chapter 9, although some of this knowledge can be obtained by actual experience in propagating plants.

Hardwood Cuttings (Deciduous Species)
This is one of the least expensive and easiest methods of vegetative propagation. Hardwood cuttings are easy to prepare, are not readily perishable, may be shipped safely over long distances if necessary, and require little or no special equipment during rooting (*50*).

The cuttings are prepared during the dormant season—late fall, winter, or early spring—from wood of the previous season's growth, although with a few species, such as the fig, olive, and certain plum varieties, two-year-old or older wood can be used. Hardwood cuttings are most often used in propagation of deciduous woody plants, although some broad-leaved evergreens, such as the olive, can be propagated by leafless hardwood cuttings. Many deciduous ornamental shrubs are started readily by this type of cutting. Some common ones are privet, forsythia, wisteria, honeysuckle, and spiraea. Rose rootstocks, such as *Rosa multiflora*, are propagated in great quantities by hardwood cuttings. A few fruit species are propagated commercially by this method—for example, fig, quince, olive, mulberry, grape, currant, gooseberry, pomegranate, and some plums. Certain trees such as the willow and poplar, are propagated by hardwood cuttings.

The propagating material for hardwood cuttings should be taken from healthy, moderately vigorous stock plants growing in full sunlight. The wood selected should not be from extremely rank growth with abnormally long internodes, or from small, weakly growing interior shoots. Wood of moderate size and vigor is the most desirable. The cuttings should have an ample supply of stored foods to nourish the developing roots and shoots until the new plant becomes self-sustaining. Tip portions of a shoot are usually low in stored foods and are discarded. Central and basal parts make the best cuttings.

Hardwood cuttings vary considerably in length—from 4 to 30 in. Long cuttings, when they are to be used as rootstocks for fruit trees, permit the insertion of the varietal bud into the original cutting following rooting, rather than into a smaller new shoot arising from the original cutting.

At least two nodes are included in the cutting; the basal cut is usually just below a node and the top cut $\frac{1}{2}$ to 1 in. above a node. However, in preparing stem cuttings of plants with short internodes, little attention is ordinarily given to the position of the basal cut, especially when quantities of cuttings are prepared and cut to length, many at a time, as by a band saw.

The diameter of the cuttings may range from $\frac{1}{4}$ in. to 1 or even 2 in., depending upon the species. Three different types of cuttings can be prepared, as shown in Figure 10-1: the "mallet," the "heel," and the straight cutting. The mallet includes a short section of stem of the older wood, whereas the heel cutting includes only a small piece of the older wood. The straight cutting, not including any of the older wood, is most commonly

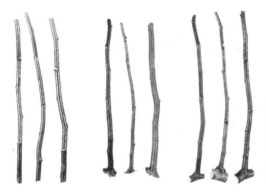

Figure 10–1   Types of hardwood cuttings. *Left:* straight—the type ordinarily used. *Center:* heel cuttings. A small piece of older wood is retained at the base. *Right:* mallet cuttings. An entire section of the branch of older wood is retained.

used, giving satisfactory results in most instances. A very old method of hardwood cutting propagation is shown in Figure 10–2.

Where it is difficult to distinguish between the top and base of the cuttings, it is advisable to make one of the cuts at a slant rather than at right angles. In large-scale operations, bundles of cutting material are cut to the desired lengths by band saws or other types of mechanical cutters rather than individually by hand (Figure 10–3). For large-scale commercial operations planting the cuttings is mechanized, using equipment as illustrated in Figure 10–4, but for planting a limited number of cuttings the method shown in Figure 10–5 would be satisfactory.

There are several methods commonly used for preparing and handling hardwood cuttings before planting:

(a) Winter callusing   During the dormant season, make the cuttings of uniform length, tie them with heavy rubber bands into convenient-sized bundles, placing the tops all one way, and store them under cool, moist conditions until spring. The bundles of cuttings may be buried out-of-doors in sandy soil, sand, or sawdust in a well-drained location. They may be placed horizontally or buried in a vertical position, but **upside down** with the basal end of the cuttings several inches below the surface of the soil. The basal

Figure 10–2   A type of hardwood cutting, called a "truncheon," is planted horizontally several inches below ground. One or more shoots may arise from latent buds; adventitious roots develop from either the new shoot or the original cutting, or both. Two- or three-year-old wood, an inch or two in diameter, is usually used. Shown here for propagation of the olive.

Figure 10-3   Sawing hardwood cuttings to length with a band saw. This method is much faster than preparing each cutting individually and in most cases gives equally good results.

Figure 10-4   Machine for large-scale planting of hardwood cuttings. Developed at the Tree Nursery Division, P.F.R.A., Indian Head, Saskatchewan, Canada, primarily for propagation of willow and poplar for shelter belt use. This 4-unit machine will plant 10 to 12 thousand cuttings per hour. Courtesy Canada Department of Regional Economic Expansion.

Figure 10-5   Steps in making and planting hardwood cuttings. *Top left:* preparing the cuttings from dormant and leafless one-year-old shoots. A common length is 6 to 8 in., and the basal cut is generally made just below a node. *Top right:* treating the cuttings with a root-promoting substance. On the left a bundle of cuttings is being dipped in a commercial talc preparation. On the right another method is illustrated. The basal ends of the cuttings are being soaked for 24 hr in a dilute solution of the chemical. With easily rooted plants such treatments are unnecessary. *Middle left:* the cuttings may be planted immediately, but with some plants it is helpful to callus the cuttings for several weeks in a box of moist shavings or peat moss before planting. *Middle right:* planting the cuttings in the nursery row. A *dibble* (a heavy, pointed, flat-bladed knife) is a useful tool for inserting the cutting and at the same time firming the soil around the previously planted cutting. *Bottom left:* the cuttings should be planted 3 or 4 in. apart and deeply enough so that just one bud shows above ground. A loose, sandy loam is best for starting hardwood cuttings. *Bottom right:* several weeks after planting, the cuttings start to grow. They must be watered frequently if rains do not occur, and weeds must be controlled.

ends are somewhat warmer and better aerated than the terminal ends. This procedure tends to promote root initiation at the base, while retarding bud development at the top. At planting time in the spring, the bundles of cuttings are dug up and the cuttings planted right side up. In regions with mild winters, the bundles of cuttings are often stored during this callusing period in large boxes of moist sand, sawdust, peat moss, or shavings, either in an unheated building or out-of-doors. This probably would not be enough protection for the cuttings, however, in regions where severe, subzero winter temperatures are experienced. A cool, but above-freezing, cellar would be satisfactory for such climates. If refrigerated rooms are available, the cuttings can be safely stored during the callusing period at temperatures of about 40° F (4.5° C) until they are ready to plant.

(b) Direct spring planting   It is often sufficient with easily rooted species to gather the cutting material during the dormant season, wrap it in heavy paper or polyethylene with slightly damp peat moss, and store at 32° to 40° F (0° to 4.5° C) until spring. The cutting material should not be allowed to dry out or to become excessively wet during storage. At planting time, the cuttings are made into proper lengths and planted in the nursery.

Stored cutting material should be examined frequently. If signs of bud development appear, lower storage temperatures should be used or the cuttings should be made and planted without delay. If the buds are far developed when the cuttings are planted, leaves will form before the roots appear, and the cuttings will die, owing to water loss from the leaves.

(c) Direct fall planting   In regions with mild winters, cuttings can be made in the autumn and planted immediately in the nursery. Callusing, and perhaps rooting, may take place before the dormant season starts, or the formation of roots and shoots may occur simultaneously the following spring. Hardwood cuttings of peach and peach × almond hybrids have been successfully rooted in the nursery by this method provided they were treated prior to planting with indolebutyric acid and captan (*15*). Fall-planted cuttings could be injured by rodents and, unless herbicides are used, weed growth may be considerable.

(d) Warm temperature callusing   Take the cuttings in the fall while the bud are in or entering the "rest period" (see p. 224), treat them with a root-promoting chemical (see p. 291), then store under moist conditions at relatively warm temperatures—65° to 70° F (18° to 21° C)—for 3 to 5 weeks to stimulate root initiation. After this, plant the cuttings in the nursery (in mild climates) or hold in cold storage (35° to 40° F; 2° to 4.5° C) until spring. Experimentally, good rooting of hardwood pear cuttings occurred when the cuttings were allowed to callus (and initiate roots) while the buds were under the "rest" influence and did not start growth and compete for food reserves in the cuttings (*1*).

(e) Bottom heat callusing   This method has been successful for difficult-to-root subjects such as some apple, pear, and plum rootstocks. Cuttings are collected in either the fall or late winter, the basal ends treated with root-promoting chemicals (IBA at 2500 to 5000 ppm) then placed upright for about 4 weeks in damp packing material over bottom heat at 65° to 70° F

(18° to 21° F), but with the top portion of the cuttings left exposed to the cool outdoor temperatures. It is best to do this in a covered open shed for protection against excessive moisture from rains. The East Malling Research Station in England has developed (23, 24, 25) commercial procedures, as shown in Figure 10–6, for propagating difficult subjects by this method. Cuttings must be transplanted before buds commence growth; this is usually done as roots begin to emerge. It is important to prevent decay in the cuttings by avoiding excessive application of water to the rooting compost. As long as the correct stimulation has been given it is not essential to await root emergence before transplanting.

This procedure is probably best suited for regions having relatively mild winters (16, 17, 23). When soil or weather conditions are not suitable for planting after roots become visible, it has been satisfactory to leave the cuttings undisturbed in the rooting bed, shut off the bottom heat, then plant them in the nursery when conditions do become suitable (4).

(f) Plastic bag storage   The hardwood cuttings are taken during the dormant season, the bases dipped into a root-promoting material—for example IBA at 2000 ppm, for a few seconds—then sealed in a polyethylene bag which is placed in the dark at a temperature of about 50° F (10° C). Studies with this technique using peach hardwood cuttings showed 85 to 100 percent rooting after about 50 days (45). While high rooting can often be obtained by this method it is often difficult to obtain survival of the cuttings following transplanting.

Hardwood Cuttings (Narrow-Leaved Evergreen Species)
Cuttings of this type have leaves and must be rooted under moisture conditions that will prevent excessive drying as they usually are slow to root, taking several months to a year. Some species root much more readily than others. In general *Chamaecyparis, Thuja,* and the low-growing *Juniperus* species root easily and the yews (*Taxus sp.*) fairly well, whereas the upright junipers, the spruces (*Picea sp.*), hemlocks (*Tsuga sp.*), firs (*Abies sp.*), and pines (*Pinus sp.*) are more difficult. In addition, there is considerable variability among the different species in these genera in regard to the ease of rooting of cuttings. Cuttings taken from young seedling stock plants root much more readily than those taken from older trees. Treatments with root-promoting substances, particularly indolebutyric acid (see p. 292) at relatively high concentrations, are usually beneficial in increasing the speed of rooting, the percentage of cuttings rooted, and obtaining heavier root systems (32, 33, 51).

Narrow-leaved evergreen cuttings ordinarily are best taken between late fall and late winter (see Figure 9–17). Rapid handling of the cuttings after the material is taken from the stock plants is important. The cuttings are best rooted in a greenhouse with relatively high light intensity and under conditions of high humidity or very light misting but without heavy wetting of the leaves. A bottom heat temperature of 75° to 80° F (24° to 26.5° C) has given good results. Dipping the cuttings into a fungicide helps prevent

Figure 10–6 Steps in propagation by hardwood cuttings using the bottom-heat technique for difficult-to-root materials. *Upper left:* removing 'M 26' apple cuttings from hard-pruned, vigorous hedges by cutting one-year shoots at their base. *Upper right:* two-foot-long cuttings inserted (after IBA treatment) to a 10 in. depth in insulated bins filled with rooting compost (one-half coarse peat and one-half grit— 3/16 in. gravel and washed sand) maintained at 70°F (21°C). Bins are situated in a cool building to retard bud development. *Lower left:* root development after six weeks (shown here for plum cuttings). *Lower right:* apple rootstock hardwood cuttings ('M 26,' 'M 106,' 'MM 111') after one season's growth in nursery. Photos courtesy East Malling Research Station, England.

Figure 10–7   Narrow-leaved evergreen cuttings. These rooted juniper cuttings are about 5 in. long.

disease attacks. Sand alone is a satisfactory rooting medium, as is 1:1 mixture of perlite and peat moss. Some individual cuttings take longer to root than others. The slower-rooting ones can be restuck in the rooting medium, and often will root eventually.

The type of wood to use in making the cuttings varies considerably with the particular species being rooted. As shown in Figure 10–7, the cuttings are made 4 to 8 in. long with all the leaves removed from the lower half of the cutting. Mature terminal shoots of the previous season's growth are usually used. In some instances, as with *Juniperus chinensis* 'Pfitzeriana,' older and heavier wood can also be used, thus resulting in a larger plant when it is rooted. On the other hand, some nurserymen use small tip cuttings, 2 to 3 in. long, placed very close together in a flat for rooting. In some species, as *Juniperus excelsa,* older growth taken from the sides and lower portion of the stock plant roots better than the more succulent tips. Cuttings of *Taxus* root best if they are taken with a piece of old wood at the base of the cutting; such cuttings seem less subject to fungus attacks (*58*). In certain of the narrow-leaved evergreen species, some type of basal wounding (see p. 290) is often beneficial in inducing rooting.

Semi-Hardwood Cuttings

Cuttings of this type are usually made from woody, broad-leaved evergreen species, but leafy summer cuttings taken from partially matured wood of deciduous plants could also be considered as semi-hardwood. Cuttings of broad-leaved evergreen species are generally taken during the summer from new shoots just after a flush of growth has taken place and the wood is partially matured. Many ornamental shrubs, such as camellia, pittosporum, euonymus, the evergreen azaleas, and holly, are commonly propagated by semi-hardwood cuttings. A few fruit species, such as citrus and olive, can also be propagated in this manner.

The cuttings are made 3 to 6 in. long with leaves retained at the upper end, as shown in Figure 10–8. If the leaves are very large, they should be reduced in size to lower the water loss and to allow closer spacing in the cutting bed. The shoot terminals are often used in making the cuttings, but the basal parts of the stem will usually root also. The basal cut is usually just below a node. The cutting wood should be obtained in the cool, early morning hours when the stems are turgid, and kept wrapped in clean moist burlap or put in large polyethylene bags. Keep out of the sun at all times until the cuttings are made.

Figure 10–8 Semi-hardwood cuttings as illustrated by the *Camellia*. *Top:* cuttings made in late summer from partially matured wood. *Below:* cuttings starting to root.

It is necessary that leafy cuttings be rooted under conditions which will keep water loss from the leaves at a minimum; commercially they are ordinarily rooted under intermittent mist sprays. Bottom heat and growth-regulator treatments are also beneficial (see pp. 22 and 291). Rooting media, such as 1:1 mixture of perlite and peat moss, or perlite and vermiculite, give satisfactory results. (Rooting procedures for semi-hardwood cuttings are shown in Figure 10–9.)

Figure 10–9 Steps in placing semi-hardwood cuttings in mist propagating beds for rooting, shown here with English holly. *Upper left:* cutting row in rooting medium with heavy knife against board. *Upper right:* selecting prepared cuttings for inserting. *Lower left:* dipping cuttings for 5 sec. in liquid rooting "hormone" (auxin) preparation. *Lower right:* sticking cuttings in rooting medium. Courtesy Klass Ellerbrook, West Oregon Nursery, Portland, Oregon.

Softwood (Greenwood) Cuttings

Cuttings prepared from the soft, succulent, new spring growth of deciduous or evergreen species may properly be classed as softwood cuttings. Many ornamental woody shrubs can be started by softwood cuttings. Typical examples are the hybrid French lilacs, forsythia, magnolia, weigela, and spiraea. Other examples are shown in Figure 10–10. Some deciduous ornamental trees, such as the maples, can also be started in this manner. Although fruit tree species are not commonly propagated by softwood cuttings, apple, peach, pear, plum, apricot, and cherry will root, especially under mist (see. p. 297).

Softwood cuttings generally root easier and quicker than the other types but require more attention and equipment. This type of cutting is always made with leaves attached. They must, consequently, be handled carefully to prevent drying, and be rooted under conditions which will avoid excessive water loss from the leaves. Temperature should be maintained during rooting at 75° to 80° F (23° to 27° C) at the base and 70° F (21° C) at the leaves for most species. Softwood cuttings produce roots in two to four or five weeks in most cases. In general, they respond well to treatments with root-promoting substances (see p. 291).

It is important in making softwood cuttings to obtain the proper type of cutting material from the stock plant. Such material will vary greatly, however, with the species being propagated. Extremely fast-growing, soft, tender shoots are not desirable, as they are likely to deteriorate before rooting. At the other extreme, older woody stems are slow to root or may just drop their leaves and not root. The best cutting material has some degree of flexibility but is mature enough to break when bent sharply. Weak, thin, interior shoots should be avoided as well as vigorous, abnormally thick, or heavy ones. Average growth from portions of the plant in full light is the most desirable to use. Some of the best cutting material is the lateral or side branches of the stock plant. Heading back the main shoots will usually force out numerous lateral shoots from which cuttings can be made. Softwood

Figure 10–10 Softwood cuttings of several ornamental species. *Top:* cuttings made in late spring from young shoots. *Below:* cuttings after rooting. Left to right: *Myrtus, Pyracantha, Oleander,* and *Veronica.*

Figure 10–11   Typical herbaceous cuttings. Left to right: chrysanthemum, begonia, and geranium. It is often necessary with large-leaved plants, such as the begonia, to trim back some of the leaves to prevent wilting and to conserve space in the propagating bench.

cuttings are 3 to 5 in. long with two or more nodes. The basal cut is usually made just below a node. The leaves on the lower portion of the cutting are removed, with those on the upper part retained. Large leaves should be reduced in size to lower the transpiration rate and to occupy less space in the propagating bed. All flowers or flower buds should be removed. In some nurseries where quantities of cuttings are prepared, bundles of cutting material are rapidly cut into uniform lengths by paper cutters.

The cutting material is best gathered in the early part of the day and should be kept moist, cool, and turgid at all times by wrapping in damp, clean burlap or placing in large polyethylene bags (kept out of the sun). Laying the cutting material or prepared cuttings in the sun for even a few minutes will cause serious damage. Soaking the cutting material or cuttings in water for prolonged periods to keep them fresh is undesirable.

Herbaceous Cuttings

This type of cutting is made from such succulent, herbaceous plants as geraniums, chrysanthemums, coleus, or carnations. They are 3 to 5 in. long with leaves retained at the proper end, as shown in Figure 10–11, or without leaves (Figure 10–12). Most florists' crops are propagated by herbaceous cuttings. They are rooted under the same conditions as softwood cuttings, requiring high humidity. Bottom heat is also helpful. Under proper conditions, rooting is rapid and in high percentages. Although root-promoting

Figure 10–12   A type of stem cutting consisting only of a leafless stem piece; used here in propagating the monocotyledonous plant, *Dieffenbachia picta*. Latent buds develop into shoots along with the formation of adventitious roots.

substances are usually not required, they are often used to gain uniformity in rooting and development of heavier root systems. Herbaceous cuttings of some plants that exude a sticky sap, such as the geranium, pineapple, or cactus, do better if the basal ends are allowed to dry for a few hours before they are inserted in the rooting medium. This practice tends to prevent the entrance of decay organisms.

Leaf Cuttings

In this type of cutting, the leaf blade, or leaf blade and petiole, are utilized in starting a new plant. Adventitious roots and an adventitious shoot (see p. 216) form at the base of the leaf; the original leaf does not become a part of the new plant.

One type of propagation by leaf cuttings is illustrated by *Sansevieria*. The long tapering leaves are cut into sections 3 to 4 in. long as shown in Figure 10–13. These leaf pieces are inserted three-fourths of their length into sand, and after a period of time a new plant forms at the base of the leaf piece, the original cutting disintegrating. The variegated form of *Sansevieria, S. trifasciata laurenti,* is an example of a periclinal chimera (see p. 191) which will not reproduce true to type from leaf cuttings; to retain its characteristics, it must be propagated by division of the original plant.

In starting plants with thick, fleshy leaves, as *Begonia rex,* by leaf cuttings, the large veins are cut on the undersurface of the mature leaf, which is then laid flat on the surface of the propagating medium. The leaf is pinned or held down in some manner, with the natural upper surface of the leaf exposed. As shown in Figure 10–14, after a period of time under humid conditions, new plants will form at the point where each vein was cut. The old leaf blade will gradually disintegrate.

Another method, sometimes used with fibrous-rooted begonias, is to cut large, well-matured leaves into triangular sections, each containing a piece

Figure 10–13   Leaf cuttings of *Sansevieria.* *Left:* the thick, leathery leaves are cut into pieces 3 or 4 in. long. To avoid trying to root upside down, the basal end can be marked by cutting on a slant as shown with two of the cuttings. *Right:* development of the plant. The original cutting does not become a part of the new plant.

Figure 10–14   Propagation of *Begonia rex* by leaf cuttings. New plants arise at wounds made in the large veins of the leaf which is laid flat on the surface of the rooting medium.

of a large vein. The thin outer edge of the leaf is discarded. These leaf pieces are then inserted upright in sand with the pointed end down. The new plant develops from the large vein at the base of the leaf piece.

A rapid method of propagating begonias is by stamping out leaf disks (2 cm diam.) with a cork borer, obtaining 40 to 50 disks from a single leaf, each disk giving a new plant *(26)*. Treatment of the disks with indolebutyric acid and kinetin has stimulated both root and shoot development. Disks are placed on moistened filter paper in covered petri dishes during the regeneration period. This method may also be useful in propagating other species with large flat leaves.

The African violet (*Saintpaulia*) is typical of leaf cuttings which can be made of an entire leaf (leaf blade plus petiole), the leaf blade only, or just a portion of the leaf blade. The new plant forms at the base of the petiole or midrib of the leaf blade. (See Figure 10–15.) An unusual type of leaf

Figure 10–15   Leaf cuttings of African violet *(Saintpaulia) (top)* and of *Peperomia (below)*. *Left:* each cutting consists of a leaf blade and petiole. *Right:* leaf cuttings after rooting. One or more new plants will form at the base of the petiole. The original leaf can be cut off and used again for rooting.

Figure 10–16   Leaf cuttings of *Kalanchoe pinnata* (*Bryophyllum pinnata*), air plant. *Left*: New plants developing from foliar "embryos" in the notches at the margin of the leaf. *Right*: Leaves ready to lay flat on the rooting medium. They should be partially covered or pegged down to hold the leaf margin in close contact with the rooting medium.

cutting is illustrated in Figure 10–16, where many new plants arise at the margins of the leaf. The leaf itself eventually deteriorates.

Leaf cuttings should be rooted under the same conditions of high humidity used for softwood or herbaceous cuttings. Root-promoting chemicals are usually helpful.

Leaf-Bud Cuttings

This type of cutting consists of a leaf blade, petiole, and a short piece of the stem with the attached axillary bud (Figure 10–17).

Such cuttings are of particular value where roots but not shoots are initiated from detached leaves, the axillary bud at the base of the petiole providing for the new shoot. A number of plant species such as the black raspberry *(Rubus occidentalis)*, blackberry, boysenberry, lemon, camellia, and rhododendron are readily started by leaf-bud cuttings, as well as many tropical shrubs and most herbaceous greenhouse plants usually started by stem cuttings. Red raspberries *(Rubus idaeus)* apparently will not reproduce in this manner.

Figure 10–17   Leaf cuttings of some plants, such as the jade plant *(Crassula argentea)*, produce roots but no shoots *(left)*. By preparing leaf-bud cuttings *(right)*, however, a new shoot develops, along with the roots.

This method is particularly valuable when propagating material is scarce, because it will produce at least twice as many new plants from the same amount of stock material as can be started by stem cuttings. Each node can be used as a cutting. Leaf-bud cuttings are best made from material having well-developed buds and healthy, actively growing leaves.

Treatment of the cut surfaces with one of the root-promoting substances should stimulate root production. The cuttings are inserted in the rooting medium with the bud $\frac{1}{2}$ to 1 in. below the surface. High humidity is essential, and bottom heat is desirable for rapid rooting. Sand, or sand and peat moss, 1:1, are satisfactory rooting media for leaf-bud cuttings.

Root Cuttings

Best results are likely to be attained if the root pieces are taken from young stock plants in late winter or early spring when the roots are well supplied with stored foods but before new growth starts. The period during the spring when the parent plant is rapidly making new shoot growth should be avoided. Root cuttings of the Oriental poppy (*Papaver orientale*) should be taken in midsummer, the dormant period for this species.

Securing cutting material in quantities for root cuttings can be quite laborious unless it can be obtained by trimming roots from nursery plants as they are dug.

It is important with root cuttings to maintain the correct polarity when planting. To avoid planting them upside down, the proximal end (nearest the crown of the plant) may be made with a straight cut and the distal end (away from the crown) with a slanting cut. The proximal end of the root piece should always be up. In planting, insert the cutting vertically so that the top is at about soil level; with many species, however, it is satisfactory to plant the cuttings horizontally 1 or 2 in. deep (Figure 10–18).

In using root cuttings to propagate chimeras with variegated foliage, such as some *Aralias* and *Pelargoniums,* the new plants will lose their variegated form (see p. 192).

Propagation by root cuttings is very simple, but the root size of the plant being propagated may determine the best procedure.

Plants with small, delicate roots   Root cuttings of such plants should be started in flats of sand or finely screened soil in the greenhouse or hotbed. The roots are cut into short lengths, 1 to 2 in. long, and scattered horizontally over the surface of the soil. They are covered with a ½-in. layer of fine soil or sand. After watering a polyethylene cover or a pane of glass

Figure 10–18   Propagation by root cuttings. *Left:* horseradish (*Armoracia rusticana*) root pieces planted horizontally. *Right:* apple (*Malus sylvestris*) root pieces set vertically. Adventitious buds arising from the root piece form the new shoot system.

Table 10–1   Some species that can be propagated by root cuttings.

Actinidia chinensis (Chinese gooseberry)	*Plumbago* sp. (leadwort)
Aesculus paviflora (bottle-brush buckeye)	*Populus alba* (white poplar)
Ailanthus altissima (tree-of-heaven)	*Populus tremula* (European aspen)
Albizia julibrissin (silk tree)	*Populus tremuloides* (quaking aspen)
Aralia spinosa (devil's walking stick)	*Prunus glandulosa* (dwarf flowering
Artocarpus altilis (breadfruit)	almond)
Broussonetia papyrifera (paper mulberry)	*Pyrus calleryana* (oriental pear)
Campsis radicans (trumpet vine)	*Rhus copallina* (shining sumac)
Celastrus scandens (American bitter sweet)	*Rhus glabra* (smooth sumac)
Chaenomeles japonica (Japanese flowering	*Rhus typhina* (staghorn sumac)
quince)	*Robinia hispida* (rose acacia)
Clerodendrum trichotomum (glory-bower)	*Robinia pseudoacacia* (black locust)
Comptonia peregrina (sweet fern)	*Rosa blanda* (rose)
Daphne genkwa (daphne)	*Rosa nitida* (rose)
Eschscholzia californica (California poppy)	*Rosa virginiana* (rose)
Koelreuteria paniculate (goldenrain tree)	*Rubus* sp. (blackberry, raspberry)
Ficus carica (fig)	*Sassafas albidum* (sassafras)
Malus sp. (apple, flowering crabapple)	*Sophora japonica* (Japanese pagoda tree)
Myrica pennsylvanica (bayberry)	*Syringa vulgaris* (lilac)
Papaver orientale (oriental poppy)	*Ulmus carpinifolia* (smooth-leaved elm)
Phlox sp. (phlox)	

should be placed over the flat to prevent drying until the plants are started. The flats are set in a shaded place. After the plants become well formed, they can be transplanted to other flats or lined-out in nursery rows for further growth.

Plants with somewhat fleshy roots   This type of root cutting is best started in a flat of sandy soil in the greenhouse or hotbed. The root pieces should be 2 to 3 in. long and planted vertically. New adventitious shoots should form rapidly, and as soon as the plants become well established with good root development they can be transplanted to their permanent location.

Plants with large roots, propagated out-of-doors   Root cuttings of this type are made 2 to 6 in. long. They are tied in bundles, care being used to keep the same ends together to avoid planting upside down later. The cuttings are packed in boxes of damp sand, sawdust, or peat moss for about three weeks and held at about 40° F (4.5° C). After this they should be planted 2 to 3 in. apart in a well-prepared nursery soil with the tops of the cuttings level with, or just below, the top of the soil.

Table 10–1 lists many of the species which can be propagated by root cuttings (*8, 12, 41, 49*).

Rooting Media
Cuttings of many species root easily in a variety of rooting media but those more difficult to root may be greatly influenced by the kind of rooting medium used, not only in the percentage of cuttings rooted, but in the quality of root system formed (*27*).

Combinations of some of the materials listed below often give better results than any one used alone. It is advisable to experiment with the plants being propagated under the actual environmental conditions at hand to determine the best rooting mixture.

Soil is ordinarily used for planting deciduous hardwood cuttings and root cuttings. A well-aerated sandy loam is preferable to a heavy clay soil, a higher percentage of the cuttings forming roots, which are usually of better quality. Also, in the lighter, sandy soils, cuttings may be planted and—after rooting—dug much sooner following rains than when the heavier soils are used. The nursery soil should be free from nematodes, verticillium, and crown gall. Nematodes can be effectively eliminated by treating the soil prior to planting with some fumigant such as D-D (dichloropropene-dichloropropane). This requires a three-week period or more following application for the fumigant to dissipate (*35*). (See page 35.) Soil is usually not considered a suitable rooting medium for the more succulent softwood and semi-hardwood types of cuttings, although some commercial nurserymen have used it successfully. Cuttings of certain easily rooted plants, such as chrysanthemums and geraniums, are sometimes started directly in small containers or plant bands, using a mixture of 2 parts coarse sand to 1 part soil. This mixture preferably should be heat-treated or fumigated before using (see p. 33).

Sand is a widely used rooting medium for cuttings. It is inexpensive and readily available. Clean, sharp plaster sand, free from organic matter and soil, as usually supplied to the building trade, is excellent. Sand is not as retentive of moisture, however, as most other rooting media, necessitating more frequent watering. The sand should be fine enough to retain some moisture around the cuttings yet coarse enough to allow water to drain freely through it. Used alone, very fine particle sand or very coarse sand does not give good results with cuttings of most woody ornamentals. As with other rooting media, it is best to use the sand only once for rooting cuttings unless it can be sterilized (see p. 33).

For evergreens, such as yews, junipers, and arborvitaes, sand is probably the most satisfactory rooting medium to use. With some species, however, cuttings rooted in sand produce a long, unbranched, brittle root system in contrast to the more desirable fibrous and branched systems developed in other media (*27*).

Peat moss (see p. 27) is often added to sand in varying proportions, mainly to increase the water-holding capacity of the mixture. This combination makes a good rooting medium for cuttings of many species. Mixtures used vary from 2 parts sand and 1 of peat moss to 1 part sand and 3 of peat moss.

Including peat moss in a rooting medium considerably increases the mixture's water-holding capacity, and consequently the danger of overwatering. High proportions of peat moss in the mixture if kept wet, as in a mist bed, will sometimes cause deterioration of the roots soon after they are formed.

Shredded sphagnum moss (see p. 27) is sometimes used as a rooting medium when mixed with an equal part of sand (*9*).

Vermiculite (see p. 27) is often used as a rooting medium. Tests (*5, 34*) have shown that cuttings of some plants root better in the larger particle sizes, whereas others do better in the smaller sizes. A mixture of equal parts of vermiculite and perlite (or a medium-grade sand) usually gives better results than either material used alone.

Perlite (see p. 28) is widely used as a rooting medium for leafy cuttings, especially under mist, owing to its good drainage properties. It may be used alone but is best when used in combination, in varying proportions, with peat moss or vermiculite.

Water can be used to root cuttings of easily propagated species. Its great disadvantage is lack of aeration. Artificially aerating water with air or oxygen, can produce excellent rooting of cuttings of some species (*46, 62*). In aerated water, the best roots are produced near the basal end of the cuttings, whereas in nonaerated water, the best roots are produced near the surface of the water where the oxygen content is higher (see Figure 9–21).

Moisture-saturated air can be used as a rooting medium by placing cuttings in closed frames in which the relative humidity is maintained by mist nozzles close to 100 percent. This method has resulted in satisfactory root formation with some plants and is especially successful with root cuttings, but does not lend itself to large-scale use.

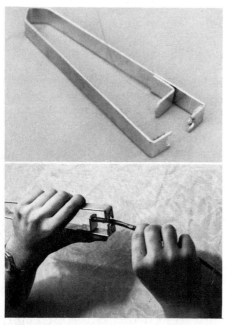

Figure 10–19   Patented tool designed for making wounding cuts in the base of cuttings to stimulate rooting. Four sharp prongs make the actual cuts as the cutting is pulled through the opening, as shown in the lower photo.

WOUNDING

Root production on stem cuttings can be promoted by wounding the base of the cutting in a number of plant species, such as juniper, arborvitae, rhododendron, maple, magnolia, and holly species (*59*). Wounds may be produced in cuttings of narrow-leaved evergreen species, such as arborvitae, by stripping off the lower side branches of the cuttings. A vertical cut with the tip of a sharp knife down each side of the cutting for an inch or two, penetrating through the bark and into the wood, may be enough. A more drastic wound is made with a razor blade device. This consists of four single-edge blades soldered together along their backs. Four wound cuts are then made simultaneously with this equipment.

Larger cuttings, as magnolias, may be more effectively wounded by removing a thin slice of bark for about an inch from the base on two sides of the cutting, exposing the cambium but not cutting deeply into the wood.

For the greatest benefit, the cuttings should be treated after wounding with one of the root-promoting compounds, either a talc or a concentrated-solution-dip preparation (see p. 246), working the material into the wounds.

The device shown in Figure 10–19 can be used for wounding cuttings rapidly and uniformly.

TREATING CUTTINGS WITH GROWTH REGULATORS

The purpose of treating cuttings with auxin-type growth regulators ("hormones") is to increase the percentage of cuttings which form roots, to hasten root initiation, to increase the number and quality of roots produced per cutting and to increase uniformity of rooting (see p. 246). Figure 10–20 shows, for example, the benefits of such materials. Plants whose cuttings root easily may not justify the additional expense and effort of using these materials. Best use of rooting hormones is with plants whose cuttings will root only with difficulty. The use of these substances, however, does not permit other good practices in cutting propagation, such as the maintenance of proper water relations, temperature, and light conditions, to be ignored. The value of these chemicals in propagation is well established, as shown by the tremendous number of reports in scientific and trade journals of tests made on the rooting of cuttings of almost all plant species of economic importance (*38, 53*). Although treatment of cuttings with root-promoting substances is useful in propagating plants, the ultimate size and vigor of such treated plants is no greater than obtained with untreated plants (*6*).

Figure 10–20 Effect of wounding and auxin treatment on the rooting of cuttings of *Juniperus sabina* 'Tamariscifolia' under intermittent mist in the greenhouse. *Top:* wounded. *Below:* not wounded. *Left:* treated with indolebutyric acid at 4000 ppm by the concentrated-dip method. *Center:* treated with indolebutyric acid in talc, at 8000 ppm. *Right:* not treated. Cuttings started March 5, dug April 23.

Materials

The synthetic root-promoting chemicals that have been found most reliable in stimulating adventitious root production in cuttings are **indolebutyric acid** and **naphthaleneacetic acid,** although there are others which can be used (see p. 246). Indolebutyric acid is probably the best material for general use, because it is nontoxic over a wide concentration range and is effective in promoting rooting of a large number of plant species. These chemicals are available in commercial preparations, dispersed in talc, or liquid formulations which can be diluted with water to the proper strength. The pure chemicals are also available from chemical manufacturers,* so it is possible for the propagator to prepare his own solutions.

Methods of Application
Commercial Powder Preparations

Complete directions come with the commercial materials, together with a list of plants which are likely to respond to the particular preparation. Woody, difficult-to-root species should be treated with the more concentrated preparations, whereas tender, succulent, and easily rooted species should be treated with lower-strength materials. Fresh cuts should be made at the base of the cuttings shortly before they are dipped into the powder. The operation is faster if a bundle of cuttings is dipped at once rather than each cutting individually, although the inner cuttings in the bundle may not receive as much powder as those on the outside. The powder adhering to the cuttings after they are lightly tapped is sufficient. If there is little or no natural moisture at the base of the cuttings, they may be pressed against a damp sponge before being dipped in the powder so that more will adhere.

It is advisable in using powder preparations to place a small portion of the stock material into a temporary container, sufficient for the work at hand, and discard any remaining portion, after use rather than dipping the cuttings into the entire stock of powder; this may lead to its early deterioration due to contamination with moisture and fungi or bacteria.

The cuttings should be inserted into the rooting medium immediately after treatment. To avoid brushing off the powder during insertion, a thick knife may be used to make a trench in the rooting medium before the cuttings are inserted. (See Fig. 10-8).

Talc preparations have the advantage of being readily available and easy to use. Uniform results may be difficult to obtain, owing to the variability in the amount of the material adhering to the cuttings. This is influenced by such factors as the amount of moisture at the base of the cutting and the texture of the stem (hairy or smooth).

Dilute Solution Soaking Method

This is an older procedure in which the basal inch of the cuttings is soaked in a dilute solution of the material for about 24 hours just before they are

---

* ICN Pharmaceuticals, Inc., Cleveland, Ohio 44128; Eastman Organic Chemicals, Rochester, New York 14650.

inserted into the rooting medium. The concentrations used vary from about 20 ppm for easily rooted species to about 200 ppm for the more difficult species.

To prepare one liter of a 100-ppm solution of a root-promoting substance, 100 mg of the pure chemical is dissolved in about 10 ml of alcohol (ethyl, methyl, or isopropyl). This is then diluted with water to make one liter. Naphthaleneacetic acid dissolves best in a few drops of ammonium hydroxide before adding to the water. The acid form of these growth substances is not directly soluble in water. The potassium salt of indolebutyric acid, which is water-soluble, is available.

An approximate 100-ppm solution of indolebutyric acid can be prepared by dissolving a level ¼ teaspoon of the chemical in a small amount of alcohol and adding to 1 gal. of water, stirring thoroughly.

During the soaking period, the cuttings should be held at about 68° F (20° C) but not placed in the sun. The amount of the chemical absorbed by the cuttings depends somewhat upon the surrounding conditions during this period, which may lead to some variation in the result obtained.

Concentrated-Solution-Dip Method
A concentrated solution varying from 500 to 10,000 ppm (0.05 to 1.0 percent) of the chemical in 50 percent alcohol is prepared, and the basal (¼ to ½ in.) of the cuttings are dipped in it for a short time (about 5 seconds); then the cuttings are inserted into the rooting medium. This method of application has a number of advantages over the others. It eliminates the necessity of providing equipment for soaking the cuttings and returning later to insert them in the rooting medium. In addition, more uniform results are likely to be obtained, because the uptake of the chemical by the cuttings is not influenced as much by surrounding conditions as is the case with the other two methods. The same solution can be reused for many thousands of cuttings, but it must be tightly sealed when not in use, because the evaporation of the alcohol will change its concentration. It is best to use only a portion of the material at a time, just sufficient for the immediate needs, discarding it after use rather than pouring it back into the stock solution. The propagator (or his pharmacist) can prepare these solutions himself, using the pure crystals, although liquid concentrates, available commercially, can be obtained and diluted according to directions.

To prepare 100 ml of a 4000-ppm solution of a root-promoting substance, weigh out 400 mg of the chemical and dissolve it in 100 ml of 50 percent alcohol (ethyl, methyl, or isopropyl).

An approximate 4000-ppm solution of indolebutyric acid can be prepared by dissolving a level ¼ teaspoon of the pure crystals in 3⅓ fluid oz of 50 percent alcohol.

Figure 10-21   Tomato leaf cuttings provide a sensitive test for the effectiveness of root-promoting substances. Left to right: no treatment; treated with indolebutyric acid in talc at 1000 ppm; at 3000 ppm; and at 8000 ppm. 'Marglobe' variety after 12 days in sand.

Growth regulators used in excessive concentrations for the species may inhibit bud development (6), cause yellowing and dropping of leaves, blackening of the stem, and eventual death of the cuttings. An effective, nontoxic concentration has been used if the basal portion of the stem shows some swelling, callusing, and profuse root production just above the base of the cutting. A concentration just below the toxic point is considered the most favorable for root promotion.

Some negative results obtained in using growth regulators as an aid in rooting cuttings may be due to the use of old or deteriorated chemicals. A simple test using tomato leaf cuttings makes it possible to determine in a few days whether the preparation planned for use has root-promoting properties (20). Tomato leaf cuttings, of the type shown in Figure 10-21, are treated with the material, then inserted in moist sand in a glass- or polyethylene-covered box together with a group of untreated cuttings for comparison. After one week the cuttings can be observed. Tomato cuttings are sensitive to growth regulators and will give a good indication of the effectiveness of the material by the extent of their root production.

Use fresh preparations whenever possible. Dilute solutions, e.g., 25 ppm, lose their activity within a few days, especially if they become contaminated with foreign material. Solutions used in the concentrated-solution-dip method of application, which contain a high percentage of alcohol, will retain their activity almost indefinitely.

Usually the base of a leafy cutting is dipped into the root-promoting preparation, but dipping the foliar portion of the cutting is effective for some species, provided a high enough concentration (2000 to 10,000 ppm) is used (29). There is evidence, too, that with hardwood cuttings of some species, dipping just the basal cut surface gives better results than dipping an inch or more of the base (25).

In treating cuttings with rooting chemicals it often is desirable to obtain measurements of the influence of such growth regulators on root initiation and development. One measurement is the *percentage of cuttings rooted* as compared to untreated controls. Some indication of the *numbers* and *length of roots* formed on the cuttings is also necessary. This can be done by actual root counts and length measurements but if quantities of cuttings are involved this may be impractical. In some experimental trials *arbitrary numerical rating systems* of the developing root mass are used. Sometimes the roots are excised and *fresh* or *dry-weight measurements* taken. A rapid and precise, nondestructive method using a "*rhizometer*" has been described *(31)*. This is a boxlike device employing a light source and a photoelectric cell. The effect of the root mass in reducing light transmission to the photoelectric cell is measured by a galvanometer. By using a previously prepared calibration curve the reduction in light transmission due to the roots can rapidly be translated into root surface area.

TREATMENT OF CUTTINGS WITH FUNGICIDES

As a precaution against fungus infection it may be advisable to give the cutting material a dip into a fungicidal preparation, such as benomyl (0.5 gm per liter; 3 oz per 50 gal.), either before or after the cuttings are made.

Dipping the cutting bases into a combination fungicide-indolebutyric acid mixture often gives better results than an IBA treatment alone. Simple preparations may be made:

(1) by mixing captan (50 percent wettable powder), 1:1 (w/w), with a commercial talc preparation containing 0.8 percent IBA to give a 25 percent captan and a 0.4 percent (4000 ppm) indolebutyric acid concentration, or

(2) by diluting benomyl (50 percent wettable powder) to a 10 percent concentration by mixing with talc (2 gm benomyl plus 8 gm talc), then mixing this, 1:1 (w/w), with a commercial talc preparation containing 0.8 percent IBA to give a mixture containing 5 percent benomyl and 0.4 percent indolebutyric acid.

If indolebutyric acid is used as a concentrated-solution-dip, after this treatment and after the cutting bases are allowed to dry, the bases can then be swirled around in a fungicidal powder, either 25 percent captan (50 percent wettable powder diluted 1:1 (w/w) with talc), or 5 percent benomyl (2 gm 50 percent wettable powder to 16 gm talc), before sticking in the rooting medium.

ENVIRONMENTAL CONDITIONS FOR ROOTING LEAFY CUTTINGS *

For the successful rooting of leafy cuttings, the essential environmental requirements are proper temperature (65° to 75° F; 18° to 27° C), an atmosphere conducive to low water loss from the leaves; ample light; and a clean, moist, well-aerated, and well-drained rooting medium. There are many possible types of equipment that are satisfactory for providing these conditions—from a simple glass jar or polyethylene cover placed over a few cuttings stuck in sand, to elaborate greenhouse benches with automatic mist control and automatically controlled electric heating cables below the cuttings. One of the simplest devices, which is satisfactory for rooting a limited

* See also p. 257.

Figure 10–22   Polyethylene plastic sheeting can be used for starting cuttings of easily rooted species. The basal ends of the cuttings are inserted in damp sphagnum or peat moss and rolled in the polyethylene as shown here. The roll of cuttings should then be set upright in a humid location for rooting.

number of cuttings, is a wooden box half-filled with the rooting medium, with a pane of glass or a sheet of polyethylene film over the top. If this is placed in a heated room in the winter next to a window, cuttings of many species can be rooted in it. Such a box can also be used successfully in the summer for rooting cuttings of some plants, if placed out-of-doors in the shade. Cuttings of many plants can be rooted successfully with only artificial light, if placed under a large fluorescent lamp fixture (see p. 255). A simple procedure for rooting a few cuttings is illustrated in Figure 10–22.

In commercial operations, large-scale rooting of leafy cuttings is done in hotbeds, cold frames, or beds in greenhouses or plastic houses. In these structures, proper environmental conditions can be established and maintained (see Chapter 2).

SANITATION
Most commercial propagators recognize the value of maintaining strict sanitary procedures during all stages of making and rooting leafy cuttings. It is much easier to prevent attacks of disease organisms than to try to stop them. Losses can be considerable from a disease attack where hundreds of thousands of cuttings are involved.

During the preparation of facilities for rooting the cuttings and while making and inserting them in the rooting medium, the following procedures will aid considerably in preventing losses from attacks by noxious organisms (*39*).

Propagating benches in the greenhouse should be washed thoroughly with water and sprayed with a copper naphthenate solution (1 part to 5 parts paint thinner *). (*Caution:* Creosote and certain phenolic materials used as wood preservatives give off fumes which are toxic to plants, and should not be used.) Flats for rooting the cuttings should be thoroughly washed, filled with the rooting medium, and then sterilized (see p. 33). The work benches

---

* A petroleum distillate.

where the cuttings are to be prepared should be washed thoroughly with water and sprayed with a Clorox (1 part to 4 parts water) solution or a formaldehyde (1 qt 38 percent formaldehyde to 5 gal. of water) solution. Tools used in preparing the cuttings should be dipped frequently in such solutions.

The cutting material itself should be free of insects and disease organisms. As the material is gathered, it can be placed on and wrapped in large sheets of clean, washed black polyethylene. These are protected from the sun and brought into the propagating room as soon as possible. The cutting material can be placed on a large raised wire rack and washed thoroughly with water. Mist nozzles should be placed over this rack to keep the material fresh while the cuttings are being prepared. After the cuttings are made, they may be soaked for about ten minutes in a mild fungicide, such as captan or benomyl. Following this, the cuttings are allowed to drain in a wire rack and are then treated with a root-promoting compound just before they are stuck into the propagating medium.

Efforts to conduct propagating operations under clean, sterilized conditions are of no avail, however, unless all components are included—the cuttings themselves, the flats, the rooting medium, the area where the cuttings are prepared, the tools used in making the cuttings, and the rooting bench. Keep the watering hose nozzle off the floor where it could pick up and later distribute harmful organisms.

PREPARING THE ROOTING FRAME AND INSERTING THE CUTTINGS

The frames or benches should preferably be raised or, if on the ground, equipped with drainage tile, so there is never any question of perfect drainage of excess water.

The frames or flats should be deep enough so that about 4 in. of rooting medium can be used. The depth should be enough so that an average-length cutting—3 to 5 in.—can be inserted up to half its total length, with the end of the cutting still an inch or more above the bottom of the frame. The rooting medium should be watered thoroughly before the cuttings are inserted, which should be as soon as possible after they are prepared. It is very important that the cuttings be protected from drying at all stages during their preparation and insertion.

After a section of the rooting bench or a flat is filled with cuttings, it should be well watered to settle the rooting medium around the cuttings.

MIST SYSTEMS FOR ROOTING CUTTINGS *

In the propagation of plants by leafy cuttings, one of the chief problems is to maintain the cuttings without wilting until roots are produced. In the past this was accomplished by keeping the relative humidity of the air sur-

* See also p. 252.

rounding the cuttings at a high level; the foliage, benches, and floors were sprinkled by hand several times a day during the rooting period.

An *intermittent-mist* water spray over the cuttings in the rooting bed is a very effective aid in rooting leafy cuttings of a great many kinds of plants. This system is widely used by propagators throughout the world. Such sprays provide a film of water over the leaves and cuttings; this lowers their temperature and increases the humidity around the leaves, thus reducing transpiration and respiration. Cuttings of certain plants, however—particularly succulent types with fleshy leaves and some others that show foliage leaching —do not do well under mist, rooting more readily in a closed-frame propagating bed.

This mist technique enables the rooting of cuttings of plants previously considered very difficult or impossible to root. This is true in some cases because it permits the use of soft, succulent, fast-growing cutting material early in the season which (in some species) is much more likely to root than older, more mature, hardened wood. In addition, intermittent mist keeps slow-rooting cuttings alive for a long period of time, giving them a chance to root before they die from dessication. By the use of mist propagation techniques, large cuttings with considerable leaf area can be rooted, permitting the production of large-size, salable plants in a short time (*54*).

Mist beds can be set up either in a greenhouse for use in summer and winter, or out-of-doors in a lath house or in open sun for use during the warmer months of the year. Over these beds, as shown in Figure 10–23, nozzles are placed which produce a fine fog-like mist spaced so as to give

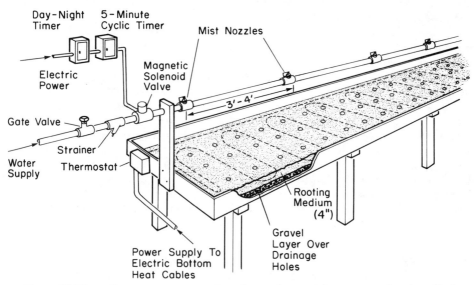

Figure 10–23   Component parts of an intermittent mist propagating installation with electric bottom-heat cable. One timer turns the mist system on in the morning and off at night. The second is a short interval timer to provide the intermittent mist cycles.

complete coverage of the bed. Figures 10–24 and 10–25 show typical mist propagation installations. Mist nozzles are also useful in experimental controlled-environment chambers (10).

Mist Nozzles

Two basic types of spray nozzles are available with several modifications of each: (a) the oil-burner, whirling action type, and (b) the deflection type (Figure 10–26).

The oil-burner nozzle produces an evenly distributed fine spray and uses a relatively small amount of water. The mist is produced in this nozzle by water passing through small grooves set at an angle to each other. Nozzles have been developed especially for mist propagation. These generally emit a flat, 160-deg. angle pattern to give a wide coverage—3½ to 4 ft—and are designed to operate well at the usual water pressure of about 30 lb per sq in. Water output is relatively low: 2½ to 5 gal. per hour.

The deflection nozzle develops a mist by a fine stream of water striking a flat surface. The larger aperture used in this type reduces clogging but uses more water. It can operate on a low water pressure more effectively than the oil-burner type. Some makes can be shut off individually, which facilitates working in the propagating beds.

There are various possible methods of placing the water pipes to which the nozzles are attached. One is to lay the main feeder pipe down the center of the bed, either below, at, or above the surface of the rooting medium, with the nozzles at the end of risers from this pipe. Another method is to place the feeder pipe well above the cuttings down the center of the bed with the nozzles directed downward. Whatever arrangement is used, the

Figure 10–24   Typical greenhouse mist propagation installations.

Figure 10–25   Commercial out-of-door mist installation for rooting cuttings. Courtesy C. W. Stuart and Co., Inc., Newark, New York.

Figure 10–26   Types of nozzles used in mist propagation installations. The two on the left are deflection nozzles. The one on the right (two views) is the whirling action, oil-burner type.

nozzles should be placed close enough together and the water pressure should be high enough so that the entire bed is completely under the mist. Unless the mist actually wets the leaves, rooting is likely to be unsatisfactory.

Controls

Intermittent mist during the daylight hours, which supplies water at intervals frequent enough to keep a film of water on the leaves, gives better results than continuous mist. Since it would be impractical to turn the mist on and off by hand at short intervals throughout the day, automatic control devices are necessary. Several types are available, all operating to control a solenoid (magnetic) valve in the water line to the nozzles (Figure 10–23).

In a mist installation, especially the outdoor type, the cuttings will be damaged if the leaves are allowed to become dry for very long. Even ten minutes without water on a hot, sunny day can be disastrous. In setting up the control system to provide an intermittent mist, every precaution should be taken to guard against accidental failure of the mist applications. This includes the use of a "normally open" solenoid value—that is, one constructed so that, if the electric power becomes disconnected, the valve is open and water passes through it. Application of electricity closes the valve and shuts off the water. If an accidental power failure occurs or any failure in the electrical control mechanism takes place, the mist remains on continuously, and no damage to the cuttings results. On the other hand, in using a "normally closed" solenoid, which requires an electric current to open it and allow the passage of water, any failure of the power would mean complete stoppage of the mist and, if not soon detected, a possible total loss of the cuttings.

Electrically operated timer mechanisms are available which will operate the mist as desired. A successful type uses two timers acting together—one turns the entire system on in the morning and off at night; the second, an interval timer, operates the system during the daylight hours to produce an intermittent mist—at any desired combination of timing intervals, such as 6 seconds ON and 90 seconds OFF. This type of control mechanism is relatively foolproof, and although it does not automatically compensate for variations in humidity conditions, in most situations it can be adjusted closely enough to give satisfactory results. Time clocks for regulating the application of water are preferred by many propagators because of their reliability (*18*).

In another type of control mechanism, the so-called "electronic leaf," a small piece of plastic containing two terminals is placed under the mist along with the cuttings (*19, 52*).

The alternate wetting and drying of the terminals makes and breaks the electric circuit which, in turn, controls the solenoid valve. There are several variations in this type of control. In one, a piece of filter paper is used as the sensing material connected between two electrodes (*3, 56*). The electronic leaf would theoretically maintain a film of water on the leaves of the cuttings at all times, automatically compensating for changes in the evaporating power of the air.

One type of control is based upon a thermostat placed with the cuttings. When the temperature at the leaf level reaches a certain point, the solenoid is activated, and mist is applied. This lowers the thermostat temperature, and the mist is shut off. A combination of such a thermostat plus a time clock has been used successfully (*14*). The timer operates the mist unless the temperature goes above a certain point, when the thermostat turns on the mist.

Another type of control is based upon the weight of water. A small stainless-steel screen is attached to a lever actuating a mercury switch. When the mist is on, water collects on the screen until its weight trips the mercury switch, shutting off the solenoid. When the water evaporates from the screen, it raises, closing the switch connection, which opens the solenoid, again turning on the mist. This type of control is best adapted to regions where considerable fluctuations in weather patterns may occur throughout the day, from warm and sunny to overcast, cool, and rainy; the unit reacts automatically to changes in the evaporating power of the air.

Controls based upon the relationship between light intensity and transpiration rate are available. These contain a photoelectric cell which conducts current in proportion to the light intensity. It activates a magnetic counter, or charges a condenser, so that after a certain period of time the solenoid valve is opened and the mist applied. The higher the light intensity the more frequently mist is applied. At dawn and dusk very little is used, and at night none. During cloudy days less mist is used than during bright, sunny days. Such a control system would not be well suited for outdoor mist beds, where transpiration is affected by wind movement as well as by light intensity (*40, 57*).

The danger of electrical shock should always be kept in mind when installing and using any electrical control unit in a mist bed where considerable water is present. The complete electrical installation should be done by a competent electrician.

Operation

Difficulties may arise in operating a mist-propagation bed. Lack of sufficient water pressure to operate the nozzles properly can be overcome by installing a small, electrically operated, rotary booster pump between the water source and the solenoid valve. If there is much sand in the water, it is advisable to install filters in the supply line, which will reduce the clogging of the strainers in the nozzles.

Algae growth often develops a green coating on and around mist-propagation installations after an extended period of operation. This coating is not particularly harmful to the cuttings but is unsightly and—because it is very slippery—can be a hazard. This slimy material is principally blue-green (*Oscillatoria, Phormidium,* and *Arthrospira*) and green (*Stichococcus* and *Chlamydomonas*) algae (*7, 11*). Allowing the water to remain off for a period each night so that the mist area can completely dry out will hold the algae growth in check. Sprinkling powdered Bordeaux Mixture on the walks and benches, or using certain proprietary algae inhibitors, as "Algae-Go 36–20," will help control the algae.

In propagating cuttings under mist, it is essential that a well-drained rooting medium be used and the bed raised, equipped with drainage tile, or otherwise provided for adequate removal of any excess water.

The quality of the water used in the mist—and in watering cuttings during rooting—can influence the rooting obtained. Water relatively high in total salts, if there is enough calcium and magnesium, may be quite satisfactory. But water high in such salts as sodium or potassium carbonates, bicarbonates, or hydroxides, can be very detrimental, especially when coupled with low levels of calcium salts and when the rooting medium contains peat (or other materials of high exchange capacity) (*36, 44*). However, with sand as the rooting medium, high sodium levels in the mist are not so detrimental (*37*). Equivalent amounts of sodium or potassium salts of nitrates, phosphates, or chlorides in the water are not so likely to cause injury. With chrysanthemum cuttings, lack of rooting in media containing peat, a condition associated with mist water having a high sodium to calcium ratio (4.3:0.6), can be overcome by adding gypsum ($CaSO_4$) to the rooting medium at the rate of 45 gms. per sq ft. Figure 10–27 shows the lack of rooting with chrysanthemum cuttings when there is high sodium in the rooting medium.

Figure 10–27 Injurious effects of excess sodium salts in a sand-peat rooting medium on rooting of chrysanthemum cuttings. Left to right: control, 23, 45, and 68 gms. per sq. ft. Courtesy R. D. Raabe (*44*).

Nutrient Mist

Mineral nutrients added to the water used for misting may improve root quality and subsequent growth of the rooted cuttings. Slow-to-root cuttings held under mist for prolonged periods may have a large amount of their nutrients leached away. A convenient solution to use for nutrient mist can be prepared from a commercial soluble fertilizer, e.g., "Ra-pid-Gro," containing nitrogen, phosphorus, and potassium (23–19–17, respectively), applied at the rate of 6 oz per 100 gals. of water. This may be made up separately in a large tank and applied to the cuttings as mist through a separate pressure pump at the rate of about 12 sec. every 2½ min. (*61*), or by a proportioning pump connected into the water supply (*48*).

Hardening-Off

Moving the rooted cuttings from under mist to a dryer environment must be carefully done. With some plants, *Prunus* sp., for example, it is important to remove the cuttings immediately from the mist once they are rooted. Otherwise, rapid defoliation and deterioration of the roots occur.

There are several ways of successfully taking the rooted cuttings from the mist conditions:

(a) The cuttings may be left in place in the mist bed but with the duration of the misting periods gradually decreased, either by lessening the ON periods and increasing the OFF periods or by leaving the misting intervals the same but gradually decreasing the time for which the mist is in operation each day.

(b) In some out-of-door operations, the rooted cuttings are left in place and allowed to send their roots on through the rooting medium to the soil beneath. The propagating frame is moved to a different location to root another set of cuttings.

(c) Another method is to root the cuttings in flats and move the flats after rooting to another mist frame where they are "hardened-off" and then potted into containers. Cuttings may be left in the rooting medium until the dormant season, when they can be dug more safely, to be either lined-out in the nursery row for further growth or potted and brought into the greenhouse. If the rooted cuttings are left in the rooting medium for a considerable time, it is advisable to water them at intervals with a nutrient solution (see p. 37).

(d) Some propagators root their cuttings directly in small containers set up in flats. Then, after rooting, the plants may be easily moved for transplanting without disturbing the roots. An alternate method is to root the cuttings in a solid, block-type rooting medium which, after rooting, permits transplanting without disturbing the roots (Figure 2–9). Several such products made from wood products and/or compressed peat, plus some added fertilizer, are available.

(e) Another method is to pot the cuttings immediately after rooting and hold them for a time in a humid, shaded location, e.g., a fog chamber, closed frame, or greenhouse.

It is a common experience to have the buds of cuttings that are rooted under mist enter an apparent dormant or physiological rest condition in which they do not continue further growth that season even though they are well rooted. This may possibly be overcome by altering the day-length conditions (see p. 255).

Anti-Transpirants

As an alternative to the use of mist it is possible to reduce water loss from leafy cuttings by dipping the leaves into one of the various anti-transpirant materials * prior to sticking. This partially seals the transpiring surfaces and reduces the unwanted rapid water loss from the tissues. Although not as effective as mist, there are instances where anti-transpirants have been successfully used for this purpose (*60*).

CARE OF CUTTINGS DURING ROOTING

Hardwood stem cuttings or root cuttings started out-of-doors in the nursery require only the usual care given to other crop plants, such as adequate soil moisture, freedom from weed competition, and insect and disease control. Best results are obtained if the nursery is established in full sun where shading and root competition from large trees or shrubbery do not occur.

Leafy softwood or semi-hardwood stem cuttings and leaf-bud or leaf cuttings being rooted under high humidity require close attention throughout the rooting period. The temperature should be controlled carefully. The cuttings must not be allowed to show wilting for any length of time. Glass-covered frames, exposed even for a few hours to strong sunlight, will build up excessively high and injurious temperatures, owing to the heat accumulating under the glass. Such equipment should always be protected by cloth screens, whitewash on the glass, or some other method of reducing the light intensity.

If bottom heat is provided, thermometers should be inserted in the rooting medium to the level of the base of the cutting and checked at frequent intervals, especially at first. A temperature of about 75° F (24° C) is desirable. Excessively high temperatures in the rooting medium, even for a short time, are likely to result in death of the cuttings.

It is important to maintain humidity conditions as high as possible in rooting leafy cuttings to reduce water loss from the leaves to a minimum. Without automatic mist equipment, syringing the leaves with a spray nozzle at frequent intervals is necessary especially during hot weather. Although more time-consuming, several light sprinklings with water each day are better than heavy soakings at less frequent intervals. A nozzle should be used that breaks the water into a fine spray. A drop of the humidity to a low level with a consequent pronounced wilting of the cuttings, if pro-

---

* *Wilt Pruf,* Nursery Specialty Products Co.; *Mobileaf,* Mobil Research & Development Co.; *Vapor Gard,* Miller Chemical & Fertilizer Co.

longed for any length of time, may so injure the cuttings that rooting will not occur, even though high-humidity conditions are subsequently resumed. However, most nursery operations use intermittent-mist propagating beds to overcome such problems.

Adequate drainage must be provided so that excess water can escape and not cause the rooting medium to become soggy and waterlogged. When peat or sphagnum moss is used as a component of the rooting medium, it is especially important to see that it does not become excessively wet.

It is also necessary to maintain sanitary conditions in the propagating frame. Leaves that drop should be removed promptly, as well as any obviously dead cuttings. Organisms find ideal conditions in a humid, closed propagating frame with low light intensity and, if not controlled, can destroy thousands of cuttings overnight.

Disease problems under mist-propagation conditions have not been serious. Probably the frequent washing of the leaves by the aerated water removes spores before they are able to germinate. The greater light intensity and air movement also decrease the incidence of diseases.

If mites, aphids, or mealy bugs appear on the leaves of the cuttings, immediate control measures are necessary.

The more sophisticated techniques, such as the use of carbon-dioxide enriched atmospheres or the application of carbonated mist (*30*), have produced faster and greater rooting by supplying added CO_2 for photosynthesis (see p. 40). However, these benefits would be found only in situations (high temperature and light intensity) where the CO_2 level surrounding the cuttings is limiting growth. Equipment is available for supplementing the natural CO_2 supply.

HANDLING CUTTINGS AFTER ROOTING
Hardwood Cuttings
Rooted hardwood cuttings in the nursery row are usually dug during the dormant season after the leaves have dropped. With fast-growing species, the cuttings may be sufficiently large to dig after one season's growth. Slower-growing species may require two or even three years to become large enough to transplant.

The digging should take place on cool, cloudy days, when there is no wind. If possible, digging should not be done when the soil is wet, especially if it has a high clay content. Most of the soil should drop readily from the roots after the plants are removed. After the plants are dug, they should be quickly heeled-in in a convenient location, placed in cold storage, or replanted immediately in their permanent location. "Heeling-in" is to place dug, bare-rooted deciduous nursery plants close together in trenches with the roots well covered. This is a temporary provision for holding the young plants until they can be set out in their permanent location. (See Figure 10–28.)

Commercial nurseries often store quantities of deciduous plants for several months in cool, dark rooms with the roots protected by damp wood shavings, shingle tow, or some similar material. Nursery stock to be kept

Figure 10–28   Temporary storage of nursery stock by "heeling-in" in raised beds filled with damp wood shavings sufficiently deep to cover the roots.

for extended periods should be held under refrigerated conditions at 32° to 35° F (0° to 2° C) (*28*).

If only a few plants are to be removed from the nursery row, they can be dug with shovels, but in large-scale nursery operations some type of mechanical digger, such as is shown in Figure 10–29, is generally used. This digger "undercuts" the plants. A sharp U-shaped blade travels 1 to 2 ft below the soil surface under the nursery row, cutting through the roots. Sometimes a horizontal, vibrating, "lifting" blade is attached to, and travels behind, the cutting blade. This slightly lifts the plants out of the soil, making them easy to pull by hand.

Unless very small, plants of evergreen species usually cannot be successfully handled bare-root, as is done with dormant and leafless deciduous plants. The presence of leaves on evergreen plants requires the continuous contact of the roots with soil. Therefore, large, salable plants of broad- or narrow-leaved evergreens, and occasionally deciduous plants, are either grown in containers or dug and sold "balled and burlapped." By the latter method, the plants are removed from the soil by carefully digging a trench around each individual. The soil mass around the roots is sometimes tapered at top and bottom, resulting in a ball of soil in which the roots are embedded. It is important that the soil be at the proper moisture level—not too wet and not too dry—otherwise it will fall apart. The ball is tipped gently onto a large square of burlap, which is then pulled tightly and sewed in place with heavy twine. If properly done, this adequately holds the soil to the roots, and the plant can then be moved safely for considerable distances and replanted successfully.

Prior root pruning is necessary to properly prepare field-grown nursery stock for transplanting, either by hand "ball and burlapping" (Figure 10–30) or by machine digging. This is done to produce a compact, fibrous root system and should be started when the young plants (*liners*) are first set in the field. Long or curled and twisted roots should be cut back to a few

Figure 10–29 Digging machines specially constructed for undercutting nursery stock in commercial nurseries. Photo on left shows large U-shaped blade which can be lowered to travel 1 to 2 ft below the soil surface to cut roots. Vibrating "lifter" behind the cutting blade facilitates the movement of the blade through the soil.

inches from the crown of the liner. Root pruning the second year should also be done, either by a tractor-powered U-blade cutter (see Figure 10–29) for large-scale operations, or with a sharp shovel if only a few plants are involved. This procedure helps confine the roots to the soil mass which will be taken with the plant during the digging and balling operation.

The production of balled and burlapped nursery stock is being replaced, especially in areas with mild winters, by the production of plants in containers, primarily because of the lower labor costs involved and the greater opportunities for mechanization.

Softwood, Herbaceous, Semi-Hardwood, Leaf-Bud, Leaf Cuttings

Cuttings such as these, rooted with leaves attached and under conditions of high humidity, require considerable care in being removed from the rooting medium. After rooting has started, the humidity should be lowered and ventilation of the bed provided. They should be dug as soon as a substantial root system with secondary roots has formed. Many propagators have experienced rapid rooting of cuttings only to have them die when they are dug and potted. Sometimes this trouble may be overcome by leaving the cuttings in the rooting bed longer, until after the first-formed primary roots have branched to develop a dense, fibrous secondary root system to which the rooting medium clings in a ball.

Figure 10–30 Steps in ball and burlapping nursery stock, the procedure being used here in digging tree roses. Plant is carefully removed with a ball of soil adhering to the roots *(top)*. Soil must be at proper moisture level so it does not fall apart. Burlap is wrapped around soil ball and tied with heavy cord *(middle)*. Plants ready to be moved and transplanted *(below)*. This method is widely used for all types of evergreen plants which have not been grown in containers.

Most of the rooting media used contain little or no mineral nutrients available to the plants, hence the cuttings are dependent upon the nutrients already stored in the stems and leaves at the time the cuttings were made. This is generally sufficient to maintain the cuttings during the period of root formation. With some species, however, it may be helpful to water the cuttings with a nutrient solution about 10 days before they are to be removed, especially if digging has been delayed until a secondary root system has developed.

If for any reason the cuttings are to be kept for a prolonged period in the rooting medium, they should be watered several times with a nutrient solution (see p. 37).

In digging, the cuttings should be lifted gently from the rooting medium with a trowel or some similar device, taking care not to break off the roots. It is desirable for the cuttings to be lifted out with a mass of the rooting medium still adhering to the roots. This can be done if a material such as peat moss or vermiculite is included in the rooting medium. The cuttings are ready for potting when most of the roots are 1 to 2 in. long. (See p. 29 for potting mixtures.)

The rooted cuttings should be watered thoroughly shortly after potting. It is very important that potted cuttings be moved gradually from the protected conditions under which they were rooted (high humidity, low light intensity) to out-of-door conditions (low humidity, high light intensity, and wind). It is often best to leave the cuttings for several days after they have been potted under the same conditions in which they were rooted. If they were rooted in mist, they must be given especially close attention and moved gradually to the dryer atmosphere. Before placing the rooted cuttings in full sun, they should be hardened-off for one or two weeks in a cold frame or lathhouse or under some partial protection from the sun (see p. 22).

COLD STORAGE OF ROOTED AND UNROOTED LEAFY CUTTINGS

Sometimes it may be convenient to take cuttings at certain times, such as when nursery plants are being sheared and shaped, for later rooting. This was done with the Kurume type of azaleas overwintered in a cold greenhouse. Softwood cuttings were taken in the spring and held in polyethylene bags at temperatures from 31° to 40° F (–0.5° to 4.5° C) for up to ten weeks, with subsequent rooting equal to that from unstored cuttings (*43*). Unrooted chrysanthemum and carnation cuttings can be stored in sealed plastic bags for several weeks at 31° F (–0.5° C) for subsequent rooting. In tests (*2*) on the effects of storage on subsequent performance of the plants, it was found that cuttings rooted after storage gave better results than those stored after rooting.

In some situations cuttings may be rooted in late summer or early fall for planting outdoors the following spring. Several studies have shown that it is possible to hold rooted cuttings of some species in polyethylene bags for prolonged periods under cold storage temperatures of 35° to 40° F (1.5° to 4.5° C) (*55*). In one test, rooted cuttings of certain plants were safely stored for about five months at temperatures of 34° to 39° F (1° to 4° C) when placed in polyethylene bags. Packing material around the roots was of no advantage (*47*). Other studies (*13*) on cold storage of rooted cuttings of 31 different species for six months at two temperatures (32° and 40° F) showed better survival with many of the species at 32° than at 40°, although no difference was noted with other species. With some species, dusting prior to storage with 5 percent captan increased survival, but in others this made no difference. Apparently some species survive cold-storage treatments much better than others.

The storage procedures for all types of nursery plants are, of course, an important consideration for commercial nurseries (*21, 28*).

REFERENCES

1 Ali, N., and M. N. Westwood, Rooting of pear cuttings as related to carbohydrates, nitrogen, and rest period, *Proc. Amer. Soc. Hort. Sci.,* 88:145–50. 1966.

2 Alstadt, R. A., and W. D. Holley, Effects of storage on the performance of carnation cuttings, *Colo. Fl. Gr's. Ass'n. Bul.,* 173:1–2. 1964.

3 Bean, G., E. S. Trickett, and D. A. Wells, Automatic mist control equipment for the rooting of cuttings, *Jour. Agr. Eng. Res.,* 2:44–48. 1957.

4 Carlson, R. F., Factors influencing root formation in hardwood cuttings of fruit trees, *Mich. Quart. Bul.,* 48:449–54. 1966.

5 Chadwick, L. C., The effect of certain mediums and watering methods on the rooting of cuttings of some deciduous and evergreen plants, *Proc. Amer. Soc. Hort. Sci.,* 53:555–66. 1949.

6 ———, and D. C. Kiplinger, The effect of synthetic growth substances on the rooting and subsequent growth of ornamental plants, *Proc. Amer. Soc. Hort. Sci.,* 36:809–16. 1938.

7 Coorts, G. D., and C. C. Sorenson, Organisms found growing under nutrient mist propagation, *HortScience,* 3(3):189–90. 1968.

8 Creech, J. L., Root cuttings, *Nat. Hort. Mag.,* 33:21–24. 1954.

9 Creech, J. L., R. F. Dowdle, and W. O. Hawley, Sphagnum moss for plant propagation, *USDA Farmers' Bul. 2085,* 1955.

10 Davis, E. A., A laboratory chamber for rooting cuttings, *Bot. Gaz.,* 129(1):86–89. 1968.

11 Durrell, L. W., and R. Baker, Algae causing clogging of cooling systems, *Colo. Flower Grs. Ass'n. Bul.,* 111. Apr.–May 1959.

12 Flemer, W., III, Propagating woody plants by root cuttings, *Proc. Plant Prop. Soc.,* 11:42–47. 1961.

13 Flint, H. L., and J. J. McGuire, Response of rooted cuttings of several woody ornamental species to overwinter storage. *Proc. Amer. Soc. Hort. Sci.,* 80: 625–29. 1962.

14 Floor, J., Verslag van proeren met stekken onder waterverneveling. (Report

on experiments with mist propagation of cuttings), *Inst. voor de Vered. van Tuinb. Wageningen. Meded.*, 188. 1957.

15 Hansen, C. J., and H. T. Hartmann, The use of indolebutyric acid and captan in the propagation of clonal peach and peach-almond hybrid rootstocks by hardwood cuttings, *Proc. Amer. Soc. Hort. Sci.*, 92:135–40, 1968.

16 Hartmann, H. T., W. H. Griggs, and C. J. Hansen, Propagation of own-rooted Old Home and Bartlett pears to produce trees resistant to pear decline, *Proc. Amer. Soc. Hort. Sci.*, 82:92–102. 1963.

17 Hartmann, H. T., C. J. Hansen, and F. Loreti, Propagation of apple rootstocks by hardwood cuttings, *Calif. Agr.*, 19(6):4–5. 1965.

18 Hess, C. E., and W. E. Snyder, A simple and inexpensive time clock for regulating mist in plant propagation procedures, *Proc. Plant Prop. Soc.*, 3:56–61. 1953.

19 ———, Interrupted mist found superior to constant mist in tests with cuttings, *Amer., Nurs.*, 100(12):11–12, 82. 1954.

20 Hitchcock, A. E., and P. W. Zimmerman. The use of green tissue test objects for determining the physiological activity of growth substances, *Contrib. Boyce Thomp. Inst.*, 9:463–518. 1938.

21 Hocking, D., and R. D. Nyland, Cold storage of coniferous seedlings, *AFRI Res. Rpt. No. 6,* Col. Forestry, Syracuse Univ., 1971.

22 Howard, B. H., Propagation techniques, *Sci. Hort.*, 23:116–26. 1971.

23 Howard, B. H., and R. J. Garner, High temperature storage of hardwood cuttings as an aid to improved establishment in the nursery, *An. Rpt. E. Malling Res. Sta. for 1964,* pp. 83–87. 1965.

24 Howard, B. H., The influence of 4 (indolyl-3) butyric acid and basal temperature on the rooting of apple rootstock hardwood cuttings, *Jour. Hort. Sci.*, 43:23–31. 1968.

25 Howard, B. H., and N. Nahlawi, Dipping depth as a factor in the treatment of hardwood cuttings with indolybutyric acid, *An. Rpt. E. Malling Res. Sta. for 1969,* pp. 91–94. 1970.

26 Lagerstedt, H. B., Propagation of begonias from leaf disks, *HortScience,* 2(1):20–22, 1967.

27 Long, J. C., The influence of rooting media on the character of the roots produced by cuttings, *Proc. Amer. Soc. Hort. Sci.*, 29:352–55. 1932.

28 Mahlstede, J. P., and W. E. Fletcher, *Storage of Nursery Stock.* Washington, D. C.: Amer. Assoc. Nurserymen, 1960, pp. 1–62.

29 McGuire, J. J., L. S. Albert, and V. K. Shutak, Effect of foliar applications of 3-indolebutyric acid on rooting of cuttings of ornamental plants, *Proc. Amer. Soc. Hort. Sci.*, 93:699–704. 1968.

30 Molnar, J. M., and W. A. Cumming, Effect of carbon dioxide on propagation of softwood, conifer, and herbaceous cuttings, *Canad. Jour. Plant Sci.*, 48:595–99. 1968.

31 Morrison, I. K., and K. A. Armson, The rhizometer—a device for measuring roots of tree seedlings, *Forestry Chron.*, 44(5):21–23. 1968.

32 Myhre, A. S., and C. D. Schwartze, Rooting evergreen cuttings with hormones, *Proc. Amer. Soc. Hort. Sci.,* 51:639–50. 1948.

33 Nelson, S. H., Mist propagation of evergreens in the greenhouse during winter, *Proc. Plant Prop. Soc.,* 9:67–76. 1959.

34 O'Rourke, F. L., and M. A. Maxon, Effect of particle size of vermiculite media on the rooting of cuttings, *Proc. Amer. Soc. Hort. Sci.,* 51:654–56. 1948.

35 Osborne, W. W., Soil sterilization and fumigation, *Proc. Plant Prop. Soc.,* 11: 57–67. 1961.

36 Paul, J. L., Water quality and mist propagation, *Proc. Inter. Plant Prop. Soc.,* 18:183–86. 1968.

37 Paul, J. L., and A. T. Leiser, Influence of sodium in the mist water on rooting of chrysanthemums, *HortScience,* 3(3):187–88. 1968.

38 Pearse, H. L., Growth substances and their practical importance in horticulture, *Commonwealth Bur. of Hort. and Plant. Crops, Tech. Comm. No. 20,* 1948.

39 Petersen, F., Current methods in the selection and production of nursery stock, *Proc. Plant Prop. Soc.,* 11:235–40. 1961.

40 Petersen, H., A photoelectric timing control for mist application, *Proc. 15th Int. Hort. Cong.,* 3:273–79. 1962.

41 Pike, A. V., Propagation by roots, *Horticulture,* 50(5):56, 57–61. 1972.

42 Pokorny, F. A., An evaluation of various equipment and media used for mist propagation and their relative costs, *Ga. Agr. Exp. Sta. Bul.,* n.s. 139. 1965.

43 Pryor, R. L., and R. N. Stewart, Storage of unrooted azalea cuttings, *Proc. Amer. Soc. Hort. Sci.,* 82:483–84. 1963.

44 Raabe, R. D., and J. Vlamis, Rooting failure of chrysanthemum cuttings resulting from excess sodium or potassium, *Phytopath.,* 56:713–17. 1966.

45 Scaramuzzi, F., Nuova technica per stimolare la radicazione delle talle legnose di ramo (A new technique for stimulating rooting in hardwood cuttings), *Riv. Ortoflorofrutticoltura Ital.,* 49:101–4. 1965.

46 Smith, P. F., Rooting of guayule stem cuttings in aerated water, *Proc. Amer. Soc. Hort. Sci.,* 44:527–28. 1944.

47 Snyder, W. E., and C. E. Hess, Low temperature storage of rooted cuttings of nursery crops, *Proc. Amer. Soc. Hort. Sci.,* 67:545–48. 1956.

48 Sorenson, D. C., and G. D. Coorts, The effect of nutrient mist on propagation of selected woody ornamentals, *Proc. Amer. Soc. Hort. Sci.,* 92:696–703. 1968.

49 Stoutemyer, V. T., Root cuttings, *Plant Prop.,* Inter. Plant Prop. Soc., 14(4): 4–5. 1968.

50 ———, Hardwood cuttings, *Plant Prop.,* Inter. Plant Prop. Soc., 15(3): 10–14. 1969.

51 Swartley, J., and L. C. Chadwick, Synthetic growth substances as aids to root production on evergreen and softwood deciduous cuttings, *Proc. Amer. Soc. Hort. Sci.,* 37:1099–1104. 1939.

52 Templeton, H. M., The phytotektor method of rooting cuttings, *Proc. Plant Prop. Soc.,* 3:51–56. 1953.

53 Thimann, K. V., and J. Behnke-Rogers, *The Use of Auxins in the Rooting of Woody Cuttings*, Harvard Forest, M. M. Cabot Foundation, Petersham, Mass., 1950.

54 Tinga, J. H., J. J. McGuire, and R. J. Parvin, The production of pyracantha plants from large cuttings, *Proc. Amer. Soc. Hort. Sci.*, 82:557–61. 1963.

55 Vanderbrook, C., The storage of rooted cuttings, *Amer. Nurs.*, 103(9):10, 76–77. 1956.

56 Vanstone, F. H., Equipment for mist propagation developed at the N.I.A.E., *An. Appl. Biol.*, 47:627–31. 1959.

57 Waxman, S., and J. H. Whitaker, A light-operated interval switch for the operation of a mist system, *Prog. Rpt. No. 40, Storrs (Conn.) Agr. Exp. Sta.*, 1960.

58 Wells, J. S., Pointers on propagation: propagation of *Taxus, Amer. Nurs.*, 96(11):13, 37–38, 43, 1952.

59 ———, Wounding cuttings as a commercial practice, *Proc. Plant Prop. Soc.*, 12:47–55. 1962.

60 Whitcomb, C. E., and L. T. Davis, Jr., Anti-transpirants—a better way to root cuttings? *Amer. Nurs.*, 132(7):9, 100–101. 1970.

61 Wott, J. A., and H. B. Tukey, Jr., Influence of nutrient mist on the propagation of cuttings, *Proc. Amer. Soc. Hort. Sci.*, 90:454–61. 1967.

62 Zimmerman, P. W., Oxygen requirements for root growth of cuttings in water, *Amer. Jour. Bot.*, 17:842–61. 1930.

SUPPLEMENTARY READING

Brooklyn Botanic Garden. Handbook on Propagation. *Plants and Gardens*. Vol. 13, No. 2 (1957).

Doran, W. L., "Propagation of Woody Plants by Cuttings," *Massachusetts Agricultural Experiment Station Bulletin 491* (1957).

International Plant Propagators' Society, *Proceedings of Annual Meetings*.

Rowe-Dutton, P., "Mist Propagation of Cuttings," *Commonwealth Agricultural Bureaux, Farnham Royal, Bucks, England, Digest No. 2* (1959).

Welch, H. J., *Mist Propagation and Automatic Watering*. London: Faber and Faber, 1970.

Theoretical Aspects of Grafting and Budding

11

Grafting is the art of joining parts of plants together in such a manner that they will unite and continue their growth as one plant. The part of the graft combination which is to become the upper portion is termed the *scion* (*cion*), and the part which is to become the lower portion or root is termed the *rootstock,* or *understock,* or just the *stock* (*169*). All methods of joining plants are properly termed *grafting,* but when the scion part is only a small piece of bark (and sometimes wood) containing a single bud, the operation is termed *budding.*

REASONS FOR GRAFTING AND BUDDING

Perpetuating Clones (see p. 182) That Cannot Be Readily Reproduced by Cuttings, Layers, Division or Other Asexual Methods

Cultivars of some groups of plants, including most fruit and nut species, many woody plants, such as eucalyptus and spruce are not propagated commercially by cuttings, because they cannot be rooted at all or in satisfactory percentages by the methods now available. Additional individuals can be started by layering or division, but for propagation in considerable quantities, it is necessary to resort to budding or grafting scions of the desired cultivar on rootstocks with which they are compatible (see p. 337).

Obtaining the Benefits of Certain Rootstocks *

Many plant cultivars selected for desirable fruit or ornamental qualities do not have comparably suitable roots, and require grafting onto other roots

---

* Detailed discussions of the rootstocks available for the various fruit and ornamental species are given in Chapters 17 and 18.

to give satisfactory plants. For many plant species, rootstocks are available which tolerate unfavorable conditions, such as heavy, wet soils, or which resist soilborne pests or disease organisms. Also, size-controlling rootstocks are available for some species which can cause the composite grafted tree to have exceptional vigor or to become very or partially dwarfed. Some stocks, particularly in citrus species, give better size and quality of the fruit of the scion cultivar than do others.

Special rootstocks for glasshouse vegetable crops have long been used in Europe to avoid root diseases such as *Fusarium* and *Verticillium* wilt. In Holland forcing types of greenhouse cucumbers are grafted onto *Cucurbita ficifolia,* and commercial tomato cultivars are grafted onto vigorous F_1 hybrid, disease-resistant rootstocks (*154*).

Rootstocks can be divided into two groups: *seedling* and *clonal* (including apomictic seedlings; see p. 61).

Seedling rootstocks have certain advantages. Production of seedlings is relatively simple and economical (see Chapter 7). Most seedling plants do not retain viruses occurring in the parent plant (although some viruses are seed-transmitted; see p. 63).

In some instances (e.g., plum rootstocks), the root system developed by seedlings tends to grow deeper and to be more firmly anchored than rootstocks grown from cuttings.

However, seedling rootstocks have the disadvantage that genetic variation among them may lead to variability in the growth and performance of the scion of the grafted plant. Such variation is most likely to occur if the seed is obtained from unknown, unselected sources. The seed source plants may be unusually heterozygous or may have been cross-pollinated with related species. Different seed sources within the same species may vary considerably in their general value for producing rootstocks for a particular kind of plant, so that an entire species need not be condemned for rootstock purposes. With careful testing, individuals in that species could perhaps be selected to be either a new clonal rootstock or a mother-tree seed source.

Variability among seedling rootstocks can be reduced by careful selection of the parental seed source as to identity, and by protection from cross-pollination. Variation can also be reduced if all trees of the same age are dug from the nursery row at one time, and small or obviously off-type seedlings or budded trees discarded. In most nurseries the trees are graded by size, all those of the same grade being sold together. The practice of retaining slow-growing seedlings or budded trees for an additional year's growth tends to perpetuate variability. Many fruit orchards in the United States grown on seedling rootstocks show no more variability due to the rootstock used than to other unavoidable environmental differences in the orchard, principally soil variability.

Clonal rootstocks have received much attention in European countries, especially England, in regard to their development and use. These are propagated vegetatively either by stool layering or as rooted cuttings; each individual rootstock plant is the same genetically as all the other plants in the clone and can be expected to have identical growth characteristics in a

Figure 11–1   Scion roots of an 'Old Home' pear grafted on quince. (A) Original quince roots. (B) Scion roots arising from the 'Old Home' pear above the graft union. These have assumed the major support of the tree. The dwarfing influence due to the quince roots has disappeared.

given environment. Clonal rootstocks are desirable not only to produce uniformity but—and this is equally important—to preserve special characteristics and specific influences on scion cultivars, such as disease resistance, growth, and flowering habit. To maintain rootstock influence, deep planting should be avoided, as illustrated in Figure 11–1. The deeper the graft union below the soil surface, the higher the incidence of scion rooting is likely to be (*25*).

The combinations of different clonal rootstocks with different scion cultivars allows much refinement in the performance of grafted trees. Each particular graft combination requires thorough testing, however, before its performance is established.

In propagating and using clonal rootstocks it is very important that disease-tested material be obtained if possible, as any diseases present in such stocks are maintained and spread, along with the rootstock material. However, this is of little value unless "clean" scion material is also available.

More than two kinds of plants can be combined together in a vertical arrangement. In addition to having a rootstock and scion, one may insert between them, by grafting, a third kind. Such a section is termed an *intermediate stem section*, an *intermediate stock*, or *interstock*. This practice is called *double-working* (see p. 413).

There are several reasons for using double-working in propagation. One is to circumvent certain kinds of incompatibility (see p. 337). A second reason is that the interstock may possess a particular characteristic (such as disease resistance or cold-hardiness) not possessed by either the rootstock or the scion, which makes it valuable for the main framework of the tree. In addition a certain scion cultivar may be required for disease resistance

Figure 11–2   A commercial application of double-working. *Left:* rubber trees in Brazil of a high-yielding, blight-susceptible East India clone budded on seedling rootstocks have been injured by attacks of leaf blight. *Right:* trees of the same high-yielding clone double-worked to use a blight-resistant top. These trees consist of seedling rootstock, high-yielding stem, and blight-resistant top. Courtesy M. H. Langford *(97)*.

(Figure 11–2). Another reason is that the interstock may have a definite influence on the growth of the tree. For example, when a stem piece of the dwarfing 'Malling 9' apple stock is inserted between a vigorous rootstock and a vigorous scion cultivar, it will stimulate flowering and reduce growth of the composite tree, in comparison with a similar tree but without the interstem piece *(136)*.

Nurserymen supplying trees on seedling or clonal rootstocks, or with a clonal interstock, should identify such stocks on the label just as they do for the scion cultivar.

Changing Cultivars of Established Plants (Topworking)

A fruit tree, or an entire orchard, may be of a cultivar that is undesirable. It could be an old cultivar no longer in demand, or an unproductive one, or one with poor growth habits, or possibly one that is susceptible to prevalent diseases or insects. As long as a compatible type is used, the top may be regrafted to a more desirable cultivar, if such exists.

Provision for adequate cross-pollination in fruit trees can be obtained by topworking scattered trees throughout the orchard to the proper pollinizing cultivar. Or, if a single tree is unfruitful owing to lack of cross-pollination, a branch of the tree may be grafted to the proper pollinizing cultivar.

A single pistillate (female) plant of a dioecious (pistillate and staminate flowers borne on separate individual plants) species, such as the hollies (*Ilex*), may be unfruitful owing to the lack of a nearby staminate (male) plant to provide proper pollination. This can be corrected by grafting a scion taken from a staminate plant on one branch of the pistillate plant.

Of interest to the home gardener is the fact that several cultivars of almost any fruit species can be grown on a single tree of that species by topworking each scaffold branch to a different cultivar. In a few cases, different species can be worked on the same tree. For example, on a single citrus rootstock it would be possible to have oranges, lemons, grapefruit, mandarins, and limes, or on a peach root system could be grafted plum, almond, apricot, and nectarine.

Hastening the Growth of Seedling Selections.
In various fruit breeding projects, the young seedling selections, if left to grow on their own roots, may take five to ten or more years to come into bearing. Sometimes, unless the juvenility influence (see p. 184) is too strong, some benefit can be obtained in shortening this period by using the seedling shoots as scions which are grafted onto large, established trees (*172*) or onto certain dwarfing rootstocks (*173*).

It is possible to graft several seedling selections on one mature tree, but this could be inadvisable owing to the danger of virus contamination coming from the old tree or, perhaps, one or more of the seedling scions, and spreading to the others.

Sometimes desirable new seedling plants produced in breeding programs never attain the ability to grow well on their own roots, but when grafted on a vigorous, compatible rootstock develop into plants of the desired form and stature (*39*). Examples of this are crosses of *Syringa laciniata* × *S. vulgaris* which produce Chinese lilac hybrids, but many of the seedlings lack sufficient vigor to live more than two or three years. However, if they are budded on the tree lilac, *Syringa amurensis japonica*, they survive and grow vigorously (*144*).

Obtaining Special Forms of Plant Growth
By grafting certain combinations together it is possible to produce unusual types of plant growth, such as "tree" roses or "weeping" cherries or birches.

Repairing Damaged Parts of Trees
Occasionally the roots, trunk, or large limbs of trees are severely damaged by winter injury, cultivation implements, certain diseases, or rodents. By the use of bridge-grafting, or inarching, such damage can be repaired and the tree saved. This is discussed in detail on pp. 396 and 399.

Studying Virus Diseases
It is characteristic of virus diseases that they can be transmitted from plant to plant by grafting. This makes possible testing for the presence of the virus in plants which may carry the disease but show few or perhaps no symptoms. By grafting scions or buds from a plant suspected of carrying the virus onto an indicator plant known to be highly susceptible, with prominent symptoms, detection is easily accomplished (*110*). This procedure is known as *indexing* (see Chapter 8).

In order to detect the presence of a latent virus in a symptomless carrier, it is not necessary to use only combinations which make a permanent, compatible graft union. For example, the 'Shirofugen' flowering cherry (*Prunus serrulata*) is used to detect viruses in peach, plum, almond, and apricot. A temporary, incompatible union is often a sufficient bridge for virus transfer (*94*).

NATURAL GRAFTING
One occasionally sees branches that have become grafted together naturally following a long period of being pressed together without disturbance (see Figure 11–3). English ivy (*Hedera helix*) forms such grafts, and detailed studies have been made of translocation in natural grafts of this species (*112*). Not so obvious but of much greater significance and occurrence, particularly in stands of forest species, such as pine, hemlock, and Douglas fir, is the natural grafting of roots (*60, 157*).

Such grafts are most common between roots of the same tree or between roots of trees of the same species. Grafts between roots of trees of different species are rare. In the forest, living stumps sometimes occur, kept alive because their roots have become grafted to those of nearby intact, living trees (*108*). The anatomy of natural grafting of aerial roots has been studied; the initial contact is established by the formation and fusion of epidermal hairs (*134*). Such natural grafting permits virus transmission from an infected tree to its neighbors. This can be important in closely set orchard and nursery plantings of fruit trees, where numerous root grafts could occur and result in the slow spread of virus throughout the planting (*92*). Natural root grafting is a potential source of error in virus indexing

Figure 11–3   Natural grafts in branches *(left)* and roots *(right)*. Such grafts sometimes occur after the tissues have been pressing tightly against each other without disturbance for a long time. Photo of roots courtesy Miller and Woods (*108*). Copyright 1965. University of Chicago.

procedures where virus-free and virus-infected trees are grown in close proximity *(63)*. In addition, fungal pathogens such as those causing oak wilt and Dutch elm disease can be spread by such natural root connections. That natural grafting does take place throws some doubt on the assumption usually held that each plant is an independent, discrete organism.

FORMATION OF THE GRAFT UNION

A number of detailed studies have been made of the healing of graft unions, mostly with woody plants *(35, 52, 106, 138, 146, 156, 162)*.

Briefly, the usual sequence of events in the healing of a graft union is as follows (Figure 11–4):

(a) Freshly cut scion tissue capable of meristematic activity is brought into secure, intimate contact with similar freshly cut stock tissue in such a manner that the cambial regions of both are in close proximity. Temperature and humidity conditions must be such as to promote growth activity in the newly exposed, and surrounding, cells.

(b) The outer exposed layers of cells in the cambial region of both scion and stock produce parenchyma cells which soon intermingle and interlock; this is called *callus tissue* (Figures 11–5 and 11–6).

(c) Certain cells of this newly formed callus which are in line with the cambium layer of the intact scion and stock differentiate into new cambium cells.

(d) These new cambium cells produce new vascular tissue, xylem toward the inside and phloem toward the outside, thus establishing vascular connection between the scion and stock, a requisite of a successful graft union.

The healing of a graft union can be considered as the healing of a wound *(13)*. Such injury to tissue as would occur if the end of a branch were split longitudinally would heal quickly if the split pieces were bound tightly together. New parenchyma cells would be produced by abundant proliferation from cells of the cambium region of both pieces, forming callus tissue. Some of the newly produced parenchyma cells differentiate into cambium cells, which subsequently produce xylem and phloem.

If, between the two split pieces, one interposed a third, detached, piece which had been cut so that a large number of its cells in the cambial region could be placed in intimate contact with cells of the cambial region of the two split pieces, proliferation of parenchyma cells from all cambial areas would soon result in complete healing, with the foreign, detached piece joined completely between the two original split pieces. A graft union is essentially a healed wound, with an additional, foreign, piece of tissue incorporated into the healed wound.

This added piece of tissue, the scion, will not resume its growth successfully, however, unless vascular connection has been established so that it may obtain water and mineral nutrients. In addition, the scion must have a terminal meristematic region—a bud—to resume shoot growth and, eventually, to supply photosynthates to the root system.

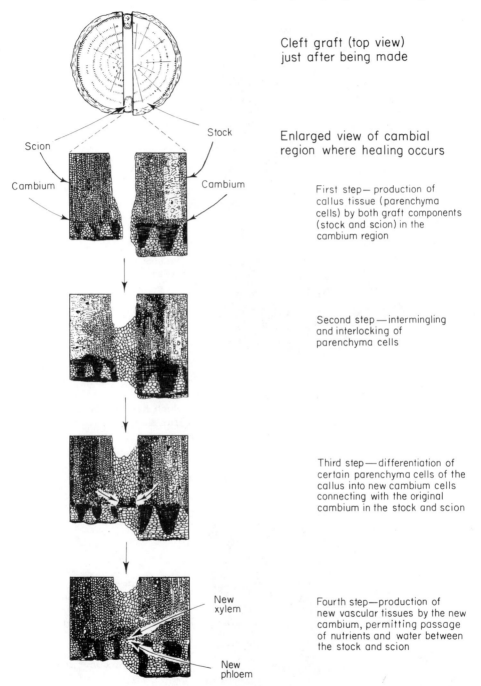

Cleft graft (top view) just after being made

Enlarged view of cambial region where healing occurs

Scion

Stock

Cambium

Cambium

First step— production of callus tissue (parenchyma cells) by both graft components (stock and scion) in the cambium region

Second step—intermingling and interlocking of parenchyma cells

Third step—differentiation of certain parenchyma cells of the callus into new cambium cells connecting with the original cambium in the stock and scion

New xylem

Fourth step—production of new vascular tissues by the new cambium, permitting passage of nutrients and water between the stock and scion

New phloem

Figure 11–4 Diagrammatic developmental sequence during the healing of a graft union as illustrated by the cleft graft.

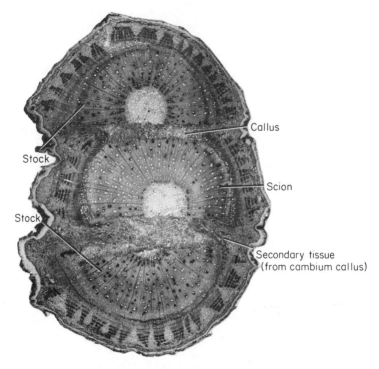

Figure 11–5   Cross section of a *Hibiscus* wedge graft showing the importance of callus development in the healing of a graft union. Cambial activity in the callus has resulted in the production of secondary tissues which have joined the vascular tissues of stock and scion. × 10. Reprinted with permission from K. Esau, *Plant Anatomy* (New York: John Wiley & Sons, Inc., 1953).

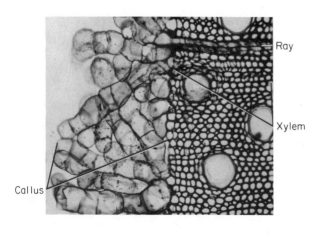

Figure 11–6   Callus production from incompletely differentiated xylem, exposed by excision of a strip of bark. × 120. Reprinted with permission from K. Esau, *Plant Anatomy* (New York: John Wiley & Sons, Inc., 1953).

In the healing of a graft union, the parts of the graft that are originally prepared and placed in close contact do not themselves move about or grow together. The union is accomplished entirely by cells which develop *after* the actual grafting operation has been made.

In addition, it should be stressed that in a graft union *there is no intermingling of cell contents.* Cells produced by the stock and by the scion each maintain their own distinct identity.

Considering in more detail the steps involved in the healing of a graft union, we may say that the first one listed below is a preliminary step, but nevertheless, it is essential and one over which the propagator has control.

(a) *Establishment of intimate contact of a considerable amount of the cambial region of both stock and scion under favorable environmental conditions* (see also p. 374).

Temperature conditions that will cause high cell activity are necessary. Usually, temperatures from 55° to 90° F (12.8° to 32° C), depending upon the species, are conducive to rapid growth. Outdoor grafting operations should thus take place at a time of year when such favorable temperatures can be expected and when the plant tissues, especially the cambium, are in a naturally active state. These conditions generally occur during the spring months. Temperature levels under greenhouse and bench grafting situations can, of course, be readily controlled, thereby permitting greater reliability of results.

The new callus tissue arising from the cambial region is composed of thin-walled, turgid cells which can easily become desiccated and die. It is important for the production of these parenchyma cells that the air moisture around the graft union be kept at a high level. This explains the necessity of thoroughly waxing the graft union or placing root grafts in a moist medium to maintain a high degree of tissue hydration.

It is important, too, that the region of the graft union be kept as free as possible from pathogenic organisms. The thin-walled parenchyma cells, under relatively high humidity and temperature conditions, provide a favorable medium for growth of fungi and bacteria, which are exceedingly detrimental to the successful healing of the union. Prompt waxing of the graft union helps prevent such infection.

It is essential that the two original graft components be held together firmly by some means, such as wrapping, tying, or nailing, or better yet, by wedging (as in the cleft or saw-kerf grafts) so that the parts will not move about and dislodge the interlocking parenchyma cells after proliferation has started.

The statement is often made that for successful grafting the cambium layers of stock and scion must be "matched." Although this is desirable, it is unlikely that complete matching of the two cambium layers is, or ever can be, attained. In fact, it is only necessary that the cambial regions be close enough together so that the parenchyma cells from both stock and scion produced in this region can become interlocked. It is in the region of the cambium that the essential callus production is the highest. Two badly matched cambial layers may delay union or, if extremely mismatched, prevent union. In studies (*119*) of grafting monocotyledonous plants, it was found that a cambium layer is not required for a successful graft union but that any meristematic tissue could be utilized for this purpose and is capable of forming a union between stock and scion.

(b) *Production and interlocking of parenchyma cells (callus tissue) by both stock and scion.*

During the grafting operation the cells cut and damaged by the grafting knife turn brown and die, forming a necrotic plate. Wound periderm may develop and the contact layers become well suberized. Underneath these dead cells new parenchyma cells (callus) arise in one to seven days from both stock and scion, coming from the parenchyma of the phloem rays and the immature parts of the xylem. The actual cambial layer itself seems to take little part in this first development of the callus (*91, 141, 150*). In grafting scions on established stocks, the stock produces most of the callus. These parenchyma cells, comprising the spongy callus tissue, fill the space between the two components of the graft (scion and stock), becoming intimately interlocked and providing some mechanical support, as well as allowing for some passage of water and nutrients between the stock and the scion. For a time, between the callus arising from the stock and that arising from the scion there is a more or less continuous brown line consisting of the above-mentioned dead and crushed cells remaining from the grafting cuts. This line of cells is gradually resorbed, however, and disappears. At the final stages of healing the cells of the outer layer of callus become suberized (*162*). Exposed dead cells (tracheids, vessels) are sealed off with a deposit of gum (*15*).

(c) *Production of new cambium through the callus "bridge."*

At the edges of the newly formed callus mass, parenchyma cells which are touching the cambial cells of the stock and scion differentiate into new cambium cells some two to three weeks after grafting. This cambial formation in the callus mass proceeds farther and farther inward away from the original stock and scion cambium, and on through the callus bridge, until a continuous cambial connection between stock and scion forms. Figure 11–7 shows the development of a cambial bridge through callus in a mango "chip bud" graft union.

(d) *Formation of new xylem and phloem from the new vascular cambium in the callus bridge.*

The newly formed cambial sheath in the callus bridge begins typical cambial activity, laying down new xylem and phloem, along with the original vascular cambium of the stock and scion, continuing this throughout the life of the plant.

In the formation of new vascular tissues following cambial continuity, it appears that the type of cells formed by the cambium is influenced by the cells of the stock adjacent to the cambium. For example, xylem ray cells are formed where the cambium is in contact with xylem rays of the stock, and xylem elements where they were in contact with xylem elements (*131*).

The new xylem tissue originates from the activities of the scion tissues rather than from that of the stock. This is shown by "ring grafting" (where a

Figure 11-7 Cambial bridge developing through callus in mango "chip bud" graft union—after 12 days. Sc = scion; St = stock; Ph = phloem; Xy = xylem; C = cambium; Ċa = callus; LC = laticiferous canal. Courtesy Dr. James Soule *(156)*.

ring of bark from a young tree is removed and replaced by a ring of bark from another tree) *(146)*. Researchers found that using bark rings of the 'Scugog' apple, which has purple xylem, subsequent xylem growth of the tree following grafting was entirely purple in color just under the 'Scugog' ring of bark, whereas in the remainder of the tree the xylem remained white *(190)*.

This production of new xylem and phloem thus permits vascular connection between the scion and the stock. Under conditions of high transpiration this must occur before much new leaf development takes place, arising from buds on the scion; otherwise, the enlarging leaf surfaces on the scion shoots will have little or no water supply to offset that lost by transpiration, and the scion will quickly become desiccated and die. Under some conditions, however, where vascular connections fail to occur, enough translocation can take place through the parenchyma cells of the callus to permit survival. In grafts of *Vanilla* orchid, a monocot, scions survived and grew for two years with only parenchyma union *(119)*.

A somewhat different developmental sequence occurs in tobacco *(37)* and cotton *(86)*. Here, xylem tracheary elements or phloem sieve tubes, or both, form directly by differentiation of callus into these vascular elements. A cambium layer subsequently forms between the two vascular elements. Apparently parenchyma cells, which make up the callus, can differentiate into tracheid-like elements with relative ease.

Buds, as would occur on the scions of grafts, are effective in inducing differentiation of vascular elements in the tissues on which they are grafted. This bud influence has been shown by inserting a bud into a piece of *Cichorium* root, consisting only of old vascular parenchyma, and observing as shown in Figure 11-8, that under the influence of auxin produced by the bud, the old parenchyma cells differentiate into groups of conducting elements (*23, 58*).

Induction of vascular tissues in callus, as would be formed in the healing of a graft union, is under the control of materials originating from growing points of shoots. Furthermore, the stimulus from such shoot apices in xylem formation can be replaced by appropriate concentrations of such hormones as auxins, cytokinins, gibberellins, as well as sugars (*165*). It has been demonstrated (*186*) that sugar concentrations of 2.5 to 3.5 percent, plus IAA or NAA at 0.5 mg/liter, will cause the induction of xylem and phloem in the callus, with a cambium in between.

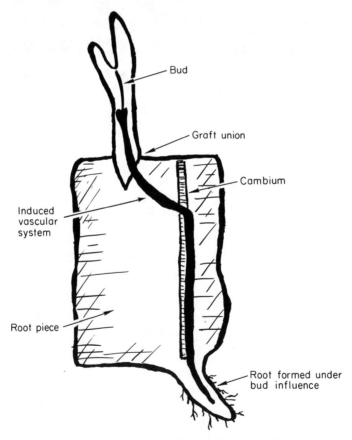

Figure 11-8   Induction of vascular tissue by grafting a bud on to a piece of *Chicorium* root tissue. Redrawn from Gautheret (*58*).

THE HEALING PROCESS IN T-BUDDING

In T-budding, the bud piece usually consists of the epidermis, cork layer, cortex, phloem, cambium, and often some xylem tissue; attached externally to this is a lateral bud subtended, perhaps, by a leaf petiole. In budding, this piece of tissue is laid against the exposed xylem and cambium of the stock.

Detailed studies of the healing process in T-budding have been made for the rose (*19*), citrus (*105*), and apple (*117*).

When the bud piece is removed from the budstick, and when the flaps of bark on either side of the "T" incision on the stock are raised, the cambium and newly formed cambial derivatives in these tissues are usually destroyed, owing to their very tender, succulent nature.

In the apple, when the bark of the stock is lifted in the budding operation the separation occurs in the young, undifferentiated xylem. The entire cambial zone remains attached to the inside of the bark flaps. Very shortly after the bud shield is inserted a necrotic plate of material is released from the cut cells. Next, after about two days, callus parenchyma cells start developing from the rootstock xylem rays and break through the necrotic plate. Some callus parenchyma from the bud scion ruptures through the necrotic area in a similar manner. As additional callus is produced it surrounds the bud shield and holds it in place. The callus originates almost entirely from the rootstock tissue, mainly from the exposed surface of the xylem cylinder. Very little callus is produced from the sides of the bud shield. Callus proliferation continues rapidly for two to three weeks until all internal air pockets are filled. Following this, a continuous cambium is established between the bud and the rootstock. The callus then begins to lignify, and isolated tracheary elements appear. Lignification of the callus is completed about 12 weeks after budding (*117, 178*).

In the rose, about three days after budding, the terminal cells of the broken xylem rays and adjacent cambial derivatives on the exposed surface of the stock begin to enlarge and divide, leading the production of callus strands. In the same manner, callus strands develop from terminal cells of broken phloem rays and adjacent young secondary phloem cells on the cut surface of the inner side of the bud piece. Within 14 days the space between the stock and the bud piece is completely filled with callus, which has developed mainly from the proliferating immature secondary xylem of the stock and the immature secondary phloem of the bud piece. During the second week, short areas of cambium cells appear in this newly developed callus tissue. By the tenth day, a completed band of cambium tissue extends over the face of the stock and is joined to the uninjured cambium on either side of the bud piece.

After cambial continuity is completed, continuity of vascular tissues soon becomes established between the bud and stock. In T-budding, then, the primary union is between the surface of the phloem on the inner face of the shield and the meristematic xylem surface of the stock. A secondary type of union may occur, however, at the edges of the shield piece, as it would in chip budding (see p. 448) (*15*).

The various stages and time intervals involved in the healing of the union in T-budding of citrus have been determined as follows (*105*):

Stage of Development	*Approximate Time after Budding*
1 First cell division	24 hours
2 First callus bridge	5 days
3 Differentiation:	
(a) in the callus of the bark flaps	10 days
(b) in the callus of the shield	15 days
4 First occurrence of xylem tracheids:	
(a) in the callus of the bark flaps	15 days
(b) in the callus of the shield	20 days
5 Lignification of the callus completed:	
(a) in the bark flaps	25 to 30 days
(b) under the shield	30 to 45 days
6 First occurrence of the meristematic layers in the callus between shield and bark flaps	15 days

FACTORS INFLUENCING THE HEALING OF THE GRAFT UNION

As anyone experienced in grafting or budding knows, the results obtained are often inconsistent, an excellent percentage of "takes" occurring in some operations, whereas in others the results are discouraging. There are a number of factors which can influence the healing of graft unions.

Incompatibility (see p. 337)

One of the symptoms of incompatibility in grafts between distantly related plants is a complete lack, or a very low percentage, of successful unions. Grafts between some plants known to be incompatible, however, will initially make a satisfactory union, even though the combination eventually fails.

Kind of Plant

Some plants are much more difficult to graft than others even when no incompatibility is involved. Difficult ones, for example, are the hickories, oaks, and beeches. Nevertheless, such plants, once successfully grafted, grow very well with a perfect graft union. In topgrafting apples and pears, even the simplest techniques usually give a good percentage of successful unions, but in topgrafting certain of the stone fruits, such as peaches and apricots, much more care and attention to details are necessary. Strangely enough, topgrafting peaches to some other compatible species, such as plums or almonds, is more successful than reworking them back to peaches. Many times one method of grafting will give better results than another, or budding may be more successful than grafting, or vice versa. For example, in topworking the native black walnut (*Juglans hindsii*) to the Persian walnut (*Juglans regia*) in California, the bark graft method is more successful than the cleft graft.

Some easily grafted plants, such as apple, form a "wound gum" plugging exposed xylem elements after the grafting operation, thus preventing excessive desiccation and death of tissues. Other plants, such as the walnut, in which graft unions heal with difficulty, form such "wound gum" very slowly, and in these desiccation and death of tissues in the area of the graft union may be extensive.

Some species, such as the Muscadine grape (*Vitis rotundifolia*), mango (*Mangifera indica*), and *Camellia reticulata,* are so difficult to propagate by the usual grafting or budding methods that "approach grafting," in which both partners of the graft are maintained for a time on their own roots, is often used. This variation among plant species and cultivars in their grafting ability is probably related to their ability to produce callus parenchyma, which is essential for a successful graft union.

Temperature, Moisture, and Oxygen Conditions During and Following Grafting

There are certain environmental requirements which must be met for callus tissue to develop.

Temperature has a pronounced effect on the production of callus tissue (Figure 11–9). In apple grafts little, if any, callus is formed below 32° F (0° C) or above about 104° F (40° C). Even around 40° F (4° C), callus development is slow and meager, and at 90° F (32° C) and higher, callus production is retarded, with cell injury becoming more apparent as the temperature increases, until death of the cells occurs at 104° F. Between 40° and 90° F, however, the rate of callus formation increases directly with the temperature (*152*). In such operations as bench grafting, callusing may be

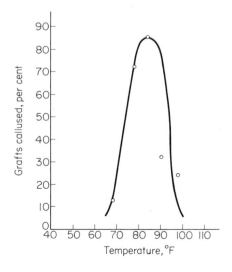

Figure 11–9   Influence of temperature on the callusing of walnut (*Juglans*) grafts. Callus formation is essential for the healing of the graft union. Maintaining an optimum temperature following grafting is very important for successful healing of walnut grafts. Adapted from data of Sitton (*153*).

allowed to proceed slowly for several months by storing the grafts at relatively low temperatures (45° to 50° F; 7° to 10° C), or if rapid callusing is desired, they may be kept at higher temperatures for a shorter time.

Following bench grafting of grapes, a temperature of 75° to 80° F (24° to 27° C) is about optimum; 85° F (29° C) or higher results in profuse formation of a soft type of callus tissue which is easily injured during the planting operations. Below 70° F callus formation is slow, and below 60° F (5° C) it almost ceases.

Grafting operations performed late in the spring when excessively high temperatures may occur, often result in failure. Tests (*69*) of walnut top-grafting in California during very hot weather in May showed that white-washing the area of the completed graft union definitely promoted healing of the union. Whitewash reflects the radiant energy of the sun, thus resulting in lower bark temperatures. In addition, in these tests, scions placed on the north and east sides of the stub survived much better than those on the south and west, probably due to the lower temperatures resulting from their shaded position.

Since the parenchyma cells comprising the important callus tissue are thin-walled and tender, with no provision for resisting desiccation, it is obvious that if they are exposed to drying air for very long, they will be killed. This was found (*152*) to be the case in studies of the effect of humidity on healing of apple grafts. Air moisture levels below the saturation point inhibited callus formation, the rate of desiccation of the cells increasing as the humidity dropped. In fact, the presence of a film of water against the callusing surface was much more conducive to abundant callus formation than just maintaining the air at 100 percent relative humidity.

Highly turgid cells are more likely to give proliferation of callus than those in a wilt condition. *In vitro* studies (*49*) of stem pieces of ash (*Fraxinus excelsior*) have shown that callus production on the cut surfaces was markedly reduced as the water potential (turgidity) decreased.

Unless a completed graft union is kept by some means at a very high humidity level, the chances of successful healing are rather remote. With most plants, thorough waxing of the graft union, which retains the natural moisture of the tissues, is all that is necessary. Often root grafts are not waxed but stored in a moist packing material during the callusing period. Damp peat moss or wood shavings are good media for callusing, providing adequate moisture and aeration.

It has been shown (*152*) that oxygen is necessary at the graft union for the production of callus tissue. This would be expected since rapid cell division and growth are accompanied by relatively high respiration, which requires oxygen. For some plants, a lower percentage of oxygen than is found naturally in air is sufficient, but for others healing of the union is better if it is left unwaxed but placed in a well-moistened medium. This may indicate that the latter plants have a high oxygen requirement for callus formation. Waxing restricts air movement and oxygen may become a limiting factor, thus inhibiting callus formation. This situation seems to be true for the grape in which, usually, the union is not covered with wax or other air-excluding materials during the callusing period.

There is evidence (*24*) that light will inhibit callus development. *In vitro* callus cultures of black cherry (*Prunus serotina*) became much larger in darkness than in light for some clones.

Growth Activity of the Stock Plant

Some propagation methods, such as T-budding and bark grafting, depend upon the bark "slipping," which means the cambium cells are actively dividing, producing young thin-walled cells on each side of the cambium. These newly formed cells separate readily from one another, so the bark "slips." Initiation of cambial activity in the spring results from the onset of bud activity since, shortly afterwards, cambial activity can be detected beneath each developing bud, with a wave of cambial activity progressing down the stems and trunk. This stimulus is due to production of auxin and gibberellins originating in the expanding buds (*179*).

In budding seedlings in the nursery in late summer it is important that they have an ample supply of soil moisture just before and during the budding operation. If they should lack water during this period, active growth is checked, cell division in the cambium stops, and it becomes impossible to lift the bark flaps to insert the bud.

There is evidence to show that callus proliferation—essential for a successful graft union—occurs most readily at the time of year just before and during "bud break" in the spring, diminishing through the summer and into fall. Increasing callus proliferation takes place again in late winter, but this is not dependent upon breaking of bud dormancy (*158*).

At certain periods of high growth activity in the spring, plants exhibiting strong root pressure (such as the walnut, maple, and grape) show excessive sap flow or "bleeding" when cuts are made preparatory to grafting. Grafts made with moisture exudation around the union will not heal, and should be done at some other stage of growth. Such "bleeding" at the graft union can be overcome by making slanting knife cuts around the tree through the bark and into the xylem to permit the exudation to take place below the graft union. When potted rootstock plants to be grafted (for example, in species of *Fagus, Betula,* or *Acer*) show excessive root pressure, they should be put in a cool place with reduced watering until the "bleeding" stops.

On the other hand, potted rootstock plants, as junipers or rhododendrons, when first brought in to a warm greenhouse in winter for grafting, are dormant, and grafting done then would be unsuccessful. Grafting should be delayed until the rootstock plants have been held for several weeks at 60° to 65° F (15° to 18° C) and new roots start to form; then the rootstock plant is physiologically active enough for the union to heal.

When the rootstock plant is physiologically overactive (excessive root pressure and "bleeding"), or underactive (no root growth being made), some form of side-graft (see p. 378), in which the rootstock top is not at first removed, should be used. In situations in which the rootstock is neither overactive nor underactive, one of the many forms of topgrafting (see p. 415), in which the top of the stock is completely removed at the time the graft is made, is likely to be successful (*50*).

Propagation Techniques

Sometimes the techniques used in grafting are so poor that only a small portion of the cambial regions of the stock and scion are brought together. Although healing occurs in this region and growth of the scion may start, after a large leaf area develops and high temperatures and high transpiration rates occur, sufficient movement of water through the limited conducting area cannot take place, and the scion subsequently dies. Other errors in grafting technique, such as poor or delayed waxing, uneven cuts, or use of desiccated scions can, of course, result in grafting failure.

Poor grafting techniques, although they may delay adequate healing for some time—weeks or months—do not in themselves cause any permanent incompatibility. Once the union is adequately healed, growth can proceed normally.

Virus Contamination, Insect Pests, and Diseases

Using virus-infected propagating materials in nurseries can reduce bud "take" as well as the vigor of the resulting plant (*127*). In stone fruit propagation the use of budwood free of ring spot virus has consistently given improved percentages of "takes" over infected wood.

Topgrafting olives in California is seriously hindered in some years by attacks of the American plum borer (*Euzophera semifuneralis*), which feeds on the soft callus tissue around the graft union, resulting in the death of the scion. In England, nurserymen are often plagued with the red bud borer (*Thomasiniana oculiperda*), which feeds on the callus beneath the bud-shield in newly inserted T-buds (*57*).

Sometimes bacteria or fungi gain entrance at the wounds made in preparing the graft or bud unions. For example, it was found that a rash of failures in grafts of *Cornus florida* 'Rubra' on *C. florida* stock was due to the presence of the fungus *Chalaropsis thielavioides* (31). Chemical control of such infections materially aids in promoting healing of the unions (*48*).

In South and Central America, rubber (*Hevea*) trees are propagated by a modification of the patch bud. A major cause of budding failures in these countries has been infection of the cut surfaces by a fungus, *Diplodia theobromae,* but control of this infection can be obtained by fungicidal treatments (*96*). In topgrafting mangos in Florida, control of the fungus diseases anthracnose and scab is essential for success. This is done by spraying the rootstock trees and the source of scionwood regularly with copper fungicides before grafting is attempted (*121*).

Growth Substances in Relation to
Healing of the Graft Union

Trials with the use of growth substances, particularly auxin, applied to tree wounds or to graft unions have not given consistent results in promoting subsequent healing; consequently such materials are not generally used for this purpose (*69, 104*). In tissue culture studies (see p. 525), however, a definite

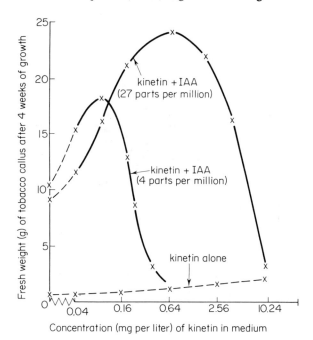

Figure 11-10 Relationship between relative concentration of auxin (indoleacetic acid) and kinetin upon callus production. Tobacco callus grown in aseptic culture medium. From original data of Murashige and Skoog *(118)* as summarized by van Overbeek *(125)*.

relationship exists between callus production (which is essential for graft healing) and the levels of certain applied growth substances, particularly kinetin and auxin *(49, 54)*. (See p. 248.) Figure 11-10 shows the relationship between auxin (indoleacetic acid) and kinetin levels in the production of callus in tobacco segments maintained under aseptic conditions *(118, 125)*. There is some evidence, too, that abscisic acid stimulates callus production, especially when applied to tissues in combination with auxin *(1)*, or with kinetin *(14)*.

Further studies may show some practical benefit from the use of combinations of auxin and kinetin or abscisic acid plus auxin in stimulating callus formation and subsequent healing of graft unions, thereby facilitating grafting of combinations considered to be difficult.

POLARITY IN GRAFTING
Proper polarity is essential if the graft union is to be permanently successful. In all commercial grafting operations correct polarity is strictly observed.

As a general rule, as shown in Figure 11-11, in grafting two pieces of stem tissue together, the morphologically proximal end of the scion should be inserted into the morphologically distal end of the stock. But in grafting a piece of stem tissue on a piece of root, as is done in root grafting, the proximal end of the scion should be inserted into the proximal end of the root piece.

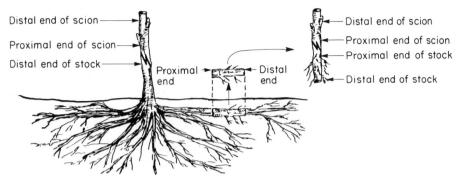

Figure 11-11   Polarity in grafting. In top grafting, the proximal end of the scion is attached to the distal end of the stock. In root grafting, however, the proximal end of the scion is joined to the proximal end of the stock.

The *proximal* end of either the shoot or the root is that nearest the stem-root junction of the plant.

The *distal* end of either the shoot or the root is that furthest from the stem-root junction of the plant and nearest the tip of the shoot or root.

Should a scion be inserted with reversed polarity—"upside down"—in bridge-grafting, for example, it is possible for the two graft unions to be successful and the scion to stay alive for a time. But, as seen in Figure 11–12, the reversed scion does not increase from its original size, whereas the scion with correct polarity enlarges normally.

As described on p. 411, in nurse-root grafting, the rootstock may purposefully be grafted to the scion with reversed polarity. Union will occur, and the root will supply water and mineral nutrients to the scion but the scion is unable to supply necessary organic materials to the rootstock and the stock eventually dies. In nurse-root grafting, the graft union is purposely set well below the ground level, and the scion itself produces adventitious roots which ultimately become the entire root system of the plant.

In T-budding or patch budding, the rule for observance of correct polarity is not as exacting. Buds can be inserted with reversed polarity and still make permanently successful

Figure 11-12   Bridge graft on a pear tree five months after grafting. Center scion was inserted with reversed polarity. Although the scion is alive it has not increased from its original size. The two scions on either side have grown rapidly.

unions. As shown in Figure 11–13, the buds start growing downward, then the shoots recurve and start upward (*143*). In the inverted bud piece, the cambium seems capable of continued functioning and growth. However, in the vessels as well as in the fibers formed from cambial activity, there is a twisting configuration which apparently keeps translocation and water conduction oriented as in the original position (*32*).

Figure 11–13   Two-year-old 'Stayman Winesap' apple budded on 'McIntosh' seedling by inverted T-buds. Note development of wide angle crotches. Courtesy Arnold Arboretum, Jamaica Plain, Mass.

LIMITS OF GRAFTING

Since one of the requirements for a successful graft union is the close matching of the callus-producing tissues near the cambium layers, grafting is generally confined to the dicotyledons in the angiosperms, and to the gymnosperms, the cone-bearing plants. Both have a vascular cambium layer existing as a continuous tissue between the xylem and the phloem. In the monocotyledonous plants of the Angiospermae, which do not have a vascular cambium, grafting is more difficult, with a low percentage of "takes." There are cases of successful graft unions between the stem parts of monocots. By making use of the meristematic properties found in the intercalary tissues (located at the base of internodes), successful grafts have been obtained with various grass species as well as the large tropical monocotyledonous vanilla orchid (*119, 120*).

Before a grafting operation is started, it should be determined that the plants to be combined are capable of uniting and producing a permanently successful union. There is no definite rule that can predict exactly the ultimate outcome of a particular graft combination except that *the more closely the plants are related botanically, the better the chances are for the graft union to be successful.*

Grafting Within a Clone
A scion can be grafted back on the plant from which it came, and a scion from a plant of a given clone can be grafted to any other plant of the same clone. For example, a scion taken from an 'Elberta' peach tree could be grafted successfully to any other 'Elberta' peach tree in the world.

Grafting Between Clones Within a Species
In the tree fruit and nut crops different clones within a species can almost always be grafted together without difficulty and produce satisfactory trees. However, in forest-tree species, notably Douglas fir (*Pseudotsuga menziesii*),

incompatibility problems have arisen in grafting together individuals of the same species, such as selected *P. menziesii* clones onto *P. menziesii* rootstock seedlings (*36*).

Grafting Between Species Within a Genus
For plants in different species but in the same genus, the situation is confused. In some cases grafting can be done successfully, in others it cannot. Grafting between most species in the genus *Citrus*, for example, is successfully and widely used commercially. Cultivars of the almond (*Prunus amygdalus*), the apricot (*Prunus armeniaca*), the European plum (*Prunus domestica*), and the Japanese plum (*Prunus salicina*)—all different species—are grafted commercially on the peach (*Prunus persica*), a still different species, as a rootstock. On the other hand, almond and apricot, both in the same genus, cannot be intergrafted successfully. The complexity of the situation is further illustrated by the fact that the 'Beauty' cultivar of Japanese plum (*Prunus salicina*) makes a good union when grafted on the almond, but another cultivar of *P. salicina*, 'Santa Rosa,' cannot be successfully grafted on the almond. Thus, compatibility between species may depend upon the particular clone or seedling being used, either for stock or scion. In another example, some almond cultivars (which are clones) are highly successful when grafted on the 'Marianna 2624' plum as a rootstock, while other cultivars are not (*93*).

In some cases a given interspecies graft is successful but the reciprocal combination is not. For instance, 'Marianna' plum (*Prunus cerasifera* × *P. munsoniana?*) on peach (*Prunus persica*) roots makes an excellent graft combination, but grafts of the peach on 'Marianna' plum roots either soon die or fail to develop normally (*2, 102*). Although many Japanese plum (*Prunus salicina*) cultivars can be successfully grafted on the European plum (*P. domestica*), the reverse, the European plum grafted on the Japanese plum, is unsuccessful (*77*).

Grafting Between Genera Within a Family
When the plants to be grafted together are in different genera but in the same family, the chances of the union being successful become more remote. Cases can be found in which such grafts are successful and used commercially, but in most instances such combinations are failures.

Trifoliate orange (*Poncirus trifoliata*) is used commercially as a dwarfing stock for various species in the genus *Citrus*. The quince (*Cydonia oblonga*) has long been used as a dwarfing rootstock for certain pear (*Pyrus communis*) cultivars. The reverse combination, quince on pear, though, is unsuccessful. Intergeneric grafts in the nightshade family, Solanaceae, are quite common. Tomato (*Lycopersicon esculentum*) can be grafted successfully on Jimson weed (*Datura stramonium*), tobacco (*Nicotiana tabacum*), potato (*Solanum tuberosum*), and black nightshade (*Solanum nigrum*). The evergreen loquat (*Eriobotrya japonica*) can be grafted on quince roots (*Cydonia oblonga*), a deciduous tree, giving a dwarfed loquat plant.

Grafting Between Families

Successful grafting between plants of different botanical families is usually considered to be impossible but there are reported instances (*94, 191*) in which it has been accomplished. These are with short-lived, herbaceous plants, though, for which the time involved is relatively short. Grafts, with vascular connections between the scion and stock, were successfully made (*123*), using white sweet clover, *Melilotus alba* (Leguminosae), as the scion and sunflower, *Helianthus annuus* (Compositae), as the stock. Cleft grafting was used, with the scion inserted into the pith parenchyma of the stock. The scions continued growth with normal vigor for over 5 months. As far as is known, however, there are no instances in which woody perennial plants belonging to different families have been successfully and permanently grafted together.

GRAFT INCOMPATIBILITY

The ability of two different plants, grafted together, to produce a successful union and to develop satisfactorily into one composite plant is termed *compatibility*. The opposite, of course, would be *incompatibility*. The distinction between a compatible and an incompatible graft union is not clearcut. On one hand, stock and scion of closely related plants, as described in the previous section, unite readily and grow as one plant. On the other hand, stock and scion of unrelated plants grafted together are likely to fail completely in uniting. Characterization of incompatibility, however, is not that distinct, for many graft combinations lie between these two extremes. They may unite initially, with apparent success (*20*), (Figures 11–14 and 11–15) but gradually develop incompatibility symptoms with time, due either to failure at the union or to the development of abnormal growth patterns. These symptoms are described further in this chapter. A strong, well-healed union and one showing incompatibility symptoms are illustrated in Figure 11–16.

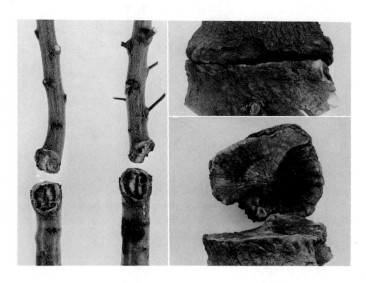

Figure 11–14   Breakage at the graft union due to incompatibility. *Left:* one-year-old nursery trees of apricot on almond seedling rootstock. *Right:* a 15-year-old 'Texas' almond tree on seedling apricot rootstock which broke off cleanly at the graft union—a case of "delayed incompatibility" symptoms.

Figure 11–15 Apple grafts (right portion of tree) five months after being grafted on a pear tree. This is an incompatible combination; the apple grafts eventually died although they initially made a strong growth.

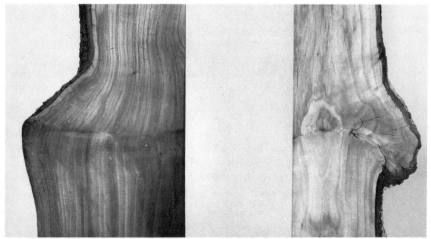

Figure 11–16 Radial sections through two fruit-tree graft unions. *Left:* a strong, well-knit compatible graft union. *Right:* a weak, poorly connected union, showing incompatibility symptoms.

Types of Incompatibility

Graft incompatibility in fruit trees has been classified by Mosse (*115*) into two types: (1) *translocated incompatibility* and (2) *localized incompatibility*. Most known cases can be placed into one or the other of these groups.

(1) *Translocated incompatibility* includes those cases in which the incompatible condition is *not* overcome by the insertion of a mutually compatible interstock because, apparently, some labile influence can move across it. This type involves phloem degeneration, and can be recognized by the development of a brown line or necrotic area in the bark. Consequently, restriction of movement of carbohydrates occurs at the graft union—accumulation above and reduction below. Reciprocal combinations may be compatible. In the various combinations in this category the range of bark tissue breakdown can extend from virtually no union at all, to a mechanically weak union with distorted tissues, to a strong union with tissues normally connected.

An example of a combination in this category is that of 'Hale's Early' peach grafted onto 'Myrobalan B' plum rootstock (*78*). This forms a weak union in which distorted tissues occur. Abnormal quantities of starch accumulate at the base of the peach scion. If the mutually compatible 'Brompton' plum is used as an interstock between the 'Hale's Early' peach and the 'Myrobalan B' rootstock, the incompatibility symptoms still persist, with an accumulation of starch in the 'Brompton' interstock. However, in other studies (*80*), in which peach/Myrobalan plum grafts were made, but using small seedlings at the cotyledonary stage, no signs of incompatibility appeared even after the grafted trees were 13 years old. In the same study all trees of this combination grafted at the usual stage in the nursery showed incompatibility symptoms one year after grafting. Possibly some factor responsible for the incompatibility symptoms is not present in the juvenile seedling tissues.

In another example of translocated incompatibility, the combination of 'Nonpareil' almond on 'Marianna 2624' plum shows complete phloem breakdown, although the xylem tissue connections are quite satisfactory. However, another cultivar, the 'Texas' almond, on 'Marianna 2624' plum produces a good, compatible combination. Inserting a 6-in. piece of 'Texas' almond as an interstock between the 'Nonpareil' almond and the 'Marianna' plum stock fails to overcome the incompatibility between these two components. Bark disintegration occurs at the normally compatible 'Texas' almond/ 'Marianna' plum graft union, owing, presumably, to some factor translocated in the phloem from the 'Nonpareil' scion above, through the 'Texas' interstock, to the 'Marianna' plum rootstock (*93*).

Virus-induced incompatibility cases can be included in translocated incompatibility. One component of the combination may carry a virus and be symptomless, but the other component may be susceptible to it (*27*). For example, at one time incompatibility was blamed for the difficulty encountered in working sweet orange (*Citrus sinensis*) on sour orange (*C. aurantium*) roots in South Africa (about 1910) and in Java (1928), even though this combination was a success commercially in other parts of the world.

The incompatibility was believed to be due to the production of some substance by the scion which was toxic to the stock (*166*). In the light of subsequent studies stimulated by the development of orange "tristeza" or "quick decline" in Brazil and California, it is clear that the toxic substance from the sweet orange scions was a virus tolerated by the sweet orange, but lethal to the sour orange roots (*11, 183*).

Incompatibility can be induced in a compatible combination by the introduction into the system of a third component. The apple, 'Cox's Pippin,' is fully compatible with *Malus toringoides* as a rootstock, but when a graft combination of these two components is grafted onto *Malus hupehensis* (*M. theifera*) roots, severe incompatibility symptoms develop at the usually normal 'Cox's Pippin'/*M. toringoides* union (*100*).

(2) *Localized incompatibility* includes combinations in which the incompatibility reactions apparently depend upon actual contact between stock and scion; separation of the components by insertion of a mutually compatible interstock overcomes the incompatibility symptoms. In the incompatible combination the union structure is often mechanically weak, with continuity of cambium and vascular tissues broken, although there are cases in which the union is strong and the tissues joined normally. Often external symptoms develop slowly, appearing in proportion to the degree of anatomical disturbance at the graft union. Root starvation eventually results, owing to translocation difficulties across the defective graft union.

An example of an incompatible combination in this category is that of 'Bartlett' ('Williams') pear grafted directly on quince rootstock. With the use of a "compatibility bridge" of 'Old Home' (or 'Beurré Hardy') pear as an interstock, the three-part combination is completely compatible, and satisfactory tree growth takes place (*114*).

Masses of parenchymatous or bark tissue, or both, rather than normally differentiated tissues, are commonly found at the union in this type of incompatibility, interrupting the normal vascular connection between stock and scion. Sometimes the conducting vessels make connections around the masses of callus tissue, whereas in other cases the ends of the vessels are separated by the masses of parenchyma. In extreme cases the vascular tissue becomes distorted, but the parenchyma layer between stock and scion is continuous. Apparently the cambial region of either scion or stock or both has failed to produce other than parenchyma cells at the line of union, resulting in a continuous sheet of such cells between scion and stock (*131, 181*).

In a study (*15*) of defective graft unions, it was concluded that incompatible combinations, such as pear on apple, differ from compatible combinations principally in the failure to maintain cambium continuity. A break in the cambium appears at the end of the growing season. "Regrafting" often occurs naturally but becomes less common as the tree grows older. Then, as the stock and scion expand, the cambium failure leaves a zone of parenchymatous, sometimes suberized, tissue between them, resulting finally in interruption of the vascular system and the development of a mechanically weak structure.

Inclusions of bark tissue may develop. This occurred in trees of the incompatible combinations of apple on pear (*132*) and plum on cherry (*131*).

A layer of bark extended almost to the point at which the cambium layers of stock and scion were placed when the graft was made. A small amount of continuous new parenchyma tissue allowed enough movement of water and nutrients to enable the tree to start growth. But the layer of bark laid down by the cambiums of both scion and stock and pinched between the xylem on either side formed a layer of mechanical weakness. Thus, by the second season, with a larger top and greater resistance to wind movement, a break could easily occur at the graft union.

Callus production in the cambial region of both stock and scion, with subsequent proliferation and interlocking to form a strong connection, would be an expected characteristic of a compatible combination. Using *in vitro* callus cultures of a wide range of woody plants this was found (54) to be true for most known compatible combinations, while with known incompatible combinations no interpenetration would occur. But in one incompatible combination, chestnut on persimmon, and vice-versa, a strong interlocking of callus did take place, with vigorous subsequent growth.

Symptoms of Incompatibilty

Graft union malformations resulting from incompatibility can usually be correlated with certain external symptoms. Below are some symptoms that have been associated with incompatible graft combinations:

(a)   Failure to form a successful graft or bud union in a high percentage of cases.

(b)   Yellowing foliage in the latter part of the growing season, followed by early defoliation. Decline in vegetative growth, appearance of shoot dieback, and general ill health of the tree.

(c)   Premature death of the trees, which may live only a year or two in the nursery.

(d)   Marked differences in growth rate or vigor of scion and stock.

(e)   Differences between scion and stock in the time at which vegetative growth for the season begins or ends.

(f)   Overgrowths at, above, or below the graft union.

An isolated case of one or more of the above symptoms does not necessarily mean the combination is incompatible; some of these symptoms can also result from unfavorable environmental conditions, such as lack of water or some essential nutrient, attacks by insects or diseases, or poor grafting or budding techniques (3).

Incompatibility is indicated by *the breaking off of trees at the point of union, particularly when they have been growing for some years and the break is clean and smooth, rather than rough or jagged.* This may occur in a year or two after the union is made—for instance, in the apricot on almond roots (see Figure 11–14).

In certain other cases, e.g., some apricot cultivars grafted on myrobalan plum roots, such breakage may not take place until the trees are full-grown and bearing crops (46, 51, 132). The occurrence of this situation in a single instance provides justification for stating that the particular combination is incompatible. Incompatibility in some other graft combinations is indicated by the appearance of a transverse brown line or necrotic area at the union (see p. 324). This can be observed by removing a narrow strip of bark through the union.

However, morphological abnormalities at the graft union are not always associated with incompatibility symptoms. In a comparison (*116*) of three-year-old double-worked trees of the combination:

(a) 'Conference' pear on 'C8' pear on 'Quince A' roots with the combination,
(b) of 'Conference' pear on 'Quince A', and 'Quince A' on 'C8' pear—

the lower unions in both cases—'C8' pear on 'Quince A,' and 'Quince A' on 'C8' pear—were very poor, the woody tissues of stock and scion separated in numerous places by masses of nonlignified, living parenchyma cells. Such badly connected and discontinuous unions would be expected to result in incompatibility symptoms, owing to the restriction of the passage of food substances between stock and scion. The second combination ('Conference' pear on 'Quince A' on 'C8' pear) did show such symptoms (sparse new growth with few leaves, which were yellowish-brown in color), obviously being very incompatible. On the other hand, trees of 'Conference' pear on 'C8' pear on 'Quince A,' with just as bad a mechanical union as the other combination, made entirely satisfactory growth.

Scion overgrowth at the graft union is often believed to be a certain indication of incompatibility. There does not seem to be justification for this if this is the only symptom that is present (*2, 15*) (Figure 11–17).

Figure 11–17   It is possible for the scion to greatly overgrow the stock and yet the combination develop into a large, strong tree. This is shown here for the Caucasian wingnut (*Pterocarya fraxinifolia*) grafted on *Pterocarya stenoptera*.

Delayed Symptoms of Incompatibility

In some instances the stock-scion combination grows in an apparently normal fashion for varying periods of time—perhaps for many years—then difficulties arise.

A good example of "delayed incompatibility" is the so-called "black-line" of walnuts (Figure 11–18); this occurs in certain Persian walnut orchards in Oregon, California, and France where cultivars of *Juglans regia* are grafted onto seedling rootstocks of *J. hindsii* (Northern California black walnut) or onto Paradox roots (*J. hindsii* × *J. regia*). Affected trees grow normally, bearing good crops, until they reach maturity—15 to 20 or more years of age. When the trouble appears, a thin layer of cambium and phloem cells dies at the graft union. The dead tissue develops at one point and gradually extends around the tree at the graft union until the tree becomes girdled; the vertical width of the dead area may reach 12 in. Such girdling kills the tree above the union, but the rootstock usually develops sprouts and remains alive. While the presence of a virus has been suspected, there has been no proof that any is involved. *(109, 148).*

There is evidence that compatibility relationships between clones may change with time *(100)*. For example, the pear variety 'Bristol Cross,' when it was introduced in 1932, was known to make a strong union when grafted directly on quince roots. Thirty years later, this variety when worked on quince roots required a mutually compatible interstock—such as 'Beurré Hardy'—in order to produce an acceptable union. Likewise, 'Conference' pear on quince in 1937 developed strong unions and vigorous trees *(29)*, but 20 years later many individual trees of this combination showed weak, incompatible unions *(187)*. There is the likelihood that such changes result

Figure 11–18   Delayed incompatibility of *Juglans regia* (Persian walnut) on *J. hindsii* rootstock. *Left:* mature tree has been completely girdled, and the top has died because of the failure of the phloem and cambium at the graft union. Many new shoots have arisen from the rootstock. *Right:* radial section through the graft union showing a transverse "black-line" (dead phloem tissue) under the outer bark just at the scion/stock junction. Courtesy Serr and Forde *(148).*

from mutations, or possibly latent viruses, which have appeared in one or more of the components (*100*).

A mutation in the opposite direction—from incompatibility to compatibility—is offered as the most likely explanation for the appearance of the Swiss 'Bartlett' ('Williams') pear clone, found to be compatible with quince (*128*). Bartlett/quince combinations have long been notorious for their incompatibility.

Causes of Incompatibility

Although incompatibility is clearly related to genetic differences between stock and scion, the mechanisms by which particular cases are expressed are not clear. With the large numbers of genetically different plant materials that can be combined by grafting, a wide range of different physiological, biochemical, and anatomical systems are brought together, with many possible interactions, both favorable and unfavorable. Several proposals have been advanced in attempts to explain incompatibility, but generally the evidence supporting most of them is inadequate and often conflicting.

One possible mechanism is that *different growth characteristics of the stock and scion occur* (*29, 182*). That is, if marked differences occur in vigor or in the time of starting or completing vegetative growth for the season, incompatibility may be expected. However, completely compatible combinations can be found that have growth differences between stock and scion that are as great as the differences found in stock and scion combinations whose incompatibility is attributed to those differences. Also, unmistakable incompatibility exists between some combinations which show little, if any, difference in growth rate or time of starting growth (*79*). Cambial activity in the spring, which originates with swelling of the scion cultivar buds, and spreads downward through the stem and through the graft union, will activate the cambium of the rootstock equally well in compatible and incompatible combinations. A study (*65*) in which growth rates of scion and rootstock were measured close to the graft union in incompatible pear-quince combinations failed to show that the incompatibility was associated with differences in growth rates or time of cambial activity. Although differences in growth rates may influence the severity of the symptoms, it is unlikely that this is a basic cause of incompatibility.

Another possible mechanism is *physiological and biochemical differences between stock and scion*. This is supported by studies with incompatible combinations of certain pear cultivars on quince rootstock (*64, 65, 66*). The experimental evidence supports the following conclusions:

(1) When certain pear cultivars are grafted onto quince roots, a cyanogenic glucoside, prunasin—normally found in quince, but not in pear, tissues—is translocated from the quince into the phloem of the pear. The pear tissues break down the prunasin in the region of the graft union, with hydrocyanic acid as one of the decomposition products. This enzymatic breakdown is hastened by high temperatures. In addition, different pear cultivars vary in their ability to decompose the glucoside.

(2)   The presence of the hydrocyanic acid leads to a lack of cambial activity at the graft union, with pronounced anatomical disturbances in the phloem and xylem at the union resulting. The phloem tissues are gradually destroyed at and above the graft union. Conduction of water and materials is seriously reduced in both xylem and phloem.

(3)   A reduction of the levels of the sugar reaching the quince roots leads to further decomposition of prunasin, liberating hydrocyanic acid and killing large areas of the quince phloem.

(4)   A water-soluble and readily diffusable inhibitor of the action of the pear enzyme (which breaks down the glucoside) occurs in the various pear cultivars, although they differ in their content of this inhibitor. This may explain why certain pear cultivars are compatible and others incompatible with quince rootstock.

Buchloh (17) in Germany studied compatible and incompatible pear-quince grafts, making electron microscope observations of cell-wall structure at the graft unions. Microspectrophotographic examination of the cell walls of compatible graft combinations showed the concentration of lignin at the line of union to be as high as that in cell walls not a part of the union. In incompatible combinations, however, the adjoining cell walls of the two components contained no lignin, with the parts either not connected or interlocked only by cellulose fibers. From these studies it was concluded that processes involved in the lignification of cell walls are involved in the formation of strong unions in pear-quince grafts. Reactions which inhibited the formation of lignin and the establishment of a mutual middle lamella between the two components resulted in weak unions. This work indicates that lignification of the adjoining cell walls at the graft union is an important basic process in producing a compatible combination. In pear-quince graft unions there is an accumulation of brown pigments in cells near the contact zone, which is attributable to hydrolysis of arbutin, a glycoside in pear, as well as the oxidation of the resulting hydroquinone. The oxidation products then inhibit cellular lignification in the union zone. Compatible cultivars are believed to reduce such oxidation products whereas incompatible pear cultivars do not (18).

Peach grafted on 'Marianna' plum is a combination which generally forms an incompatible union. Although peach buds unite readily with this plum stock, and grow satisfactorily the first season, later an enlargement appears above the graft union, followed by wilting of the peach leaves and death of the tree. Anatomical studies (102) showed this to be a case of incompatibility in which good xylem connections develop at the graft union but the phloem tissues fail to unite. Death of the roots results, with subsequent wilting and death of the peach top. However, if leafy shoot growth was retained on the 'Marianna' plum stock to nourish the roots, the trees could be kept alive indefinitely, a situation similar to other incompatibility cases (184).

Another possible mechanism is that abnormalities in incompatible graft combinations may result from viruses which have arisen directly as a result of the grafting; that is, by invasion of cells of one of the graft partners by proteins of the other (38, 42). Latent (symptomless) viruses, virus complexes,

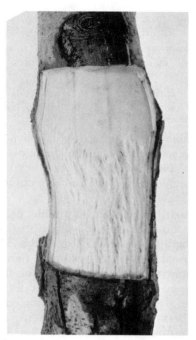

Figure 11-19   Latent viruses in the scion portion of a graft combination may cause symptoms to appear in a susceptible rootstock following grafting. Here "stem-pitting" virus symptoms have developed in the sensitive 'Virginia Crab' apple rootstock. The wood of the scion variety—above the graft union—is unaffected. Courtesy H. F. Winter *(188)*.

and mycoplasmalike bodies do occur; and in particular graft combinations, in which a susceptible cultivar is introduced to such a latent organism in its graft partner, rapid decline of the entire plant takes place *(107)*. This is a serious situation since it is well known that many clones are widely infected with latent viruses *(47)*. Figure 11–19 shows an occurrence in apple.

The pear decline disease, occurring first in Italy and later in western North America, killed hundreds of thousands of pear trees in California alone *(122, 149)*. Early studies *(6)* showed that the trouble was related to the rootstock used and indicated it to be an "induced incompatibility," with the causal factor unknown. 'Bartlett' trees on oriental pear roots, such as *Pyrus pyrifolia,* were highly susceptible to the disease, while trees on *P. communis* roots were not. Subsequent research showed that pear decline was not due to an inherent stock-scion incompatibility but was associated with what was thought to be a virus *(89)*, transmitted by an insect vector, *Psylla pyricola.* The 'Bartlett' scion cultivar and the *P. communis* roots were observed to be resistant to decline, but the rootstock, *P. pyrifolia,* was highly susceptible, showing a phloem degeneration just below the graft union. Later studies *(82, 83)* indicated the infective agent to be a mycoplasmalike organism rather than a virus. For the decline condition to appear, the concurrent presence of the infective bodies, pear psylla, and a susceptible rootstock was necessary. The mycoplasmalike organisms are sensitive to tetracycline antibiotics; thus by tree injections control measures are possible *(124)*.

Predicting Incompatible Combinations

To predict in advance, by some reliable test, whether or not a certain scion-rootstock combination would be incompatible or not would be of considerable economic value. This could be particularly important in the development of new clonal rootstocks. Several methods have been tried for making such predictions, none with complete success.

Various laboratory methods have been developed for evaluating stock-scion compatibility in young nursery trees without growing them on to maturity (*53*). These include water conductivity measurements through the graft union, macroscopic evaluation of the external appearance of the graft union, microscopic evaluation of the graft union, and breaking-strength tests.

One biochemical method (*140*) is based upon the pear-quince incompatibility due to the presence of the glucoside, prunasin (p. 344). Quince rootstocks known for their relatively good unions with pear, e.g., 'Provence C51,' had a low prunasin content (0.33 ppm), while the quince stock 'Portugal E,' known for its weak unions, had a relatively high prunasin content (3.8 ppm). A determination of the prunasin content in quince would thus give an indication of its compatibility chances with pear.

In one study (*98*), five different characteristics to be used as an index of defective union structure in two-year-old apricot scions on peach seedlings were correlated with subsequent tree performance. It was concluded that in this case macroscopic examination for discontinuity of inner bark tissue at the graft union in the young trees was the simplest and, since it did not destroy the tree, the most practical method of determining incompatibility status. The bark discontinuity appeared as a pitting of the inner bark, and could easily be seen by slicing off a portion of the bark covering the graft union.

Correcting Incompatible Combinations

If a graft combination known to be incompatible has been made and is discovered before the tree dies or breaks off at the union, it is possible, in graft combinations of the localized type, to correct this condition by bridge grafting (see p. 399) with a mutually compatible stock, if such exists. If a tree has been mistakenly propagated on a rootstock known to show symptoms of incompatibility with the scion cultivar, and with the probability that the tree will eventually break off at the graft union, it is also possible to correct this condition by inarching (see p. 396) with seedlings of a compatible rootstock. If breakage does not occur until the inarches are strong enough to support the tree, it may thus be possible to save it, the inarched seedlings becoming the main root system.

SCION-STOCK (SHOOT-ROOT) RELATIONSHIPS

Combining two (or more, in the case of interstocks) genotypes into one plant by grafting—one part producing the top and the other part the root—can produce unusual growth patterns which may be different from what would have occurred if each component part were grown separately. Some of these have major horticultural value. Others are detrimental and are to be avoided. Certain of these growth patterns may result (a) from incompatibility reactions as described in the previous section. Others may develop

(b) because one of the graft partners possesses one or more specific characteristics not found in the other. Examples are resistance to certain diseases, insects, or nematodes—or tolerance of certain adverse weather or soil conditions. Other growth patterns following grafting may result (c) from specific interactions between the stock and the scion which would alter size, growth, productivity, fruit quality, or other horticultural attributes. In practice it may be difficult to separate which of the three kinds of influencing factors is dominant in any given graft combination growing in a particular environment.

Effects of the Rootstock on the Scion Cultivar
Size and Growth Habit

Size control, and sometimes an accompanying change in tree shape, is one of the most significant rootstock effects. This seems to be largely because the rootstock causes the vigor of a given scion cultivar to be altered. By proper rootstock selection in apples, the complete range—from very dwarfed to very large—of tree size has been obtained with a given scion cultivar grafted to the rootstock series.

That specific rootstocks can be used to influence the size of trees has been known since ancient times. Theophrastus—and later the Roman horticulturists—made use of dwarfing apple rootstocks that could be easily propagated. The name "Paradise," which refers to a Persian park or garden—"pairidaeza"—was applied to dwarfing apple stocks about the end of the fifteenth century (22).

A wide assortment of size-controlling rootstocks has now been developed for certain of the major tree fruit crops (16). Most notable is the series of vegetatively propagated apple rootstocks collected and developed at the East Malling Research Station in England, beginning in 1912. These were classified into four groups according, primarily, to the degree of vigor imparted to the scion variety: *dwarf, semi-dwarf, vigorous,* and *very vigorous* (72, 113, 129). Similarly, the size-controlling effects of the rootstock on sweet cherry (*Prunus avium*) scion cultivars has been known since the early part of the eighteenth century (81). Mazzard (*P. avium*) seedling rootstocks produce large, vigorous, long-lived trees, whereas *P. mahaleb* seedlings, as a rootstock, tend to produce smaller trees of shorter life (45). However, individual seedlings of these species, increased and maintained as clones, can produce distinct rootstock effects different from that of the whole species. Rootstock effects on tree size and vigor are also recognized in citrus, pear, and other species (9, 55). A discussion of specific rootstocks for the various fruit and nut crops is given in Chapter 17.

Prediction of rootstock effects cannot be made with certainty without considering the entire system in which it is used. This includes the particular cultivar used as the scion top which can modify the rootstock influence. Each combination of scion and rootstock must be tested before any conclusion regarding the behavior of the composite tree can be reached (71). The environment in which the particular combination is to be grown must also be taken into account. If grafted plants are grown under optimum conditions the differences in performance of those on strong growing stocks vs. those on weaker stocks may not be as great as when such plants are grown under less favorable conditions. Good soil and cultural conditions are required for successful tree performance when very dwarfing rootstocks are used.

Often, alterations in the normal shape of the tree are associated with the dwarfing effect due to certain rootstocks. A low and spreading, rather than upright, form may develop. Such effects may possibly be due to changes in the auxin levels in the tree. This is illustrated by grafts of the 'McIntosh' apple on the apomictic, semi-dwarfing rootstock *Malus sikkimensis (143)*.

There occur in plants symptomless viruses which, in themselves, may exert a dwarfing influence. Some dwarfing effects could be due to the presence of a virus, and, if no harmful symptoms appeared, perhaps useful dwarfing could be produced in this manner (62). It is possible that removing viruses from dwarfing clonal stocks by heat treatment may decrease their dwarfing influence.

It would be useful in developing new clonal rootstocks from seedlings to be able to predict whether such stocks would be dwarfing or invigorating. Studies (7, 103) in England indicate that apple rootstocks known to produce dwarf trees have a high proportion of bark to wood in the lateral roots, whereas stocks causing increased vigor in the scion cultivar have a lower proportion of bark to wood. Also, much of the functional wood tissue of roots of dwarfing apple stocks is composed of living cells, whereas in non-dwarfing, vigorous rootstocks, the wood consisted of a relatively large amount of lignified tissue without living cell contents. Similar studies (164) in Italy with grafted apple and pear trees did not show such a relationship.

Efforts have been made to relate the concentration of extracted electrolytes to vigor imparted by clonal rootstocks by measuring electrical resistance of an ethanol extract from the shoot tips. In these studies, dwarfing stocks generally yielded an extract of high electrolytic content, as shown by high specific conductivity, whereas shoots from more vigorous rootstocks yielded fewer electrolytes, as shown by low specific conductivity (87). Further studies (161) showed, too, that electrical conductance of a water extract of shoots of dwarfing apple rootstocks is considerably higher than that of extracts taken from invigorating rootstocks. If shoots of dwarfing stocks contain proportionately more living cells, more living protoplasm, and thus more electrolytes than vigorous stocks (7), this could account for their greater electrical conductivity. Another method has been used which relates electrical conductivity of the living shoot itself to rootstock vigor (160). It was determined that shoots of dwarfing stocks exhibited low electrical resistance whereas those of vigorous stocks showed high resistance.

Fruiting Precocity, Fruit Bud Formation, Fruit Set, and Yield
Fruitfulness of a tree can be influenced by the rootstock used. In general, fruiting precocity is associated with dwarfing rootstocks and slowness to start fruiting with vigorous rootstocks. Long-term yield studies, conducted in England, involving several apple rootstocks showed that the results varied

according to tree age and tree spacing (*129, 130*). Trees on 'Malling 9' roots planted 12 x 12 feet apart showed highest accumulated yield per tree up to ten years of age because of their early bearing. By the tenth year this was surpassed by trees on the moderately vigorous 'Malling 4' roots. By the fifteenth year this was surpassed by trees on the vigorous 'Malling 1' roots, but by the twentieth year yield of all trees was superseded by trees grown on the very vigorous 'Malling 16' roots. On a yield-per-acre basis, for the first 20 years dwarf trees on the 'Malling 9' rootstock, planted 12 x 12 feet were the highest yielding of any group; after that time large trees on the vigorous 'Malling 16' roots gave the highest per-acre yields. It is probable that in cases such as this, yield is the product of a balance between invigorating effects (fruit-bearing surface) from a strong-growing rootstock, e.g., 'Malling 16,' and a flower-forming stimulus that is strong in a stock, such as 'Malling 9.'

Vigorous, strongly growing rootstocks in some cases result in a larger and more vigorous plant which produces greater crops over a long period of years. On the other hand, trees on dwarfing stocks may be more fruitful, and if closely planted, produce higher yields per acre, especially in the early years of bearing.

The presence of the graft union itself may stimulate earlier and perhaps heavier bearing. For instance, in studies with citrus (*85*) five rootstocks—sour orange, sweet orange, trifoliate orange, grapefruit, and rough lemon—all came into bearing two seasons earlier when budded to themselves than when unbudded, although in each case the trees were about the same size.

If there is an imperfect graft union, as with a partial incompatibility as occurs in some combinations, a reduction in translocation at the graft union can have a girdling effect and thus lead to increased fruitfulness.

Rootstock influence can vary greatly with different kinds of plants. In growing Oriental persimmon (*Diospyros kaki*) cultivars the rootstock seems to have a direct effect on flower production and fruit set. In tests (*147*) using the 'Hachiya' persimmon, trees on *D. lotus* roots produced more flowers but matured fewer fruits than similar trees on *D. kaki* roots, while trees on *D. virginiana* roots produced so few flowers that crops were very poor.

In grapes, where yield is dependent upon the vigor of the current season's growth, the rootstock used can be a strong influencing factor (*70*). Large yield increases of certain American types (*Vitis labrusca*) of grapes were obtained (*174*) when they were grafted on vigorous rootstocks, in comparison with own-rooted plants. Over a six-year period, yield of 'Concord' vines was increased from 30 to 150 percent, depending upon the rootstock used. On the other hand, with *V. vinifera* scions, use of 'Dog Ridge' (*V. champini*), an extremely vigorous rootstock cultivar, on fertile soils can lead to such strong growing vines that they become unproductive.

Size, Quality, and Maturity of Fruit

There is considerable variation among plant species in regard to the effect of the rootstock on fruit characteristics on the scion cultivar. No carry-over of characteristics of the fruit which the stock would produce is encountered in fruit of the scion cultivar. For example, quince, commonly used as a pear rootstock, has fruits with a pronounced tart and astringent flavor, yet this flavor does not appear in the pear fruits. The peach is often used as a rootstock for apricot, yet there is no indication that the apricot fruits have taken on any characteristics of peach fruits.

Although there is no intermingling of fruit characteristics between the stock and the scion, certain rootstocks can affect fruit quality of the scion cultivar. An outstanding example of this is the "black-end" defect of pears. 'Bartlett,' 'Anjou,' and some other pear cultivars on several rootstocks often produce fruits which are abnormal at the calyx end. The injury consists of blackened flesh, which in severe cases cracks open. Sometimes the calyx end of the fruit is hard and protruding. Such fruit is worthless commercially. It has been shown (*43, 76*) that this trouble develops when the trees are propagated on certain rootstocks, such as *Pyrus pyrifolia*, but only rarely when the French pear, *P. communis,* is used. This trouble affects only the fruit; no symptoms of adverse tree growth appear.

The development of black-end fruits disappears if trees on *P. pyrifolia* rootstock are inarched with *P. communis* seedlings, and—after the inarches are able to support the tree—the original *P. pyrifolia* roots are cut away. Black-end fruits continue to appear unless the original connection of the 'Bartlett' top with the *P. pyrifolia* roots is broken. Whether the *P. pyrifolia* roots are producing substances toxic to the fruit or whether there is some other interaction has not been established.

In citrus, striking effects of the rootstock appear in fruit characteristics of the scion cultivar (*10*). If sour orange (*Citrus aurantium*) is used as the rootstock, fruits of sweet orange, tangerine, and grapefruit are smooth, thin-skinned, and juicy, with excellent quality, and they store well without deterioration. Sweet orange (*C. sinensis*) rootstocks also result in thin-skinned, juicy, high-quality fruits. Citrus fruits on grapefruit (*C. paradisi*) stocks are usually excellent in size, grade, and quality if heavy fertilization is provided. But when rough lemon (*C. limon*) is used as the stock, the fruits are often thick-skinned, somewhat large and coarse, inferior in quality, and low in both sugar and acid.

Fruit size of both 'Washington Navel' and 'Valencia' orange is strongly influenced by the rootstock. The largest navel orange fruits are produced on sour orange stocks and the smallest on the Palestine sweet lime. The largest 'Valencia' oranges are associated with the dwarfing trifoliate orange stock, whereas sweet orange rootstocks produce the smallest fruits.

While many such tests have shown that the various characteristics of citrus fruit are affected by the rootstock used, the underlying physiological mechanisms remain unknown.

Tomato (*Lycopersicon esculentum*) grafted on Jimson weed (*Datura stramonium*) roots had been used at one time in the southern part of the United States, owing to the resistance of these roots to nematodes. As it was known that Jimson weed contains poisonous alkaloids, concern was felt about whether such alkaloids might not be translocated from the rootstock to the tomato fruits. Tests showed that this was the case. The alkaloid content of tomato fruits from grafted plants ranged from 1.77 to 13.00 mg per pound of fresh fruit, whereas ungrafted control plants had zero alkaloid content *(99)*.

Similar to this are other examples of translocation of compounds between stock and scion in intergeneric grafts in Solanaceae. In a series of reciprocal grafts between tomato and tobacco, nicotine was present in tomato scions when they were grown on tobacco rootstocks. When tobacco scions were grown on tomato rootstocks, tobacco alkaloid production was greatly reduced, most of the alkaloid synthesis being at the graft union and in the tobacco stem immediately above the union. Such localization was not found in tomato scions on tobacco roots *(44, 155)*.

Miscellaneous Effects of the Rootstock on the Scion Cultivar—
Winter-Hardiness, Disease Resistance, and Time of Fruit Maturity

In citrus the rootstock used can affect the cold-hardiness of the scion cultivar. During killing freezes in the winter of 1950–51 in the Texas Rio-Grande Valley, young grapefruit trees on 'Rangpur' lime roots survived much better than those on rough lemon or sour orange, whereas trees on 'Cleopatra' mandarin were the most severely damaged *(33)*. Survival following a severe winter freeze in Florida in 1962 showed that there was a wide range in tree hardiness in oranges and grapefruit which could be attributed to the rootstock used *(55)*.

Different rootstocks respond differently to certain soil conditions, thus resulting in an altered effect on the behavior of the scion cultivar. This is illustrated in Figure 11–20. Almond and myrobalan plum roots tolerate excess boron in the soil better than 'Marianna' plum roots. Thus, in this case, fairly good growth of the French prune occurred on almond and myrobalan plum roots under conditions in which it was severely injured when worked on 'Marianna' plum or apricot roots *(68)*.

The four rootstocks commonly used for stone fruits—plum, peach, apricot, and almond—differ markedly in their response to adverse soil conditions *(46)* and consequently can affect growth of the scion cultivar. For example, trees with myrobalan plum as the rootstock would be the most tolerant of excessive soil moisture, followed by peach or apricot roots, almond being the most susceptible to injury from such conditions.

In citrus, considerable variability exists in the tolerance of the various rootstocks to adverse soil conditions. The choice of rootstock upon which to work the scion cultivar is very important. In Texas, for example, the severity of lime-induced chlorosis symptoms on calcareous soils is greatly influenced by the rootstock used. Tests of grapefruit on 36 different rootstocks showed that with 4 of the stocks no chlorosis appeared, but severe chlorosis developed with 13 of the rootstocks *(34)*.

These examples illustrate cases in which the behavior of the scion cultivar was affected by the rootstock used. This was, in turn, traced to reactions of the rootstock to certain soil conditions. This can be extended further. It is known that some rootstocks are more tolerant than others to adverse soil situations, such as the presence of nematodes (*Meloidogyne* sp.) or oak root fungus (*Armillaria mellea*). The growth of the scion cultivar would subse-

Figure 11–20   The tolerance of fruit trees to toxic amounts of certain elements may be influenced by the rootstock used. This is illustrated here by 'French' prune trees on four different rootstocks which were irrigated with water containing different levels of boron. Rootstocks were: *(top left)* almond seedlings; *(top right)* myrobalan plum seedlings; *(below left)* apricot seedlings; *(below right)* 'Marianna' plum cuttings. Irrigation water containing the following five concentrations of boron were used for each rootstock (left to right): 1/2 ppm (tap water); 2 ppm; 3 ppm; 5 ppm; 10 ppm. Courtesy C. J. Hansen.

quently be influenced by the rootstock through the latter's relative ability to withstand such adverse conditions.

Effects of the Scion Cultivar on the Rootstock

Although there is a tendency to attribute all cases of dwarfing or invigoration of a grafted plant to the rootstock, the effect of the scion on the behavior of the composite plant may be fully as important as that of the rootstock. Unquestionably, however, the scion, the interstock, the rootstock, and the graft union itself all interact to influence each other and determine the over-all behavior of the plant. In certain combinations, however, a particular member of the combination could have a marked influence no matter what part of the plant it becomes. For instance, a dwarfing stock will exert a dwarfing influence on the entire plant whether used as rootstock, intermediate stock, or scion.

Effect of the Scion Cultivar on the Vigor of the Rootstock

This is the major influence of the scion on the stock, just as it was in the case of rootstock effect on scion cultivar. If a strongly growing scion cultivar is grafted on a weak rootstock, the growth of the rootstock will be stimulated so as to become larger than it would have been if left ungrafted. Conversely if a weakly growing scion cultivar is grafted on a vigorous rootstock, the growth of the rootstock will be lessened from what it might have been if left ungrafted. In citrus, for example, when the scion cultivar is less vigorous than the rootstock cultivar, it is the scion cultivar rather than the rootstock which determines the rate of growth and ultimate size of the tree *(84)*.

That the scion influences the growth of the rootstock was recognized at least as early as the middle of the nineteenth century *(59)*. It has long been known, particularly with apples, that the size, nature, and form of the root system which develops from the seedling rootstocks of grafted trees can be affected by the cultivar of the scion *(151, 170)*. Different scion cultivars may cause a characteristic root growth pattern to develop in the rootstock. For example, if apple seedlings are budded with the 'Red Astrachan' apple, a very fibrous root system with few taproots develops. If other similar seedlings are budded with 'Oldenburg' or 'Fameuse,' the subsequent root system is not fibrous but has a two- or three-pronged deep taproot system *(75)*. In fact, nurserymen propagating apples often can identify many of the scion cultivars by the appearance of the root system of the grafted rootstock.

This effect of the scion cultivar on the type of root system developed by the stock has been noted principally with seedling stocks. The vegetatively propagated clonal stocks have not usually shown such an effect *(159)*. For instance, the root systems of 'Malling 9' or '12,' when each is grafted with different scion cultivars, seem to retain their own distinct morphological characters regardless of the scion cultivar used, although the scion cultivar may have a marked influence on the quantity of roots.

Effect of the Scion Cultivar on Cold-Hardiness of the Rootstock

In some species at least, the cold-hardiness of a particular rootstock can be influenced by the particular scion cultivar grafted on it. This effect is not due necessarily to a winter-hardy scion imparting hardiness to the rootstock. Rather, it is probably related to the degree of maturity attained by the rootstock, certain scion cultivars tending to prolong growth of the roots long into the fall so that insufficient maturity of the root tissues is reached by the time killing low winter temperatures occur. The rootstock, if left ungrafted, or grafted to a scion cultivar which stops growth in early fall, may mature its tissues sufficiently early so as to develop adequate winter-hardiness.

Cold-hardiness of citrus roots is affected by the scion cultivar. Sour orange seedlings budded to 'Eureka' lemon suffered much more from winter injury than unbudded seedlings. The lemon tops were killed, as well as a portion, several inches deep, of the sour orange stock, but unbudded sour orange seedlings suffered only slight foliage injury *(182)*.

Effects of an Intermediate Stock on Scion and Rootstock

The ability of certain dwarfing clones, inserted as an interstock between a vigorous top and vigorous root, to produce a dwarfed and early-bearing fruit tree has been known for centuries. It is reported (*126*) that one of the earliest records (*95*) of such procedures was given in 1681, in England, advocating the use of the Paradise apple as an interstock to induce precocity in apple trees grown on crabapple rootstocks. Figure 11–21 shows, for example, the degree of dwarfing which can be induced by interstocks.

Tests (*163*) made to determine whether intermediate stocks of various apple cultivars inserted between the apple rootstock 'Malling 2' and the scion cultivar 'Jonathan' or 'Delicious' had any effect on the behavior of the scion showed that in every case a depression of growth occurred in comparison with cases in which the scion cultivar itself was used as the interstock.

An intermediate dwarfing stem piece seems to have a built-in mechanism that causes reduced growth in the rootstock as well as in the scion top (*126*).

Figure 11–21   Effect of interstock on size of six-year-old Cox's Orange Pippin apple on 'MM 104' roots. *Top left:* Cox/M.9/MM.104. *Top right:* Cox/M.27/MM.104. *Lower left:* Cox/MM.104/MM.104. *Lower right:* Cox/M.20/MM.104. Courtesy M. S. Parry, W. S. Rogers, and the Editors, *Journal of Horticultural Science* (*126*)

Repeated comparisons of the influence exerted on the scion cultivar by the rootstock and an interstock show that although both have an influence the rootstock's is greater (*167, 176*).

Dwarfing of apple trees by the use of a dwarfing interstock, such as 'Malling 7,' 'Malling 8' ('Clark Dwarf'), or 'Malling 9', has been widely used commercially for many years. This method has the advantage of allowing the use of well-anchored, vigorous seedlings as the rootstock rather than a brittle, poorly anchored dwarfing clone. However, excessive suckering from the roots may occur due to the dwarfing interstock, even in rootstock types that normally do not sucker freely.

This interstock effect could, in some cases, be due to the introduction of an additional graft union with the possibility of translocation restrictions. Imperfect graft unions are indicated as a cause of the dwarfing exerted on orange trees by a lemon interstock. In contrast to the dwarfing situation in apples with a 'Malling 9' interstock, the lemon itself is strong-growing.

On the other hand, there is evidence that the observed effects of the interstock are due directly to an influence of the interstock piece rather than to abnormalities at the graft union (*145*). The dwarfing effect of 'Malling 9' would seem to be due to something more than restrictions at the graft union, since this is an early-bearing, dwarf tree itself.

Roberts (*136*) has shown quantitatively that the initial response obtained from a 'Malling 9' interstock in composite 'Starking Delicious'/'M. 9'/'M. 16' trees was early and heavy flowering. This was followed—as a result of the heavy cropping—by a reduction in tree size. The degree of the response obtained was proportional to the length of the 'Malling 9' interstock. This supports earlier evidence that the interstock effect is due to a direct influence of the stock, since increasing the length of the interstock intensifies its effect (*40, 61*). However, studies with Old Home pear as an interstock between 'Bartlett' and quince showed no effects due to different interstock lengths (*185*).

Possible Mechanisms for the Effects of Stock on Scion and Scion on Stock

The nature of the rootstock-scion relationship is very complex and probably differs among genetically different combinations. The fundamental mechanism by which stock and scion influence each other has yet to be adequately determined. Some of the explanations offered for the observed effects are speculative, often conflicting, and sometimes not well substantiated. Several theories have been advanced as possible explanations for the interaction between stock and scion.

One suggested mechanism is that the rootstock influences are the result of *translocation effects* rather than the *absorbing ability* of the root system. That the stem portion of the tree has, to some extent at least, a definite influence is shown by experiments (*12*) in which commonly used rootstock materials were used as intermediate stocks between a vigorous root system and the scion cultivar. The expected effects were still present, although to a lesser degree, even though the materials were used as interstocks rather than as the entire root-absorbing system. This same influence on the tree was noted if the intermediate stock tissue was reduced to just a ring of bark *(137)*.

By this reasoning, the uniformity of trees which are produced on vegetatively propagated rootstocks is due to the uniformity of the *stems* of such stocks, whereas the variable stems of seedlings are responsible for the variation encountered in the growth of the scion cultivar. This is especially so if they are grafted or budded high on the stem so that there is more opportunity for stem influence to be exerted. But if the grafting is done on the roots of seedlings (those with relatively passive influence)—as is done in root grafting—this variability does not appear, owing to the absence of influence contributed by the stem section. Therefore a fairly uniform group of trees could be produced, even though they were on seedling rootstocks. This same situation has been noted with cherries. Orchards consisting of nursery trees which were low-worked on Mazzard seedlings were quite uniform. Orchards from nursery trees budded high on such seedlings were quite variable.

However, some workers in England concluded from their experiments *(73, 176)* that the *root system* itself, rather than the *stem* of the rootstock, plays the major role in rootstock effects on the scion. Beakbane and Rogers *(8)*, after experiments conducted with apple trees for 19 years, concluded that the characteristic rootstock influence was due to the roots themselves. This effect did not depend upon the presence of a piece of rootstock stem, although it did tend to increase the rootstock influence.

Ramirez and Tabuenca *(133)* may have resolved such contradictions in stem and root influence by showing that if the scion or interstock is to influence root morphology, it must be a type that has dominating characteristics.

Chandler *(28)* and Gardner, Bradford, and Hooker *(56)*, discussing the subject in general terms, asserted that the effects of stock on scion and scion on stock can be explained by *physiological factors,* chiefly the influences due to *changes in vigor.* Chandler pointed out that when the scion is the more vigorous part of the combination, the carbohydrate supply to the roots should be greater. And since certain roots supply and are supplied by certain branches, it would be expected that the branching habit of the top would influence the branching habit of the roots, thus explaining the different root types obtained in using different scion cultivars. If trees of different cultivars were pruned to exactly the same number and distribution of branches, then the difference in rootstock growth associated with the different scion cultivars might not occur.

Differences in growth rate were suggested by Vyvyan *(175, 177)* as a possible explanation of some of the observed reciprocal effects of stock and scion, especially the influence on vigor. He pointed out that for a given scion cultivar on a given soil, the stem/root ratio in any given scion-stock combination is remarkably constant regardless of tree size or age. This constant ratio between stem and root means that both grow at the same rate. If two different cultivars with markedly different growth rates are combined as stem and root by grafting into one plant which would subsequently have a constant growth rate, there must be some alteration in the growth rate of the components. Either the slower growing cultivar must speed up or the faster growing one must decrease its rate. Thus the stock and the scion would be expected to mutually affect each other's growth rate.

Rootstock effects are not invariably related to vigor. Such things as flowering, fruit setting, fruit size, and fruit color or quality may be affected by the rootstock used even on trees showing an equal amount of vigor. For example, the marked effect of the rootstock on the development of black-end in pears (see p. 351) is certainly not due to an alteration in vigor.

Tukey and Brase (*170*) concluded from their studies that no one part of a grafted tree could be considered to have complete control, but that all—rootstock, interstock, and scion—influenced the growth of the whole, although generally the rootstock had the dominant role.

Although no completely satisfactory explanation of how the three genetically different components of a grafted plant—rootstock, interstock, and scion—interact to influence the growth, flowering, and fruiting responses of the composite plant, three approaches can be considered: (1) *nutritional uptake and utilization*, (2) *translocation of nutrients and water*, and (3) *alterations in endogenous growth factors*.

Nutrition

It could possibly be that dwarfing rootstocks cause small trees by starvation effects. This has not been the case, however. Dwarf trees often contain higher concentrations of organic and mineral nutrients than vigorous ones (*30, 135*). The fruitful condition existing in young apple trees worked on the dwarfing 'Malling 9' roots was found to be associated with accumulation of starch in the shoots early in the season (*30*). Such an early starch storage would be expected to be favorable for the initiation of flower bud primordia. Nonfruitful trees on the vigorous 'Malling 12' roots failed to show such a starch accumulation. The increased supply of water and nutrients from the vigorous roots would stimulate production of new growth rather than retard growth and would not allow for the accumulation of carbohydrates, as would be the case with weaker, dwarfing rootstocks.

Tests (*41*) have shown that trees with a dwarfing interstock had a higher reducing sugar percentage in the leaves than did comparable trees without such an interstock, indicating an alteration in utilization.

Mineral absorption by the various rootstocks which is made available for use by the scion cultivar can explain certain rootstock influences. For example, the very vigorous 'Shalil' peach root system grown at low nutrient levels was able to pick up and furnish the scion cultivar with a greater supply of nutrients and water than the less vigorous 'Lovell' rootstock. Under these conditions, the scion cultivar showed better growth on 'Shalil' than on 'Lovell' roots. But when the salt level was higher and included toxic chloride ions, the greater accumulation of such salts, subsequently translocated to the scion cultivar, caused injury and growth depression. In this case, the difference in vigor of two rootstocks caused opposite effects under two different soil conditions—a low-salt and a high-salt condition (*74*).

When the nutrient status of a tree is judged by leaf analysis, it is apparent that all components—rootstock, interstock, and scion cultivar—can affect the mineral nutrient levels. In one study with apples, the scion had the greatest influence and the rootstock the least (*171*).

Studies (*4, 5*) in England with 'Malling 9' and the even more dwarfing 'Malling 27' have provided evidence that trees on these stocks are smaller because they are restricted to fewer growing points, which continue growth for a shorter period. The even slower growth rate of the root system so limits the trees that they are unable to completely make use of photosynthates in continued growth.

Translocation

The fact that interstocks of such dwarfing apple clones as 'Malling 9' or 'Clark Dwarf' will cause a certain amount of dwarfing could indicate that translocation is involved, owing either (1) to partial blockage at the graft unions or (2) to a reduction in movement of water or nutrient materials (or both) through the interstock piece itself.

One explanation advanced (*180*) for the dwarfing effect of certain rootstocks is supported by studies concerning the efficiency of water conductivity of the graft union. There are indications that the graft union does introduce an additional resistance to the flow of water. This resistance was greater in unions of which the 'Malling 9' stock was one of the components. Certain of the growth characteristics of trees on 'Malling 9' roots, such as small leaves, short internodes, and the early cessation of seasonal shoot growth, are those generally associated with a slight water deficit in the tree. As mentioned before, however, although a restricted graft union may contribute to the dwarfing influence of certain rootstocks, the primary influence probably lies in the nature of the growth characteristics of such stocks.

In a study of the translocation of radioactive phosphorus (P^{32}) and calcium (Ca^{45}) from the roots to the tops of one-year 'McIntosh' apple trees grown in solution culture, it was shown that over three times as much of both elements was found in the scion top when the vigorous 'Malling 16' root was used in comparison with the dwarfing 'Malling 9' (*21*). This may indicate a superior ability of the vigorous stock to absorb and translocate mineral nutrients to the scion in comparison with the dwarfing stock. Or it may only mean that the 'Malling 9' roots, with their high percentage of living tissue, formed a greater "sink" for these materials, retaining them in the roots.

Endogenous Growth Factors

The idea was proposed by Sax (*142, 145*) that some of the growth alterations noted when certain interstocks are used, or when a ring of bark in the trunk of young trees is inverted, may be due to interference with the normal translocation of natural growth substances and nutrients from the leaves to the roots. It is known that certain factors do exist (e.g., vitamin B_1) that are necessary for root growth, and anything that would stop or reduce the flow of such substances would limit growth of the rootstock and subsequently dwarf the entire tree.

Dwarfing rootstocks may exhibit their characteristic effects due to their own low production of endogenous growth promoters or to their inability to conduct or utilize such substances produced by the scion. Young trees on dwarfing stocks can make vigorous growth for a year or two in the nursery (*168*) when growth-regulating materials may still be present in sufficient amounts, but several years later dwarfing develops, possibly due to an increasing lack of growth promoters to give normal root or top growth.

There is evidence (*67*) to show that the amount of indoleacetic acid (a growth-promoting auxin) that is destroyed by root and shoot bark tissue of various apple rootstocks is correlated inversely with the scion vigor induced by the rootstock. In a study (*111*) of leaf extracts from various size-controlling apple rootstocks, it was found that those giving the greatest dwarfing contained materials which stimulated oxidative breakdown of indoleacetic acid.

In a comparison of endogenous growth-regulating factors in the bark of the dwarfing 'Malling 9' apple rootstock with those in the bark of the invigorating 'Malling 16', it was noted (*101*) that 'Malling 9' contained lower

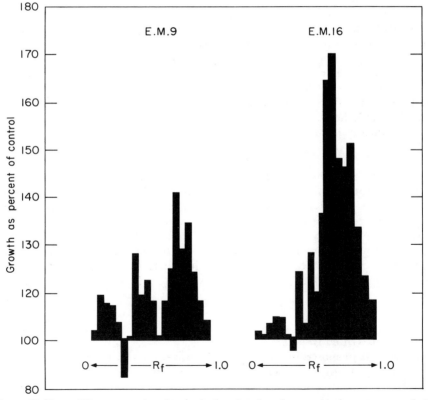

Figure 11–22   Histogram showing relative levels of growth promoters and inhibitors in *(left)* the leaves of a dwarfing apple rootstock 'Malling 9,' and *(right)* an invigorating apple stock 'Malling 16,' as determined by the wheat coleoptile bioassay. After Martin and Stahly *(101)*.

amounts of growth-promoting materials—but more growth inhibitors—than did 'Malling 16.' Figure 11–22 shows the comparative amounts of growth promoters and inhibitors found in leaf extracts of 'Malling 9' and 'Malling 16.' Tissue extracts from four size-controlling apple rootstocks 'Malling 16' (invigorating), 'Malling 1' (semi-invigorating), and 'Malling 7' (semi-dwarfing), and 'Malling 9' (dwarfing), showed increasing levels of an endogenous inhibitor—in the same order—which apparently was abscisic acid (*189*).

There is the possibility, too, that different levels of endogenous gibberellin, as a growth promoter, may account, in part, for the size-controlling characteristics of various rootstocks. It is known that roots do produce gibberellin and that the amounts of gibberellin in the transpiration stream are sufficient to have a decided growth-controlling influence (*90*). Extracts of the dwarfing 'Malling 9' apple rootstocks were found (*88*) to contain lower amounts of gibberellic acid-like substances than extracts of the more invigorating 'Malling 1' and 'Malling 25.' Such low gibberellin levels in the more dwarfing stock could be due either to lower production or to more rapid destruction. The differential dwarfing due to different rootstocks thus could possibly be accounted for, in part, by different levels of gibberellin translocated from the roots to the shoot system (*26*). Injections of gibberellic acid into grafted apple trees have given increasing stimulation to the top growth as rootstock vigor decreased, while injections of the growth inhibitor, abscisic acid, caused shoot growth to cease (*139*).

The various dwarfing and invigorating types of rootstocks apparently contain different amounts of naturally occurring growth-promoting and growth-inhibiting materials, varying during the year, which could control not only their own growth, but could pass through the graft union and similarly affect growth of the scion top. This seems to offer the best explanation advanced so far for the mechanisms involved in size-controlling rootstocks.

REFERENCES

1 Altman, A., and R. Goren, Promotion of callus formation by abscisic acid in citrus bud cultures, *Plant Physiol.*, 47:844–46. 1971.

2 Amos, J., T. N. Hoblyn, R. J. Garner, and A. Witt, Studies in incompatibility of stock and scion. I. Information accumulated during twenty years of testing fruit tree rootstocks with various scion varieties at East Malling, *Ann. Rpt. East Malling Res. Sta. for 1935*, pp. 81–99. 1936.

3 Argles, G. K., A review of the literature on stock-scion incompatibility in fruit trees, with particular reference to pome and stone fruits, *Imp. Bur. of Fruit Prod., Tech. Comm. No. 9*, 1937.

4 Avery, D. J., Comparisons of fruiting and deblossomed maiden apple trees and of non-fruiting trees on a dwarfing and an invigorating rootstock, *New Phytol.*, 68:323–36. 1969.

5 ———, Effects of fruiting on the growth of apple trees on four rootstock varieties, *New Phytol.*, 69:19–30. 1970.

6 Batjer, L. P., and H. Schneider, Relation of pear decline to rootstocks and sieve tube necrosis, *Proc. Amer. Soc. Hort. Sci.*, 76:85–97. 1960.

7 Beakbane, A. B., Anatomical structure in relation to rootstock behavior, *Rpt. 13th Int. Hort. Cong.*, Vol. 1, pp. 152–58. 1953.

8 ———, and W. S. Rogers, The relative importance of stem and root in determining rootstock influence in apples, *Jour. Hort. Sci.*, 31:99–110. 1956.

9 Bitters, W. P., Citrus rootstocks for dwarfing, *Calif. Agr.*, 4(2):5–14. 1950.

10 ———, Physical characters and chemical composition as affected by scions and rootstocks. Chap. 3 in W. B. Sinclair (ed.), *The Orange: Its Biochemistry and Physiology*. Berkeley: University of California Division of Agricultural Science, 1961.

11 Bitters, W. P., and E. R. Parker, Quick decline of citrus as influenced by top-root relationships, *Calif. Agr. Exp. Sta. Bul. 733*, 1953.

12 Blair, D. S., Rootstock and scion relationship in apple trees, *Sci. Agri.*, 19:85–94. 1938.

13 Bloch, R., Wound healing in higher plants, *Bot. Rev.*, 18:655–79. 1952.

14 Blumenfield, A., and S. Gazit, Interaction of kinetin and abscisic acid in the growth of soybean callus, *Plant Physiol.*, 45:535–36. 1969.

15 Bradford, F. C., and B. G. Sitton, Defective graft unions in the apple and pear, *Mich. Agr. Exp. Sta. Tech. Bul. 99*, 1929.

16 Brase, K. D., and R. D. Way, Rootstocks and methods used for dwarfing fruit trees, *N.Y. State Agr. Exp. Sta. Bul. 783*, 1959.

17 Buchloh, G., The lignification in stock-scion junctions and its relation to compatibility, in J. B. Pridham (ed.), *Phenolics in Plants in Health and Disease*. Long Island City, N.Y.: Pergamon Press, Inc., 1960.

18 ———, Verwachsung und Verwachsungsstorungen als Ausdruck des Affinitatsgrades bei Propfungen von Birnenvarietaten auf *Cydonia oblonga, Beit. Biol. Pfl.*, 37:183–240. 1962.

19 Buck, G. J., The histological development of the bud graft union in roses, *Proc. Amer. Soc. Hort. Sci.*, 62:497–502. 1953.

20 ———, and B. J. Heppel, A bud-graft incompatibility in *Rosa, Jour. Amer. Soc. Hort. Sci.*, 95(4):442–46. 1970.

21 Bukovac, M. J., S. H. Wittwer, and H. B. Tukey, Effect of stock-scion interrelationships on the transport of P^{32} and Ca^{45} in the apple, *Jour. Hort. Sci.*, 33:145–52. 1958.

22 Bunyard, E. A., The history of the Paradise stocks, *Jour. Pom.*, 2:166–76. 1920.

23 Camus, G., Recherches sur le role des bourgeons dans les phénomènes de morphogénèse, *Revue Cytol. et Biol. Veg.*, 11:1–199. 1949.

24 Caponetti, J. D., G. C. Hall, and R. E. Farmer, Jr., In vitro growth of black cherry callus: effects of medium, environment, and clone, *Bot. Gaz.* 132(4):313–18. 1971.

25 Carlson, R. F., The incidence of scion-rooting of apple cultivars planted at different soil depths, *Hort. Res.*, 7(2):113–15. 1967.

26　Carr, D. J., D. M. Reid, and K. G. M. Skene, The supply of gibberellins from the root to the shoot, *Planta,* 63:382–92. 1964.

27　Cation, D., and R. F. Carlson, Determination of virus entities in an apple scion/rootstock test orchard, *Quart. Bul. Mich. Agr. Exp. Sta., Rpt.* I, 43(2):435–43, 1960. Rpt. II, 45(17):159–66. 1962.

28　Chandler, W. H., *Fruit Growing.* Boston: Houghton Mifflin Company, 1925.

29　Chang, Wen-Tsai, Studies in incompatibility between stock and scion, with special reference to certain deciduous fruit trees, *Jour. Pom. and Hort. Sci.,* 15:267–325. 1938.

30　Colby, H. L., Stock-scion chemistry and the fruiting relationships in apple trees, *Plant Phys.,* 10:483–98. 1935.

31　Collins, R. P., and S. Waxman, Dogwood graft failures, *Amer. Nurs.,* 108(8):12. 1958.

32　Colquhoun, T. T., Polarity in *Casuarina paludosa, Trans. and Proc. Roy. Soc. South Australia,* 53:353–58. 1929.

33　Cooper, W. C., Influence of rootstock on injury and recovery of young citrus trees exposed to the freezes of 1950–51 in the Rio Grande Valley, *Proc. 6th Ann. Rio Grande Valley Hort. Inst.,* pp. 16–24. 1952.

34　————, and E. O. Olson, Influence of rootstock on chlorosis of young Red Blush grapefruit trees, *Proc. Amer. Soc. Hort. Sci.,* 57:125–32. 1951.

35　Copes, D. A., Graft union formation in Douglas-fir, *Amer. Jour. Bot.,* 56(3):285–89. 1969.

36　————, Initiation and development of graft incompatibility symptoms in Douglas-fir, *Silvae Genet.,* 19:101–7, 1970.

37　Crafts, A. S., Phloem anatomy in two species of *Nicotiana,* with notes on the interspecific graft union, *Bot. Gaz.,* 95:592–608. 1934.

38　Crane, M. B., Origin of viruses, *Nature,* 155:115–16. 1945.

39　————, and E. Marks, Pear-apple hybrids, *Nature,* 170:1017. 1952.

40　Dana, M. N., H. L. Lantz, and W. E. Loomis, Effects of interstock grafts on growth of Golden Delicious apple trees, *Proc. Amer. Soc. Hort. Sci.,* 81:1–11. 1962.

41　————, Studies on translocation across dwarf interstocks, *Proc. Amer. Soc. Hort. Sci.,* 82:16–24. 1963.

42　Darlington, C. D., Heredity, development, and infection, *Nature* 154:164–69. 1944.

43　Davis, L. D., and W. P. Tufts, Black end of pears III, *Proc. Amer. Soc. Hort. Sci.,* 33:304–15. 1936.

44　Dawson, R. F., Accumulation of nicotine in reciprocal grafts of tomato and tobacco, *Amer. Jour. Bot.,* 29:66–71. 1942.

45　Day, L. H., Cherry rootstocks in California, *Calif. Agr. Exp. Sta. Bul.* 725. 1951.

46　————, Rootstocks for stone fruits, *Calif. Agr. Exp. Sta. Bul.* 736, 1953.

47　Dimalla, G. G., and J. A. Milbrath, The prevalence of latent viruses in Oregon apple trees, *Plant Dis. Rpt.,* 49(1):15–17. 1965.

48 Doesburg, J. van, Use of fungicides with vegetative propagation, *Rpt. XVI Int. Hort. Cong.,* pp. 365–72. 1962.

49 Doley, D., and L. Leyton, Effects of growth regulating substances and water potential on the development of wound callus in *Fraxinus, New Phytol.,* 69:87–102. 1970.

50 Dorsman, C., Grafting of woody plants in the glasshouse, *Proc. XVII Int. Hort. Cong.,* I:366. 1966.

51 Eames, A. J., and L. G. Cox, A remarkable tree-fall and an unusual type of graft union failure, *Amer. Jour. Bot.,* 32:331–35. 1945.

52 Evans, G. E., and H. P. Rasmussen, Anatomical changes in developing graft unions of *Juniperus, Jour. Amer. Soc. Hort. Sci.,* 97(2):228–32. 1972.

53 Evans, W. D., and R. J. Hilton, Methods of evaluating stock/scion compatibility in apple trees, *Canad. Jour. Plant Sci.,* 37:327–36. 1957.

54 Fujii, T., and N. Nito, Studies on the compatibility of grafting of fruit trees. I. Callus fusion between rootstock and scion, *Jour. Jap. Soc. Hort. Sci.,* 41(1):1–10. 1972.

55 Gardner, F. E., and G. H. Horanic, Cold tolerance and vigor of young citrus trees on various rootstocks. *Proc. Fla. State Hort. Soc.,* 76:105–10. 1963.

56 Gardner, V. R., F. C. Bradford, and H. D. Hooker, Jr., *Fundamentals of Fruit Production,* 2nd ed. New York: McGraw-Hill, 1939.

57 Garner, R. J., and D. H. Hammond, Studies in nursery technique. Shield budding. Treatment of inserted buds with petroleum jelly, *Ann. Rpt. East Malling Res. Sta. for 1938,* pp. 115–17. 1939.

58 Gautheret, R. J., La culture des tissus vegetaux, *Proc. 6th Inter. Cong. Exp. Cytol.,* pp. 437–49. 1947.

59 Goodale, S. L., Influence of the scion upon the stock, *Hort.,* 1:290. 1846.

60 Graham, B. F., Jr., and F. H. Bornmann, Natural root grafts, *Bot. Rev.,* 32(3):255–92. 1966.

61 Grubb, N. H., The influence of intermediate stem pieces in double-worked apple and pear trees, *Sci. Hort.,* 7:17–23. 1939.

62 Guengerich, H. W., and D. F. Milliken, Bud transmission of dwarfing in sweet cherry, *Plant Dis. Rpt.,* 50:367–68. 1966.

63 ———, Root grafting, a potential source of error in apple indexing, *Plant Dis. Rpt.,* 49:39–41. 1965.

64 Gur, A., The compatibility of the pear with quince rootstock, *Spec. Bul. No. 10,* pp. 1–99, Agr. Res. Sta., Rehovot (Israel). 1957.

65 ———, and R. M. Samish, The relation between growth curves, carbohydrate distribution, and compatibility of pear trees grafted on quince rootstocks, *Hort. Res.,* 5:81–100. 1965.

66 ———, and E. Lifshitz, The role of the cyanogenic glycoside of the quince in the incompatibility between pear cultivars and quince rootstocks, *Hort. Res.,* 8:113–34. 1968.

67 Gur, A., and R. M. Samish, The role of auxins and auxin destruction in the vigor effect induced by various apple rootstocks, *Beitr. Biol. Pflanz.,* 45:91–111. 1968.

68 Hansen, C. J., Influence of the rootstock on injury from excess boron in French (Agen) prune and President plum, *Proc. Amer. Soc. Hort. Sci.,* 51:239–44. 1948.

69 ———, and H. T. Hartmann, Influence of various treatments given to walnut grafts on the percentage of scions growing, *Proc. Amer. Soc. Hort. Sci.,* 57:193–97. 1951.

70 Harmon, F. N., Comparative value of thirteen rootstocks for ten vinifera grape varieties in the Napa Valley in California, *Proc. Amer. Soc. Hort. Sci.,* 54:157–62. 1949.

71 Hartmann, H. T., Rootstock effects in the olive, *Proc. Amer. Soc. Hort. Sci.,* 72:242–51. 1958.

72 Hatton, R. G., The influence of different rootstocks upon the vigor and productivity of the variety budded or grafted thereon, *Jour. Pom. and Hort. Sci.,* 6:1–28. 1927.

73 ———, The influence of vegetatively raised rootstocks upon the apple, with special reference to the parts played by the stem and root portions in affecting the scion, *Jour. Pom. and Hort. Sci.,* 9:265–77. 1931.

74 Hayward, H. E., and E. M. Long, Vegetative responses of the Elberta peach on Lovell and Shalil rootstocks to high chloride and sulfate solutions, *Proc. Amer. Soc. Hort. Sci.,* 41:149–55. 1942.

75 Hedrick, U. P., Stocks for fruits, *Rpt. N. Y. State Fruit Grow. Ass'n.,* pp. 84–94. 1915.

76 Heppner, M. J., Pear black-end and its relation to different rootstocks, *Proc. Amer. Soc. Hort. Sci.,* 24:139. 1927.

77 ———, and R. D. McCallum, Grafting affinities with special reference to plums, *Calif. Agr. Exp. Sta. Bul. 438,* 1927.

78 Herrero, J., Incompatibilidad entre patrón e injerto. II, Efecto de un intermediario en la incompatibilidad entre melocotonero y mirobalán, *An. Aula Dei,* 4:167–72. 1955.

79 ———, Studies of compatible and incompatible graft combinations, with special reference to hardy fruit trees, *Jour. Hort. Sci.,* 26:186–237. 1951.

80 ———, and M. V. Tabuenca, Incompatibilidad entre patron e injerto. X. Comportamiento de la combinacion melocotonero/mirobalan injertado en estado cotiledonor, *An. Estac, exp. Aula Dei,* 10:937–45. 1969.

81 Hesse, H., *Teutscher Gartner.* Leipsig, 1710.

82 Hibino, H., and H. Schneider, Mycoplasma-like bodies in sieve tubes of pear trees affected with pear decline, *Phytopathology,* 60:499–501. 1970.

83 Hibino, H., G. H. Kaloostian, and H. Schneider, Mycoplasma-like bodies in the pear psylla vector of pear decline. *Virology,* 43:34–40. 1971.

84 Hodgson, R. W., Some instances of scion domination in citrus, *Proc. Amer. Soc. Hort. Sci.,* 43:131–38. 1943.

85 ———, and S. H. Cameron, On bud union effect in citrus, *Calif. Citrog.,* 20(12):370. 1935.

86 Homes, J., Histogenesis in plant grafts, in P. R. White and A. R. Groves (eds.), *Proc. Int. Conf. Plant Tissue Cult.,* Amer. Inst. Biol. Sci., pp. 553, 1965.

87 Hutchinson, A., Note on the ethanolic conductivity method for the estimation of apple rootstock vigor, *Canad. Jour. Plant Sci.*, 42:527–29. 1962.

88 Ibrahim, I. M., and M. N. Dana, Gibberellin-like activity in apple rootstocks, *HortScience*, 6(6):541–42. 1971.

89 Jensen, D. D., W. H. Griggs, C. Q. Gonzales, and H. Schneider, Pear decline virus transmission by pear psylla, *Phytopath.*, 54:1346–51. 1964.

90 Jones, O. P., and H. J. Lacey, Gibberellin-like substances in the transpiration stream of apple and pear trees, *Jour. Exp. Bot.*, 19:526–31. 1968.

91 Juliano, J. B., Callus development in graft union, *Philippine Jour. Sci.*, 75:245–51. 1941.

92 Keane, F. W. L., and J. May, Natural root grafting in cherry and spread of cherry twisted-leaf virus, *Canad. Plant Dis. Survey*, 43(2):54–60. 1963.

93 Kester, D. E., C. J. Hansen, and C. Panetsos, Effect of scion and interstock variety on incompatibility of almond on Marianna 2624 rootstock, *Proc. Amer. Soc. Hort. Sci.*, 86:169–77. 1965.

94 Kunkel, L. O., Contact periods in graft transmission of peach viruses, *Phytopath.*, 28:491–97. 1938.

95 Langford, G. T., *Plain and Full Instructions to Raise All Sorts of Fruit Trees That Prosper in England.* London: Printed by J. M. at the Rose and Crown, 1681.

96 Langford, M. H., et al., *Hevea* diseases of the Western Hemisphere, *Plant Disease Rpt. Suppl. 225*, May 15, 1954.

97 Langford, M. H., and C. H. T. Townsend, Jr., Control of South American leaf blight of *Hevea* rubber trees, *Plant Disease Rpt. Suppl. 225*, May 15, 1954.

98 Lapins, K., Some symptoms of stock-scion incompatibility of apricot varieties on peach seedling rootstock, *Canad. Jour. Pl. Sci.*, 39:194–203. 1959.

99 Lowman, M. S., and J. W. Kelley, The presence of mydriatic alkaloids in tomato fruit from scions grown on *Datura stramonium* rootstock, *Proc. Amer. Soc. Hort. Sci.*, 48:249–59. 1946.

100 Luckwill, L. C., New developments in the study of graft incompatibility in fruit trees, *Adv. in Hort. Sci. and their Appl.*, Vol. II. Long Island City, N.Y.: Pergamon Press, Inc., 1962, pp. 23–27.

101 Martin, G. C., and E. A. Stahly, Endogenous growth regulating factors in bark of EM IX and XVI apple trees, *Proc Amer. Soc. Hort. Sci.*, 91:31–38. 1967.

102 McClintock, J. A., A study of uncongeniality between peaches as scions and the Marianna plum as a stock, *Jour. Agr. Res.*, 77:253–60. 1948.

103 McKenzie, D. W., Rootstock-scion interaction in apples with special reference to root anatomy, *Jour. Hort. Sci.*, 36:40–47. 1961.

104 McQuilkin, W. E., Effects of some growth regulators and dressings on the healing of tree wounds, *Jour. Forest.*, 48(9):423–28. 1950.

105 Mendel, K., The anatomy and histology of the bud-union in citrus, *Palest. Jour. Bot.*, (R), 1(2):13–46. 1936.

106 Mergen, F., Anatomical study of slash pine graft unions, *Quart. Jour. Fla. Acad. Sci.*, 17:237–45. 1954.

107 Milbraith, J. A., and S. M. Zeller, Latent viruses in stone fruits, *Science*, 101:114–15. 1945.

108 Miller, L., and F. W. Woods, Root grafting in Loblolly pine, *Bot. Gaz.*, 126:252–55. 1965.

109 Miller, P. W., The etiology of blackline in grafted Persian walnuts, *Plant Dis. Rpt.*, 49:954. 1965.

110 ———, Technique for indexing strawberries for viruses by grafting to *Fragaria vesca, Plant Disease Rpt. 36*, pp. 94–96. 1952.

111 Miller, S. R., Growth inhibition produced by leaf extracts from size controlling apple rootstocks, *Canad. Jour. Pl. Sci.*, 45(6):519–24. 1965.

112 Millner, M. E., Natural grafting in *Hedera helix, New Phytol.*, 31:2–25. 1932.

113 Montgomery, H. B. S., Fruit tree raising, *Bul. 135, Minist. Agr., Fish. and Foods*, London. 1963.

114 Mosse, B., Further observations on growth and union structure of double-grafted pear on quince, *Jour. Hort. Sci.*, 33:186–93. 1958.

115 ———, Graft-incompatibility in fruit trees, *Tech. Comm. No. 28, Comm. Bur. Hort. and Plant. Crops*, East Malling, England, 1962.

116 Mosse, B., and J. Herrero, Studies on incompatibility between some pear and quince grafts. *Jour. Hort. Sci.*, 26:238–45. 1951.

117 Mosse, B., and M. V. Labern, The structure and development of vascular nodules in apple bud unions, *Ann. Bot.*, 24:500–507. 1960.

118 Murashige, T., and F. Skoog, A revised medium for rapid growth and bioassays with tobacco tissue cultures, *Physiol. Plant.*, 15:473–97. 1962.

119 Muzik, T. J., Role of parenchyma cells in graft union in *Vanilla* orchid, *Science*, 127:82. 1958.

120 ———, and C. D. LaRue, Further studies on the grafting of monocotyledonous plants, *Amer. Jour. Bot.*, 41:448–55. 1954.

121 Nelson, R., S. Goldweber, and F. J. Fuchs, Top-working for mangos, *Fla. Grower and Rancher*, p. 45, Jan., 1955.

122 Nichols, C. W., H. Schneider, H. J. O'Reilly, T. A. Shalla, and W. H. Griggs, Pear decline in California, *Bul. Calif. State Dept. Agr.*, 49:186–92. 1960.

123 Nickell, L. G., Heteroplastic grafts, *Science,* 108:389. 1948.

124 Nyland, G., and W. J. Moller, Control of pear decline with a tetracycline, *Pl. Dis. Rpt.*, 57:634–37. 1973.

125 Overbeek, J. van, Plant hormones and regulators, *Science*, 152:721–31. 1966.

126 Parry, M. S., and W. S. Rogers, Dwarfing interstocks: their effect on the field performance and anchorage of apple trees, *Jour. Hort. Sci.*, 43:133–46. 1968.

127 Posnette, A. F., Virus diseases of fruit plants, *Proc. XVII Int. Hort. Cong.*, Vol. 3, pp. 89–93. 1966.

128 ———, and R. Cropley, Further studies on a selection of Williams Bon Chrétien pear compatible with Quince A rootstocks, *Jour. Hort. Sci.*, 37:291–94. 1962.

129 Preston, A. P., Apple rootstock studies: thirty-five years' results with Cox's Orange Pippin on clonal rootstocks, *Jour. Hort. Sci.*, 33:194–201. 1958.

130 ———, Apple rootstock studies: thirty-five years' results with Lane's Prince Albert on clonal rootstocks, *Jour. Hort. Sci.*, 33:29–38. 1958.

131   Proebsting, E. L., Further observations on structural defects of the graft union, *Bot. Gaz.*, 86:82–92. 1928.

132   ———, Structural weaknesses in interspecific grafts of *Pyrus, Bot. Gaz.*, 82: 336–38. 1926.

133   Ramirez, D., and M. C. Tabuenca, The reciprocal effects of M. IX and M. XVI apples (English summary), *Ann. Estac. Exp. Aula Dei*, 7:164–74. 1964.

134   Rao, A. N., Developmental anatomy of natural root grafts in *Ficus globosa, Austral. Jour. Bot.*, 14:269–76. 1966.

135   Rao, Y. V., and W. E. Berry, The carbohydrate relations of a single scion grafted on Malling rootstocks IX and XIII. A contribution to the physiology of dwarfing, *Jour. Pom.*, 18:193–225. 1940.

136   Roberts, A. N., and L. T. Blaney, Qualitative, quantitative, and positional aspects of interstock influence on growth and flowering of the apple, *Proc. Amer. Soc. Hort. Sci.*, 91:39–50. 1967.

137   Roberts, R. H., Theoretical aspects of graftage, *Bot. Rev.*, 15:423–63. 1949.

138   Robitaille, R. H., and R. F. Carlson, Graft union behavior of certain species of *Malus* and *Prunus, Jour. Amer. Soc. Hort. Sci.*, 95(2):131–34. 1970.

139   Robitaille, H., and R. F. Carlson, Response of dwarfed apple trees to stem injections of gibberellic and abscisic acids, *HortScience*, 6(6):539–40. 1971.

140   Samish, R. M., Physiological approaches to rootstock selection, *Adv. in Hort. Sci. and their Appl.*, Vol. II. Long Island City, N.Y.: Pergamon Press, 1962, pp. 12–17.

141   Sass, J. E., Formation of callus knots on apple grafts as related to the histology of the graft union, *Bot. Gaz.*, 94:364–80. 1932.

142   Sax, Karl, The control of tree growth by phloem blocks, *Jour. Arn. Arb.*, 35:251–58. 1954.

143   ———, Dwarf trees, *Arnoldia*, 10:73–79. 1950.

144   ———, The effect of the rootstock on the growth of seedling trees and shrubs, *Proc. Amer. Soc. Hort. Sci.*, 56:166–68. 1950.

145   ———, Interstock effects in dwarfing fruit trees, *Proc. Amer. Soc. Hort. Sci.*, 62:201–4. 1953.

146   ———, and A. Q. Dickson, Phloem polarity in bark regeneration, *Jour. Arn. Arb.*, 37:173–79. 1956.

147   Schroeder, C. A., Rootstock influence on fruit set in the Hachiya persimmon, *Proc. Amer. Soc. Hort. Sci.*, 50:149–50. 1947.

148   Serr, E. F., and H. I. Forde, Blackline, a delayed failure at the union of *Juglans regia* trees propagated on other *Juglans* species, *Proc. Amer. Soc. Hort. Sci.*, 74:220–31. 1959.

149   Shalla, T. A., and L. Chiarappa, Pear decline in Italy, *Bul. Calif. State Dept. Agr.*, 50:213–17. 1961.

150   Sharples, A., and H. Gunnery, Callus formation in *Hibiscus rosasinensis L.* and *Hevea brasiliensis* Mull. Arg., *Ann. Bot.*, 47:827–39. 1933.

151   Shaw, J. K., The root systems of nursery apple trees, *Proc. Amer. Soc. Hort. Sci.*, 12:68–72. 1915.

152 Shippy, W. B., Influence of environment on the callusing of apple cuttings and grafts, *Amer. Jour. Bot.,* 17:290–327. 1930.

153 Sitton, B. G., Vegetative propagation of the black walnut, *Mich. Agr. Exp. Sta. Tech. Bul. 119,* 1931.

154 Smith, J. W. M., and P. Proctor, Use of disease resistant rootstocks for tomato crops, *Exp. Hort.,* 12:6–20. 1965.

155 Solt, M. L., and R. V. Dawson, Production, translocation, and accumulation of alkaloids in tobacco scions grafted on tomato rootstocks, *Plant Phys.,* 33:375–81. 1958.

156 Soule, J., Anatomy of the bud union in mango *(Mangifera indica* L.), *Jour. Amer. Soc. Hort. Sci.,* 96(3):380–83. 1971.

157 Stone, E. L., J. E. Stone, and R. C. McKittrick, Root grafting in pine trees, *Food and Life Sci. Quart.,* 6(2):19–21. 1973.

158 Sussex, I. M., and Mary E. Clutter, Seasonal growth periodicity of tissue explants from woody perennial plants *in vitro, Science,* 129:836–37. 1959.

159 Swarbrick, T., and R. H. Roberts, The relation of scion variety to character of root growth in apple trees, *Wis. Agr. Exp. Sta. Res., Bul. 78,* 1927.

160 Taper, C. D., and R. S. Ling, Estimation of apple rootstock vigor by the electrical resistance of living shoots, *Canad. Jour. Bot.,* 39:1585–89. 1961.

161 ——, and A. Hutchinson, Note on the estimation of apple rootstock vigor by the electrical conductivity of water extract, *Canad. Jour. Plant Sci.,* 43:228–30. 1963.

162 Thiel, K., Untersuchungen zur Frage der Unvertraglichkeit bei Birnenedelsorten auf Quitte A *(Cydonia* E. M. A), *Gartenbauwiss,* 1(19):127–59. 1954.

163 Thomas, L. A., Stock and scion investigations. X. Influence of an intermediate stem-piece upon the scion in apple trees. *Jour. Hort. Sci.,* 29:150–52. 1954.

164 Tomaselli, R., and E. Refatti, Sull' inesistenza di relazioni costanti tra anatomia radicale e vigoria in meli e peri innestati. (The nonexistence of constant relationship between root anatomy and vigor in grafted apple and pear trees), *Atti Ist. bot., Univ. Pavia,* Ser. 5, 18:130–40. 1960.

165 Torrey, J. G., D. E. Fosket, and P. K. Hepler, Xylem formation: a paradigm of cytodifferentiation in higher plants, *Amer. Sci.,* 59:338–52. 1971.

166 Toxopeus, H. J., Stock-scion incompatibility in citrus and its cause, *Jour. Pom. and Hort. Sci.,* 14:360–64. 1936.

167 Tukey, H. B., The dwarfing effect of an intermediate stem-piece of Malling IX apple, *Proc. Amer. Soc. Hort. Sci.,* 42:357–64. 1943.

168 ——, Similarity in the nursery of several Malling apple stock and scion combinations which differ widely in the orchard, *Proc. Amer. Soc. Hort. Sci.,* 39: 245–46. 1941.

169 ——, Stock and scion terminology, *Proc. Amer. Soc. Hort. Sci.,* 35:378–92. 1937.

170 ——, and K. D. Brase, Influence of the scion and of an intermediate stem-piece upon the character and development of roots of young apple trees, *N. Y. (Geneva) Agr. Exp. Sta. Tech. Bul. 218,* 1933.

171 Tukey, R. B., R. Langston, and R. L. Cline, Influence of rootstock, body-stock, and interstock on the nutrient content of apple foliage, *Proc. Amer. Soc. Hort. Sci.*, 80:73–78. 1962.

172 Tydeman, H. M., Experiments on hastening the fruiting of seedling apples, *Ann. Rpt. East Malling Res. Sta. for 1936*, pp. 92–99. 1937.

173 ———, and F. H. Alston, The influence of dwarfing rootstocks in shortening the juvenile phase of apple seedlings, *Ann. Rpt. East Malling Res. Sta. for 1964*, pp. 97–98. 1965.

174 Vaile, J. E., The influence of rootstocks on the yield and vigor of American grapes, *Proc. Amer. Soc. Hort. Sci.*, 35:471–74. 1938.

175 Vyvyan, M. C., The distribution of growth between roots, stems and leaves in a young apple tree and its possible bearing on the problem of stock effect on tree vigor, *Ann. Rpt. East Malling Res. Sta. for 1933*, pp. 122–31. 1934.

176 ———, The relative influence of rootstock and of an intermediate piece of stock stem in some double-grafted apple trees, *Jour. Pom. and Hort. Sci.*, 16:251–73. 1938.

177 ———, and D. H. Maggs, Progress in the study of rootstock-scion interaction, *Ann. Rpt. East Malling Res. Sta. for 1953*, pp. 141–44. 1954.

178 Wagner, D. F., Ultrastructure of the bud graft union in *Malus*, Ph.D. Diss., Iowa State University, Ames, 1969.

179 Wareing, P. F., C. E. A. Hanney, and J. Digby, The role of endogenous hormones in cambial activity and xylem differentiation, in Zimmermann, M. H. (ed.), *The Formation of Wood in Forest Trees*. New York: Academic Press, 1964.

180 Warne, L. G. G., and Joan Raby, The water conductivity of the graft union in apple trees, with special reference to Malling rootstock No. IX, *Jour. Pom. and Hort. Sci.*, 16:389–99. 1939.

181 Waugh, F. A., The graft union, *Mass. Agr. Exp. Sta. Tech. Bul. 2*, 1904.

182 Webber, H. J., Rootstock reactions as indicating the degree of congeniality, *Proc. Amer. Soc. Hort. Sci.*, 23:30–36. 1926.

183 ———, The "Tristeza" disease of sour orange rootstock, *Proc. Amer. Soc. Hort. Sci.*, 43:160–68. 1943.

184 Wellensiek, S. J., Het voorkómen van entings-incompatibiliteit door eigen bladann de onderstam (The prevention of graft-incompatibility by own foliage on the stock), *Meded. Landbouwhoogesch. Wageningen*, 49:255–72. 1949.

185 Westwood, M. N., and H. O. Bjornstad, Length of Old Home interstem makes little growth difference, *Ore. Orn. Nurs. Dig.*, 16(1):3–4. 1972.

186 Wetmore, R. H., and J. P. Rier, Experimental induction of vascular tissues in callus of angiosperms, *Amer. Jour. Bot.*, 50:418–30. 1963.

187 Williams, R. R., and A. I. Campbell, Rosetting and incompatibility of pears on Quince A, *Ann. Rpt. Long Ashton Res. Sta.*, pp. 51–56. 1956.

188 Winter, H. F., Prevalence of latent viruses in Ohio apple trees, *Ohio Farm and Home Res.*, 48:58–59. 63. 1963.

189 Yadava, U. L., and D. F. Dayton, The relation of endogenous abscisic acid to the dwarfing capability of East Malling apple rootstocks, *Jour. Amer. Soc. Hort. Sci.*, 97(6):701–5. 1972.

190 Yeager, A. F., Xylem formation from ring grafts, *Proc. Amer. Soc. Hort. Sci.*, 44:221–22. 1944.

191 Zebrak, A. R., Intergeneric and interfamily grafting of herbaceous plants, *Timirjazey Seljskohoz Akad.*, 2:115–33. 1937. *(Herb. Abst. 9:675. 1939).*

SUPPLEMENTARY READING

Argles, G. K., "A Review of the Literature on Stock-Scion Incompatibility in Fruit Trees with Particular Reference to Pome and Stone Fruits," *Imperial Bureau of Fruit Production, Technical Communication No. 9* (1937).

Beakbane, A. B., "Possible Mechanisms of Rootstock Effect," *Annals of Applied Biology,* Vol. 44 (1956), pp. 517–21.

Brase, K. D., and R. D. Way, "Rootstocks and Methods used for Dwarfing Fruit Trees," *New York Agricultural Experiment Station Bulletin 783* (1959).

Chang, W. T., "Studies in Incompatibility Between Stock and Scion with Special Reference to Certain Deciduous Fruit Trees," *Journal of Pomology and Horticultural Science,* Vol. 15 (1938), pp. 267–325.

Daniel, L., *Etudes sur la greffe.* Rennes, Paris: Imprimerie Oberthur, 1927, 1929, Vols. I and II.

Gardner, V. R., F. C. Bradford, and H. D. Hooker, *Fundamentals of Fruit Production,* 2nd ed., Sect. VI, *Propagation.* New York: McGraw-Hill Book Company, 1939.

Hatton, R. G., "The Relationship Between Scion and Rootstock with Special Reference to the Tree Fruits," *Journal of the Royal Horticultural Society,* Vol. 55 (1930), pp. 169–211.

Mosse, B., "Graft Incompatibility in Fruit Trees," *Commonwealth Agricultural Bureaux, Technical Communication No. 28* (1962).

Nelson, S. H., "Incompatibility Survey among Horticultural Plants," *Proceedings International Plant Propagators' Society,* Vol. 18 (1968), pp. 343–93.

Roberts, R. H., "Theoretical Aspects of Graftage," *Botanical Review,* Vol. 15 (1949), pp. 423–63.

Rogers, W. S., and A. B. Beakbane, "Stock and Scion Relations," *Annual Review of Plant Physiology,* Vol. 8 (1957), pp. 217–36.

Tubbs, F. R., "Tree Size Control Through Dwarfing Rootstocks," *Proceedings XVII International Horticultural Congress,* Vol. 3 (1966), pp. 43–56.

Techniques of Grafting 12

HISTORY

The origins of grafting can be traced back to ancient times. There is evidence that the art of grafting was known to the Chinese at least as early as 1000 B.C. Aristotle (384–322 B.C.) discussed grafting in his writings with considerable understanding. During the days of the Roman Empire grafting was very popular, and methods were precisely described in the writings of that era. Paul the Apostle, in his Epistle to the Romans, discussed grafting between the "good" and the "wild" olive trees (Romans 11:17–24). The Renaissance period (1300–1500 A.D.) saw a renewed interest in grafting, as well as in many others fields. Large numbers of new plants from foreign countries were imported into European gardens and maintained by grafting. By the sixteenth century the cleft and whip grafts were in widespread use in England and it was realized that the cambium layers must be matched, although the nature of this tissue was not then understood and appreciated. Propagators were handicapped by a lack of a good grafting wax; mixtures of wet clay and dung were used to cover the graft unions. In the seventeenth century many orchards were planted in England, the trees all being propagated by budding and grafting. Early in the eighteenth century Stephen Hales, in his studies on the "circulation of sap" in plants, approach-grafted three trees and found that the center tree stayed alive even when severed from its roots. Duhamel, about the same time, studied wound healing and the uniting of woody grafts. The graft union at that time was considered to act as a type of filter changing the composition of the sap flowing through it. Thouin (30) in 1821 described 119 methods of grafting and discussed changes in growth habit due to grafting. Vöchting (32) in the late nineteenth century continued Duhamel's earlier work on the anatomy of the graft union.

TERMINOLOGY

In constructing a successful graft, the goal is to connect or fit two pieces of living plant tissue together in such a way that they will unite and subsequently behave as one plant, as illustrated in Figures 12–1 and 12–2. As any technique that will accomplish this could be considered a method of grafting, it is not surprising that there are innumerable procedures for grafting described in the literature on this subject. Through the years several distinct methods have become established that enable the propagator to cope with almost any grafting problem at hand. These are described with the realization that there are many variations of each and that there are other, somewhat different, forms which could give the same results.

Definition of Terms

Scion (Cion) is the short piece of detached shoot containing several dormant buds, which, when united with the stock, comprises the upper portion of the graft and from which will grow the stem or branches, or both, of the grafted plant. It should be of the desired cultivar and free from disease. (See p. 407 for information on selection of the proper type of scion wood.)

Stock (Rootstock, Understock) is the lower portion of the graft, which develops into the root system of the grafted plant. It may be a seedling, a rooted cutting, or a layered plant. If the grafting is done high in a tree, as in topworking, the stock may consist of the roots, trunk, and scaffold branches.

Interstock (Intermediate stock, Interstem) is a piece of stem inserted by means of two graft unions between the *scion* and the *rootstock*. An inter-

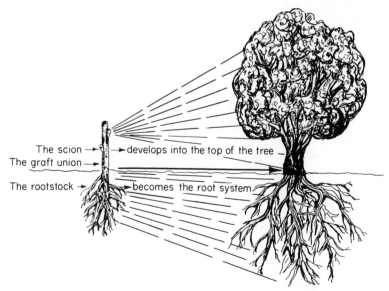

The scion → develops into the top of the tree
The graft union →
The rootstock → becomes the root system

Figure 12–1   In grafted plants the entire shoot system consists of growth arising from one (or more) buds on the scion. The root system consists of an extension of the original rootstock. The graft union remains at the junction of the two parts throughout the life of the plant.

Figure 12-2   Cultivars of the Persian (English) walnut *(Juglans regia)* grafted on *J. hindsii* rootstocks. The characteristics of these two species remain distinctly different after grafting, exactly to the junction of the graft union.

stock is used for several reasons, such as to avoid an incompatibility between the stock and scion, to make use of a winter-hardy trunk, or to take advantage of its growth-controlling properties.

Cambium is a thin tissue of the plant located between the bark (phloem) and the wood (xylem). Its cells are meristematic: they are capable of dividing and forming new cells. For a successful graft union, it is essential that the cambium of the scion be placed in close contact with the cambium of the stock.

Callus is a term applied to the mass of parenchyma cells that develops from and around wounded plant tissues. It occurs at the junction of a graft union, arising from the living cells of both scion and stock. The production and interlocking of these parenchyma (or callus) cells constitute one of the important steps in the healing process of a successful graft.

For any successful grafting operation there are five important requirements:

(a) *The stock and scion must be compatible.* They must be capable of uniting. Usually, but not always, plants closely related, such as two apple cultivars, can be grafted together. Distantly related plants, such as an oak tree and an apple tree, cannot be grafted together. (See Chapter 11 for a detailed discussion of this factor.)

(b) *The cambial region of the scion must be in intimate contact with that of the stock.* The cut surfaces should be held together tightly by wrapping, nailing, or some other such method. Rapid healing of the graft union is necessary so that the scion may be supplied with water and nutrients from the stock by the time the buds start to open. (See p. 320 for details on how the two parts of the graft unite.)

(c) *The grafting operation must be done at a time when the stock and scion are in the proper physiological stage.* Usually this means that the scion buds are dormant. For deciduous plants, dormant scion wood is collected during the winter and kept inactive by storing at low temperature (see p. 407 for the proper method of collecting and storing scion wood). The rootstock plant may be dormant or in active growth, depending upon the grafting method used.

(d) *Immediately after the grafting operation is completed, all cut surfaces must be carefully protected from desiccation.* This is done either by covering with grafting wax or by placing the grafts in moist material or in a covered grafting frame. (See p. 403 for a discussion of types of grafting wax.)

(e) *Proper care must be given the grafts for a period of time after grafting.* Shoots coming from the stock below the graft will often choke out the desired growth from the scion. Or, in some cases, shoots from the scion will grow so vigorously that they break off unless staked and tied or cut back. (See p. 419 for details on how to care for grafted trees.)

METHODS OF GRAFTING
Whip, or Tongue, Grafting
This method, as shown in Figures 12–3 and 12–4, is particularly useful for grafting relatively small material, ¼ to ½ in. in diameter. It is highly successful if properly done because there is considerable cambial contact. It heals quickly and makes a strong union. Preferably, the scion and stock should be of equal diameter. The scion should contain two or three buds with the graft made in the smooth internode area below the lower bud.

The cuts made at the top of the stock should be exactly the same as those made at the bottom of the scion. First, a long, smooth, sloping cut is made, 1 to 2½ in. long. The longer cuts are made when working with large material. This first cut should be made, preferably with one single stroke of the knife, so as to leave a smooth, flat surface. To do this, the knife must be sharp. Wavy, uneven cuts will not result in a satisfactory union.

On each of these cut surfaces, a reverse cut is made. It is started downward at a point about one-third of the distance from the tip and should be about one-half the length of the first cut. To obtain a smooth-fitting graft, this second cut should not just split the grain of the wood but should follow along under the first cut, tending to parallel it.

The stock and scion are then inserted into each other, with the tongues interlocking. It is extremely important that the cambium layers match along at least one side, preferably along both sides. The lower tip of the scion should not overhang the stock, as there is a likelihood of the formation of large callus knots. In some species, such callus overgrowths are often mistaken for crown gall knots, which are caused by bacteria. The use of scions larger than the stock should be avoided for the same reason. If the scion is smaller than the stock, it should be set at one side of the stock so that the cambium layers will be certain to match along that side. If the scion is much smaller than the stock, the first cut on the stock consists only of a slice taken off one corner. (See also Root Grafting, p. 409.)

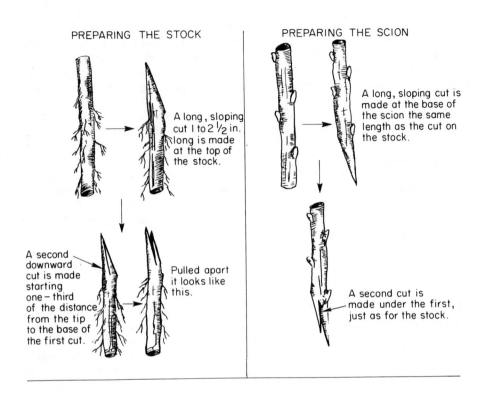

PREPARING THE STOCK

A long, sloping cut 1 to 2 1/2 in. long is made at the top of the stock.

A second downward cut is made starting one-third of the distance from the tip to the base of the first cut.

Pulled apart it looks like this.

PREPARING THE SCION

A long, sloping cut is made at the base of the scion the same length as the cut on the stock.

A second cut is made under the first, just as for the stock.

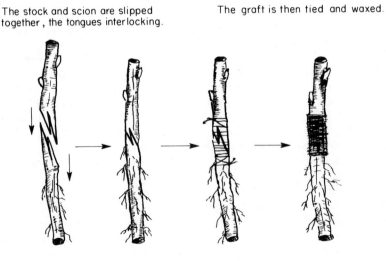

The stock and scion are slipped together, the tongues interlocking.

The graft is then tied and waxed.

Figure 12-3   The whip, or tongue, graft. This method is widely used in grafting small plant material and is especially valuable in making root grafts as illustrated here.

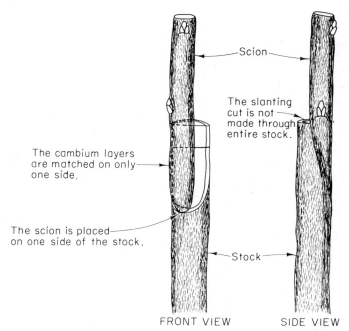

Scion

The slanting cut is not made through entire stock.

The cambium layers are matched on only one side.

The scion is placed on one side of the stock.

Stock

FRONT VIEW   SIDE VIEW

Figure 12–4   Method of making a whip, or tongue, graft when the scion is considerably smaller than the stock.

After the scion and stock are fitted together, they should be held securely in some manner until the pieces have united. There are a number of possible ways of doing this.

(a) If the unions are very well made with a tight, snug fit, it is possible that no additional wrapping or tying is needed, but it is safer to provide some type of wrapping. If not wrapped, the grafts must be protected from drying by burying in moist sand, peat moss, or sawdust until the union has healed. Or they may be planted directly in the nursery with the union below soil level. If the whip graft is used in topworking, the exposed union must be protected in some manner.

(b) With a secure fit it may be sufficient to omit tying and merely cover the union with hot grafting wax, which will secure the pieces to some extent and give good protection against drying. This is not recommended for inexperienced grafters.

(c) A common method is to wrap the union with budding rubbers or possibly raffia or waxed string, such as No. 18 knitting cotton. After wrapping, the whole union can be covered with grafting wax. Waxing may be omitted if the grafts are to be protected from drying by burying in moist sand or peat moss, or if the grafts are planted immediately with the union below the soil surface.

Grafts wrapped with budding rubbers and covered with soil should be inspected later, since the rubber decomposes very slowly below ground and may cause a constriction at the graft union.

(d) A practice widely used is to wrap the grafts with some type of adhesive tape. A special nurseryman's tape is available. The tape is drawn tightly around the graft union with the edges slightly overlapping. This holds the parts together very well and prevents drying, thus eliminating the need for waxing. If just one thickness of tape is used, it will decompose sufficiently fast (if the union is below ground) that no constriction of growth will develop. If used above ground, the tape should be cut after the graft has healed. The use of tight wrapping material such as this is especially recommended when difficulty is encountered with the formation of excessive callus.

(e) Plastic tapes are available for wrapping grafts. They are used just as adhesive tape, although they are not adhesive. The final turn of the tape is secured by slipping it under the previous turn. This tape has some elasticity. Also, it deteriorates more slowly below ground than above.

The whip graft can also be used in topworking young trees. The stocks in this case would be small, pencil-sized branches, well distributed around the tree. The grafts are usually tied with nurseryman's adhesive tape or with string, the latter covered with a coat of grafting wax. After the parts of the graft have united, the tying material must be cut; otherwise, the branch may be constricted as growth commences.

Splice Grafting

This method is the same as the whip, or tongue, graft except that the second, or "tongue," cut is not made in either the stock or scion. A simple slanting cut of the same length and angle is made in both the stock and the scion. These are placed together and wrapped or tied as described for the whip graft. The splice graft is simple and easy to make. It is particularly useful in grafting plants that have a very pithy stem or which have wood that is not flexible enough to permit a tight fit when a tongue is made as in the whip graft.

Side Grafting

There are numerous variations of the side graft; three of the most useful are described. As the name suggests, the scion is inserted into the side of the stock, which is generally larger in diameter than the scion.

Stub Graft

This method is useful in grafting branches of trees that are too large for the whip graft yet not large enough for other methods such as the cleft or bark graft. For this type of side graft, the best stocks are branches about 1 in. in diameter. An oblique cut is made into the stock branch with a chisel or heavy knife at an angle of 20 to 30 degrees. The cut should be about 1 in. deep, and at such an angle and depth that when the branch is pulled back the cut will open slightly but will close when the pull is released.

The scion should contain two or three buds and be about 3 in. long and relatively thin. At the basal end of the scion, a wedge about 1 in. long is

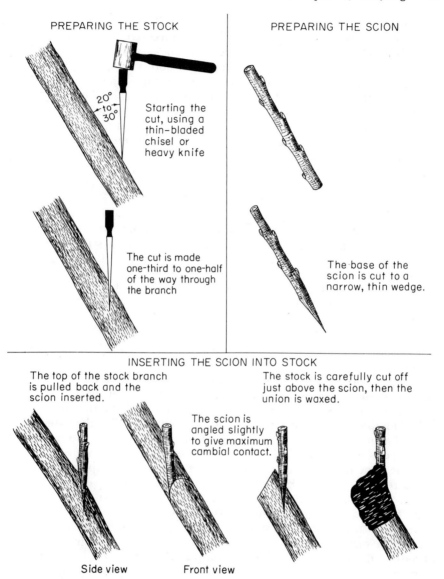

PREPARING THE STOCK

Starting the cut, using a thin-bladed chisel or heavy knife

20° to 30°

The cut is made one-third to one-half of the way through the branch

PREPARING THE SCION

The base of the scion is cut to a narrow, thin wedge.

INSERTING THE SCION INTO STOCK

The top of the stock branch is pulled back and the scion inserted.

The stock is carefully cut off just above the scion, then the union is waxed.

The scion is angled slightly to give maximum cambial contact.

Side view   Front view

Figure 12–5 Steps in preparing the side, or stub, graft. A thin-bladed chisel as illustrated here is ideal for making the cut, but a heavy butcher knife could be used satisfactorily.

made. The cuts on both sides of the scion should be very smooth, each made by one single cut with a sharp knife. It is best to insert the scion into the stock at an angle as shown in Figure 12–5 so as to obtain maximum contact of the cambium layers. The grafter inserts the scion into the cut while the upper part of the stock is pulled backward, using care to obtain the best cambium contact. Then the stock is released. The pressure of the stock should grip the scion tightly, making tying unnecessary, but if desired, the scion can be further secured by driving two small flat-headed wire nails (20 gauge, ⅝ in. long) into the stock through the scion. Wrapping the stock and scion at the point of union with nurseryman's tape may also be helpful. After the graft is completed the stock may be cut off just above the union. This must be very carefully done or the scion may become dislodged. The entire graft union must be thoroughly covered with grafting wax, sealing all openings. The end of the scion should also be covered with wax.

Under some conditions, healing will be more rapid and certain if the stock is left intact and cut off above the scion later. In grafting citrus seedlings by use of side grafting methods, it is a common practice to bend over the top of the stock about 6 in. above the graft. Some time later, after growth starts from the scion, the top is completely removed just above the graft.

This method may be used to provide a new branch at a position in a tree where it is particularly needed. To force the new scion into active growth, it may be necessary to prune back rather severely the top of the stock branch above the graft.

This grafting method is recommended (*31*) as useful in the nursery for spring grafting young nursery trees which were fall-budded but in which the bud failed to grow.

Side-Tongue Graft

This type of side graft, shown in Figure 12–6, is useful for small plants, especially some of the broad- and narrow-leaved evergreen species. The stock plant should have a smooth section in the stem just above the crown of the plant. The diameter of the scion should be slightly smaller than that of the stock. The cuts at the base of the scion are made just as for the whip graft. Along a smooth portion of the stem of the stock, a thin piece of bark and wood, the same length as the cut surface of the scion, is completely removed. Then a reverse cut is made downward in the cut on the stock, starting one-third of the distance from the top of the cut. This second cut in the stock should be the same length as the reverse cut in the scion. The scion is then inserted into the cut in the stock, the two tongues interlocking, and the cambium layer(s) matching. The graft is wrapped tightly, using one of the methods (c, d, or e) described for the whip graft on p. 377.

The top of the stock is left intact for several weeks until the graft union has healed. Then it may be cut back above the scion gradually or all at once. This forces the buds on the scion into active growth.

Side-Veneer Graft (Spliced Side Graft)

This variation of side grafting, shown in Figure 12–7, is widely used, especially for grafting small potted plants, such as seedling evergreens. A shallow downward and inward cut from 1 to 1½ in. long is made in a smooth area

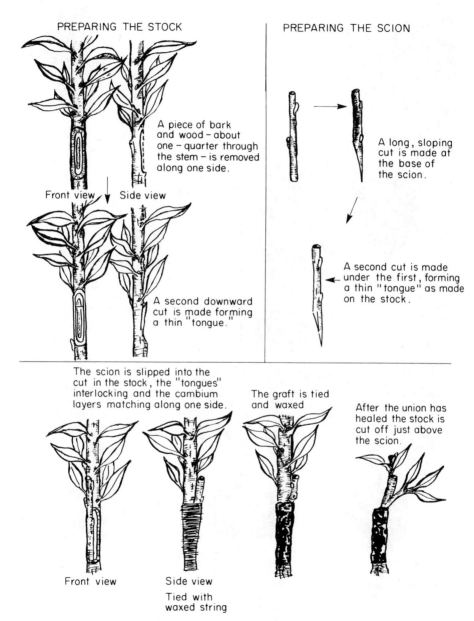

PREPARING THE STOCK

A piece of bark and wood – about one – quarter through the stem – is removed along one side.

Front view   Side view

A second downward cut is made forming a thin "tongue."

PREPARING THE SCION

A long, sloping cut is made at the base of the scion.

A second cut is made under the first, forming a thin "tongue" as made on the stock.

The scion is slipped into the cut in the stock, the "tongues" interlocking and the cambium layers matching along one side.

The graft is tied and waxed

After the union has healed the stock is cut off just above the scion.

Front view   Side view
Tied with waxed string

Figure 12–6   The side-tongue graft. This method is very useful for grafting broad-leaved evergreen plants.

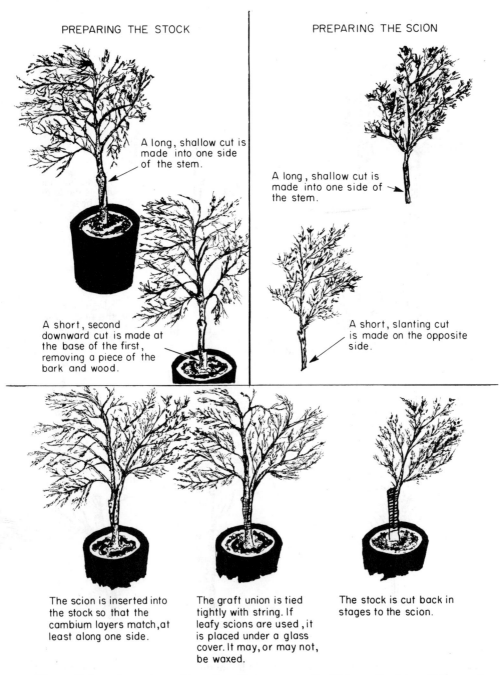

PREPARING THE STOCK

A long, shallow cut is made into one side of the stem.

A short, second downward cut is made at the base of the first, removing a piece of the bark and wood.

PREPARING THE SCION

A long, shallow cut is made into one side of the stem.

A short, slanting cut is made on the opposite side.

The scion is inserted into the stock so that the cambium layers match, at least along one side.

The graft union is tied tightly with string. If leafy scions are used, it is placed under a glass cover. It may, or may not, be waxed.

The stock is cut back in stages to the scion.

Figure 12–7   Steps in making the side-veneer graft. This method is widely used in propagating narrow-leaved evergreen species that are difficult to start by cuttings.

just above the crown of the stock plant. At the base of this cut, a second short inward and downward cut is made, intersecting the first cut, so as to remove the piece of wood and bark. The scion is prepared with a long cut along one side and a very short one at the base of the scion on the opposite side. These scion cuts should be the same length and width as those made in the stock so that the cambium layers can be matched as closely as possible.

After inserting the scion, the graft is tightly wrapped with waxed or paraffined string or with ordinary budding rubbers. The graft may or may not be covered with wax, depending upon the species. A common practice in side grafting small potted plants of some of the woody ornamental species is to plunge the grafted plants into a damp medium, such as peat moss, so that it just covers the graft union. The newly grafted plants may be placed for healing in a mist propagating house or set in grafting cases. The latter are closed boxes with a transparent cover which permits retention of a high humidity around the grafted plant until the union has healed. The grafting cases are kept closed for a week or so after the grafts are put in, then gradually opened over a period of several weeks; finally, the cover is taken off completely.

After the union has healed, the stock can be cut back above the scion either in gradual steps or all at once.

Cleft Grafting

This is one of the oldest and most widely used methods of grafting, being especially adapted to topworking trees, either in the trunk of a small tree or in the scaffold branches of a larger tree (Figures 12–8 and 12–9). Cleft grafting is also useful for smaller plants, as in crown grafting established grape vines or camellias. In topworking trees, this method should be limited to stock branches about 1 to 4 in. in diameter and to species with fairly straight-grained wood which will split evenly. Although cleft grafting can be done any time during the dormant season, the chances for successful healing of the graft union are best if the work is done in early spring just when the buds of the stock are beginning to swell but before active growth has started. If cleft grafting is done after the tree is in active growth, it is likely that the bark of the stock will separate from the wood, causing difficulties in obtaining a good union. When this occurs, the loosened bark must be firmly nailed back in place. The scions should be made from dormant, one-year-old wood. Unless the grafting is done early in the season (when the dormant scions can be collected and used immediately), the scion wood should be collected in advance and held under refrigeration until time for use. (See p. 407 for details on collecting and storing scion wood.)

In sawing off the branch for this and other topworking methods, the cut should be made at right angles to the main axis of the branch. The proper method of sawing the branches is shown on p. 417.

In making the cleft graft, a heavy knife, such as a butcher knife, or one of several special cleft grafting tools, is used to make a vertical split for a distance of 2 to 3 in. down the center of the stub to be grafted. This is done

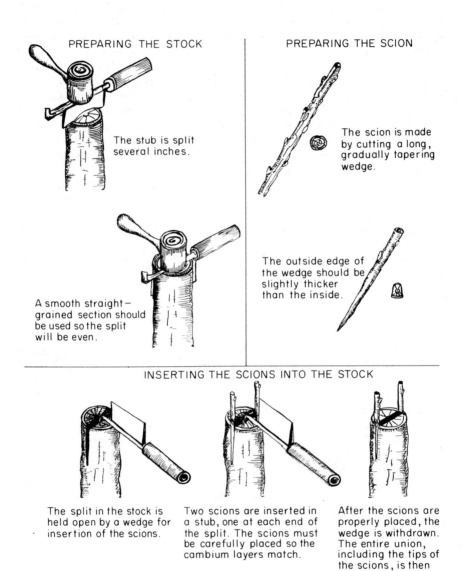

PREPARING THE STOCK

The stub is split several inches.

A smooth straight-grained section should be used so the split will be even.

PREPARING THE SCION

The scion is made by cutting a long, gradually tapering wedge.

The outside edge of the wedge should be slightly thicker than the inside.

INSERTING THE SCIONS INTO THE STOCK

The split in the stock is held open by a wedge for insertion of the scions.

Two scions are inserted in a stub, one at each end of the split. The scions must be carefully placed so the cambium layers match.

After the scions are properly placed, the wedge is withdrawn. The entire union, including the tips of the scions, is then thoroughly covered with grafting wax.

Figure 12–8   Steps in making the cleft graft. This method is very widely used and is quite successful if the scions are inserted so that the cambium layers of stock and scion match properly.

by pounding the knife in with a hammer or mallet. It is very important to have the branch sawed off in such a position that the end of the stub which is left is smooth, straight-grained, and free of knots for at least 6 in. Otherwise, when the split is made it may not be straight, or the wood may split one way and the bark another. The split should be in tangential rather than radial direction in relation to the center of the tree. This permits better placement of the scions for their subsequent growth. Sometimes the cleft is made by a longitudinal saw cut rather than by splitting. After a good, straight split is made, a screwdriver, chisel, or the wedge part of the cleft-grafting tool is driven into the top of the split to hold it open.

Two scions are usually inserted, one at each side of the stock where the cambium layer is located. The scions should be 3 or 4 in. long and have two

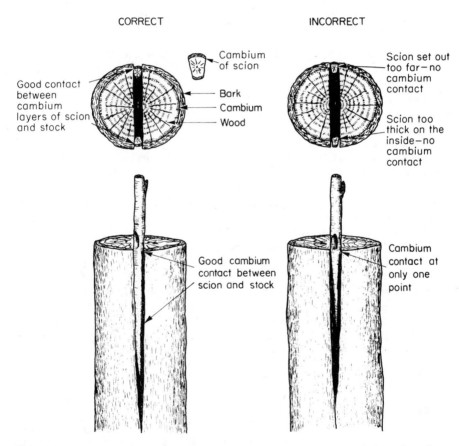

CORRECT   INCORRECT

Good contact between cambium layers of scion and stock

Cambium of scion

Bark

Cambium

Wood

Scion set out too far—no cambium contact

Scion too thick on the inside—no cambium contact

Good cambium contact between scion and stock

Cambium contact at only one point

Figure 12–9   In making the cleft graft, the proper placement of the scions is very important. The correct way of doing this is shown on the left. Scions inserted as shown on the right probably would not grow.

or three buds. The basal end of each scion should be cut into a long, gently sloping wedge—about 2 in. long. It is not necessary that the end of the wedge come to a point. The side of the wedge which is to go to the outer side of the stock should be slightly wider than the inside edge. Thus, when the scion is inserted and the tool is removed, the full pressure of the split stock will come to bear on the scions at the position where the cambium of the stock touches the cambium layer on the outer edge of the scion. Since the bark of the stock is almost always thicker than the bark of the scion, it is usually necessary for the outer surface of the scion to set slightly in from the outer surface of the stock in order to match the cambium layers.

In all types of grafting the scion must be inserted right side up. That is, the points of the buds on the scion should be pointing away from the stock. Failure to observe this means ultimate failure of the graft to grow.

The long, sloping wedge cuts at the base of the scion should be smooth, made by a single cut on each side with a very sharp knife. Both sides of the scion wedge should press firmly against the stock for their entire length. A common mistake in cutting scions for this type of graft is to make the cut on the scion too short with too abrupt a slope, so that the only point of contact is just at the top. Shaving slightly the sides of the split in the stock will often permit a smoother contact.

After the scions are properly made and inserted, the tool is carefully withdrawn, using care not to disturb the scions. They should be held so tightly by the pressure of the stock that they cannot be pulled loose by hand. No further tying or nailing is needed unless very small stock branches have been used, in which case the top of the stock can be wrapped tightly with string or adhesive tape to hold the scions in place more securely.

Thorough waxing of the completed graft is essential. (See p. 403 for a discussion of grafting waxes.) The top surface of the stub should be entirely covered, permitting the wax to work into the split in the stock. The sides of the grafted stub should be well covered with wax as far down the stub as the split has gone. The tops of the scions should be waxed but not necessarily the bark or buds of the scion. Two or three days later all the grafts should be inspected and rewaxed where openings appear. Lack of thorough and complete waxing in this type of graft is almost certain to result in failure.

Saw-Kerf (Notch or Wedge) Grafting

This can be used in place of the cleft graft. (See Figure 12–10.) It is especially useful in topworking trees with branches 2 to 4 in. or more in diameter. It does not have the disadvantage of a deep split in the stock (as does the cleft graft), which may permit the entrance of decay-producing organisms. Saw-kerf grafting can be done over a long period of time—two to three months— before growth of the stock starts in the spring. If the work is to be performed late in the season, at the time active growth is starting, it is necessary to collect the scion wood earlier and hold it under refrigeration to keep it dormant until the grafting is to be done. Curly grained stock branches, which would not split evenly for the cleft graft, can be worked readily by the saw-kerf graft. The saw-kerf graft is somewhat more difficult for beginners to perform than most of the other types, but in the hands of experienced workers it can be done rapidly and is highly successful, especially with certain hard-to-graft species, such as the peach.

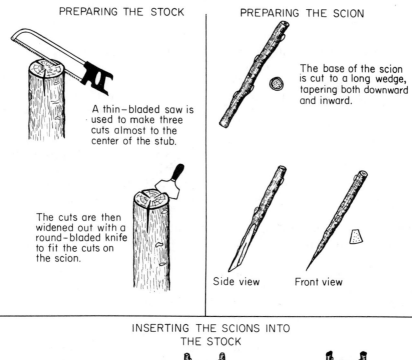

PREPARING THE STOCK

A thin-bladed saw is used to make three cuts almost to the center of the stub.

The cuts are then widened out with a round-bladed knife to fit the cuts on the scion.

PREPARING THE SCION

The base of the scion is cut to a long wedge, tapering both downward and inward.

Side view   Front view

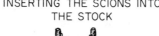

INSERTING THE SCIONS INTO THE STOCK

If the scions match the cuts in stock, they will be held securely just by tapping into place. No nailing or tying is necessary.

Three scions are usually put into each stub. The cambium layers of the scion and stock must match.

The graft union and the ends of the scion are finally covered with grafting wax.

Figure 12–10   Steps in making the saw-kerf (notch) graft.

Usually, three scions are inserted into each stock branch. A cut with a thin-bladed, fine-toothed saw is made into the stub for each scion. This cut should extend 1 to 1½ in. toward the center of the stub and about 4 in. down the side of the stub. Then, using a very sharp "round knife," the grafter widens this saw cut to fit the scion. The knife should be placed at the bottom of the saw cut and brought upward and inward to cut out thin slices of wood. Care should be taken not to get the cut in the stock too wide for use with the available scions.

The scions should be 4 or 5 in. long and contain two or three buds. The basal end of the scion is cut to a wedge shape with the outside edge of the wedge somewhat thicker than the inside edge so as to conform to the general shape of the widened saw cut as made in the stock. The wedge cut on the scions should be 1½ to 2 in. long and, as in all grafting, the cut areas should be perfectly smooth, with no wavy surfaces. The scion should be tried for a good fit as the groove in the stock is widened to the right size. It is usually easier to enlarge the cut in the stock to fit the scion rather than trying to fit the scion to the cut in the stock. After the cuts on the scion and stock are completed, the scion is tapped firmly into place. The cambium layers should cross to insure contact. Since the bark of the scion is usually thinner than that of the stock, at the correct position the scion will set slightly in from the outside of the stock. If this graft is properly made, the scion should be held very securely by tapping it into place. No further nailing or tying is necessary but, as with all exposed grafts, the cut surfaces must be thoroughly waxed.

The round knife is a useful tool in making this graft. It may be adapted from a leather-worker's knife or a cook's mincing knife. The edge of the knife should be sharpened so that it is flat on one side and beveled on the other, being used so that the flat side is held against the wood in widening out the saw cut made in the stock. If a round knife of this type is not available, a large-bladed grafting knife may be substituted.

Another simpler method of preparing this type of graft is to make a V-cut about 1 in. long straight into the side of the stock to a depth equal to the diameter of the scion. This is done by making two cuts with a heavy knife pounded in with a mallet or plastic tip hammer. Then a screwdriver is pounded vertically downward into the stock to knock out the wedge piece of wood resulting from the two V cuts. The base of the scion can then be prepared as a simple V-shaped wedge which is then tapped downward into the opening in the stock with the cambium layers of scion and stock matching. If properly made the scion should fit tightly without nailing. Finally, all exposed cut surfaces should be thoroughly waxed.

Bark Grafting

This method is rapid, simple, readily performed by amateurs, and if properly done, gives a high percentage of "takes." It requires no special equipment and can be performed on branches ranging from 1 in. up to a foot or more in diameter. The latter size is not recommended as it is difficult to heal over such large stubs before decay-producing organisms get started. The bark graft, since it depends on the bark separating readily from the wood, can only be done after active growth of the stock has started in the spring. As dormant scions must be used, it is necessary to gather the scion wood for deciduous species during the dormant season and hold it under refrigeration until the grafting operation is done. (See p. 407 for instructions on storing scion wood.) For evergreen species, freshly collected scion wood can be used. Scions are not as securely attached to the stock as in some of the other methods and are more susceptible to wind breakage during the first year even though healing has been satisfactory. Therefore the new shoots arising from the scions probably should be staked during the first year, or cut back to about half their length, especially in windy areas. After a few years' growth, the bark graft union is as strong as the unions formed by other methods.

There are several modifications of the bark graft. Three important types are described.

Bark Graft (Method No. 1)

Several scions are inserted into each stub. For each scion, a vertical knife cut about 2 in. long is made at the top end of the stub through the bark to the wood. The bark is then lifted slightly along both sides of this cut, in preparation for the insertion of the scion. The scion should be of dormant wood, 4 or 5 in. long, containing two or three buds, and be $1/4$ to $1/2$ in. in thickness. One cut about 2 in. long is made along one side at the base of the scion. With large scions, this cut extends about one-third of the way into the scion, leaving a "shoulder" at the top. The purpose of this shoulder is to reduce the thickness of the scion to minimize the separation of bark and wood after insertion in the stock. The scion should not be cut too thin, however, or it will be mechanically weak and break off at the point of attachment to the stock. If small scions are used, no shoulder is necessary. On the side of the scion opposite the first long cut, a second, shorter cut is made, as shown in Figure 12–11, thereby bringing the basal end of the scion to a wedge shape. The scion is then inserted between the bark and the wood of the stock, centered directly under the vertical cut through the bark. The longer cut on the scion is placed against the wood, and the shoulder on the scion is brought down until it rests on top of the stub. The scion is then ready to be fastened in place. A satisfactory method is to nail the scion into the wood, using two nails per scion. Flat-headed nails $5/8$ to 1 in. long, of 19 or 20 gauge wire, depending on the size of the scions, are satisfactory. The bark on both sides of the scion should also be securely nailed down, or it will tend to peel back from the wood. Another method commonly used with soft-barked trees, such as the avocado, is to insert all the scions in the stub and then hold them in place by wrapping with string, adhesive tape, or waxed cloth around the stub. This is more effective than nailing in preventing the scions from blowing out but probably does not give as tight a fit. Both nailing and wrapping are advisable for maximum strength. If a wrapping material is used, it may be necessary to cut this later to prevent constriction.

In cases in which the bark of the stock is quite thick and small scions are used, it may not be necessary to make the vertical slit in the bark. The scions,

PREPARING THE STOCK

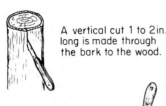

A vertical cut 1 to 2 in. long is made through the bark to the wood.

The bark on both sides of the cut is slightly separated from the wood.

PREPARING THE SCION

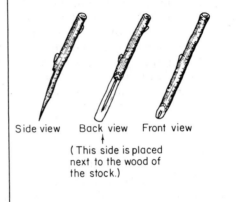

The scion is cut as shown below, a long cut with a shoulder on one side, and a shorter cut on the opposite side.

Side view   Back view   Front view

(This side is placed next to the wood of the stock.)

INSERTING THE SCIONS INTO THE STOCK

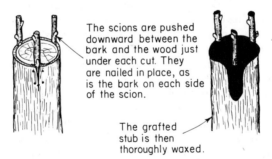

The scions are pushed downward between the bark and the wood just under each cut. They are nailed in place, as is the bark on each side of the scion.

The grafted stub is then thoroughly waxed.

Figure 12–11   Steps in preparing the bark graft (Method No. 1). In grafting some thick-barked plants the vertical cut in the bark is unnecessary, the scion being inserted between the bark and wood of the stock.

cut as described before, can be pushed into place between the bark and the wood. They can then be nailed or tied securely with string or waxed cloth.

After the stub has been grafted and the scions fastened by nailing or tying, all cut surfaces, including the end of the scions, should be thoroughly covered with grafting wax.

Bark Graft (Method No. 2)
This type of graft (Figure 12–12) is similar to method No. 1, except that the scion is not inserted centered on the vertical cut in the bark of the stock.

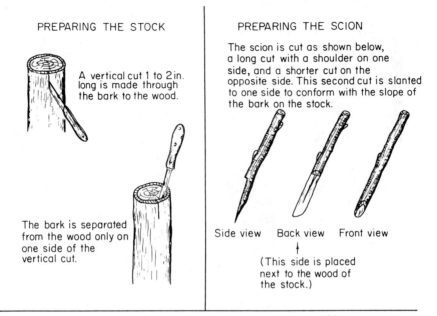

PREPARING THE STOCK

A vertical cut 1 to 2 in. long is made through the bark to the wood.

The bark is separated from the wood only on one side of the vertical cut.

PREPARING THE SCION

The scion is cut as shown below, a long cut with a shoulder on one side, and a shorter cut on the opposite side. This second cut is slanted to one side to conform with the slope of the bark on the stock.

Side view   Back view   Front view

(This side is placed next to the wood of the stock.)

INSERTING THE SCIONS INTO THE STOCK

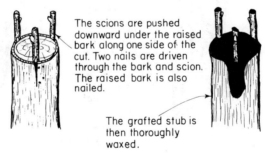

The scions are pushed downward under the raised bark along one side of the cut. Two nails are driven through the bark and scion. The raised bark is also nailed.

The grafted stub is then thoroughly waxed.

Figure 12–12   Steps in making the bark graft (Method No. 2).

The bark is lifted only along one side of the vertical cut; the bark on the other side is not disturbed. The scion is inserted under the raised bark and held in place with two nails driven through the bark of the stock, through the scion, and into the wood of the stock. The scion is cut with a shoulder, just as described for method No. 1. However, the short cut on the back of the scion, rather than being parallel to the first longer cut, is slanted to one side slightly to conform to the slope of the bark under which the scion is inserted. As in method No. 1, the raised bark near the scion should be fastened securely in place with two or more nails or wrapped with tape to prevent it from peeling back. This modification of the bark graft has been widely used with good results. It has an advantage over method No. 1, in that one of the edges of the scion is placed against undisturbed bark with

PREPARING THE STOCK | PREPARING THE SCION

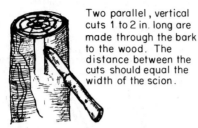

Two parallel, vertical cuts 1 to 2 in. long are made through the bark to the wood. The distance between the cuts should equal the width of the scion.

The scions are made with a long sloping cut on one side and a shorter cut on the opposite side.

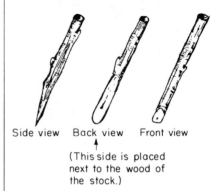

A horizontal cut is made between the two vertical cuts and most of the piece of bark is removed. A small flap is left at the bottom.

Side view   Back view   Front view

(This side is placed next to the wood of the stock.)

INSERTING THE SCION INTO THE STOCK

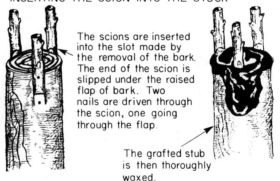

The scions are inserted into the slot made by the removal of the bark. The end of the scion is slipped under the raised flap of bark. Two nails are driven through the scion, one going through the flap.

The grafted stub is then thoroughly waxed.

Figure 12–13   Steps in making the bark graft (Method No. 3).

its intact cambium cells. This tends to promote more rapid healing of the union. In method No. 1, where the bark is lifted away from the scion on both sides, intact cambium cells are some distance from the scion and healing is usually slower.

Bark Graft (Method No. 3)

In this method, as shown in Figures 12–13 and 12–14, *two* knife cuts about 2 in. long are made through the bark of the stock down to the wood, rather than just one as in the other two methods. The distance between these two cuts should be exactly the same as the width of the scion. The piece of bark

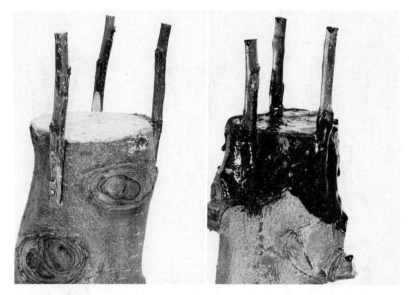

Figure 12–14 Limbs grafted by the bark graft method (No. 3). *Left:* grafting just completed. *Right:* completed graft following waxing.

between the cuts should be lifted and the terminal two-thirds cut off. The scion is prepared with a smooth slanting cut along one side at the basal end completely through the scion. This cut should be about 2 in. long but *without* the shoulder, in contrast to the other two methods. On the opposite side of the scion, a cut about ½ in. long is made, forming a wedge at the base of the scion. The scion should fit snugly into the opening in the bark with the longer cut inward and with the wedge at the base slipped under the flap of remaining bark. Rapid healing can be expected because both sides of the scion are touching undisturbed bark and cambium cells, which is not the case in the other two types of the bark graft.

The scion should be nailed into place with two nails, the lower nail going through the flap of bark covering the short cut on the back of the scion. If the bark along the sides of the scion should accidentally become disturbed, it must be nailed back into place.

Method No. 3 is well adapted for use with thick-barked trees, such as walnuts, on which it is not feasible to insert the scion under the bark.

Approach Grafting

The distinguishing feature of approach grafting is that two independent, self-sustaining plants are grafted together. After a union has occurred, the top of the stock plant is removed above the graft and the base of the scion plant is removed below the graft. Sometimes it is necessary to sever these parts gradually rather than all at once. Approach grafting provides a means of establishing a successful union between certain plants which are difficult

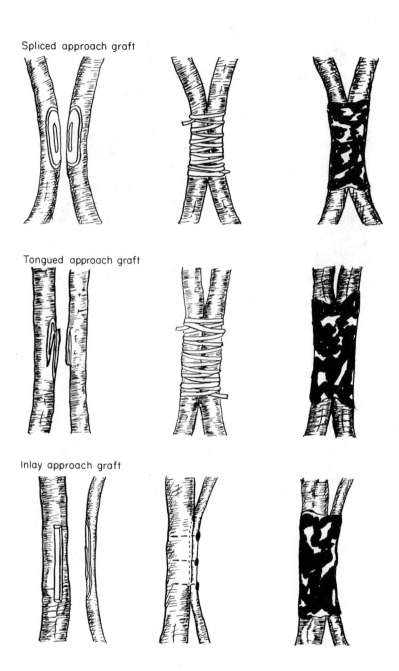

Spliced approach graft

Tongued approach graft

Inlay approach graft

Figure 12–15   Three methods of making an approach graft.

Figure 12–16   Approach grafting used in obtaining a desirable scion cultivar on a seedling camellia. *Top:* seedling plant in container is set close to a large plant of the desired cultivar. The graft union is made and tightly wrapped with adhesive tape. *Below:* after the graft union has healed, which may take several weeks or longer, the stock plant is cut off above the graft union, and the scion is severed from the parent plant just below the graft union. Approach grafting is sometimes necessary for plants very difficult to graft by other methods.

to graft together otherwise. It is usually performed with one or both of the plants to be grafted growing in a container. Rootstock plants in containers may be placed adjoining an established plant which is to furnish the scion part of the new, grafted plant (see Figure 12–16).

This type of grafting can be done at any time of the year, but healing of the union is more rapid if it is performed at a season when growth is active. As in other methods of grafting, the cut surfaces should be securely fastened together, then covered with grafting wax to prevent drying of the tissues.

Three useful methods of making approach grafts are described below and illustrated in Figure 12–15.

Spliced Approach Graft

Preferably the two stems should be approximately the same size. At the point where the union is to occur, a slice of bark and wood 1 to 2 in. long is cut from both stems. This cut should be the same size on each so that identical cambium patterns will be made. The cuts must be perfectly smooth and as nearly flat as possible so that when they are pressed together there will be close contact of the cambium layers. The two cut surfaces are then bound tightly together with string, raffia, or nurseryman's tape. The whole union should then be covered with grafting wax. After the parts are well united, which may require considerable time in some cases, the stock above the union and the scion below the union are cut, and the graft is then completed. It may be necessary to reduce the leaf area of the scion if it is more than the root system of the stock can sustain. Figure 12–16 shows the use of this method with camellias.

Tongued Approach Graft
This is the same as the spliced approach graft except that after the first cut is made in each stem to be joined, a second cut—downward on the stock and upward on the scion—is made, thus providing a thin tongue on each piece. By interlocking these tongues a very tight, closely fitting graft union can be obtained.

Inlay Approach Graft
This method may be used if the bark of the stock plant is considerably thicker than that of the scion plant. A narrow slot, 3 or 4 in. long, is made in the bark of the stock plant by making two parallel knife cuts and removing the strip of bark between. This can only be done when the stock plant is actively growing and the bark "slipping." The slot should be exactly as wide as the scion to be inserted. The stem of the scion plant, at the point of union, should be given a long, shallow cut along one side, of the same length as the slot in the stock plant and deep enough to go through the bark and slightly into the wood. This cut surface of the scion branch should be laid into the slot cut in the stock plant and held there by nailing with two or more small, flat-headed wire nails. The entire union must then be thoroughly covered with grafting wax. After the union has healed, the stock can be cut off above the graft and the scion below the graft.

Inarching
This method is similar to approach grafting in that both stock and scion plants are on their own roots at the time of grafting; it differs in that the top of the new rootstock plant usually does not extend above the point of the graft union as it does in approach grafting. Inarching is generally considered to be a form of "repair grafting," being used in cases in which the roots of an established tree have been damaged by such things as cultivation implements, rodents, or disease. It can be used to very good advantage in saving a valuable tree or improving its root system (Figure 12–17).

Seedlings (or rooted cuttings) planted beside the older, damaged tree, or suckers arising near its base, are grafted into the trunk of the tree to provide a new root system to supplant the damaged roots. The seedlings to be inarched into the tree should be spaced about 5 or 6 in. apart around the cir-

Figure 12–17   Inarching can be used for invigorating existing trees by replacing a weak rootstock with a more vigorous one. Here a Persian walnut tree has been inarched with vigorous Paradox hybrid *(J. hindsii × J. regia)* seedlings.

cumference of the tree if the damage is extensive. The tree will usually stay alive for some time after the injury occurs unless it is very severe. A satisfactory procedure for inarching is to plant seedlings of a compatible species or cultivar (see p. 337) around the tree during the dormant season. Then, as active growth commences in early spring, the grafting operation can be done.

Inarching old, weakly growing trees with strong, vigorous seedling rootstocks has on some occasions *(15)* proved beneficial in promoting renewed active growth of the old trees.

The seedling plants to provide the new root system are usually considerably smaller than the tree to be repaired. As illustrated in Figure 12–18, the graft union is made in a manner similar to that described for method No. 3 of the bark graft. The upper end of the seedling, which should be $\frac{1}{4}$ to $\frac{1}{2}$ in. thick, is given a long shallow cut along the side for 4 to 6 in. This cut should be on the side next to the trunk of the tree and made deep enough to remove some of the wood, thus exposing two strips of cambium tissue. At the end of the seedling another shorter cut, about $\frac{1}{2}$ in. long, is made on the side opposite the long cut; this makes a sharp, wedge-shaped end on the seedling stem. A long slot is made in the trunk of the older tree by removing a piece of bark the exact width of the seedling and just as long as the cut surface made on the seedling. A small flap of bark is left at the upper end of the slot, under which the wedge end of the seedling is inserted. Then the seedling is nailed into the slot with four or five small, flat-headed wire nails. The nail at the top of the slot should go through the flap of bark and through the end of the seedling. If any of the bark of the tree along the sides of the seedling should accidentally be pulled loose, it is necessary to nail it back in place. After nailing, the entire area of the graft union should be thoroughly waxed.

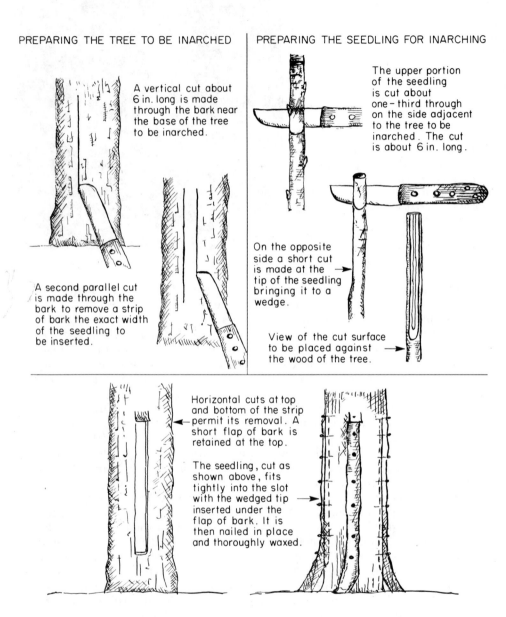

PREPARING THE TREE TO BE INARCHED

A vertical cut about 6 in. long is made through the bark near the base of the tree to be inarched.

A second parallel cut is made through the bark to remove a strip of bark the exact width of the seedling to be inserted.

PREPARING THE SEEDLING FOR INARCHING

The upper portion of the seedling is cut about one-third through on the side adjacent to the tree to be inarched. The cut is about 6 in. long.

On the opposite side a short cut is made at the tip of the seedling bringing it to a wedge.

View of the cut surface to be placed against the wood of the tree.

Horizontal cuts at top and bottom of the strip permit its removal. A short flap of bark is retained at the top.

The seedling, cut as shown above, fits tightly into the slot with the wedged tip inserted under the flap of bark. It is then nailed in place and thoroughly waxed.

Figure 12–18   Steps in inarching a large plant with smaller ones planted around its base.

In inarching some species, such as the walnuts, better healing of the union takes place if the top of the seedling is retained and allowed to extend above the graft union for a period of time, finally being cut off just above the union.

Owing to the food materials translocated from the larger tree, the seedlings grow very rapidly and soon provide a considerable number of new roots for the grafted tree. If shoots arise from the inarches, they should be suppressed by cutting off their tips. The shoots should finally be removed entirely as the union becomes well established.

Bridge Grafting

This is a form of repair grafting, and is used when the root system of the tree has not been damaged but there is injury to the bark of the trunk. Sometimes cultivation implements, rodents, disease, or winter injury will damage a considerable trunk area, often girdling the tree completely. If the damage to the bark is extensive, the tree is almost certain to die, because the roots will be deprived of their food supply from the top of the tree. Trees of some species, such as the elm, cherry, and pecan, can heal over extensively injured areas by the development of callus tissue. But trees of most species which have had the bark of the trunk severely damaged should be bridge grafted if they are to be saved, as illustrated in Figure 12–19.

The bridge grafting operation is best performed in early spring just as active growth of the tree is beginning and the bark is slipping easily. The scions to be used should be taken when dormant from one-year-old growth,

Figure 12–19   Injured trunk of a cherry tree successfully bridge grafted by a modification of the bark graft.

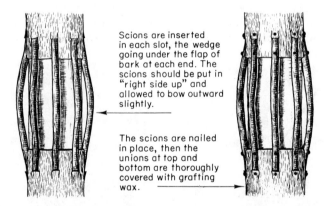

PREPARING THE STOCK

All dead and damaged bark around the wound is trimmed back to live healthy tissue.

Cuts are made in the bark at top and bottom of the wound, just as for Method 3 of the bark graft. The slots in the bark should be the same width as the scions to be inserted.

PREPARING THE SCIONS

One long, slanting cut is made at each end of the scion, with both cuts on the same side.

A second, short, slanting cut is made on the back side of the scion, bringing the ends to a sharp wedge. Buds can be trimmed off the scions if desired.

Scions are inserted in each slot, the wedge going under the flap of bark at each end. The scions should be put in "right side up" and allowed to bow outward slightly.

The scions are nailed in place, then the unions at top and bottom are thoroughly covered with grafting wax.

Figure 12–20   A satisfactory method of making a bridge graft, using a modification of Method No. 3 of the bark graft.

¼ to ½ in. in diameter, of the same or compatible species, and held under refrigeration until the grafting work is to be done. (See p. 407 for information on collecting and storing scion wood.) In an emergency, one may successfully perform bridge grafting late in the spring, using scion wood whose buds have already started to grow. Remove the developing buds or new shoots.

The first step in bridge grafting is to trim the wounded area back to healthy, undamaged tissue by removing dead or torn bark. Then every 2 or 3 in. around the injured section a scion is inserted, attached at both the upper and lower ends into live bark. It is important that the scions be inserted right side up. If they are put in reversed, they may make a union and stay alive for a year or two, but the scions will not grow and enlarge in diameter as they would if inserted correctly (see Figure 11–12).

Figure 12–20 shows the details of making a satisfactory type of bridge graft. It is essentially the same as method No. 3 of the bark graft. Just above and below the injured area, a slot 2 to 3 in. long and exactly the width of the scion is cut in the bark of the trunk for each scion. The piece of bark is removed, with the exception of a flap about ½ in. long which is left at the end of the slot. The scions are cut to fit into these slots at each end of the wound, and long enough to bow outward slightly. This bow allows for good contact at each end and permits some swaying of the trunk in the wind without tearing the scions loose. To prepare the scions, a cut is made along both ends on the side which is to fit into the slot. The cut should be the same length as the slot and deep enough to remove some bark and wood and expose two strips of cambium tissue. Then, on the opposite side of the scion, at each end, another, shorter cut, about ½ in. long, is made to form the end of the scion into a wedge shape. The ends of the scion should be inserted under the flaps of bark and nailed in place, using ¾-in. 20 gauge flat-headed wire nails. One nail should go through the flap of bark at each end of the scion with enough additional nails to hold the scion securely.

After all the scions have been inserted, the cut surfaces must be thoroughly covered with grafting wax, particular care being taken to work the wax around the scions, especially at the graft unions. The exposed wood of the injured section may also be covered with grafting wax to prevent the entrance of decay organisms and to prevent excessive drying out of the wood, which is important, being the path for upward movement of water and nutrients in the tree.

The buds on the scions will often push into growth if the grafts are successful. These shoots should be removed, because no branches would be desired in this position. The scions will rapidly enlarge in size and completely heal over the wound in a few years.

Bracing

The same type of graft union practiced in inarching and bridge grafting can often be used to establish a "natural" brace in young trees (Figure 12–21). This method is useful in supporting branches that may be in danger of breaking off or where there is a weak crotch. A small branch, about pencil size or a little larger, coming a foot or so above the weak crotch is grafted into the adjacent branch to be supported. It should be wrapped spirally and upward partly around the branch. In the region

Figure 12–21   Brace graft after two seasons' growth.

where the graft union is to be, the bark on the large branch is cut just under the small branch to its exact width. This should extend for a length of 6 in. or more, and the spiral piece of bark should be removed. Grafting such as this could only be done at a time of year when there is active growth and the bark is slipping easily. Early spring, just as the new growth is starting, is the preferable time.

After the slot in the bark is ready, the "scion"—the small branch—should be smoothly cut on the lower side through about a third of its thickness and for the length of the slot—6 in. or more. If the cuts are well made, the small branch should fit snugly into the slot. It is helpful to trim the end of the small branch to a wedge point by cutting it on the top side so that it can be inserted under the bark at the top end of the slot.

The next step is to nail the branch in place in the slot with small, flat-headed wire nails (5/8 or 3/4 in., 20 gauge) placed at 1- or 2-in. intervals. The graft union should then be thoroughly waxed to prevent drying. It is advisable to tie the two branches together temporarily with a strong cord to prevent whipping by the wind, which might pull the graft union loose.

TOOLS AND ACCESSORIES FOR GRAFTING

Special equipment needed for any particular method of grafting has been illustrated along with the description of the method. There are some pieces of equipment, however, which are used in all types of grafting.

Knives

For propagation work, the two general types used are the budding knife and the grafting knife (Figure 12–22). Where a limited amount of either budding or grafting is done, the budding knife can be used satisfactorily for both operations. The knives have either a folding or a fixed blade. The fixed-blade type is stronger, and if a holder of some kind is used to protect the cutting edge, it is probably the most desirable. A well-built, sturdy knife of high-quality steel is essential if much grafting work is to be done. The knife must be kept very sharp in order to do good work.

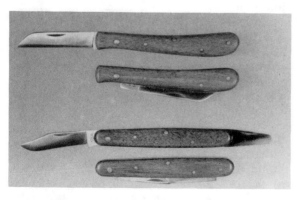

Figure 12–22   Types of folding knives used in plant propagation. *Top:* grafting knife. *Below:* budding knife. The blunt part on the right of this knife is used in T-budding to open the bark flap for insertion of the shield bud piece.

HOW TO SHARPEN A KNIFE PROPERLY

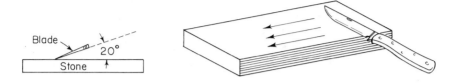

The initial grinding may be done with a fairly coarse stone, but a hard, fine-grained stone should be used for the final grinding. Do not use a carborundum stone, because it is too abrasive and will grind off too much metal. Some prefer to use knives beveled on both sides, whereas others prefer a knife beveled on one side only, the back side being flat. In sharpening the knife, hold it so that only the edge of the blade touches the stone in order that a stiff edge for cutting can be obtained. Use the whole width of the stone so that its surface will remain flat. A correctly sharpened knife of high-quality steel should retain a good edge for several days' work, with only occasional stropping on a piece of leather.

Grafting Waxes

Grafting wax has two chief purposes: (a) It seals over the graft union, thereby preventing the loss of moisture and death of the tender, exposed cells of the cut surfaces of the scion and stock. These cells are essential for callus production and healing of the graft union. (b) It prevents the entrance of various decay-producing organisms which may lead to wood rotting.

A good grafting wax should adhere well to the plant surfaces, not be washed off by rains, not be so brittle as to crack and chip during cold weather, and not be so soft that it will melt and run off during hot days, but still be pliable enough to allow for swelling of the scion and growth enlargement of the stock without cracking.

Although a great many different materials have been used as grafting waxes, various formulations of the so-called *hot waxes* or *cold waxes* are generally preferred.

Hot Waxes

The ingredients used in this type of wax are resin, beeswax, either raw linseed oil or tallow, and lampblack or powdered charcoal. The proportions of each ingredient may vary considerably without affecting the results appreciably. The purpose of the lampblack is to give some color to the otherwise colorless wax so that it will show more readily when the grafts have been well covered. Lampblack also imparts a more workable consistency to the wax, eliminating some of the stickiness and stringiness. After application, the dark-colored wax may absorb more heat and thus remain soft and pliable. During hot weather, however, this could be undesirable, because the wax may become too soft.

Directions for Making Hot Grafting Wax (Hard Type)
Ingredients:

Resin	5 lb
Beeswax	¾ lb
Raw linseed oil	½ pt
Lampblack	1 oz
Fish glue	1½ oz

Heat the glue in a double boiler with just enough water to dissolve it. Melt the other ingredients in another container and allow the mixture to cool but still remain fluid. Add the glue slowly to the partly cooled mixture, stirring continually. Pour out into shallow greased pans or wooden boxes lined with greased paper. Allow to harden. To use, chip or break into small lumps. Reheat in a grafting wax melter and apply with a small paint brush.

The hard type of hot grafting wax solidifies upon cooling and must be reheated just before being applied to the graft union. It is important that this wax be at the right temperature when it is used. If the wax is heated until it is bubbling and boiling, it will injure the plant tissues. At the other extreme, if the wax is too cool, it will not flow easily into all the crevices in the bark, thus leaving openings for the entrance of air. *The wax should be hot enough to flow easily yet not be at the bubbling stage.*

For heating the wax, any small burner is satisfactory. Two types are illustrated in Figure 12–23. A brush is used to apply the wax, but provision should be made to suspend it from the side, for if the brush rests at the bottom of the burner the heat will burn the bristles. Homemade grafting wax melters can easily be constructed (27).

Hand Waxes

Hand wax, which does not require heating, is soft and pliable and is applied to the graft union by pressing in and around the graft with the fingers. It is sticky, however, and unpleasant to use. Hand wax is made by heating together 4 parts of resin, 2 parts of beeswax, and 1 part of tallow. (An alternative formula is 4 parts resin, 1 part beeswax, and 2 parts tallow or linseed oil.) The tallow is melted first, then the beeswax is added, and finally the resin. When the ingredients have melted together, the mixture is poured into a large vessel of cold water. After cooling it is removed, then pulled and worked with the hands until it is light yellow. It is then made into small balls and wrapped in waxed paper for storing.

This soft hand wax is suitable for small operations, but for general use the hard wax is more satisfactory. Hand waxes are somewhat more difficult and time-consuming to apply than the more fluid hot and cold waxes.

Cold Waxes

A commercially prepared type of grafting wax consisting of an emulsion of asphalt and water is available. It has proved quite satisfactory and is widely

Figure 12-23 Two types of wax pots for melting hot grafting wax. Alcohol is commonly used as the fuel for the burners. *Left:* units disassembled. *Right:* burners ready to use.

used (*29*). This material is about 50 percent water, which evaporates after application, leaving a coat of asphalt over the graft. For the remaining wax to be thick enough to adequately protect the graft, the original application should be fairly heavy. Since this material is water soluble until it dries, rains occurring within a day after application are likely to wash it off. It would then, of course, be necessary to rewax the grafts immediately. If grafting is done during rainy weather, it is advisable to use the hot type of wax, which is not affected by rains.

The emulsions of this type of cold wax are broken down by freezing, so it is very important that the containers be stored in a warm place during cold weather. *Do not confuse these materials with roofing compounds, which have a similar appearance but are entirely unsuitable for use as grafting wax.*

For small-scale operations, grafting waxes are available in aerosol applicator cans. Several repeated applications of the wax are generally necessary to give sufficient coverage for adequate protection.

Tying and Wrapping Materials

Some of the grafting methods, particularly the whip or tongue, require that the graft union be held together by tying until the parts unite. This can be done in several ways—the simplest would be merely tying with ordinary string and covering with grafting wax. For large-scale operations, waxed string is convenient because it will adhere to itself and to the plant parts, without tying. It should be strong enough to hold the grafted parts together yet weak enough to be broken by hand.

A good waxed string for this purpose can be prepared by soaking balls of No. 18 knitting cotton for about 10 minutes in a mixture of the following ingredients which have been melted together: resin, 2 lbs; beeswax, 1 lb; tallow, ½ lb; linseed oil, ½ lb; paraffin, ¼ lb.

A special nurseryman's adhesive tape is manufactured which is similar to surgical adhesive tape except that it is lighter in weight and is not sterilized. It is more convenient to use than waxed cloth tape. Adhesive tape is useful for tying and sealing whip grafts. When using any kind of tape or string for wrapping grafts, it is very important not to wrap too many layers or the material may eventually girdle the plant unless it is cut. When this type of wrapping is covered with soil it usually will rot and break before damage can occur, unless it has been wrapped in too many layers.

Plastic polyethylene or polyvinyl chloride (PVC) grafting and budding tapes are available. They are not adhesive, so they must be secured by folding the end of the tape under the last turn. These plastic tapes are slightly elastic and will allow for some diameter growth of the grafts, but must eventually be removed.

Grafting Machines

Several machines or devices have been developed to make graft and bud unions, and some are widely used. Bench grafting of grapes has been quite successful with the use of an electrically driven circular saw which cuts square notches in the ends of the stock and scion. This enables them to be pushed together, giving a very tight and secure fit (*1, 2*). (See Figures 12–24 and 12–25.)

The hand-operated device shown in Figure 12–26, which is manufactured in France, can be used to rapidly prepare a wedge graft. It was developed primarily for grafting grapes but has given satisfactory results in making tree fruit root grafts.

Figure 12–24   Machine for cutting notches in ends of grape scions and stocks for making the type of graft illustrated in Figure 12–25. Blades on left make cuts for stock; that on right makes cuts for scion.

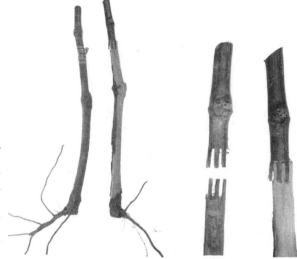

Figure 12-25 Grapes propagated by machine grafting. Small, one-budded scions are grafted on rooted cuttings during the dormant season. The graft union is wrapped with a budding rubber, then allowed to callus before planting.

SELECTION AND STORAGE OF SCION WOOD

Since almost all grafting of deciduous species takes place in late winter or early spring, the use of scion wood that grew the previous summer is necessary. It is important that suitable wood be obtained and properly cared for until it is used. Storage of scions should not be attempted if succulent, herbaceous plants are being grafted; such scions should be obtained at the time of grafting and used immediately. In grafting certain of the broad-leaved evergreen species, such as olives or citrus, it is also unnecessary to collect the scion wood previously and hold it in storage. Grafting of these species is done in the spring before much active growth starts; the scion wood is taken directly from the tree as needed, using wood on the basal part of the shoots containing dormant, axillary buds. The leaves are removed at the time of collection.

In selecting scion material of woody plants, the following points should be observed.

Figure 12-26 A type of wedge grafting as done by a French grafting device. *Top left:* appearance of graft as made by the cutting blades. (After this, graft must be tightly wrapped with grafting tape.) *Top right:* apple graft union after one year's growth in the nursery. *Below:* grafting device in use. Grafts can be made with this device much faster than by the whip, or tongue, method.

(a) For most species, the wood should be one year old or less (current season's growth). Avoid including older growth, although in some species, such as the fig or olive, two-year-old wood is satisfactory, or even preferable, if it is of the proper size.

(b) Healthy, well-developed vegetative buds should be visible. Avoid wood with flower buds. Usually, vegetative buds are narrow and pointed, whereas flower buds are round and plump (see Figure 13–1).

(c) The best type of scion material is vigorous, but not overly succulent, well-matured, and hardened shoots from the upper part of the tree, which have made 2 to 3 feet of growth the previous summer. Such growth develops on relatively young, well-grown, vigorous plants; high production of scion material can be promoted by pruning the plant back severely the previous winter. Watersprouts from older trees sometimes make satisfactory scion wood but suckers arising from the base of grafted trees should not be used. A satisfactory size is from $\frac{1}{4}$ to $\frac{1}{2}$ in. in diameter.

(d) The best scions are obtained from the center portion or from the basal two-thirds of the shoots. The terminal sections are likely to be too succulent and pithy, and low in stored carbohydrates. Well-matured wood with short internodes should be selected.

(e) Scion wood should be taken from stock trees where there is good evidence that it is pathogen-free and genetically true-to-type. Virus diseases, undesirable sports, chimeras, or virus-like genetic disorders must be avoided. The best procedure is to collect from scion orchards where trees have originated from virus-indexed stock taken from known source trees which have been carefully inspected for trueness to cultivar and which have been progeny tested for performance. Various certification schemes are available in many countries to regulate such propagative material. Where this is not available, one should carefully inspect the source tree or trees, preferably in the growing season prior to collection. Avoid source plants that show evidence of not being typical. For fruit-bearing species it may be desirable to collect scion material from a bearing rather than a nonbearing tree so that trueness to name can be accurately determined.

(f) For deciduous plants which are to be grafted in early spring, the scion wood can be collected almost anytime during the winter season when they are fully dormant (4). In climates with severe winters, the wood should be gathered at a time when it is not frozen. Any wood that shows freezing injury should not be used. Where considerable winter injury is likely, it is best to collect the scion wood and put it in storage after leaf fall but before the onset of the winter season.

Scion wood collected prior to grafting must be properly stored. It should be kept slightly moist and at a temperature which will prevent development of the buds. A common method is to wrap the wood, in bundles of 25 to 100 sticks, in heavy, waterproof paper. A small amount of some clean, slightly moist material, such as sawdust, wood shavings, or peat moss, should be sprinkled through the bundle. Sand should not be used, because it will adhere to the scion wood and dull the edge of the grafting knife during the grafting operation. Packing material should not be wet, or various fungi will develop and damage the buds, even at low storage temperatures. The packing should be barely moist; scion wood is more apt to be damaged by being too wet than by being too dry. If it is to be stored for a prolonged period, the bundle should be examined every few weeks to see that the wood is not becoming either dried out or too wet. When the buds show signs of swelling, the wood should either be used or be placed in a lower storage temperature.

Polyethylene plastic sheeting is a good material for wrapping scion wood. It allows the passage of oxygen and carbon dioxide, which are exchanged during the respiration process of the stored wood, but retards the passage of water vapor. Therefore, if this type of wrapping is tightly sealed, no moist packing material is needed; the natural moisture in the wood will be sufficient, since little will be lost. Polyethylene bags are useful for storing small quantities of scion wood. All bundles should be accurately labeled.

The temperature at which the wood is stored is important. If it is to be kept for just a short time before grafting—two or three weeks—the temperature of the home refrigerator—40° to 50° F (5° to 10° C)—is satisfactory. If stored for a period of 1 to 3 months, scion wood should be held at about 32° F (0° C) (*4*) to keep the buds dormant. However, buds of some species, such as the almond and sweet cherry, will start growth after about three months even at this temperature.

Do not store scion wood in a home freezer where the very low temperatures—about 0° F (−18° C)—may injure the buds.

If cold storage facilities are not available, the scion wood can be kept for a period of time in cold winter regions by burying the bundles in the ground 12 to 18 in. deep, below frost level, on the north side of a building or tall hedge. Drainage should be good so that excess water does not remain around the scions and cause the buds to deteriorate.

Attempting to use scion wood in which the buds are starting active growth is almost certain to result in failure. In such cases, after grafting, the buds will quickly leaf out before the graft union has occurred; consequently the leaves, by transpiration, will withdraw water from the scions and cause them to die.

In top-working pecans (*3*) it was found that good results could be obtained by using precut scions; that is, scions were cut in advance by skilled persons at a convenient time, then stored in polyethylene bags for periods up to nine days before inserting in the graft unions. Grafting success was reduced slightly by the use of precut scions.

GRAFTING CLASSIFIED ACCORDING TO PLACEMENT

Grafting may be classified according to the part of the plant on which the scion is placed—a root, the crown (the junction of the stem and root at the ground level), or various places in the top of the plant.

Root Grafting

In this class of grafting the rootstock seedling, rooted cutting, or layered plant is dug up, and the roots are used as the stock for the graft. The entire root system may be used (*whole-root graft*—Figure 12–27), or the roots may be cut up into small pieces and each piece used as a stock (*piece-root graft*). Both methods give satisfactory results. As the roots used are relatively small (¼ to ½ in. in diameter), the whip, or tongue, graft (p. 375) is generally used. Root grafting is usually performed indoors during the late winter or early spring. The scion wood collected previously (see p. 407) is held in storage, while the rootstock plants are also dug in the late fall and stored under cool (40° to 50° F; 4° to 10° C) and moist conditions until the grafting is done.

The term *bench grafting* is sometimes given to this process, because it is often performed at benches by skilled grafters as a large-scale operation. A number of plants are propagated commercially by root grafting—apples, pears, grapes, and such ornamentals as the wisteria and rhododendron.

In making root grafts, the root pieces should be 3 to 6 in. long and the scions about the same length, containing two to four buds. After the grafts are made and properly tied, they are bundled together in groups of 50 to 100 and stored for callusing in damp sand, peat moss, or other packing material. They may be placed in a cool cellar or under refrigeration at approximately 45° F (7° C) for about two months. The callusing period for apples can be shortened to around 30 days if the grafts are stored at a temperature of about 70° F (21° C) and at a high humidity. To use this higher callusing temperature the material should be collected in the fall and the grafts made before any cold weather has overcome the rest period of the scion buds. After the unions are well healed, the grafts must be stored at cool temperatures— 35° to 40° F (2° to 4° C)—to overcome the "rest period" of the buds and to hold them dormant until planting (*17*). The grafts are lined-out in early spring in the nursery row directly from the low temperature storage conditions. For general callusing purposes, temperatures from 45° F (7° C) to 70° F (21° C) are the most satisfactory. By the proper regulation of temperature, callusing processes may be accelerated by increased temperature or retarded by decreased temperature so that, within reasonable limits, a desired degree of callus formation may be had within a given length of time. Provision should be made for adequate aeration of the callusing grafts (*26*).

Figure 12–27   Whole root apple grafts made by the whip or tongue method and wrapped with adhesive nursery tape.

As soon as the ground can be prepared in the spring, the grafts are lined out in the nursery row 4 to 6 in.

apart. They should be planted before growth of the buds or roots begins. If this starts before the grafts can be planted, they should be moved to lower temperatures (30° to 35° F; − 1° to 2° C). The grafts are usually planted deep enough so that the graft union is just below the ground level, but if the roots are to arise only from the rootstock, the graft should be planted with the union well above the soil level. It is very important to prevent scion rooting where certain definite influences, such as dwarfing or disease resistance, are expected from the rootstock. (See Figure 11–1.)

After one summer's growth, the grafts should be large enough to transplant to their permanent location. If not, the scion may be cut back to one or two buds, or headed-back somewhat to force out scaffold branches, and then allowed to grow a second year. With the older root system a strong, vigorous top is obtained the second year.

Nurse-Root Grafting
Under certain conditions, it is desired to have a stem cutting of a difficult-to-root species on its own roots. One way this can be done is by making a root graft, using the plant to be grown on its own roots as the scion and a root of a compatible species as the stock. The scion may be made longer than usual and the graft planted deeply with the major portion of the scion below ground. In some cases scion-rooting is promoted by rubbing a rooting stimulant, such as indolebutyric acid into several vertical cuts made through the bark at the base of the scion, just above the graft union. This is done just before planting, and the grafts are set deeply so that most of the scion is covered with soil (20, 21). After one or two seasons of growth many of the scions will have roots. The temporary nurse rootstock is then cut off and the top reduced in proportion to the root system. The rooted scion is replanted to grow on its own roots.

Scion rooting is often better if the nurse root is buried deeply enough so that a new shoot of the current season's wood, developing from a bud on the scion, will be rooted, rather than trying to root the older, lignified tissue of the original scion. This is essentially a form of mound layering (see p. 464). As the new scion shoot grows, soil is gradually mounded up around it to a height of 5 or 6 in., although the terminal leaves are at no time covered (22).

Several methods of handling, as illustrated in Figure 12–28, eliminate the necessity of digging up the graft and cutting off the rootstock. The rootstock piece will eventually die if it is grafted onto the scion in an inverted position (22). The inverted stock piece sustains the scion until scion roots are formed, but the stock fails to receive food from the scion and eventually dies, thus leaving the scion on its own roots. In another method an incompatible rootstock is used. Hence, if the graft is planted deeply, scion roots will gradually become more important in sustaining the plant, and the incompatible rootstock will finally cease to function.

In a third method the base of the scion, just above the graft union, is bound with some type of wrapping material to eventually girdle and cut off the rootstock (Figure 12–29). Excellent results have been obtained with ordinary budding rubber strips (0.016 gauge) (5). Budding rubbers disintegrate

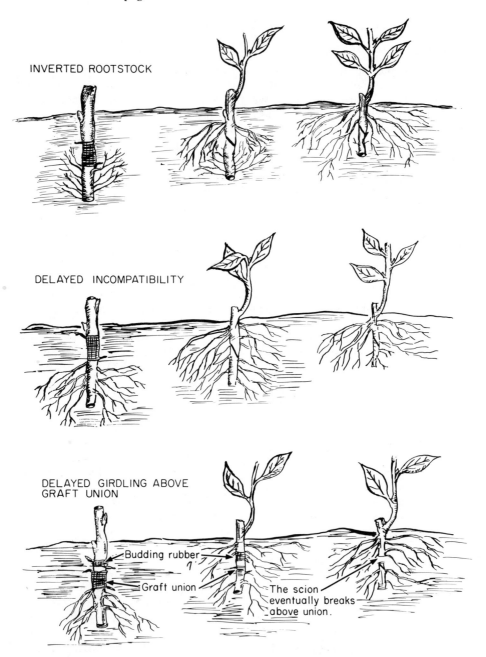

INVERTED ROOTSTOCK

DELAYED INCOMPATIBILITY

DELAYED GIRDLING ABOVE GRAFT UNION

Budding rubber

Graft union

The scion eventually breaks above union.

Figure 12–28   The "nurse-root" graft is a temporary graft used to induce the scion to develop its own roots. The nurse root sustains the plant until the scion roots form; then it dies. Three methods are shown here for preparing a nurse-root graft.

within a month when exposed to sun and air; when buried in the soil, they will last as long as two years, allowing sufficient time for the scion to become rooted. Yet, owing to the slow deterioration of the rubber below ground, the rootstock is finally girdled and cut off.

Crown Grafting

A graft union made at the root-stem transition region—the "crown" of the plant—on an established rootstock is termed a "crown graft." Several methods of grafting are commonly used in crown grafting, such as the whip, side, cleft, saw-kerf, or bark graft, depending upon the species and size of the root-stock.

Crown grafting of deciduous plants is best done in late winter or early spring, shortly before new growth starts. The scions should be prepared from well-matured, dormant wood of the previous season's growth.

Since the operation is performed just below, at, or just above the soil level, it is possible to cover the graft union, or even the entire scion, with soil and thus eliminate the necessity

Figure 12–29   Nurse-root graft. A 'Malling 9' apple scion was root-grafted to an apple seedling nurse root. Just above the graft union the scion was wrapped with a budding rubber strip (see arrow). After two years in the nursery, vigorous scion roots were produced. The budding rubber has effectively constricted development of the seedling nurse root, which can now be broken off and discarded. Courtesy D. S. Brown.

for waxing. The union should be tied securely with string or tape to hold the grafted parts together until healing takes place.

Double-Working

A double-worked tree has three parts, usually all different genetically—the rootstock, the interstock, and the scion, or fruiting, top (*11*). Such a tree has two graft unions, one between the rootstock and interstock, and one between the interstock and the scion. The interstock may be less than an inch in length or extensive enough to include the trunk and secondary scaffold branches. Double-working is used for various purposes, such as (a) overcoming graft incompatibility between a desired top cultivar and the rootstock, (b) providing a cold- or disease-resistant trunk, (c) obtaining a dwarfing effect from the use of certain intermediate stocks, or (d) obtaining the strong trunk or crotch systems of certain cultivars.

Examples of double-working are (1) the propagation of 'Bartlett' pears on quince as a dwarfing rootstock by using a mutually compatible interstock,

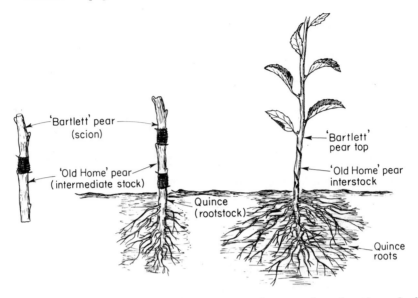

'Bartlett' pear (scion)

'Old Home' pear (intermediate stock)

'Bartlett' pear top

'Old Home' pear interstock

Quince (rootstock)

Quince roots

Figure 12–30   There are three distinct parts and two graft unions in a double-worked plant, illustrated here by a 'Bartlett' pear grafted on an 'Old Home' pear which in turn is grafted on quince as a rootstock. The 'Old Home' pear, in this case, is the intermediate stock, or interstock.

such as 'Old Home' or 'Hardy' pear, (Figure 12–30) and (2) the propagation of dwarfed apple trees consisting of the scion variety grafted onto a 'Malling 9' interstem piece which is grafted onto either an 'MM 106' or 'MM 111' rootstock (7).

Several methods are used for developing double-worked nursery trees. The grafting in these techniques can be done by machine grafting devices as illustrated in Figures 12–24 and 12–26 or by use of the whip or tongue graft.

(a) Rootstock "liners", either seedlings, clonal rooted cuttings, or rooted layers, are set out in the nursery row in early spring. These are then fall-budded with the interstock buds, growth from which, a year later, is fall-budded with the scion cultivar buds. Three years are required to produce a nursery tree by this method.

(b) The interstock piece is bench-grafted onto the rooted rootstock—either a seedling or a clonal stock—in late winter. After callusing, the grafts are lined-out in the nursery row in the spring. These are then fall-budded to the scion cultivar. By this method the nursery tree is propagated in two years.

(c) A variation of method (b) is to prepare, by bench grafting, two graft unions —the scion grafted to the interstock, and the interstock grafted to the rooted rootstock. After callusing, the completed graft, with two unions, is lined-out in the nursery row. Depending upon growth rate, a nursery tree can be obtained in one or two years.

(d) The scion piece is bench-grafted onto the interstem piece in late winter or early spring; then, after callusing, this component is grafted—as shown in Figure

12–30—onto rootstocks which have been grown in place in the nursery row. A nursery tree is obtained in one or two years by this method.

(e)   Double-shield budding, in which the double-working is done in one operation by budding, is illustrated in Figures 13–16 and 13–17. A nursery tree is produced in one or, if growth is slow, two years after budding.

Top-Grafting (Topworking)

One of the principal uses of grafting is to change the cultivar of an established plant–tree, shrub, or vine (Figures 12–31 and 12–32). If budding is used, then the process is termed *top-budding* (see p. 448).

In some cases pulling out the existing trees and planting new nursery trees may be more economical than topworking. This is especially true if the existing trees are old or diseased or if the species is relatively short-lived, as is the peach. Long-lived species, such as apples and pears, are worth top-

Figure 12–31   Topworking a 55-year-old 'Glou Morceau' pear tree to the 'Comice' cultivar. *Left:* just after grafting. The scions are well placed in the proper-size branches, but they should be shorter. *Center:* the same tree one year later. *Right:* the same tree six years later. Courtesy W. P. Tufts, California Agricultural Experiment Station.

Figure 12-32   Proper method of topworking trees. *Left:* scions, which have been inserted into fairly small branches, are starting to grow. Tree has been whitewashed to prevent sunburn injury. *Right:* same tree about six weeks later. Stakes have been nailed to branches to which the shoots developing from scions are tied to prevent their breaking off in winds.

working if the tree is in a healthy condition. Also to be considered in planning for topworking is the hazard that viruses are present in the scion wood. In the peach, for example, virus diseases such as "stubby twig," "ringspot," or "necrotic leaf spot" may be introduced by infected scions into an otherwise healthy orchard. Care should be taken to obtain scion wood for top-grafting, if possible, from known virus-tested sources. Of equal significance is the possibility that the stock trees may be infected with viruses or part of a virus complex. The value of scion wood from virus-tested sources can be nullified if the stock is virus-infected.

Preparation for Top-Grafting
Any of the methods of grafting described earlier in this chapter—whip, side, cleft, saw-kerf, or bark—can be used for top-grafting, which is usually done in the spring, shortly before new growth starts. The exact time depends upon the method to be used. The cleft, side, whip, and saw-kerf graft can be done before the bark is slipping. The bark graft must be done when the bark is slipping, preferably just as the buds of the stock tree are starting growth.

It is usually advisable to obtain an ample amount of good-quality scion wood prior to grafting and store it under the proper conditions (see p. 407).

In preparing the stock tree for topworking, one must decide, for each individual tree, which and how many scaffold branches (usually three to five) should be used. If the work is done high in the tree in the smaller, secondary scaffold branches, an earlier return to bearing will result than when fewer

Figure 12-33 Sawing off a branch in preparation for top-grafting so that no bark-tearing occurs. *Top left:* the first cut is made in a smooth area starting on the under side of the branch and continuing about one-third of the distance through. *Top right:* the second cut is made starting from the upper part of the branch and cutting downward. It should be back 1 or 2 in. on the branch from the first cut. *Below left:* after the branch breaks off and falls, the second cut is continued. *Below right:* final smooth cut, ready for grafting.

and larger limbs are grafted lower in the tree. The branches to be grafted should be well distributed around the tree and up and down the main trunk, avoiding branches with weak, narrow crotches. All others can be removed unless one or more nurse branches are used. Figure 12–33 shows the proper method of preparing the branches for grafting.

Topworking is an extremely severe pruning operation for the tree and results in a considerable imbalance between the root system and the top. However, deciduous trees soon recover and the new growth from the scions and from latent buds on the stock restores a balance without damage, providing the tree is healthy and vigorous and the grafting is done when it is dormant or shortly after growth starts in the spring. However, under some situations, not always well understood, the tree will be adversely affected by the heavy cutting back, and will show intense leaf burning and may even fail to survive. This situation is particularly noticeable when the grafting is delayed until after new spring growth is well underway, where all foliage and new shoots are stripped off the stock plants, or where high temperatures prevail shortly after grafting.

Also when all branches of the tree are cut back and grafted, considerable energy goes into the scion growth, producing vigorous, succulent shoots that may break out if not properly pruned or supported. In addition, such rank, succulent growth is very susceptible to winter freezing damage in cold climates.

To avoid the problems cited above it is advisable to retain some of the foliage on the stock plant as well as small lower branches and even large limbs to be held temporarily as *"nurse branches."* However, such nurse limbs should be pruned fairly severely, otherwise the grafted scions may not make adequate growth.

It is essential that nurse limbs be left when topworking broad-leaved evergreen trees, as citrus or olives. If they are retained on the south and west parts of the tree they will shade the grafted portions and reduce the chances of sunburn injury to the exposed branches.

A practice which is often recommended, especially for older trees, is to topwork them during a two-year period, grafting perhaps two main branches on the northeast side of the tree the first year and retaining two branches on the southwest as nurse branches. The second year these are topworked.

Topworking is most successful when done on relatively young trees where the branches to be grafted are not larger than 3 or 4 in. in diameter and are relatively close to the ground. When attempting to topwork large, old trees, it is often necessary to go high up in the trees to find branches with a diameter as small as 4 in. If the grafting is done on such branches, the new top is inconveniently high for the various orchard operations, such as thinning and harvesting. The other alternative is cutting off the branches or main trunk close to the ground and inserting the scions into wood one or more feet in diameter. Although many scions can be inserted around the tree between the bark and the wood by the bark-graft method and may grow well for several years, it is quite likely, as shown in Figure 12–34, that wood rot will develop in the center of the stub before the growth of the scions can heal it over. Also, some scions are not mechanically held in place securely and may be blown out by strong winds after they reach considerable size.

It is important that the branch to be topworked be cut off in such a location that the region just below the cut is smooth and free from knots or small branches, so that there will be a satisfactory place for inserting the scions. The branches are best cut off about 9 to 12 in. from the main trunk to keep the tree headed low. The branch should not be cut off more than a few hours before the grafting is to be done.

In Preparation for Topworking (Top-Grafting)

a Do the work in the spring when the trees are dormant or shortly after growth starts.

b Select for grafting three to five well-placed scaffold branches which are not larger than about 4 in. in diameter and which are conveniently close to the ground.

c Retain nurse branches for broad-leaved evergreen trees and for deciduous trees where the winters are severe.

d Cut off the branches properly so that the bark is not torn down the trunk.

Grafting on a cool, overcast day with no wind blowing offers the most protection from drying of the cut surfaces of the scion and stock until they can be covered with grafting wax. Grafting on hot, sunny, and windy days should be avoided. During grafting, the scion wood must not dry out by being exposed to the sun. It should be kept moist and cool in some container or be wrapped in moist burlap. (See p. 407 for information on collecting scion wood.)

Immediately after each stub is grafted it should be thoroughly covered with grafting wax as described on p. 386. The need for prompt and thorough coverage of all cuts, including the tip end of the scions, cannot be stressed too strongly. The wax should be worked into the bark of the stock, sealing all small cuts or cracks where air could penetrate in and around the cut surfaces where healing tissue is expected to develop. Waxes containing bees-wax will attract bees, which may remove the wax, necessitating rewaxing.

Figure 12–34   An improper method of topworking trees. This walnut tree was cut off close to the ground, and a number of scions were inserted around its circumference by the bark-graft method. Two of the scions grew successfully, but the remainder failed. Most of the original tree is dead; healing of such a large cut is almost impossible. A better method of topgrafting trees is shown in Figure 12–32.

Subsequent Care of Topworked Trees

After the actual top-grafting (or top-budding) operation is finished, much important work needs to be done before the topworking is successfully completed. A good grafting job can be ruined by improper care of trees.

If the grafting has been done in late spring when growth is active, trees of some species, such as the walnut, will "bleed" to a considerable extent from the grafted stub, even though it has been covered with grafting wax. This flow of sap around the scions can be so heavy as to interfere with the normal healing processes at the graft union. If this condition appears, it can often be corrected by making several slanting cuts around the tree with a knife or saw through the bark, into the water conducting tissues, in the trunk of the tree several feet below the grafted stubs. The bleeding will then take place at these cuts rather than around the graft union. This extensive sap flow is not particularly harmful to the tree and will usually stop within a few days. Boring a series of random holes into the trunk around the tree will also accomplish the same purpose.

If certain bacterial diseases, such as crown gall, are present, this practice may spread the bacteria. In such cases, dipping the knife, saw, or bit into a disinfectant, such as 10 percent Clorox, is advisable.

In three to five days after grafting, the trees should be carefully inspected and the graft unions rewaxed if cracks or holes appear in the wax.

It is essential to prevent sunburn on the trunk and large branches which are exposed to the sun by the removal of the protecting top foliage. This is especially important if the grafting has been done late in the season when hot weather can be expected, and when no protecting nurse branches have been retained. The energy from the sun absorbed by the dark-colored bark can raise the temperature of the living cells below the bark to a lethal level.

It is advisable to whitewash the trunk, branches and scions of the grafted trees. Various cold-water paints, some made especially for this purpose, are available. The white color reflects a considerable portion of the sun's radiant energy, thus keeping the temperature of the living tissues within safe limits. Interior, water-base white house paints (both latex and acrylic) mixed with water, 1:1, prevent sunburn for one season. Exterior paints give longer protection but are more likely to cause injury to the tree (*23*).

To Prepare Whitewash for Grafted Trees

Formula I.   Quicklime 5 lb; salt, ½ lb: sulfur, ¼ lb. Add water to the lime to start it slaking. Then add the salt and sulfur. This mixture should be prepared in a crockery or wood container. Allow to age for several days, then dilute to a consistency just thick enough to apply with a paint brush. Dilute to a thinner consistency for spray application.

Formula II.   Hydrated lime, 25 lb; zinc sulfate, 2 lb; water, 50 gal. Add water first to a power sprayer with agitator running, then add other components. Spray mixture on trees.

When the grafting has been done late in the season, scions can be protected from the sun's heat by covering the end of the grafted stub—as well as the scions—with a large paper bag, with the corners cut off to permit ventilation. The bag is tied securely onto the grafted stub with heavy string. However, whitewashing the scions and the grafted stub has given better protection in comparative tests (*13*).

Another help in preventing sunburn is to retain some of the watersprouts which soon start growth along the trunk and branches of the grafted tree. They must be kept under control or they will quickly shade out the developing scions. Rather than removing the watersprouts completely, the grafter can head them back to several inches in length, and they will shade the bark underneath. An additional benefit is that the food manufactured by this leaf area will help sustain the tree until the new scions develop sufficient foliage to do so.

Trees just grafted should be amply supplied with water so that the tissues are in a high state of turgidity. This is necessary in order to obtain good callus production, which is essential for healing of the graft union.

During the first summer after topworking, new shoots arising from below the grafted branches must be kept pruned back so as not to interfere with

Figure 12-35   Problems encountered in handling limbs that have been top-grafted. *Left:* where both scions "take," one is cut back heavily to allow the other to become predominant so as not to form a weak crotch. It is retained for two or three years, however, to assist in healing the grafting wound, but will eventually be removed. *Right:* where only one scion in a fairly large branch "takes," difficulty may occur in rapid healing of the wound. The wood under the dead scion should be cut off at an angle, then waxed; this will give a minimum cut surface for healing.

the growth of the scions. Nitrogen fertilizers should be withheld for a year or two after grafting because no stimulation of growth is usually needed. The water requirement of the trees will be less, owing to the removal of a considerable amount of leaf area when the tops were cut back for grafting.

Two to four scions are generally inserted in each stub. If all the scions grow, they should all be retained during the first year, because they will help heal over the stub. However, just one branch, from the best placed and strongest growing scion, should be retained permanently. Growth from the remaining scions should be retarded by rather severe pruning, keeping them alive to help heal the branch, but allowing the permanent scion to become the dominant one. To keep two or more scions for permanent branches at one point will undoubtedly result in a weak crotch, which will eventually break. The first year, then, the best practice is to retain all scions to help heal the stub but, by pruning, to retard the growth of all but the best one. Such a procedure is illustrated in Figure 12-35. If two shoots arise from the scion to be retained permanently, select the best of these and remove the other. After the stub has healed over, the temporary scions can be removed completely. This may be in the second or third year.

If the permanent scion grows rather vigorously, the danger arises of its becoming topheavy and breaking off during winds. This may be handled in two ways—either by retarding the growth of the permanent branch by pruning it back or by nailing a lath or other type of stick onto the tree and tying the new branch to this stick. *When tying a cord around a branch, always make a loop so that there is no chance of the branch being girdled as it grows.*

Should only one scion grow at each stub, the problem of securing adequate healing of the stub on the side opposite the living scion may prove to be serious. Healing may be helped by sawing off the stub at an angle away from the surviving scion. The cut surface of the stub can be covered with grafting wax to retard wood decay. It may take a number of years for the single scion to heal over the stub, and it is possible that wood decay may start before it can do so.

When none of the scions grow in a stub, there are still some possibilities for getting it topworked. One is by allowing several well-placed water-sprouts arising just below the cut surface to grow, then top-budding them during the summer. Or the watersprouts may be allowed to grow so as to keep the branch alive and healthy, then the grafting operation repeated the following year, making a fresh cut a foot or so below the original cut.

If nurse branches have been used, they should be cut back and away from the scions at intervals throughout the summer so that scion shoots are always fully exposed to the sun and have sufficient space to grow. Nurse branches should be removed entirely or grafted by the beginning of the second or third year.

When the top of the tree has been finally worked over to the new cultivar, it will grow vigorously for a few years. Good pruning practices are needed to prevent badly placed branches from developing.

Frameworking

In this method of changing the cultivar of fruit trees, all the main scaffold branches are retained on the tree but most of the small laterals throughout the top of the tree are replaced by a large number of scions of the replacement cultivar. These are inserted by the side graft method (see p. 378) with the cuts made very close to the origin of lateral branches (about 1 in. in diameter) arising from the secondary scaffolds. Long scions having seven or eight buds are often used. In frameworking the tree quickly returns to fruiting, but considerable labor is needed for grafting, and much aftercare is required in keeping growth from the original tree below the scions cut away so that only the replacement cultivar is maintained.

HERBACEOUS GRAFTING

Grafting herbaceous types of plants is used for various purposes, such as studying virus transmission, stock-scion physiology, and grafting compatibility, as well as for the commercial greenhouse production of certain cucurbitaceous crops, particularly in Europe. Usually such grafts are made while the plants are quite small, the stock being grafted shortly after seed germination. Such material is generally very soft, succulent, and susceptible to injury. In one technique (see Figure 12–36), a simple splice graft is used with a diagonal cut made through the seedling stock, just above the cotyledons (16). A piece of thin-walled polyethylene tubing of the proper size to give a snug fit is slipped over the cut end of the stock. The basal end of the scion receives a diagonal cut similar in length and angle to that given the stock. The scion is then slipped down into the plastic tubing so that the two

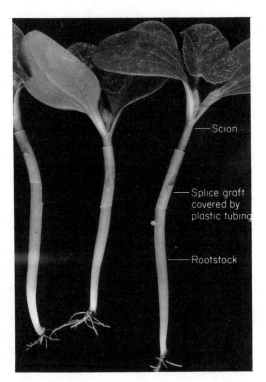

Scion

Splice graft covered by plastic tubing

Rootstock

Figure 12–36 Grafting young, herbaceous plant material together by using a splice graft. A single, slanting cut is made in the upper part of the stock. A piece of transparent plastic tubing is slipped over this to give a snug fit. A similar slanting cut is made in the lower portion of the scion, which is then slipped into the plastic tubing so that the cut surfaces match. After the graft has healed, the tubing is cut away with a razor blade.

cut surfaces make intimate contact. The tubing holds the graft in place until healing occurs—about 12 days after grafting. Then the tubing may be slipped off over the scion if there are no leaves and if the bud has not expanded; otherwise, it may be cut off with a razor blade.

In another procedure (*8*) used in grafting older herbaceous stock plants, several leaves are retained below the graft union. The cleft graft is used (but with only one scion), and the graft union is bound with raffia, budding rubbers, or adhesive latex tape. To prevent drying out, the entire plant—following grafting—is covered with a supported polyethylene bag. The grafted plant is then set in the shade until the graft has healed; then the plastic cover can be removed.

A method of establishing *Vitis vinifera* grapes on phylloxera or nematode-resistant rootstocks is by "greenwood" grafting, in which the grafting is done in the spring, placing the scion—taken from new growth—on a new green shoot which has developed from the rootstock plant. A simple splice graft (see p. 378) is made with the sloping cuts 1 to 1½ in. long. The stock and scion pieces must be the same diameter. The scion has only one bud. The cuts are matched as closely as possible, and the graft union is completely covered by wrapping with a budding rubber. The graft union should be healed and the bud growing by two weeks after grafting. Since these grafts are quite fragile, they must be tied to a stake (*6, 14*).

Figure 12–37   Rooted chestnut grafts 19 days after grafting by the nurse-seed grafting method *(18)*. Courtesy Connecticut Agricultural Experiment Station.

NURSE-SEED
GRAFTING *(18, 19, 24, 25)*

Germination of some large-seeded woody species, such as the chestnut, is hypogeal (see p. 112) (that is, the cotyledons remain below ground in the seed coats with the shoot tip appearing above ground). In a grafting procedure using certain species with seeds of this type, when the seeds have just germinated, the petioles of the cotyledons are cut off transversely just at the seed, leaving the cotyledons inside. A knife point is inserted into the seed between the cut petioles, making an opening for the scion. The scion, which is prepared from dormant wood of the previous season's growth, is cut to a wedge shape at the base, as for a cleft graft. The scion is inserted into the cut between the cotyledons so that the exposed cambium surfaces of the scion are in close contact with the cut surfaces of the cotyledons. The "seed grafts" are then lined-out in the rooting medium with the union about $1\frac{1}{2}$ in. below the surface. A graft union takes place, and roots arise from the cut cotyledon petioles. This type of grafting is done in early spring, using properly stratified seed (see p. 157) and dormant scion wood. Chestnuts, avocados, and camellias have been grafted successfully by this method; in fact, grafted camellia plants of the desired cultivar can be obtained in the same length of time it would take for the seedlings to become large enough to graft by the conventional cleft graft method. (See Figure 12–37.)

In a modification of the nurse-seed graft *(24)* which has worked well in propagating chestnuts, pecans, walnuts, and oaks, the actual graft union is made into the seedling hypocotyl at the point of attachment of the cotyledons (see Figure 12–38). The scion is cut to a thin wedge and inserted into a split made in the hypocotyl. After insertion of the scion, the graft is wrapped firmly with rubber strips or waxed string.

CUTTING-GRAFTS

In this type of graft, a leafy scion is grafted onto a leafy, unrooted stem piece (which is to become the rootstock), and the combination is then placed in a rooting medium under intermittent mist (see p. 297) for simultaneous healing of the graft union and rooting of the stock. This procedure

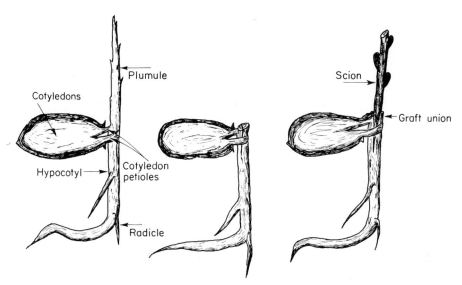

Figure 12–38   Modified nurse-seed graft. *Left:* seedling plant, ready for grafting. *Center:* hypocotyl cut open longitudinally, ready for insertion of the scion. *Right:* scion inserted; the graft is now ready for tying *(24)*. Redrawn from Moore.

was utilized many years ago in studying stock-scion physiology in citrus *(12)*. More recently it has been used in commercial propagation of various types of citrus on clonal dwarfing rootstocks *(9)*. It is also of value in propagating certain difficult-to-root conifers *(28)* and rhododendrons *(10)*.

For citrus a simple splice graft (see p. 378) is used. The slope of the cut is at a 30-deg. angle, ½ to ¾ in. long; the union is tied with a rubber band. The base of the stock is dipped into a root-promoting material, such as indolebutyric acid (see p. 293), and then the grafts are placed under mist, or in a closed case, in flats of the rooting medium over bottom heat. After healing of the union and rooting of the stock, the grafts are allowed to harden by discontinuing the mist and bottom heat for about two weeks. Then the grafts are ready for planting in 1–gal. cans or other containers.

In using this method for conifers the difficult-to-root scion is grafted in early spring to the unrooted (but easily rooted) stock by a side graft (see p. 378). The graft union is made about 2 in. above the base of the stock. The stock and scion are both 10 to 12 in. long. The base of the stock is treated with an indolebutyric acid rooting compound (see p. 293), and then the combination is inserted into a flat of a a well-drained rooting medium, such as perlite, so that the medium covers the graft union. The flats containing the grafts are kept in a cool greenhouse under intermittent mist. About 8 weeks later the cuttings should be rooted and the graft union healed. Then the top of the stock is cut off just above the graft union, and the grafted plant is transferred to an appropriate-sized container of soil.

REFERENCES

1 Alley, C. J., Mechanized grape grafting, *Calif. Agr.*, 11(6):3, 12. 1957.

2 ———, Can grafting be mechanized? *Proc. Inter. Plant Prop. Soc.*, 20:244–48. 1970.

3 Anonymous, Pre-cut scions, *Agr. Res.*, 17(6):11. 1968.

4 Bhar, D. S., R. J. Hilton, and G. C. Ashton, Effect of time of cutting and storage treatment on growth and vigor of scions of *Malus pumila* cv. McIntosh, *Canad. Jour. Plant Sci.*, 46:69–72. 1966.

5 Brase, K. D., The nurse-root graft, an aid in rootstock research, *Farm Research*, 17(1):16. 1951.

6 Carlson, V., How to green graft grapes, *Calif. Agr. Ext. Pub. AXT. 115*, 1963.

7 Cummins, J. N., Systems for producing multiple-stock fruit trees in the nursery, *Plant Prop.*, Inter. Plant Prop. Soc., 19(4):7–9. 1973.

8 Denna, D. W., A simple grafting technique for cucurbits, *Proc. Amer. Soc. Hort. Sci.*, 81:369–70. 1962.

9 Dillon, D., Simultaneous grafting and rooting of citrus under mist, *Proc. Inter. Plant Prop. Soc.*, 17:114–17. 1967.

10 Eichelser, J., Simultaneous grafting and rooting techniques as applied to rhododendrons, *Proc. Inter. Plant Prop. Soc.*, 17:112. 1967.

11 Garner, R. J., Studies in nursery technique: The production of double worked pear trees, *Ann. Rpt. East Malling Res. Sta. for 1939*, pp. 84–86. 1940.

12 Halma, F. F., and E. R. Eggers, Propagating citrus by twig-grafting, *Proc. Amer. Soc. Hort. Sci.*, 34:289–90. 1936.

13 Hansen, C. J., and H. T. Hartmann, Influence of various treatments given to walnut grafts on the percentage of scions growing, *Proc. Amer. Soc. Hort. Sci.*, 57:193–97. 1951.

14 Harmon, F. N., and E. Snyder, Some factors affecting the success of green-wood grafting of grapes, *Proc. Amer. Soc. Hort. Sci.*, 52:294–98. 1948.

15 Hearman, J., et al., The reinvigoration of apple trees by the inarching of vigorous rootstocks, *Jour. Pom. and Hort. Sci.*, 14:376–90. 1936.

16 Holt, J., A simple way of grafting herbaceous plants, *Gard. Chron.*, 143:332. 1958.

17 Howard, G. S., and A. C. Hildreth, Introduction of callus tissue on apple grafts prior to field planting and its growth effects, *Proc. Amer. Soc. Hort. Sci.*, 82:11–15. 1963.

18 Jaynes, R. A., Nurse seed grafts of chestnut species and hybrids, *Proc. Amer. Soc. Hort. Sci.*, 86:178–82. 1965.

19 Jaynes, R. A., and G. A. Messner, Four years of nut grafting chestnut, *Proc. Inter. Plant Prop. Soc.*, 17:305–11. 1967.

20 Jones, F. D., Hormone on root graft, *Amer. Nurs.*, 72(11):6–7. 1950.

21 Kerr, W. L., A simple method of obtaining fruit trees on their own roots, *Proc. Amer. Soc. Hort. Sci.*, 33:355–57. 1936.

22 Lincoln, F. B., Layering of root grafts—a ready method for obtaining self-rooted apple trees, *Proc. Amer. Soc. Hort. Sci.*, 35:419–22. 1938.

23 Micke, W. C., J. A. Beutel, and J. A. Yeager, Water base paints for sunburn protection of young fruit trees, *Calif. Agr.*, 20(7):7. 1966.

24 Moore, J. C., Propagation of chestnuts and camellias by nurse seed grafts, *Proc. Inter. Plant Prop. Soc.*, 13:141–43. 1963.

25 Park, Kyo S., Studies on juvenile tissue grafting of some special use trees, II. *Korean Jour. Bot.*, 11(3): 88–97. 1968.

26 Shippy, W. B., Influence of environment on the callusing of apple cuttings and grafts, *Amer. Jour. Bot.*, 17:290–327. 1930.

27 Sitton, B. G., and E. P. Akin, Grafting wax melter, *USDA Leaflet 202*, 1940.

28 Teuscher, H., Speeding production of hard-to-root conifers, *Amer. Nurs.*, 116(7):16. 1962.

29 Thompson, L. A., and C. O. Hesse, Some factors which may affect the choice of grafting compounds for top-working trees, *Proc. Amer. Soc. Hort. Sci.*, 56:213–16. 1950.

30 Thouin, A., Monographie des Greffes, ou description Technique (in Royal Hort. Soc. Lib., London), 1821.

31 Upshall, W. H., The stub graft as a supplement to budding in nursery practice, *Proc. Amer. Soc. Hort. Sci.*, 47:187–89. 1946.

32 Vöchting, H., Ueber Transplantation am Pflanzenkörper, 1892.

SUPPLEMENTARY READING

Baltet, C., *The Art of Grafting and Budding*, 6th ed. London: Crosby, Lockwood, 1910.

Banta, E. S., "Fruit Tree Propagation," *Ohio Agricultural Extension Bulletin 481* (1967).

Chandler, W. H., *Deciduous Orchards,* 3rd ed. Philadelphia: Lea & Febiger, 1957, Chap. 13.

————, *Evergreen Orchards,* 2nd ed. Philadelphia: Lea & Febiger, 1958.

Garner, R. J., *The Grafter's Handbook,* 3rd ed. London: Faber & Faber, Ltd., 1968.

"Grafting Fruit Trees," *Ministry of Agriculture, Fisheries, and Food Advisory Leaflet 326,* London (1965).

Hansen, C. J., and H. T. Hartmann, "Propagation of Temperate Zone Fruit Plants," *California Agricultural Experiment Station Circular 471* (rev.) (1966).

Snyder, J. C., and R. D. Bartram, "Grafting Fruit Trees," *Pacific Northwest Extension Publication (Idaho, Oregon, and Washington)* (1965).

Spangelo, L. P. S., R. Watkins, and E. J. Davies, "Fruit Tree Propagation," *Canada Department of Agriculture Publication 1289,* Ottawa (1968).

Way, R. D., F. G. Dennis, and R. M. Gilmer, "Propagating Fruit Trees in New York," *New York Agricultural Experiment Station Bulletin 817* (1967).

Techniques of Budding 13

In contrast to grafting, in which the scion consists of a short detached piece of stem tissue with several buds, budding utilizes only one bud and a small section of bark, with or without wood. Budding is often termed "bud grafting," since the physiological processes involved are the same as in grafting.

The commonly used budding methods depend upon the bark's *"slipping."* This term indicates the condition in which the bark can be easily separated from the wood. It denotes the period of year when the plant is in active growth, when the cambium cells are actively dividing, and newly formed tissues are easily torn as the bark is lifted from the wood. Beginning with new growth in the spring, this period should last until the plant ceases growth in the fall. However, adverse growing conditions, such as lack of water, defoliation, or low temperatures, may lead to a tightening of the bark and can seriously interfere with the budding operation. Getting the stock plants in the proper condition for budding is an important consideration. Of the commonly used methods described here, only one—the chip bud—can be done when the bark is not slipping.

The budding operation, particularly T-budding, can be performed more rapidly than the simplest method of grafting, some rose budders inserting as many as 2000 to 3000 or more T-buds a day if the tying is done by helpers. If performed under the proper conditions, the percentage of successful unions in T-budding is very high—90 to 100 percent. Budding is widely used in producing nursery stock of rose and fruit tree cultivars, where hundreds of thousands of individual plants are propagated each year. Therefore, for propagation operations involving large numbers of plants, where speed and low mortality are essential, budding upon selected rootstocks is likely to be chosen.

The use of budding is confined generally to young plants or the smaller branches of large plants where the buds can be inserted into shoots which are from $\frac{1}{4}$ to 1 in. in diameter. Topworking young trees by top-budding is quite successful. Here the buds are inserted in small, vigorously growing branches in the upper portion of the tree (see p. 448).

Budding may result in a stronger union, particularly during the first few years, than is obtained by some of the grafting methods, and thus the shoots are not as likely to blow out in strong winds. Budding makes more economical use of propagating wood than grafting, each bud potentially being capable of producing a new plant of the desired cultivar. This may be quite important if propagating wood is scarce. In addition, the techniques involved in budding are simple and can be easily performed by the amateur.

ROOTSTOCKS FOR BUDDING

In propagating nursery stock of the various fruit and ornamental species by budding, a rootstock plant is used. It should have the desired characteristics of vigor, growth habit, and disease resistance, as well as being easily propagated. This rootstock plant may be a rooted cutting, a rooted layer, or, more commonly, a seedling (see Chapter 7). Usually, one year's growth in the nursery row before budding is to be done is sufficient to produce a rootstock plant large enough to be budded, but seedlings of slow-growing species may require two seasons.

To produce nursery trees free of harmful pathogens (as viruses, fungi, or bacteria) it is essential that the rootstock plant, as well as the budwood, be free of such organisms. (See p. 315.)

TIME OF BUDDING—FALL, SPRING, OR JUNE

The important budding methods are used at seasons of the year when the stock plant is in active growth and the cambial cells are actively dividing so that the bark separates readily from the wood. It is also necessary that well-developed buds of the desired cultivar be available at the same time. These conditions exist for most plant species at three different times during the year. In the Northern Hemisphere, these periods are late July to early September (*fall budding*), March and April (*spring budding*), and late May and early June (*June budding*). In the Southern Hemisphere, similar periods would be late January to early March (*fall budding*), September and October (*spring budding*), and late November and early December (*"June" budding*).

Fall Budding

This is the most important time of budding in the propagation of fruit tree nursery stock, although actually the budding is done mostly in late summer rather than fall. The rootstock plants are usually large enough by late summer to accommodate the bud, and the plants are still actively growing, with the bark slipping easily. Once growth has stopped and the bark adheres tightly to the wood, budding can no longer be done.

In fall budding, the budsticks, consisting of the current season's shoots, are obtained at or near the time of budding. They should be vigorous and should contain vegetative or leaf buds (Figure 13–1). Short, slowly growing shoots on the outer portion of the tree should be avoided, because they may have chiefly flower buds rather than vegetative buds. Flower buds are usually round and plump, whereas leaf buds are smaller and pointed. Some species have mixed buds, the node containing both vegetative and flower buds. These are satisfactory for use in budding.

Every effort should be made to make sure that the trees from which the budsticks are obtained are free of any bacterial, fungus, or virus diseases. *Using infected budsticks can infect every budded nursery tree with the disease* (see p. 197).

As the budsticks are selected the leaves should be removed immediately, leaving only a short piece of the leaf stalk or petiole attached to the bud; this will aid in handling the bud later on. The budsticks should be kept from drying by wrapping in some material such as clean, moist burlap and keeping them in a cool, shady location until they are needed. The budsticks should be used promptly after cutting, although they can be stored for a short time if kept cool and moist. It is best, if possible, when a considerable amount of budding is being done, to collect the budsticks as they are being used, a day's supply at a time.

The best buds to use on the stick are usually those in the middle and basal portions. Buds on the succulent terminal portion of the shoot should be discarded. In certain species, such as the sweet cherry, buds on the basal portion of the shoots are flower buds, which of course should not be used.

In fall budding, after the buds have been inserted, there is nothing more to be done until the following spring. *Although eventually the rootstock is to be cut off above the bud, in no case should this be done im-*

Figure 13–1   In budding it is important that vegetative rather than flower buds be used. Vegetative buds are usually small and pointed, while flower buds are larger and more plump. Differences between vegetative and flower buds in three fruit species are illustrated here. *Left:* almond. The shoot on the left has primarily flower buds and should not be selected for budding. The shoot on the right has buds which are more suitable. *Center:* peach. The shoot on the right has excellent vegetative buds while those on the left shoot are mostly flower buds. *Right:* pear. All the buds on the shoot at the left are flower buds. Buds on the shoot at the right are good vegetative buds, suitable for budding.

mediately after the bud has been inserted. Healing of the bud piece to the stock is greatly facilitated by the normal movement of water and nutrients up and down the stem of the rootstock. This would, of course, be stopped if the top of the rootstock were cut off above the bud.

If the budding operation is done properly, the bud piece should unite with the stock in two to three weeks, depending upon the growing conditions. If the leaf stalk or petiole drops off cleanly next to the bud, this is a good indication that the bud has united, especially if the bark piece retains its normal light brown or green color and the bud stays plump. On the other hand, if the leaf petiole does not drop off cleanly but adheres tightly and starts to shrivel and darken, while the bark piece also commences to turn black, it is likely that the operation has failed. If the bark of the rootstock is still slipping easily and budwood is still available, there may be time for the budding to be repeated.

Even though the bud union has healed, in most deciduous species the bud usually does not grow or "push out" in the fall, since it is in a physiological rest period. It remains just as it is until spring, at which time the chilling winter temperatures have overcome the rest influence and the bud is ready to grow. There are some exceptions to this; for example, in fall budding of maples, roses, honey locust, and certain other plants, some of the buds may start growth in the fall. In northern areas, if such fall-forced buds do not start early enough for the shoots to mature before cold weather starts, they are likely to be winter-killed.

In the spring, just before new growth begins, the rootstock is cut off immediately above the bud. It is desirable to make a sloping cut, slanting away from the bud. Although this cut may be waxed, it is usually not essential unless the stock is large in diameter. Cutting back the rootstock forces the inserted bud into growth. In citrus budding, it is a common practice to partially cut the stock above the bud and to lop or bend it over away from the bud. The leaves of the stock still furnish the roots with some nutrients, but the partial cutting forces the bud into growth. After the new shoot from the bud is well established, the top is completely removed.

In northern regions, fall-inserted buds are sometimes covered with soil during the winter until danger of frost has passed and are then uncovered and topped-back in late spring.

Where strong winds occur, and in species in which the new shoots grow vigorously, support for the newly developing shoot may be necessary. One practice sometimes followed is to cut off the rootstock several inches above the bud, using this projecting stub as a support on which to tie the tender young shoot arising from the bud. This stub is removed after the shoot has become well established. In another procedure, stakes may be driven into the ground next to the stock to which the developing shoot is tied at intervals during its growth.

Cutting back to force the main bud to grow also forces many latent buds on the rootstock into growth. These must be rubbed off as soon as they appear, or they will soon choke out the desired inserted bud. It may be necessary to go over the budded plants several times before these

Figure 13–2   Row of fall-budded nursery trees about one year after budding. The inserted buds have grown through the spring and summer. The roots have grown through two seasons. The tops of the rootstocks were cut off above the inserted buds the spring following budding.

"sprouts" stop appearing. Nurserymen refer to this procedure as "suckering."

As shown in Figure 13–2, the shoot arising from the inserted bud becomes the top portion of the plant. After one season's growth in the nursery with favorable conditions of soil, water and nutrients, temperature, and insect and disease control, this shoot will have developed sufficiently to enable the plant to be dug and moved to its permanent location during the following dormant season. Such a tree would have a one-year-old top and a two- or perhaps three-year-old root, but it is still considered a "yearling" tree. If the top makes insufficient growth the first year, it can be allowed to grow a second year, and is then known as a two-year-old tree. However, abnormal, slow-growing trees should be discarded.

Spring Budding

This method is similar to fall budding except that insertion of the bud takes place the following spring as soon as active growth of the rootstock begins and the bark separates easily from the wood. The period for successful spring budding is limited, and budding should be completed before the rootstocks have made much new growth.

Budsticks are chosen from the same type of shoots—in regard to vigor of growth and type of buds—that would be used in fall budding, except that they are not collected until the dormant season the following winter. The leaves would, of course, have fallen by this time, and the buds would have experienced sufficient chilling to overcome their rest period. Budwood must be collected while it is still dormant—before there is any evidence of the buds swelling. Since the buds must be dormant when they are inserted, and since the rootstocks must be in active growth, it is necessary in spring budding that the budsticks be gathered some time in advance of the time of budding and stored at temperatures (32° to 40° F; to 0° to 4° C) to hold the buds dormant. The budsticks should be wrapped in bundles with damp peat moss or some similar material to prevent drying out.

In spring budding, the actual budding operation should be done just as soon as the bark on the rootstock slips easily. Then, about two weeks after budding, when the bud unions have healed, the top of the stock must be cut off above the bud to force the inserted bud into active growth. At the same time, latent buds on the rootstock begin to grow and should be removed. Sometimes it is helpful to permit such shoots from the rootstock to develop to some extent to prevent sunburn and help nourish the plant. They must be held in check, however, and eventually removed.

Although the new shoot from the inserted bud gets a later start in spring budding than in fall budding, spring buds will usually develop rapidly enough, if growing conditions are favorable, to make a satisfactory top by fall. Fall budding, however, is to be preferred for several reasons: the higher temperatures at that time promote more certain healing of the union, the budding season is longer, there is no necessity to store the budsticks, the inserted buds start growth earlier in the spring, and the pressure of other work is usually not so great for nurserymen in the late summer as in the spring. Spring budding is used sometimes on rootstocks which were fall-budded but on which the buds failed to take.

June Budding

June budding is used to obtain a "one-year-old" budded tree in a single growing season. Budding is done in the early part of the growing season and the inserted bud forced into growth immediately during the same season. As a method of nursery propagation, June budding is confined to regions which have a relatively long growing season—in the U.S. this includes California and the southern states. In the propagation of fruit trees, June budding is used mostly in producing such stone fruits as peaches, nectarines, apricots, almonds, and plums. Peach seedlings are generally used as the rootstock, but almond seedlings and plum hardwood cuttings can also be used. Budding is done by the T-bud method. If seeds are planted in the fall, or stratified seeds as early as possible in the spring, the seedlings usually attain sufficient size (12 in. high and at least ⅛ in. in diameter) to be budded by mid-May or early June (in the Northern Hemisphere). Preferably, June budding should not be done much after mid-June, or a nursery tree of satisfactory size will not be obtained by fall. June-budded trees are not as large by the end of the growing season as those propagated by fall or spring budding, but they are of sufficient size (about ⅜ in. to ⅝ in. caliper and 3 to 5 ft tall) to produce entirely satisfactory trees.

Budwood used in June budding consists of current season's growth, that is, of new shoots which have developed since growth started in the spring.

By late May or early June, these shoots will usually have grown sufficiently to have a well-developed bud in the axil of each leaf. At this time of year these buds will not have entered the rest period, so when they are used in budding they continue their growth on through the summer, producing the top portion of the budded seedling.

For June-budded trees handling subsequent to the actual operation of budding is somewhat more exacting than for fall- or spring-budded trees. The rootstocks are smaller and have less stored food than those used in fall or spring budding. The object behind the following procedures—shown in Figure 13–3—is to keep the rootstock (and later the budded top) actively and continuously growing so as to allow no check in growth, while at the same time changing the seedling shoot to a budded top. The bud should be inserted high enough (5 to 6 in.) on the stem so that a number of leaves— at least three or four—can be retained below the bud. The method of T-budding with the "wood out" described on p. 437 should be used. Healing of the inserted bud should be very rapid at this time of year, since temperatures are relatively high, and rapidly growing, succulent plant parts are used. By four days after budding, healing should have started, and the top of the rootstock can be cut back somewhat—2 to 5 in. above the bud —leaving at least one leaf above the bud and several below it. This opera-

Figure 13-3   It is important in June budding that the stock be cut back to the bud properly. *Far left:* the bud is inserted high enough on the stock so that there are several leaves below the bud. *Center left:* three or four days after budding, the stock is partially cut back—2 to 5 in. above the bud. *Center:* ten days to two weeks after budding, the stock is completely removed just above the bud. *Center right:* this forces the bud and other buds on the stock into growth; the latter must subsequently be removed. *Far right:* appearance of the budded tree after the new shoot has made considerable growth.

tion will force the inserted bud into growth and will check terminal growth of the rootstock. It will also stimulate shoot growth from basal buds of the rootstock, which will produce additional leaf area. This continuous leaf area is necessary so that there always will be enough leaves to keep manufacturing food for the small plant. Ten days to two weeks after budding, the rootstock can be cut back to the bud, which should be starting to grow. If the budding rubber has not broken, it should be cut at this time. Other shoots arising from the rootstock should be headed back to retard their growth. After the inserted bud grows and develops a substantial leaf area, it can supply the plant with the necessary nutrients. By the time the shoot from the inserted bud has grown 8 to 12 in. high, it should have enough leaves so that all other shoots and leaves can be removed. Later inspections should be made to remove any shoots arising from the rootstock below the budded shoot.

Another method which works well is to partially cut the stock just above the bud and break it over. Nutrients are still able to pass from the top of the stock to the roots, but this partial blocking forces the inserted bud into growth.

June budding is often of considerable value to nurserymen who find that their regular supply of fall- or spring-budded trees is insufficient to meet their expected demand. By this method they can, as late as the end of June, still propagate trees to be ready by fall, provided that they have a supply of rootstock plants large enough to bud (5).

The steps in fall, spring, and June budding are compared in Figure 13-4.

METHODS OF BUDDING

T-Budding (Shield Budding)

This method is known by both names, the "T-bud" designation arising from the T-like appearance of the cut in the stock, whereas the "shield bud" name is derived from the shield-like appearance of the bud piece when it is ready for insertion in the stock.

T-budding is by far the most common method of budding and is widely used by nurserymen in propagating nursery stock of most fruit tree species, roses, and some ornamental shrubs. Its use is generally limited to stocks which are about $\frac{1}{4}$ to 1 in. in diameter, with fairly thin bark, and which are actively growing so that the bark will separate readily from the wood. If the bark is so tight on the wood that it has to be forcibly pried loose, the chances of the bud healing successfully are rather remote. The operation should then be delayed until the bark is slipping easily.

The bud is inserted into the stock 2 to 10 in. above the soil level in a smooth bark surface. There are different opinions as to which is the proper side of the stock in which to insert the bud. If extreme weather conditions are likely to occur during the critical healing period just following budding, it may be desirable to place the bud on the side of the stock on which as much protection as possible may be obtained. Some

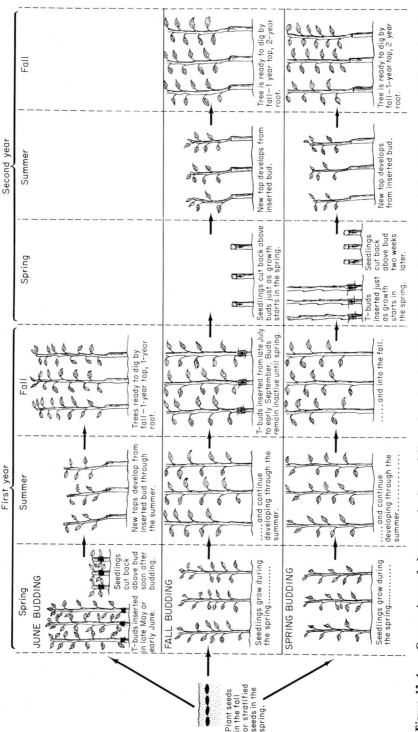

Figure 13-4 Comparison of the steps in June, fall, and spring budding. The actual techniques in budding are not difficult, but it is very important that the various operations be done at the proper time.

believe that if the bud is placed on the windward side, there is less chance of the young shoot breaking off. Otherwise, it probably makes little difference where the bud is inserted, the convenience of the operator and the location of the smoothest bark being the controlling factors. When rows of closely planted rootstocks are budded, it is more convenient to have all the buds on the same side for later inspection and manipulations.

The cuts to be made in the stock plant are illustrated in Figure 13–5. There are various modifications of this technique; most budders prefer to make the vertical cut first, then the horizontal crosscut at the top of the T. As the horizontal cut is made, the knife is given a twist to throw open the flaps of bark for insertion of the bud. It is important that neither the vertical nor horizontal cut be made longer than necessary, because this requires additional tying later to close the cuts.

After the proper cuts are made in the stock and the incision is ready to receive the bud, the shield piece is cut out of the budstick.

To remove the shield of bark containing the bud, a slicing cut is started at a point on the stem about ½ in. below the bud, continuing under and about an inch above the bud. The shield piece should be as thin as possible but still thick enough to have some rigidity. A second horizontal cut is then made ½ to ¾ in. above the bud, thus permitting the removal of the shield piece.

There are two methods of preparing the shield—with the "wood in" or with the "wood out." This refers to the little sliver of wood just under the bark of the shield piece and which will remain attached to it if the second horizontal cut is deep and goes through the bark and wood, joining the first slicing cut. Some professional budders believe it is best to remove this sliver of wood, but others retain it. In budding certain species, however, such as maples and walnuts, much better success is usually obtained with "de-wooded" buds. If it is desired to prepare the shield with the wood out, the second horizontal cut should be just deep enough to go through the bark and not through the wood. Then if the bark is slipping easily, the bark shield can be snapped loose from the wood (which still remains attached to the budstick) by pressing it against the budstick and sliding it sideways. A small core of wood comprising the vascular tissues supplying the bud is present, and this should remain in the bud, rather than adhering to the wood and leaving a hole in the bud. If the shield is pulled outward rather than being slid sideways from the wood, this core usually pulls out of the bud, eliminating the chances of success. In June budding of fruit trees, the shield piece is usually prepared with the wood out. In most other instances, however, the wood is left in. In spring budding, using dormant budwood, this sliver of wood is tightly attached to the bark and cannot be removed.

The next step is the insertion of the shield piece containing the bud into the incision in the stock plant. The shield is pushed under the two raised flaps of bark until its upper horizontal cut matches the same cut on the stock. The shield should fit snugly in place, well covered by the two flaps of bark, but with the bud itself exposed.

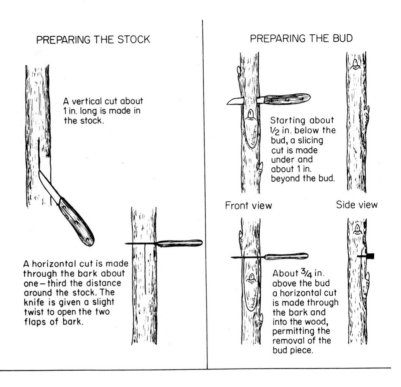

PREPARING THE STOCK

A vertical cut about 1 in. long is made in the stock.

A horizontal cut is made through the bark about one – third the distance around the stock. The knife is given a slight twist to open the two flaps of bark.

PREPARING THE BUD

Starting about 1/2 in. below the bud, a slicing cut is made under and about 1 in. beyond the bud.

Front view

Side view

About 3/4 in. above the bud a horizontal cut is made through the bark and into the wood, permitting the removal of the bud piece.

INSERTING THE BUD INTO THE STOCK

The shield piece is inserted by pushing it downward under the two flaps of bark

—until the horizontal cuts on the shield and the stock are even.

The bud union is then tightly tied with some wrapping material.

Figure 13–5   Steps in making the T-bud (shield bud). There are several variations of the procedure shown here.

Figure 13–6 Wrapping materials used in budding. *Left:* nurserymen's adhesive tape. *Center:* plastic tape. *Top right:* raffia. *Below right:* budding rubber strips.

No waxing is necessary, but the bud union must be wrapped, using materials such as those shown in Figure 13–6, to hold the two components firmly together until healing is completed. Rubber budding strips, especially made for wrapping, are widely used for this purpose (Figure 13–7). Their elasticity provides sufficient pressure to hold the bud securely in place. The rubber, being exposed to the sun and air, usually deteriorates, breaks, and drops off after several weeks, at which time the bud should be healed in place. If the budding rubber is covered with soil, the rate of deterioration will be much slower. This material has the advantage of eliminating cutting the wrapping ties, which can be a costly operation if many thousands of plants have been budded. The rubber will expand as the rootstock grows, and thus there is little danger of constriction.

In tying the bud, the ends of the budding rubbers are held in place by inserting them under the adjacent turn. The bud itself should not be

Figure 13–7 Steps in the development of a T-bud. *Left:* bud after being inserted and wrapped. *Center:* bud has healed in place, the budding rubber has dropped off, and the stock has been cut back above the bud. *Right:* shoot development from the inserted bud. All buds arising from the stock have been rubbed off.

covered. The amount of tension given the budding rubber is quite important. It should not be too loose, or there will be too little pressure holding the bud in place. On the other hand, if the rubber is stretched extremely tight, it may be so thin that it will deteriorate rapidly and break too soon—before the bud union has taken place. Often the tying is done from the top down to avoid forcing the bud out through the horizontal cut.

Raffia (fiberlike leaf segments of certain *Raphia* species) has been widely used for wrapping buds. This material is soaked in water overnight before it is used so that it will be flexible. Raffia must be cut later—about ten days after budding—to prevent constriction at the bud union as the plant grows. If such nonelastic ties are not promptly cut, the resultant constriction can have a very adverse affect on subsequent growth of the bud (7).

Plastic ties of polyvinyl chloride (PVC) film, $3/8$ in. wide, are quite useful for budding. Such material is moisture-proof and elastic, and since it is transparent, it permits inspection of the buds after covering (*1*). The next step, the cutting back of the rootstock, is described on p. 429.

Inverted T-Budding

In rainy localities, water running down the stem of the rootstock is believed to enter the T-cut, soak under the bark, and cause decay of the shield piece. Under such conditions an inverted T-bud may give better results, since it is more likely to shed excess water. In citrus budding, the inverted T method is widely used, even though the conventional method also gives excellent results. In species which bleed badly during budding, such as chestnuts, the inverted T-bud allows better drainage and better healing. Strong supporters of both conventional and inverted T-budding can be found, and in a given locality the usage of either with a given species tends to become traditional.

The techniques of the inverted T-bud method are the same as those already described, except that the incision in the stock has the transverse cut at the bottom rather than at the top of the vertical cut, and in removing the shield piece from the budstick the knife starts above the bud and cuts downward below it. The shield is removed by making the transverse cut $1/2$ to $3/4$ in. below the bud. The shield piece containing the bud is inserted into the lower part of the incision and pushed upward until the transverse cut of the shield meets that made in the stock.

It is important in using this inverted T-bud method that a normally oriented shield bud piece should not be inserted into an inverted incision in the stock. The bud would then have a reversed polarity. Although such upside down buds do live and grow, at least in some species, their use may not result in the expected shoot development (see p. 335).

Patch Budding

The distinguishing feature of patch budding and related methods is that a rectangular patch of bark is removed completely from the stock and replaced with a patch of bark of the same size containing a bud of the cultivar to be propagated.

Patch budding is somewhat slower and more difficult to perform than T-budding, but it is widely and successfully used on thick-barked species, such as walnuts and pecans, in which T-budding gives poor results, presumably owing to the poor fit around the margins of the bud. Patch budding, or one of its modifications, is also extensively used in propagating various tropical species, such as the rubber tree (*Hevea brasiliensis*).

Patch budding requires that the bark of both the stock and budstick be slipping easily. It is usually done in late summer or early fall, but can also be done in the spring. In propagating nursery stock, the diameter of the rootstock and the budstick should preferably be about the same, from ½ to 1 in. Although the budstick should not be much larger than about 1 in. in diameter, the patch can be inserted successfully into stocks as large as 4 in. in diameter, although adequate healing of such large stubs may be a problem.

Special knives (see Figure 13–8) have been devised to remove the bark pieces from the stock and the budstick. Some type of double-bladed knife that will make two transverse parallel cuts 1 to 1⅜ in. apart is necessary. These cuts, about an inch in length, are made through the bark to the wood in a smooth area of the rootstock several inches above the ground. Then the two transverse cuts are connected at each side by vertical cuts made with a single-bladed knife.

The patch of bark containing the bud is cut from the budstick in the same manner in which the bark patch is removed from the stock. Using the same two-bladed knife, the budder makes two transverse cuts through the bark, one above and one below the bud. Then two vertical cuts are

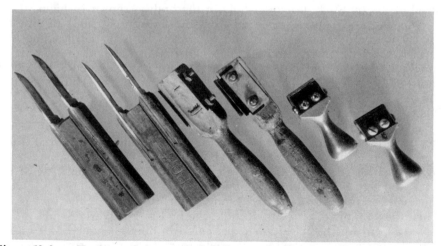

Figure 13–8   Tools used in patch budding and related methods. *Left:* manufactured double-bladed knives. *Center:* double-bladed knives made with two razor blades. *Right:* manufactured patch-budding tools consisting of four rectangular cutting blades.

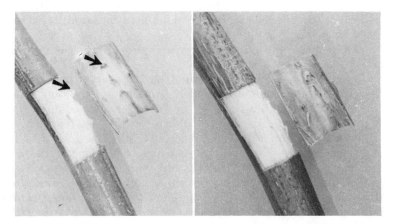

Figure 13-9   Removing the bud patch from the budstick in patch budding. *Left:* incorrect. The core of wood in the bud, comprising the vascular tissues, has broken off leaving a hole in the bud. Such a bud is not likely to grow. *Right:* correct. The patch was pushed off sideways, and the core of wood has remained inside the bud.

made on each side of the bud so that the bark piece will be about an inch wide. The bark piece containing the bud is now ready to be removed. It is important that it be slid off sideways rather than being lifted or pulled off. There is a small core of wood, the bud trace, which must remain inside the bud if a successful "take" is to be obtained. By sliding the bark patch to one side, this core is broken off, and it stays in the bud. If the bud patch is lifted off, this core of wood is likely to remain attached to the wood of the budstick, leaving a hole in the bud, as shown in Figure 13-9.

After the bud patch is removed from the budstick, it must be inserted immediately on the stock, which should already be prepared, needing only to have the bark piece removed. The patch from the budstick should fit snugly at the top and bottom into the opening in the stock, since both transverse cuts were made with the same knife. It is more important that the bark piece fit tightly at top and bottom than that it fit along the sides. These procedures are illustrated in Figures 13-10 and 13-11.

The inserted patch is now ready to be wrapped. Often the bark of the stock will be thicker than the bark of the inserted bud patch so that, upon wrapping, it is impossible for the wrapping material to hold the bud patch tightly against the stock. In this case it is necessary that the bark of the stock be pared down around the bud patch so that it will be of the same thickness, or preferably slightly thinner, than the bark of the bud patch. Then the wrapping material will hold the bud patch tightly in place.

In cases in which difficulty is experienced in obtaining successful unions with the patch bud method, it may help to make the four cuts of the rectangle in the stock one to three weeks ahead of the time when the actual budding is to be done. This bark patch is not removed from the stock, however, until the patch containing the bud is ready to be inserted. When

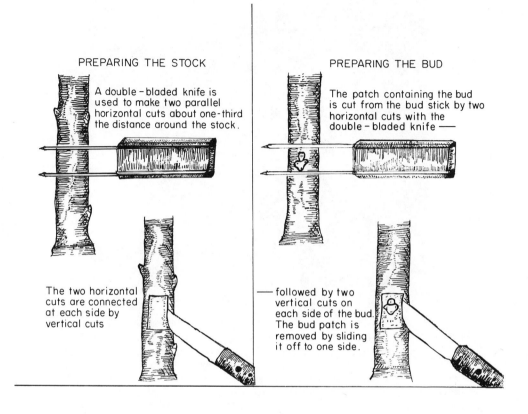

PREPARING THE STOCK

A double-bladed knife is used to make two parallel horizontal cuts about one-third the distance around the stock.

The two horizontal cuts are connected at each side by vertical cuts

PREPARING THE BUD

The patch containing the bud is cut from the bud stick by two horizontal cuts with the double-bladed knife —

— followed by two vertical cuts on each side of the bud The bud patch is removed by sliding it off to one side.

INSERTING THE BUD INTO THE STOCK

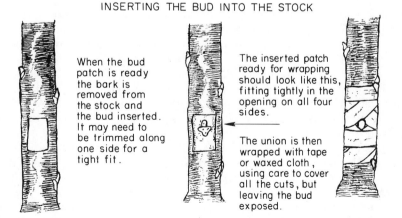

When the bud patch is ready the bark is removed from the stock and the bud inserted. It may need to be trimmed along one side for a tight fit.

The inserted patch ready for wrapping should look like this, fitting tightly in the opening on all four sides.

The union is then wrapped with tape or waxed cloth, using care to cover all the cuts, but leaving the bud exposed.

Figure 13–10   Steps in making the patch bud. This method is widely used for propagating thick-barked plants.

Figure 13–11   Steps in the development of a patch bud. *Top left:* patch bud after being inserted and wrapped with tape. *Top right:* after about ten days, the tape is slit along the back side to release any constricting pressure. *Below left:* the tape is completely removed after about three weeks, at which time the patch should be healed in place. *Below right:* cutting back the stock above the bud forces it into growth.

the cuts are made ahead of time, the wounding causes the callusing process to start, so that when the new bark patch is inserted, it heals very rapidly.

In wrapping the patch bud, a material should be used which not only will hold the bark tightly in place but will cover all the cut surfaces to prevent the entrance of air under the patch, with subsequent drying and death of the tissues. The bud itself must not be covered during wrapping. The most satisfactory material is nurserymen's adhesive tape. Another method is to tie the bud patch with heavy cotton string and cover the cut edges with grafting wax.

It is important in patch budding, especially with walnuts, that the wrapping not be allowed to cause a constriction at the bud union. When the stock is rapidly growing, it is necessary to cut the tape about ten days after budding. A single vertical knife cut on the side opposite the bud is sufficient, but care should be taken not to cut into the bark. Sometimes an extra strip of tape is placed vertically before wrapping to protect the bark when the tape is cut. The cut tape should not be pulled off.

Patch budding is best performed in late summer when both the seedling stock and the source of budwood are growing rapidly and their bark slipping easily. The budsticks for patch budding done at this time should have the leaf blades cut two to three weeks before they are taken from the tree. The petiole or leaf stalk is left attached to the base of the bud, but by the time the budstick is taken this petiole has dropped off or is easily removed.

Patch budding can also be done in the spring after new growth has appeared on the stocks and it has been determined that the bark is slipping. There is a problem, however, in obtaining satisfactory buds to use at this time of year, since it is necessary that the bark of the budstick separate readily from the wood. At the same time, the buds should not be starting to swell. There are two methods by which satisfactory buds can be obtained for patch budding in the spring. In one, the budsticks are selected during the dormant winter period and stored at low temperatures (about 36° F; 2° C) and wrapped in moist sphagnum or peat moss to prevent their drying out. Then, about three weeks before the time the spring budding is to be done, they are brought into a warm room and set with their bases in a container of water, or they may be left packed in damp peat moss. The increased temperature will cause the cambium layer to become active, and soon the bark will slip sufficiently for the buds to be used. Although a few of the more terminal buds on each stick may start swelling in this time and cannot be used, there should be a number of buds in a satisfactory condition.

The second method of obtaining buds for spring patch budding is to take them directly from the tree which is the source of the budwood, but at the time the budding is to be done. If the trees are inspected carefully, it will be seen that not all of the buds start pushing at once. The terminal

Figure 13–12   *Left:* shoot development from patch budding operation shown in Figure 13–11 after two years' growth. *Right:* strong, smooth union developed after 14 years. *Juglans regia* 'Hartley' on Paradox rootstock *(Juglans regia* × *J. hindsii).*

ones are usually more advanced than the basal ones. There is a period when the bark is slipping easily throughout the shoot containing the desired buds, but when only a few of the buds have developed so far that they cannot be used. The remaining buds, which are still dormant but upon bark that can be removed readily, may be taken and used immediately for budding. It is easier to obtain suitable buds from young trees which made vigorous shoot growth the previous year than it is from old trees. When the budding is done will be governed then by the stage of development of the buds. This will vary considerably with the species and cultivar being used. Stage of development of the rootstock is not as critical. Its bark must be slipping well, and the budding should be done before the stock plant has made much new growth.

Flute Budding

Flute budding is similar to patch budding, except that the patch of bark removed from the rootstock almost completely encircles it, leaving a narrow connection (about one-eighth of the circumference) between the upper and lower parts of the stock. In taking the bud patch from the budstick, a two-bladed knife is used, with the two transverse cuts completely encircling the budstick. A single vertical cut connects the two horizontal cuts, permitting the bud patch to be removed. In fitting this bud patch to the stock, it may be necessary to shorten its circumference by a vertical cut to remove the surplus amount of bark. If the bud patch fails to unite, the narrow connecting strip of bark on the stock keeps the top of the stock alive (Figure 13–13).

Ring or Annular Budding

By this method, a complete ring of bark is removed from the stock and a complete ring from the budstick. In order for the two to match, the size of the stock and that of the budstick should be about the same. Since the stock is completely girdled, if the bud patch fails to heal in, the stock above the ring may eventually die. This method is not as widely used as the ordinary patch bud, since it has no particular advantages and is rather cumbersome to perform (Figure 13–13).

I-Budding

In this type of budding, the bud patch is cut just as for patch budding, that is, in the form of a rectangle or square. Then, with the same parallel-bladed knife, two transverse cuts are made through the bark of the stock. These are joined at their centers by a single vertical cut to produce the figure I. The two flaps of bark can then be raised for insertion of the bud patch beneath them. It may make a better fit to slant the side edges of the bud patch. In tying the I-bud, care should be taken to see that the bud patch does not buckle upward and fail to touch the stock.

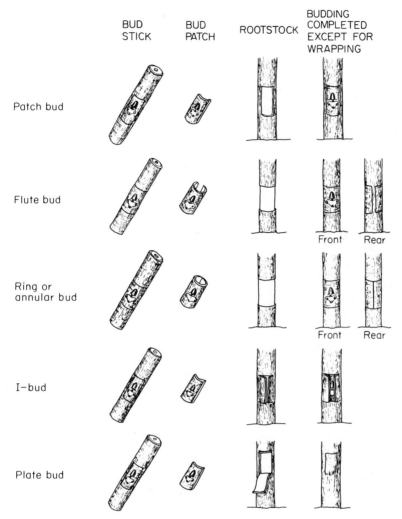

	BUD STICK	BUD PATCH	ROOTSTOCK	BUDDING COMPLETED EXCEPT FOR WRAPPING

Patch bud

Flute bud — Front — Rear

Ring or annular bud — Front — Rear

I–bud

Plate bud

Figure 13–13   There are many variations of the patch bud, some of which are shown here. The naming of these types is somewhat confused; the most generally accepted names are given here.

This method should be considered for use when the bark of the stock is much thicker than that of the budstick. In such cases, if the patch bud were used, considerable paring down of the bark of the stock around the patch would be necessary. This operation is not necessary in the I-bud method (Figure 13–13).

Seedling growth of some trees, such as the pecan, is so slow that several growing seasons may be required before the trees are large enough to be patch budded. It has been shown experimentally that several repeated applications of gibberellin in a lanolin paste to the stems beginning ten days after seedling emergence so stimulated stem growth that budding could be done three months later (*8*).

Chip Budding

This method can be used at times when the bark is not slipping, i.e., early in the spring before growth starts or during the summer when active growth stops prematurely owing to lack of water or some other cause. It is generally used with small material, $\frac{1}{2}$ to 1 in. in diameter. Chip budding is not as fast or simple as T-budding and is not likely to be used if conditions are favorable for T-budding. Chip budding in the fall has given excellent results in budding grape cultivars on phylloxera or nematode-resistant rootstocks (*3, 4*). It is not commonly used on deciduous tree fruits.

As illustrated in Figure 13–14, a chip of bark is removed from a smooth place between nodes near the base of the stock and replaced by another chip of the same size and shape from the budstick which contains a bud of the desired cultivar. The chips in both stock and budstick are cut out in the same manner. The first cut is made just below the bud and down into the wood at an angle of about 45 deg. The second cut is started about $\frac{1}{2}$ in. above the bud and goes inward and downward behind the bud until it intersects the first cut. The order of making these two cuts may be reversed. The chip is removed from the stock and replaced by the one from the budstick. If they have both been cut to the same size and shape—as they should be—a good fit is obtained. It is important that the cambium layer of the bud piece be placed so as to concide with that of the stock, preferably on both sides of the stem, but at least on one side.

There are no protective flaps of bark to prevent the bud piece from drying out, as there are in T-budding. It is very important, then, that the chip bud be wrapped to seal the cut edges as well as to hold the bud piece tightly into the stock. Nurserymen's adhesive tape works very well for this, although string can be used if all the cut edges are covered with grafting wax. If the bud is inserted into the stock close to the ground level, as it is in grape propagation, it is sufficient to wrap the bud with budding rubber and cover the whole bud union immediately with several inches of finely pulverized moist soil, which can be removed after the bud has united. Budding rubbers in the soil are slow to disintegrate, so they may need to be cut or removed before constriction occurs.

In chip budding, as in the other methods, the stock is not cut back above the bud until the union is completed. If the chip bud is inserted in the fall, the stock is cut back just as growth starts the next spring. If the budding is done in the spring, the stock is cut back about ten days after the bud has been inserted.

TOP-BUDDING

In young trees when there is an ample supply of vigorous shoots at a height of 4 to 6 ft, top-budding provides a fast and certain method of top-

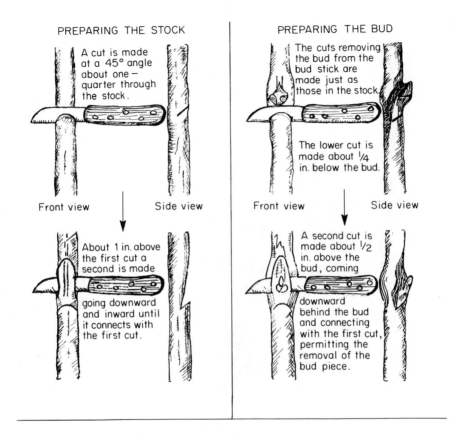

PREPARING THE STOCK

A cut is made at a 45° angle about one – quarter through the stock.

Front view   Side view

About 1 in. above the first cut a second is made going downward and inward until it connects with the first cut.

PREPARING THE BUD

The cuts removing the bud from the bud stick are made just as those in the stock.

The lower cut is made about ¼ in. below the bud.

Front view   Side view

A second cut is made about ½ in. above the bud, coming downward behind the bud and connecting with the first cut, permitting the removal of the bud piece.

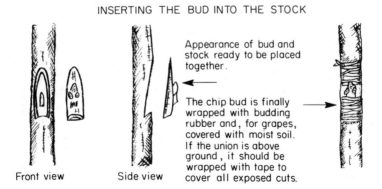

INSERTING THE BUD INTO THE STOCK

Appearance of bud and stock ready to be placed together.

The chip bud is finally wrapped with budding rubber and, for grapes, covered with moist soil. If the union is above ground, it should be wrapped with tape to cover all exposed cuts.

Front view   Side view

Figure 13–14   The chip bud is really a form of grafting, a variation of the side-veneer graft, with the scion reduced to a small piece of wood containing only a single bud. Chip budding is widely used in propagating Vinifera grapes on nematode and phylloxera resistant rootstocks.

Figure 13–15 Topworking a young tree by top-budding. T-buds were inserted in the positions shown by arrows and have grown for one season. All other shoots have been removed. From L. H. Day, "Apple, Quince, and Pear Rootstocks in California," *Calif. Agr. Exp. Sta. Bul. 700.*

working (Figure 13–15). It can be used in older trees, too, if they are cut back rather severely the year before to provide a quantity of vigorous watersprout shoots fairly close to the ground.

Depending upon the size of the tree, 10 to 15 buds are placed in vigorously growing branches $\frac{1}{4}$ to $\frac{3}{4}$ in. in diameter in the upper portion of the tree—about shoulder height. Although a number of buds could be placed in a single branch, usually only one will be saved to develop into secondary branches, which will then form the permanent new top of the tree. The T-bud method is used on thin-barked species, and the patch bud on those with thick bark.

Top-budding is usually done in midsummer, as soon as well-matured budwood can be obtained and while the stock tree is still in active growth with the bark slipping easily. Orchard trees generally stop growth earlier in the season than young nursery trees; therefore the budding must be done earlier. When top-budding is done at this time of year, the buds usually remain inactive until the following spring. At that time, just as vegetative growth starts, the stock branches are cut back just above the buds. This forces the buds into active growth, and they should develop into good-sized branches by the end of the summer. At the time the shoots are cut back to the buds, all other unbudded branches should be removed at the trunk. It is important that the trees be inspected carefully through the summer and any shoots removed that are arising from any but the inserted buds.

Top-budding can also be done in the spring just as the tree to be top-worked is starting active growth and the bark is slipping easily. The techniques for spring budding of nursery stock, described on p. 432, are used. Although not commonly done, June budding could be used for top-budding with the methods described on p. 433.

DOUBLE-WORKING BY BUDDING

In propagating nursery trees, some of the budding methods can be used in developing double-worked trees (see p. 413). As shown in Figure 13–16, the intermediate stock can be budded on the rootstocks; then the following year the desired cultivar is budded on the interstock. Although quite effective, this is a rather lengthy process, taking three years. As illustrated in Figures 13–17 and 13–18, it is possible to develop a double-worked tree in one operation in one year by the double-shield bud method. A T-bud is used, but just under and below it a budless shield piece of the desired interstock is inserted.

MICRO-BUDDING

This type of budding is used successfully in propagating citrus trees and probably could be utilized also for other tree and shrub species. It has been of commercial importance in the citrus districts of southeastern Australia (9). Micro-budding is similar to ordinary T-budding, except that the bud piece is reduced to a very small size. The leaf petiole is cut off just above the bud, and then the bud is removed from the budstick by a flat cut just underneath the bud, with a razor-sharp knife. Only the bud itself and a small piece of wood under it are used. In the stock an inverted

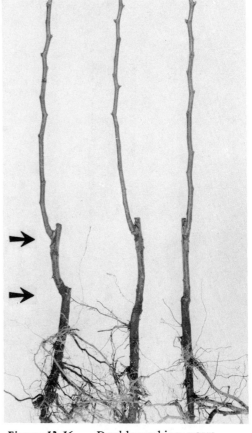

Figure 13–16   Double-working pears by budding. At the bottom arrow, the 'Old Home' pear was T-budded on a rooted quince cutting. A year later, at the top arrow, the 'Bartlett' cultivar was budded on the 'Old Home.' Three years were required to produce these nursery trees.

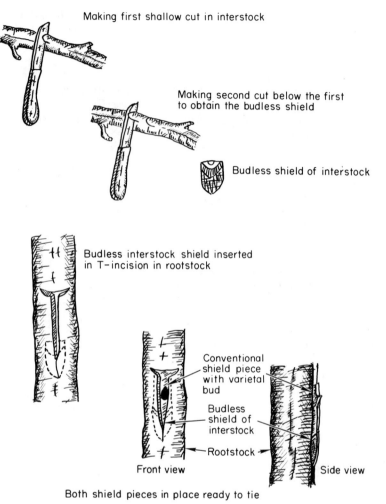

Making first shallow cut in interstock

Making second cut below the first to obtain the budless shield

Budless shield of interstock

Budless interstock shield inserted in T–incision in rootstock

Conventional shield piece with varietal bud

Budless shield of interstock

Rootstock

Front view

Side view

Both shield pieces in place ready to tie

Figure 13–17   Double-shielding budding is double-working by budding in one operation. In Garner's method *(2)*, as illustrated here, the intermediate stock is reduced to a small, budless shield piece inserted under and just below the regular shield piece used in T-budding.

T-cut is made, and the micro-bud is slipped into this, right side up. The entire T-cut, including the bud, is covered with polyvinyl chloride (PVC) plastic budding tape (3/8 in. in width and 0.002 in. thick). The tape is allowed to remain for 10 to 14 days for spring budding and three weeks for fall budding, after which it is removed by cutting with a knife. By this time the buds should have healed in place; subsequent handling is the same as for conventional T-budding.

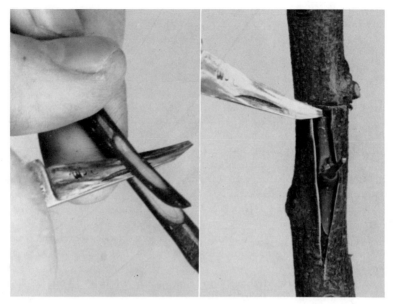

Figure 13-18   Steps in the double-budding technique of Nicolin *(6). Left:* cutting out the oval bark piece of the mutually compatible interstock variety. *Right:* placing the cultivar bud shield on top of the previously inserted oval bark piece. The budding operation is now ready for tying with a budding rubber. From K. D. Brase and R. D. Way, "Rootstocks and Methods Used for Dwarfing Fruit Trees," *N. Y. State Agr. Exp. Sta. Bul. 783.* Courtesy Prof. Karl Brase.

REFERENCES

1   Bryden, J. D., Use of plastic ties in fruit tree budding, *Agr. Gaz. New S. Wales (Austral.),* 68:87–88. 1957.

2   Garner, R. J., Double-working pears at budding time, *Ann. Rpt. East Malling Res. Sta. for 1952,* pp. 174–75. 1953.

3   Harmon, F. N., and J. H. Weinberger, The chip-bud method of propagating Vinifera grape varieties on rootstocks, *USDA Leaflet 513,* 1969.

4   Lider, L. A., Field budding and the care of the budded grapevine, *Calif. Agr. Ext. Ser. Leaflet 153,* 1963.

5   Mertz, W., Deciduous June-bud fruit trees, *Proc. Int. Plant Prop. Soc.,* 14:255–59. 1964.

6   Nicolin, P., Nicolieren, a new method of grafting, *Deutsche Baumschule,* 5:186–87. 1953.

7   Smith, N. G., R. J. Garner, and W. S. Rogers, Delayed growth of apple scions in relation to early budding, bud constriction, and some other factors, *An. Rpt. East Malling Res. Sta. for 1961,* pp. 51–56. 1962.

8 Taylor, R. M., Influence of gibberellic acid on early patch budding of pecan seedlings, *Jour. Amer. Soc. Hort. Sci.,* 97(5):677–79. 1972.

9 Wishart, R. D. A., Microbudding of citrus, *S. Austral. Dept. Agr. Leaflet 3660,* 1961.

SUPPLEMENTARY READING

Baltet, C., *The Art of Grafting and Budding,* 6th ed. London: Crosby, Lockwood, 1910.

Brase, K. D., "Propagation of Fruit Trees by Budding," *Farm Research* (New York Agricultural Experiment Station), Vol. 18. No. 3 (1952).

Carlson, R. F., and A. E. Mitchell, "Budding and Grafting Fruit Trees," *Michigan Agricultural Extension Bulletin 508* (1971).

Garner, R. J., *The Grafter's Handbook,* 3rd ed. London: Faber & Faber, Ltd., 1968.

Hansen, C. J., and H. T. Hartmann, "Propagation of Temperature Zone Fruit Plants," *California Agricultural Experiment Station Circular 471* (revised) (1966).

Webber, H. J., "Nursery Methods," in *The Citrus Industry,* Vol. II. Berkeley: University of California Press, 1948.

Layering 14

Layering is a propagation method by which adventitious roots are caused to form on a stem while it is still attached to the parent plant. The rooted, or layered, stem is then detached to become a new plant growing on its own roots. Layering may be regarded as a preparation for subsequent division. This may be a natural means of reproduction, as in black raspberries and trailing blackberries, or it may be induced by the "artificial" methods described in this chapter.

Water and minerals are continually supplied to the layered shoot, because the stem is not severed and the xylem remains intact. Thus, layering does not depend upon the length of time that a severed shoot (cutting) can be maintained before rooting occurs. This is an important reason why layering is more successful with many plants than propagation by cuttings.

FACTORS AFFECTING THE REGENERATION OF PLANTS BY LAYERING

Root formation during layering is stimulated by various stem treatments (as shown in Figure 14–1) which cause an interruption in the downward translocation of organic materials—carbohydrates, auxin, and other growth factors—from the leaves and growing shoot tips. These materials accumulate near the point of treatment, and rooting occurs in this general area even though the stem is still attached to the parent plant.

A step common to all methods of layering is to eliminate light from the part required to form roots. It is accomplished in mound layering and trench layering by covering the newly developing shoots as they grow so that the basal part of the layered shoot is never exposed to light. If the part is already in existence, as is a shoot on a stool, or a tree branch or shoot, a blanching process follows. Blanching is quite different from etiola-

455

BEFORE ROOTING   AFTER ROOTING

Shoot bent to a sharp "v"

Shoot cut or broken on lower side

Shoot cut on the upper side; terminal end brought upright by twisting at the cut.

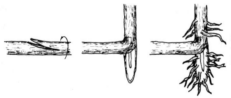

Girdling is accomplished by removing a strip of bark from around stem.

Girdling is accomplished by wrapping copper wire around stem.

Figure 14–1   Treatments used to stimulate rooting during layering.

tion, in which shoots are produced entirely in complete darkness and are leafless. Shoots produced in light and then covered may be blanched but are not truly etiolated. A large measure of the success with which shy-rooting plants are rooted by layering apparently results from etiolation and blanching.

Applying root-promoting substances, such as indolebutyric acid, during layering is sometimes beneficial, as it is with cuttings, although the method of application may be somewhat different (*6, 14, 25, 26, 30*). Applying the material to girdling cuts as a powder, in lanolin, or as a solution in 50 percent alcohol can be utilized effectively (see p. 291).

Root formation on layers depends upon continuous moisture, good aeration, and moderate temperatures in the rooting zone. These conditions are best provided by a rooting medium such as a mixture of light soil and saw-dust. Prolonged dry spells and compact, heavy soils hinder root development, particularly during the initial stages of rooting. The addition of granulated peat moss to the soil mounded around apple and quince stock plants has promoted rooting (*29*). Excessively high temperatures in the upper layers of soil during the spring and summer may reduce the moisture content and cause compaction, not only inhibiting rooting but injuring the shoots as well (*28*). In winter, uncovered stool beds with layers removed may be damaged by sunscald in mild climates. Likewise, winter exposure can lead to serious winter-killing (*21*). This damage has been prevented by temporarily remounding until spring.

CHARACTERISTICS AND USES OF LAYERING

The principal advantage of layering is the success with which stems will develop roots by this method. Many clones whose cuttings will not root easily can be propagated by layering, enabling the plant to be established on its own roots. Most methods of layering are relatively simple to perform and can be practiced out-of-doors in the nursery or garden. When small numbers of plants are involved, layering can give a high degree of success with somewhat less skill, effort, and equipment than is necessary with cuttings. With those few kinds of plants in which layering occurs naturally, it is a simple and economical method of propagation.

In some cases, a larger plant can be produced in a shorter time than if it was started as a cutting. However, since transplanting becomes increasingly difficult as the size of the layer increases, special precautions are necessary to establish successfully the larger plant on its own roots.

On the other hand, layering is an expensive method of propagation since it requires considerable hand labor and does not lend itself to the large-scale techniques of mechanization. A layered plant requires a certain amount of individual attention, depending upon the particular method in use, even though the operations involved are in themselves simple. Also, the number of salable plants from a given number of stock plants is smaller than when cuttings, buds, or scions are taken. Layering methods tend to be cumbersome, and the stock plants take up a considerable area which is difficult to cultivate and maintain free of weeds.

Layering is usually limited by American nurserymen to those plants which propagate naturally in this manner—e.g., black raspberries, trailing blackberries, gooseberries, and currants—and those plants too difficult or impossible to propagate by other methods, yet of sufficient value to justify the cost. For instance, the filbert (*Corylus* sp.), Muscadine grape (*Vitis rotundifolia*), and litchi (*Litchi chinensis*) are propagated commercially in this manner. Layering is used in propagating certain clonal rootstocks, such as Malling apple stocks (see Chapter 17), which are not easily rooted as cuttings or which require special equipment, such as mist installations.

Some ornamental trees and shrubs are propagated by layering (5, 7), particularly in Europe (33). Established blocks of plants to be used for layering have been in production for many years in some cases, and nurserymen are skilled in their management.

Layering is perhaps best utilized by the amateur horticulturist who wishes to propagate a relatively small number of plants, or by specialists involved with reproducing certain kinds of plants. In these cases, expense per plant and the individual attention necessary would not be factors in the choice of method.

PROCEDURES IN LAYERING
Tip Layering
In tip layering, rooting takes place near the tip of the current season's shoot which is bent to the ground. The shoot tip begins to grow downward into the soil but recurves to produce a sharp bend in the stem, from which roots develop. This natural method of reproduction is characteristic of trailing blackberries, dewberries, and black and purple raspberries.

Stems of these plants are biennial in that the canes are vegetative during the first year, fruitful the second, and pruned out after fruiting. In the nursery it is advisable to set aside stock plants solely for propagation. Healthy young plants are set 12 ft. apart to give room for subsequent layering. The plants are cut down to within 9 in. of the ground as soon as planted. Vigorous new canes which arise are "summer topped" by pinching off 3 to 4 in. of the tip after growth of 18 to 30 in. This encourages lateral shoot production, and will increase the number of potential tip layers, and also next year's fruit crop. By late summer, the canes begin to arch over, and their tips assume a characteristic appearance in that the terminal ends become elongated and the leaves small and curled to give a "rat-tail" appearance. The best time for layering is when only part of the lateral tips have attained this appearance. If the operation is done too soon, the shoots may continue to grow instead of forming a terminal bud. If it is done too late, the root system will be small.

The tips are preferably layered by hand, a spade or trowel being used to make a hole with one side vertical and one sloping slightly towards the parent plant. The tip is placed in the hole with the shoot lying along the sloping side and the returned soil is pressed firmly against it. Placed thus, the tip cannot continue to grow in length and becomes "telescoped," soon forming an abundant root system and developing a vigorous young vertical shoot.

The plants are ready for digging by the end of the same season. The rooted tip consists of a terminal bud, a large mass of roots, and 6 to 8 in. of the old cane to serve as a "handle" and to mark the location of the new plant (Figure 14–2). Since the tip layers are tender, easily injured, and subject to drying out, digging should be done preferably just before replanting. The remainder of the layered shoots attached to the parent plant are cut back to 9 in. as in the first year. Economical quantities of shoots are produced annually for as long as ten years. Rooted tip layers are planted in the late fall or early spring. New canes develop rapidly during the first season.

Figure 14–2   Tip layer of boysenberry.

Simple Layering

Simple layering (Figure 14–3 and 14–4) is performed by bending a branch to the ground and covering it partially with soil or rooting medium, but leaving the terminal end exposed. The end of the branch is sharply bent to an upright position about 6 to 12 in. back from the tip. The sharp bending of the shoot may be all that is necessary to induce rooting, although additional benefit may be gained by twisting to loosen the bark (*17*). Cutting or notching the underside of the stem is often practiced. Other methods are illustrated in Figure 14–1. The bent part of the shoot is next inserted into the soil so that it can be covered to a depth of 3 to 6 in. A wooden peg, bent wire, or stone may be used to hold the layer in place, and a vertical wooden stake can be inserted beside the layer to support the exposed shoot and hold it upright. If the branch is relatively inflexible and hard to bend, the tension may be lessened by notching the upper side of the shoot on the highest part of the bend back from the layer itself.

The usual time for layering is in the early spring, and dormant, one-year-old shoots are used. Low, flexible branches of the plant which can be bent easily to the ground are chosen. In some cases, the suckers produced near the crown of the plant may serve as a source of shoots for layering. Layering could also be delayed until later in the growing season after the current season's shoots have attained sufficient length and become hardened. This timing would be used, perhaps, with some broad-leaved evergreens such as rhododendron and magnolia. As a general rule, shoots older than one year are not satisfactory for layering.

Shoots layered in the spring will usually be adequately rooted by the end of the first growing season and can be removed either in the fall or in the next spring before growth starts. Mature shoots layered in summer should

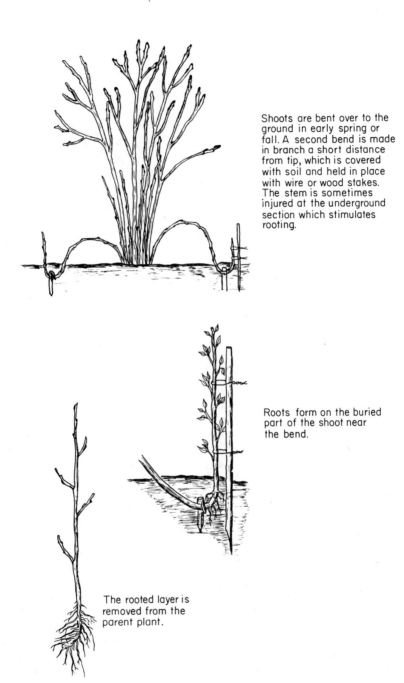

Shoots are bent over to the ground in early spring or fall. A second bend is made in branch a short distance from tip, which is covered with soil and held in place with wire or wood stakes. The stem is sometimes injured at the underground section which stimulates rooting.

Roots form on the buried part of the shoot near the bend.

The rooted layer is removed from the parent plant.

Figure 14–3   Steps in propagation by simple layering.

Figure 14–4 Propagation of two Dieffenbachia plants by simple layering. Leggy stems were curved and placed into containers of soil. After several months, strong root systems formed at the curved portion of stems *(lower left);* new plants are then severed from the mother plant for independent growth.

be left through the winter and either removed the next spring before growth begins or left until the end of the second growing season. When the rooted layer is removed from the parent plant, it is treated essentially in the same way as a rooted cutting of the same plant. Evergreen plants should be potted and kept humid and cool for a time. It should be possible to plant a well-rooted layer of a dormant deciduous plant directly into a nursery row or permanent location if the top is reduced to a size corresponding to the root system. Rooted layers removed in the fall can be planted in a cold frame or shaded greenhouse in a peat-sand mixture. Although defoliation may occur, root activity often continues for several weeks, and by spring the root system may be well developed (*8*).

A supply of rooted layers can be produced over a period of years by establishing a stool bed composed of stock plants far enough apart to allow room for all shoots to be layered. This procedure has been used commercially to propagate certain hard-to-root shrubs (*7, 17*).

Compound or Serpentine Layering

Compound layering is essentially the same as simple layering, except that the branch is alternately covered and exposed along its length. Generally, the stem is injured or girdled at the lower part of the stem and covered in the same manner as for simple layering. Roots develop at each of these buried sections. The exposed part of the stem should have at least one bud to develop a new shoot. After rooting takes place or at the end of the growing season, the branch is cut in sections made up of the new shoot and the portion containing roots. Several new plants are thus possible from a single branch.

This method is used for propagating plants which have long, flexible shoots, such as the Muscadine grape. Ornamental vines, such as *Wisteria* and *Clematis* can also be propagated this way, although this method is used more by amateurs than by commercial propagators.

Air Layering (Chinese Layering, Pot Layerage, Circumposition, Marcottage, Gootee)

In air layering, roots form on the aerial part of the plant where the stem has been girdled or slit at an upward angle. The injured portion is enclosed at the point of injury with a rooting medium which is maintained continuously moist. The procedure is most successful in regions of high humidity or in greenhouses. Enclosures used to surround the rooting medium have included metal or wooden boxes, split flower pots, waxed-paper cones, and rubber sheeting. Wrapping with sphagnum moss alone may be adequate if humidity is very high or daily syringing is practiced. Polyethylene film, which has high permeability to gases (carbon dioxide and oxygen), low transmission of water vapor, and sufficient durability to withstand long periods of weathering, has largely replaced other materials for covering the rooting medium (*10, 13, 32*). However, better rooting occurred in *Pinus radiata* layers with aluminum foil replacing sphagnum moss and plastic film (*3*).

Air layering is used to propagate a number of tropical and subtropical trees and shrubs (*11, 20, 31*), including the litchi (*13*) and the Persian lime (*Citrus aurantifolia*) (27). With polyethylene film for wrapping the layers, it is possible to extend the method of air layering out-of-door plants from the tropics to the temperate zones. Its greatest application may be for the amateur horticulturist or the propagator of selected plants for cultural or for scientific purposes rather than for the commercial nurseryman.

Air layers are made in the spring on wood of the previous season's growth or, in some cases, in the late summer with partially hardened shoots. Wood older than one year can be used in some cases, but rooting is less satisfactory and the larger plants produced are somewhat more difficult to handle after rooting. The presence of numerous active leaves on the layered shoot speeds root formation. With tropical greenhouse plants, layering should be done after several leaves have developed during a period of growth.

The first step in air layering is to girdle or cut the bark of the stem at a point 6 to 12 in. or more from the tip end (see Figure 14–5). A strip of

Figure 14–5 Steps in making an air layer on a *Ficus elastica* plant using polyethylene film. *(Top)* The stem should be girdled for a distance of about 1 in. to induce adventitious root formation above the cut. A ball of slightly damp sphagnum moss is placed around the girdled section *(middle)*. A wrapping of polyethylene film is placed around the sphagnum moss and tied at each end *(below)*.

bark ½ to 1 in. wide, depending upon the kind of plant, is completely removed from around the stem. Scraping the exposed surface to insure complete removal of the phloem and cambium is desirable to retard healing. Another procedure is to make a slanting cut about 2 in. long up and to the center of the stem, keeping the two surfaces apart by sphagnum or a piece of wood. Application of a root-promoting material, such as indolebutyric acid (see p. 291), to the exposed wound has been beneficial. Increasing concentration, up to 4 percent, of IBA in talc has increased rooting and survival in pecan air layers (26). About two handfuls of *slightly* moistened sphagnum moss are placed around the stem to enclose the cut surfaces. If the moisture content of the sphagnum moss is too high, stem tissue decay may occur.

A piece of polyethylene film 8 to 10 in. square is wrapped carefully about the branch so that the sphagnum moss is completely covered. The ends of the sheet should be folded (as in wrapping meat) with the fold placed on the lower side. The two ends must be twisted to make sure that no water can seep inside. Adhesive tape, such as electricians' waterproof tape, serves well to wrap the ends; the winding should be started well above the end of the plastic film to enclose the ends, particularly the upper one, securely. Budding rubbers and florist's ties are other materials which can be used for this purpose. Following girdling or partial cutting, and enclosing in moist material, the air layer should be supported by tying to a neighboring untreated branch or to a stick or cane attached to the parent plant. Otherwise it may break off.

The time for removal of the layer from the parent plant is best determined by observing root formation through the transparent film. In some plants, rooting occurs in two to three months or less. Layers made in spring or early summer are best left until the shoots become dormant in the fall, and are removed at that time. Holly, lilac, azalea, and magnolia should be left for two seasons (8). In general, it is desirable to remove the layer for transplanting when it is not actively growing.

Since the difficulties of transplanting increase in proportion to the size of the top, pruning is usually advisable to balance the top with the roots, but may not be imperative if the following precautions are followed. The rooted layer should be potted into a suitable container and placed under cool, humid conditions, such as an enclosed frame, where the plants can be frequently syringed. If done in the fall, a sufficiently large root system may develop by spring to permit successful growth in the open. Placing the rooted layers under light mist (see p. 297) for several weeks, followed by gradual hardening-off, is probably the most satisfactory procedure (22).

Mound (Stool) Layering

Mound or stool layering involves cutting a plant to the ground during the dormant season and mounding soil or other media around the base of the newly developing shoots in the spring to encourage roots to form on them. Covering with soil blanches the shoots and encourages root formation. Plants with stiff branches that do not bend easily and which are capable of producing an abundance of shoots from the crown year after year are particularly adapted to this method. Plants commonly propagated in this manner include the clonal apple stocks, quince, currants, and gooseberries (See Figures 14–6 and 14–7).

A stool bed is established on loose, fertile, well-drained soil one year before the propagation is to begin. The mother plants should be set 12 to 15 in. apart in the row, but the spacing between rows varies with different conditions and types of nursery equipment used. Width between rows should be sufficient to allow for cultivation and hilling operations during spring and summer. In England $3\frac{1}{2}$-ft rows are recommended (12), but in New York a minimum of 8 ft was found desirable (1). Planting in a shallow trench will permit the shoots to arise from the crown at a low level. Plants

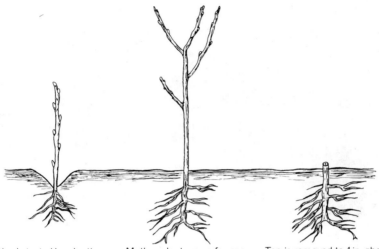

Stool bed started by planting a rooted layer in a small trench.

Mother plant grows for one season to become established.

Top is removed to 1 in. above ground just before growth begins.

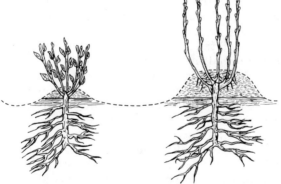

When new shoots are 3 to 5 in. high soil or sawdust is added to half their height. Soil is then added at intervals until it is 6 to 8 in. deep.

At end of season roots have formed at base of covered shoots.

Rooted layers are cut off as closely as possible to the base and are lined out in nursery row.

Mother stool with layers removed at the beginning of the next season. Additional new shoots will be layered.

Figure 14–6   Steps in propagation by mound (stool) layering.

Figure 14–7   *Left:* rooting medium of wood shavings and soil pulled away showing root production by end of summer. *Right:* stool bed use for propagating 'MM. 106' apple rootstocks.

are cut back to 15 to 18 in. from the ground and left to grow unchecked for a season, the space between the rows being kept cultivated.

Before new growth starts the following spring, all plants are cut back to an inch above the ground level. Two to five new shoots usually develop from the crown the second year, more in later years. When these shoots have grown 3 to 5 in., loose soil, sawdust, or a soil-sawdust mixture is drawn up around each shoot to one-half its height. When the shoots have grown to a total height of 8 to 10 in., a second hilling operation takes place. Additional rooting medium is again added, to be mounded around the bases of the shoots but not to more than half their total height. In England, with a spacing of 3½ ft between the rows, a ridging plow is drawn between the rows so that two ridges of soil are produced about 15 in. apart. Each row of plants will thus have a ridge on each side. The trench made by the plow is kept cultivated, and the loose soil is shoveled around and between the shoots. The shoots arise close together and, to prevent crowding, should be spread apart by placing a shovelful of soil in the middle of a cluster of shoots. A third and final hilling operation is made in midsummer when the shoots have developed to a total length of approximately 18 in. The base of the shoots will then have been covered with soil to a depth of 6 to 8 in. A depression is left in the center of the mound to facilitate water penetration.

Layered shoots of easily propagated plants should have rooted sufficiently by the end of the growing season to be separated from the parent stool for lining out in the nursery row. This operation should be delayed until plants are dormant, since much root development takes place during moist autumn months. In severe climates, the operation should be delayed until danger of winter injury is past; after removing the layers, the mother plant must be re-covered by several inches of soil. The rooted layers are cut close to their base to keep the height of the stool plant low. After cutting away the shoots, the mother stool remains exposed until new shoots have grown 3 to 5 in., when the hilling-up is begun for the next year. Shoots which do not root or which root poorly can be cut off and treated as hardwood cuttings (*1*).

A stool bed can be used for 15 to 20 years with proper handling, providing it is maintained in a vigorous condition, with disease, insect, and weed control. Overhead irrigation may be essential in some areas to maintain proper moisture conditions in the rooting zone (*4*). Soil must be added while the base of the shoots are still soft and succulent and not when they have hardened.

Girdling the bases of the shoots by wiring about six weeks after they begin to grow will stimulate rooting in many plants (*10, 16, 18, 20, 24*). The size of the root system in apple layers has been increased on shoots growing through the spaces of a galvanized screen ($\frac{3}{16}$ in. square) laid in an 18 in. strip down the row over the top of the cut-back stumps. New shoots growing through this screen gradually become girdled as the season progresses (*15*).

Budded plants of apple and citrus (*9, 19*) have been produced by budding the layer in place in the stool bed. This may be done in the middle of the growing season whereupon the bedded layer is transplanted to the nursery in the fall for an additional season's growth.

Trench Layering

Trench layering (etiolation method) consists of growing a plant or a branch of a plant in a horizontal position in the base of a trench and filling in soil around the new shoots as they develop, so that the shoot bases are etiolated. Roots develop from the bases of these new shoots (see Figure 14–8).

The first step in this procedure involves the establishment of the mother bed which, as in mound layering, can be used over a period of years. Rooted layers or one-year-old nursery-budded or grafted trees are planted 18 to 30 in. apart at an angle of 30 to 45 deg. down a row. The rows should be 4 to 5 ft apart—wide enough to allow for cultivation and to draw soil up around the plants to a height of 6 in. The plants are then cut back to a uniform length—18 to 24 in.—and left to grow one season. In some cases (in layering walnuts, for instance) plants can be placed horizontally in the trench and the developing shoots layered the first year.

Before the beginning of growth in the spring, the parent layers are bent over and laid flat on the bottom of a trench dug along the row, about 2 in. deep and wide enough to receive the entire layer. Weak lateral branches are cut to $\frac{1}{2}$ in. in length and strong laterals tipped back. Wooden pegs may be used to hold the mother layer in place. Short lengths of wire bent to form a U, or a longer single wire rolled on the end and set at an angle, will be useful to hold down the small shoots. It is important that all parts of the shoot be completely flat on the floor of the trench.

Before the buds swell, the entire layer is covered with 1 to 2 in. of fine soil or other rooting medium such as peat moss, sawdust, or wood shavings. Another inch of medium is added when the developing shoots have pushed through the first soil layer but before their leaves have expanded. Several more additions of soil are made the first two or three weeks of the growing season to ensure the etiolation of the basal 2 to 3 in. of the shoot. When the shoots have grown an additional 3 to 4 in., soil is again added to half the height of the exposed shoot. Further additions are made at intervals until

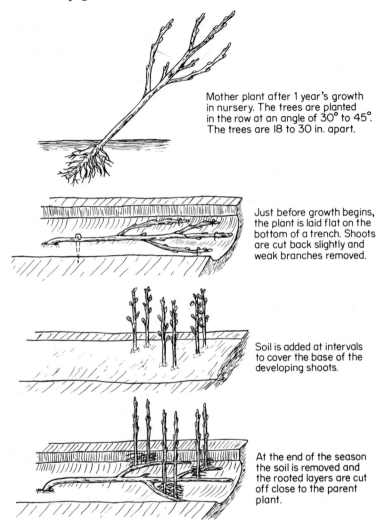

Mother plant after 1 year's growth in nursery. The trees are planted in the row at an angle of 30° to 45°. The trees are 18 to 30 in. apart.

Just before growth begins, the plant is laid flat on the bottom of a trench. Shoots are cut back slightly and weak branches removed.

Soil is added at intervals to cover the base of the developing shoots.

At the end of the season the soil is removed and the rooted layers are cut off close to the parent plant.

Figure 14–8   Steps in propagation by trench layering.

the bases of the shoots are finally covered to a depth of 6 to 8 in. by mid-summer. Roots form from the etiolated base of the current season's shoot.

At the end of the growing season after the plants have become dormant—or the next spring—the soil is removed from around the layered shoots, and rooted layers are cut off from the original layered stock as close to the base as possible. Unrooted shoots can be left to be pegged down the next year to fill in gaps or to rejuvenate an old bed. New shoots may arise from adventitious buds along the old original layer underground. However, some of the vigorous shoots growing from the mother plant can be pegged down each year to produce a new supply of rooted layers. If all the shoots are rooted, care must be taken to leave at least one shoot every foot or so for future layering.

Trench layering is primarily a nursery method for propagating certain fruit tree rootstocks difficult to propagate by other methods. It could also be practiced on established shrubs or trees by bending long, flexible shoots or vines to the ground as is done in simple layering, but laying them flat in trenches. The shoot is covered along its entire length, but the tip is left exposed. New shoots which develop from the buds along the stem grow upward through the soil, with roots forming at their base. This latter procedure is sometimes known as *continuous layering*.

PLANT MODIFICATIONS SUITABLE FOR NATURAL LAYERING

Some plants exhibit modifications of their vegetative structure or method of growth which lead to their natural vegetative increase. Those listed below could be considered natural forms of layering and are often utilized for propagation.

Runners

A *runner* is a specialized stem which develops from the axil of a leaf at the crown of a plant, grows horizontally along the ground, and forms a new plant at one of the nodes. The strawberry is a typical plant propagated in this way (Figure 14–9). Other plants propagated by runners include bugle (*Ajuga*), the strawberry geranium (*Saxifraga sarmentosa*), and the ground cover, *Duchesnea indica*. Plants of these species grow as a typical rosette or crown.

In most strawberry cultivars, runner formation is related to the length of day and temperature. Runners are produced in long days of 12 to 14 hours

Figure 14–9 Runners arising from the crown of a strawberry plant. New plants are produced at every second node. The daughter plants, in turn, produce additional runners and runner plants.

or more with high midsummer temperatures. New plants are produced at alternate nodes. These take root but remain attached to the mother plant. New runners are in turn produced by the daughter plants. The connecting stems die in the late fall and winter, and each daughter plant becomes separate from the others.

In propagating by runners, the rooted daughter plants are dug when they have become well rooted, and then transplanted to the desired locations. (See p. 572 for strawberry propagation.)

Stolons

Stolons are special modified stems, produced by some plants, that grow horizontal to the ground. These may be prostrate or sprawling stems growing above-ground as found in some woody species, such as *Cornus stolonifera*. The term also describes the horizontal stem structure occurring in Bermuda grass, *Ajuga*, mint (*Mentha*) and *Stachys* (23). Stolon-like underground stems are involved in tuberization (p. 495); they are the stems that develop into a potato tuber.

The stolon can be treated as a naturally occurring rooted layer and can be cut from the parent plant and planted.

Offsets

An *offset* is a characteristic type of lateral shoot or branch which develops from the base of the main stem in certain plants. This term is applied generally to a shortened, thickened stem of rosette-like appearance. Many bulbs (see Chapter 15 for details) reproduce by producing typical offset bulblets from their base. The term offset (or *offshoot*, as is sometimes used) also applies to lateral branches arising on stems of monocotyledons. The date palm produces lateral shoots from the base of the plant by which it is propagated (Figure 14–10). The pineapple is also propagated by offsets, although in commercial culture these are termed ratoons, suckers, or slips, depending upon the location on the plant where they are produced. Details of date and pineapple propagation are given in Chapter 17. Lateral shoots arising from rhizomes, as in the banana or orchid, are also offsets or offshoots.

Offsets are removed by cutting them close to the main stem with a sharp knife. If it is well rooted, the offset can be potted as is done with any rooted cutting. If insufficient roots are present, the shoot is placed in a favorable rooting medium and treated as a leafy stem cutting.

In cases in which offset development is meager, cutting off the main rosette may stimulate the development of offsets from the old stem just as removing the terminal bud stimulates lateral shoots in any other type of plant. For instance, in *Echeveria* the main stem may elongate so that the plant becomes a fleshy rosette borne on top of a fleshy, bare stem. The rosette, with a short piece of stem, may be removed and rooted, and new offsets will develop from buds at the base. It is desirable that this operation be carried out while the stem is somewhat soft and succulent, rather than allowing it to become hard and woody (2).

Figure 14–10   Removing a date offshoot by using chisel and sledge hammer. From R. W. Nixon, "Date Culture in the United States," *USDA Cir. 728.*

Offshoots of the date palm do not root readily if separated from the parent plant. They are usually layered for a year prior to removing (see p. 553).

Suckers

A *sucker* is a shoot which arises on a plant from below ground, as shown in Figure 14–11. The most precise use of this term is to designate a shoot which arises from an adventitious bud on a root. However, in practice, shoots which arise from the vicinity of the crown are also referred to as suckers even though originating from stem tissue. Nurserymen generally designate any shoot produced from the rootstock below the bud union of a budded tree as a sucker and refer to the operation of removing them as "suckering." In contrast, a shoot arising from a latent bud of a stem several years old, as, for instance, on the trunk or main branches, should be termed a *watersprout.*

The tendency to "sucker" is a characteristic possessed by some plants and not by others (Figure 14–12). The ability of a plant to sucker and the ability of a plant to grow from root cuttings are closely related. In such cases, however, nurserymen prefer to use root cuttings rather than depend on naturally produced suckers. Since suckers arise from adventitious buds, they may show juvenile characteristics (see p. 184) and regenerate roots more readily than shoots in the adult phase.

Suckers are dug out and cut from the parent plant. In some cases part of the old root may be retained, although most new roots arise from the base of the sucker. It is important to dig the sucker out rather than pull it, to avoid injury to its base. Suckers are treated essentially as a rooted layer or as a cutting, in case few or no roots have formed. They are usually dug during the dormant season.

Figure 14–11   Suckers arising as adventitious shoots from the roots of a red raspberry plant. After they are well rooted the suckers may be cut from the parent plant and transplanted to their permanent location.

Figure 14–12   Suckering in Coast Redwood *(Sequoia sempervirens)*. A commercial method of propagating this important lumber tree is by allowing suckers from roots to develop *(left)* after the original tree was logged. Some 100 years later *(right)* the second growth *(sucker)* trees can be cut for lumber. Mendocino county, California.

Crowns

The term *crown* as generally used in horticulture designates that part of a plant at the surface of the ground from which new shoots are produced. In trees or shrubs with a single trunk, the crown is principally a point of location near the ground surface marking the general transition zone between stem and root. In herbaceous perennials, the crown is the part of the plant from which new shoots arise annually. The crown of many herbaceous perennials consists of many branches, each being the base of the current season's stem, which originated from the base of the preceding year's branch. These lateral shoots are stimulated to grow from the base of the old stem as it dies back after blooming. Adventitious roots develop along the base of the new shoots. These new shoots eventually flower either the same year they are produced or the following year. As a result of the annual production of new shoots and the dying back of old shoots, the crown may become extensive within a period of a relatively few years.

In certain plants, for example, the strawberry or the African violet (*Saintpaulia*), the stem is a short and thickened structure from which the leaves are produced in a rosette-like arrangement. The entire body of the plant is often referred to as the crown. Lateral shoots or offsets are produced from the base of the crown. An old plant may be composed of a number of "crowns" or "crown divisions" which have been produced in this manner.

Multibranched woody shrubs may develop extensive crowns. Although an individual woody stem may persist for a number of years, new, vigorous shoots are continuously produced from the crown, and eventually crowd out the older shoots. If left undisturbed, such shrubs develop into extensive thickets. Under normal handling, the older shoots are regularly removed by pruning to give way to the younger, more vigorous shoots.

Division of the crown is an important method of propagation for herbaceous perennials, and to some extent for woody shrubs, because of its simplicity and reliability. Such characteristics make this method particularly useful to the amateur or professional gardener who is generally interested in only a modest increase of a particular plant. Many herbaceous perennials must be divided every two to three years to prevent the plants from becoming overcrowded.

Crowns of outdoor herbaceous perennials are usually divided in the spring just before growth begins or in late summer or autumn at the end of the growing season. As a general rule, those plants which bloom in the spring and summer and produce new growth after blooming should be divided in the fall. Those which bloom in summer and fall and make little or no new growth until spring should be divided in early spring. Potted plants are divided when they become too large for the particular container in which they are growing.

In crown division, plants are dug and cut into sections with a knife. In herbaceous perennials, as the Shasta daisy or day lily (see Figure 14–13), where an abundance of new rooted offshoots are produced from the crown, each may be broken from the old crown and planted separately, the older part of the plant clump being discarded. If a larger clump is desired, then a

Figure 14–13   Propagation by crown division illustrated by division of day lily (*Hemerocallis*) clump.

section of the old crown bearing a number of new shoots from its base may be used. In the case of plants in which the crown consists of a number of rosettes or offsets, as occurs in the African violet, division should be made between each of the rosettes, but roots should be present on each section.

Shrubs may be divided in the same manner with a shovel or hatchet. Such an operation should be carried out at a time when the plant is dormant. The top should be cut back and the roots trimmed at the time of division, and each section planted as a new shrub. If crowns are old or otherwise extensive, and if there is no shortage of the cultivar, the sub crowns farthest from the center of the parent plant should be used, the older middle crowns being discarded.

REFERENCES

1 Brase, K. D., and R. D. Way, Rootstocks and methods used for dwarfing fruit trees, *New York Agr. Exp. Sta. Bul. 783*, 1959.

2 Butterfield, H. M., The propagation of Echeverias and related succulents, *Jour. Calif. Hort. Soc.*, 15:30–35. 1954.

3 Cameron, R. J., The leaching of auxin from air layers, *New Zealand Jour. Bot.*, 6(2):237–39. 1968.

4 Carlson, R. F., and H. B. Tukey, Cultural practices in propagating dwarfing rootstocks in Michigan, *Mich. Agr. Exp. Sta. Quart. Bul.* 37:492–97. 1955.

5 Chase, H. H., Propagation of oriental magnolias by layering, *Proc. Int. Plant Prop. Soc.*, 14:67–69. 1964.

6 Ching, F., C. L. Hamner, and F. Widmoyer, Air-layering with polyethylene film, *Mich. Agr. Exp. Sta. Quart. Bul. 39:* 3–9. 1956.

7 Congdon, M. L., Mass production of deciduous shrubs by layering, *Proc. Plant. Prop. Soc.*, 4:39–45. 1954.

8   Creech, J. L., Layering, *Nat. Hort. Mag.*, 33:37–43. 1954.

9   Duarte, O., and C. Medina., Propagation of citrus by improved mound layering, *HortScience*, 6:567. 1971.

10   Du Preez, D., Propagation of guavas, *Farming in South Africa*, 29:297–99. 1954.

11   Feilden, G. St. C., and R. J. Garner, Vegetative propagation of tropical and subtropical plantation crops, *Imp. Bur. Hort. and Plant. Crops, Tech. Comm. No. 13,* 1940.

12   Garner, R. J., *The Grafter's Handbook,* 2nd ed. London: Faber & Faber, Ltd., 1958.

13   Grove, W. R., Wrapping air layers with rubber plastic, *Proc. Fla. State Hort. Soc.,* 60:184–89. 1947.

14   Hanger, F. E. W., V. M. H., and A. Ravenscroft, Air layering experiments at Wisley, *Jour. Roy. Hort. Soc.,* 79:111–16. 1954.

15   Hogue, E. J., and R. L. Granger, A new method of stool bed layering, *Hort. Science,* 4:29–30. 1969.

16   Hostermann, G.. Versuche zur vegetativen Vermehrung von Gehozen nach dem Dahlemer Drahtungsverfahren, *Ber. Deutsch. Bot. Ges.,* 48:66–70. 1930.

17   Knight, F. P., The vegetative propagation of flowering trees and shrubs, *Jour. Roy. Hort. Soc.,* 70:319–30. 1945.

18   Maurer, K. J., Möglichkeiten der vegetativen Vermehrung der Walnuss, *Schweiz. Z. Obst. V. Weinb.,* 59:136–37. 1950.

19   Medina, C., and O. Duarte, Propagating apples in Peru by an improved mound layering method, *Jour. Amer. Soc. Hort. Sci.,* 96:150–51. 1971.

20   Mowry, H., L. R. Toy, and H. S. Wolfe, Miscellaneous tropical and subtropical Florida fruits, *Fla. Agr. Ext. Serv. Bul. 156,* pp. 1–110. 1953.

21   Modlibowska, I., and C. P. Field, Winter injury to fruit trees by frost in England (1939–1940), *Jour. Pom. Hort. Sci.,* 19:197–207. 1942.

22   Nelson, R., High humidity treatment for air layers of lychee, *Proc. Fla. State Hort. Soc.,* 66:198–99. 1953.

23   Nitsch, J. P., Perennation through seeds and other structures, in F. C. Steward (ed.), *Plant Physiology,* Vol VIA. New York: Academic Press, 1971, pp. 413–79.

24   Oppenheim, J. E., A new system of citrus layers, *Hadar,* 5:2–4. 1932.

25   Singh, L. B., *The Mango.* London: Leonard Hill, 1960.

26   Sparks, D., and J. W. Chapman, The effect of indole-3-butyric acid on rooting and survival of air-layered branches of the pecan, *Carya illinoensis* Koch, cv. 'Stuart,' *HortScience,* 5(5):445–46. 1970.

27   Sutton, N. E., Marcotting of Persian limes, *Proc. Fla. State Hort. Soc.,* 67: 219–20. 1954.

28   Thomas, L. A., Stock and scion investigations. II. The propagation of own-rooted apple trees. *Jour. Counc. Sci. Industr. Res. Org., Austral.,* 11:175–79. 1938.

29   Tukey, H. B., and K. Brase, Granulated peat moss in field propagation of apple and quince stocks, *Proc. Amer. Soc. Hort. Sci.,* 27:106–13. 1930.

30   Vieitez, E., Estudios sobre la reproduciion vegetativa del castano. I. Enraiza-

miento en al acodo alto mediante el empleo de fitohormonas. *An. Edaf. Fis. Veg. Madrid,* 12:337–56. 1953.

31   Watkins, J. V., Propagation of ornamental plants, *Fla. Agr. Exp. Sta. Bul. 150,* pp. 1–15. 1952.

32   Wyman, D., Air layering with polyethylene films, *Jour. Roy. Hort. Soc.,* 77:135–40. 1952.

33   ———, Layering plants in Holland, *Arnoldia,* 13:25–28. 1953.

SUPPLEMENTARY READING

Garner, R. J., "Propagation by Cuttings and Layers. Recent Work and Its Application, with Special Reference to Pome and Stone Fruits," *Imp. Bur. Hort. Plant. Crops, Tech. Commun.* 14. 1944.

Ministry of Agriculture, Fisheries and Food (London) "Fruit Tree Raising—Rootstocks and Propagation," *Bulletin 135.* 5th ed. 1969.

Tukey, H. B., *Dwarfed Fruit Trees.* New York: The Macmillan Company, 1964.

Propagation by Specialized Stems and Roots

15

This chapter deals with propagation by specialized vegetative structures— *bulbs, corms, tubers, tuberous roots, rhizomes,* and *pseudobulbs* (see Table 15–1). The primary function of these modified plant parts is food storage for the plant's survival. Plants possessing them are generally herbaceous perennials in which the shoots die down at the end of a growing season, and the plant survives in the ground as a dormant, fleshy organ which bears buds to produce new shoots the next season. Such plants are well suited to withstanding periods of adverse growing conditions in their yearly growth cycle. The two principal climatic cycles for which such performance is adapted are the warm-cold cycle of the temperate zones and the wet-dry cycle of tropical and subtropical regions.

The second function of these specialized organs is vegetative reproduction. The propagation procedure which utilizes the production of naturally detachable structures, such as the bulb and corm, is generally spoken of as *separation.* In cases in which the plant is cut into sections, as is done with the rhizome, stem tuber, and tuberous root, the process is spoken of as *division.*

Table 15–1   Flowering plants grown from specialized stems or roots

Bulbs	Corms	Tubers	Tuberous Roots and Stems	Rhizomes
I *Hardy; spring-flowering; planted in the fall*				
Camassia Chionodoxa (Glory-of-the-snow) Erythronium (Trout lily) Fritillaria (Guinea hen flower) Galanthus (Snowdrop) Hyacinthus (Hyacinth) Iris (bulbous) Leucojum (Snowflake) Muscari (Grape hyacinth) Narcissus (Daffodil) Scilla (Squill) Tulipa (Tulip)	Crocus		Anemone (Wind flower) Arisaema (Jack-in-the-pulpit) Claytonia (Spring beauty) Dicentra (Bleeding heart) Eranthis (Winter aconite) Eremurus (Desert candle)	Convallaria (Lily-of-the-valley) Iris (bearded types)
II *Semi-hardy to hardy; summer- and fall-flowering; planted in the fall*				
Lilium (Lily) Lycoris (Hardy amaryllis)	Colchicum (Autumn crocus) Crocus (some species) Sternbergia	Begonia evansiana (aerial)		
III *Tender; summer- and fall-flowering; dug in the fall and stored over winter; planted in the spring when danger of freezing is over; these can be kept outdoors over winter in mild climates*				
Amaryllis belladonna Galtonia (Summer hyacinth) Hymenocallis Polianthes (Tuberose) Tigridia (Tiger flower) Watsonia Zephyranthes (Zephyr lily)	Freesia Gladiolus Ixia Tritonia	Caladium	Begonia (tuberous) Dahlia Gloriosa	Canna

Table 15–1 Continued

Bulbs	Corms	Tubers	Tuberous Roots and Stems	Rhizomes
IV *Tender; mostly grown in greenhouses or as house plants; grown outdoors in mild climates*				
Clivia (Kafir lily) *Crinum* *Eucharis grandiflora* (Amazon lily) *Haemanthus* (Blood lily) *Hippeastrum* (Amaryllis) *Nerine* *Oxalis* (bulbous types) *Vallota* (Scarborough lily) *Veltheimia*		*Caladium*	*Agapanthus* (Lily-of-the-Nile) *Arum* (Black calla) *Colocasia* (Elephant's ear) *Cyclamen* *Ranunculus* (florist's types) *Sinningia* (Gloxinia)	*Achimenes* *Zantedeschia* (Calla lily)

BULBS
Definition and Structure

A *bulb* is a specialized underground organ consisting of a short, fleshy, usually vertical stem axis (*basal plate*) bearing at its apex a growing point or a flower primordium enclosed by thick, fleshy scales. (See Figure 15–1.) Bulbs are produced by monocotyledonous plants in which the usual plant structure is modified for storage and reproduction.

Most of the bulb consists of *bulb scales,* which morphologically are the continuous, sheathing leaf bases (see Figure 15–2). The outer bulb scales are generally fleshy and contain reserve food materials, whereas the bulb scales toward the center function less as storage organs and are more leaf-like. In the center of the bulb, there is either a vegetative meristem or an unexpanded flowering shoot. Meristems develop in the axil of these scales to produce miniature bulbs, known as *bulblets,* which when grown to full size are known as *offsets.* In various species of lilies, bulblets may form in the leaf axils either on the underground portion or on the aerial portion of the stem. The aerial bulblets are called *bulbils,* while the underground organs are called *stem bulblets.* There are two types of bulbs:

(a) *Tunicate (laminate)* bulbs, represented by the onion, daffodil, and tulip. These bulbs have outer bulb scales which are dry and membranous. This covering, or *tunic,* provides protection from drying and mechanical injury to the bulb. The

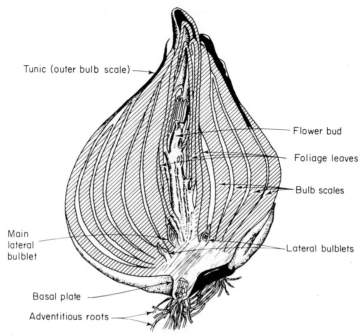

Figure 15-1 The structure of a tulip bulb—an example of a tunicated laminate bulb. Longitudinal section representing stage of development shortly after the bulb is planted in the fall. Redrawn from Mulder and Luyten *(37)*.

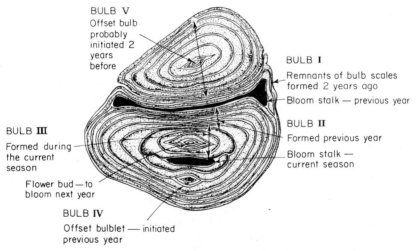

Figure 15-2 Cross section of a daffodil bulb. The continuous, concentric leaf scales found in the laminate bulb are shown. Also shown is the perennial nature of the daffodil bulb, which continues to grow by producing a new bulb annually at the main meristem. Lateral offset bulbs are also produced; parts of five individual, differently aged bulbs are shown here. Redrawn from Huisman and Hartsema *(24)*.

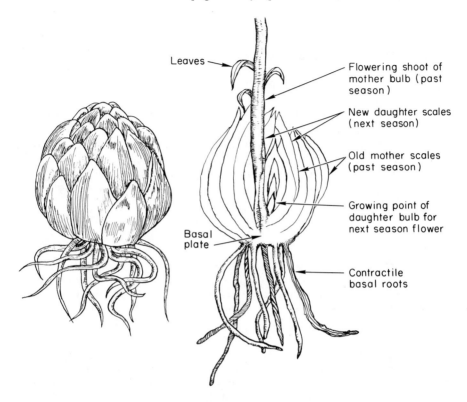

Leaves

Flowering shoot of
mother bulb (past
season)

New daughter scales
(next season)

Old mother scales
(past season)

Growing point of
daughter bulb for
next season flower

Basal
plate

Contractile
basal roots

Figure 15–3   *Left:* outer appearance of a scaly bulb of lily *(Lilium hollandicum).*
Right: Longitudinal section of bulb of *L. longiflorum* 'Ace,' after flowering stage,
showing old mother bulb scales and new daughter bulb scales. Bulb obtained in
fall near digging time *(10).*

fleshy scales are in continuous, concentric layers, or *lamina,* so that the structure
is more or less solid.

(b)   *Non-tunicate (scaly)* bulbs, represented by the lily (Figure 15–3). These bulbs
do not possess the enveloping dry covering. The scales are separate and attached
to the basal plate. In general, the non-tunicate bulbs are easily damaged and
must be handled more carefully than the tunicate bulbs; they must be kept continuously moist because they are injured by drying. Root primordia are present
at the time of harvesting. They do not elongate until planted under proper conditions and at the proper time. They occur in a narrow band around the outside
edge on the bottom of the basal plate. In the non-tunicate lily bulb new roots
are produced in midsummer or later, and persist through the following year *(43).*
In most lily species, roots also form on the stem above the bulb.

   Thickened *contractile* roots are produced in many species that shorten
to pull the bulb to a given level in the ground *(51).*

Growth Pattern

An individual bulb goes through a characteristic cycle of development, beginning with its initiation as a meristem and terminating in flowering and seed production. This general developmental cycle is composed of two stages: (a) vegetative and (b) reproductive. In the vegetative stage the bulblet grows to flowering size and attains its maximum weight. The subsequent reproductive stage includes the induction of flowering, differentiation of the floral parts, elongation of the flowering shoot, and finally flowering and (sometimes) seed production. Various bulb species have specific environmental requirements for the individual phases of this cycle which determine their seasonal behavior, environmental adaptations and methods of handling. They can be grouped into classes according to their time of bloom and method of handling.

Spring-Flowering Bulbs

Important commercial crops included in this group are the tulip, daffodil, hyacinth, and bulbous iris, although other kinds are grown in gardens.

(a) *Bulb formation*. The vegetative stage begins with the initiation of the bulblet on the basal plate in the axil of a bulb scale. In this initial period, which usually occupies a single growing season, the bulblet is insignificant in size, since it is present within another growing bulb and can be observed only if the bulb is dissected. Its subsequent pattern of development and the time required for the bulblet to attain flowering size are somewhat different for different species. The bulbs of the tulip and the bulbous iris, for instance, disintegrate upon flowering and are replaced by a cluster of new bulbs and bulblets which were initiated the previous season. The largest of these may have attained flowering size at this time, but smaller ones require additional years of growth (see Figures 15–1 and 15–4). The flowering bulb of the daffodil, on the other hand, continues to grow from the center year by year, producing new offsets which may remain attached for several years (see Figures 15–2 and 15–4). The hyacinth bulb also continues to grow year by year, but because the number of offsets produced is limited, artificial methods of propagation are usually used.

The size and quality of the flower are directly related to the size of the bulb. Only bulbs greater than a certain minimum size are capable of initiating flower primordia. Commercial value is largely based on bulb size (2), although the condition of the bulb and freedom from disease are also important quality factors.

The greatest increase in size and weight of the developing bulb takes place in the period during and (mostly) after flowering, as long as the foliage remains in good condition (9). Cultural operations that include irrigation; weed, disease, and insect control; and fertilization encourage vegetative growth. The benefit, however, is to the next year's flower, because larger bulbs are produced. Conversely, adverse situations, such as poor growing conditions, removal of foliage, and premature digging of the bulb, result in smaller bulbs and reduced flower production.

Figure 15–4   Propagation by offset bulbs. *Top left:* bulbous iris. The old bulb disintegrates, leaving a cluster of bulbs. *Top right:* daffodil. Bulb continues to grow from the inside each year, but continuously produces lateral bulbs which eventually split away. *Below left:* daffodil. Three types of bulbs: "split," or "slab," bulb; "round" bulb; and "double-nose" bulb. *Below right:* hyacinth. Lateral bulblets produced, but old bulb continues to develop from inside.

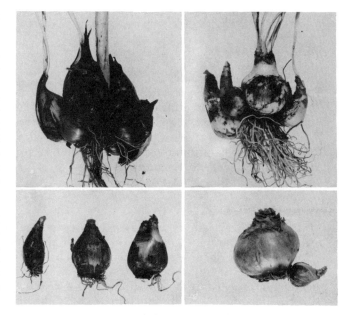

Moderately low temperatures tend to prolong the vegetative period, whereas higher temperatures may cause the vegetative stage to cease and the reproductive stage to begin. Thus a shift from cool to warm conditions early in the spring, as occurs in mild climates, will shorten the vegetative period, result in smaller bulbs, and consequently produce inferior blooms the following year (*40*). Commercial bulb-producing areas for hárdy spring-flowering bulbs are largely in regions of cool springs and summers, such as The Netherlands and the Pacific Northwest of the United States.

The relative length of photoperiod apparently is not an important factor affecting bulb formation in most species. It has been shown, however, to be significant in some *Allium* species, such as onion and garlic (*23, 33*).

(b) *Flower bud formation and flowering.* The beginning of the reproductive stage and the end of the vegetative stage is indicated by the drying of the foliage and the maturation of the bulb. From then on, no additional increase in size or weight of the bulb takes place. The roots disintegrate, and the bulb enters a seemingly "dormant" period. However, important internal changes take place, and in some species the vegetative growing point undergoes transition to a flowering shoot. In nature, all bulb activity takes place underground during this period; in horticultural practice, the bulbs are dug, stored, and distributed during this three- to four-month period.

Temperature controls the progression from the vegetative stage to flowering (*22, 23*). Differentiation of flower primordia occurs at moderately warm temperatures in late summer or early fall either in the ground or in storage. Subsequent exposure to lower but above-freezing temperatures is required to promote flower stalk elongation. As temperatures increase in the spring the flower stalk elongates and the bulb plant subsequently flowers.

The optimum temperature—determined by the shortest time in which the bulb would flower—has been established for these developmental phases for the important bulb species, such as tulip (*50*), hyacinth (*8*) or daffodil (*50*), depending also upon cultivar. Such information is important in establishing forcing schedules for retail florist sales (*4*). Holding the bulbs continuously at high temperatures (86° to 90° F; 30° to 32° C; or more), or at temperatures near freezing, will inhibit or retard floral development and can be used to lengthen the period required for flowering. With a shift to favorable temperatures, flower bud development will continue. This treatment (below or above optimum) can be used when shipping bulbs from Northern- to Southern-Hemisphere countries.

Lilies

Lilies have a growth cycle geared to the seasonal pattern of the summer-winter cycles (annual thermoperiodism) of the temperate zone. Their non-tunicate bulbs do not go "dormant" in late summer and fall as does the tunicate type of bulb but have unique characteristics that must be understood for proper handling. Although different lily species have somewhat different methods of reproduction (*43, 52*) the pattern, as determined for the Easter lily (*Lilium longiflorum*), can serve as a model (*5, 10*).

Lilies flower in the late spring or early summer at the apex of the leaf-bearing *stem axis*. The flower-producing bulb is known as the *mother bulb* made up of the basal plate, fleshy scales, and the flowering axis (Figure 15–3). Prior to flowering, a new daughter bulb(s) is developing within the mother bulb. It had been initiated the previous fall and winter from a growing point in the axil of a scale at the base of the stem axis. During spring the daughter bulb initiates new scales and leaf primordia at the growing point. Natural chemical inhibitors in the daughter scales prevent elongation of the daughter axis, which remains dormant—but it can be promoted to grow by exposure to high temperature (100° F; 37.5° C), to low temperature (40° F; 4.5° C), or by treatment with gibberellic acid (*47*).

After flowering of the mother bulb, no more scales are produced by the daughter bulb, but it increases in size (circumference) and weight until it equals the weight of the mother bulb surrounding it. Inhibitory effects of the daughter scales decrease, as does the response to dormancy-breaking treatments (*47*). Fleshy basal roots persist on the mother bulb through the fall and winter, and new adventitious roots develop in late summer or fall from the basal plate. Warm temperatures promote root formation (*11*). Bulbs should be dug for transplanting after they "mature" in the fall. The top may or may not have died down. Bulbs should be handled carefully to avoid injury and to prevent drying. The commercial value of the bulb depends on size (transverse circumference) and weight at the time of digging (*31*); and on the condition of the fleshy roots and the freedom of the bulb from disease.

Transition of the meristem to a flowering shoot does not take place until after first, the stem axis has protruded through the "nose" of the bulb (*4*), and then the bulb has been subjected to chilling temperatures (*29*). The critical temperatures are 60° to 65° F (15.5° to 18.5° C) or less, and the chilling effect becomes more and more effective down to 35° to 40° F (2° to 4.5° C). Storing bulbs at warm (70° F; 21° C, or more), or low (31° F; −0.5° C), temperatures will keep bulbs dormant and delay blooming (46). Moisture content of the storage medium is important; if too dry the bulb will deteriorate and if too wet, it will decay.

Following the flower induction stage, and with the onset of higher temperatures, the stem elongates, initiating first leaves and then flowers. The outer scales of the "old" mother bulb rapidly disintegrate early in the spring as the new mother bulb produces the flower.

Stem bulblets may develop in the axils of the leaves underground or, in some species, bulbils may develop above ground. These appear about the time of flowering.

Flowering time and the size and quality of the Easter lily bloom can be closely regulated by manipulating the temperature at various stages following the digging of bulbs in the fall (10). Day length also influences development, but to a lesser extent (48). Control of flower induction is referred to as the "programming" phase and the subsequent handling as the "greenhouse phase." In programming, a bulb may be given natural cooling by planting immediately in pots and placing out-of-doors—but with frost protection—until brought into the greenhouse. A bulb may be precooled (PC) at 35° to 40° F (2° to 4.5° C) while packed in moist peat. This latter procedure will produce the most rapid shoot elongation and shortest time to flower after placing the plant at growing temperatures. Particular cultivars, such as 'Ace' or 'Nellie White,' may be most effectively handled by controlled temperature forcing (CTF) in which the nonprecooled bulb is placed at 60° to 65° F (15.5° to 17° C) for two weeks and subsequently at 35° to 40° F (2° to 4.5° C) for six to seven weeks before shifting to the greenhouse.

Tender, Winter-Flowering Bulbs

There are a number of flowering bulbs from tropical areas whose growth cycle is related to a wet-dry, rather than a cold-warm climatic cycle. The amaryllis (*Hippeastrum vittata*) is an example (22, 49). This bulb is a perennial, growing continuously from the center with the outer scales disintegrating. New leaves are produced continuously from the center during the vegetative period extending from late winter to the following summer. In the axil of every fourth leaf (or scale) that develops, a meristem is initiated. Thus throughout the vegetative period a series of vegetative offsets is produced. By fall the leaves mature and the bulb becomes dormant, during which time the bulb should be dry. In this period, the *fourth* growing point from the center and any external to it differentiate into flower buds, and the shoot begins to elongate slowly. After two or three months of dry storage, the bulbs can be watered, causing the flowering shoots to elongate rapidly, with flowering taking place in midwinter. Maximum foliage development and bulb growth are essential to produce a bulb large enough to form a flowering shoot.

Propagation

Offsets

Offsets provide a simple and reliable method for propagating many kinds of bulbs. This method is sufficiently rapid for the commercial production of tulip, daffodil, bulbous iris, and grape hyacinth, but in general it is too slow for the lily, hyacinth, and amaryllis.

If undisturbed, the offsets may remain attached to the mother bulb for several years. They can also be removed at the time the bulbs are dug and replanted into beds or nursery rows to grow into flowering-sized bulbs. This may require several growing seasons, depending upon the kind of bulb and size of the offset.

(a) **Tulip** (*8, 15*). Bulb planting takes place in the fall. Two systems of planting are used: the bed system, used extensively in Holland, and the row or field system, used mostly in the United States and England. Beds are usually 3 ft. wide and separated by 12- to 18-in. paths. The soil is removed to a depth of 4 in., the bulbs set in rows 6 in. apart, and the soil replaced. In the other system, single or double rows are placed wide enough apart to permit the use of machines. To improve drainage, two or three adjoining rows may be planted on a ridge. Bulbs are spaced one to two diameters apart, with small bulbs scattered along the row. A mulch may be applied after planting but removed the following spring before growth.

Planting stock consists principally of those bulbs of the minimum size for flowering (9 to 10 cm or smaller in circumference). Since the time required to produce flowering sizes varies with the size of the bulb, the planting stock is graded so that all those of one size can be planted together. For instance, an 8-cm or larger bulb normally requires a single season to become of flowering size; a 5- to 7-cm bulb, two seasons; and those 5 cm or less, three years (*15*).

During the flowering and subsequent bulb growing period of the next spring, good growing conditions should be provided so that the size and weight of the new bulbs will be at a maximum. Foliage should not be removed until it dries or matures. Important cultural operations include removal of competing weed growth, irrigation, fungicidal sprays to control *Botrytis blight* (*13*), and fertilization. Beds should be inspected for disease early in the season and for trueness-to-cultivar at the time of blossoming. All diseased or off-type plants should be rogued out (*20*). It is desirable to remove the flower heads at blooming time, because they may serve as a source of *Botrytis* infection and can lower bulb weight (*1*).

Bulbs are dug in early to midsummer when the leaves have turned yellow or the outer tunic of the bulb has become dark brown in color. In the Pacific Northwest of the United States, where summer temperatures are cool and the leaves remain green for a longer period, digging may take place before the leaves dry. If the bulbs are dug too early or if warm weather causes early maturation, the bulbs may be small in size. The bulbs are dug by machine or by hand with a short-handled spade. After the loose soil is shaken from the bulbs, they are placed in trays in well-ventilated storage houses for drying, cleaning, sorting, and grading. General storage temperatures are 65° to 68° F (18° to 20° C). To force early flowering, the bulbs should be held at

68° F for three to five weeks and then placed at 48° F (9° C) for eight weeks. Later flowering can be produced by holding bulbs at 72° F (22° C) for ten weeks. For shipment from Northern to Southern Hemisphere countries, the bulbs can be held at 31° F (−1° C) until late December, when they are shifted to a higher temperature (78° F; 25.5° C) (*8*).

(b) **Daffodil** (*16*). Daffodil bulbs are perennial and produce a new meristem growing point at the center every year. Offsets are produced which grow in size for several years until they break away from the original bulb, although they are still attached at the basal plate. An offset bulb, when it first separates from the mother bulb, is known as a "split," "spoon," or "slab," and can be separated from the mother bulb and planted. Within a year it becomes a "round," or "single-nose," bulb containing a single flower bud. One year later a new offset should be visible, enclosed within the scales of the original bulb, indicating the presence of two flower buds. At this stage the bulb is known as a "double-nose." By the next year the offsets split away, then the bulb is known as a "mother bulb." Grading of daffodil bulbs is principally by age, that is, as splits, round, double-nose, and mother bulbs. The grades marketed commercially are the round and the double-nose bulb. The mother bulbs are used as planting stock to produce additional offsets, and only the surplus is marketed. Offsets, or splits, are replanted for additional growth.

Storage should be at 55° to 60° F (13° to 16° C) with a relative humidity of 75 percent. To force earlier flowering, they can be stored at 48° F (9° C) for eight weeks. To delay flowering, store at 72° F (22° C) for 13 to 15 weeks. For shipment from the Northern to the Southern Hemisphere, the bulbs can be held at 86° F (30° C) until October, then stored at 31° F (−1° C) until late December, and then at 77° F (25° C) (*8*).

Hot water treatment plus a fungicide for stem and bulb nematode control is important (*20*). A three- or four-hour treatment at 110° F (43° C) is used, but that temperature must be carefully maintained or the bulbs may be damaged.

(c) **Lilies.** Lilies increase naturally, but except for a few species this increase is slow and of limited propagation value except in home gardens (*18, 43, 52*). Several methods of bulb increase are found among the different species. For instance, *Lilium concolor, L. hansonii, L. henryi,* and *L. regale* increase by bulb splitting. Two to four lateral bulbets are initiated about the base of the mother bulb, which disintegrates during the process, leaving a tight cluster of new bulbs. *Lilium bulbiferum, L. canadense, L. pardalinum, L. parryi, L. superbum,* and *L. tigrinum* multiply from lateral bulbets produced from the rhizome-like bulb. This is sometimes called "budding-off."

Bulblet Formation on Stems

Underground stem bulblets are used to propagate the Easter lily (*Lilium longiflorum*) and some other lily species (Figure 15–5). In the field, flowering of the Easter lily occurs in early summer. Bulblets form and increase in size from spring throughout summer (*5, 42*). Between mid-August and mid-

Figure 15–5   Two methods of lily propagation. *Left:* bulblets are produced at the base of individual bulb scales; this method of propagation is possible for nearly all lily species. *Right:* underground stem bulblets; these are produced only by some lily species.

September in the Northern Hemisphere the stems are pulled from the bulbs and stacked upright in the field. Periodic sprinkling keeps the stems and bulblets from drying out. Similarly, the base of the stem can be "heeled in" the ground at an angle of 30 to 45 deg. or laid horizontally in trays at high humidity.

About mid-October the bulblets are planted in the field 4 in. deep and an inch apart in double rows which are spaced 36 in. apart. Here they remain for the following season. They are dug in September as yearling bulbs and again replanted, this time 6 in. deep and 4 to 6 in. apart in single rows. At the end of the second year, they are dug and sold as commercial bulbs.

Digging is done in September after the stem is pulled. The bulbs are graded, packed in peat moss, and shipped. Commercial bulbs range in size from 7 to 10 in. in circumference. Lily bulbs must be handled carefully so they will not be injured and must be kept from drying out. The fleshy roots should also be kept in good condition. For long-term storage that will prevent flowering, the bulbs should be packed on polyethylene-lined cases with peat moss at 30 to 50 percent moisture and stored at 31° F (−1° C) *(46)*.

Control of viruses, fungus diseases, and nematodes during propagation is important in bulb production. Methods of control include using pathogen-free stocks for propagation *(3, 42)*, growing plants in pathogen-free locations with good sanitary procedures, and treating the bulbs with fungicides.

Aerial stem bulblets, commonly known as *bulbils,* are formed in the axil of the leaves of some species, such as *Lilium bulbiferum, L. sargentiae, L. sulphureum,* and *L. tigrinum.* Bulbils develop in the early part of the season and fall to the ground several weeks after the plant flowers. They are harvested shortly before they fall naturally and are then handled in essentially the same manner as underground stem bulblets. Increased bulbil production can be induced by disbudding as soon as the flower buds form. Likewise, some lily species which do not form bulbils naturally can be induced to do so by pinching out the flower buds and a week later cutting off the upper half of the stems. Species which respond to the latter procedure include *Lilium candidum, L. chalcedonicum, L. hollandicum, L. maculatum* and *L. testaceum (43)*.

Stem Cuttings

Lilies may be propagated by stem cuttings. The cutting is made shortly after flowering. Instead of roots and shoots forming on the cutting, as would occur in other plants, bulblets form at the axils of the leaves and then produce roots and small shoots while still on the cutting.

Leaf-bud cuttings, made with a single leaf and a small heel of the old stem, may be used to propagate a number of lily species. A small bulblet will develop in the axil of the leaf. It is handled in the same manner as for the other methods described here.

Bulblet Formation on Scales (Scaling)

In *scaling,* individual bulb scales are separated from the mother bulb and placed in growing conditions so that adventitious bulblets form at the base of each scale (Figure 15–5). Three to five bulblets will develop from each scale. This method is particularly useful for rapidly building up stocks of a new cultivar or to establish pathogen-free stocks. Almost any lily species can be propagated by scaling (*18, 19*).

Scaling is done soon after flowering in midsummer, although it might be done in late fall or even in midwinter. The bulbs are dug, the outer two layers of scales are removed and the mother bulb is replanted for continued growth. It is possible to remove the scales down to the core, but this will reduce subsequent growth of the mother bulb. The scales should be kept from drying and handled so as to avoid injury. Scales with evidence of decay should be discarded and the remaining ones dusted with a fungicide. Naphthaleneacetic acid (1 ppm) will stimulate bulblet formation.

Scales are field planted in beds or frames no more than $2\frac{1}{2}$ inches deep. Bulblets form on the scale during the first year to produce "yearlings." These are replanted for a third year to produce "commercials." Another procedure is to place scales in trays or flats of moist sand, peat moss, sphagnum moss, or vermiculite for six weeks at 65° to 70° F (18° to 21° C). The scales are inserted vertically to about half their length. Small bulblets and roots should form at the base within three to six weeks. The scales are transplanted either into the open ground or into pots or flats of soil, and then planted in the field the following spring. Subsequent treatment is the same as described for underground bulblets.

A simple method of propagating lilies by scales, after removing them from the bulb, is to dust them with a fungicide then place the scales so they are not touching in damp vermiculite in a polyethylene bag. The bag is closed and tied then set for six to eight weeks where the temperature is fairly constant at about 70° F (21° C). After bulblets are well developed at the base of the scales, the bag with the scales still inside should be refrigerated at 35° to 40° F (2° to 4.5° C) for at least eight weeks to overcome dormancy. The small bulblets can then be potted and placed in the greenhouse or out-of-doors for further growth.

Figure 15–6   Basal cuttage. Hyacinth bulb which has been scored. Note the bulblets starting to appear.

Basal Cuttage

The hyacinth is the principal plant propagated by this method, although others such as *Scilla* can be handled in this way. Specific methods include "scooping" and "scoring" (*8, 17*). Mature bulbs which have been dug after the foliage has died down and are 17 to 18 cm or more in circumference are used. In *scooping,* the entire basal plate is scooped out with a special curve-bladed scalpel, a round-bowled spoon, or a small-bladed knife. Adventitious bulblets develop from the base of the exposed bulb scales. Depth of cutting should be enough to destroy the main shoot. In *scoring,* three straight knife cuts are made across the base of the bulb, as shown in Figure 15–6, each deep enough to go through the basal plate and the growing point. Growing points in the axils of the bulb scales grow into bulblets.

To combat decay that may develop during the later incubation period, any infected bulbs should be discarded; the tools disinfected frequently with an alcohol, formalin, or mild carbolic acid solution; and the cut bulbs dusted with a fungicide. Most important is to callus the bulbs at about 70° F (21° C) for a few days to a few weeks in dry sand or soil or in open trays, cut side down. After callusing, the bulbs are incubated in trays or flats, in dark or diffuse light, at 70° F, which is increased to 85° to 90° F (29.5° to 32° C) over a two-week period and held at high humidity (85 percent) for 2½ to 3 months.

The mother bulbs are planted about 4 in. deep in nursery beds in the fall. The next spring bulblets produce leaves profusely. Normally the mother bulb disintegrates during the first summer. Annual digging and replanting of the graded bulblets is required until they reach flowering sizes. Bulbs for greenhouse forcing should be 17 cm (6¾ in.) or more in circumference; bulbs for bedding should be 14 to 17 cm (5½ to 6¾ in.) in circumference (2). On the average, a scooped bulb will produce 60 bulblets, but four to five years will be required to produce flowering sizes; and a scored bulb will produce 24, requiring three to four years (*8*).

Leaf cuttings

This method is successful for blood lily (*Haemanthus*), grape hyacinth (*Muscari*), hyacinth, and Cape cowslip (*Lachenalia*) (*12*), although the range of species is probably wider.

Leaves are taken at a time when they are well developed and green. An entire leaf is cut from the top of the bulb and may in turn be cut into two or three pieces. Each section is placed in a rooting medium with the basal end several inches below the surface, as described for rooting cuttings. The leaves should not be allowed to dry out, and bottom heat is desirable. Within two to four weeks small bulblets form on the base of the leaf and roots develop. At this stage the bulblets are planted in soil.

Bulb Cuttings

Among plants which respond to this method of propagation are the *Albuca, Chasmanthe, Cooperia, Haemanthus, Hippeastrum, Hymenocallis, Lycoris, Narcissus, Nerine, Pancratium, Scilla, Sprekelia* and *Urceolina* (12).

A mature bulb is cut into a series of eight to ten vertical sections, each containing a part of the basal plate. These sections are further divided by sliding a knife down between each third or fourth pair of concentric scale rings and cutting through the basal plate. Each of these fractions makes up a bulb cutting consisting of a piece of basal plate and segments of three or four scales.

The bulb cuttings are planted vertically in a rooting medium, such as peat moss and sand, with just their tips showing above the surface. The subsequent technique of handling is the same as for ordinary leaf cuttings. A moderately warm temperature, slightly higher than for mature bulbs of that kind, is required. New bulblets develop from the basal plate between the bulb scales within a few weeks, along with new roots. At this time they are transferred to flats of soil to continue development.

CORMS

Definition and Structure

A *corm* is the swollen base of a stem axis enclosed by the dry, scale-like leaves. In contrast to the bulb, which is predominantly leaf scales, a corm is a solid stem structure with distinct nodes and internodes. The bulk of the corm consists of storage tissue composed of parenchyma cells. In the mature corm, the dry leaf bases persist at each of these nodes and enclose the corm. This covering, known as the *tunic,* protects it against injury and water loss. At the apex of the corm is a terminal shoot which will develop into the leaves and the flowering shoot. Axillary buds are produced at each of the nodes. In a large corm, several of the upper buds may develop into flowering shoots, but those nearer the base of the corm are generally inhibited from growing. However, should something prevent the main buds from growing, these lateral buds would be capable of producing a shoot. (See Figures 15–7 and 15–8.)

Two types of roots are produced from the corm: a fibrous root system developing from the base of the mother corm and enlarged, fleshy contractile roots developing from the base of the new corm. The latter roots apparently develop in response to the fluctuating temperatures near the soil surface and with exposure of the leaves to light. At lower soil depths temperature fluctuations decrease (26) and contraction ceases once the corm is at a given depth.

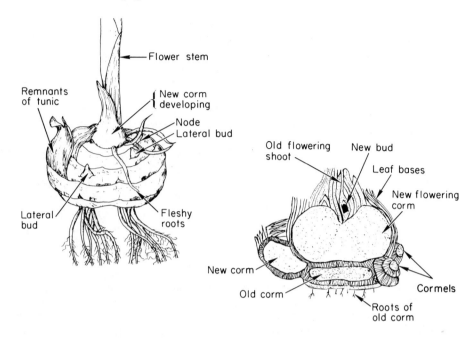

Figure 15-7   Gladiolus corm. *Left:* external appearance. *Right:* longitudinal section showing solid stem structure.

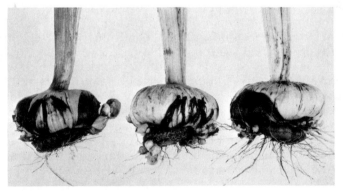

Figure 15-8   Stage of development of gladiolus corms during the latter part of the growing season. In each case, the remains of the originally planted corm can still be seen just below the newly formed corm. Many small cormels have also been produced.

Growth Pattern

Gladiolus and crocus are typical cormous plants. The gladiolus is semi-hardy to tender and, in areas with severe winters, the corm must be stored over winter and replanted in the spring. At the time of planting, the corm is a vegetative structure (*21, 39*). New roots develop from its base, and one or more of the buds begin to develop leaves. Floral initiation takes place within a few weeks after the shoot begins to grow. At the same time the base of the shoot axis thickens, and a new corm for the succeeding year begins to form above the old corm. Stolon-like structures bearing miniature corms or *cormels* on their tip develop from the base of the new corm. The new corm continues to enlarge, and the old corm begins to shrivel and disintegrate as its contents are utilized in flower production. After flowering, the foliage continues to manufacture food materials, which are stored in the new corm. At the end of the summer, when the foliage dries, there are one or more new corms and perhaps a great number of little cormels. The corms are dug and stored over winter until they are planted the following spring.

Propagation

New Corms

Propagation of cormous plants is principally by the natural increase of new corms. Flower production in corms, as in bulbs, depends upon food materials stored in the corm the previous season, particularly during the period following bloom. In gladiolus, cool nights and long growing periods are favorable for production of very large corms. Fertilization and other good management practices during bloom have their greatest effect on the next year's flowers. Plants are left in the ground for two months following blooming, or until frost kills the tops. After digging, the plants are placed in trays with a screen or slat bottom arranged to allow air to circulate between them, and cured at 90° to 92° F (32° to 33° C) or lower at 80 to 85 percent relative humidity. A few hours at 95° F (35° C) may be helpful. Then the new corms, old corms, cormels, and tops can be easily separated. The corms are graded according to size, sorted to remove the diseased ones, treated with a fungicide, and returned to a 95° F temperature for an additional week. This curing process suberizes the wounds and helps combat *Fusarium* infection. The corms are then stored at 40° F (5° C) with a relative humidity of 70 to 80 percent in well-aerated rooms to prevent excessive drying. It may also be desirable to treat them with a suitable fungicide (*32*) immediately before planting.

Cormels

These are miniature corms which develop between the old and the new corms. One or two years' growth is required for them to reach flowering size. Shallow planting of the corms, only a few inches deep, results in greater production of cormels; increasing the depth of planting reduces cormel production.

Cormels are separated from the mother corms and stored over winter for planting in the spring. Dry cormels become very hard and may be slow to start growth the following spring, but if they are stored at about 40° F (5° C) in slightly moist peat moss, they will stay plump and in good condition. Soaking dry cormels in cool running water for one to two days and holding them moist until planting at first sign of root development will hasten the onset of growth.

Disease-free cormels can be obtained by hot-water treatments which should be done between two and four months after digging. Holding cormels at room temperatures to keep them dormant will increase tolerance to this treatment. The cormels are soaked in water at air temperature for two days, then placed in a 1:200 dilution of commercial 37 percent formaldehyde for four hours, and then immersed in a water bath at 135° F (57° C) for 30 minutes. The temperature should be maintained within 1° F, plus or minus, of this temperature. At the end of the treatment, the cormels are cooled quickly, dried immediately, and stored at 40° F (5° C) in a clean area with good air circulation.

The cormels are planted in the field in furrows about 2 in. deep in the manner of planting large seeds. Only grass-like foliage is produced the first season. The cormel does not increase in size but produces a new corm from the base of the stem axis, in the manner described for full-sized corms. At the end of the first growing season, the beds are dug and the corms separated by size. A few of the corms may attain flowering size, but most require an additional year of growth.

Size grades in gladiolus are determined by diameter. There are seven grades, the smallest ⅜ to ½ in. in diameter, the largest 2 in. or more (2).

Division of the Corm

Large corms can be cut into sections, retaining a bud with each section. Each of these should then develop a new corm. Segments should be dusted with a fungicide because of the great likelihood of decay of the exposed surfaces.

TUBERS

Definition and Structure

A *tuber* is a modified stem structure which develops below ground as a consequence of the swelling of the subapical portion of a stolon (see p. 470) and subsequent accumulation of reserve materials.

The most notable example of a plant which produces tubers by which it is propagated is the Irish potato (*Solanum tuberosum*). (See Figure 15–9.) The *Caladium,* grown for its striking foliage, is also propagated by tubers, as is the Jerusalem artichoke (*Helianthus tuberosus*).

A tuber has all the parts of a typical stem. The "eyes," present in regular order over the surface, represent nodes, each consisting of one or more small

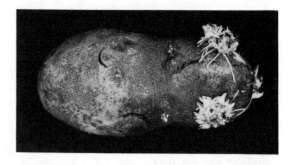

Figure 15-9   Potato tuber. Note "eyes" (axillary buds). Those on terminal (distal) end of tuber are starting to grow into new shoots which are already producing adventitious roots. Basal (proximal) end of tuber, which was attached to a stolon, is to the left.

buds subtended by a leaf scar. The arrangement of the nodes is a spiral, beginning with the terminal bud on the end opposite the scar resulting from the attachment to the stolon. The terminal bud is at the apical end of the tuber, oriented farthest (distally) from the crown of the plant. Consequently it shows the same apical dominance as any stem.

Growth Patterns

A tuber is a storage and propagative organ that is produced in one growing season, remains dormant during the winter, and then produces new shoots the following spring to start a new cycle. Most information on growth cycles in tubers is from studies of the potato (*6, 14, 45*). At the base of the main shoot adventitious roots are initiated, and lateral buds grow out horizontally to become the *stolons,* as shown in Figure 15–10. Continued growth of the stolon takes place during long photoperiods and is associated with the presence of auxin and a high gibberellin level. Tuberization begins with inhibition of terminal growth and the initiation of cell enlargement and division in the subapical region of the stolons. This stage is associated with short or intermediate day lengths, reduced temperatures (particularly at night),

Figure 15-10   Tubers of Irish potato showing their development from stolons arising from stem tissue. Note adventitious root system originating from main plant stem. Tuber is attached to stolon at the tuber's morphological basal (proximal) end.

high light intensity, low mineral content, and a reduction in gibberellin levels in the plant.

Tuberization has been explained as being caused by the production of a tuber-inducing substance that is produced in the leaves and the mother tuber (*38*). It seems to be necessary for the stolon tip to have attained a particular physiological age. Continued tuber enlargement is dependent on a continuing adequate supply of photosynthate. Conditions which favor rapid and luxurious plant growth above ground, such as an abundance of nitrogen, or high temperatures, are not conducive to tuber production (*34*). In the fall, the tops of the plants die down and the tubers are dug. At this time, the buds of potato tubers are dormant for six to eight weeks. This condition must disappear before sprouting will take place.

Propagation
Division

Propagation by tubers can be done either by planting the tubers whole or by cutting them into sections, each containing one or more buds or "eyes." These small pieces of tuber to be used for propagation of the potato are commonly referred to as "seed." The weight of the tuber piece should be 1 to 2 oz to provide sufficient stored food for the new plant to become well established.

Division of tubers is done with a sharp knife shortly before planting. The cut pieces should be stored at warm (68° F; 20° C) temperatures and relatively high humidities (90 percent) for two to three days prior to planting. During this time the cut surfaces heal (*suberization*) and the "seed" piece is effectively protected against drying and decay. Treatment of potato tubers prior to cutting for the control of *Rhizoctonia* and scab may be desirable (*34*). Caladium tubers are produced commercially in Florida (*44*). The tubers are cut into sections, usually two buds per piece. These are planted 3 to 4 in. deep, 4 to 6 in. apart in rows 18 to 24 in. apart. Harvest begins in November. After harvest, the tubers are dried in open sheds for six weeks or artificially dried for 48 hours. Further storage should be at temperatures above 60° F (16 ° C).

TUBERCLES

Begonia evansiana and the cinnamon vine (*Dioscorea batatas*) produce small aerial tubers, known as *tubercles,* in the axils of the leaves. These tubercles may be removed in the fall, stored over winter, and planted in the spring (*12*). Short days induce tuberization (*38*).

TUBEROUS ROOTS AND STEMS
Definition and Structure

Certain species of herbaceous perennials produce thickened underground structures which contain large amounts of stored food. In certain plants,

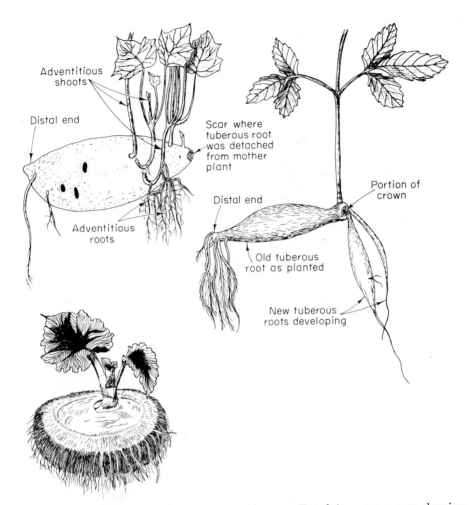

Figure 15–11   Types of tuberous roots and stems. *Top left:* sweet potato showing adventitious shoots. *Top right:* dahlia during early stages of growth. The old root piece will disintegrate in the production of the new plant; the new roots can be used for propagation. *Below left:* tuberous begonia, showing its vertical orientation. This type continues to enlarge each year.

such as sweet potato (*Ipomoea batatus*) and *Dahlia,* these thickened structures are true tuberous roots with external and internal structures of roots. They lack nodes and internodes; buds are present only on the crown or stem (*proximal*) end; fibrous roots are commonly produced only on the opposite (*distal*) end. Polarity is the reverse of that of the true tuber. (Figure 15–11.)

Other plants, as tuberous begonia, cyclamen, or gloxinia, have thickened structures that have arisen from enlarged hypocotyl (stem) tissue (*25, 41*). These have a vertical arrangement and may show features of stems.

Growth Behavior

Secondary tuberous roots produce swollen sections on lateral roots. In the dahlia, these swollen sections are borne in a cluster, each tuberous root attached to the crown of the plant. These roots are biennial. They are produced in one season, after which they go dormant as the herbaceous shoots die. The following spring, buds from the crown produce new shoots which utilize the food materials from the old root during their initial growth. The old root then distintegrates, and new roots are produced, which in turn maintain the plant through the following dormant period.

In the tuberous begonia, buds are produced at the proximal end (the crown). Fibrous roots are produced from the basal portion. This tuberous stem is perennial and lives for a number of years, continuing to increase in size and produce additional buds from the crown.

Initiation of tuberization and growth of tuberous roots occur apparently as a result of continuing exposure to short days (less than 12 hours) (*14, 30, 36*).

Propagation

Adventitious Shoots

The fleshy roots of a few species of plants such as sweet potato have the capacity to produce adventitious shoots if subjected to the proper conditions. The roots are laid in sand so that they do not touch one another and are covered to a depth of about 2 in. The bed is kept moist. The temperature should be about 80° F (27° C) at the beginning and about 70° to 75° F (21° to 24° C) after sprouting has started. As the new shoots, or *slips,* come through the covering, more sand is added so that eventually the stems will be covered for 4 to 5 in. Adventitious roots develop from the base of these adventitious shoots. After the slips are well rooted, they are pulled

Figure 15–12   Propagation of sweet potato. Adventitious shoots ("slips") develop when the mother root is placed under warm, moist conditions. *Top:* heat treatment (100°F; 34.5°C) for 26 hr before planting *(left)* strongly increases slip production by overcoming proximal dominance, as shown in unheated tuber on right. Photo courtesy Welch and Little *(49). Bottom:* after they are well rooted, the slips are removed and planted.

from the parent plant and transplanted into the field (*34*). If sweet potato roots are cut in half and the pieces are subjected to 110° F (43° C) for about 26 hours, slip production increases. This treatment overcomes the apical dominance and also controls nematodes and fungus diseases (*49*). (See Figure 15–12.) (Vine cuttings from established beds can also be used in sweet potato propagation.)

Division

In addition to the method described above for propagating the sweet potato, certain cultivars can be propagated by dividing the tuberous root into 20- to 25-gm pieces, treating with a fungicide, then giving a pre-sprouting treatment for four weeks of 80° F (26.5° C) and 90 percent relative humidity before planting (*7*).

However, most plants with fleshy roots must be propagated by dividing the crown so that each section bears a shoot bud. This is necessary in the dahlia, for example. The plant is dug with its cluster of roots intact, dried for a few days, and stored at 40° to 50° F (4° to 10° C) in sawdust or vermiculite. Open storage may result in shriveling. The root cluster is divided in the late winter or spring shortly before planting. In warm, moist conditions the buds begin to grow, and division can be done with assurance that each section will have a bud. (See Figure 15–13.)

The perennial tuberous root of the tuberous begonia can also be divided as long as each section has a bud. To combat decay, the cut surface should be dusted with a fungicide and each section dried for several days after cutting and before placing in a moist medium.

Vegetative propagation in such plants as these can often be done better with stem, leaf, or leaf-bud cuttings. The cuttings will develop tuberous

Figure 15–13 Propagation of dahlia. To produce a new plant each separate tuberous root must have a section of the crown bearing a shoot bud, as shown by the detached root on the left.

roots at their base. This process can be stimulated if the stem cutting initially includes a small piece of the fleshy root.

RHIZOMES
Structure
A *rhizome* is a specialized stem structure in which the main axis of the plant grows horizontally at or just below the ground surface. A number of economically important plants, such as bamboo, sugar cane, banana, and many grasses, as well as a number of ornamentals, such as rhizomatous *Iris* and lily-of-the-valley, have rhizome structures. Most are monocotyledons, although a few dicotyledons—for example, low bush blueberry (*Vaccinium angustifolium*)—have analagous underground stems classed as rhizomes. Many ferns and lower plant groups have rhizomes or rhizome-like structures.

Figure 15–14 shows structural features of a rhizome (*53*). The stem appears segmented because it is composed of nodes and internodes. A leaf-like *sheath* is attached at each node; it encloses the stem and, in an expanded form, becomes the foliage leaves. When the leaves and sheaths disintegrate, a scar is left at the point of attachment identifying the node and giving a segmented appearance. Adventitious roots and lateral growing points develop in the vicinity of the node. Upright-growing, above-ground shoots and flowering stems (*culms*) are produced either terminally from the rhizome tip or from lateral branches.

Two general types of rhizomes are found (*35*). The first (the *pachymorph*) is illustrated by rhizomatous *Iris* in Figure 15–15 and by ginger in Figure 15–16. The rhizome is thick, fleshy, and shortened in relation to length. It appears as a many-branched clump made up of short, individual sections. It is determinate; that is, each clump terminates in a flowering stalk, growth continuing only from lateral branches. The rhizome tends to be oriented horizontally with roots arising from the lower side.

The second type (the *leptomorph*) is illustrated by the lily-of-the-valley in Figure 15–14. The rhizome is slender with long internodes. It is indeterminate; that is, it grows continuously in length from the terminal apex and from lateral branch rhizomes. The stem is symmetrical and has lateral buds at most nodes, nearly all remaining dormant. This type does not produce a clump but spreads extensively over an area.

Intermediate forms between these two types also exist. These are called *mesomorphs* (*35*).

Growth Pattern
Rhizomes grow by elongation of the growing points produced at the terminal end and on lateral branches. Length also increases by growth in the intercalary meristems in the lower part of the internodes. As the plant continues to grow and the older part dies, the several branches arising from one plant may eventually become separated to form individual plants of a single clone.

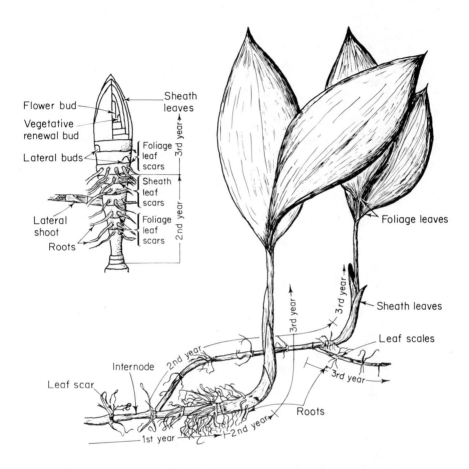

Figure 15–14 Structure and growth cycle of lily-of-the-valley *(Convallaria majalis)*. *Right:* section of rhizome as it appears in late spring or early summer with one-, two-, or three-year-old branches. A new rhizome branch begins to elongate in early spring and terminates in a vegetative shoot bud by the fall. The following spring the leaves of the bud unfold; food materials manufactured in the leaves by photosynthesis are accumulated in the rhizome. Growth the second season is again vegetative. Early in the third season a flower bud begins to form, and at the same time a vegetative growing point forms in the axil of the last leaf. *Top left:* section of the three-year-old branch showing terminal flower bud and lateral shoot bud enclosed in leaf sheaths. Such a section is sometimes known as a *pip* or *crown* and is forced for spring bloom. In the early spring the flowering shoot expands, blooms, and then dies down, the shoot bud beginning a new cycle of development. Redrawn from Zweede *(53)*.

Figure 15–15   Structure of an iris (rhizomous type) plant as it appears about the time of flowering. A two-year-old section which had flowered the previous year and is now dying back is shown at D. The lateral branch arising from it consists of the one-year-old vegetative section (B); the current-season, lateral vegetative branches (A); and the current-season, terminal flowering shoot (C). The vegetative shoot (A) will flower the following year.

Figure 15–16   Tuberous rhizome of the tropical ginger plant (*Zingiber officinale*). This is easily propagated by division of the thick rhizome, which is the source of commercial ginger.

Rhizomes exhibit consecutive vegetative and reproductive stages, but the growth cycle differs somewhat in the two types described. In the pachymorph rhizome of *Iris* (Figure 15–15), a growth cycle begins with the initiation and growth of a lateral branch on a flowering section. The flowering stalk dies, but these new lateral branches produce leaves and grow vegetatively during the remainder of that season. Continued growth of the underground stem, storage of food, and the production of a flower bud at the conclusion of the vegetative period are dependent upon photosynthesis. Consequently foliage should not be removed during this period. A flowering stalk is produced the following spring and no further terminal growth can take place. In general, plants with this structure flower in the spring and grow vegetatively during the summer and fall.

Plants with a leptomorph habit as a general rule (with exceptions) grow vegetatively during the beginning of the growth period and flower later in the same period. The length of time during which an individual rhizome remains vegetative varies with different kinds of plants. An individual branch in the lily-of-the-valley in Figure 15–14, for instance, is vegetative three years before a flower bud forms. Some bamboo species remain vegetative for many years, but then they change abruptly and the entire plant produces flowers.

In some rhizomatous plants, such as blueberry, rhizome development is increased by higher temperatures and long photoperiod, and is correlated with vigorous above-ground growth (27).

Propagation
Division of Clumps and Rhizomes
Division is the usual procedure for propagating plants with a rhizome structure, but the procedure may vary somewhat with the two types. In pachymorph rhizomes, individual sections (or culms) are cut off at the point of attachment to the rhizome, the top is cut back, and the piece is transplanted to the new location. Leptomorph rhizomes can be handled in essentially the same way by removing a single lateral "offshoot" from the rhizome and transplanting it. The tip of the lily-of-the-valley rhizome bearing a flower bud (see Figure 15–14) called a "pip" is removed along with the rooted section below and transplanted.

Division is usually carried out at the beginning of a growth period (as in early spring) or at or near the end of a growth period (i.e., in late summer or fall).

Propagation is carried out by cutting the rhizome into sections, being sure that each piece has at least one lateral bud, or "eye"; it is essentially a stem cutting. Bananas, for instance, are propagated in this way. This general method works well for the leptomorph rhizomes, in which a dormant lateral growing point is present at most nodes. The rhizomes are cut or broken into pieces, and adventitious roots and new shoots develop from the nodes. Rhizome-producing turf grasses, for instance, are cut up into sections and the individual "sprigs" transplanted. New plants can be established readily by this method.

Figure 15–17   Cattleya orchid showing rhizome structure and upright elongated pseudobulbs as basal part of shoots.

Culm Cuttings

In large rhizome-bearing plants, such as bamboos, the aerial shoot, or culm, may be used as a cutting. These may be whole culm cuttings, in which the entire aerial shoot is laid horizontally in a trench. New branches arise at the nodes. A stem cutting of three- or four-node sections planted vertically in the ground may be used.

PSEUDOBULBS

Definition and Structure

A *pseudobulb* (literally "false bulb") is a specialized storage structure, produced by many orchid species, consisting of an enlarged, fleshy section of the stem made up of one to several nodes (see Figure 15–17). In general the appearance of the pseudobulb varies with the orchid species. The differences can be used to identify species.

Growth Pattern

These pseudobulbs arise during the growing season on upright growths which develop laterally or terminally from the horizontal rhizome. Leaves and flowers form either at the terminal end or at the base of the pseudobulb, depending upon the species. During the growth period, they accumulate stored food materials and water and assist the plants in surviving the subsequent dormant period.

Propagation

Offshoots

In a few orchids, such as the *Dendrobium* species, the pseudobulb is long and jointed, being made up of many nodes. Offshoots develop at these nodes. From the base of these offshoots roots develop. The rooted offshoots are then cut from the parent plant and potted.

Division

Most important commercial species of orchids, including the *Cattleya, Laelia, Miltonia,* and *Odontoglossum,* may be propagated by dividing the rhizome into sections, the exact procedure used depending upon the particular kind of orchid. Division is done during the dormant season and preferably just before the beginning of a new period of growth. The rhizome is cut with a sharp knife back far enough from the terminal end to include four to five pseudobulbs in the new section, leaving the old rhizome section with a number of old pseudobulbs, or "back bulbs," from which the leaves have dehisced. The section is then potted, whereupon growth begins from the bases of the pseudobulbs and at the nodes. The removal of the new section of the rhizome from the old part stimulates new growth, or "back breaks," to occur from the old parts of the rhizome. These new growths grow for a season and can be removed the following year.

An alternate procedure is to cut partly through the rhizome and leave it for one year. New back breaks will develop which can be removed and potted.

Back Bulbs and Green Bulbs

"Back bulbs" (i.e., those without foliage) are commonly used to propagate clones of *Cymbidium*. These are removed from the plant, the cut surface is painted with a grafting compound, and they are placed in a rooting medium for new shoots to develop. When

Figure 15–18   A "back bulb" of a *Cymbidium* orchid was removed from the parent plant and placed in a rooting medium; the offshoot shown above then developed. This offshoot is now ready for removal and potting. When this is done, a second offshoot, or "break," should appear. Courtesy A. Kofranek.

the stage shown in Figure 15–18 is reached, the shoot can be removed from the bulb and potted. This back bulb can be repropagated and a second shoot developed from it.

"Green bulbs" (i.e., those with leaves) can also be used in *Cymbidium* propagation. Treatment with indolebutyric acid, either by soaking or by painting with a paste, has been shown to be beneficial *(28)*.

REFERENCES

1 Allen, R. C., Factors affecting the growth of tulips and narcissi in relation to garden practice, *Proc. Amer. Soc. Hort. Sci.*, 35:825–29. 1937.

2 Amer. Assoc. Nurserymen, Inc., Comm. on Hort. Stand., *American Standard for Nursery Stock*. Washington, D.C.: Amer. Assoc. Nurs., Inc., 1956.

3 Baker, K. F., and P. A. Chandler, Development and maintenance of healthy planting stock, Sect. 13 in *Calif. Agr. Exp. Sta. Man. 23*, 1957.

4 Ball, V. (ed.), *The Ball Red Book*, 12th ed. Chicago: Geo. J. Ball, Inc., 1972.

5 Blaney, L. T., and A. N. Roberts, Growth and development of the Easter lily bulb *Lilium longiflorum* Thunb. 'Croft,' *Proc. Amer. Soc. Hort. Sci.*, 89:643–50. 1966.

6 Booth, A., The role of growth substances in the development of stolons, in J. D. Ivins and F. L. Milthorpe (eds.), *The Growth of the Potato*. London: Butterworth & Co., Ltd., 1963, pp. 99–113.

7 Bouwkamp, J. C., and L. D. Scott, Production of sweet potatoes from root pieces, *HortScience*, 7(3):271–72. 1972.

8 Crossley, J. H., Hyacinth culture; Narcissus culture; Tulip culture, *Handbook on Bulb Growing and Forcing*, Northwest Bulb Growers Assoc., pp. 79–84, 99–104, 139–44. 1957.

9 Curtis, A. H., Growth studies of King Alfred narcissus bulbs, *Proc. Amer. Soc. Hort. Sci.*, 36:781–82. 1938.

10 De Hertogh, A. A., A. N. Roberts, N. W. Stuart, R. W. Langhans, P. G. Linderman, R. H. Lawson, H. F. Wilkins, and D. C. Kiplinger, A guide to terminology for the Easter lily, *HortScience*, 6:121–23. 1971.

11 De Hertogh, A. A., and N. Blakely, The influence of temperature and storage time on growth of basal roots of nonprecooled and precooled bulbs of *Lilium longiflorum* Thunb. cv. 'Ace,' *HortScience*, 74:409–10. 1972.

12 Everett, T. H., *The American Gardener's Book of Bulbs*. New York: Random House. 1954.

13 Gould, C. J., Blights of lilies and tulips, *Plant Diseases. USDA Yearbook of Agriculture, 1953*. Washington, D.C.: U.S. Govt. Printing Office, pp. 611–16.

14 Gregory, L. E., Physiology of tuberization in plants (tubers and tuberous roots), in *Encyclopedia of Plant Physiology*, Vol. 15. Berlin: Springer-Verlag, 1965, pp. 1328–54.

15 Griffiths, D., The production of tulip bulbs, *USDA Bul. 1082*, 1922.

16 ———, Daffodils, *USDA Cir. 122*, 1930.

17 ———, The production of hyacinth bulbs, *USDA Cir. 112*, 1930.

18 ———, The production of lily bulbs, *USDA Cir. 102*, 1930.

19 ———, Artificial propagation of the lily, *Proc. Amer. Soc. Hort. Sci.*, 29:519–21. 1932.

20 Harrison, A. D., Bulb and corm production, *Bul. 62, Minist. Agr., Fish. and Foods*, London. 1964, pp. 1–84.

21 Hartsema, A. M., Periodieke ontwikkeling van *Gladiolus hybridum* var. Vesuvius, *Verh. Koninkl. Ned. Akad. van Wet.*, 36(3):1–34. 1937.

22 ———, Influence of temperatures on flower formation and flowering of bulbous and tuberous plants, in *Encyclopedia of Plant Physiology*, Vol. 16. Berlin: Springer-Verlag, 1961, pp. 123–67.

23 Heath, O. V., and M. Holdsworth, Morphologenic factors as exemplified by the onion plant, *Symposia for the Soc. Exp. Biol.*, II:326–50. 1948.

24 Huisman, E., and A. M. Hartsema, De periodieke ontwikkeling van *Narcissus pseudonarcissus* L., *Meded. Landbouwhoogesch., Wageningen*, DL. 37 (Meded. No. 38, Lab. v. Plantenphys. onderz., Wageningen). 1933.

25 Jacobi, E. F., *Plantkunde voor tuinbouwscholen*. Zwolle, The Netherlands: W. E. J. Tjeenk Willink, 1950.

26 Jacoby, B., and A. H. Halevy, Participation of light and temperature fluctuations in the induction of contractile roots of gladiolus. *Bot. Gaz.,* 131(1):74–77. 1970.

27 Kender, W. J., Rhizome development in the lowbush blueberry as influenced by temperature and photoperiod, *Proc. Amer. Soc. Hort. Sci.,* 90:144–48. 1967.

28 Kofranek, A. M., and G. Barstow, The use of rooting substances in *Cymbidium* green bulb propagation, *Amer. Orch. Soc. Bul.,* 24(11):751–53. 1955.

29 Langhans, R. W., and T. C. Weiler, Vernalization in Easter lilies, *HortScience,* 3:280–82. 1968.

30 Lewis, C. A., Some effects of daylength on tuberization, flowering, and vegetative growth of tuberous-rooted begonias, *Proc. Amer. Soc. Hort. Sci.,* 57:376–78. 1951.

31 Lin, P. C., and A. N. Roberts, Scale function in growth and flowering of *Lilium longiflorum,* Thunb. 'Nellie White,' *Jour. Amer. Soc. Hort. Sci.,* 95(5): 559–61. 1970.

32 Magie, R. O., Some fungi that attack gladioli, *Plant Diseases. USDA Yearbook of Agriculture, 1953.* Washington, D.C.: U.S. Govt. Printing Office, pp. 601–7.

33 Mann, L. K., Anatomy of the garlic bulb and factors affecting bulb development, *Hilgardia,* 21:195–251. 1952.

34 MacGillivray, J. H., *Vegetable Production.* New York: Blakiston, 1953.

35 McClure, F. A., *The Bamboos: A Fresh Perspective.* Cambridge, Mass.: Harvard University Press, 1966.

36 Moser, B. C., and C. E. Hess, The physiology of tuberous root development in Dahlia, *Proc. Amer. Soc. Hort. Sci.,* 93:595–603. 1968.

37 Mulder, R., and I. Luyten, De periodieke ontwikkeling van der Darwin tulip. *Verh. Koninkl. Ned. Akad. van Wet.,* 26:1–64. 1928.

38 Nitsch, J. P., Perennation through seeds and other structures, in F. C. Steward (ed.), *Plant Physiology,* Vol. VIA. New York: Academic Press, 1971, pp. 413–79.

39 Pfieffer, N. E., A morphological study of *Gladiolus, Contrib. Boyce Thomp. Inst.,* 3:173–95. 1931.

40 Rees, A. R., *The Growth of Bulbs.* New York: Academic Press, 1972.

41 Reinders, E., and R. Prakken, *Leerboek der Plantkunde.* Amsterdam: Scheltema & Holkema N. V., 1964.

42 Roberts, A. N., and L. T. Blaney, Easter lilies. Culture, *Handbook on Bulb Growing and Forcing,* Northwest Bulb Growers Assoc., pp. 35–43. 1957.

43 Rockwell, F. F., E. C. Grayson, and J. de Graaf, *The Complete Book of Lilies.* Garden City, N.Y.: Doubleday & Company, Inc., 1961.

44 Sheehan, T. J., Caladium production in Florida, *Fla. Agr. Ext. Circ. 128,* 1955.

45 Slater, J. W., Mechanisms of tuber initiation, in J. D. Ivins and F. L. Milthorpe (eds.), *The Growth of the Potato.* London: Butterworth & Co., Ltd., 1963, pp. 114–20.

46 Stuart, N. W., Moisture content of packing medium, temperature and duration of storage as factors in forcing lily bulbs, *Proc. Amer. Soc. Hort. Sci.,* 63:488–94. 1954.

47 Wang, S. Y., and A. N. Roberts, Physiology of dormancy in *Lilium longiflorum* Thunb. 'Ace' *Jour. Amer. Soc. Hort. Sci.,* 95(5):554–58. 1970.

48 Weiler, T. C. and R. W. Langhans. Growth and flowering responses of *Lilium longiflorum* Thunb. 'Ace' to different day lengths, *Jour. Amer. Soc. Hort. Sci.,* 97(2):176–77. 1972.

49 Welch, N. C., and T. M. Little, Heat treatment and cutting for increased sweet potato slip production, *Calif. Agr.,* 21(5):4–5. 1967.

50 Went, F. W., Thermoperiodicity, in A. E. Murneck and R. O. Whyte (eds.), *Vernalization and Photoperiodism.* Waltham, Mass.: Chronica Botanica, 1948.

51 Wilson, K., and J. N. Honey, Root contraction in *Hyacinthus orientalis, Ann. Bot.,* 30:47–61. 1966.

52 Woodcock, H. B. D., and H. T. Stearn, *Lilies of the World.* New York: Charles Scribner's Sons, 1950.

53 Zweede, A. K., De periodieke ontwikkeling van *Convallaria majalis, Verh. Koninkl. Ned. Akad. van Wet.,* 27:1–72. 1930.

SUPPLEMENTARY READING

Crockett, J. V., *Bulbs.* New York: Time-Life Books, 1971.

"Daffodil Handbook," *American Horticultural Magazine,* Vol. 45, No. 1 (1966), 227 pp.

Everett, T. H., *The American Gardener's Book of Bulbs.* New York: Random House, 1954.

Genders, R., *Bulbs, A Complete Handbook.* New York: The Bobbs-Merrill Company, 1973.

Gould, C. J., "The Flower Bulb Industry," *Washington Agricultural Experiment Station Circular 318* (1959), pp. 1–59.

Harrison, A. D., "Bulb and Corm Production," *Ministry of Agriculture, Fisheries, and Food Bulletin No. 62* (London) (1964), pp. 1–84.

Hartsema, A. M., "Influence of Temperatures on Flower Formation and Flowering of Bulbous and Tuberous Plants," in W. Ruhland (ed.), *Encyclopedia of Plant Physiology,* Vol. 16. Berlin: Springer-Verlag, 1961, pp. 123–67.

Ivins, J. D., and F. L. Milthorpe (eds.), *The Growth of the Potato.* London: Butterworth & Co., Ltd., 1963.

Kiplinger, D. C., and R. W. Langhans (eds.), *Easter Lilies, The Culture, Diseases, Insects and Economics of Easter Lilies.* Columbus: Ohio State Univ.; and Ithaca, N. Y.: Cornell Univ., 1967.

N. A. Gladiolus Council, *The World of the Gladiolus.* Edgewood, Md.: Edgewood Press, 1972.

Rees. A. R., *The Growth of Bulbs.* London: Academic Press, Ltd., 1972.

———, "The Physiology of Ornamental Bulbous Plants," *Botanical Review,* Vol. 32 (1966), pp. 1–22.

Aseptic Methods of Micro-Propagation **16**

Micro-propagation is the development of new plants in an artificial medium under aseptic conditions from very small pieces of plants, such as embryos, seeds, stems, shoot tips, root tips, callus, single cells and pollen grains. For propagation new roots and/or shoots, or small embryos (embryoids) must develop to produce an entire plant. For research a tissue, such as callus, or an organ, such as a root, may be maintained in culture indefinitely without an entire plant developing. The kind of growth pattern that develops depends upon the genetic potential of the plant cultured and upon the chemical and physical environment to which the plant part is subjected. Potentially, plants of all species could be reproduced by these techniques if their nutritional, hormonal, and cultural requirements were sufficiently known. Of those species that have been investigated, responses and degrees of success have varied considerably.

Micro-propagation has been used in research laboratories throughout the world since the first (but unsuccessful) attempt by Haberlandt in 1902 (9) to isolate and grow plant tissue in aseptic culture.

By 1934 P. P. White was able to grow tomato roots continuously *in vitro* by supplying them with yeast extract, the essential ingredients turning out to be certain B vitamins, notably thiamin. In 1939 three investigators, Nobecourt and Gautheret in France, and White in the United States, reported independently the indefinite culture of plant callus tissue in a synthetic medium (58). Since these early pioneering efforts, the technique of tissue and organ culture has been utilized extensively as a standard research tool in many plant physiology, plant pathology, and genetics laboratories (6, 19, 55, 56).

However, this method has also become practical for plant propagation, *(10, 28, 30, 33)*. In 1922 Knudson *(24, 25)* described culture techniques for germinating orchid seeds, a procedure now in standard horticultural use. Embryos whose germination is delayed because they are dormant within the intact seed, or which abort if they remain until maturity, can be excised from the seed and successfully germinated by *embryo culture (26, 37, 45, 47, 51)*. The successful production of whole haploid and subsequently diploid tobacco plants from pollen grains in aseptically cultured anthers has significant applications in plant breeding *(3, 38)*. Even more significant is the production of hybrid tobacco plants created by uniting in culture cells which were previously isolated from two separate tobacco species *(4)*.

Vegetative (clonal) propagation of specific plants can be achieved by a variety of techniques described in this chapter, including production of vegetative embryos *(embryoids)*, and variations of *tissue culture* and *shoot-tip* (meristem) *culture*. The small piece of plant used to begin a culture has been referred to as an *explant*.

The principal applications of these techniques to plant propagation are: (a) to establish pathogen-free plants (see p. 197), (b) to isolate genetically unique cells or cell lines that can become new plant variants, and (c) to enable rapid multiplication under conditions that maintain freedom from disease. These techniques are particularly significant for providing new methods of rapid vegetative propagation for plants where normal reproduction is very slow (e.g., orchids) or not possible.

GENERAL TECHNIQUES FOR PREPARING CULTURES FOR MICRO-PROPAGATION

Plant parts invariably have various microorganisms (bacteria, fungi, yeasts, etc.) on their external surface, although internally or within such closed structures as buds such organisms are generally not found. Sterile culture media are usually so favorable for growth of microorganisms that the excised plant part would quickly be destroyed if contaminating organisms were not eliminated before its transfer to the culture medium.

Preparation Room

Micro-propagation should be performed in a clean room, free of dust and convection currents that carry spores of microorganisms. Satisfactory freedom from contamination can sometimes be achieved on a table or laboratory bench in a relatively open room if the work is done rapidly, if careful techniques are followed, and if room openings, such as doors and windows, are closed or are properly filtered.

Best results are achieved by doing the operations in an enclosed transfer chamber under positive filtered air pressure, sterilized inside, well lighted, and suitably ventilated. Filling the chamber with steam, as from a pressure cooker, prior to using the chamber will help remove dust particles that may carry contaminating organisms. The chamber may be a large walk-in type with a chair and bench or it may be an enclosed box in which only the operator's hands are inserted.

Ultra-violet (UV) germicidal lamps may be used to sterilize the interior of the chamber. These are turned on about two hours prior to using the chamber but must be turned off during operations. The light should be directed inward and not be looked at directly because of danger to the eyes. UV light will not penetrate glass.

Equipment Used in Micro-Propagation

Although a wide diversity of procedures and equipment is used by different propagators, certain basic materials are required.

Autoclave or pressure cooker   Such equipment is used to sterilize the media, containers, petri dishes, and the various instruments required in the transfer operations. In an autoclave, temperatures of 250° F (120° C) at 15 lbs pressure can be obtained. Twenty to 30 minutes at this temperature is usually used.

Containers for growing cultures   These include test tubes of various sizes, bottles, flasks, etc. They should be Pyrex so that they can be sterilized and the openings flamed during the transfer operations. Closures should prevent recontamination and allow aeration but reduce moisture loss. Non-absorbent cotton plugs, screwcap bakelite covers, various kinds of metal or plastic covers, and closures made of polyurethane foam, parafilm, and polypropylene are available from commercial supply houses. *Petri dishes* (shallow, flat, covered containers made of Pyrex glass or plastic) are useful either for growing cultures or to hold material during the transfer operations. They can be sterilized in an autoclave (30 minutes or more) in metal containers or simply wrapped in paper bags. They may also be sterilized dry in an oven (338° F (170° C) for three to four hours.

Alcohol lamps   Small glass or metal laboratory lamps with a wick are used to sterilize instruments or to flame the opening of test tubes and other containers.

Disinfectants   Ethyl alcohol (70 to 90 percent) is very useful to wipe surfaces of working areas, to rinse hands or gloves, and to dip instruments, with or without subsequent flaming. Alcohol exaporates quickly and is nontoxic to hands. *Because of its flammability, extreme care should always be used with alcohol around a flame.* Calcium or sodium hypochlorite (see p. 164) are useful for sterilizing equipment and surfaces of working areas but, more important, to disinfect plant tissues and seeds without injuring them. Mercuric chloride (1:1000) is useful for disinfecting working surfaces but may be somewhat toxic to the skin. Merthiolate, diluted 1:1 with either water or alcohol, is useful for disinfesting some seeds (7, 47).

Sterile water   This is needed to wash plant pieces after disinfestation. It should be autoclaved for periods up to an hour depending upon volume. Water can be sterilized in a large flask (Pyrex) closed by a nonabsorbent cotton plug. It is dispensed through Pyrex glass siphon and rubber or plastic tubing closed with a clamp.

Tweezers, forceps, and scalpels   These can be obtained from scientific supply houses and come in various sizes, shapes and models. In some cases, special microtools may have to be made *(43)*. For instance, for excising shoot tips, a piece of razor blade may be broken off and cemented to a wooden handle. These instruments can be sterilized before use in an autoclave or they can be dipped in alcohol and flamed during the transfer operation, or exposed to UV light.

Rubber gloves, aprons, and laboratory coats   These are desirable to help maintain an antiseptic atmosphere. Hands should be washed with alcohol or Clorox, or gloves may be worn.

Labels and marking pencils   China marking pencils made of wax and useful for writing on glass surfaces. Various kinds of gummed labels are available including those that are color coded and which change color to signal change with time.

Fire extinguisher   A CO_2 type extinguisher should be available where alcohol lamps are used and flaming is practiced.

Media Preparation

Nutrients needed in the culture medium vary with the kind of plant and the purposes for producing the culture. Specific media and techniques have been developed empirically for certain plants and may not apply to others. Most tissue and organ cultures are grown on a semi-solid agar medium which gives good support to the culture. Purified powdered agar is used at a concentration of 0.5 to 1.0 percent, although for some embryo cultures the upper part of this range is too concentrated and can reduce growth *(50)*. Agar dissolves in water when heated but solidifies when cool into a semi-solid gel. Cultures may also be grown on a liquid medium, but this requires some method of aeration. A filter-paper insert, cut into a narrow strip, bent, and placed within the tube in a manner to extend above the medium will provide support and act as a wick for nutrients. This technique is useful if rooted plantlets are to be transplanted, since better roots with more root hairs are produced than if on agar *(21, 22)*. For the purpose of continuously bathing the plant tissue with the solution, machinery to rotate the flasks or to shake them vigorously are available or can be constructed.

The various kinds of nutrients to be added to the agar solution can be grouped into the following categories:

A. Minerals   A supply of inorganic elements—nitrogen, phosphorus, potassium, calcium, magnesium, and sulfur—is essential for all plant tissues and organs growing *in vitro*. Various combinations of these have been used by different individuals and there is considerable latitude in the kinds and amounts of chemicals utilized to supply the elements. Several well-known nutrient combinations that have been successful are listed below. Each material is first made up as a stock solution at 10 or 100 times the concentration required, autoclaved and stored, preferably in a refrigerator, until use.

Micro-nutrient elements may or may not be required but are usually added routinely. The following stock solution (*20*) will provide the required materials. One milliliter of this solution is added per liter of culture medium. Similar mixtures may also be used (*6, 18, 36, 55*).

$MnSO_4 \cdot 4H_2O$	1.81 g	$CuSO_4 \cdot 5H_2O$	0.08 g
H_3BO_3	2.86 g	$(NH_4)_2MoO_4$	0.09 g
$ZnSO_4 \cdot 7H_2O$	0.22 g	Distilled water	995 ml

Iron is usually essential and can be supplied in several ways: iron tartrate (1 ml of a 1 per cent stock solution) (*20, 57*); inorganic iron ($FeCl_3 \cdot 6H_2O$, 1 mg/1) or $FeSO_4$, 2.5 mg/1 (*55*); or chelated iron (*36, 57*). Chelated iron can be supplied as NaFeEDTA, 25 mg/1, or by mixing, Na_2EDTA and $FeSO_4 \cdot 7H_2O$ in equimolar concentrations to give 0.1 mM Fe.

B. Sugar   This is required for plant tissues and organs (except for mature embryos). Sucrose, 2 to 4 percent, is usually used, but glucose is also satisfactory.

C. Vitamins and other materials   Thiamin (0.1 to 1 mg/1), nicotinic acid (0.5 mg/1), and pyridoxine (0.5 mg/1) are required for some plant tissues and can be routinely added. Other materials sometimes beneficial include inositol (100 mg/1), pantothenic acid (0.1 mg/1), and biotin (0.1 mg/1). These materials are mostly soluble in water and should be made in stock solutions of 100 × final concentration. They should be stored in a refrigerator.

D. Growth regulators (**auxins and cytokinins**)   Synthetic auxins widely used are naphthaleneacetic acid (NAA) (0.1 to 10 mg/1) and 2,4-dichlorophenoxyacetic acid (2,4-D) (0.05 to 0.5 mg/1), or sometimes indoleacetic acid (IAA) (1 to 50 mg/1). Auxin materials should be dissolved in a small amount of alcohol before being added to water to make up a stock solution. Kinetin (0.01 to 10.0 mg/1) is required for tissue culture of many species. A stock solution is prepared by first dissolving it in a small amount of HCl and then diluting with water.

E. Organic complexes   The most useful of these materials is coconut milk at a rate of 10 to 15 percent by volume. It comes from either young (preferred) or ripe coconuts, but there is considerable variation in its effect. If immature embryos are being grown, the milk should be sterilized by filtering rather than by heating (*53*). Yeast extract (*55*) or amino acid complexes, e.g., casein hydrolysate (*44*) have sometimes been used. A banana puree has been beneficial in orchid seed cultures (*15*).

The following are examples of some widely used media:

Murashige and Skoog basic medium (*36*)   This medium was first used to grow tobacco callus but has been widely used for other tissues.

NH_4NO_3	400 mg/l	IAA	2.0 mg/l
$Ca(NO_3)_2 \cdot 4H_2O$	144	kinetin	0.04-0.2
KNO_3	80	thiamin	0.1
KH_2PO_4	12.5	nicotinic acid	0.5
$MgSO_4 \cdot 7H_2O$	72	pyridoxine	0.5
KCl	65	glycine	2.0
$NaFe-EDTA$	25	myo-inositol	100
H_3BO_3	1.6	Casein hydrolysate	1000
$MnSO_4 \cdot 4H_2O$	6.5	sucrose	2%
$ZnSO_4 \cdot 7H_2O$	2.7	powdered purified agar	1%
KI	0.75		

Murashige and Skoog high salt medium *(36)*   This was developed to produce maximum growth from tobacco callus but may have too high a salt concentration for some kinds of tissue.

NH_4NO_3	1650 mg/l	H_3BO_3	6.2 mg/l
KNO_3	1900	$MnSO_4 \cdot 4H_2O$	22.3
$CaCl_2 \cdot 2H_2O$	440	$ZnSO_4 \cdot 4H_2O$	8.6
$MgSO_4 \cdot 7H_2O$	370	KI	0.83
KH_2PO_4	170	$Na_2MoO_4 \cdot 2H_2O$	0.25
$Na_2EDTA(1)$	37.3	$CuSO_4 \cdot 5H_2O$	0.025
$FeSO_4 \cdot 7H_2O(1)$	27.8	$CoCL_2 \cdot 6H_2O$	0.025

(1) 5 ml/l of a stock solution containing 5.57 gm $FeSO_4 \cdot 7H_2O$ and 7.45 gm Na_2-EDTA per liter of water.

Organic constituents are the same as added to the basic medium.

White's medium   This medium was first developed for root cultures and has been used for routine culturing of many plant tissues *(55)*. Very often it has been modified for specific plants and has been supplemented by additional organic materials.

KNO_3	80 mg/l	KCl	65 mg/l
$Ca(NO_3) \cdot 4H_2O$	300	$Fe(SO_4)_3$	2.5
$NaH_2PO_4 \cdot 4H_2O$	19	glycine	3.0
Na_2SO_4	200	nicotinic acid	0.5
sucrose	2%	thiamin	0.1
agar	0.5 to 1 %	pyridoxine	0.1
$MgSO_4 \cdot 7H_2O$	750 mg/l		

A micro-nutrient solution is usually added. (See p. 513.)

Knudson C *(25)*   This medium is widely used for orchid seed and for shoot-tip (meristem) cultures.

$Ca(NO_3)_2 \cdot 4H_2O$	1000 mg/l	$MgSO_4 \cdot 7H_2O$	250 mg/l
$(NH_4)_2SO_4$	500	$MnSO_4 \cdot 4H_2O$	0.0075 gm/l
KH_2PO_4	250	$FeSO_4 \cdot 7H_2O$	0.025 gm/l

To this is added 2% sucrose and 1.5% agar.

[This medium has been modified *(15)* by replacing the KH_2PO_4 with 18 ml of a potassium phosphate buffer prepared by combining 97.5 ml of 0.1 M KH_2PO_4 solution (13.6 gm in 1 liter water) and 2.5 ml of 0.1 M KH_2PO_4 solution (17.4 gm in 1 liter water). This maintains the pH at 5.3 without further adjustment. To this, 1 ml of a micro-nutrient solution (see p. 513) is added.]

Knop's solution *(55)*   This basic nutrient solution was first developed by Knop in 1865 but has been modified by various others *(20, 45)*. It is useful for growing embryos.

KNO_3 .... 125–200 mg/l
$Ca(NO_3)_2 \cdot 4H_2O$ 500–800
K_2HPO_4 125–200
$MgSO_4 \cdot 7H_2O$ 125–200

To this solution is added a micro-nutrient stock solution and iron. Various organic supplements and sucrose could be added.

Preparation of the medium requires a supply of pipettes, volumetric flasks, and containers to measure accurately the required materials. To prepare a nutrient solution, start with a portion of the required water, which should be distilled, or preferably redistilled, or de-ionized (particularly if critical studies on nutrition are being made). Add to this the various aliquots of nutrients from the stock solutions with an appropriate pipette. Sugar is added to the solution and allowed to dissolve. The agar may be either heated separately to combine with the remainder of the solution or simply added to the solution to be heated all together. Water is then added to make up the solution to the required volume. The mixture is heated to dissolve the agar. Care must be taken not to overheat, since the agar solution tends to boil, or it may burn on the bottom of the flask. The hot solution is mixed to insure uniformity and immediately distributed into culture vessels.

Ten to twenty-five milliliters of medium are used per culture flask. The usual pH range of the medium for effective growth is 5 to 6. This can be determined with a pH indicator and adjusted by adding acid (e.g., HCl) or base (e.g., KOH) accordingly. Culture tubes and medium should be sterilized at 250° F (120° C) for 15 minutes in an autoclave or pressure cooker. For materials that are affected by high temperature, sterilization should be by microbial filters and the solution added to the remainder of the medium, which has been previously autoclaved.

Most of the chemicals, instruments, containers, and the like listed in this section are standard equipment in scientific laboratories. They can be obtained from scientific supply houses. Ready-mixed culture media is also available from some of these companies. The concentrations of materials given are those that have been used in various experimental activities. The required materials and the optimum concentrations should be determined for each situation.

PROCEDURES FOR CULTURING
VARIOUS TISSUES AND ORGANS
Embryo Culture

An embryo culture is prepared by aseptically removing the embryo from the seed and placing it on a sterilized culture medium to germinate. Because the nutritional requirements of a developing embryo become less complex as it approaches maturity the culture medium used will depend upon the stage of development (see p. 56). Embryos excised in their last stage of development, morphologically full-size although possibly low in weight, require a very simple medium—inorganic salts and sugar (A and B) or only salts. However, in their early embryonic stages the requirements are complex, and little success has been achieved until the early heart-shaped stage is attained (37, 41), in which the cotyledons begin to elongate. Nutritional requirements at this stage include inorganic salts (A), sugar (B), possibly vitamins (C), and sometimes compounds of group E. Unautoclaved coconut milk (53, 54), casein hydrolysate (44), or a high (8 percent) sucrose supply (42) have been beneficial. Embryos part way in their second stage of growth, where they are increasing in size, have been cultured more or less successfully.

An agar medium is usually most useful for embryo culture. To prepare a culture, surface sterilize the fruit for 15 minutes with a disinfectant, such as carbolic acid (5 percent for five minutes), alcohol, calcium or sodium hypochlorite or merely wash thoroughly. Cut open the fruit if soft or crack in a vise if hard. Extract the seed (or ovule) with a sterile tweezers and place into a sterile Petri dish. Remove the seed coverings or cut open the ovule to extract the embryo. Place it on the surface of the medium or in the case of large seeds insert the radicle end one-third to half-way down. All of these operations should normally be done in a sterilized transfer room.

If chilled for several months the immature embryos of some plants, such as *Prunus* and other fruit tree species, respond by better survival and normal growth (47). One month at 41° F (5° C) and four months at 34° F (1° C) is desirable for *Prunus* embryos.

Merthiolate has been used to treat mature *Prunus* seeds in preparation for aseptic stratification (7, 47). Dry seeds are soaked for five minutes in a 1:2000 solution of merthiolate in 50 percent alcohol in an Erlenmeyer flask. They are then washed for five minutes in sterile water (autoclaved for one hour at 250° F; 120° C), with a siphon made from rubber attached to a glass tube extending from the flask of sterile water. The seeds are left to soak for 24 hours in the water, the excess poured off, and the flasks—with just enough water to half-cover the seeds—placed at the stratification temperature.

Flasks or tubes containing planted embryos (Figure 16–1) should be placed at room temperature in diffuse light, protected from the direct rays of the sun. When roots and shoots have developed, the seedlings are transplanted to a sterilized growing medium (see Chapter 7). These seedlings are unusually tender and particular care must be given to protect them from strong light, high temperatures, and drying when they are transplanted from the container.

Figure 16–1 Seedlings growing aseptically on nutrient agar medium. *Right:* peach seedlings. *Top:* orchid seedlings. Courtesy Shaffer Tropical Gardens, Santa Cruz, California.

Orchid Seed Cultures (15, 48)

Orchid seeds are extremely tiny and it is estimated that some 30,000 seeds are produced in a single Cattleya pod. The seeds may be extracted from a green pod before it has opened naturally (about 60 percent mature) without seed sterilization, except for cleaning the pod with disinfectant.

In a second method, the seeds are removed from the fruit and treated with disinfestant. Orchid seeds are placed in a small vial or flask and covered with five to ten times their volume of sodium or calcium hypochlorite solution plus a drop or two of a wetting agent. Seeds are left for five minutes (sometimes longer), shaken periodically. At the end of this time, seeds will have sunk to the bottom of the tube. Gently pour off the disinfestant and add a portion of sterile water to half the vial. After shaking a few times, pour off the water containing seeds into the germination flask. An alternate procedure is to dip out the seeds with a platinum wire loop, as used for bacteriological studies, or a spatula.

The culture medium mostly used for orchids is Knudson's C (see p. 514) with 1.0 to 1.5 percent agar. Germination is apparent within a few months and by the fourth to sixth month the small plants must be transplanted to a new flask. About a year after sowing, the young plants are moved to containers (*15, 48*).

Pollen Grain Culture

Whole plants have been produced from pollen grains of tobacco (*38*) (Figure 16–2). Unopened flower buds are sterilized in a 1:3 Clorox:water solution for 10 to 20 minutes. The flower buds are then opened aseptically and the anthers removed and planted in a culture solution containing inorganic materials, sucrose, and vitamins (A, B, C). IAA may be helpful and iron seems essential (*51*). Within four to six weeks small plantlets emerge that are

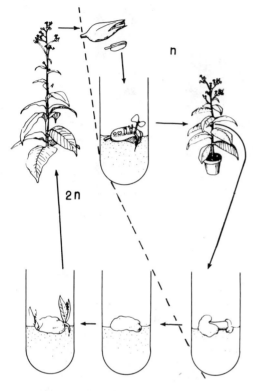

Figure 16–2   Procedure for obtaining haploid plants from *Nicotiana tabacum*, then diploid plants from the haploid ones. A flower bud at the right stage is excised from a flowering plant *(upper left)*. The immature stamen is removed and planted aseptically on the proper nutrient medium. Pollen grains at the uni-nucleated, microspore stage develop into haploid embryos which germinate and form plantlets which are transplanted into pots in the greenhouse. The haploid plants flower abundantly but do not set seed as they contain only *1n* number of chromosomes per cell. In a second step, stem sections of haploid plants are surface sterilized then planted aseptically on a nutrient medium which favors proliferation of a callus *(lower right)*. This callus can be transferred to the same medium in order to get rid of the initial explant and to let the process of endomitosis produce diploid cells. The callus is then transferred to a new medium which favors the formation of adventitious shoots *(lower left)* from which whole plants can be raised, the majority of which are diploid and capable of setting seed. Courtesy J. P. Nitsch *(38)*.

then transferred to another culture medium with 1 percent sucrose and no IAA. Eventually these grow sufficiently to be transplanted to soil. It is essential that the pollen grains be at the correct stage or only callus is produced; this is after the tetrad stage but before they have matured or developed starch. At this time they are individual, with a single nucleus.

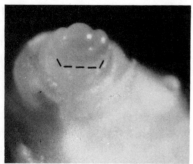

Figure 16–3   Shoot tip of carnation stem with outer leaves removed, showing the apical and lateral meristems (growing points). Part of shoot tip to be excised for culturing is indicated by lines. Courtesy W. P. Hackett.

Plants produced by this method are haploid (i.e., with only a single set of chromosomes). Chromosome numbers may be doubled by treating with colchicine solution *(3, 38)*, thus producing a homozygous diploid plant.

Shoot-Tip Culture
This technique is used to produce pathogen-free clones (see p. 197) of such plants as carnation, chrysanthemum, garlic, potato, strawberry, dahlia, lily, iris, freesia, gladiolus, and orchid *(22)*. Variations of this procedure are used as a rapid method of asexually multiplying orchids, as shown in Figure 16–5. Orchids nor-

mally reproduce very slowly by vegetative means. Such techniques may be used to propagate other plants, as bromeliads (*28*).

In shoot-tip culture the apical meristem and the first 1 or 2 leaf primordia below it (Figure 16–3) are excised from the shoot tip and placed in conditions to form roots at their base, produce a small plantlet. The meristem itself is not able to grow independently unless some leaf primordia are retained.

The shoot-tip enclosed in developing leaves is normally free of contaminating organisms, but one should avoid overhead watering of source plants because water may carry bacteria into leaf axils. For many plants, such as carnation (*21*) and chrysanthemum (*2*), a growing shoot several centimeters long is cut from the plant and the larger, lower leaves stripped away.

The stem piece is laid on a sterile surface and the remaining leaves near the tip removed with a sterile needle or scalpel. The exposed apical meristem appears as a small dome of tissue, as shown in Figure 16–4. A segment of the tip, 0.2 to 0.5 mm in length, consisting of the shoot-tip and the one to four subtending leaves, is removed and transplanted to the culture medium.

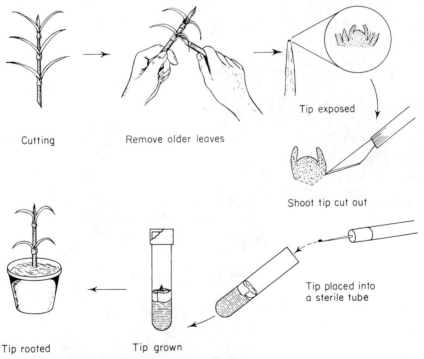

Cutting   Remove older leaves

Tip exposed

Shoot tip cut out

Tip placed into a sterile tube

Tip rooted   Tip grown

Figure 16–4 Shoot-tip culture of carnation. Following the arrows: the carnation cutting is obtained, the larger leaves are stripped away, and then the small, enclosing leaves at the extreme tip are removed to expose the growing point. The tip and the next subtending leaf primordia are removed with a scalpel and placed on the surface of a paper wick in a test tube with nutrient media. The shoot tip grows in the tube until large enough to be transplanted to a container. A full-size plant is shown at bottom left. Redrawn from Holley and Baker (*21*).

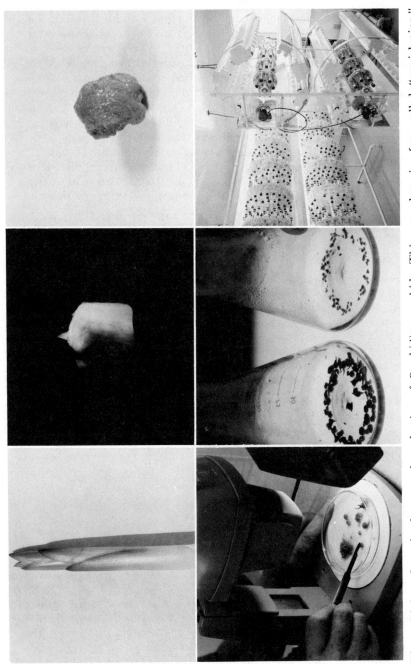

Figure 16–5 Steps in the asexual production of *Cymbidium* orchids. This procedure is often called "mericloning." *Top row* (left to right: growing shoot–source of meristem; excised shoot apical meristem; protocorm-like bodies, after enlarging in nutrient medium and before dividing. *Second row* (left to right): dividing each enlarged protocorm-like structure into several pieces; before and after dividing; rotating wheels with divided mericlones in nutrient solutions for enlargement and prevention of root-shoot polarity.

Figure 16–5 (continued). *Third row* (left to right): protocorm-like material placed on nutrient agar for polarizing growth and development of roots and shoots; greenhouse filled with flasks for growing on into plants; well developed plants large enough for flatting. *Bottom row* (left to right): well developed orchid plants, ready for potting; potted *Cymbidium* orchid, originating from a divided protocorm-like body; orchid house with plants coming into flowering. Courtesy Richard Smith, Rod McLellan Co., South San Francisco (48).

An additional or alternate procedure is to surface sterilize the stem by dipping into alcohol and then soak in a hypochlorite solution for ten minutes followed by washing with sterile water.

The medium needed for shoot-tip culture is usually simple. The basic requirements include inorganic salts and sugar. For some plants, materials in group C—particularly thiamin—appear necessary. Naphthaleneacetic acid (1 ppm) is recommended for root initiation of some plants—e.g., carnation. When roots begin to appear the stem piece is transferred to a medium without NAA. Agar has been used as a support, but filter paper wicks may be more satisfactory because a better root system is produced.

The shoot-tip cuttings are placed in moderate light and temperature. Adventitious roots should develop in three to four weeks. After the plantlet has grown enough to produce some small leaves, it can be transplanted into a soil medium.

Variations of this basic technique have been developed, illustrated by methods outlined below for several kinds of plants.

Orchids *(29, 31, 48, 58)*   As shown in Figure 16–5, an actively growing vegetative shoot is excised, the outer leaves stripped away, the bulb-like section dipped in alcohol, soaked in hypochlorite solution (15 to 20 minutes), then washed in sterile water. The remaining leaves are removed to expose the shoot apex, which is then excised and placed (usually) on Knudson's medium with 2 percent sucrose. It may be on agar, filter paper, or liquid medium. Within four to six weeks a small, round protocorm develops from this tip and will eventually produce a root and a shoot in the same way that an orchid seed germinates. However, before a shoot develops, the protocorm is removed from the flask and cut into sections which are replanted separately into new media. At monthly intervals the process is repeated as additional protocorms continue to develop. By monthly transfers and consecutive sectioning this multiplication process can build up large culture stocks of any one clone within a relatively short time.

Alternate procedures are to place the shoot tip into a liquid medium where it is continuously agitated on a shaking device or rotated on a circulating wheel. Under these conditions, new lateral protuberances continue to develop and are recultured every three to four weeks after slicing. Production of new protocorms will continue as long as the shaking or rotating continues and/or they are resliced and recultured every three to four weeks. When new plants with roots and shoots are desired, individual pieces are placed onto an agar medium, the shaking discontinued, and plants allowed to develop without reculturing. Within eight months leaves and roots are well developed and the young plants are removed from the flasks and planted.

Proliferation cultures of carnation *(11)*   A shoot-tip with four leaf primordia is prepared in the usual manner. Cuts are made in each leaf to mutilate it and the shoot-tip is placed in Murashige and Skoog high salt medium with naphthaleneacetic acid (2 ppm). Under these conditions a mass of callus with dark green areas is produced (Figure 16–6). Subcultures are made by dividing the mass into five to six pieces (including a green area) and placing the pieces in high salt media and 1 ppm NAA. In six to eight weeks cultures,

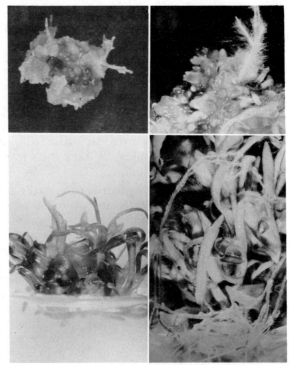

Figure 16–6 Proliferation cultures of carnation shoot apices. *Top left:* callus growth from mutilated shoot apices grown on high salt medium and an auxin (NAA) at 2 mg/l. *Top right:* subculture from top left growing on a medium with high salt but with NAA at 1 mg/l. *Bottom left:* subculture from top right but with low salt medium and no NAA. *Bottom right:* more advanced stage than that of bottom left, showing shoots and roots *(11)*. Courtesy W. P. Hackett.

as shown in Figure 16–6 are produced. This culture is subcultured every six weeks. If small pieces (about 50 mg) are transferred to fresh medium with lower concentrations of inorganic salts and no NAA, small rooted plantlets develop, as in normal shoot-tip cultures; these are eventually transplanted to soil. Unrooted tips are treated as cuttings and handled accordingly.

Chrysanthemum (2) Stem apices are removed without sterilization, retaining the apex and two leaf primordia about 0.2 mm below the apex. This is cultured in flasks on Murashige and Skoog medium plus 10 percent coconut milk, or 25 ppm inositol plus 0.8 ppm kinetin and 0.5 ppm IAA on agar. The growth produced is intermediate between normal shoots and leafy callus. After six weeks the explants are transferred to a rotating vessel using the same medium but without the agar. After three weeks the explants are transferred to larger culture flasks with 250 ml of liquid; subcultures are made monthly using explants of intermediate growth. The material is filtered through two layers of cheesecloth and the most differentiated segments selected for transfer. Eventually cells with shoots are removed and placed in a stationary flask on layers of cheesecloth partially submerged in liquid. As shoots elongate, roots form; after three months this medium is replaced by one with only inorganic materials. After hardening, the plants are transplanted to soil in a greenhouse.

Asparagus (17, 35, 60) Stock plants are produced *in vitro* by removing lateral growing points taken from spears 15 to 20 cm long previously sterilized in a hypochlorite solution for ten minutes, followed by washing. The outer scale leaves are removed. Murashige and Skoog medium is used plus NAA 0.3 ppm, kinetin 0.1 ppm, thiamin 1.0 ppm and sucrose 2.5 to 3 percent. Helpful but not required are myo-inositol 100 ppm, nicotinic acid 0.5 ppm, pyridoxine 0.1 ppm, adenine 40 ppm, and malt 500 ppm. Relatively low light intensity (1000 to 3000 lux; 93 to 280 ft.c.) is important, as is warm night temperature (84° F; 29° C). Spears and adventitious roots are formed in three to six weeks. These can be used as stock plants that are cultured by cutting into single bud segments every six weeks. To produce transplantable stock the plants are recultured in media without NAA and placed at higher light intensity (10,000 lux; 930 ft.c.), then transplanted into soil when the roots are about 8 cm long.

Gladiolus (61)   Young inflorescence stalks 250 to 300 mm long with five to six fully developed leaves are severed from the plant. Each stalk has the outer leaves removed and the remaining ones wiped with alcohol; the stalk is then taken into a sterile room. Leaves are rinsed with sterile water and removed to expose the inner stem. It is cut off below the first floret and the remainder sterilized in a hypochlorite solution for 15 minutes, then washed with sterile water. The terminal 50 mm piece is cut into disks 3 to 4 mm thick which are placed for sectioning on sterile filter paper soaked with 1 percent ascorbic acid to prevent tissue browning.

Murashige and Skoog medium is modified by replacing KH_2PO_4 with NaH_2PO_4 at 300 ppm and adding nicotine acid 5 ppm, pyridoxine 1 ppm, thiamine 0.5 ppm, and adenine 160 ppm. NAA at 10 ppm and kinetin at 0.5 ppm is added to the agar medium to grow the original explant. Best results occur by placing the explants upside down on the medium. Within six weeks roots and small structures resembling cormlets are then produced on the top of the section along with a thin layer of callus. The explants are reversed in orientation and recultured on the basal medium but with NAA at 0.5 ppm and kinetin at 0.5 ppm. With redividing every 6 weeks there is continuous production of small cormlets and roots. Such plants are very sensitive at transplanting and must be hardened-off gradually.

Gloxinia (16)   Shoots that are 4 to 5 cm long are cut from tubers, then rinsed with tapwater. A 10- to 15-mm-long trimmed shoot is placed into an antioxidant solution and then disinfected with hypochlorite solution for five minutes, followed by rinsing with sterile antioxidant. The 1 to 2 mm section of shoot-tip is excised and placed in a culture tube containing a Murashige and Skoog high salt medium. Low light intensity (300 ft c.) at 16 hours photoperiod is important. With IAA at 0.1 ppm and kinetin at 30.0 ppm, maximum numbers of shoots, tubers, and roots are produced. These can be recultured and redivided for multiplication in three to four weeks. A single node or shoot can be removed and grown on a medium with IAA 1 ppm for rooting. After one month these explants can be transferred to soil.

Shoot-Apex Grafting (34)

A method of producing virus-free citrus clones is to micro-graft virus-free shoot apices onto pathogen-free seedling rootstock plants growing in sterile culture. The seed is sterilized and germinated in a test tube on a sterile medium. At three weeks it is removed, placed in a sterile Petri dish lined with sterile filter paper. The seedlings are decapitated 3 cm below the apex. Terminal shoots 2- to 3-cm long are removed from a mature "scion" citrus tree, surface sterilized, and transferred onto moist filter paper in a sterile Petri dish.

The shoot apices, with primordial leaves, are excised under a dissecting microscope using micro-scalpels made from pieces of razor blades. The apices are inserted onto the exposed end of a decapitated rootstock. The grafted plant is then transferred to a filter paper support in a test tube; a liquid medium with a mineral solution (Group A) is used and the grafts are then grown in the light. The grafts are transferred to a fresh medium monthly and all rootstock shoots must be continuously removed. In five months successful grafts are sufficiently large to be transferred to soil.

Tissue Cultures

A tissue culture is a mass of more or less unorganized callus (parenchyma) growing on an artificial medium, separately from the plant from which it originated. Size increases by continuous cell division. After a period of time—four to six weeks—the mass is sufficiently large to divide into sections which are recultured to produce additional tissue cultures. Although used considerably in research, the value of tissue cultures as a plant propagation procedure depends upon the capacity of the callus either (a) to differentiate roots and shoots, or (b) to develop small embryo-like structures (*embryoids*) from individual cells and cell clumps. Root and shoot differentiation is more easily done with certain herbaceous plants, for example, carrot (49), tobacco (33), and endive (52), less successfully with woody plants, but has been done with aspen (62) and citrus (8), for example.

One problem with tissue cultures is the variation among cells and cell lines that has been found to occur (19). Some of these variations may have been present in the plant from which the cultures were made, such as differences in virus content (5) or differences in chromosome numbers (33). Other changes evidently occur during the growth of the cultures after they have been made, resulting from instability within the cultures and variability in the plants that are produced (39). For instance, asparagus plants have been successfully propagated by callus cultures but many of the plants produced have been tetraploid rather than diploid (27). In contrast, asparagus plants propagated by the shoot-tip method do not show such variability (35).

A tissue culture can be started from a variety of plant parts which have cells capable of dividing. Usually tissues near the vascular area of stems and roots proliferate most satisfactorily, but cultures have been started from fruits, endosperm, pollen, and embryos.

Excision procedures, as outlined below, have been used successfully.

(a) A fleshy root, such as carrot, is cut open aseptically (following surface sterilization). Pieces of root tissue are removed with a cork borer, cut transversely into disks, and then planted on an agar surface.

(b)   Thick herbaceous stems, as found in tobacco, are cut into sections about 15 cm long, stripped of leaves, cleansed with 95 per cent alcohol and cut into 5-cm sections. Cylinders of the inner stem (pith) are removed with a cork borer, cut transversely into disks, and planted.

(c)   Immature shoots of woody plants can be cut into 10 to 15 mm sections, soaked in sodium hypochlorite (Clorox-water 1:9) for 15 to 20 minutes and then washed with sterile water. The two ends are cut off and discarded and the remainder cut into short segments and planted on the medium. Leaf petiole segments can be prepared in this way. Mature stems can also be handled in a similar manner, but these are more easily split in half than cut transversely. The cut side should be up. It may be desirable to remove buds of such pieces prior to treatment to facilitate sterilization.

(d)   Bark and cambium sections can be removed from woody plants aseptically after the outer roughened layer is removed and the part below surface-sterilized with alcohol or other material.

(e)   Developing embryos, germinating seeds, or shoot-tips can also be used as starting points.

Tissue culture media usually include material listed in the A,B,C, and D groups (see p. 512). An auxin, such as NAA or 2,4-D, is almost always required and, for many plants, kinetin is essential. Complex materials (group F) such as coconut milk have been stimulative and necessary for some plants. Darkness generally produces the best callus growth but light seems to be necessary for differentiation of shoots to occur. For those plants such as tobacco where differentiation of shoots and roots will take place, a proper balance of particular nutrients and growth regulators in the culture medium appears to be essential *(46)*. High auxin seems to promote root initiation whereas high kinetin or adenine tends to promote shoot initiation. The presence of each in proper balance has resulted in the production of both shoots

Figure 16–7   Nurse-tissue procedure used to grow single cells of a tobacco hybrid *(Nicotinana tobacum × N. glutinosa)*. *Far left:* small piece of filter paper is placed on top of a well-growing piece of callus tissue. *Center left:* a single cell is placed on top of filter paper, where it divides and grows into a small mass of callus. *Center right:* this callus piece continues to increase in size because it absorbs materials through the filter paper from the nurse tissue below. *Far right:* after this callus piece has attained some size, it can be removed from the filter paper; it will then grow independently. The minimum size for removal is somewhere between the two center stages. Courtesy A. C. Hildebrandt *(19)*.

Figure 16–8 Development from a single cell to a complete plant in tobacco. *Left top:* undifferentiated callus tissue which developed from a single pith cell produced by the nurse-culture technique. *Left below:* adventitious roots develop on the callus culture when it is grown on a medium high in auxin (IAA) and low in kinin (kinetin). *Center:* stems and leaves are produced when the callus is grown on a medium low in IAA and high in kinetin. *Right:* with the proper balance of IAA and kinetin, stems, leaves, and roots are produced, and a complete plant eventually develops. The plant on the right started from a single tetraploid cell; that on the left developed from a normal diploid cell. Courtesy T. Murishige.

and roots (see p. 220). The special nutrient requirements and environmental conditions must be established for each kind of plant being cultured. For instance, geranium callus forms shoots and later roots on low auxin and high kinetin and with 12 to 16 hours light period (*40*). Aspen cultures have formed shoots in culture with 0.1 to 0.24 ppm 6-benzylaminopurine (BAP) (a cytokinin) and no auxin, and have formed roots when transferred to a medium with no cytokinin but with auxin—i.e., NAA 0.25 to 1 ppm (*59*).

Single cell culture (23) For research purposes single cell cultures have been produced from tissue cultures which have been subjected to continuous shaking or which have been grown in special glass containers continuously rotated. Under these conditions, small cell clumps, or individual cells—particularly if the callus is soft and friable—slough-off into the culture medium. In general, friability is favored by a relatively high auxin medium or one containing coconut milk (*49*). Single cells may be removed with a micropipette under a dissecting microscope and transferred to a special growing environment. This growing environment can be provided by placing the cell on a small piece of filter paper that has been inserted on top of a well-growing callus "nurse" tissue (Figure 16–7 and Figure 16–8) of the same kind (*32*), or by growing the cell in a special "microchamber" in a "prepared" medium—that is, one in which these cells or similar tissue cultures were previously grown (*5*).

Production of embryoids (12, 13, 49) This procedure has been developed using carrot tissue. Tissue cultures are developed and maintained in a liquid medium containing a relatively low salt concentration, such as White's medium supplemented with unautoclaved coconut milk or casein hydrolysate

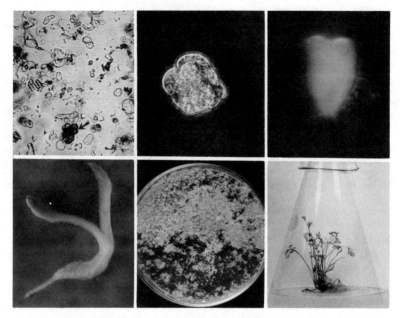

Figure 16–9   Embryoid production in carrot *(49)*. *Top left:* highly magnified view of suspended cells and cell clumps derived from tissue cultures of carrot growing in a liquid medium. Arrow points to a clump of cells—the beginning stage for an embryoid. *Top center:* single cell clump (higher magnification) illustrating the globular stage of development. *Top right:* a more advanced heart-shaped stage of embryoid development. *Below left:* mature embryoid that developed continuously from the globular stage, through the heart and torpedo (not shown, see text) stages. *Below center:* thousands of carrot plantlets that developed as embryoids. *Below right:* single carrot plant that grew from a single embryoid transplanted from previous stage. Courtesy F. C. Steward and M. O. Mapes.

and NAA 2 ppm on a rotating wheel. A small amount of the liquid containing a suspension of cells and cell clumps is removed with a small sterile pipette and transferred to a petri dish or test tube on a liquid or an agar medium.

The new medium must have no auxin and should contain ammonium nitrogen *(1, 14)*. The Murashige and Skoog medium is particularly useful in growing the plantlets in this second stage *(1)*. Small embryoids develop in this culture that can eventually become normal plants, as shown in Figure 16–9.

REFERENCES
1 Ammirato, P. V., and F. C. Steward, Some effects of environment on the development of embryos from cultured free cells, *Bot. Gaz.*, 132:149–58. 1971.

2 Ben-Jaacov, J., and R. W. Langhans, Rapid multiplication of chrysanthemum plants by stem-tip proliferation, *HortScience,* 7(3):289–90. 1972.

3 Burk, L. G., G. R. Gywnn, and J. F. Chaplin, Diploidized haploids from aseptically cultured anthers of *Nicotiana tabacum, Jour. Hered.,* 63(6):355–60. 1972.

4 Carlson, P. S., H. H. Smith, and Rosemarie Dearing, Parasexual interspecific plant hybridization. *Proc. Nat. Acad. Sci.* U.S.A., 69:2292–94. 1972.

5 Chandra, N., and A. C. Hildebrandt, Growth in microculture of single tobacco cells infected with tobacco mosaic virus, *Science,* 152:789–91. 1966.

6 Chaturvedi, H. C., and G. C. Mitra, Clonal propagation from somatic callus cultures, *HortScience,* 9(2): 118–20. 1974.

7 Gilmore, A. E., A technique for embryo culture of peaches, *Hilgardia,* 20:147–169. 1950.

8 Grinblat, U. H., Differentiation of *Citrus* stem in *vitro, Jour. Amer. Soc. Hort. Sci.,* 97:599–603, 1972.

9 Haberlandt, G., Kulturversuche mit isolierten Pflanzenzellen. *Sitzungsber. Akad. der Wiss. Wien, Math.-Naturwiss. Kl.,* 111:69–92. 1902.

10 Hackett, W. P., Applications of tissue culture to plant propagation, *Proc. Int. Plant Prop. Soc.,* 16:88–92. 1966.

11 ———, and J. M. Anderson, Aseptic multiplication and maintenance of differentiated carnation shoot tissue derived from shoot apices, *Proc. Amer. Soc. Hort. Sci.,* 90:365–69. 1967.

12 Halperin, W., Alternative morphogenetic events in cell suspensions, *Amer. Jour. Bot.,* 53:443–53. 1966.

13 ———, and D. F. Wetherell, Adventive embryony of the wild carrot *Daucus carota, Amer. Jour. Bot.,* 51:274–83. 1964.

14 ———, Ammonium requirements for embryogenesis *in vitro, Nature,* 205: 519–20. 1965.

15 Harrison, C., and J. Arditti, Growing orchids from seed, *The Orchid Digest,* 7:199–204. 1970.

16 Haramaki, C., Tissue culture of Gloxinia, *Proc. Inter. Plant Prop. Soc.,* 21: 442–48. 1971.

17 Hasegawa, P. M., T. Murashige, and F. H. Takatori, Propagation of asparagus through shoot apex culture. II. Light and temperature requirements, transplantability of plants and cyto-histological characteristics, *Jour. Amer. Soc. Hort. Sci.,* 98(2):140–42, 1973.

18 Heller, R., Some aspects of the inorganic nutrition of plant tissue cultures in *Plant Tissue Culture.* Berkeley, Calif.: McCutchan Publishing Corp., 1965, pp. 1–18.

19 Hildebrandt, A. C., Tissue and single cell cultures of higher plants as a basic experimental method, in H. F. Linskens and M. V. Tracey (eds.), *Modern Methods for Plant Analysis.* Berlin: Springer-Verlag, 1962, pp. 383–421.

20 Hoagland, D. R., and D. E. Arnon, The water-culture method for growing plants without soil, *Calif. Agr. Exp. Sta. Cir. 347,* 1950.

21 Holley, W. D., and R. Baker, *Carnation production,* Dubuque, Iowa: William C. Brown & Co., 1963.

22 Hollings, M., Disease control through virus-free stock, *Ann. Rev. Phytopath.,* 3:367–96. 1965.

23 Jones, L. E., A. C. Hildebrandt, A. J. Riker, and J. H. Wu, Growth of somatic tobacco cells in microculture, *Amer. Jour. Bot.,* 47:468–75. 1960.

24 Knudson, L. Nonsymbiotic germination of orchid seeds, *Bot. Gaz.,* 73:1–15. 1922.

25 ———, Nutrient solutions for orchid seed germination, *Amer. Orch. Soc. Bul.,* 12:77–79. 1943.

26 Lammerts, W. E., Embryo culture an effective technique for shortening the breeding cycle of deciduous trees and increasing germination of hybrid seed, *Amer. Jour. Bot.,* 29:166–71. 1942.

27 Malnassy, P., and J. H. Ellison, Asparagus tetraploids from callus tissue, *HortScience,* 5:444–45. 1970.

28 Mapes, Marion O., Tissue culture of bromeliads, *Proc. Inter. Plant Prop. Soc.,* 23: 47–55. 1973.

29 Marden, L., The exquisite orchids, *Nat. Geog. Mag.,* 139(40):485–512. 1971.

30 Marston, M. E., Plant propagation using *in vitro* culture techniques, *Proc. Inter. Plant Prop. Soc.,* 20:359–63. 1970.

31 Morel, G. M., Tissue culture—a new means of clonal propagation of orchids. *Amer. Orch. Soc. Bul.,* 33:473–78. 1964.

32 Muir, W. H., A. C. Hildebrandt, and A. J. Riker, The preparation, isolation, and growth in culture of single cells from higher plants, *Amer. Jour. Bot.,* 45:589–97. 1958.

33 Murashige, T., Plant propagation through tissue cultures. *Ann. Rev. Plant Physiol.* 25:135–66. 1974.

34 ———, W. P. Bitters, T. S. Rangan, E. M. Nauer, C. N. Roistacher, and P. B. Holliday, A technique of shoot apex grafting and its utilization towards recovering virus-free *Citrus* clones, *HortScience,* 7:118–19. 1972.

35 Murashige, T., M. N. Shabde, P. M. Hasegawa, F. H. Takatori, and J. B. Jones, Propagation of asparagus through shoot apex culture. I. Nutrient medium for formation of plantlets. *Jour. Amer. Soc. Hort. Sci.,* 97(2):158–61. 1972.

36 Murashige, T., and F. Skoog, A revised medium for rapid growth and bioassays with tobacco tissue cultures, *Physiol. Plant.,* 15:473–97. 1962.

37 Narayanaswami, S., and K. Norstog, Plant embryo culture, *Bot. Rev.,* 30:587–628. 1964.

38 Nitsch, J. P., and C. Nitsch, Haploid plants from pollen grains, *Science,* 163:85–87. 1969.

39 Partanen, C. R., Plant tissue culture in relation to developmental cytology, *Int. Rev. Cytol.,* 15:215–43. 1963.

40 Pillai, S. K., and A. C. Hildebrandt, Induced differentiation of geranium plants from undifferentiated callus in vitro, *Amer. Jour. Bot.,* 56:52–58. 1969.

41 Raghavan, V., and J. G. Torrey, Effects of certain growth substances on the growth and morphogenesis of immature embryos of *Capsella* in culture, *Plant Phys.,* 39:691–99. 1964.

42 Rietsema, J., S. Satina, and A. F. Blakeslee, The effect of sucrose on the growth of *Datura stramonium* embryos *in vitro, Amer. Jour Bot.,* 40:538–48. 1953.

43 Romberger, J. A., R. J. Varnell, and C. A. Tabor, Culture of apical meristems and embryonic shoots of *Picea abies*—approach and techniques, *U.S. Dept. Agr. Tech. Bul. No. 1409.* 1970.

44 Sanders, M. E., and P. R. Burkholder, Influence of amino acids on growth of *Datura* embryos in culture, *Proc. Natl. Acad. Sci. (U.S.),* 34:516–26. 1948.

45 Sanders, M. E., and N. K. Ziebur, Artificial culture of embryos, in P. Maheshwari (ed.), *Recent Advances in the Embryology of Angiosperms.* Delhi: Int. Soc. Plant Morph., Univ. of Delhi, 1963, pp. 297–325.

46 Skoog, F., and C. O. Miller, Chemical regulation of growth and organ formation in plant tissues cultured *in vitro, Symp. Soc. Exp. Biol.,* 11:118–31. 1957.

47 Smith, C. A., C. H. Bailey, and L. F. Hough, Methods for germinating seeds of some fruit species with special reference to growing seedlings from immature embryos, *New Jersey Agr. Exp. Sta. Bul. 823:*1–29. 1969.

48 Smith, R. J., Orchid propagation by *in vitro* culture techniques, *Proc. Inter. Plant Prop Soc.,* 22:174–77. 1972.

49 Steward, F. C., M. O. Mapes, A. E. Kent, and R. D. Holstein, Growth and development of cultured plant cells, *Science,* 143:20–27. 1964.

50 Stoltz, L. P., Agar restriction of the growth of excised mature iris embryos, *Jour. Amer. Soc. Hort. Sci.,* 96:681–84. 1971.

51 Vasil, I. K., Plants: Haploid tissue cultures. In P. F. Kruse, Jr., and M. K. Patterson, Jr. (eds.), *Tissue Culture Methods and Applications.* New York: Academic Press, 1973.

52 Vasil, I. K., and A. C. Hildebrandt, Variations in morphogenetic behavior in plant tissue cultures. I. *Cichorium endivia;* II. *Petroselinum hortense, Amer. Jour. Bot.,* 53:860–68, 869–74. 1966.

53 van Overbeek, J., Cultivation *in vitro* of small *Datura* embryos, *Amer. Jour. Bot.,* 29:472–77. 1942.

54 ———, R. Siu, and A. J. Haagen-Smit, Factors affecting the growth of *Datura* embryos *in vitro, Amer. Jour. Bot.,* 31:219–24. 1944.

55 White, P. R., *The Cultivation of Animal and Plant Cells,* 2nd ed. New York: The Ronald Press Company. 1963.

56 ———, and A. R. Grove (eds.), *Proceedings of an International Conference on Plant Tissue Culture.* Berkeley, Calif.: McCutchan Publishing Corp., 1965.

57 Wilmar, J. C., A. C. Hildebrandt, and A. J. Riker, Iron nutrition for growth and chlorophyll development of some plant tissue cultures, *Nature* 202:1235. 1964.

58 Wimber, D. E., Clonal multiplication of Cymbidiums through tissue culture of the shoot meristem, *Amer. Orch. Soc. Bul.,* 32:105–7. 1963.

59 Wolter, K. E., Root and shoot initiation in aspen callus cultures, *Nature,* 219:509–10. 1968.

60 Yang, Hsu-Jen, and W. J. Clore, Rapid vegetative propagation of asparagus through lateral bud culture, *HortScience,* 8(2):141–42. 1973.

61 Ziv, M., A. H. Halevy, and R. Shilo, Organs and plantlets regeneration of *Gladiolus* through tissue culture, *Ann. Bot.,* 34:671–76. 1970.

SUPPLEMENTARY READING

Gautheret, R. J., *La culture des tissus végétaux*. Paris: Masson et Cie, 1959.

Haissig, B. E., "Organ Formation *in vitro* as Applicable to Forest Tree Propagation," *Botanical Review*, Vol. 31, (1965), pp. 607–26.

Kruse, P. F., Jr., and M. K. Patterson, Jr. (eds.), *Tissue Culture Methods and Applications*. New York: Academic Press, 1973.

Murashige, T., "Plant Propagation Through Tissue Cultures," *An. Rev. Plant Physiol*, Vol 25 (1974), pp. 135–166.

Steward, F. C. (ed.), *Plant Physiology* Vol. VB: *Analysis of Growth: The Responses of Cells and Tissues in Culture*. New York: Academic Press, 1969.

Street, H. E., *Plant Tissue and Cell Culture*. Berkeley, Calif.: Univ. of California Press, 1973.

White, P. R., *The Cultivation of Animal and Plant Cells*, 2nd ed. New York: The Ronald Press Company, 1963.

————, and A. R. Grove (eds.), *Proceedings of an International Conference on Plant Tissue Culture*. Berkeley, Calif.: McCutchan Publishing Corp., 1965.

Winton, L. L., Bibliography of somatic callus cultures from deciduous trees. *Genetics and Physiology Notes no. 17*. Appleton, Wisc.: The Institute of Paper Chemistry, 1972.

Propagation Methods and Rootstocks for the Important Fruit and Nut Species

17

Few fruit or nut crop cultivars reproduce true when propagated by seed. It is necessary, therefore, that they be propagated by some asexual method. Most tree fruit and nut species are propagated by budding or grafting on roostocks—seedlings, rooted cuttings, or layered plants. In many cases, cuttings would be the simplest and easiest method to use, but for other than a few species such as the grape, fig, olive, quince, currant, and gooseberry, fruit and nuts are difficult to propagate by cutting, so other methods are used.

ACTINIDIA CHINENSIS   See Gooseberry, Chinese.

ALMOND (*Prunus amygdalus* Batsch)   The almond is propagated by T-budding on seedling rootstocks by fall, spring, or June budding. Root and stem cuttings have given but slight success.

Rootstocks for Almond   Traditionally, almond seedlings have been used. Peach seedlings are favored for irrigated orchards under intensive cultivation or where nematodes are a problem. The clonal rootstock, 'Marianna 2624' plum (see p. 569), is of value in certain situations. Almond × peach hybrids are a class of almond rootstocks developed in recent years.

Almond (P. amygdalus)   Almond seedlings are quite satisfactory in deep and well-drained soils. Seeds of the bitter types, or certain commercial cultivars such as 'Texas' ('Mission'), are commonly used. Almond seeds require stratification for three or four weeks before planting. In poorly drained soils almond roots are often unsatisfactory, owing to their susceptibility to infection by crown rot (*Phytophthora* sp.) and crown gall (*Bacterium tumefasciens*). Their deep-rooting tendency is an advantage in orchards grown on unirrigated soils or where drought conditions occur. Almond seedlings are tolerant of high-lime soils and, of all of the rootstocks available for almonds, are the least affected by excess boron salts. Almond seedlings are susceptible to root-knot and root-lesion nematodes as well as to oak root fungus (*Armillarea mellea*).

533

Peach (P. persica)   Peach seedlings are widely used as rootstocks for almond where irrigation is practiced and where nematodes are present. Any of the peach stocks are satisfactory, but 'Lovell' and 'Nemaguard' (for nematode resistance) are most commonly used. The graft unions often show a scion overgrowth but are quite acceptable. Peach roots are not as susceptible to crown gall or crown rot as almond roots. Almond on peach roots in irrigated soils will grow faster for the first several years and bear heavier crops during the first 15 to 20 years than those on almond roots, but trees on almond roots tend to live longer and may eventually outgrow those on peach.

'Marianna 2624' plum (see p. 569)   This clonal rootstock is used to a limited extent in heavy, wet soils or where oak root fungus or root-knot nematodes (to which it is immune) are present. Almond trees on this stock are about one-third smaller than those on the other available rootstocks. Not all almond cultivars are compatible with this stock *(91);* those that cannot be used include 'Nonpareil,' 'Davey,' 'Milow,' and 'Kapareil.' A clonal rootstock selection of *Prunus insititia,* known as 'Haven's 2B,' has been used as an interstock between 'Nonpareil' and 'Marianna 2624' with partial success in overcoming the incompatibility.

Almond-peach hybrids (92)   These are first generation offspring of peach and almond that can be produced by natural crossing between adjoining trees of the two species. Hybrid plants are identified in the nursery row by their high vigor and intermediate appearance between the parents. Combinations produced include 'Titan' almond × 'Nemaguard' peach and 'Nonpareil almond × 'Nemaguard' peach. Clonal selections 'GF-677' and 'GF-557' rooted by leafbud cuttings under mist have been made in France. Rooting of selected clones is possible by hardwood cuttings directly planted in the nursery, as described on page 277. These stocks are noted for their vigor and excellent compatibility with scion cultivars but experience is too limited to establish their range of adaptation.

ANACARDIUM OCCIDENTALE   *See* Cashew.

ANANAS COMOSUS   *See* Pineapple.

ANNONA CHERIMOLA   *See* Cherimoya.

APPLE *(Malus sylvestris Mill.)*   T-budding is used either as fall budding or spring budding on seedling or clonal vegetatively propagated rootstocks. Root grafting is used in some places, usually as whole-root grafts. Propagation by hardwood cuttings, with the exception of certain clonal rootstocks, is little used. Rooting of softwood cuttings under mist with the aid of root-promoting substances can be done, but this method is not used commercially *(83).*

All apple rootstocks, either seedling or clonal, are in the genus *Malus,* although the apple will grow for a time and even come into bearing grafted on pear *(Pyrus communis)* roots. Most apple cultivars will not grow much longer than a year on quince *(Cydonia oblonga)* roots. However, 'Winter Banana' seems to be successful on the quince, at least for a time, living as long as 20 years, although dwarfed. Apples have been grown also on roots of hawthorn *(Crataegus)* and mountain ash *(Sorbus)* species.

Rootstocks for Apple *(20, 181)*   **Seedling Stocks.** Apples on seedling roots generally make a larger tree than when any of the clonal stocks are used as roots. French Crab seedlings *(Malus sylvestris)* from France were widely used both in North America and Europe for many years in propagating apples. Since 1930, U.S. quarantine regulations have prohibited importation of seedlings into the U.S. Seeds can be imported, however, and produce a preferred type of seedling, resistant to crown gall and hairy root.

Seedlings of 'Delicious,' 'Golden Delicious,' 'McIntosh,' 'Winesap,' 'Yellow Newtown,' 'Rome Beauty,' (particularly 'Delicious'), are widely and successfully used as rootstocks. Such seedlings are quite uniform, and no incompatibility problems have arisen. However, in purchasing seed it is often difficult to determine the seed source. Grading nursery trees tends to eliminate weak or off-type plants. In the colder portions of the U.S.—the Dakotas and Minnesota—the hardier Siberian crabapple *(Malus baccata)* and seedlings of such cultivars as 'Antonovka,' containing some *M. baccata* parentage, are used. In Poland 'Antonovka' seedlings are the chief apple rootstock. Some nurseries in British Columbia, Canada, use 'McIntosh' seedlings for their winter-hardiness, upright growth in the nursery, and early fall shedding of leaves, although they tend to be somewhat susceptible to hairy root, a form of crown gall, and to powdery mildew. The resulting trees tend to be variable in size and performance.

Apples with the triploid number of chromosomes, such as 'Gravenstein,' 'Baldwin,' 'Stayman Winesap,' 'Arkansas,' 'Rhode Island Greening,' 'Bramley's Seedling,' and 'Tompkins King,' produce seeds that are of low viability and are not recommended as a seed source. Seeds of 'Wealthy,' 'Jonathan,' or 'Hibernal' have given unsatisfactory results.

The principal source of seeds is the pomace from processed apples. For spring sowing, seeds require stratification for 60 to 90 days at 35° to 45° F (2° to 7° C) to germinate. Some nurseries fall-plant seeds to receive the natural winter chilling. To avoid seedlings which become crooked in breaking through the crust, soil must be raked over in the spring. To obtain a branched root system, the seedlings may be undercut while still small to prevent the development of a taproot; a straight root may be preferred, however, when bench grafting is to be done. Seedlings that do not grow to a satisfactory size in one year should be culled out.

Various Asiatic species of *Malus* have been tried as dwarfing or semi-dwarfing rootstocks or interstocks for apple cultivars *(151, 152)*. Some of these, *Malus hupehensis, M. toringoides, M. sargenti,* and *M. sikkimensis,* are apomictic. These stocks are moderately hardy and are resistant to crown gall. *Malus sikkimensis* seedlings are uniform, and scion cultivars worked on this stock are restricted in growth and start bearing early. However, such trees are relatively unproductive compared to those on the dwarfing Malling stocks.

Apple roots are resistant to root-knot and root-lesion nematodes, moderately resistant to oak root fungus, and highly resistant to verticillium wilt.

Rootstocks for Apple   Clonal Stocks. Numerous clonal, asexually propagated apple rootstocks have been developed. It is important in using these clonal stocks to obtain only virus-tested material, if possible.

'Northern Spy' (77)   This is resistant, with a notable exception in South Africa *(46)*, to woolly aphids *(Eriosoma lanigera)*, an insect causing serious injury to apple trees by infesting the roots, particularly in areas having mild winters. A woolly aphid resistant tree can be obtained by propagating 'Northern Spy' nursery trees by layering, or by root grafting onto a 'Northern Spy' root piece so that they are on their own roots, then working them to the desired cultivar. Because of its resistance to woolly aphid, 'Northern Spy' has been used as a rootstock, primarily in New Zealand and Australia *(25)*. It produces trees of moderate vigor but does not do well in low-fertility soils and is not suited for replant situations.

'Hibernal'   This very winter-hardy cultivar was imported into the U.S. by the USDA from Russia in 1870. It formerly was widely used in the Midwestern states for its winter-hardy properties in developing the root, trunk, and primary scaffold system. In the orchard the 'Hibernal' was then topworked by budding or grafting

to the desired cultivar *(188)*. However, some trees on 'Hibernal' roots are poorly shaped, unbalanced, and lean with the wind.

'Alnarp 2'   This rootstock was developed in Sweden and is widely used there for its winter-hardy properties. The roots are well anchored and give vigorous trees, comparable to those on 'Malling 25' roots. Where apple mildew *(Podosphaera leucotricha)* is prevalent, 'Alnarp 2' itself has shown susceptibility.

'Robusta No. 5 (M. robusta [M. baccata × *M. prunifolia])*   This vigorous, very hardy clonal apple rootstock, propagated by stooling or stem cuttings, was developed at the Central Experimental Farm, Ottawa, Canada. It is apparently resistant to fire blight and crown rot and seems to be compatible with most apple cultivars. It is extensively used as an apple rootstock in eastern Ontario and Quebec, as well as in the New England states. It is the best stock for use where extreme winter-hardiness is required, but in the Pacific Northwest it has shown the unfavorable habit of starting growth too early in the spring following three or four warm days in late winter. It is not a dwarfing stock.

'Virginia Crab'   This stock originated as a chance seedling in a nursery in Iowa in 1862 and was propagated by vegetative methods in order to retain its desirable root and trunk characteristics. 'Virginia Crab' was long used for developing the root system, trunk, and scaffold branches. However, it is no longer propagated due to its susceptibility to a stem pitting defect caused by a virus latent in many scion cultivars (see Figure 11–20) *(184, 199)*.

Malling Series   Beginning in 1912 the East Malling Research Station in England selected and classified a series of vegetatively propagated apple rootstocks which ranged from very dwarfing to very invigorating in their effect on the scion cultivar *(201)*. This size-controlling influence is modified by the scion cultivar being used. The dwarfing influence of these various rootstocks does not extend to the fruit, however; fruit size, especially on young dwarfed trees, is often larger than on standard-size trees.

The Malling stocks apparently give completely compatible graft unions with most apple cultivars. Trees on this series of rootstocks have been planted in varying amounts in many parts of the world. These stocks have proved to be hardy except in regions with extremely severe winters, such as the northern U.S. and Canada. They have done well on both heavy and light soils but none are resistant to woolly aphid.

Malling-Merton Series (138)   The John Innes Horticultural Institution and the East Malling Research Station began work jointly in 1928 on breeding a new series of apple rootstocks to provide resistance to woolly aphids and to give a range in tree vigor. Of this group four stocks, 'MM 104,' 'MM 106,' 'MM 111,' and 'M 25,' have been most widely used. All except 'Malling 25' show resistance to woolly aphids. However, all these stocks have been severely attacked by this insect in South Africa, presumably due to the presence there of a different strain of woolly aphid *(46)*.

Other improved tree characteristics associated with these stocks include high yield, induced precocity of flowering (with some stocks), well-anchored trees (with some stocks), freedom from suckering, and good propagation qualities.

All the Malling and Malling-Merton stocks are readily propagated, mostly by stool-bed layering *(21)*. While trench, or simple pegged-down, layering provides high rootstock production quickly, the high amount of labor required generally precludes its use (but see 'MM 106,' below). Root cuttings taken from young, vigorously growing nursery stock are also sometimes used. A number of rootstock clones, notably 'Malling 26,' 'MM 106,' and 'MM 111,' respond very well to propagation by the hardwood cutting method described on page 279.

A summary of the characteristics of the most useful of these two groups of clonal rootstocks follows, with the rootstocks grouped according to their effect on the vigor of the scion cultivar. However, the particular scion cultivar used has a definite influence, too, on the size of the composite tree.

Dwarfing Stocks (191)

'Malling 27' (139)   This is by far the most dwarfing of all the Malling stocks, producing trees only about four feet tall, about half the size of those on 'Malling 9.' It is a cross between 'Malling 13' and 'Malling 9.' Fifteen years' trials show it may have use in high-density plantings. Virus-tested propagating material was released by the East Malling Research Station, England, in 1970.

'Malling 8' ('French Paradise,' 'Clark Dwarf')   This stock is considered in England to be inferior to 'Malling 9,' which has similar dwarfing characteristics. Trees on this stock require staking because the root system is weak and brittle and gives poor anchorage. It is best used as an intermediate stock, giving a dwarfing effect without the limitation of its poor root system. The amount of dwarfing seems to be proportional to the length of the interstock. It has been known in the United States for many years as 'Clark Dwarf' (14). This stocks seems to be quite winter-hardy.

'Malling 9' ('Jaune de Metz')   This originated as a chance seedling in France about 1878 and has been widely used in Europe as an apple rootstock. It is a dwarfed tree itself and is a valuable dwarfing rootstock much in demand for producing small trees for the home garden or for commercial high-density plantings. Such trees are seldom over nine feet tall when mature and usually start bearing in the first year or two after planting.

'Malling 9' has numerous thick, fleshy, brittle roots and requires a fertile soil; the trees require staking or trellising for support. It is moderately hardy and resistant to crown rot *(Phytophthora)*, but susceptible to crown gall. This stock seems to grow better at relatively low soil temperatures (below about 60° F) than at higher soil temperatures (121). It is propagated by stooling.

When used as an intermediate stock in double-working it will cause dwarfing of the scion cultivar but not as much as when it is used as the rootstock (152, 183).

'Malling 26'   This stock was introduced in 1959 at the East Malling Research Station, originating from a cross between 'Malling 16' and 'Malling 9' (146). It produces a tree somewhat larger and more sturdy than 'Malling 9' though less so than 'Malling 7 or 'MM 106,' but still requires staking. It is easily propagated by softwood cuttings under mist or by hardwood cuttings (75), but is reported to be a poor producer in stool beds. It is quite winter-hardy. Some breakage has occurred close above the union in nursery trees worked to 'Cox's Orange Pippin' and 'Laxton's Superb,' particularly following rapid growth in a wet season. Older, otherwise healthy trees on 'Malling 26' sometimes exhibit fluted trunks.

Semi-Dwarfing Stocks

'Malling 7'   This stock produces trees somewhat larger than those on "Malling 26' roots. 'Malling 7' has a stronger, deeper, root system than '9' and produces an early-bearing, semi-dwarf tree. It is tolerant of excessive soil moisture but is susceptible to crown gall and is not very well anchored, requiring staking for the first few years. It seems to make good growth over a wide soil temperature range (111). 'Golden Delicious' on 'Malling 7,' however, has been reported in Illinois (167) to show abnormal phloem development and sieve tube necrosis, with eventual death of the trees. This stock also has the undesirable characteristic of suckering badly, and the trees are not very winter-hardy. 'Malling 7' is easily propagated by stooling or by leafy cuttings under mist.

'*Malling-Merton 106*'   This stock produces trees about half the size of those on seedling rootstocks and about the same as those on 'Malling 7,' but more productive with earlier cropping. It has been very successful in the U.S. *(120)* and in England it is the most popular apple rootstock for bush trees and well-spaced hedgerows. On good soils wtih some scion cultivars it can produce a standard-size tree. The roots are well anchored and do not sucker, as does 'Malling 7.' It is promising as a replacement for '7' and is considered by some to be the best of the semi-dwarfing stocks. In some areas 'MM 106' has shown susceptibility to crown rot; this may be its chief weakness. It grows well in the nursery but drops its leaves later in the fall than most other understocks and is not resistant to early fall freezes. Hardwood and softwood cuttings root easily and stool beds are quite productive, but there is a tendency for the stool to lose vigor after a few years. This has encouraged nurserymen to replant their 'MM 106' stool beds every eight years or so on fresh ground.

Vigorous Stocks

'*Malling 2*' ('*Doucin*')   This has been the most commonly used apple under-stock in England but is being replaced by 'MM 111.' Trees on 'Malling 2' tend to be vigorous and fruitful, come into bearing early, and are smaller than trees on seedling roots. It is moderately susceptible to crown gall but resistant to crown rot. If the desired cultivar is budded high on this stock (10 to 12 in. above soil level) so that the tree can be planted deeply, satisfactory root anchorage can be obtained. 'Malling 2' grows considerably better at low (below 60° F) than at high soil temperatures *(121)*. It is somewhat difficult to propagate by cuttings and the layers on young stool plants root rather sparsely until the plant carries a good number of shoots of limited vigor.

'*Malling-Merton 111*'   Trees on this stock are comparable in vigor to those on 'Malling 2,' but grow better when young, start bearing earlier, and survive periods of drought better. It does well on a wide range of soil types. Stool beds are highly productive with heavy root systems developing. Hardwood cuttings root well with proper treatment *(75)* and softwood cuttings under mist root well. 'MM 111' is slightly more winter-hardy than 'MM 104' and much more so than 'Malling 7' or 'MM 106.'

'*Malling-Merton 104*'   Trees on this stock are drought-resistant and well anchored (depending upon the scion cultivar), but its lack of tolerance for poorly drained soils and its susceptibility to collar rot are bringing it into disfavor. The trees bear heavily and are more vigorous than those on 'Malling 2.' Root development is somewhat one-sided. Trees on 'MM 104' are almost as large as those on seedling roots. It is difficult to propagate by hardwood cuttings *(75)*. 'MM 111' is generally preferred to 'MM 104.'

Very Vigorous Stocks

'*Malling 16*' ('*Ketziner Ideal*')   Trees on this rootstock are large, well anchored, and will come into bearing about the same time, or later than, those on seedling roots. 'Malling 16' seems to do well over a wide range of soil temperatures *(121)*. It is usually propagated by stooling, but can also be started by root cuttings. It is very susceptible to woolly aphid and is not so widely used as in the past.

'*Malling-Merton 109*'   Trees on this stock are about the same size as those on seedling stocks. Its chief advantage is its resistance to woolly aphid and its stimulation of early bearing. Trees on this stock sometimes tend to lean badly and are not tolerant of waterlogged soils. Roots of 'MM 109' produced very poorly anchored 'Delicious' trees in Oregon, but in an eight-year trial in New Zealand it has produced strong-growing, high-yielding trees *(111)*.

'Malling 25' Although not as resistant to woolly aphid as the Malling-Merton stocks, it is more resistant than 'Malling 16.' It produces large, well-anchored, vigorous trees, induces early fruit-bud formation, and gives excellent fruit set and yields. It may be the best stock in this group.

APRICOT *(Prunus armeniaca* L.) Apricot cultivars are propagated commercially by T-budding on various seedling rootstocks in the genus *Prunus*. Fall budding is the usual practice, but spring and June budding may be used.

Rootstocks for Apricot *(124, 174)* Three stocks are commercially suitable—apricot seedlings, peach seedlings, and in some cases, myrobalan plum seedlings. Seeds of all these species require low-temperature (41° F; 5° C) stratification before planting in the spring—three to four weeks for the apricot and about three months for the peach and myrobalan plum. On good, well-drained soils, apricot seedlings are the recommended rootstock for apricot cultivars.

Apricot (P. armeniaca) Apricot seeds can be obtained from drying yards and canneries. Seeds of 'Royal' or 'Blenheim' produce excellent rootstock seedlings in California. Since the apricot root is almost immune to the root-knot nematode (*Meloidogyne* sp.), it should be used where this pest is present. In addition, it is somewhat resistant to the root-lesion nematode. It is susceptible to crown rot (*Phytophthora* sp.), and is not tolerant of poor soil-drainage conditions. Apricot roots are not as susceptible to crown gall *(Agrobacterium tumefaciens)* as are peach and plum roots. Apricot seedlings are susceptible to oak root fungus and highly susceptible to verticillium wilt.

Peach (P. persica) In California peach seedlings are satisfactory as a rootstock for apricot cultivars but sometimes the union is enlarged or rough. Although the peach itself is short-lived, apricot trees 85 years old growing satisfactorily on peach have been known. In unirrigated orchards or where drought conditions prevail, apricots on peach seedling roots make better growth than those on apricot roots. Peach roots are not tolerant of wet soils, growing better on light or well-drained soils. For trees to be planted in a location formerly occupied by peach roots, some stock other than peach should be used, because peach roots often grow poorly on soils formerly occupied by them. In some regions apricot cultivars show definite incompatibility on peach seedlings, so it cannot be assumed that all cultivars will do well on peach roots *(19, 96)*.

Myrobalan plum (P. cerasifera) Although there are successful high-yielding apricot orchards grown on this rootstock, it cannot be unqualifiedly recommended. In a few instances the trees have broken off at the graft union in heavy winds, and die-back conditions have been noted. Nurserymen often have trouble getting apricots started on myrobalan roots, some of the trees failing to grow rapidly and upright or else having weak or rough unions. After these weaker trees are culled out, the remaining ones seem to grow satisfactorily. In older trees of this combination, the myrobalan root usually grows much larger than the apricot trunk, giving a "churn bottom" tree. This stock is useful for apricot when the trees are to be planted in heavy soils or under excessive soil moisture conditions, which the myrobalan root will tolerate. An alternative to myrobalan plum seedlings is the related vegetatively propagated 'Marianna 2624' plum (see p. 569), on which apricots seem to do well. In apricot rootstock trials in Australia *(85)* trees on Myrobalan seedlings, 'Myro 29C,' and Marianna plum far outyielded those on peach or apricot roots over an eight-year period.

Western sand cherry (P. besseyi). Although it is not used commercially, this rootstock will produce a semi-dwarf tree.

ASIMINA TRILOBA *See* Papaw.

AVOCADO (Persea sp.) *(134, 149)*   Nursery trees of the avocado are propagated commercially in California by T-budding and tip grafting (whip or splice graft) *(179)* avocado seedlings. In Florida and some of the Caribbean countries T-budding is occasionally done, but the usual nursery practice is to graft mature tip scions, either as side grafts—or sometimes the side-veneer—or cleft grafts on young succulent seedlings. In South Africa, tip grafting when the seedlings are ¼ in. in diameter is recommended. The grafted trees are raised in containers from which they are transplanted to the orchard *(109)*. T-budding one-year-old seedlings in the fall is the most widely used method in Israel *(150)*.

Seeds are generally planted shortly after removal from the fruit, care being taken not to allow them to dry out, although they can be stored six to eight months if packed in dry peat moss and held at 41° F (5° C) with 90 percent relative humidity. To eliminate infection from *Phytophthora cinnamoni* (avocado root rot), seeds should be dipped in hot water (120° to 125° F; 49° to 52° C) for 30 minutes before planting *(202)*. Germination of stored seed is hastened by removing the brown seed coats or by cutting a thin slice off the seed before planting. The seed coats can be removed by wetting the seeds and allowing them to dry in the sun. Seeds of the Mexican race, which ripens its fruit in the fall, may be planted in beds in late fall or early winter. They should be placed with the large, basal end down, just deep enough to cover the tips. If grown in a warm area, the sprouted seeds will be ready to line-out in the nursery row the following spring. By summer or fall the seedlings will usually be large enough to permit T-budding; if not, they can be budded the following spring.

It is important to select the budwood properly. The best buds are usually near the terminal ends of completed growth cycles with fully matured, leathery leaves. To prevent drying, the leaves should be removed when the budwood is taken.

Four to six weeks after budding, the seedling rootstocks should be cut off 8 to 10 in. above the bud, or bent over a few inches above the bud. The remaining portion of the seedling above the bud is not cut off until the bud shoots have completed a cycle of growth. The new shoots are usually staked and tied. In digging, the nursery trees are usually "balled and burlapped" for removal to their permanent location. This is done just following the first or second growth cycle of the bud or sometimes just as the first flush starts.

The seedlings are sometimes grown in soil in containers made from 12- by 16-in. sheets of felt building paper, each stapled to form an open-end cylinder about 6 in. in diameter. These are set in greenhouse benches and filled with soil. One seed is planted in each pot in the fall, and by late winter the seedlings are large enough to graft, using the whip graft method. The scion consists of a shoot 1 to 6 in. long, with several buds, which is grafted near the base of the seedling. The scions are tied with rubber or plastic budding tape. The grafted trees are ready to set out in their permanent location at the end of their first growth cycle, about three months after grafting.

In Florida, side grafting on young, succulent West Indian seedlings is used *(103)*. The seedlings, grown in gallon containers, are grafted when they are 6 to 10 in. high and ¼ to ⅜ in. in diameter. The scions are shoot terminals, 2 to 3 in. long, with a plump terminal bud, taken just as it resumes growth.

Stem cuttings of young Mexican and Guatemalan avocado seedlings can be rooted, and cuttings taken from mature Mexican and 'Fuerte' trees have also been rooted, but with more difficulty. It is difficult to root cuttings taken from mature trees of the Guatemalan race, but they can be rooted if the basal portion of the leafy shoot to be made into the cutting is etiolated, that is, allowed to grow only in complete darkness. The terminal portion of such a shoot develops in the light until three to five leaves have formed. Those shoots with etiolated bases can then be detached and rooted in a propagating case *(42, 43)*. Use of auxin-type rooting materials has not been beneficial. Avocados started as rooted cuttings eventually make satisfactory trees but, in general, grow poorly in the initial stages.

Rootstocks for Avocado

Mexican race (P. drymifolia)   Seedlings of this race are preferred in California for their cold-hardiness and their resistance to lime-induced chlorosis and the diseases *Dothiorella* and *Verticillium*. They are, however, susceptible to injury from high salinity. In Florida, where seedlings of large diameter are preferred for grafting, the Mexican types are little used, owing to their thin shoots.

Guatemalan race (P. americana)   These are occasionally used in California when there is a scarcity of Mexican seeds. Guatemalan seedlings are often initially more vigorous than Mexican seedlings but are more susceptible to diseases and to injury from cold.

West Indian race (P. americana)   These seedlings are too liable to frost injury to be used commercially under California conditions but are widely used in Florida. The large seed produces a large, pencil-size shoot suitable for side grafting in two to four weeks after germination.

West Indian × *Guatemalan hybrids*   Seedlings of this hybrid are occasionally used in Florida.

BANANA *(Musa sp. (148, 166)*   The banana "tree" is not a tree but a large perennial herb. The "stem" is not a stem but consists of compressed, curved leaf stalk bases arranged spirally in strips. The bases of the leaf stalks are attached to the true stem, a rhizome (a horizontal, underground stem which develops into a "corm" structure). New "suckers" grow from buds on the corm and soon develop their own roots and a base as large as the parent plant, which dies and deteriorates shortly after the fruit bunch is harvested. A banana planting may live for a considerable time but it really is a succession of new plants, each arising as a sucker from a rhizomatous bud; any given sucker fruits only once, then dies.

Since the edible types rarely produce seeds, commercial propagation of the banana is asexual, consisting of division of the rhizome and replanting of the pieces or the suckers. A large rhizome is cut into pieces weighing 7 to 10 lb, depending on the cultivar, which are termed "heads." These should contain at least two buds capable of growing into suckers. Each sucker produces two branches in the first crop. Fairly large "sword" suckers 3 to 6 ft high with well-developed roots are also used, but the leaves must be shortened considerably to reduce water loss after the sucker is cut from the parent plant. These suckers are removed with a sharp cutting tool inserted vertically about half way between the parent stalk and stem of the sucker. These sword suckers only produce one bunch of fruit in the first crop, but they are often preferred, owing to the large size of the bunch.

BLACKBERRY *(Rubus sp.)*   The upright type of blackberry produces suckers readily, its propagation consisting of digging up suckers, with an attached root piece, during early spring and replanting in a new location. The suckers are often grown an additional year in the nursery to develop stronger plants before being set out in a new planting.

The trailing type of blackberry, such as the youngberry, boysenberry, loganberry, or dewberry, does not produce many suckers. This type is propagated by tip layering.

All blackberries can be propagated by root cuttings, but some thornless forms, such as the thornless youngberry and thornless loganberry, revert to the thorny type if propagated in this manner (see p. 286).

Both the upright and the trailing types of blackberries may be started easily by stem cuttings or leaf-bud cuttings (see p. 286) taken from leafy shoots and rooted under high humidity, particularly in mist-propagating beds. Treatment of such cuttings with root-promoting chemicals is beneficial.

BLUEBERRY, HIGHBUSH (*Vaccinium corymbosum* L. and *V. australe* Small) *(29, 32, 35, 106, 156)*   Stem cuttings, both dormant hardwood and leafy softwood, are used. The blueberry can also be started by leaf-bud cuttings. Most blueberry cuttings are grown in the nursery for one year after the year of rooting and then sold as two-year plants for setting in their permanent location.

Blueberry cultivars can be propagated by T-budding in mid-summer to seedling plants or rooted cuttings. The highbush blueberry is successfully worked onto the rabbiteye blueberry (*V. ashei*) as a rootstock to take advantage of the latter's wide soil adaptability and vigor *(44)*.

Hardwood Cuttings   The blueberry is difficult to propagate by hardwood cuttings, but good results can be obtained. Cutting material—consisting of vigorous, firm, unbranched shoots of the previous season's growth, about pencil size—should be taken from dormant plants in late winter or early spring before the buds start to swell. Only vegetative wood, without fruit buds, should be used. The cuttings should contain three or four buds and be 4 to 5 in. long. A polyethylene-covered frame in a lathhouse is a suitable rooting structure. Bottom heat is helpful, but root-promoting chemicals have generally failed to improve rooting. A mixture of half sand and half ground sphagnum peat moss is a satisfactory rooting medium.

Cuttings should be spaced about 2 in. apart in the rooting bed and set with the top bud just showing. When leaves appear, the frame should be raised slightly to allow for ventilation. Watering must be frequent enough to maintain a high humidity. Roots start to form in about two months.

Softwood Cuttings   Leafy cuttings are taken in spring from shoots about 5 in. long after the first flush of growth has fully expanded leaves. Intermittent mist and a peat moss-perlite rooting medium gives good rooting conditions.

Rooted blueberry cuttings are best handled by allowing them to remain in the rooting frame until early the following spring, when they are lined-out in the nursery row. Fertilizing the cutting beds with nitrogen after roots start forming will produce larger and thriftier plants.

BLUEBERRY, LOWBUSH (*Vaccinium angustifolium* Ait.) *(106)*   This is probably best propagated by leafy softwood cuttings under intermittent mist with bottom heat, using sand and peat moss (1:1) as a rooting medium *(90)*. Cuttings taken in late spring and early summer from actively growing shoots root well, some clones giving almost 100 percent rooting. Transfer rooted cuttings to peat pots for further growth and overwintering. Cuttings made from rhizomes can also be rooted. These are best taken in early spring or late summer and fall, avoiding the midsummer rest period of the rhizome buds.

Seed propagation is sometimes used. Seeds are removed from ripe berries, then spread over a well-drained, acid type soil mix containing one-third peat moss. Cover with a layer of finely ground sphagnum moss which is kept moist until the seeds germinate, usually in three to four weeks. Seedlings $3/4$ in. tall are transferred to peat pots.

BLUEBERRY, RABBITEYE (*Vaccinium ashei* Reade)   This can be propagated by leafy softwood cuttings taken in midsummer and rooted under intermittent mist with a 1 per cent indolebutyric acid in talc treatment. Added lights to give a 16 hr. daylength may improve root production. *(30)*.

BUTTERNUT (*Juglans cinerea* L.)   There are several butternut cultivars and these are propagated by grafting onto *Juglans nigra* seedlings, using a form of the bark graft or any method successful with the other nut species. Excessive sap "bleeding" is a problem in graft union healing and necessary steps must be taken to overcome it (see p. 419).

CACAO (*Theobroma cacao* L.) *(185)*   Almost all commercial cacao plantings consist of seedling trees, which are highly variable. Cacao can be propagated vegetatively, however, and with the development of new, superior clones and the use of modern techniques for rooting softwood cuttings, it is likely that more vegetatively propagated plantings will be made.

In the large cacao-producing areas of West Africa and South America, emphasis is on seedling propagation with seed taken from selected clones. Vegetative propagation is used to produce the seed bearers. Cross-pollination between high-yielding clones and use of the resultant seeds to produce bearing trees is an important propagation method.

Seedling Propagation   Freshly harvested mature seeds should be planted immediately. Cacao seeds quickly deteriorate after harvesting, normally losing all capacity to germinate within a week after removal from the pod. Prevention of drying plus storage at 75° to 85° F (24° to 29° C) prolongs seed life *(7)*.

A common practice is to plant three or four seeds in a hole. If they all germinate they are allowed to grow, the plants being treated as branches of one tree. Another method is to start the seedlings in a nursery bed and transplant the small trees to their permanent location. Alternatively, the seedlings may be started in baskets, bamboo or paper cylinders, or clay pots, from which they are removed later and planted. A germination temperature of about 80° F (26° C) should be used.

Asexual Propagation   After young seedling trees have attained sufficient size, they can be used as rootstocks on which superior clones are budded. The patch bud is most succesfsul, but T-budding may be used *(36)*. In areas where unfavorable soil conditions occur, superior clones are sometimes grafted on resistant clonal rootstocks. In the world industry, however, budding and grafting are done on a negligible scale. Air layering is quite successful.

Leafy softwood cuttings can be rooted in high percentages under intermittent mist (five minutes in each hour) in about 50 percent shade if treated with indolebutyric acid plus a fungicide, such as captan. Cuttings will root in a closed plastic bag with a moist rooting medium at the base and with the bags suspended under a cloth shade at a low light intensity *(37)*

CARICA PAPAYA   *See* Papaya.

CAROB (*Ceratonia siliqua* L.)   This subtropical evergreen tree is usually propagated by seeds which germinate without difficulty. Transplanting of bare root seedlings gives poor results so the seeds are planted in their permanent location or started in containers for later transplanting. Seedlings are best budded to selected cultivars; this is most successful in late spring. Cuttings can be rooted, but with difficulty. Air layering in late summer is quite successful.

CARYA ILLINOENSIS   *See* Pecan.

CARYA OVATA   *See* Hickory.

CASHEW (*Anacardium occidentale* L.) *(3, 116)*   This tender tropical evergreen tree, grown principally in India, is usually propagated by seed; two or three are planted directly in place in the orchard, since seedlings transplant with difficulty.

They are later thinned to one tree per location. There are no seed dormancy problems, but seeds should be tested for viability by placing in water; those that float should be discarded *(142)*. Germination takes place in 15 to 20 days. For transplanting, the seeds are started in some type of container which will disintegrate, and the container—with the seedling—is set in the ground. Cashew seedlings show great variability in growth habit, yield, and nut quality. There are few named cultivars but efforts are being made to select superior, high-yielding types and propagate them by asexual methods, particularly by approach-grafting *(141)* and air layering *(123)*. Other methods, such as T-budding, patch budding, and rooting leafy cuttings, have also been successfully used in limited trials.

CASTANEA MOLLISSIMA   *See* Chestnut, Chinese.

CERATONIA SILIQUA   *See* Carob.

CHERIMOYA *(Annona cherimola* Mill.)   This is propagated by cleft grafting or T-budding selected cultivars on seedlings of cherimoya or the related sugar apple *(Annona squamosa),* which gives a dwarf plant, or on custard apple *(Annona reticulata).* The latter two species should not be used as rootstocks in cold areas or in poorly drained soils where they are subject to root rot.

Some of the seedling forms developed in Mexico and South America come nearly true from seed, and in many regions seed propagation is used exclusively. The seeds retain their viability for many years if kept dry, but germinate in a few weeks after planting. After the young seedlings are 3 or 4 in. high they should be transferred from flats to small pots, and when about 8 in. high, to larger pots or to open ground.

CHERRY *(Prunus avium* L., *P. cerasus* L.)   Cherry trees are propagated by T-budding the desired cultivar on a seedling rootstock. Rootstock seedlings are often grown closely planted in a seed bed for one year, then are lined-out about 4 in. apart in the nursery and grown a second year before budding. If growing conditions are good, seeds may be planted directly in the nursery in early spring, and the seedlings will be large enough for budding by late summer or early fall.

Rootstocks for Cherry (33, 125, 145)   The two most common stocks are Mazzard *(Prunus avium)* and Mahaleb *(P mahaleb)* seedlings. A third stock, the vegetatively propagated 'Stockton Morello' *(P cerasus)* is occasionally used. These three rootstocks are used for sweet cherry *(P. avium)* cultivars. Sour cherry *(P. cerasus)* and Duke cherry (hybrids between sweet and sour cherries) cultivars are propagated on either Mazzard or Mahaleb roots. All three of these cherry rootstocks are susceptible to verticillium wilt. *P. avium* is moderately resistant to oak root fungus where *P. mahaleb* and 'Stockton Morello' are susceptible.

Mazzard (P. avium)   Mazzard seedlings used in the U.S. are available from sources in Oregon and Washington. There is considerable variation among the sources of Mazzard seeds, some undoubtedly being better than others. Seeds from trees indexed for freedom from ring spot virus should be used if possible. In germinating Mazzard seeds it is beneficial to soak them in water changed daily for about eight days prior to stratification. A warm, moist stratification period at 70° F (21° C) for three weeks prior to cold stratification may improve germination. Finally a cold (35° to 40° F; 2° to 4° C) stratification period of 120 days is used. When a fair percentage of the seeds show cracking of the endocarp with root tips emerging, they should be removed and planted *(203)*.

The clonal Mazzard stock, 'Malling 12/1,' developed at the East Malling Research Station in England, produces trees which are vigorous, uniform, and resistant to bacterial canker. Propagation is by trench or mound layering. In Oregon this stock has worked well for the root, trunk, and primary scaffold system onto which the scion cultivar is worked.

Sweet cherry cultivars make an excellent graft union with Mazzard roots, giving trees that are vigorous and long-lived, but under ideal conditions they often become so large that harvest costs may be excessive. Mazzard roots are not particularly suitable for heavy, poorly aerated, wet soils, but will tolerate such conditions better than Mahaleb. Under dry, unirrigated, drought conditions Mahaleb is more likely to survive than Mazzard, due presumably to the deep, vertical rooting habit of Mahaleb in contrast to Mazzard's shallow, horizontal root system.

Mazzard roots are immune to one species of root-knot nematode *(Meloidogyne incognita)* and resistant to *M. javanica,* but susceptible to the root-lesion nematode *(Pratylenchus vulnus).*

Mahaleb (P. mahaleb)   The seeds should be soaked in water for 24 hours, then stratified for about 100 days at 40° F (4° C). Graft unions between most cherry cultivars and Mahaleb roots are strong but not as smooth as with Mazzard roots. Leafy Mahaleb cuttings are easily rooted under intermittent mist if treated with indolebutyric acid, thus permitting the establishment of clonal Mahaleb stocks *(69).* This is the principal rootstock in the U.S. for 'Montmorency,' the leading sour cherry cultivar. Mahaleb roots may produce a somewhat dwarfed tree, particularly if the budding is done high (15 to 20 in.) on the rootstock; there is evidence, however, that in good soils the trees may become as large as those on Mazzard roots. Trees on Mahaleb roots are resistant to the buckskin virus. Mahaleb-rooted trees do not grow satisfactorily in heavy, wet soils with high water tables, being susceptible to *Phytophthora* root rot. This rootstock should be used for nonirrigation or drought conditions. Trees on Mahaleb roots have more cold-hardiness than those on Mazzard or 'Stockton Morello.' Sweet cherries often grow faster for the first few years on Mahaleb than on Mazzard roots, and some cultivars start heavy bearing rather early, which may result in some dwarfing of the tree. There is evidence, especially in England, that trees on Mahaleb roots are relatively short-lived, but in the U.S. good, productive trees over 50 years old on this stock are known. Mahaleb roots are more resistant to root-lesion nematode *(Pratylenchus vulnus),* than Mazzard or 'Stockton Morello' *(33).* They are also resistant to one species of the root-knot nematode *(Meloidogyne incognita)* but susceptible to *M. javanica.*

'Stockton Morello' (P. cerasus) (82)   This clonal stock is particularly useful for growing cherries in heavy, wet soils or for giving some reduction in tree size. It is not suitable for sandy soils. Sweet cherry tends to overgrow the 'Stockton Morello' root at the graft union, but the union is strong. 'Royal Anne' grows poorly on 'Stockton Morello' and 'Chapman' is reported to be incompatible with it, requiring the use of an intermediate stock. Trees on this stock sucker badly.

'Stockton Morello' is propagated by leafy softwood cuttings from virus-tested sources, using intermittent mist and root-promoting chemicals. Cuttings should be made in spring or summer, using current season's growth. Terminal cuttings root better than subterminal *(69).*

'Stockton Morello' is immune to the root-knot nematode, *Meloidogyne incognita,* but susceptible to root-lesion nematodes.

Mongolian cherry (Prunus fruticosa) roots can be used to obtain dwarfing of both sweet and sour cherry cultivars, but this cannot be recommended as a commercially useful stock.

CHESTNUT, CHINESE *(Castanea mollissima* Blume.) *(87)*   This blight-resistant species has usually been propagated by seed, many orchards consisting of seedling trees. Clonal selections have been made, however, and named cultivars are available. Seeds should be prevented from drying. The nuts are gathered as soon as they drop and either planted in the fall or kept in moist storage one or two months at 32° to 36° F (0° to 2° C) for spring planting. Seed nuts are satisfactorily stored in tight tin cans, with one or two very small holes for ventilation, at 32° F (0° C) or slightly higher; this storage temperature also aids in overcoming embryo dormancy *(110)*. Weevils in the nuts, which will destroy the embryo, can be killed by hot-water treatment at 120° F (49° C) for 30 minutes.

After one year's growth, the seedlings should be large enough to transplant to their permanent location or be grafted to the desired cultivar *(132)*. Although the chestnut is difficult to graft or bud, bark grafting and inverted T-budding have given good results. In regular T-budding, the buds tend to "drown," owing to excessive bleeding. The splice and whip grafts are also widely used.

Only *C. mollissima* seedlings should be used as rootstocks. It is likely that many graft union failures are due to the use of hybrid seedling rootstocks. The union may become defective and the tree die, even after it has been bearing for some years. Infection with blight fungus *(Endothia parasitica)* at the graft union can also cause graft failures.

A method for obtaining chestnut clones on their own roots is the "buried inarch" technique. Dormant unrooted scions of the desired cultivar are inarched into a young seedling tree. The base of the scion is buried in the soil several inches deep with at least one bud above ground. After the inarched scion becomes well-rooted, it can be detached at the graft union and subsequently grown as an independent plant *(86)*.

CITRUS *(Citrus sp.) (128, 144)*   Propagation methods are the same for all species of citrus. Members of this genus are readily intergrafted and can also be grafted to other closely related genera such as *Fortunella* (kumquat) and *Poncirus* (trifoliate orange). Citrus cultivars are propagated commercially by T-budding on seedling rootstocks. In all types of citrus propagation it is very important to use true-to-type material free of transmissible pathogens *(93)*.

Many of the citrus species can be propagated by rooting leafy cuttings, or by leaf-bud cuttings, although nursery trees are not commonly propagated in this manner *(40, 42, 57)*. Except for the psorosis virus, transmissible diseases do not appear in seedlings unless they become infected from insect vectors or by budding.

In Florida the Persian lime *(C. aurantifolia)* is propagated to some extent by air layering, as is the pummelo *(C. grande)* in southeast Asian countries.

Probably the most rapid method of obtaining a citrus cultivar worked on a given rootstock is the use of "cutting-grafts," (see p. 424) to produce dwarf plants. Most citrus nursery trees are grown in field nurseries, but there is increasing interest in container-growing *(115)*, particularly for dooryard trees and for replacement of nursery trees in the orchard.

Growing Citrus Nursery Stock   It is important to avoid using soils infested with citrus nematodes *(Tylenchulus semipenetrans)* or burrowing nematodes *(Radopholus similis)*, or soilborne diseases, although citrus is resistant to verticillium wilt. For nurseries it is preferable to use virgin soil or at least a soil that has not been formerly planted to citrus. For small operations, raised seedbeds enclosed by 12-in. boards can be used. A soil mixture of ¾ sandy loam and ¼ peat moss is satisfactory. Treating the soil with a fumigant such as DD (dichloropropane-dichloropropene) at the rate of 700 to 1000 lb per acre will minimize the chances of nematode infestations. To reduce the hazard of fungus infection, methyl bromide is often used to fumigate the seedbed and nursery site. Following treatment, planting should be delayed for six to ten weeks to allow the fumigant to dissipate. Some soils in California, however, have remained toxic to citrus for a year following such treatment. The seedbed should be in a lathhouse, or some other provision should be made for screening the young seedlings from the full sun.

Since there is considerable variation in the performance of seedlings taken from different trees, it is best to select the seeds from healthy, virus-tested old trees, known to produce vigorous, uniform seedlings which develop into satisfactory orchard trees after being budded to the desired cultivar.

Citrus seeds generally have no dormancy condition but are injured by being allowed to dry; they may be planted immediately after being extracted from the ripe fruit. Certain species, such as the trifoliate orange or its hybrids, mature their fruits in the fall. If the seeds are to be planted at that time, they should be held in moist storage at 30° to 40° F (−1° to 4° C) for at least four weeks before planting.

Seeds may be stored in polyethylene bags at a low temperature (40° F; 4° C); before storage they should be soaked for ten minutes in water at 120° F (49° C) to aid in eliminating seedborne diseases. Treatment of the seed with a fungicide, such as thiram, is also beneficial.

The best time to plant the seed is in the spring after the soil has warmed (above 60° F; 15° C). Seeds should be planted in rows 2 to 3 in. apart, and 1 in. apart in the row. They are pressed lightly in the soil and covered with a ½- to ¾-in. layer of clean, sharp river sand. This prevents crusting and aids in the control of "damping-off" fungi. To further aid in controlling these organisms, apply aluminum sulfate to the surface of the seedbed at the rate of 1¼ oz per sq ft just before the seeds are planted. The material may be either scattered over the surface and raked in or dissolved in water and sprinkled over the seedbed. The soil should be kept moist at all times until the seedlings emerge. Either extreme, allowing the soil to become dry and baked or overly wet, should be avoided. Electric soil-heating cables placed below the seedbed to maintain a temperature of 80° to 85° F (27° to 29° C) will hasten germination. By this method, seeds may be planted in the winter months, and the seedlings will be large enough to line-out in the nursery that spring. Many can be budded by fall or the following spring. This often shortens the propagation time by 6 to 12 months.

After the seedlings are 8 to 12 in. tall, they are ready to be transplanted from the seedbed to the nursery row. This is best done in the spring after danger of frost has passed. The seedlings are dug with a spading fork after the soil has been wet thoroughly to a depth of 18 in. They can then be loosened and removed with little danger of root injury. All stunted or off-type seedlings or those with crooked, misshapen roots should be discarded.

The nursery site should be in a frost-free, weed-free location on a medium textured, well-drained soil at least 24 in. deep and with irrigation water available. Old citrus soils should be avoided unless heavily fumigated with DD or methyl bromide before planting. The seedlings should be planted at the same depth as in the seedbed and spaced 10 to 12 in. apart in 3- to 4-ft rows.

Citrus seedlings are usually budded in the fall in Florida and California, starting in mid-September, early enough so that warm weather will insure a good bud union, yet late enough so that bud growth does not start and the wound callus does not grow over the bud itself.

Most commercial citrus-producing areas of the world have programs to determine the presence of virus and virus-like diseases in trees which are used as budwood sources. There are about 12 known viruses and the mycoplasma-like organism causing "stubborn" disease which can infect citrus *(93)*. Good indexing procedures have been worked out for most of them *(17).* Budwood should be taken only from known high-producing, disease-free trees. It is desirable to select the budsticks from a single tree, avoiding any off-type "sporting" branches. The best type of budwood is that next to the last flush of growth, or the last flush after the growth hardens. A round budstick gives more good buds than an angular one. The best buds are those in the axils of large leaves. The budsticks are usually cut at the time of budding, the leaves removed, and the budsticks protected against drying. Budsticks may be stored for several weeks if kept moist and held under refrigeration at 40° to 55° F (4° to 13° C).

The T-bud method is quite satisfactory for citrus. The bud piece is cut to include a sliver of wood beneath the bud. Fall buds are unwrapped in six to eight weeks after budding, spring buds in about three weeks. In California and Texas the buds are usually inserted at a height of 12 to 18 in., but in Florida the buds are inserted very low on the stock—1 or 2 in. above the soil. Such low budding is often necessitated by profuse branching in rough lemon and sour orange seedlings, which is caused by partial defoliation due to scab and anthracnose spot.

Buds inserted in the fall are forced into growth in the spring by "lopping" the top of the seedlings 2 or 3 in. above the bud. This is done just before spring growth starts, and consists of partly severing the top, allowing it to fall over on the ground. The top thus continues to nourish the seedling roots, but the bud is forced into growth. Lopping of spring buds is done when the bud wraps are removed—about three weeks after budding. If possible, the "lops" should be left until late summer, at which time they are cut off just above the bud union (Figure 17-1). Although lopping is satisfactory, it may make irrigation and cultivation difficult. An alternative practice is to first cut the seedling completely off 12 to 14 in. above the bud, then later cut it back immediately above the bud. However,

Figure 17-1 Budded citrus seedlings are often "lopped over" just above the inserted bud. This forces the bud into growth, even though the top continues to provide nourishment for the plant. Courtesy Department of Pomology, University of California, Davis.

this practice does not force the bud as well as lopping or cutting the seedling just above the bud.

Young citrus nursery trees may be dug "balled and burlapped" or bare-root. Bare-rooted trees should have the tops pruned back severely before digging. Transplanting of such trees is best done in early spring, but balled trees can be moved any time during spring before hot weather starts.

A marked variation occurs among the several citrus species in regard to the vigor and growth habit of their seedlings. Sweet orange and grapefruit seedlings are somewhat bushy and difficult to handle, as are those of the 'Sampson' tangelo which are also slow growing. Under California conditions, rough lemon, sour orange, 'Cleopatra' mandarin, and 'Troyer' citrange seedlings are vigorous, rapid growers with a uniform, upright growth habit which makes them easy to use.

Polyembryony is high in seeds of most citrus species used as rootstocks. The sexual seedling is often weak and makes a poor rootstock. The other seedlings arising from the nucellus are asexual and have the same characteristics as the seed-bearing plant. Consequently they are uniform, and make good rootstocks if the parent tree is desirable. Nucellar seedlings are vigorous, thorny, and upright-growing, but slow to start bearing if used as orchard trees. These undesirable qualities (thorniness and delayed bearing) are less pronounced in nursery trees propagated from budwood taken from the upper part of old nucellar seedling trees. Nucellar cultivars have been developed for all the commercial citrus cultivars, and are used because of their increased vigor, tree size, and yields, and their usual freedom from viruses *(18)*.

Rootstocks for Citrus *(8, 129)*.

Sweet orange (C. sinensis).   This is a good rootstock for all citrus cultivars, producing large, vigorous trees. Sweet orange is adapted to well-drained, light- to medium-loam soils, but owing to its susceptibility to gummosis (*Phytophthora* sp.), it is not suited to poorly drained, heavy soils. It produces standard-size fruits that are thin-skinned, juicy, and of fairly high quality. The seeds germinate readily, but the seedlings, which are 70 to 90 percent nucellar, are relatively slow growers and tend to produce low-branched, bushy trunks in the nursery.

Sour orange (C. aurantium)   This is an excellent stock for most citrus species, owing to its vigor, hardiness, deep root system, resistance to gummosis diseases, and to the high quality, smooth, thin-skinned, and juicy fruit produced by cultivars worked on it. However, it is subject to "tristeza," a virus disease that is transmitted by an insect vector or by using infected budwood *(11, 58)*. The scion top itself may be quite tolerant to the virus, but in combination with sour orange root the stock is affected, owing to death of the phloem tissues in the bud union area, which then results in starvation of the root. Grapefruit and mandarin, as well as orange cultivars, on sour orange are subject to tristeza. In California sour orange is no longer recommended as a rootstock for orange or grapefruit cultivars. In Florida, however, almost one-third of the orange trees are on sour orange roots, but tristeza has not been serious there, except in certain limited areas. Presumably the strains of tristeza in Florida are less virulent than those in California and other citrus areas. In Texas, also, sour orange rootstock is widely used. In Australia strains of sour orange—originating in Israel—have been tested that are tolerant to tristeza *(177)*.

Rough lemon (C. limon)   This stock is well adapted to sandy soils, and because of this about 60 percent of all citrus trees in Florida are on rough lemon roots. Trees on this rootstock outyield those on all other stocks, although the fruit produced is of lower quality. Its use in California is restricted to the lighter soil areas, particularly in the desert regions. Its susceptibility to *Phytophthora* precludes its use on heavy soils. Both trees and fruit are more susceptible to cold injury on this stock than on most other commercial stocks *(45)*. Fruit from trees on rough lemon roots are early maturing, but have a thick rind, are low in both sugar and acid, and are often coarse-textured compared with fruit from trees on other rootstocks.

Rough lemon produces numerous seeds which germinate well. Ninety to 100 percent of the seedlings are nucellar. They grow upright with single, unbranched trunks which are easy to bud and handle in the nursery. Rough lemon also can be easily propagated by cuttings. Sweet orange cultivars are tolerant to quick decline when worked on rough lemon roots. The chief advantages of this stock are its vigor and its ability to produce a bearing tree quickly, particularly on light, sandy soils. Some orange cultivars on rough lemon develop an incompatible bud union.

Grapefruit (C. paradisi)   This has been used occasionally as a rootstock with conflicting reports on its behavior. Good-quality fruits of excellent size are produced on this stock. It is unsatisfactory for light, sandy soils and is less cold-resistant than sour orange and sweet orange. Because sweet orange on this stock is susceptible to quick decline, and because production is variable, grapefruit finds limited use as a rootstock.

Trifoliate orange (Poncirus trifoliata)   This dwarfing citrus rootstock has been used to some extent for many years. In northern Florida and along the Gulf coast to Louisiana, it has long been used as a stock for Satsuma oranges and kumquats, for which it is excellent. *Poncirus trifoliata*, as a rootstock, does best on medium-textured soils. It will tolerate heavy soil but grows quite slowly. On light, sandy soil its growth may be so poor that trees on this stock are worthless. Trifoliate orange is commonly used as a stock for ornamental citrus and in home orchards for dwarfed trees. Trees on this stock yield heavily and produce high-quality fruits. Trifoliate orange is a deciduous species noted for its winter-hardiness and its *Phytophthora* and nematode resistance. Both the tree and its fruits, when worked on this stock, are more resistant to cold than other rootstock-top combinations, making it particularly adaptable to the colder citrus-growing regions. A citrus cultivar worked on trifoliate orange is one of the few examples of an evergreen top on a deciduous rootstock.

Trees on trifoliate roots are often affected by exocortis, or "scaly butt." This may be avoided by using uncontaminated nucellar buds or by taking buds from trees on trifoliate stock that are free of the disease, as evidenced by the lack of scaly butt symptoms. This stock is quite resistant to gummosis, but it is very susceptible to citrus canker *(Phytomonas citri)* and moderately so to scab. It is susceptible to lime-induced chlorosis but tolerant of excess boron. 'Eureka' lemon and 'Troyer' citrange are reported to develop a graft union disorder when worked on this stock, but this can be overcome, at least partially, by using a sweet orange interstock.

Trifoliate orange fruits produce large numbers of plump seeds which germinate easily. The upright-growing, thorny seedlings, about 60 percent of which are nucellar, are easy to bud and handle in the nursery, but their slow growth often necessitates an extra year in the nursery before salable trees are produced.

'Cleopatra' mandarin (C. reticulata)   This is widely used, particularly in Florida, as a stock for other mandarin types, and has come into use in California and Texas as a replacement for sour orange. Its resistance to gummosis, comparative salt tolerance, and resistance to tristeza seemingly justify its greater use. In

addition, trees on this stock show good yields of high-quality fruit, although fruit size is somewhat smaller than average. Its chief disadvantages are the slow growth of the seedlings, slowness in coming into bearing, and susceptibility to *Phytophthora parasitica* root rot. (It is, however, resistant to *P. citrophthora*.) About 80 percent of the seedlings are nucellar. Some cases of incompatibility, especially with 'Eureka' lemon, have appeared.

Citranges (Trifoliate orange × *sweet orange* hybrids)   There are many named cultivars—'Savage,' 'Morton,' 'Troyer,' and 'Carrizo,' and so on—some of which are proving useful both in California and in Florida.

'Savage' is especially suitable as a dwarfing stock for grapefruit and is also satisfactory for the mandarins. It is resistant to gummosis, and trees on this stock are more cold-hardy than those on many of the other rootstocks *(10, 45)*.

'Morton' citrange is not particularly dwarfing. Trees worked on it are heavy producers of excellent-quality fruit. It produces so few seeds, however, that to get quantities of nursery trees started is difficult. In addition, orange cultivars on this stock appear to be susceptible to quick decline.

Sweet orange on 'Troyer' citrange is vigorous, cold-hardy, and resistant to gummosis, with high-quality fruit. This stock is more widely used than any other for oranges in California. In replanting citrus on old citrus soils, trees on the 'Troyer' citrange have shown outstanding vigor. 'Troyer' itself is relatively fruitful and produces 15 to 20 plump seeds per fruit, facilitating the propagation of nursery trees. It is also readily propagated by leafy cuttings, especially if they are taken from young vigorous trees, and if root-promoting chemicals are used. 'Troyer' citrange is not generally recommended as a stock for 'Eureka' lemon but may be used for the 'Lisbon' lemon, at least in the interior valley of California *(189)*. As with most other citrus rootstocks, it is not adaptable to soils with a high salt content or where the irrigation water is high in boron. Young trees are very sensitive to wet soil conditions.

Seedlings of citrange cultivars are mostly nucellar and develop strong, single trunks, easily handled in the nursery. As with trifoliate orange, only exocortis-free buds should be used on citrange rootstocks; otherwise, dwarfing—and eventual low production—will result.

Rangpur lime (C. aurantifolia × *C. reticulata)*   This is the most widely used citrus rootstock in Brazil. It produces vigorous, fruitful trees which are resistant to tristeza, but it is highly susceptible to exocortis. In Texas it has been reported to be more salt-tolerant than other citrus rootstocks. Some strains of Rangpur lime are susceptible to *Phytophthora*.

Citrus macrophylla.   This is widely used in California as a rootstock for lemons in high-boron areas, owing to its boron tolerance. It is susceptible to tristeza when sweet orange scion cultivars are used or when infected before budding to lemons.

COCONUT (*Cocos nucifera* L.) *(23, 135)*   Trees are propagated only by seed, although there is a great need to develop methods of clonal propagation from superior trees. There are some named seedling cultivars which maintain their characteristics quite dependably by seed propagation. It is important to select seeds from trees that produce large crops of high-quality nuts.

The nuts are usually germinated in seedbeds. Seedlings that develop rapidly and have strong, vigorous shoots are selected. The nuts, still in the husk, are set at least 12 in. apart in the bed and laid on their sides with the stem end containing the "eyes" slightly raised. The sprout emerges through the eye on the side that has the longest part of the triangular hull. As soon as this occurs (about a month after planting), the sprout sends roots downward through the hull and into the soil. In 6 to 18 months the seedlings are large enough to transplant to their permanent location.

COCOS NUCIFERA.   *See* Coconut.

COFFEE (*Coffea arabica* L. *(192)*   The most common method of propagation is by seed, preferably obtained from selected superior trees. Coffee seeds lose viability quickly and are subject to drying through the seed coverings. Seeds held at a moisture content of 40 to 50 percent and at 40° to 50° F (4° to 10° C) will keep for several months. There are no dormancy problems. Seeds are usually planted in seedbeds under shade *(178)*. Sometimes the seedlings are started in soil in containers formed from leaves, or in polyethylene bags, to facilitate transplanting. Germination takes place in four to six weeks. When the first pair of true leaves develop, the seedlings are transplanted to the nursery and set 12 in. apart. After 12 to 18 months in the nursery, by which time they have formed six to eight pairs of laterals, the young trees are ready to set out in the plantation.

Coffee can be propagated asexually by almost all methods, but leafy cuttings probably hold the most promise for commercial use. Cutting material should be taken only from upright-growing shoots in order to produce the desired upright-growing tree. Leafy cuttings of partially hardened wood can be rooted fairly easily, especially if treated with a root-promoting chemical. High-humidity conditions, such as in intermittent mist, must be maintained, and the cuttings should be partially shaded during rooting *(54, 171.)*

COLA NITIDA   *See* Kola.

CORYLUS AVELLANA   *See* Filbert.

CRABAPPLE   Siberian crabapple (*Malus baccata* Borkh.), Western crabapple (*M. ioensis* Britt.), and other *Malus* species. The usual propagation method is to bud or graft the desired cultivars on seedling rootstocks, either one of the crabapple species or the common apple, *Malus sylvestris*. In areas where winter-hardiness is important, *M. baccata* seedlings should be used.

Crabapples can be propagated, although with difficulty, by softwood or hardwood cuttings, especially if root-promoting chemicals are used.

CRANBERRY (*Vaccinium macrocarpon* Ait.) *(31, 56)*   This vine type of evergreen plant produces trailing runners upon which are numerous short upright branches. Propagation is by cuttings made from either runners or upright branches. Cutting material is obtained by mowing the vines in early spring before new growth has started. The cuttings are then set directly in place in their permanent location without previous rooting at distances of 6 to 18 in. apart each way. Two to four cuttings are set in sand in each "hill." The cuttings are 5 to 10 in. long and set deep enough so that only an inch is above ground. A more rapid method of starting a cranberry bog is to scatter the cuttings over the ground and work them into the soil with a special disk-type planter. This is justifiable when there is an abundance of cutting material and a scarcity of labor for setting the cuttings by hand. Water is applied to the bog immediately after planting. The cuttings root during the first year and make some top growth, but the plants do not start bearing until three or four years.

CURRANT (*Ribes* sp.)   Currants are readily propagated by hardwood cuttings prepared from well-matured wood of the previous summer's growth. Cuttings 8 to 10 in. long are made in late fall, stored in moist sand, sawdust, or peat moss at about 35° F (2° C), then planted in early spring. The plants can be transplanted to their permanent location in one or two years, depending upon their growth. Currants can also be propagated by mound layering.

CYDONIA OBLONGA   *See* Quince.

DATE (*Phoenix dactylifera* L.) (*1, 2, 122*)   Propagation of the date is either by seed or by offshoots. This is a monocotyledonous plant having no continuous cambial cylinder, so it cannot be propagated by budding or grafting. In commercial plantings, most of the trees are female, but a few male trees are necessary for pollination purposes.

In seed propagation about half of the trees produced are males, whereas the seedling female trees produce fruits of variable and generally inferior types. The commercial grower is therefore little interested in seedling trees, preferring the superior named cultivars, even though the latter must be propagated by the vegetative offshoot method. Date seeds germinate readily.

Offshoots arise from axillary buds near the base of the tree. If they are near the ground level, they will develop roots in the soil after three to five years on the parent tree. Large, well-rooted offshoots, weighing 40 to 100 lb, are more likely to grow than smaller ones. Offshoots higher on the trunk can be induced to root if moist rooting medium is held against the base of the offshoot by means of a box or a polyethylene tube. Unrooted offshoots arising higher on the stem can be cut off and rooted in the nursery, but in relatively low percentages.

Considerable skill is required to properly cut off a date offshoot. Soil is dug away from the rooted offshoot, but with a ball of moist earth as thick as possible attached to the roots. The connection with the parent tree should be exposed on each side by removing loose fiber and old leaf bases. A special chisel, having a blade flat on one side and beveled on the other, is used to sever the offshoot. The first cut is made to the side of the base of the offshoot close to the main trunk. The beveled side of the chisel is toward the parent tree, which gives a smooth cut on the offshoot. A single cut may be sufficient, but usually one or more cuts from each side are necessary to remove the offshoot. The offshoot should never be pried loose; it should be cut off cleanly. After removal, it should be handled carefully and replanted as soon as possible, care being taken to prevent the roots from drying out (see Figure 14–10).

DEWBERRY   *See* Blackberry.

DIOSPYROS sp.   *See* Persimmon.

ERIOBOTRYA JAPONICA   *See* Loquat.

FEIJOA (*Feijoa sellowiana* Berg.)   Propagation is mainly by seeds, which germinate without difficulty. They should be started in flats of soil and later transplanted to the nursery row. Cultivars can be grafted on feijoa seedlings, although it is difficult to obtain a high percentage of successful unions. Leafy softwood cuttings treated with root-promoting substances and started under closed frames can sometimes be rooted.

FICUS CARICA   *See* Fig.

FIG (*Ficus carica* L.) (*27, 94*)   The fig is easily propagated by hardwood cuttings. Two- or three-year-old wood or basal parts of vigorous one-year shoots with a minimum of pith are suitable for cuttings. These are grown for one or two seasons in the nursery, then transplanted to their permanent location. A common method in European countries is to plant long cuttings (3 to 4 ft) their full length in the ground where the tree is to be located permanently; sometimes two cuttings are set in one location to increase the chance of having one grow.

The fig can be budded, using either T-buds inserted in vigorous one-year-old shoots on heavily pruned trees, or patch buds on older shoots.

Fig roots are resistant to oak root fungus *(Armillaria mellea)* and verticillium wilt but quite susceptible to both root-knot *(Meloidogyne* sp.) and lesion nematodes *(Pratylenchus vulnus) (34)*.

Seed propagation is used only for breeding new cultivars. The small seeds can be germinated easily in flats of well-prepared soil. Fertile seeds should first be separated from the sterile ones by placing all of them in water; fertile seeds sink, but sterile ones float.

Figs can be air layered. One-year-old branches, if layered in early spring, are usually well rooted by midsummer.

FILBERT *(Corylus avellana* L.) *(112, 155)*   Simple layering is the usual method of commercial filbert propagation. The suckers arising from the base of vigorous young trees four to eight years old are layered in early spring. After one season's growth, a well-rooted tree 2 to 6 ft tall may be obtained, ready to set out in the orchard. Old orchard trees are not suitable to use for layering. Sometimes suckers arising from the roots are dug and grown in the nursery row for a year or, if well-rooted, planted directly in place in the orchard.

Leafy tip cuttings of some cultivars taken in late spring to midsummer will root under mist if treated with indolebutyric acid, but it is often difficult to induce shoot growth from the rooted cuttings. Hardwood cuttings of some cultivars can be rooted utilizing the bottom heat methods described on p. 279. Other cultivars are very difficult to propagate by cuttings.

Budding or grafting filbert cultivars on seedling stocks is rarely practiced, because of the difficulty in obtaining successful unions.

Filbert seeds are easily germinated but require a stratification period of several months at 32° to 40° F (0° to 4° C).

FRAGARIA *sp.*   *See* Strawberry.

GOOSEBERRY *(Grossularia* or *Ribes* sp.)   Mound layering is used commercially. Layered shoots of American cultivars usually root well after one season. They are then cut off and transferred to the nursery row for a second season's growth before they are set out in their permanent location. The slower-rooting layers of European cultivars may have to remain attached to the parent plant for two seasons before they develop enough roots to be detached from the parent plant. Some cultivars, such as the 'Houghton,' 'Poorman,' and 'Van Fleet,' can be started fairly easily by hardwood cuttings *(140)*.

GOOSEBERRY, CHINESE. KIWI FRUIT *(Actinidia chinensis* Planch.) *(39)*   Both male and female vines of this dioecious subtropical fruit must be planted to ensure fruiting.

Seed   Seed propagation can be used but the sex of the vine cannot be determined until fruiting at seven years or more. Select seed from soft, well-ripened fruit; dry and store at 41° F (5° C). After at least two weeks at this temperature subject seed to fluctuating temperatures—50° F (10° C) night and 68° F (20° C) day for two or three weeks before planting.

Cuttings   Softwood tip cuttings taken from late spring to midsummer may be rooted under mist if wounded, then treated with an 0.8 percent indolebutyric acid in talc powder preparation. Dipping the bases of the cuttings in captan is helpful. Plants can also be started by root cuttings.

Grafting and Budding   Seedlings can be grafted by the whip or tongue method in late winter using dormant scion wood. T-budding in late summer is also successful. Plants grafted on seedlings tend to be more vigorous than those started as rooted cuttings.

GRAPE (*Vitis* sp.) *(198)*   Grapevines are propagated by seeds, cuttings, layering, budding, or grafting. Seeds are used in breeding programs to produce new cultivars. Grape propagation methods have been modernized by the use of virus-indexed, "clean" planting stock, mist-propagation techniques for leafy cuttings, and rapid machine-grafting procedures. Most commercial propagation is by dormant hardwood cuttings. For types difficult to root, such as the Muscadine *(Vitis rotundifolia)*, layering or the use of leafy cuttings under mist is necessary *(163)*. Budding or grafting on rootstocks is used occasionally to increase vine life, plant vigor, and yield. Where noxious soil organisms, such as phylloxera *(Dactylosphaera vitifoliae)*, or root-knot nematodes *(Meloidogyne* sp.), are present, and cultivars of susceptible species such as *V. vinifera* are to be grown, it is necessary to graft or bud on a resistant rootstock.

Root-knot nematodes can be eradicated from grapevine rootings by dipping them in hot water (125° to 130° F; 51.5° to 54.5° C) for five to three min., respectively *(97)*.

Seeds   Grape seeds are not difficult to germinate. Best results with vinifera grape seeds are obtained after a moist stratification period at 33° to 40° F (0.5° to 4° C) for about 12 weeks before planting *(63)*.

Dormant Cuttings   Most grape cultivars are traditionally propagated by dormant hardwood cuttings, which root readily. Cutting material should be collected during the winter from healthy, vigorous, mature vines. Well-developed current season's canes should be used; they should be of medium size and have moderately short internodes. Cuttings $\frac{1}{3}$ to $\frac{1}{2}$ in. in diameter and 12 to 16 in. long are generally used. One season's growth in the nursery should produce plants large enough to transplant to the vineyard.

Root-promoting auxin-type chemicals have not been particularly helpful in rooting hardwood grape cuttings *(61)*.

Leafy Cuttings   Leafy greenwood grape cuttings root profusely under mist in about ten days if given relatively high (80 to 85° F; 26.5 to 29.5° C) bottom heat and if treated with indolebutyric acid. Scarce planting stock (as virus-indexed material) can be increased very rapidly by using one-budded stem cuttings (see Figure 17-2), then consecutively taking additional such cuttings from the shoot arising from the bud on the rooted cutting, and so on.

Figure 17-2   Leafy, one-node grape stem cuttings rooted under mist. In the grape, roots arise readily from the internodes.

Layers   Grape cultivars difficult to start by cuttings can be propagated by layering, using either simple, trench, or mound layering (see Chapter 14).

Grafting   Bench grafting *(64)* is widely used; scions are grafted on either rooted or unrooted disbudded rootstock cuttings by the whip graft or, better, by machine grafting (see Figures 12–23 and 12–24). The grafts are made in late winter or early spring from completely dormant scion and stock material. The stocks are cut to 12 to 14 in., with the lower cut just below a node and the top cut an inch or more above a node. All buds are removed from the stock to prevent subsequent suckering. Scion wood should be selected which has the same diameter at the stock.

After grafting, using a one-bud scion, the union is stapled together or wrapped with budding rubber. The grafts should be held for two to four weeks in well-aerated, moist wood shavings or peat moss at about 80° F (26.5° C) for callusing. Plant the grafts in the nursery as soon as the unions have healed, before there is much shoot or root growth, but not before the ground has warmed.

The grafts are planted so that the unions are just above the soil level. Immediately after planting, the graft is covered with a wide ridge of soil so that the scions are covered to a depth of 2 or 3 in. After the grafts are growing vigorously and the shoots are 8 to 12 in. high, the bench grafts are uncovered for the first time. Each one should be examined carefully. Roots arising from the scion should be removed, although some scion roots are not undesirable for a time to help the graft get started. Removing the scion roots too late will retard root growth from the stock. Suckers from the stock should also be removed. At this time, the tying material should be cut if it has not already deteriorated. The grafts must be covered back to the same depth.

Variations developed *(190)* on the above procedures are illustrated in Figure 17–3.

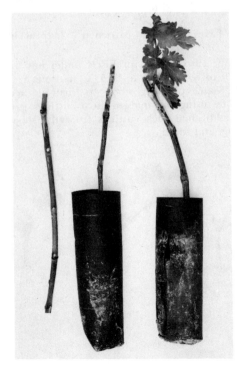

Figure 17–3   Steps in grafting a grape cultivar on to a resistant rootstock. *Left:* a one-budded scion grafted on an unrooted rootstock cutting using a French grafting device (see Figure 12–26). *Center:* after callusing, followed by paraffining scion and graft union, the stock is planted in a tube of felt building paper (or a plastic-coated cardboard cylinder) and then set in a protected place, as a greenhouse or lathhouse. *Right:* several weeks later the scion bud starts growth. Graft union has healed and the rootstock cutting is well rooted, so that after seven to ten days hardening-off, the plant can be set out (still in the tube) directly in the vineyard.

Greenwood Grafting   Greenwood grafting is a simple and rapid procedure for propagating vinifera grapes on resistant rootstocks *(62)*. A one-budded greenwood scion is splice-grafted during the active growing season on new growth arising either from a one-year-old rooted cutting or from a cutting during midseason of the second year's growth.

Budding *(100)*   A satisfactory method of establishing grape cultivars on resistant rootstocks is by field budding on rapidly growing, well-rooted cuttings which were planted in their permanent vineyard location the previous winter or spring. A form of chip budding is performed in late summer or early fall just as soon as fresh mature buds from wood with light brown bark can be obtained and before the stock begins to go dormant. In areas where mature buds cannot be obtained early in the fall, growers may store under refrigeration budsticks collected in the winter and bud them late spring or early summer.

The bud is inserted in the stock 2 to 4 in. above the soil level, preferably on the side adjacent to the supporting stake. It is tied in place with budding rubber, but is not waxed. The bud is then covered with 5 to 10 in. of well-pulverized, *moist* soil to prevent drying. In areas of extremely hot summers, or in soils of low moisture, variable results are likely to be obtained, and bench or nursery grafted vines should be used.

T-budding is not practiced with grapes, owing to the relatively large size of the buds or "eyes."

Rootstocks for American Grapes *(11, 102)*

'*Salt Creek*' (*Ramsay*), '*Dog Ridge*,' '*Champanel*,' '*Lukfata*' (*V. champini*) These stocks have been very effective in the southern coastal region of the U.S. in increasing yields and prolonging vine life of the "bunch" type of grapes.

V. rupestris '*Constantia*,' '*Couderc 3309*,' '*V. cordifolia* × *V. riparia 125-1*,' '*Cynthiana*,' '*Wine King*,' '*Lenoir*'   These stocks have been useful in the inland states of the U.S. in increasing yields, plant vigor, and evenness of fruit ripening.

'*Couderc 3309*,' '*SO-4*' ('*Oppenheim #4*') and '*Teleki 5-A*' are recommended in New York State for their resistance to root parasites, especially in replant situations.

Rootstocks for Vinifera Grapes *(89, 98, 99)*

'*St. George*' (*V rupestris*)   This vigorous, phylloxera-resistant stock, noted for its drought tolerance, is especially suitable for shallow, nonirrigated soils; it is not resistant to root-knot nematodes. It is readily propagated by cuttings and easily grafted. It tends to sucker profusely, so disbudding before planting should be carefully done, and suckering continued for the first three or four years.

'*Ganzin No. 1*' ('*A* × *R. #1*')   This phylloxera-resistant rootstock is highly recommended for fertile, irrigated soils. Vines on this stock under such conditions are generally more productive than those on '*St. George*.' It is susceptible to root-knot nematodes and does not do well on dry, hillside soils. Cuttings root easily but it does not bench graft well, as it calluses poorly.

'*Couderc 1613*'   This stock is the one most widely used in the interior valley grape growing region of California. It is resistant to root-knot nematodes. Although suitable for fertile, irrigated, sandy loam soils, it produces weak, unproductive vines in nonirrigated or sandy soils of low fertility. Cuttings are easily rooted.

'*Harmony*'   This rootstock developed by the USDA is a cross between a '*Dog Ridge*' seedling and a '*1613*' seedling. It has more vigor and more phylloxera and nematode resistance than '*1613*.' Cuttings root readily and it buds and grafts easily. It is recommended as a replacement for '*1613*,' and especially for table grape vineyards.

'Dog Ridge' and 'Salt Creek' ('Ramsay') (V. champini)   These closely related stocks are moderately resistant to root-knot nematodes and are extremely vigorous. They should be used only in low-fertility soils. In fertile soils, the vines are often so vigorous that they are unproductive. Both are moderately resistant to phylloxera. Cuttings of these stocks are difficult to root, especially *'Salt Creek.'* On better soils, *'Salt Creek'* is usually preferred to *'Dog Ridge'* because the latter's extreme vigor causes poor fruit set. These two stocks have performed best for raisin and wine grape vineyards.

GRAPEFRUIT   *See* Citrus.

GROSSULARIA sp.   *See* Gooseberry.

GUAVA *(Psidium cattleianum* Sabine—the Cattley or strawberry guava—and *Psidium guajava* L.—the common, tropical, or lemon guava) *(103)*   The Cattley guava has no named cultivars, and most nursery plants are propagated by seed. This species comes nearly true from seed, large-fruited, superior trees being used as the seed source. It is difficult to propagate by vegetative methods.

The common guava has several cultivars, but they are not grown extensively, because of difficulties in asexual propagation. Most trees of this species are propagated by seeds, which germinate easily and in high percentages, but do not come true to type. Seeds should be taken from the best type of tree available, and the flowers to produce the seeds should be self-pollinated, which will reduce seedling variability.

Seedlings are somewhat susceptible to damping-off organisms, and should be started in sterilized soil or otherwise protected by fungicides. When about 1½ in. high, the seedlings should be transplanted into individual containers. In six months the plants should be about 12 in. high and can be transplanted to their permanent location.

For large-scale propagation of guava cultivars, grafting or budding is necessary. Chip budding has been successful, done any time during the summer using greenwood buds from selected cultivars inserted into seedling stocks about 5 mm in thickness. Plastic wrapping tape is used to cover the buds. The stock is cut off above the inserted bud after about three weeks *(84)*.

Cuttings are difficult to root, best results being obtained by rooting succulent shoots under mist after treating them with about 200 ppm of indolebutyric acid *(133)*.

Air layering gives good results. Simple and mound layering are effective methods of starting new plants; the layers may be tightly wrapped with wire just below the point where roots are wanted, or can be ringed, with indolebutyric acid (500 ppm) in lanolin rubbed into the cuts to promote rooting *(107)*.

HICKORY *(Carya ovata* Koch—Shagbark hickory)   Hickory cultivars are propagated by grafting or budding on seedlings of the Shagbark hickory *(C ovata)*, or pecan *(C. illinoensis) (130)*. However, owing to grafting and transplanting difficulties and slow growth of the trees, there is little planting of grafted trees.

To avoid transplanting failure caused by the long taproot, several stratified nuts are planted in the spring where the trees are to be located permanently. The best tree is saved and topworked, usually by bark grafting, to the desired cultivar *(105)*. If grafted nursery trees are to be used, the taproot should be cut a year previous to digging or the tree transplanted in the nursery once or twice to force out lateral roots.

Hickory nuts will germinate without stratification but should be kept in moist, cool storage until planting. Fall planting is successful if the soil in cold climates is well mulched to prevent excessive freezing and thawing.

Patch budding is used by nurserymen in the commercial propagation of hickories, usually performed in late summer. The seedling stocks are grown for two years or more before they are large enough to bud.

JUGLANS sp. *See* Walnut. *See* Butternut.

JUJUBE *(Zizyphus jujuba Mill.)*—Chinese date   The leading cultivars—'Lang' and 'Li'—are propagated by budding or grafting on jujube seedlings. The seeds should be stratified at about 40° F (4° C) for several months before planting. Jujubes may also be propagated by root cuttings or hardwood stem cuttings.

KIWI FRUIT   *See* Gooseberry, Chinese.

KOLA *(Cola nitida* Vent.)   Propagation of the kola, the tropical caffein-containing nut used to make refreshing beverages, is largely by seed planted directly in the field or in nurseries for later field planting. Seeds for planting should be harvested only when completely mature. A pronounced juvenility pattern (see p. 184) occurs in kola; seven to nine years are required for the seedlings to flower so that asexual propagation using mature growth is essential for early production.

Rooting of terminal leafy cuttings in polyethylene-covered frames can be done but root-promoting "hormones" have not been successful *(186)*. Air layering is quite successful in obtaining early bearing, commercial production taking place 12 to 18 months after planting the rooted layers. IBA treatments aid in rooting the layers. Patch budding is also a successful method of vegetative propagation *(5)*.

LEMON   *See* Citrus.

LIME   *See* Citrus.

LITCHI CHINENSIS   *See* Litchi.

LOGANBERRY   *See* Blackberry.

LOQUAT *(Eriobotrya japonica* [Thunb.] Lindl.)   This is propagated by T-budding or side grafting superior cultivars on loquat seedlings or by top-grafting older, established trees, using the cleft graft. The quince can be used as a rootstock, producing a dwarfed tree. Air layering is successful; indolebutyric acid in lanolin at 250 ppm rubbed into the layered surface will increase rooting *(169)*. For ornamental use, seedling trees are satisfactory.

LITCHI *(Litchi chinensis* Sonn.)   Seed-produced litchi plants are inferior, so the many cultivars of this species are propagated by asexual methods, principally air layering. This can be done at any time of the year, but with best results in spring or summer *(103)*. Large limbs air layer easier than small ones.

Tip cuttings from a flush of growth in the spring have been rooted in fairly high percentages under mist in the full sun. Use of root-promoting chemicals are beneficial. Hardwood cuttings from an active flush of new growth have rooted more readily than those from dormant hardwood cuttings.

Litchi seeds germinate in two to three weeks if planted immediately upon removal from the fruit, but they lose their viability in a few days if not planted. Seedling trees are rarely grown for their fruit; they require 10 to 15 years to start bearing, and the fruit is likely to be of low quality *(24)*.

MACADAMIA *(Macadamia integrifolia* Maiden & Betche, and *M. tetraphylla* L. Johnson) *(147, 175)*   Trees of this subtropical evergreen nut tree are grown as seedlings, but vegetative propagation from a few superior selections by grafting is usually practiced. Macadamia is highly resistant to *Phytophthora cinnamomi* and will tolerate heavy clay soil.

Fresh seeds should be planted in the fall as soon as they mature, either directly in the nursery or in sand boxes in a lathhouse, then transplanted to the nursery after the seedlings are 4 to 6 in. tall. It is important not to crack the seeds, because they are readily attacked by fungi. Only seeds that sink when placed in water should be used *(59)*. Seeds retain viability for about 12 months at 40° F (4° C), but at room temperature viability starts decreasing after about four months. Scarifying or soaking the seed in hot water hastens germination.

Selected clones are propagated by one of the side graft methods. Leaves should be retained for a time on the rootstock. Rapid healing of the union is promoted if the rootstock is checked in growth prior to grafting by water or nitrogen deficiency to permit carbohydrate accumulation. Also, ringing the branches that are to be the source of the scions several weeks before they are taken increases their carbohydrate content and promotes healing of the union. Budding has generally been unsuccessful.

Macadamia can be propagated by rooting leafy, semi-hardwood cuttings of mature, current season's growth; 3- to 4-in. tip cuttings are best. Treatment with indolebutyric acid at 8000 to 10,000 ppm is beneficial (see p. 291). Cuttings should be placed in a closed propagating frame or under intermittent mist for rooting. Bottom heat at 75° F (24° C) is beneficial. There are pronounced cultivar differences in ease of rooting. *M. tetraphylla* cuttings root more readily than those of *M. integrifolia.*

Air layering is successful. Some believe that a better root system develops when grafted seedlings are used than when propagation is by cutting or air layering.

MALUS BACCATA   *See* Crabapple.

MALUS IOENSIS   *See* Crabapple.

MALUS SYLVESTRIS   *See* Apple.

MANDARIN   *See* Citrus.

MANGO *(Mangifera indica* L.) *(170)*   Most plantings of this evergreen tropical fruit are seedlings, although many superior selections are maintained vegetatively. Polyembryonic (see p. 62) cultivars commonly occur in mango. The seedlings may be sexual or nucellar in origin, either condition occurring in the seed, or both conditions occurring simultaneously. However, growth of several shoots from one seed does not necessarily indicate the presence of nucellar embryos, since, in certain cultivars shoots develop from below ground, arising in the axils of the cotyledons of one embryo, which may or may not be of zygotic origin *(4)*. Monoembryonic cultivars should not be propagated by seed, as they do not come true.

Mango seeds are used either to produce a true-to-type nucellar seedling of some superior clone or a rootstock on which is budded or grafted the desired clone. The seeds should be planted as soon as they are mature, although they can be stored in the fruits or in polyethylene bags at about 70° F (21° C) for at least two months. Low-temperature (below 50° F; 10° C) storage and excessive drying should be avoided. Removing the tough endocarp which surrounds the seed, followed by planting in a sterilized medium, should result in good germination in two to three weeks. Soon after the seedlings start to grow, they should be transplanted to pots or to the nursery.

Mangos are commonly propagated in Florida by a type of veneer grafting or by chip budding. A week after budding, the stock is cut off two to three nodes above the bud, with a final removal to the bud when the bud shoot is 3 or 4 in. long. The best budwood is prepared from terminal growth, $\frac{1}{4}$ to $\frac{3}{8}$ in. in diameter, that have been hardened. The leaves are removed, with the exception of two or three terminal ones. The buds will swell in two or three weeks, and are then ready to use. If the buds are to be used on stocks older than three weeks, ringing the base of the shoots from which the buds are to be taken about ten days before they are used increases their carbohydrate supply and seems to promote healing. Budding is best done when the rootstock seedlings are two to three weeks old—in the succulent red stage. Four to six weeks after budding, the inserted bud should start growth *(103)*. T-budding has also been successful.

Approach grafting, termed "inarching" in India, has been used since ancient times in propagating the mango. Veneer grafting is also successful, and its use is being encouraged *(118)*. Rootstocks are mostly from assorted seedlings, although selected monoembryonic rootstocks can be multiplied by air layering or by cuttings, and subsequently clonally propagated by stooling *(117)*.

Air layering is successful, especially when etiolated shoots are treated with indolebutyric acid at 10,000 ppm; such treatments have given 100 percent rooting with 90 percent survival *(119)*. The mango is difficult to propagate by cuttings, but it can be done by using leafy cuttings under mist along with indolebutyric acid treatments.

MORUS sp.   *See* Mulberry.

MULBERRY (*Morus* sp.)   The mulberry is readily started by hardwood cuttings (see p. 273) 8 to 12 in. long, made from wood of the previous season's growth and planted in early spring.

MUSA sp.   *See* Banana.

NECTARINE *(Prunus persica)*   *See* Peach.

OLEA EUROPAEA   *See* Olive.

OLIVE (*Olea europaea* L.) *(73)*   Olives are propagated by budding or grafting on seedling or clonal rootstocks, by hardwood or semi-hardwood cuttings, or by suckers from old trees.

Seeds of small-fruited cultivars germinate more easily than those of large-fruited ones. Often germination is prolonged over one or two years. The usual practice is to plant many more seeds than will be needed as seedlings to offset the low germination percentage. The seed is enclosed in a hard endocarp or "pit." Removing, clipping or cracking this endocarp materially hastens germination which, however, is still slow and erratic.

The seedlings grow slowly, and they may take a year or two to become large enough to be grafted or budded. Several methods have been used successfully, including T-budding, patch budding, whip grafting, and side-tongue grafting. In Italy, a widely used method is to bark graft small seedlings in the nursery row in the spring. The stocks are cut off several inches above ground, and one small scion is inserted in each seedling, followed by tying and waxing. After grafting or budding, one or two more years are required before a tree large enough for transplanting to the orchard is produced.

Hardwood cuttings may be made from two- or three-year-old wood about an inch in diameter and 8 to 12 in. long. All leaves are removed. It is often helpful to soak the basal ends of the cuttings in a solution of indolebutyric acid (15 ppm for 24 hours), followed by storage in moist sawdust at 60° to 70° F (15° to 21° C) for a month preceding spring planting in the nursery *(66)*.

Semi-hardwood cuttings from vigorous, one-year-old wood about $\frac{1}{4}$ in. in diameter can be successfully rooted. They should be 4 to 6 in. long with two to six leaves retained on the upper portion of the cutting. It is best to take the cuttings in early summer or midsummer and root them under high-humidity *(72)*. Olive cuttings root well under intermittent mist and respond markedly to treatment with indolebutyric acid at about 4000 ppm for five seconds. There is considerable variability among cultivars in ease of rooting.

An ancient propagation method is to use the characteristic swellings, sometimes called "ovuli," found on the trunks of old trees, which are cut off and planted in early spring. These contain both adventitious root initials and dormant buds, hence new root and shoot systems can be regenerated. This practice is damaging to the parent tree, however, and is not widely used.

Another method of propagating olives is to saw large branches, 3 or 4 in. in diameter, into pieces about 12 in. long and plant them horizontally several inches below the surface of the soil. Sometimes they are split into several pieces, called truncheons. (See Figure 10–2.) Several shoots, each with an accompanying root system, will grow from each piece; the truncheon may then be cut apart to form several small plants, which are grown separately another year in the nursery row to produce a tree.

Rootstocks for Olives   *Olea europaea* seedlings are often used, with budding or grafting, although considerable variation in tree vigor and size may result from this practice *(68)*. A more suitable rootstock, giving uniform trees, is rooted cuttings of a strong-growing cultivar. Olive trees are very susceptible to *Verticillium* wilt; where this occurs resistant clonal rootstocks should be used *(76)*.

ORANGE   *See* Citrus.

PAPAYA *(Carica papaya* L.) *(60)*   Propagation by seed is the most practical method. They can be sown in flats of soil or in seedbeds in the open with germination in two to three weeks. The seeds do equally well if taken from fresh fruit or stored dry. The best seeds are from controlled pollinations between superior trees. When the seedlings are about 4 in. tall they are transplanted. This is generally done once or twice before they are put in their permanent location.

A method of starting seedlings without disturbing the roots is to plant four to eight seeds in a container and then thin to two to four of the strongest when they are about 4 in. tall. They can then be set in the field without disturbing the root system. Young papaya seedlings are very susceptible to damping-off organisms, so the soil in which they are started should be sterilized if possible.

In Florida, the usual practice is to plant seeds in midwinter and set the young plants in the field by early spring. They grow during spring and summer, and usually mature their first fruits by fall, the plants bearing all winter and the following season.

Papaya cuttings are not difficult to root if bottom heat at 85° F (29° C) is maintained and entire branches with the basal swelling are used *(182)*. Owing to the nonbranching growth habit of the plant, the amount of cutting material is limited. Scions from particularly good plants may be grafted on seedlings, but this is not done commercially.

PAPAW *(Asimina triloba.* Dun.)   Propagation is by seed, which should be removed from the fruits as soon as they mature. Seeds are either planted in the fall or given a rather long stratification period and planted in early spring. They are slow to germinate, some starting in the second year. Young seedlings are slow growing and, because they are quite sensitive to strong sunlight, are usually started under partial shade. The seedlings are somewhat difficult to transplant successfully after the first year, owing to their long taproot *(49)*.

PASSIFLORA EDULIS   *See* Passion Fruit.

PASSION FRUIT *(Passiflora edulis* Sims.)   This subtropical tender evergreen fruit is propagated chiefly by seed, which germinates two to three weeks after planting. Propagation by cuttings of mature wood is also practiced occasionally. To avoid attacks of *Fusarium* wilt, which is a serious problem, it is advisable to graft the purple-fruited form onto seedlings of golden passion fruit *(12)* or *P. flavicarpa (180),* which are resistant to this disease.

PEACH AND NECTARINE *(Prunus persica* Batsch)   Nursery trees of these fruits are propagated by T-budding on seedling rootstocks. Fall budding is most common, but spring budding or—in regions with long growing seasons—June budding is also done.

Some peach cultivars can be propagated by leafy, succulent softwood cuttings taken in the spring, treated with a root-promoting material, and rooted in a mist propagating bed but this is not done commercially *(71).* In areas with mild winters, some peach cultivars can be started from hardwood cuttings if they are treated with indolebutyric acid (4000 ppm for 5 sec.), then set out in the nursery in the fall.

Rootstocks for Peach (126)   Most peach cultivars are propagated on peach seedlings. Apricot and almond seedlings are sometimes used, as well as (in Europe) 'Brompton' and 'Common Mussel' plum clones. Peach seeds must be stratified at about 40° F (4° C) for three to four months, and almond and apricot for four weeks, before they are planted.

Peach (P. persica)   Peach seedlings are the most satisfactory rootstock for peach and should be used unless certain special conditions warrant other stocks. Seeds of 'Elberta,' 'Halford,' or 'Lovell' are usually used, since they germinate well and produce vigorous seedlings. Seedlings of these cultivars are not resistant to root-knot nematodes, however, and where this is a problem, resistant peach stocks (see below) should be considered. Seeds from peach cultivars whose fruits mature early in the season should not be used because their germination percentage is usually low. It is best to obtain seeds from the current season's crops, since viability decreases with each year of storage.

Peach roots are quite susceptible to the root-knot nematode *(Meloidogyne* sp.), especially in sandy soils. They are also susceptible to the root-lesion nematode, *Pratylenchus vulnus.* There is much interest therefore in nematode-resistant types. Seedlings of the 'S-37' peach rootstock are resistant to attacks of *Meloidogyne incognita,* but do not show such resistance to *M. javanica,* another widely distributed nematode species. 'Nemaguard,' a *Prunus persica* × *P. davidiana* hybrid rootstock introduced by the USDA in 1959, produces seedlings which are uniform and resistant to both *M. incognita* and *M. javanica,* and which form strong, well-anchored trees, although certain peach and plum varieties have not done well on it and it has shown susceptibility to bacterial canker *(165).*

A very winter-hardy peach rootstock, 'Siberian C,' withstanding 12° F (−11° C) soil temperatures, was introduced in 1967 by the Canadian Department of Agriculture Research Station at Harrow, Ontario. It does well on light, sandy soils and will give about 15 percent dwarfing to the scion cultivar. It is, however, susceptible to root-knot and root-lesion nematodes *(95).* 'S-37' and 'Nemaguard' may not be hardy in the colder growing areas.

Nursery trees on peach roots often make unsatisfactory growth when planted on soils previously planted to peach trees. Peach roots are susceptible to oak root fungus, crown rot, crown gall, and verticillium wilt.

Apricot (P. armeniaca)   Apricot seedlings are occasionally used as a rootstock for the peach. The graft union is not always successful, but numerous trees and commercial orchards of this combination have produced fairly well for many years. Seedlings of the 'Blenheim' apricot seem to make better rootstocks for peaches than those of 'Tilton.' The apricot root is highly resistant to root-knot but not to root-lesion nematodes.

Almond (P. amygdalus)   Almond seedlings have been used with limited success as a rootstock for peaches. There are trees of this combination growing well, but in general this is not a satisfactory combination. The trees are often dwarfed and tend to be short lived.

Western sand cherry (P. besseyi)   When this was used in limited experiments as a dwarfing rootstock for several peach cultivars, it was found that although the bud unions were excellent about 40 percent of the nursery trees failed to survive. The remainder grew well, however, and developed into typical dwarf trees with healthy dark green foliage. The trees bore normal-size fruit in the second or third year after transplanting to the orchard *(15)*.

Nanking cherry (P. tomentosa)   This may be suitable for some peach cultivars as a dwarfing rootstock.

'Brompton' and 'St. Julien A' plum   In England these rootstocks have been compatible with all peach and nectarine cultivars worked on them, and produce medium to large trees. Peach seedlings are rarely used as peach rootstocks in England.

PEAR *(Pyrus communis* L.)   Pears are commonly propagated by fall budding, using the T-bud method, on either seedling pear rootstocks or rooted quince cuttings. Pear trees are also started by whole-root grafting, using the whip or tongue method. Pear seeds must be stratified for 60 to 100 days at about 40° F (4° C). They are then planted thickly about ½ in. deep in a seedbed, where they are allowed to grow during one season. The following spring they are dug, the roots and top are cut back, and then they are transplanted to the nursery row, where they are grown a second season, ready for budding in the fall.

Some pear cultivars, such as 'Old Home' and 'Bartlett,' can be propagated by hardwood cuttings or by leafy cuttings under mist if treated with indolebutyric acid *(70, 74, 193)*. In pear rootstock test plantings, own-rooted 'Bartlett' trees have shown excellent production, are resistant to pear decline *(74)* (see below) and with age become partially dwarfed, a desirable attribute for high-density plantings.

Rootstocks for Pear (53, 65, 194)   In using rootstocks from seedlings of one of the several *Pyrus* species available, the seed source is very important if a rootstock of a certain species is expected. The various *Pyrus* species hybridize freely, some bloom at the same time, and cross-pollination is necessary for seeds to develop. It is best to use as the seed source an isolated group of trees of a given known species and to avoid collecting seeds from a planting containing trees of several different species, seeds from which are likely to produce hybrid seedlings.

"Pear decline," believed to be due to mycoplasma-like bodies *(80)* and known to be spread by pear psylla *(Psylla pyricola)*, devastated pear orchards, first in Italy *(162)* and later in western North America. Pear decline is associated with the rootstock used *(9, 13, 153)* so it is an important factor to be considered in rootstock selection. For plantings in decline-prevalent areas only rootstocks known to be resistant to decline should be used.

Seedlings from cultivars which may have resulted from pollination by a decline-susceptible species, such as *Pyrus pyrifolia* or *P. ussuriensis*, should not be used. Where pear psylla exist, and where they may be carrying the mycoplasma-like bodies causing pear decline, seedlings of the Oriental pears—*P. pyrifolia* or *P. ussuriensis*—should *not* be used, as trees with these as rootstocks are highly susceptible to decline.

French pear (P. communis)   These are seedlings of the common pear, taking their name from the fact that in earlier days seeds were imported into the U.S. in large quantities from France. French pear seedlings as used now are generally grown from seeds of 'Winter Nelis' or 'Bartlett.' This rootstock is vigorous and winter-hardy, produces uniform trees with a strong, well-anchored root system, and is resistant to pear decline. It forms an excellent graft union with all pear cultivars and will tolerate relatively wet (but not waterlogged) and heavy soils. French pear roots are resistant to verticillium wilt, oak root fungus *(Armillaria mellea)*, to root-knot and root-lesion nematodes, and to crown gall.

The two serious defects of French pear roots are their susceptibility to pear root aphid, *Eriosoma pyricola,* and to fire blight (pear blight), *Erwinia amylovora.* Millions of pear trees on French pear roots have died from fire blight, owing to the high susceptibility of this stock.

Seedlings of the 'Kieffer' pear—a hybrid between *P. communis* and *P. pyrifolia* —have been satisfactorily used for many years in Australia as a pear rootstock *(25).*

Blight-resistant French pear rootstocks   Pear cultivars resistant to blight, such as 'Old Home' and 'Farmingdale' *(143),* are sometimes used as an intermediate stock budded or grafted on seedling rootstocks. Topworking with the desired cultivar takes place after the trunk and primary scaffold branches develop. If a blight attack occurs in the top of the tree, it will stop at the resistant body stock, which can subsequently be regrafted after the blight has been cut out. Since 'Old Home' can be propagated by cuttings *(74),* it is possible to have the root, trunk, and primary scaffold system of this blight-resistant cultivar. Such trees are vigorous and highly resistant to pear decline *(52, 74).* Only 'Old Home' that is free of "pear bark measles" should be used *(113).*

Japanese pear [P. pyrifolia (P. serotina)]   This was widely used in the U.S. as a pear rootstock from about 1900 to 1925, but is no longer considered of value, owing to its high susceptibility to pear decline and to the physiological defect, "black-end" or "hard-end," which may occur in 'Bartlett,' 'Anjou,' 'Winter Nelis,' and other cultivars when they are propagated on it *(78).* Fruits affected with black-end are unusable because of hardening and cracking of the flesh at the blossom end, often with the development of blackened areas.

P. calleryana   This stock is blight resistant and produces vigorous trees with a strong graft union; fruit quality is good, with no black-end. 'Bartlett' orchards with this rootstock have produced well in California, and it is popular in the southern part of the U.S. for 'Kieffer' and other hybrid pears. In areas with cold winters it lacks winter-hardiness. Where good pear psylla control has been practiced, trees on *P. calleryana* roots are resistant to pear decline. However, trees with *P. calleryana* roots show less resistance to oak root fungus than those with *P. communis* roots. Seedlings of a selection of this species, known as 'D-6,' are widely used as a pear rootstock in Australia *(51).*

P. ussuriensis   This Oriental stock has been used in the past to some extent; many pear cultivars on this root develop black-end, although not to the extent found with *P. pyrifolia* roots. It also produces small trees which are susceptible to pear decline, and should not be used in areas where this may occur.

P. betulaefolia   This species has vigorous seedlings, resistance to leaf spot and pear root aphid, tolerance to alkali soils, an adaptability to a wide range of climatic conditions, good resistance to pear decline and produces large trees. It is, however, susceptible to fire blight.

Quince (Cydonia oblonga) (195)   This has been used for centuries as a dwarfing stock for pear. Some cultivars, however, fail to make a strong union directly on the quince, hence double-working, with an intermediate stock such as 'Old Home,' or 'Hardy,' is necessary. Cultivars which require such a compatible interstock when worked on quince roots include 'Bartlett,' 'Bosc,' 'Winter Neils,' 'Seckel,'

'Easter,' 'Clairgeau,' 'Guyot,' 'Clapp's Favorite,' 'and 'El Dorado.' A selection of 'Bartlett,' originating in Switzerland, which is compatible with 'Quince A' rootstock has been found, arising presumably as a bud mutation from an incompatible form *(136, 137)*. The following pears appear to be compatible when worked directly on quince: 'Anjou,' 'Old Home,' 'Hardy,' 'Packham's Triumph,' 'Gorham,' 'Flemish Beauty,' 'Duchess,' and 'Maxine.'

Quince roots are resistant to pear root aphids and nematodes, but are susceptible to oak root fungus, fire blight, excess lime, and are not winter-hardy in areas where extremely low temperatures occur. In some areas trees on quince roots have developed pear decline, but in others they have not. This situation may possibly be explained, however, by the use of different quince stocks. The black-end trouble has not developed with pears on quince roots. There are a number of quince cultivars, most of which are easily propagated by hardwood cuttings or layering. 'Angers' quince is commonly used as a pear rootstock, because its cuttings root readily, it grows vigorously in the nursery, and does well in the orchard.

The East Malling Research Station in England has selected several clones of quince suitable as pear rootstocks, and designated them as 'Quince A', 'B,' and 'C.' 'Quince A' ('Angers') has proved to be the most satisfactory stock. 'Quince B' (Common quince) is somewhat dwarfing, whereas 'Quince C' produces very dwarfed trees. It is important to use only virus-tested quince stock.

PECAN (Carya illinoensis Koch*) (6, 164, 168)*   Pecans are propagated by budding or grafting on seedling rootstocks *(16)*. Leafy, softwood cuttings will root under mist. Plants have also been started by root cuttings and by trench and air layering, the latter with the aid of indolebutyric acid treatments.

Pecan seeds sometimes start growing in the hulls even before the nuts are harvested. The seeds lose their viability, however, in warm, dry storage. To prevent this, they should be placed at $32°$ F ($0°$ C) immediately after harvest until planted. Midwinter planting, with seedling emergence in the spring, is a successful procedure. If the nuts have become dry, it is helpful to soak them in water for several days before planting. Only deep, well-drained, sandy soil should be used for growing pecan nursery trees. Young seedlings are tender and should be shaded against sunburn. In the summer, toward the end of the second growing season, the seedlings are large enough to bud to the desired cultivar. Patch budding, or sometimes ring budding, is the usual method employed. After the new top grows for one or sometimes two seasons, the nursery tree is ready to transplant to its permanent location. Young pecan trees have a long taproot, which must be handled carefully in digging and replanting.

Sometimes the seedlings are changed to the desired cultivar by crown grafting in late winter or early spring, using the whip, or tongue, method. Large pecan seedlings, growing in place in the orchard can be topworked to the desired cultivar by bark grafting (see p. 388) limbs $1\frac{1}{2}$ to $3\frac{1}{2}$ in. in diameter.

Rootstocks for Pecans   Commercially, pecan cultivars are propagated almost entirely on pecan *(C. illinoensis)* seedlings. Wild seedling trees that are large and vigorous can be used as a source of the seeds. Seeds from the 'Moore,' 'Waukeenah,' 'Elliott,' 'Curtis,' and 'Halbert' cultivars are reported to produce excellent seedlings. As a possible rootstock for wet soils, seedlings of one of the hickory species, *C. aquatica,* have been used experimentally. Although pecan scions will grow on hickory species, the nuts generally do not attain normal size. Pecan roots are susceptible to verticillium wilt.

PERSEA sp.   See Avocado.

PERSIMMON *(Diospyros* sp.) *(50)*   **American persimmon** *(D. virginiana* L.) *(38).* The named cultivars of this species are commonly propagated by budding or grafting on *D. virginiana* seedlings. Root cuttings are also successful.

Seed germination is rather slow, owing to the seed's slow rate of water absorption. Fall sowing, or stratification at about 50° F (10° C) for 60 to 90 days, is advisable. In transplanting any but small seedlings, the taproot should be cut 12 in. below ground a year before moving to force out lateral roots.

Oriental persimmon *(D. kaki* L.)   The cultivars of this fruit are propagated by grafting on seedling rootstocks. Crown grafting by the whip, or tongue, method in early spring when both scion wood and stock are still dormant is a common practice, although bench grafting is also used. Budding can be done but it is less successful than grafting.

The usual practice in germinating seeds of both *D. lotus* and *D. kaki* is to stratify the seeds from fall until early spring. If the seeds have dried out, they should be soaked in warm water for two days before stratification at 50° F (10° C) for 120 days. Excessive drying of the seed is harmful, especially for *D. kaki.* The seeds are planted either in flats or in the nursery row. Young persimmon seedlings require shading.

Rootstocks for Oriental Persimmon (81)

Diospyros lotus   This has been widely used in California; it is very vigorous and drought-resistant, and produces a rather fibrous type of root system which transplants easily. This stock is quite susceptible to crown gall and verticillium and will not tolerate poorly drained soils but is highly resistant to oak root fungus. 'Hachiya' does not produce well on *D. lotus* stock, owing to the excessive shedding of fruit in all stages *(154).* 'Fuyu' scions usually do not form a good union with *D. lotus,* although 'Fuyu' topworked on a *D. kaki* cultivar which has been established on *D. lotus* roots makes a satisfactory tree.

D. kaki   This stock is the one most favored in Japan and is probably best for general use, because it develops a good union with all cultivars and is resistant to crown gall and oak root fungus but susceptible to verticillium. Trees worked on it grow well and yield satisfactory commercial crops. The seedlings have a long taproot with few lateral roots, making transplanting somewhat difficult.

D. virginiana   Seedlings of this species are utilized in the southern part of the U.S. and seem to be adapted to a wide range of soil conditions. It has not proven satisfactory in some localities, however. In California, 'Hachiya' on this stock is distinctly dwarfed and yields poorly, owing to the sparse bloom *(154).* Most oriental cultivars make a good union with this stock, but there are diseases carried by *D. kaki* scions that will move into the *D. virginiana* roots, causing them to die. This stock seems to be quite tolerant either of drought or of excess soil moisture conditions. It produces a fibrous type of root system easy to transplant, but tending to sucker badly.

PHOENIX DACTYLIFERA   *See* Date.

PINEAPPLE *(Ananas comosus* Merr.) *(26, 108)*   All commercial propagation of the pineapple is by asexual methods. As shown in Figure 17–4, there are three main types of planting material—*suckers, slips,* and *crowns.* Suckers coming from below ground are rarely used as planting material. Those above ground are cut from the mother plant about one month after the peak fruit harvest. Slips are taken from the plant two to three months following the peak harvest. When used in propagation, crowns are taken either before or at the time of harvest.

Crown

Slips

Suckers

Peduncle

Main stem of mother plant

Ground sucker

Figure 17–4   Parts of the pineapple plant showing the three major types of asexual planting material—crowns, suckers, and slips. In Hawaii slips are, by far, the type most commonly used. Courtesy Pineapple Research Institute of Hawaii.

Slips are the most popular type of planting material for commercial use. More of these are produced by the plant than either suckers or crowns. After removal from the mother plant, slips can be stored for a relatively long time and still retain sufficient vigor for replanting. If the slips are all about the same size and age when planted, they will flower and fruit at approximately the same time.

Slips will produce fruit 18 to 20 months after planting, suckers at about 15 months, and crowns at about 22 months. Because of these differences, the three types of planting material are never mixed in one field.

In South Africa, "stumps" are also used as propagating material. These are fruit stems which have borne and have been removed from the parent plant during pruning.

Before planting, all types of planting material must be cured or dried for one to several weeks after they are cut from the mother plant. This allows a callus layer to develop over the cut surface, reducing losses from decay organisms after they are planted.

PISTACHIO (*Pistacia vera* L.) *(88, 196)*   The pistachio nut is usually propagated by T-budding on seedling rootstocks of *Pistacia* species. Budding is possible over a considerable period of time, but if done before midspring, when sap flow may be excessive, the percentage of takes is apt to be low. A marked improvement in bud union occurs as budding is extended through the summer and fall. Best results are sometimes obtained by planting seedling rootstocks in their permanent orchard location, then T-budding them two feet or more above ground after the seedlings are well established. To collect seeds for rootstocks obtain fruits when the hulls turn blue-green. The hulls should be removed, as they apparently contain a germination inhibitor. *P. terebinthus* seed should be held in moist sand at 41° to 50° F (5° to 10° C) for six weeks before sowing, then germinated at 68° F (20° C) or slightly below. With early planting of seed and adequate irrigation and fertilization through the summer, the more vigorous seedlings may reach budding size by fall. Buds of *P. vera* cultivars are quite large, requiring a fairly large seedling to accommodate them. Because of their long taproot, bare-root nursery trees should be transplanted to their permanent planting site as early as possible. A satisfactory method of starting pistachio rootstock trees in containers—to avoid transplanting loss—is to plant the seeds either in tubes made of black felt sheathing paper 4 in. in diameter and 8 to 24 in. tall or in large pressed-fiber containers which are not removed when the trees are planted.

Rootstocks for Pistachio   *Pistacia atlantica* and *P. terebinthus* are recommended rootstocks; both are susceptible to verticillium wilt but are somewhat resistant to root-knot nematodes. Only the most vigorous seedlings should be used.

PLUM *(Prunus* sp.)   Plums are almost always propagated by T-budding in the fall on seedling rootstocks or, with certain stocks, on rooted cuttings or layers. T-budding in the spring can also be done. Some plums can be propagated by hardwood cuttings *(70)* and some by leafy, softwood cuttings under intermittent mist *(71)*.

Rootstocks for Plum *(47, 127)*   *Myrobalan plum (P. cerasifera).* This is the most widely used plum rootstock, being particularly desirable for the European plums, *P. domestica* (which includes the commercially important prune cultivars). It is also a satisfactory stock for the Japanese plums, *P. salicina.* Some kinds of plums—'President,' 'Kelsey,' 'Stanley,' and 'Robe de Sergeant'—are not entirely compatible with this stock, however. Cultivars which are Japanese-American hybrids *(P. salicina* × *P. americana)* are best worked on American plum seedlings.

Myrobalan roots are adapted to a wide range of soil and climatic conditions. They will endure fairly heavy soils and excess moisture, and are resistant to crown rot but susceptible to root-knot nematodes and oak root fungus. They grow well on light sandy soils.

Myrobalan seeds require stratification for about three months at 36° to 40° F (2° to 4° C). They may then be planted thickly in a seedbed for one season, then transplanted to the nursery and grown for a second season before budding. The seeds may also be planted directly in the nursery row and grown there for one season, the seedlings being budded in late summer or fall.

Certain very vigorous myrobalan selections are propagated by hardwood cuttings. One of these, 'Myro C,' is immune to root-knot nematodes. In England a selection, 'Myrobalan B,' was developed at the East Malling Research Station and is propagated by hardwood cuttings. It is particularly valuable in producing vigorous trees, although there are cultivars not completely compatible with it *(47)*. Most plum cultivars in England belong to the *P. domestica* species.

Marianna plum (P. cerasifera × *P. munsoniana?)*   This is a clonal rootstock which originated in Texas as an open-pollinated cross between the myrobalan plum and, supposedly, *P. munsoniana.* It is propagated by hardwood cuttings *(67, 70)*. Some plums have grown well on it; others have not. An exceptionally vigorous seedling selection of the parent 'Marianna' plum, made by the California Agricultural Experiment Station in 1926, is widely used under the identifying name, 'Marianna 2624.' It is adaptable to heavy, wet soils and is immune to root-knot nematodes, resistant to crown gall, oak root fungus, and verticillium wilt, but is susceptible to bacterial canker. It is not as deep-rooted as Myrobalan plum seedlings, and trees on 'Marianna' sometimes blow over in strong winds.

Peach (P. persica)   Many plum and prune orchards in California are on peach seedling rootstocks. This stock has proved satisfactory for light, well-drained soils. However, peach rootstocks should be avoided if the trees are to be planted on a site formerly occupied by a peach orchard. In some areas plum on peach roots tends to overbear and develop a die-back condition. Peach is not satisfactory as a stock for some plum cultivars, including 'Sugar' prune and 'Robe de Sergeant.'

Apricot (P. armeniaca)   Apricot seedlings can be used as a plum stock in nematode-infested sandy soils for those cultivars which are compatible with apricot. Japanese plums tend to do better than European plums on apricot roots.

Almond (P. amygdalus)   Some plum cultivars can be grown successfully on almond seedlings. The 'French' prune does very well on this stock, the trees growing faster and bearing larger fruit than when myrobalan roots are used. Plum cultivars on almond roots tend to overbear, sometimes to the detriment of the tree. Except where plantings are to be made on well-drained, sandy soils, high in lime or boron, this stock probably should not be used for plums.

'Brompton,' and 'Common' plum (P. domestica)   These two clonal plum stocks are used chiefly in England. 'Brompton' seems to be compatible with all plum cultivars and tends to produce medium to large trees. 'Common' plum, which produces small to medium trees, has shown incompatibility with some cultivars *(47, 114)*. Propagation is by hardwood cuttings.

'St. Julien,' 'Common Mussel,' 'Damas,' and 'Damson' (P. insititia)   The first three stocks are used mostly in England. 'Damas C' produces medium to large trees, whereas 'Common Mussel' generally produces small to medium trees. The latter stock seems to be compatible with all plums. 'St. Julien A' produces small to medium trees for all compatible scion cultivars. Results have been variable with these stocks, since they vary widely in type. At the East Malling Research Station in England, the more promising types of 'St. Julien,' 'Mussel,' and 'Damas' have been selected and kept true-to-type by vegetative propagation, and listed as clones A, B, C, D, and so forth. 'St. Julien' has been used to some extent as a plum stock for the *P. domestica* cultivars in the United States.

In England the following clonal stocks are recommended for plums: for *vigorous* trees, 'Myrobalan B'; for *semi-vigorous* trees, 'Brompton'; for *intermediate* trees, 'Marianna,' 'Pershore'; for *semi-dwarf* trees, 'Common' plum, 'St. Julien A'; for *dwarf* trees, 'St. Julien K.'

Florida sand plum (P. angustifolia)   This may be useful as a dwarfing stock for compatible plum cultivars. In California, after 16 years, 'Giant,' 'Burbank,' and 'Beauty' on this stock were healthy and very productive, with a dwarf type of growth.

Japanese plum (P. salicina)   Seedlings of this species are used in Japan as plum stocks but apparently not elsewhere. European plums *(P. domestica)*, when topworked on Japanese plum stocks, result in very shortlived trees; the reverse combination, though, produces compatible unions.

Western sand cherry (P. besseyi)   This stock has produced satisfactory dwarfed plum trees of the Japanese and European types, but poor bud unions and shoot growth developed when it was used as a stock for cultivars of *P. insititia (15)*.

POMEGRANATE *(Punica granatum* L.)   The pomegranate is easily propagated by hardwood cuttings. After one season's growth, the plants are usually large enough to be moved to their permanent location. Softwood cuttings taken during the summer are also easily rooted if maintained under high-humidity conditions. The pomegranate forms suckers readily, and these may be dug up during the dormant season with a piece of root attached.

PRUNUS AMYGDALUS   *See* Almond.

PRUNUS ARMENIACA   *See* Apricot.

PRUNUS PERSICA   *See* Peach.

PSIDIUM sp.   *See* Guava.

PUNICA GRANATUM   *See* Pomegranate.

PYRUS COMMUNIS   *See* Pear.

QUINCE *(Cydonia oblonga* Mill.)   The quince is usually propagated by hardwood cuttings, which root readily, "heel" cuttings more so than those made entirely from one-year-old wood. Cuttings from two- or three-year-old wood also root easily. The burrs or knots found on this older wood are masses of adventitious root initials. Cuttings made from the basal portion of one-year-old wood root better than those made from the terminal end. The cuttings usually make sufficient growth in one season to transplant to their permanent location. Quince can also be propagated by mound layering.

Occasionally, commercial quince cultivars are T-budded on rooted cuttings (such as 'Angers') or sometimes on seedlings.

RASPBERRY, BLACK *(Rubus occidentalis* L.) and **PURPLE** *(R. occidentalis* × *R. idaeus)*   Tip layering is the usual method of propagating these species, the black rooting more easily than the purple. The black raspberry can be propagated also by leaf-bud cuttings, roots forming in about three weeks. Cuttings should be taken in early summer and rooted under high-humidity conditions or in a mist-propagating frame *(176)*.

RASPBERRY, RED *(Rubus strigosus* Michx., *R. idaeus* L.)   The red raspberry is easily propagated by removing suckers of one-year-old growth. These are dug in early spring, care being taken to leave a piece of the old root attached. Also, young, green suckers of new wood may be dug in the spring shortly after they appear above ground. When these plants are dug, they should have some new roots starting. A piece of the old root is taken with them. Such suckers should be dug and transplanted during cool, cloudy weather and irrigated well after resetting, otherwise they will not survive. Sucker production can be stimulated by inserting a spade deeply at intervals in the vicinity of old plants to cut off roots, each root piece then sending up a shoot. Production of suckers can also be stimulated by mulching with straw or sawdust.

The red raspberry can be propagated by root cuttings. For thick root pieces 6 in. root lengths are best; for thin roots use 2 in. pieces. Shallow (1/2 in.) planting is advised *(79)*. If given good care, strong nursery plants will be produced in one year.

Leafy softwood cuttings made in early spring from young sucker shoots just emerging from the soil, with a 1- or 2-in. etiolated section from below the surface, will give almost 100 percent rooting when placed in a propagating frame. *(197)*. Shoot cuttings taken directly from the canes are almost impossible to root.

RIBES sp.   *See* Currant, Gooseberry.

RUBUS sp.   *See* Blackberry; Raspberry, Black; Raspberry, Red.

STRAWBERRY *(Fragaria* sp.) *(187)*   Strawberries are propagated by runners or, in certain everbearing types—which produce few runners—by crown division. Seed propagation is used only in breeding programs for the development of new cultivars. Treatment of seed with sulfuric acid for 15 to 20 min. before planting will give good germination. *(157)*.

The nursery field is set out in early spring, and runner plants are ready for digging the following fall and winter. They may be dug any time after they have had sufficient winter chilling to overcome the rest period of the buds. The digging should be completed in the spring before the plants start to grow, particularly if they are to be held in cold storage. Special machine diggers handle great quantities of plants (Figure 17–5). Immediately after digging, the plants are hauled to packing sheds, where all leaves and petioles are removed. Plants are packed in vented cardboard or wood boxes for storing or shipping. A thin polyethylene liner is used to keep the plants from drying out *(200)*. They are usually handled in 1000-

Figure 17–5   Equipment used for large-scale digging of strawberry nursery plants. The entire row is scooped up and elevated to a revolving trommel, which separates the plants from the soil. The plants are then packed in wet burlap bags and taken to sheds for trimming and packing. Courtesy Wheeler's Nursery, Los Molinos, California.

or 2000-plant units, depending upon plant size. The plants are placed firmly in the boxes with roots to the center. Such plants can be stored and kept in good condition for one year if held between 28° and 30° F (−2° to −1° C) *(55)*. Maintaining this temperature is critical.

Everbearing cultivars that produce few runners are propagated by crown division. Certain cultivars, such as the 'Rockhill,' may produce 10 to 15 strong crowns per plant by the end of the growing season. In the spring, such plants are dug and carefully cut apart; each crown may then be used as a new plant.

Certification programs for strawberries have been established, hence nursery sources of pathogen-free, true-to-name plants of most cultivars are available *(41)*. "Meristem subclones" of the important strawberry cultivars free of all detectible viruses have been developed following heat treatment of the source plants. *(28, 173)*. Strawberry plants can be freed of certain nematodes by hot-water treatments, dormant plants being immersed for two minutes at 127° F (53° C) *(48)*.

TANGERINE   *See* **Citrus.**

THEOBROMA CACAO   *See* **Cacao.**

VACCINIUM CORYMBOSUM   *See* **Blueberry, Highbush.**

VACCINIUM MACROCARPON   *See* **Cranberry.**

VITIS sp.   *See* **Grape.**

WALNUT (*Juglans* sp.) (**Persian, English, Carpathian**) walnut (*J. regia* L.)   Walnut cultivars are propagated by either patch budding (or one of its variations) or whip grafting one-year-old seedling rootstocks, or by topworking seedling trees, one to four years old or more, planted in place in the orchard. Bark grafting in late spring or patch budding in the spring or summer works well for the latter method.

Rootstocks for Persian Walnut (160, 161)   Nuts of most of the *Juglans* species used as seedling rootstocks should be either fall-planted or stratified for about three months at 36° to 40° F (2° to 4° C) before they are planted in the spring, to obtain good germination. Although Persian walnut seeds will germinate without any cold treatment, a stratification period hastens germination. It is better to plant the seeds before they start sprouting in the stratification boxes, but with care sprouted seeds can be successfully planted. At the end of one season, the seedlings should be large enough to bud.

Northern California black walnut (J. hindsii)   This is the stock most commonly used in California. The seedlings are vigorous, and make a strong graft union. They are resistant to verticillium, oak root fungus (*Armillaria mellea*), and root-knot nematode (*Meloidogyne* sp.), but are susceptible to crown rot (*Phytophthora* sp.), crown gall, the root-lesion nematode (*Pratylenchus vulnus*), and, in some areas, to a serious defect known as "black-line." The latter has appeared in some walnut districts of California and Oregon and in France, affecting grafted Persian walnut trees on this rootstock after they are 15 to 30 years old. It is characterized by a breakdown of the tissues in the cambial region at the graft union, leading to a girdling effect on the trees (see Figure 11–19). The cause of this defect has not yet been established, but presumably it is a delayed incompatibility, modified by climatic influences *(159)*. Symptoms are an unhealthy, yellowish-green leaf color, reduction in leaf area and crop, black lesions at the graft union, and a heavy production of watersprouts or suckers from the black walnut stock below the graft union.

Southern California black walnut (J. californica)   Seedlings of this species were once popular as rootstocks but have been little used in recent years. It is highly susceptible to crown rot; the crown below the union usually becomes greatly enlarged; and the roots sucker badly. It is not as vigorous as *J. hindsii*. Sometimes this stock is mistakenly used in place of *J. hindsii*, thus producing trees very susceptible to crown rot troubles. "Black-line" cases have appeared in trees worked on this stock.

Eastern black walnut (J. nigra)   Rootstocks of this species produce good trees, but they grow slowly when young, delaying the age when the trees reach full bearing. This stock is ordinarily used in the eastern part of the United States and in Europe for the hardy Carpathian walnut cultivars. Its long taproot tends to make transplanting nursery trees difficult. No incompatibility problems have appeared.

Persian walnut (J. regia)   Seedlings of this species produce good trees with an excellent graft union and are highly resistant to crown rot. The roots are susceptible to crown gall, oak root fungus, and salt accumulations in the soil, and are not as resistant to root-knot nematodes as *J. hindsii*. Nurserymen object to the slow initial growth of the seedlings. Trees on this stock should be quite satisfactory in soils free of oak root fungus and salt accumulation. Its use is recommended in some localities as a means of avoiding the black-line trouble, to which it is presumably immune. In Oregon, where black-line is very serious, seedlings of the 'Manregian' clone (of the Manchurian race of *J. regia*), imported by the USDA as P.I. No. 18256, are used extensively as rootstocks. They are vigorous, cold-hardy, and—in Oregon—no more susceptible to oak root fungus than *J. hindsii* seedlings *(131)*.

Paradox walnut (J. hindsii × J. regia)   This is a first-generation (F_1) hybrid stock, the seedlings being obtained from seed taken from *J. hindsii* trees, whose pistillate (female) flowers have been wind-pollinated with pollen from nearby *J. regia* trees. When seeds from such a *J. hindsii* tree are planted, some of the seedlings may be the hybrid progeny. The amount varies widely, from 0 to almost 100 percent, depending upon the individual mother tree. The hybrids are easily distinguished by their large leaves in comparison with the smaller-leaved, self-pollinated *J. hindsii* seedlings (see Figure 4–4). Seedlings from Paradox trees themselves should not be used for producing rootstocks, owing to their great variability in all characteristics. Although first-generation (F_1) seedlings are variable in some characteristics, most of them exhibit hybrid vigor and make excellent rootstocks for the Persian walnut. They are resistant to root-lesion nematodes and crown rot, and are tolerant of heavy, wet soils. Generally, Paradox seedlings are more susceptible to oak root fungus and crown gall than those of *J. hindsii*. They are just as susceptible to the black-line trouble as *J. hindsii* seedlings. Trees on Paradox rootstocks grow and yield as well as, or better than, those on *J. hindsii* roots, and may produce large-sized nuts with better kernel color *(160)*. In very heavy or low-fertility soils, trees on Paradox roots will grow faster than those on *J. hindsii*.

Since it is difficult to secure Paradox seeds in quantity, vegetative propagation methods would be very desirable in order to establish superior Paradox clones from which large numbers of rootstock plants could be obtained. Propagation by leafy cuttings under mist, hardwood cuttings, and trench layering can be done but with difficulty *(104, 158)*. Piece root grafting has also been used successfully. Trials with intermediate stem pieces of *J. nigra* or *J. sieboldiana cordiformis* between *J. regia* tops and Paradox roots have given substantially dwarfed trees. *(161)*.

Royal walnut (J. nigra × J. hindsii or J. californica)   This is used to a limited extent as a rootstock for Persian walnuts. It is more tolerant of excessive soil moisture than *J. hindsii*, but probably not more than *J. nigra* or Paradox.

WALNUT, BLACK *(Juglans nigra L.)*   The several named black walnut cultivars are propagated by patch or ring budding and grafting (especially the modified cleft

Figure 17–6   Steps in the modified cleft graft used in propagating the Eastern black walnut. Left to right: scion cut to a wedge. Stock cut diagonally with the scion inserted. Waxed cloth placed over cut surfaces and the union tied with raffia. Scion and stub after waxing *(105)*.

graft) on eastern black walnut seedlings *(J. nigra)*, which is a satisfactory rootstock *(22, 105, 172)*.

The modified cleft graft, as illustrated in Figure 17–6, is suitable for seedling stocks up to about 1 in. in diameter. The scion is cut to a wedge, slightly thicker on one side than on the other. The stock is cut diagonally rather than being split. The scion is inserted into this cleft, the thicker part out, with the cambium of stock and scion matching.

In growing Eastern black walnut for timber production, seedlings are started in nurseries and only the strongest, most vigorous trees are set out in the plantation. Careful planting is necessary to obtain the essential rapid, early growth.

ZIZYPHUS JUJUBA   *See* Jujube.

REFERENCES

1 Albert, D. W., Propagation of date palms from offshoots, *Ariz. Agr. Exp. Sta. Bul. 119,* 1926.

2 Aldrich, W. W., G. H. Leach, and W. A. Dollins, Some factors influencing the growth of date offshoots in the nursery row, *Proc. Amer. Soc. Hort. Sci.,* 46:215–21. 1945.

3 Argles, G. K., *Anacardium occidentale* (cashew). Ecology and botany in relation to propagation, *F.A.O. Conf. on Prop. of Trop. and Sub-trop. Fruits,* London, Item 7. 1969.

4 Arndt, C. H., Mango polyembryony and other multiple shoots, *Amer. Jour. Bot.,* 22:26. 1935.

5 Ashiru, G. A., and T. Quarcoo, Vegetative propagation of kola *(Cola nitida),* *Trop. Agr.,* 48(1):85–92. 1971.

6 Bailey, J. E., and J. G. Woodroof, Propagation of pecans, *Ga. Agr. Exp. Sta. Bul. 172.* 1932.

7 Barton, L. V., Viability of seeds of *Theobroma cacao* L., *Contr. Boyce Thomp. Inst.*, 23(4):109–22. 1965.

8 Batchelor, L. D., and W. P. Bitters, High quality citrus rootstocks, *Calif. Agr.*, 6(9):3–4. 1952.

9 Batjer, L. P., and H. Schneider, Relation of pear decline to rootstocks and sieve-tube necrosis, *Proc. Amer. Soc. Hort. Sci.*, 76:85–97. 1960.

10 Bitters, W. P., Rootstocks with dwarfing effect, *Calif. Agr.*, 4(2):5, 14. 1950.

11 ———, and E. R. Parker, Quick decline of citrus as influenced by top-root relationships, *Calif. Agr. Exp. Sta. Bul. 733*, 1953.

12 Blacker, G. W. I., Grafting passion fruit vines, *New Zeal. Jour. Agr.*, 101: 401–5. 1960.

13 Blodgett, E. C., H. Schneider, and M. D. Aichele, Behavior of pear decline disease on different stock-scion combinations, *Phytopath.*, 52:679–84. 1962.

14 Brase, K., Similarity of the Clark Dwarf and East Malling Rootstock VIII, *Proc. Amer. Soc. Hort. Sci.*, 61:95–98. 1953.

15 ———, and R. D. Way, Rootstocks and methods used for dwarfing fruit trees, *N. Y. Agr. Exp. Sta. Bul. 783*, 1959.

16 Bryant, M. D., Propagating the pecan tree, *New Mex. State Univ. Pl. Sci. Guide*, 400. H-604. 1969.

17 Calavan, E. C., E. F. Frolich, J. B. Carpenter, C. N. Roistacher, and D. W. Christiansen, Rapid indexing for exocortis of citrus, *Phytopath.*, 54:1359–62. 1964.

18 Cameron, J. W., and R. K. Soost, Nucellar lines of citrus, *Calif. Agr.*, 7(1):8, 15, 16. 1953.

19 Carlson, R. F., Growth and incompatibility factors associated with apricot scion/rootstock in Michigan, *Mich. Quart. Bul.*, 48(1):23–29. 1965.

20 ———, Rootstocks in relation to apple cultivars, in W. H. Upshall (ed.), *North American Apples: Varieties. Rootstocks. Outlook.* East Lansing, Mich.: Michigan State University Press, 1970.

21 Carlson, R. F., and H. B. Tukey, Cultural practices in propagating dwarfing rootstocks in Michigan, *Mich. Agr. Exp. Sta. Quart. Bul.* 37(4):492–97. 1955.

22 Chase, S. B., Budding and grafting eastern black walnut, *Proc. Amer. Soc. Hort. Sci.*, 49:175–80. 1947.

23 Child, R., *Coconuts.* London: Longmans, Green & Company, Ltd., 1964.

24 Cobin, M., The lychee in Florida, *Fla. Agr. Exp. Sta. Bul. 546*, 1954.

25 Cole, C. E., The fruit industry of Australia and New Zealand, *Proc. XVII Int. Hort. Cong.* Vol. IV, pp. 321–68. 1966.

26 Collins, J. L., *The Pineapple—History, Cultivation, Utilization.* London: Leonard-Hill, 1960.

27 Condit, I. J., *Ficus, the Exotic Species*, Berkeley. Univ. California Division of Agricultural Sciences, 1969.

28 Converse, R. A., The propagation of virus-tested strawberry stocks, *Proc. Inter. Plant Prop. Soc.*, 22:73–76. 1972.

29 Coorts, G. D., and J. W. Hull, Propagation of highbush blueberry *(Vaccinium Australe* Small) by hard and softwood cuttings, *The Plant Prop.*, Inter. Plant Prop. Soc., 18(2):9–11. 1972.

30 Couvillon, G. A., and F. A. Pokorny, Photoperiod, indolebutyric acid, and type of cutting wood as factors in rooting of rabbiteye blueberry *(Vaccinium ashei Reade)*, cv. Woodward, *HortScience,* 3(2):74–75. 1968.

31 Cross, C. E., I. E. Demoranville, K. H. Deubert, R. M. Devlin, J. S. Norton, W. E. Tomlinson, and B. M. Zuckerman, Modern cultural practice in cranberry growing, *Mass. Agr. Exp. Sta. and Ext. Serv. Bul. No.* 39, 1969.

32 Darrow, G. M., and J. N. Moore, Blueberry growing, *USDA Farmers' Bul.* 1951 (rev.). 1966.

33 Day, L. H., Cherry rootstocks in California, *Calif. Agr. Exp. Sta. Bul. 725,* 1951.

34 ———, and E. F. Serr, Comparative resistance of rootstocks of fruit and nut trees to attack by root lesion or meadow nematode, *Proc. Amer. Soc. Hort. Sci.,* 57:150–54. 1951.

35 Doehlert, C. A., Propagating blueberries from hardwood cuttings, *N. J. Agr. Exp. Sta. Cir. 551,* 1953.

36 Evans, H., Investigations on the propagation of cacao, *Trop. Agr.,* 28:147–203. 1951.

37 ———, Physiological aspects of the propagation of cacao from cuttings, *Rpt. 13th Int. Hort. Cong.,* Vol. 2, 1179–90. 1953.

38 Fletcher, W. F., The native persimmon, *USDA Farmers' Bul. 685,* 1942.

39 Fletcher, W. A., Growing Chinese gooseberries, *New Zealand Dept. Agr. Bul. 349,* 1971.

40 Ford, H. W., A method of propagating citrus rootstock clones by leaf-bud cuttings, *Proc. Amer. Soc. Hort. Sci.,* 69:204–7. 1957.

41 Frazier, N. F., and A. F. Posnette, Relationships of the strawberry viruses of England and California, *Hilgardia,* 27(17):455–513. 1958.

42 Frolich, E. F., Rooting citrus and avocado cuttings, *Proc. Int. Plant Prop. Soc.,* 16:51–54. 1966.

43 ———, and R. G. Platt, Use of the etiolation technique in rooting avocado cuttings, *Calif. Avoc. Soc. Yearbook, 1971–72,* pp. 97–109. 1972.

44 Galletta, G. J., and A. S. Fish, Jr., Interspecific blueberry grafting, a way to extend *Vaccinium* culture to different soils, *Jour. Amer. Soc. Hort. Sci.,* 96(3): 294–98. 1971.

45 Gardner, F. E., and G. E. Horanic, Cold tolerance and vigor of young citrus trees, *Proc. Fla. State Hort. Soc.,* 76:105–10. 1963.

46 Giliomee, J. H., D. K. Strydom, and H. J. Van Zyl, Northern Spy, Merton, and Malling-Merton rootstocks susceptible to wooly aphid, *Eriosoma lanigerum,* in the Western Cape, *S. Afr. Jour. Agr. Sci.,* 11:183–86. 1968.

47 Glenn, E. M., Plum rootstock trials at East Malling, *Jour. Hort. Sci.,* 36:28–39. 1961.

48 Goheen, A. C., and J. R. McGrew, Control of endoparasitic root nematodes in strawberry propagation stocks by hot water treatments, *Plant Dis. Rept.,* 38: 818–26. 1954.

49 Gould, H. P., The native papaw, *USDA Leaflet 179,* 1939.

50 ———, The Oriental persimmon, *USDA Leaflet 194,* 1940.

51 Greenhalgh, W. J., A pear rootstock trial at Bathurst Experiment Farm, *Agr. Gaz. N.S.W. (Aust.),* 74:350–53. 1963.

52 Griggs, W. H., and H. T. Hartmann, Old Home pears show resistance to decline when on own roots, *Calif. Agr.,* 14:8, 10. 1960.

53 Griggs, W. H., D. D. Jensen, and B. T. Iwakiri, Development of young pear trees with different rootstocks in relation to psylla infestation, pear decline, and leaf curl, *Hilgardia,* 39:153–204. 1968.

54 Guiscafre-Arrillage, J., The propagation of coffee *(Coffea arabica* L.) by cuttings, *Proc. Amer. Soc. Hort. Sci.,* 48:279–90. 1946.

55 Guttridge, C. G., D. T. Mason, and E. G. Ing, Cold storage of strawberry runner plants at different temperatures, *Exp. Hort.,* 12:38–41. 1965.

56 Hall, I. V., Growing cranberries, *Canad. Dept. Agr. Publ.,* 1282 (rev.), 1969.

57 Halma, F. F.. Own-rooted and budded lemon trees, *Proc. Amer. Soc. Hort. Sci.,* 50: 172–76. 1947.

58 ———, K. M. Smoyer, and H. W. Schwalm, Quick decline associated with sour rootstocks, *Calif. Citrogr.,* 29:245. 1944.

59 Hamilton, R. A., A study of germination and storage life of macadamia seed, *Proc. Amer. Soc. Hort. Sci.,* 70:209–12. 1957.

60 Harkness, R. W., Papaya growing in Florida, *Fla. Agr. Exp. Sta. Cir. S-180,* 1967.

61 Harmon, F. N., Influence of indolebutyric acid on the rooting of grape cuttings, *Proc. Amer. Soc. Hort. Sci.,* 42:383–88. 1943.

62 ———, A modified procedure for greenwood grafting of vinifera grapes, *Proc. Amer. Soc. Hort. Sci.,* 64:255–58. 1954.

63 Harmon, F. N., and J. H. Weinberger, Effects of storage and stratification on germination of *Vinifera* grape seeds, *Proc. Amer. Soc. Hort. Sci.,* 73:147–50. 1959.

64 Harmon, F. N., and J. H. Weinberger, Bench grafting trials with Thompson seedless grapes on various rootstocks, *Proc. Amer. Soc. Hort. Sci.,* 83:379–83. 1963.

65 Hartman, Henry, Historical facts pertaining to root and trunkstocks for pear trees, *Ore. Agr. Exp. Sta. Misc. Paper 109,* 1961.

66 Hartmann, H. T., Further studies on the propagation of the olive by cuttings, *Proc. Amer. Soc. Hort. Sci.,* 59:155–60. 1952.

67 ———, Auxins for hardwood cuttings, *Calif. Agr.,* 9(4):7, 12, 13. 1955.

68 ———, Rootstock effects in the olive, *Proc. Amer. Soc. Hort. Sci.,* 72:242–51. 1958.

69 ———, and R. M. Brooks, Propagation of Stockton Morello cherry rootstocks by softwood cuttings under mist, *Proc. Amer. Soc. Hort. Sci.,* 71:127–34. 1958.

70 Hartmann, H. T., and C. J. Hansen, Rooting pear, plum rootstocks, *Calif. Agr.,* 12(10):4, 14–15. 1958.

71 ———, Rooting of softwood cuttings of several fruit species under mist, *Proc. Amer. Soc. Hort. Sci.,* 66:157–67. 1955.

72 Hartmann, H. T., and F. Loreti, Seasonal variation in rooting leafy olive cuttings under mist, *Proc. Amer. Soc. Hort. Sci.,* 87:194–98. 1965.

73 Hartmann, H. T., and K. Opitz, Olive production in California, *Calif. Agr. Exp. Sta. Cir. 540,* 1966.

74 Hartmann, H. T., W. H. Griggs, and C. J. Hansen, Propagation of own-rooted Old Home and Bartlett pears to produce trees resistant to pear decline, *Proc. Amer. Soc. Hort. Sci.,* 82:92–102. 1963.

75 Hartmann, H. T., C. J. Hansen, and F. Loreti, Propagation of apple root-stocks by hardwood cuttings, *Calif. Agr.,* 19(6):4–5. 1965.

76 Hartmann, H. T., W. C. Schnathorst, and J. Whisler, Oblonga, a clonal olive rootstock resistant to verticillium wilt, *Calif. Agr.,* 25(6): 12–15. 1971.

77 Hearman, J., The Northern Spy as a rootstock, *Jour. Pom. and Hort. Sci.,* 14:246–75. 1936.

78 Heppner, M. J., Pear black-end and its relation to different rootstocks, *Proc. Amer. Soc. Hort. Sci.,* 24:139–42. 1927.

79 Heydecker, W., and Margaret Marston, Quantitative studies on the regeneration of raspberries from root cuttings, *Hort. Res.,* 8(2):142–46. 1968.

80 Hibino, H., G. H. Kaloostian, and H. Schneider, Mycoplasma-like bodies in the pear psylla vector of pear decline, *Virology,* 43:34–40. 1971.

81 Hodgson, R. W., Rootstocks for the Oriental persimmon, *Proc. Amer. Soc. Hort. Sci.,* 37:338–39. 1940.

82 Howard, W. L., The "Stockton" Morello cherry, *Proc. Amer. Soc. Hort. Sci.,* 21:320–23. 1925.

83 Hsu, C. S., and H. A. Hinrichs, Rooting response of dwarf apple cuttings under intermittent mist, *Proc. Amer. Soc. Hort. Sci.,* 72:15–22. 1958.

84 Jaffee, A., Chip grafting guava cultivars, *The Plant Prop.,* Inter. Plant Prop. Soc., 16(2):6. 1970.

85 Jager, L. A., Apricot rootstock trials—Scoresby, *Victorian Hort. Dig. (Austral.),* 14(1):22–23. 1970.

86 Jaynes, R. A., Buried-inarch technique for rooting chestnut cuttings, *North. Nut Grow. Assoc. 52nd Ann. Rpt.,* pp. 37–39. 1961.

87 ———, and A. H. Graves, Connecticut hybrid chestnuts and their culture, *Conn. Agr. Exp. Sta. Bul. 657,* 1963.

88 Joley, L. E., and K. Opitz, Further experiences with propagation of *Pistachia,* *Proc. Inter. Plant Prop. Soc.,* 21:67–74. 1971.

89 Kasimatis, A. N., and L. Lider, Grape rootstock varieties, *Calif. Agr. Ext. Serv. AXT–47.* 1962.

90 Kender, W. J., Some factors affecting the propagation of lowbush blueberries by softwood cuttings, *Proc. Amer. Soc. Hort. Sci.,* 86:301–6. 1965.

91 Kester, D. E., C. J. Hansen, and C. Panetsos, Effect of scion and interstock variety on incompatibility of almond on Marianna 2624 rootstock, *Proc. Amer. Soc. Hort. Sci.,* 86:169–77. 1965.

92 Kester, D. E., and R. A. Asay, Almond Breeding, in J. M. Moore and J. Janick (eds.), *Advances in Fruit Breeding.* Lafayette, Ind.: Purdue University Press, 1975.

93 Klotz, L. J., E. C. Calavan, and L. G. Weathers, Virus and virus-like diseases of citrus, *Calif. Agr. Exp. Sta. Cir. 559,* 1972.

94 Krezdorn, A. H., and G. W. Adriance, Fig growing in the south, *USDA Agr. Handbook 196,* 1961.

95 Layne, R. E. C., Peach rootstock research at Harrow, *Canada Agric.,* Summer edition, pp. 20–21. 1971.

96 Lapins, K., Some symptoms of stock-scion incompatibility of apricot varieties on peach seedling rootstock, *Canad. Jour. Plant Sci.,* 39:194–203. 1959.

97 Lear, B., and L. Lider, Eradication of root-knot nematodes from grapevine rootings by hot water, *Plant Dis. Rpt.,* 43:314–17. 1959.

98 Lider, L., Phylloxera-resistant grape rootstocks for the coastal valleys of California, *Hilgardia,* 27:287–318. 1958.

99 ———, Vineyard trials in California with nematode-resistant grape rootstocks, *Hilgardia,* 30:123–52. 1960.

100 ———, Field budding and the care of the budded grapevine, *Calif. Agr. Ext. Serv. Leaf. 153,* 1963.

101 ———, and N. Shaulis, Resistant rootstocks for New York vineyards, *N. Y. Agr. Exp. Sta. Res. Cir. No. 2,* 1965.

102 Loomis, N. H., Rootstocks for grapes in the south, *Proc. Amer. Soc. Hort. Sci.,* 42:380–82. 1943.

103 Lynch, S. J., and R. O. Nelson, Current methods of vegetative propagation of avocado, mango, lychee, and guava in Florida, *Cieba* (Escuela Agricola Panamericana, Tegucigalpa, Honduras), 4:315–77. 1956.

104 Lynn, C., and H. T. Hartmann, Rooting cuttings under mist—Paradox walnut, *Calif. Agr.,* 11(5):11, 15. 1957.

105 MacDaniels, L. H., Nut growing, *Cornell Ext. Bul. 701,* 1952.

106 Mainland, C. M., Propagation and planting, Chap. 6 in P. Eck, and N. F. Childers (eds.), *Blueberry Culture.* New Brunswick, N. J.: Rutgers University Press, 1966.

107 Majumdar, P. K., and S. K. Mukherjee, Guava: a new vegetative propagation method, *Indian Hort.,* Jan.–Mar., 1968.

108 Malan, E. F., Pineapple production in South Africa, *South Afr. Dept. Agr. Bul. 339,* 1954.

109 ———, and A. van der Meulen, Propagation of avocados, *Fmg. South Afr.,* 29:499–502. 1954.

110 McKay, J. W., and H. L. Crane, Chinese chestnut. A promising new orchard crop, *Econ. Bot.,* 7:228–42. 1953.

111 McKenzie, D. W., Apple rootstock trials: Jonathan on East Malling, Merton, and Malling-Merton rootstocks, *Jour. Hort. Sci.,* 39:69–77. 1964.

112 McMillan-Browse, P. D. A., Vegetative propagation of *Corylus, Proc. Inter. Plant Prop. Soc.,* 20:356–58. 1970.

113 Millecan, A. A., C. W. Nichols, and W. M. Brown, Jr., Pear bark measles and its association in California with Old Home interstocks, *Phytopath.,* 52:363. 1962.

114 Montgomery, H. B. S., Fruit tree raising: rootstocks and propagation, *Minist. Agr. Fish. Foods (London) Bul. 135.* 1963.

115   Moore, P. W., Propagation and growing of citrus nursery trees in containers, *Proc. Int. Plant Prop. Soc.,* 16:54–62. 1966.

116   Morton, J. F., The cashew's brighter future, *Econ. Bot.,* 15:57–78. 1961.

117   Mukherjee, S. K., and P. K. Majumdar, Standardization of rootstock of mango. I. Studies on the propagation of clonal rootstocks by stooling and layering, *Indian Jour. Hort.,* 20:204–9. 1963.

118   ———, Effect of different factors on the success of veneer grafting in mango, *Indian Jour. Hort.,* 21:46–50. 1964.

119   ———, and N. N. Bid, Propagation of mango. II. Effect of etiolation and growth regulator treatment on the success of air layering, *Indian Jour. Agr. Sci.,* 35:309–14. 1965.

120   Mullins, C. A., Twenty years' research with size-controlling apple rootstocks on the Cumberland plateau, *Tenn. Agr. Exp. Sta. Bull. 488,* 1972.

121   Nelson, S. H., and H. B. Tukey, Root temperature affects the performance of East Malling apple rootstocks, *Quart. Bul. Mich. Agr. Exp. Sta.,* 38:46–51. 1955.

122   Nixon, R. W., Growing dates in the United States, *USDA Agr. Inform. Bul. 207,* 1969.

123   Northwood, P. J., Vegetative propagation of cashew *(Anacardium occidentale)* by the air layering method, *E. Afr. Agr. For. Jour.,* 30:35–37. 1964.

124   Norton, R. A., C. J. Hansen, H. J. O'Reilly, and W. H. Hart, Rootstocks for apricots in California, *Calif. Agr. Ext. Serv. Leaf. 156,* 1963.

125   ———, Rootstocks for sweet cherries in California, *Calif. Agr. Ext. Leaf. 159,* 1963.

126   ———, Rootstocks for peaches and nectarines in California, *Calif. Agr. Ext. Serv. Leaf. 157,* 1963.

127   ———, Rootstocks for plums and prunes in California, *Calif. Agr. Ext. Serv. Leaf. 158.* 1963.

128   Opitz, K. W., R. G. Platt, and E. F. Frolich, Propagation of citrus, *Calif. Agr. Ext. Cir. 546,* 1968.

129   Opitz, K. W., and R. G. Platt, Citrus growing in California, *Univ. Calif. Agr. Exp. Sta. Man. 39,* 1969.

130   O'Rourke, F. L., Propagation of the hickories, *44th Ann. Rpt. North. Nut Growers' Assoc.,* pp. 122–27. 1953.

131   Painter, J. H., Producing walnuts in Oregon, *Ore. Agr. Ext. Bul. 795,* 1961.

132   Pease, R. W., Growing chestnuts from seed, *W. Va. Agr. Exp. Sta. Cir. 90,* 1954.

133   Pennock, W., and G. Maldonado, The propagation of guavas from stem cuttings, *Jour. Agr., Univ. Puerto Rico,* 47:280–90. 1963.

134   Platt, R. G., and E. F. Frolich, Propagation of avocados, *Calif. Agr. Exp. Sta. Cir. 531,* 1965..

135   Popenoe, J., Coconut varieties and propagation, in J. G. Woodroof (ed.), *Coconuts, Production, Processing, Products.* Westport, Conn.: Avi Publishing Co., 1970.

136   Posnette, A., and R. Cropley, A selection of Williams Bon Chrétien pear compatible with quince rootstocks, *Ann. Rpt. E. Malling Res. Sta. for 1958*, pp. 91–94. 1959.

137 ———, Further studies on a selection of Williams Bon Chrétien pear compatible with Quince A rootstocks, *Jour. Hort. Sci.*, 37:291–94. 1962.

138 Preston, A. P., Apple rootstock studies: fifteen years' results with Malling-Merton clones, *Jour. Hort. Sci.*, 41:349–60. 1966.

139 ———, Apple rootstock 3431 (M. 27), *Ann Rpt. E. Malling Res. Sta. for 1970*, pp. 143–47. 1971.

140 Rake, B. A., The propagation of gooseberries. I. Some factors influencing the rooting of hardwood cuttings, *Ann. Rpt. Long Ashton Res. Sta. for 1953*, pp. 79–88. 1954.

141 Rao, V. N. M., and I. K. S. Rao, Studies on the vegetative propagation of cashew *(Anacardium occidentale* L.). Approach grafting with and without the aid of plastic film wrappers, *Indian Jour. Agr. Sci.*, 27:267–75. 1957.

142 ———, and M. V. Hassan, Studies on seed viability in cashew, *Indian Jour. Agri. Sci.*, 27:289–94. 1957.

143 Reimer, R. C., Blight resistance in pears and characteristics of pear species and stocks, *Ore. Agr. Exp. Sta. Bul. 214.* 1925.

144 Reuther, W. (ed.), *The Citrus Industry*, Vol. I, 1967; Vol. II, 1968; Vol. III, 1973. Berkeley, Calif: Univ. California, Division of Agricultural Sciences.

145 Roberts, A. N., Cherry rootstocks, *Rpt. Ore. Hort. Soc.*, 54:95–98. 1962.

146 Rogers, W. S., Malling 26, a new semi-dwarfing apple rootstock, *Ann. Rpt. E. Malling Res. Sta. for 1957*, A41:48–49. 1958.

147 Rosedale, D. O., Growing macadamia nuts in California, *Calif. Agr. Ext. Ser. Pub. AXT–103*, 1963.

148 Ruehle, G. D., Growing bananas in Florida, *Univ. Fla. Agr. Ext. Serv. Cir. 178*, 1958.

149 ———, The Florida avocado industry, *Fla. Agr. Exp. Sta. Bul. 602*, 1958.

150 Samish, R. M., and A. Gur, Experiments with budding avocado, *Proc Amer. Soc. Hort. Sci.*, 81:194–201. 1962.

151 Sax, K., The use of *Malus* species for apple rootstocks, *Proc. Amer. Soc. Hort. Sci.*, 53:219–20. 1949.

152 ———, Interstock effects in dwarfing fruit trees, *Proc. Amer. Soc. Hort. Sci.*, 62:201–4. 1953.

153 Schneider, H., Graft transmission and host range of the pear decline causal agent, *Phytopathology*, 60:204–7. 1970.

154 Schroeder, C. A., Rootstock influence on fruit set in the Hachiya persimmon, *Proc. Amer. Soc. Hort. Sci.*, 50:149–50. 1947.

155 Schuster, C. E., Filberts in Oregon, *Ore. Agr. Ext. Bul. 628*, 1961.

156 Scott, D. H., A. D. Draper, and G. M. Darrow, Commercial blueberry growing, *U.S.D.A. Farmers' Bul. 2254.* 1973.

157 Scott, D. H., and D. P. Ink, Germination of strawberry seed as affected by scarification treatment with sulfuric acid, *Proc. Amer. Soc. Hort. Sci.*, 51:299–300. 1948.

158 Serr, E. F., Rooting Paradox walnut hybrids, *Calif. Agr.,* 8(5):7, 1954.

159 Serr, E. F., and H. I. Forde, Blackline, a delayed failure at the union of *Juglans regia* trees propagated on other *Juglans* species, *Proc. Amer. Soc. Hort. Sci.,* 74:220–31. 1959.

160 Serr, E. F. and A. D. Rizzi, Walnut rootstocks, *Univ. Calif. Agr. Ext. Publ. AXT–120.* 1964.

161 Serr, E. F., Persian walnuts in the western states, Chap. 18 in R. A. Jaynes (ed.), *Handbook of North American Nut Trees.* Knoxville, Tenn.: Northern Nut Growers' Assn., 1969.

162 Shalla, T., and L. Chiarappa, Pear decline in Italy, *Bul. Calif. Dept. Agr.,* 50:213–17. 1961.

163 Sharpe, R. H., Rooting of muscadine grapes under mist, *Proc. Amer. Soc. Hort. Sci.,* 63:88–90. 1954.

164 Sharpe, R. H., and N. Gammon, Jr., Pecan growing in Florida, *Fla. Agr. Exp. Sta. Bul. 601,* 1958.

165 Sharpe, R. H., C. O. Hesse, B. F. Lownsbery, V. G. Perry, and C. J. Hansen, Breeding peaches for root-knot nematode resistance, *Jour. Amer. Soc. Hort. Sci.* 94(3):209–12. 1969.

166 Simmonds, N. W., *Bananas,* 2nd ed., London: Longmans Green & Company, Ltd., 1966.

167 Simons, R. K., Abnormal phloem development and sieve-tube necrosis associated with Golden Delicious on EM VII rootstock, *Proc. Amer. Soc. Hort. Sci.,* 89:14–22. 1966.

168 Simpson, R. H., Growing pecan trees, in J. G. Woodroof (ed.), *Tree Nuts: Production, Processing, Products.* Vol. II, pp. 19–30. Westport, Conn.: Avi Publishing Co., 1967.

169 Singh, S., and D. V. Chugh, Marcotting with some plant regulators in loquat *(Eriobotrya japonica* Lindl.), *Indian Jour. Hort.,* 18:123–29. 1961.

170 Singh, L. B., *The Mango: Botany, Cultivation, and Utilization.* New York: Interscience Publishers, Inc., 1960.

171 Singh-Dhaliwal, T., and A. Torres-Sepulveda, Recent experiments on rooting coffee stem cuttings in Puerto Rico, *Univ. P. R. Agr. Exp. Sta. Tech. Pap. 33,* 1961.

172 Sitton, B. G., Vegetative propagation of the black walnut, *Mich. Agr. Exp. Sta. Tech. Bul. 119,* 1931.

173 Smith, S. H., R. E. Hilton, and N. W. Frazier, Meristem culture for elimination of strawberry viruses, *Calif. Agr.,* 24(8):8–10. 1970.

174 Spiegel-Roy, P., Experience with rootstocks for the apricot and stock-scion compatibilities in Israel, *Proc. 16th Int. Hort. Cong.,* pp. 587–92. 1962.

175 Storey, W. B., The macadamia in California, *Proc. Fla. State Hort. Soc.,* 70: 333–38. 1957.

176 Stoutemyer, V. T., T. J. Maney, and B. C. Pickett, A rapid method of propagating raspberries and blackberries by leaf-bud cuttings, *Proc. Amer. Soc. Hort. Sci.,* 30:278–82. 1934.

177   Stubbs, L. L., Tristeza-tolerant strains of sour orange, *F. A. O. Plant Prot. Bul.*, 11:8–10. 1963.

178   Suarez de Castro, F., Semilleros o germinadores de cafe (Coffee seed beds or germinators), *Agri. Trop.* (Bogota), 17:317–24. 1961.

179   Teague, C. P., Avocado tip-grafting, *Proc. Int. Plant Prop. Soc.*, 16:50–51. 1966.

180   Teulon, J., Propagation of passion fruit *(Passiflora edulis)* on a fusarium resistant rootstock. *The Plant Prop.*, Inter. Plant Prop. Soc.. 17(3):4–5. 1971.

181   Tiscornia, J. R., and F. E. Larsen, Rootstocks for apple and pear—a literature review, *Wash. Agr. Exp. Sta. Cir. 421*, 1963.

182   Traub, H. P., and L. C. Marshall, Rooting of papaya cuttings, *Proc. Amer. Soc. Hort. Sci.*, 34:291–94. 1937.

183   Tukey, H. B., and K. D. Brase, The dwarfing effect of an intermediate stempiece of Malling IX, *Proc. Amer. Soc. Hort. Sci.*, 42:357–64. 1943.

184   Tukey, R. B., R. L. Klackle, and J. A. McClintock, Observations on the uncongeniality between some scion varieties and Virginia Crab stocks, *Proc. Amer. Soc. Hort. Sci.*, 64:151–55. 1955.

185   Urquhart, D. H., *Cocoa,* 2nd ed. London: Longmans Green & Company, Ltd., 1961.

186   van Eignatten, C. L. M., Propagation of kola *(Cola nitida),* Commun. NEDERF (Amsterdam), Sept. 1969.

187   Waldo, G. F., R. S. Bringhurst, and V. Voth, Commercial strawberry growing in the Pacific Coast States, *USDA Farmers' Bul. 2236,* 1968.

188   Waring, J. H., and M. T. Hilborn, Hardy stocks in the apple orchard, *Me. Agr. Ext. Bul. 355,* 1947.

189   Weathers, L. G., E. C. Calavan, J. M. Wallace, and D. W. Christiansen, Lemon on Troyer citrange root, *Calif. Agr.*, 9(11):11–12. 1955.

190   Weinberger, J. H., and N. H. Loomis, A rapid method for propagating grapevines on rootstocks, *USDA Agr. Res. Serv. ARS-W-2,* 1972.

191   Weiss, G. M., and D. V. Fisher, Growing apple trees on dwarfing rootstocks, *Canad. Dept. of Agr. Res. Br.*, S.P. 15. 1960.

192   Wellman, F. L, *Coffee: Botany, Cultivation, Utilization.* London: Leonard-Hill, 1961.

193   Westwood, M. N., and L. A. Brooks, Propagation of hardwood pear cuttings, *Proc. Inter. Plant Prop. Soc.*, 13:261–68. 1963.

194   Westwood, M. N., H. R. Cameron, P. B. Lombard, and C. B. Cordy, Effects of trunk and rootstock on decline, growth, and performance of pear, *Jour. Amer. Soc. Hort. Sci.*, 96(2):147–50. 1971.

195   Westwood, M. N., and A. N. Roberts, Quince root used for dwarf pears, *Ore. Orn. and Nurs. Dig.*, 9(1):1–2. 1965.

196   Whitehouse, W. E., The pistachio nut—a new crop for the western United States, *Econ. Bot.*, 11:281–321. 1957.

197   Williams, M. W., and R. A. Norton, Propagation of red raspberries from softwood cuttings, *Proc. Amer. Soc. Hort. Sci.*, 74:401–6. 1959.

198   Winkler, A. J., J. A. Cook, L. A. Lider, and W. M. Kliewer, *General Viticulture,* 2nd ed. Berkeley: University of California Press, 1974, Chap. 9, Propagation.

199   Winter, H. F., Prevalence of latent viruses in Ohio apple trees, *Ohio Farm Home Res.,* 48:58–59, 1963.

200   Worthington, T., and D. H. Scott, Strawberry plant storage using polyethylene liners, *Amer. Nurs.,* 105(9):13, 56–57. 1957.

201   Zeiger, D., and H. B. Tukey, An historical review of the Malling apple rootstocks in America, *Mich. State Univ. Cir. Bul.* 226, 1960.

202   Zentmyer, G. A., A. O. Paulus, and R. M. Burns, Avocado root rot, *Calif. Agr. Exp. Sta. Cir.* 511, 1962.

203   Zielinski, Q. B., Some factors affecting seed germination in sweet cherries, *Proc. Amer. Soc. Hort. Sci.,* 72:123–28. 1958.

SUPPLEMENTARY READING

Brase, K. D., and R. D. Way, "Rootstocks and Methods Used for Dwarfing Fruit Trees," *New York Agricultural Experiment Station Bulletin 783,* 1959.

Carlson, R. F., "Fruit Trees—Dwarfing and Propagation," *Michigan State University Horticultural Report No. 1* (rev.), 1971.

Chandler, W. H., *Deciduous Orchards.* 3rd ed. Philadelphia: Lea & Febiger, 1957.

———, *Evergreen Orchards,* 2nd ed. Philadelphia: Lea & Febiger, 1958.

Day, L. H., "Apple, Quince, and Pear Rootstocks in California," *California Agricultural Experiment Station Bulletin 700,* 1947.

———, "Rootstocks for Stone Fruits," *California Agricultural Experiment Station Bulletin 736,* 1953.

Feilden, G. S., and R. J. Garner, "Vegetative Propagation of Tropical and Sub-Tropical Plantation Crops," *Imperial Bureau of Horticulture and Plantation Crops, Tech. Comm. No. 13,* 1940.

Hansen, C. J., and H. T. Hartmann, "Propagation of Temperate-Zone Fruit Plants," *California Agricultural Experiment Station Circular 471* (rev.), 1966.

Hutchinson, A., "Rootstocks for Fruit Trees," *Ontario (Canada) Department of Agriculture and Food. Publication 334,* 1969.

Jaynes, R. A. (ed), *Handbook of North American Nut Trees.* Knoxville, Tenn.: Northern Nut Growers' Association, 1969.

Roach, F. A., "Fruit Tree Raising—Rootstocks and Propagation," *Ministry of Agriculture, Fisheries, and Food (London) Bulletin 135, 5th ed.* 1969.

Ruehle, G. D., "Miscellaneous Tropical and Sub-Tropical Fruit Plants," *Florida Agricultural Extension Bulletin 156A.* 1958.

Tukey, H. B., *Dwarfed Fruit Trees.* New York: The Macmillan Company, 1964.

Warner, R. M., "Propagation of tropical crop plants," *Proc. Inter. Plant Prop. Soc.,* 22:181–90. 1972.

Propagation of Certain Ornamental Trees, Shrubs, and Woody Vines

18

This chapter is limited to propagation methods for those woody perennial ornamentals in which there seems to be the greatest interest. It is beyond the scope of this book to discuss propagation of all plants used as ornamentals.

ABELIA (*Abelia* sp.)   Leafy cuttings from partially matured current season's growth can be rooted easily under glass or mist in summer or fall, and respond markedly to treatments with indolebutyric acid. Hardwood cuttings may also be rooted in fall or spring.

ABIES sp.   *See* Fir.

ABUTILON (*Abutilon* sp.)   Flowering maple, Chinese bell flower. Started by leafy cuttings rooted under glass or mist in the spring or fall.

ACACIA (*Acacia* sp.)   This is uusally propagated by seeds. The impervious seed coats must be softened before planting by soaking in concentrated sulfuric acid for 20 minutes to two hours. Another method is to pour boiling water over the seeds and allow them to soak for 12 hours in the gradually cooling water. Leafy cuttings of partially matured wood can be rooted. All but very young plants are difficult to transplant because of a pronounced taproot.

ACER sp.   *See* Maple.

AESCULUS sp.   *See* Buckeye.

AILANTHUS (*Ailanthus altissima* [Mill] Swingle)   Tree of Heaven. Seed propagation is easy, and self-sowing usually occurs when both male and female trees are grown close together. Embryo dormancy apparently is present in freshly harvested seed. Stratification at about 40° F (4° C) for two months will aid germination. Seed propagation produces both types of trees, but planting male trees should be avoided, since the staminate flowers produce an obnoxious odor. Propagation of the more desirable female trees can be done by root cuttings planted in the spring.

ALBIZZIA (*Albizia julibrissin* Durazz.)   Silk tree. This is started by seeds with some treatment to overcome the impermeable seed coat, such as soaking in sulfuric acid for half an hour, plus thorough washing, before planting. Stem cuttings do not root, but root cuttings several inches long and ½ in. or more in diameter taken and planted in early spring are successful *(41)*.

ALDER *(Alnus* sp.)   This is propagated by seed which should be cleaned thoroughly and planted in late fall.

AMELANCHIER ALNIFOLIA   See Service-Berry.

ARALIA *(Aralia* sp.)   Seeds should be treated with sulfuric acid for 30 or 40 min., then planted in early fall. Do not allow seeds to become dry.

ARAUCARIA HETEROPHYLLA   See Norfolk Island pine.

ARBORVITAE (American [*Thuja occidentalis* L.] and Oriental [*T. orientalis* L.])

Seeds   Germination is relatively easy but stratification of seeds for 60 days at about 40° F (4° C) may be helpful.

Cuttings

Thuja occidentalis   These may be taken in midwinter and rooted under mist (about eight seconds every ten minutes) in the greenhouse. Best rooting is often found with cuttings taken from older plants no longer making rapid growth. The cuttings should be about 6 in. long and may be taken either from succulent, vigorously growing terminals or from more mature side growths several years old. Wounding and treating with root-inducing chemicals are beneficial. No shading should be used *(90)*.

Cuttings may also be made in midsummer, rooted out-of-doors in a shaded, closed frame. They should be several inches long and of current season's growth with somewhat matured wood at the base. They should be rooted by fall.

Thuja orientalis   Cuttings of this species are often more difficult to root than those of *T. occidentalis*. Small, soft cuttings several inches long, taken in late spring, can be rooted in mist beds if treated with a root-promoting chemical (see p. 291) *(123)*.

Grafting   The side graft is used in propagating selected clones of *T. orientalis*, with two-year-old potted *T. orientalis* seedlings as the rootstock. Grafting is done in late winter in the greenhouse. After making the grafts, the potted plants are set in open benches filled with moist peat moss just covering the union. The grafted plants should be ready to set out in the field for further growth by midspring.

ARBUTUS MENZIESII   See Madrone.

ARCTOSTAPHYLOS   See Manzanita.

ASH *(Fraxinus* sp.)   Seeds of most species germinate readily if stratified for two to four months at about 40° F (4° C). Seeds of *F. excelsior, F. nigra,* and *F. quadrangulata* should have one to three months moist storage at room temperature, followed by five to six months at about 40° F (4° C).

ASPEN   See Poplar.

AZALEA *(Rhododendron* sp.) *(9, 65, 72, 107)*   **Evergreen and Semi-evergreen Types.** Although evergreen azaleas can be propagated by seeds, grafting, and layering, almost all nursery plants are started by cuttings.

Seeds   The seed capsules should be gathered in the fall after they turn brown, then stored at room temperature in a container that will hold the seed when the capsules open. The seeds have no dormancy problems but will lose viability if held for extended periods in open storage; sealed, low-temperature storage is preferable. The seeds germinate satisfactorily in a thin layer of screened sphagnum moss over an acid soil mixture. Seed planting is usually done in the greenhouse from midwinter to early spring. Optimum germinating temperatures are about 70° F (21° C) during the day and 55° F (13° C) at night. Germination usually occurs within a month. In areas having hard water, the seedlings should be watered with distilled or rain water because the alkaline salts will soon cause injury.

Cuttings   Azalea cuttings of the evergreen and semi-evergreen types are not difficult to propagate, roots forming in three to four weeks under proper conditions. They are best taken in midsummer after the new current season's growth has become somewhat hardened but before the wood has turned red or brown. Root-promoting growth regulators are often beneficial. Azaleas do very well under mist if the medium is well drained. After roots form the mist should be reduced and finally discontinued.

After rooting, the cuttings may be potted in equal parts of leaf mold, sand, and sphagnum or peat.

Deciduous Types

Seeds   Deciduous azaleas are often propagated by seed, since it is inexpensive and rooting cuttings is difficult. The same germination procedures can be used as described for the evergreen types. Growth of the seedlings is sometimes slow, but with early spring germination and early transplanting to a good growing medium, it is possible to have plants large enough to set flower buds in their second year.

In a different procedure, which will bring the seedlings into bloom sooner, seeds are sown in late summer (from seed harvested in midwinter) on top of screened peat moss in flats covered with glass. After the seeds germinate, the day length is extended by added light to keep them growing well. The seedlings are transplanted in midfall into flats to 10 percent sand and 90 percent peat moss, where they are left (in the greenhouse) until midsummer *(51)*.

Cuttings *(10, 16)*   Deciduous azaleas are difficult to propagate by cuttings, but by taking soft, succulent material in the spring, particularly from stock plants brought into the greenhouse and forced, cuttings can be rooted. Timing is very important; it is much better to take the cuttings in spring than in summer.

Wounding, plus the use of root-promoting chemicals, is important in obtaining good rooting. Treatment with indolebutyric acid (75 ppm for 15 hours, or IBA in talc at 0.8 percent) has been found to be effective. Closed frames or intermittent mist with bottom heat (68° F; 20° C) and a rooting medium of sphagnum peat or a sand-peat-perlite mixture provide good rooting conditions.

Not only is it difficult to root deciduous azalea cuttings, but it is also a problem to get them to survive through the winter even after they are rooted. Following rooting, the cuttings should be set in a cold frame, where they will make some growth during the summer, although some cuttings may go dormant. Increased day length may prevent this. They are best left in the cold frame throughout the winter, then transplanted the following spring. There is also a problem in getting the buds to start in the spring on rooted cuttings. Increasing the day length by artificial light—after the cuttings are rooted—from late summer to midwinter is effective in inducing "bud-break" in the spring.

Grafting   This is the chief propagation method for most Ghent and Mollis hybrids. *Rhododendron luteum (Azalea pontica)* seedlings, two or three years old, are used as the understock. This method has the disadvantages of undesirable suckering from the understock and a general lack of plant vigor, perhaps owing to an unsatisfactory graft union or to a lack of adaptability of the seedling stocks to some climates. The seedlings are potted in early spring, the grafting being done during late summer. A side-veneer graft may be used. The grafted plants are set in closed frames in the greenhouse during healing of the union *(76).* Tree-type azaleas can be produced by grafting high on rooted cuttings of *R. concinnum.*

Layering   Azaleas are easily layered. It is a method worth using for obtaining a limited number of new plants, especially of the deciduous types difficult to start by cuttings. Simple, mound, trench, and air layering have all been used satisfactorily.

BAMBOO *(Arundinaria, Bambusa, Phyllostachys, Sasa* sp.) *(80)*   Bamboo is easily propagated vegetatively by division of the clumps or young rhizomes. This is best done just before the annual period of active bud growth. Use of one- or two-year-old growth taken from the periphery of the clump gives best results. Prevention of drying during transplanting is important. Best results are obtained from rhizomes taken and planted in late winter or early spring before the buds begin to push *(30).*

BARBERRY *(Berberis* sp.)   These are propagated without difficulty by fall-sowing or by spring-sowing seeds which have been stratified for two to six weeks at about 40° F (4° C). It is important to remove all pulp from seeds. Leafy cuttings taken from spring to fall can be rooted under mist. Indolebutyric acid at 5000 ppm aids rooting *(35).* Greenhouse grafting of some selected types is also practiced, and layering is done occasionally.

BEECH *(Fagus* sp.)   Seeds germinate readily in the spring from fall planting or after being stratified for thre months at about 40° F (4° C). Seeds should not be allowed to dry out. Selcted clones are grafted by either the cleft, the whip, or the side-veneer method, on seedlings of *F. sylvatica,* the European beech. Frequent transplanting or root pruning of nursery trees is necessary to prevent development of a single long tap root, which, if present, makes subsequent transplanting difficult *(28).*

BERBERIS sp.   *See* Barberry.

BETULA sp.   *See* Birch.

BIRCH *(Betula* sp.)   Seeds should be either fall-planted or spring-planted following stratification at about 40° F (4° C) for one to three months. *Betula nigra* matures its seeds in the spring, and if planted promptly the seed will germinate at once without treatment. Some of the selected, weeping forms are grafted on European white birch *(B. verrucosa)* seedlings. *B. papyrifera* seeds will germinate best if held at below freezing temperature for six weeks before planting. Birch is difficult to propagate by cuttings, but leafy cuttings will root in summer under glass if treated with indolebutyric acid at about 50 ppm for 24 hours. Low-growing species can be propagated by layering.

BITTERSWEET *(Celastrus* sp.)   These twining vines have male flowers on one plant and female on another, and the two types must be near each other for production of the attractive berries. Seeds must be removed from the berries. They can be fall-planted or stratified for about three months at 40° F (4° C). Some nurseries propagate the bittersweet asexually and supply plants of known sex. Softwood cuttings taken in midsummer root readily under glass. Treatments with indolebutyric acid will hasten rooting. Hardwood cuttings are also easily rooted and benefit by treatment with root-promoting substances.

BOTTLE BRUSH *(Callistemon* sp.)   Although seeds germinate wtihout difficulty, seedlings should be avoided because many of them will prove worthless as ornamentals. The preferred method of propagation is by leafy cuttings of partially matured wood, which root under glass quite easily.

BOUGAINVILLEA (*Bougainvillea* sp.)   This showy evergreen vine, growing outdoors only in mild climates, is propagated by hardwood cuttings taken at any time of the year and rooted in sand or other rooting media. Some cultivars, difficult to start, should be rooted as leafy cuttings under mist after treatment with a root-promoting substance.

BOXWOOD (*Buxus sempervirens* L. and *B. microphyllas* Sieb. and Zucc.)   Cuttings are commonly used—either softwood taken in spring or summer, or semi-hardwood taken in the fall. In the latter method, which is ordinarily used, the cuttings are rooted in a cool greenhouse or cold frame during the winter and spring or under mist at any time. The rooted cuttings are ready for digging at the end of summer. Seeds are rarely used, owing to the very slow growth of the seedlings. Young plants should always be transplanted with a ball of soil around their roots.

BROOM (*Cytisus* sp.)   Seeds of many of the species germinate satisfactorily if gathered as soon as mature and treated with sulfuric acid to soften the hard seed coat before planting. The seeds are best germinated in a warm location and then transferred to a cooler place when the seedlings are several inches high. The various *Cytisus* species crossbreed readily, so stock plants for seed sources should be isolated.

Cuttings can be rooted rather easily under mist in midsummer if treated with indolebutyric acid and given bottom heat.

BUCKEYE (*Aesculus* sp.) *(84)*   Horse chestnut. This may be propagated by seeds, but prompt sowing or stratification after gathering in the fall is necessary. If the seeds lose their waxy appearance and become wrinkled, their viability will be reduced. For best germination, seeds should be stratified for four months at about 40° F (4° C) immediately after collecting. In propagating the low-growing species, simple layering either in spring or fall is often used. T-budding or bench grafting, using the whip graft, can be used for selected cultivars worked on *A. hippocastanum* as the rootstock. The dwarf buckeye, *A. parviflora*, is propagated by underground stem pieces. Seeds of buckeye are poisonous if eaten.

BUCKTHORN (*Rhamnus* sp.)   This can be propagated by fall planting seed out-of-doors. Macerate fruits and clean seeds. Those of some species may germinate better if given a 20 min. treatment with sulfuric acid before planting (see p. 156).

BUTTERFLY BUSH (*Buddleia* sp.)   Softwood cuttings can be taken in the summer or fall and rooted under glass. Seeds started in the greenhouse in early spring will provide flowering plants by fall, although reproduction is not genetically true by seed.

BUDDLEIA sp.   *See* Butterfly Bush.

BUXUS sp.   *See* Boxwood.

CALLISTEMON sp.   *See* Bottle Brush.

CALLUNA VULGARIS   *See* Heather.

CAMELLIA (*Camellia* sp.) *(60)*   Camellias can be propagated by seed, cuttings, grafting, and layering. To perpetuate selected and named cultivars, cuttings, grafts, or layers must be used. Seedlings are used in breeding new cultivars, as rootstocks for grafting, or in growing hedges where foliage is the only consideration.

Seeds   In the fall when the capsules begin to turn reddish-brown and split, they should be gathered before the seeds became scattered. The seed should not be allowed to dry out before planting; best germination is obtained if they are planted before the seed coat hardens. If the seeds must be stored for long periods they will keep satisfactorily mixed with ground charcoal and stored in an airtight container placed in a cool location. After the hard seed coat develops, germination is hastened by pouring boiling water over the seeds and allowing them to remain in the cooling water for 24 hours. For germination, a well-drained acid soil high in organic matter should be used. It takes four to seven years to bring a camellia into flowering from seed.

Cuttings   Most *C. japonica* and *C. sasanqua* plants are produced commercially from cuttings which are not difficult to root. *Camellia reticulata* cuttings do not root easily, however, and this species is generally propagated by either cleft or approach grafting.

Cuttings are best taken in midsummer from the spring flush of growth after the wood has matured somewhat and changed from green to light brown in color. Tip cuttings are used, 3 to 6 in. long, with two or three terminal leaves. Rooting will be much improved if the cuttings are treated with indolebutyric acid used at 20 ppm for 24 hours. Wounding the base of the cuttings before they are treated is also likely to improve rooting. The cuttings root best under high humidity, either in a closed frame or under mist.

Camellias may also be started as leaf-bud cuttings, which are handled as stem cuttings. In this case, excessive concentrations of a root-promoting substance should be avoided because this may inhibit development of the single bud.

Grafting   Camellias are frequently grafted, not only to produce nursery plants but to change cultivars of older established plants. Vigorous seedlings or rooted cuttings of either *C. japonica* or *C. sasanqua* can be used as stocks for grafting. Any of the side graft methods is suitable.

Both cleft and bark graft methods are used for grafting larger, established plants. This is best done in the spring, two or three weeks before new vegetative growth starts. The stock and the scion should be dormant at the time of grafting. Scions several inches long from terminal shoots containing one or two leaves and several dormant buds are used, inserted into the stock plant, which is cut off 2 or 3 in. above the soil level. In cleft grafting, stocks smaller than ½ in. should be wrapped securely with string to hold the scions firmly in place. This is not necessary with larger stocks. After the graft is completed, the union is usually waxed, but sometimes not. In either case, it should be covered with well-pulverized, moist soil up to the base of the leaves on the scion. A gallon can with both ends removed may be set around the graft, which will hold the soil in place more securely. Following this, a large glass jar should be set inverted over the leafy scion, but it should be shaded to prevent excessively high temperatures inside. After the buds on the scions start growth, the jar is gradually lifted and finally removed.

The "nurse-seed" method of grafting has been successfully used for cultivars hard to root *(87)*.

Layering   To obtain a few additional plants from a single mother plant, simple layering, performed in the spring, may be done. To do this, branches must be present close to the ground which, preferably, are young and not over about ½ in. in diameter. One or two years may be required for enough roots to form before the layer can be removed.

CAMPSIS sp.   *See* Trumpet Creeper.

CAPE JASMINE   *See* Gardenia.

CATALPA (*Catalpa* sp.)   Seeds germinate readily without any previous treatment. They are stored dry over winter at room temperature and planted in late spring. For ornamental purposes, *C. bignonioides* 'Nana' is often budded or grafted high on stems of *C. speciosa,* giving the "umbrella tree" effect. A strong shoot is forced from a one-year-old seedling rootstock, which is then budded with several buds in the fall at a height of about 6 ft. Catalpa species can also be propagated in summer by softwood cuttings rooted under glass.

CEANOTHUS (*Ceanothus* sp.)   Propagation is by seed, cuttings, layering, and sometimes grafting. Seeds must be gathered shortly before the capsules open or they will be lost. Those of *C. arboreus, C. cuneatus, C. jepsoni, C. megacarpus, C. oliganthus, C. rigidis,* and *C. thyrsiflorus* have only seed coat dormancy. Placing the seeds in hot water (180° to 190° F; 82° to 87° C) and allowing them to cool for 12 to 24 hours, as described on p. 156, or even boiling in water for five minutes, will aid germination. To obtain germination in other *Ceanothus* species, which have both seed coat and embryo dormancy, the seeds should be immersed in hot water (as described above), then stratified at 35° to 40° F (2° to 4° C) for two to three months.

Leafy cuttings can be rooted under mist at any time from spring to fall, especially if treated with a root-promoting substance. Terminal softwood cuttings from vigorously growing plants in containers give good results.

Ceanothus americanus seedlings are often used as rootstocks for grafting selected clones.

CEDAR (*Cedrus* sp.)   Seeds germinate readily if not permitted to dry out. No dormancy conditions occur, but soaking the seeds in water several hours before planting is helpful. Cuttings do not root easily, but if taken in late summer or fall, treated with a root-promoting hormone and placed over bottom heat under a plastic covered frame, some rooting may be obtained *(40)*. Side-veneer grafting of selected forms on one- or two-year-old potted seedling stocks may be done in the spring. Scions should be taken from vigorous terminal growth of current season's wood rather than from lateral shoots. *Cedrus atlantica* selections are grafted on *Cedrus deodara* seedlings as a rootstock *(28)*.

CELASTRUS sp.   *See* Bittersweet.

CELTIS sp.   *See* Hackberry.

CERCIS sp.   *See* Redbud.

CHAENOMELES sp.   *See* Quince, Flowering.

CHAMAECYPARIS (*Chamaecyparis* sp.)   False cypress. After collection in the fall, the seeds should be carefully dried in a warm room or in a kiln at 90° to 110° F (32° to 43° C). Stratification at about 40° F (4° C) for two to three months will aid germination. Cuttings of most species are not difficult to root, particularly if juvenile forms are used (see p. 239). They may be taken in fall or winter and rooted in a closed frame in the greenhouse, using lateral shoots of current season's wood. Indolebutyric acid treatments are helpful *(67, 90)*.

CHERRY, FLOWERING (*Prunus* sp.)   Cultivars of *P. serrulata, P. sargentii, P. sieboldi, P. yedoensis, P. campanulata,* and *P. subhirtella* comprise the flowering cherries. Seedlings of *P. avium,* the Mazzard cherry, or *P. serrulata* are suitable as rootstocks upon which these ornamental forms may be T-budded, either in the fall or in the spring. *Prunus dropmoreana* is a suitable stock for *P. serrulata* 'Kwanzan' *(39)*. If cross-pollination can be avoided, *P. sargentii, P. campanulata,* and *P. yedoensis* will reproduce true from seed. Leafy cuttings of some of the flowering cherry species can be rooted under mist in high percentages if treated with indolebutyric acid, but subsequent survival and overwintering are sometimes difficult.

CHIONANTHUS sp.   *See* Fringe Tree.

CLEMATIS *(Clematis* sp.)   Seeds of some clematis species have embryo dormancy, so stratification for one to three months at about 40° F (4° C) is likely to aid germination. Clematis is probably best propagated by cuttings, which root under mist in about five weeks. Young wood with short internodes in the spring often gives satisfactory results, but partially matured wood taken in late spring to late summer is more commonly used. Leaf-bud cuttings taken in midsummer will also root readily under mist. Treatment with indolebutyric acid is helpful. The large-flowering hybrids are generally root-grafted by the cleft or side-veneer graft or a similar method *(102)* in the spring on *C. flammula, C. vitalba,* or *C. viticella* seedlings. The grafts are planted deeply with scion roots forming, the rootstock acting as a nurse-root. When only a few plants are needed, layering the long canes gives satisfactory results.

CORNUS sp.   *See* Dogwood.

COTINUS COGGYGRIA   *See* Smoke Tree.

COTONEASTER *(Cotoneaster* sp.) *(42)*   Seeds should be soaked for about 90 minutes in concentrated sulfuric acid and then stratified for three to four months at about 40° F (4° C). As a substitute for the acid treatment, a moist, warm (60° to 75° F; 15 to 24° C) stratification treatment for three to four months may be used. This must be followed by the cold stratification treatment. Leafy cuttings of most species taken in spring or summer will root under mist without much difficulty. Treatments with indolebutyric acid (see p. 291) are helpful. Cotoneaster can be budded high onto a pear nursery tree to porduce a "tree" cotoneaster. A blight-resistant pear, as 'Old Home,' should be used. Simple layering can be done.

COTTONWOOD   *See* Poplar.

CRABAPPLE, FLOWERING *(Malus* sp.) *(127)*   Four species of crabapples—*M. toringoides, M. hupehensis, M. sikkimensis,* and *M. florentina*—will reproduce true from seed. Selected forms of all other crabapple species, such as *M. sargentii, M. floribunda,* and *M.* 'Dolgo,' should be propagated by asexual methods. Nursery trees are commonly propagated either by root grafting, using the whip, or tongue, method, or by T-budding seedlings in the nursery row; the latter is done either as spring or as fall budding. Fall budding is considered by most nurserymen to be the fastest and most desirable method of propagating crabapples. In addition, older *Malus* trees may be topworked to the desired flowering crab cultivar. Various seedling rootstocks are used, such as *M. sylvestris* (common apple), *M. baccata, M. ionensis,* and *M. coronaria,* as well as a number of the Malling series of apple rootstocks *(39).* The flowering 'Bechtel's Crab' is said to show delayed incompatibility, which appears in 10 to 15 years, when worked on *M. sylvestris* seedlings. Dwarf trees of the 'Carmine' crabapple *(M. atrosanguinea)* have been produced by grafting onto *Cotoneaster divaricata.*

CRAPE MYRTLE *(Lagerstroemia indica* L.)   This is propagated by leafy cuttings under glass or mist in the summer. Transplanting is somewhat difficult, so all but very small plants should be moved with a ball of soil.

CRATAEGUS sp.   *See* Hawthorn.

CRYPTOMERIA, JAPANESE *(Cryptomeria japonica* [L.f] D. Don).   This can be propagated either by seeds or by cuttings. Seeds should not dry out. Cuttings, 2 to 6 in. long, should be taken from green wood at a stage of maturity at which it breaks with a snap when bent. Root in sand over bottom heat; keep the cuttings shaded and cool. After roots start to form, in about two weeks, give more light; transplant to pots when roots are about ½ in. long. Indolebutyric acid treatments promote rooting.

CUPRESSUS sp.   *See* Cypress.

CYPRESS (*Cupressus* sp.)   Seeds have embryo dormancy, so stratification for about four weeks at 35° to 40° F (2° to 4° C) will improve germination. Cuttings can be rooted if taken during winter months. Treatments with indolebutyric acid at 60 ppm for 24 hours aid rooting. Side-veneer grafting of selected forms on seedling *Cupressus* rootstocks in the spring is often practiced.

CYTISUS sp.   *See* Broom.

DAPHNE (*Daphne* sp.) *(18)*   This is best propagated by leafy cuttings in sand and peat moss (2:1) under glass; cuttings are taken in summer from partially matured current season's growth. Root-promoting substances (see p. 291) have often been helpful. Some daphne species can be started by root cuttings, e.g., *D. genkwa* and *D. mezereum*. Layering is successful, using the previous season's shoots layered in spring and removed the following spring. The daphnes do not transplant easily and should be moved only when young. Berries are very poisonous if eaten.

DEUTZIA (*Deutzia* sp.)   This is easily propagated either by hardwood cuttings lined-out in the nursery row in spring or by softwood cuttings under glass in summer.

DOGWOOD (*Cornus* sp.)   Seeds have various dormancy conditions; those of the popular flowering dogwood *(C. florida)* requires either fall planting or a stratification period of about four months at 40° F (4° C). Best germination is obtained if the seeds are gathered as soon as the fruit starts to color, and sown or stratified immediately. It is important to remove seeds from the fruit and soak in water if allowed to dry out. Other species require, in addition, a treatment to soften the seed covering. Two months in moist sand at diurnally fluctuating temperatures (70° to 85° F; 21° to 30° C), followed by four to six months at 32° to 40° F (0° to 4° C), is effective. With some species, the warm stratification period may be replaced by mechanical scarification or soaking in sulfuric acid.

Some dogwoods are not difficult to start by cuttings. Those of *C. florida* are best taken in late spring or early summer from new growth after flowering, then rooted under mist. Treatments with indolebutyric acid at 20,000 ppm have given good rooting. Hardwood cuttings taken in the spring are successful with certain species, such as *C. alba*.

Selected types, such as the red flowering dogwood, *C. florida* 'Rubra,' and the weeping forms, are difficult to start by cuttings and are ordinarily propagated by T-budding in late summer, or by whip grafting in the greenhouse in winter on *C. florida* seedlings *(23)*.

Cuttings of *C. florida* 'Rubra' have been successfully rooted by taking them in early summer after the second flush of growth and rooting under mist in one part peat moss and three parts sand *(53)*. To insure survival through the following winter in cold climates, the potted cuttings should be kept in heated cold frames to hold the temperature between 32° and 45° F (0° and 7° C).

DOUGLAS FIR (*Pseudotsuga menziesii* [Mirb.] Franco) *(4)*   Seeds of this important lumber and Christmas tree species exhibit varying degrees of embryo dormancy. For prompt germination, it is best to sow the seeds in the fall or stratify them for two months at about 40° F (4° C).

Douglas fir cuttings are rather difficult to root, but by taking them in late winter, treating them with indolebutyric acid, and rooting them in a sand–peat moss mixture, it is possible to obtain fairly good rooting. Cuttings from young trees root more easily than those from old, and cuttings from some trees are easier rooted than those from others.

ELAEAGNUS *(Elaeagnus* sp.)   Russian olive. Silverberry. Seeds planted in the spring germinate readily following a stratification period of three months at 40° F (4° C). Removal of the pit (endocarp) from silverberry seeds *(E. commutata)* resulted in about 90 percent germination of unstratified seeds, a germination inhibitor apparently being present in the pit *(25)*. Seeds of the Russian olive, *E. angustifolia,* should be treated with sulfuric acid for 30 to 60 min. before fall planting or stratification. This species can also be started by hardwood cuttings planted in spring. Root cuttings or layering are also successful. Leafy cuttings of the evergreen, *E. pungens,* will root if taken in the fall and started under glass.

ELDERBERRY *(Sambucus* sp.)   Seed propagation is difficult, owing to complex dormancy conditions involving both the seed coat and embryo. Probably the best treatment is a warm (70° to 85° F; 21° to 30° C), moist stratification period for two months, followed by a cold (40° F; 4° C) stratification period for three to five months. These conditions could be obtained naturally by planting the seed in late summer, after which germination should occur the following spring.

Since softwood cuttings root easily if taken in spring or summer, this method is generally practiced.

ELM *(Ulmus* sp.)   Seed propagation is commonly used. Elm seed loses viability rapidly if stored at room temperature, but it can be kept for several years in sealed containers at 32° to 40° F (0° to 4° C). Seeds ripening in the spring should be sown immediately, germination usually taking place promptly. For those species which ripen their seed in the fall, either fall planting or stratification for two months at about 40° F (4° C) should be used. To eliminate seedling variation when trees are to be propagated for street planting, some nurserymen bud selected clones on seedlings of the same species. Twig budding (a short twig substituted for a bud), has been successful in elm propagation. Such a bud is T-budded into a seedling rootstock in the spring.

Softwood cuttings of several elm species have been successfully rooted under mist when taken in the spring. Treatments with indolebutyric acid at 50 ppm for 24 hours have been beneficial *(97)*.

ENGLISH IVY *(Hedera helix* L.)   This is readily propagated by rooting cuttings of the juvenile (nonfruiting, lobed-leaf) form. It is also sometimes grafted onto *Fatshedera lizei (Fatsia japonica* × *Hedera helix)* as a rootstock. *Acanthopanax sieboldianus* is a suitable rootstock where root rot from excessive soil moisture is a problem.

ERICA sp.   See Heath.

ESCALLONIA *(Escallonia* sp.)   This is easily started by leafy cuttings taken after a flush of growth. Cuttings root well under mist and respond markedly to treatment with indolebutyric acid.

EUCALYPTUS *(Eucalyptus* sp.) *(95)*   This is almost entirely propagated by seeds planted in the spring. Mature capsules are obtained just before they are ready to open. No dormancy conditions occur in most species, the seed being able to germinate immediately following ripening. Seeds of some species—for example, *E. niphophila* and *E. pauciflora*—however, require stratification for about two months at 40° F (4° C) for best germination. Eucalyptus seedlings are very susceptible to damping-off, so they should be planted in flats of sterilized soil placed in a shady location. From this they are transplanted into small pots, from which they are later lined-out in nursery rows. The roots of young trees will not tolerate drying, so the young plants should be handled as container-grown stock.

Eucalyptus is difficult to start from cuttings, but 65 percent rooting of red gum *(E. camaldulensis)* has been obtained *(37)*. Leafy cuttings were taken in early spring from shoots arising from the base of young trees. These were wounded (see p. 290), treated with a 4000-ppm solution of a root-promoting mixture (IBA + NAA, 1:1) and rooted in perlite under mist over bottom heat at 70° F (21° C).

Eucalyptus ficifolia has been successfully grafted by a side-wedge method, using young, vigorous *Eucalyptus* seedling rootstocks growing in containers and placed under very high humidity following grafting. Use of scions taken from shoots which had been girdled, at least a month previously, increased success *(101)*.

EUONYMUS *(Euonymus* sp.)   Stratification for three to four months at 32° to 50° F (0° to 10° C) is required for satisfactory seed germination. Remove seeds from fruit and prevent drying. Euonymus is easily started by cuttings, hardwood in early spring for the deciduous species and leafy semi-hardwood under glass, after a flush of growth has partially matured, for the evergreen types. Layering is also successful.

EUPHORBIA PULCHERRIMA   *See* Poinsettia.

FAGUS sp.   *See* Beech.

FALSE CYPRESS   *See* Chamaecyparis.

FIR *(Abies* sp.)   Seed propagation is not difficult, but fresh seed should be used; those of most species lose their viability after one year in ordinary storage. Embryo dormancy is generally present, fall planting or stratification at about 40° F (4° C) for one to three months required for good germination. Fir seedlings are very susceptible to damping-off attacks. They should be given partial shade during the first season, since they are injured by excessive heat and sunlight.

Fir cuttings are difficult to root, but if taken in winter and treated with indolebutyric acid, good percentages can be obtained. White fir *(A. concolor)* and red fir *(A. magnifica),* the California "Silver Tip," are important Christmas tree species.

FIRETHORN   *See* Pyracantha.

FORSYTHIA *(Forsythia* sp.)   This is easily propagated by hardwood cuttings set in the nursery row in early spring or by leafy softwood cuttings under high-humidity conditions during late spring or summer.

FRAXINUS sp.   *See* Ash.

FRINGE TREE *(Chionanthus virginicus* L. and *C. retusus* L.) *(110)*   Seed propagation can be used, but it is very slow. Embryo dormancy, as well as some endosperm inhibition, seems to be present. Probably the best practice is stratification for 30 days or more at room temperature, followed by one or two months stratification at about 40° F (4° C). With fall-planted seed, germination may not occur until the second spring.

Cutting propagation of the Chinese fringe tree *(C. retusus)* has generally been considered almost impossible, but by taking softwood cuttings in late spring, and rooting under mist, using treatments with indolebutyric acid, plus a mixture of sand and vermiculite as the rooting medium, excellent rooting percentages can be obtained.

GARDENIA *(Gardenia jasminoides* Ellis)   Cape jasmine. Leafy terminal cuttings can be rooted in the greenhouse under glass or mist from fall to spring. A mixture of sand and peat moss, 1:1, is a good rooting medium. Gardenias are difficult to transplant and should be moved only when small.

GINKGO *(Ginkgo biloba* L.) *(73)*   Seed propagation is often used for this "living fossil," shown by records in rocks to have existed on earth 150 million years ago. A satisfactory procedure is to collect the "fruits" in midfall, remove the pulp, and pack the cleaned seeds in layers of moist sand for ten weeks at temperatures of 60° to 70° F (15° to 21° C) to permit the embryos to finish developing. After this the seeds require a stratification period of several months at about 40° F (4° C) for good germination. Seedlings produce either male or female trees. The plum-like "fruits" on the female trees have a very disagreeable odor, so only male trees are used for ornamental planting. For this reason, propagation by cuttings taken from male trees is advisable. Softwood cuttings in midsummer can be rooted under glass or intermittent mist, but sometimes it is difficult to get the rooted cuttings to make satisfactory growth.

GLEDITSIA TRIACANTHOS   *See* Honeylocust.

GOLDENRAIN TREE (*Koelreuteria* sp.)   This is usually propagated by seed, but root cuttings or softwood cuttings of new growth can be rooted under glass in the spring. The seeds have double dormancy, germinating best if the seed coats are softened by soaking for about 60 minutes in concentrated sulfuric acid, followed by stratification for about 90 days at 35° to 40° F (2° to 4° C) to overcome the embryo dormancy.

HACKBERRY (*Celtis* sp.) *(81)*   Seeds are ordinarily used, sown either in the fall or stratified for two or three months at about 40° F (4° C) and planted in the spring. Prior to stratification, treatments to soften the seed coat, such as soaking in concentrated sulfuric acid, may hasten germination. There is a noticeable variation in the behavior of seedling trees, especially when they are young. A better propagation method is to use cuttings taken from selected vigorous trees. At least two species, *C. occidentalis* and *C. laevigata* (sugarberry) can be started by cuttings. Grafting and chip budding have also been used.

HAMAMELIS   *See* Witch Hazel.

HAWTHORN (*Crataegus* sp.) *(27)*   These tend to reproduce true by seed. Pronounced seed dormancy is present, owing to a combination of an impermeable seed coat and embryo conditions. Probably the best procedure for rapid germination is stratification of freshly collected and cleaned seed in moist peat moss for three or four months at 70° to 80° F (21° to 27° C) (or treatment with sulfuric acid), followed by stratification for five months at about 40° F (4° C). Planting the seed in early summer will provide these conditions naturally, resulting in germination the following spring. Seeds of some *Crataegus* sp. do not have an impermeable seed coat, so the initial high-temperature storage period is unnecessary. Untreated seed may require two or three years to germinate. Since the hawthorn develops a long taproot, transplanting is successful only with very young plants.

Selected clones may be T-budded or root grafted on seedlings of *C. oxyacantha, C. arnoldiana,* or *C. mollis.*

HEATH (*Erica* sp.) and **HEATHER** (*Calluna vulgaris* Hull)   The propagation of these two closely related genera is about the same. Seeds may be germinated in flats in the greenhouse in winter or in a shaded outdoor cold frame in spring. Leafy, partially matured cuttings taken at almost any time of year, but especially in early summer, root readily under glass. Treatment with indolebutyric acid at about 50 ppm for 24 hours is helpful.

HEDERA HELIX   *See* English Ivy.

HEMLOCK (*Tsuga* sp.)   Hemlocks are propagated by seed without difficulty. Seed dormancy is variable, some lots exhibiting embryo dormancy whereas others do not. To insure good germination it is advisable to stratify the seeds for two to four months at about 40° F (4° C). Fall planting outdoors will generally give satisfactory germination in the spring. The seedlings should be given partial shade during the first season. Hemlock cuttings are somewhat difficult to root, but success has been reported with cuttings taken at all times of the year. They seem to respond to treatments with root-promoting substances. Layering is also successful.

HIBISCUS (*Hibiscus syriacus* L.)   Shrub-althea. Rose of Sharon. This is easily propagated, either by hardwood cuttings in the nursery row in spring, or by soft-wood cuttings in midsummer under glass. Lateral shoots make good cutting material. Softwood cuttings respond well to treatment with indolebutyric acid.

HIBISCUS, CHINESE (*Hibiscus rosa-sinensis* L.) *(118)*   Seeds, cuttings, budding or grafting, division, and air layering can be used.

    Cuttings   These are not difficult to root; terminal shoots of partially matured wood of most cultivars taken in spring or summer usually form roots in about six weeks. Leaf-bud cuttings can also be used. Rooting should take place under high humidity, as in a glass-covered frame.

    Grafting   Strongly growing cultivars which are resistant to soil pests and can be started easily by cuttings, such as 'Single Scarlet,' 'Dainty,' 'Euterpe,' or 'Miami Lady,' are used as rootstocks. Some clones develop into much better plants when grafted on these rootstocks than when they are on their own roots, propagated by cuttings. Whip grafting in the spring or side grafting in late spring or early summer is successful. Scions of current season's growth, about pencil size, are grafted on rooted cuttings of about the same size.

    Budding   T-budding, using an inverted T, is sometimes practiced, generally in the spring, although it is successful at any time during the year when the bark is slipping.

    Air Layering   This is practiced during the spring or summer, particularly for cultivars difficult to start by cuttings. Roots will usually form in six to eight weeks.

HOLLY (*Ilex* sp.) *(48, 94)*   Holly can be propagated by seeds, cuttings, grafting, budding, layering, and division.

Most hollies are dioecious. The female plants produce the very desirable decorative berries if male plants are nearby for pollination. In seed propagation, both male and female plants are produced in ratios of one female to three, or sometimes up to ten, male plants. Sex cannot be determined, however, until the seedlings start blooming, at 4 to 12 years.

    Seeds   Germination of holly seed is very erratic; those of some species, *I. crenata, I. cassine, I. glabra, I. vomitoria, I. amelanchier,* and *I. myrtifolia,* germinate promptly and should be planted as soon as they are gathered. Seeds of other species, *I. aquifolium, I. cornuta, I. vericilliata, I. decidua,* and most *I. opaca,* do not germinate until a year or more after planting even though stratified, owing probably to rudimentary embryos at time of harvest.

Seeds of *Ilex aquifolium* (English holly), *I. opaca* (American holly), and *I. cornuta* (Chinese holly) should be collected and cleaned as soon as the fruit is ripe in the fall and then stored at about 40° F (4° C) until spring in a mixture of moist sand and peat moss. Germination in these species generally does not occur until a year later, and then growth is very slow, two seasons being required to bring seedlings of *I. opaca* to a size large enough to be used for grafting.

    Cuttings (111, 124)   This is the method most used by commercial nurserymen, permitting large-scale production of choice clones. Semi-hardwood tip cuttings from well-matured current season's growth produce the best plants. Cuttings taken from flat, horizontal branches of *Ilex crenata* tend to produce plants having this type of growth, and those from upright growth produce upright plants.

Timing is important; best rooting is usually obtained from mid- to late summer, but cuttings may be successfully taken on into the following spring. Wounding the base of the cuttings helps induce root formation. The wounding induced by stripping off the lower leaves may be sufficient.

The use of a root-promoting chemical, particularly indolebutyric acid at relatively high concentrations (8000 to 20,000 ppm), is essential in obtaining rooting of some cultivars, whereas in others this is not needed. Boron, at 50 to 200 ppm, in combination with IBA, has increased root quality in English holly cuttings *(120)*. Bottom heat at 70° to 75° F (21° to 24° C) is beneficial. The maintenance of a high relative humidity is essential. The use of intermittent mist in a greenhouse, where high temperatures can be maintained, provides good rooting conditions. A perlite-peat moss (1:1) rooting medium is satisfactory.

The cuttings root in one to three months and can then be potted. It is very important that they be gradually hardened-off during removal from the rooting bed to outdoor conditions.

Grafting and Budding   Hollies are easily grafted, the cleft, whip, and side grafts being used. The operation is best performed during the dormant season for field grafting. Grafting is often done on greenhouse potted stock. T-budding is also suitable, being done in late summer or early spring. *Ilex opaca* is a satisfactory stock for its own cultivars and those of *I. aquifolium,* but probably the best stocks for English holly are *I. aquifolium* and *I. cornuta* 'Burfordii.' The latter has also been found to be a good rootstock for cultivars of *I. opaca (43).*

Air Layering   This is successful for a number of *Ilex* species. Layers are best started in early summer; after 10 to 14 weeks, plants 1 to 2 ft high are produced.

HONEYLOCUST, COMMON (*Gleditsia triacanthos* L.)   This is readily propagated either by seeds planted in the spring or by cuttings. The thornless honey locust, *G. triacanthos,* var. *inermis,* and the thornless and fruitless patented 'Moraine' locust are usually propagated by grafting on seedlings of the thorny type. In seed propagation, soaking the seed in sulfuric acid for one hour, followed by stratification at 36° F (2° C) for three months, gives good germination. Hardwood cuttings planted in the spring root successfully.

HONEYSUCKLE (*Lonicera* sp.)   Seeds show considerable variation in their dormancy conditions, some species having both seed coat and embryo dormancy, some only embryo dormancy, and some no dormancy. This variability also occurs among different lots of seeds of the same species. In *L. tatarica* some lots have no seed dormancy. In general, however, for prompt germination, stratification for two to three months at about 40° F (4° C) is recommended. Seeds of *L. hirsuta* and *L. oblongifolia* should have two months of warm stratification (70° to 85° F; 21° to 30° C), followed by two to three months stratification at about 40° F (4° C).

Most honeysuckle species are propagated easily by either hardwood cuttings in the spring or leafy softwood cuttings of summer growth under glass. Layering of vine types, such as 'Hall's' honeysuckle, is very easy, roots forming wherever the canes touch moist ground.

HORSE CHESTNUT   See Buckeye.

HYDRANGEA (*Hydrangea* sp.) *(79)*   This is easily rooted by leafy, softwood cuttings taken in midspring; they do well under mist and respond markedly to indolebutyric acid treatments. Leaf-bud cuttings can be used if propagating material is scarce. Hardwood cuttings planted in early spring are often used in propagating *H. paniculata* 'Grandiflora.'

HYPERICUM sp.   *See* St. Johnswort.

INCENSE CEDAR *(Calocedrus decurrens* [Torr.] Florin)   This is propagated by seed, which requires a stratification period of about eight weeks at 32° to 40° F (0° to 4° C) for good germination.

JACARANDA *(Jacaranda acutifolia* Humb. & Bonpl.) *(119)*   This is easily propagated by seed taken from capsules after blooming. Vegetative propagation is not used.

JASMINE *(Jasminum* sp.)   This is propagated without difficulty by leafy semi-hardwood cuttings taken in late summer and rooted under glass. Layers and suckers can also be used.

JUNIPER *(Juniperus* sp.) *(56, 57, 108)*   The junipers are generally propagated by cuttings, but in some cases difficult-to-root types are grafted on seedlings. The low-growing, prostrate forms are easily layered.

Seeds   Seedlings of the red cedar, *Juniperus virginiana,* or of *J. chinensis,* are ordinarily used as stocks for grafting ornamental clones.

Seeds should be gathered in the fall as soon as the berry-like cones become ripe. For best germination, seeds should be removed from the fruits, then treated with sulfuric acid for 30 minutes before being stratified for about four months at 40° F (4° C). Rather than the acid treatment, two to three months of warm (70° to 85° F; 21° to 30° C) stratification or summer planting could be used. As an alternative for cold stratification, the seed may be sown in the fall. Germination is delayed at temperatures above 60° F (15° C). Viability of the seeds varies considerably from year to year and among different lots, but it never is much over 50 percent. Planting of treated seed is usually done in the spring, either in ourdoor beds or in flats in the greenhouse. Two or three years are required to produce plants large enough to graft.

Cuttings   The spreading, prostrate types of junipers are more easily rooted than upright kinds. Cuttings are made 2 to 6 in. long from new lateral growth tips stripped off older branches. A small piece of old wood—a heel—is thus left attached to the base of the cutting. Some propagators believe this to be advantageous. In other cases, good results are obtained when the cuttings are just clipped without the heel from the older wood. Terminal growth of current season's wood also roots well.

Cuttings to be rooted in the greenhouse can be taken at any time during the winter *(69, 90)*. Exposing the stock plants to subfreezing temperatures seems to give better rooting. For propagating in an outdoor cold frame, cuttings are taken in late summer or early fall. Lightly wounding the base of the cuttings is sometimes helpful, and the use of root-promoting chemicals, especially indolebutyric acid is also beneficial. A medium-coarse sand or a 1:1 mixture of perlite and peat moss is a satisfactory rooting medium. A greenhouse temperature of about 60° F (15° C) is best for the first four to six weeks. Maintenance of a humid environment without excessive wetting of the cuttings is desirable, as is a relatively high light intensity. A light, intermittent misting can be used. Bottom heat of about 80° F (27° C) will aid rooting.

Grafting *(96)*   Vigorous seedling understocks with straight trunks, about pencil size, are dug in the fall from the seedling bed and potted in small pots set in peat moss in a cool, dry greenhouse. Seedlings potted earlier—in the spring—may also be used. After about 30 days, the greenhouse is heated and the plants kept well watered. This stimulates growth activity so that after one or two weeks the plants resume root activity and are in a suitable condition for grafting.

The scions should be selected from current season's growth taken from vigorous, healthy plants and preferably of the same diameter as the stock to be grafted. Scion material can be stored at 30° to 40° F (−1° to 4° C) for several weeks until used if kept in a saturated atmosphere.

Side-veneer or side graft methods are ordinarily used. The unions are best tied with budding rubber strips. The grafted plants are set deep enough in a greenhouse bench filled with peat moss to keep the union covered. The temperature around the graft union should be kept as constant as possible at 75° F (24° C) with a relative humidity of 85 percent or more around the tops of the plants. A lightly shaded greenhouse should be used to avoid burning the grafts. Adequate healing will take place in two to eight weeks, after which the temperature and humidity can be lowered. The stock plant is then cut off above the graft union, allowing the scion to develop.

KOELREUTERIA sp.   *See* Goldenrain Tree.

LAGERSTROEMIA INDICA   *See* Crape Myrtle.

LARCH (*Larix* sp.) (*17, 50*)   Most of these deciduous conifers are propagated easily by fall-planted seeds. Cones should be collected prior to drying and opening on the tree. Several species have empty or improperly developed seed. Seeds of some species have a slight embryo dormancy, so for spring planting, stratification for one month at about 40° F (4° C) is advisable. Cutting propagation is best done by rooting leafy-tip softwood cuttings in late summer under mist. Sand is a suitable rooting medium, and indolebutyric acid at 0.8 percent in talc promotes rooting. Cutting material should be taken from young trees only.

LIBOCEDRUS DECURRENS   *See* Incense Cedar.

LIGUSTRUM sp.   *See* Privet.

LILAC, FRENCH HYBRID (*Syringa vulgaris* cults.) (*20, 24, 31, 64*)   Grafting or budding on California privet (*Ligustrum ovalifolium*) or Amur privet (*L. amurense*) cuttings, or on lilac or green ash seedlings, is used commercially in lilac propagation. Cuttings can be rooted, however, if attention is given to proper timing. Layering or division of old plants is quite satisfactory when only a few new plants are needed.

Seeds   Seedlings are used mostly as an understock for grafting. Lilac cultivars will not reproduce true from seed. Seeds require fall-planting out-of-doors or a stratification period of 40 to 60 days at about 40° F (4° C) for good germination.

Cuttings   Ordinarily, good rooting of lilacs can be obtained only with terminal leafy cuttings taken within a narrow period shortly after growth commences in the spring. When the new, green shoots have reached a length of 4 to 6 in., they should be cut off and trimmed into cuttings. Since they are very succulent at this stage, it is difficult to prevent wilting. In a mist propagating bed rooting takes place in three to six weeks; a well-drained rooting medium must be used. Rooting can also be obtained in a polyethylene-covered bed in the greenhouse with bottom heat. Sprays with Captan (2 tsp per gal. of water) will help avoid fungus attack. The use of indolebutyric acid at about 80 ppm for 24 hours has given good rootings, as has dipping in an 0.8 percent IBA in talc preparation. Cuttings can be rooted quite well in out-of-door mist beds (*59*).

Grafting and Budding   Owing to the difficulty in rooting lilac cuttings and the fact that they must be taken at a definite time in the spring, often at the peak of the nurseryman's busy season, many propagators practice grafting.

When privet or green ash is used as the understock, the lilac may show incompatibility symptoms, but if the grafts are planted deeply, scion roots rapidly develop from the lilac and soon become the predominant root system of the plant. *Syringa vulgaris* seedlings are sometimes used as the stock, but if suckers arise from this stock, there is difficulty in distinguishing them from the selected hybrid lilac top. The plant thus becomes a mixture of growth from the scion and the rootstock. Should privet understock produce suckers, however, they can be easily identified and removed. It is best in any case, to plant lilacs "on their own roots," either as rooted cuttings or with the privet or lilac "nurse root" already removed.

Grafting is done during the winter season, using rootstock plants which have been dug and brought inside. Vigorous, one-year-old scion wood is used, taken from plants that have been heavily pruned and well fertilized to induce such growth. Different grafting methods can be used, such as the side or the whip graft. Cleft grafting on pieces of privet root is also practiced and tends to eliminate subsequent suckering from the rootstock.

T-Budding   T-budding in late summer or early fall is sometimes practiced, the lilac buds being inserted below ground on one-year-old privet cuttings or seedlings. The following spring the privet stock is cut off above the bud and soil mounted around the shoot as it develops so as to encourage subsequent rooting from the lilac. Unless this is done, the plant will be short-lived.

Layering   Simple layering of one-year shoots arising from the base of plants on their own roots provides an easy propagation method where only a few plants are needed. Air-layering of one- or two-year-old branches is also successful.

LINDEN *(Tilia americana)*   Basswood. The seeds have a dormant embryo and an impermeable seed coat which, in some species, is surrounded by a hard, tough pericarp. Such seeds are slow and difficult to germinate. Removing the pericarp, either mechanically or by soaking the seeds in concentrated nitric acid for one-half to two hours, rinsing thoroughly and drying, then soaking the seeds for about 15 minutes in concentrated sulfuric acid to etch the seed coat, followed by stratification for four months at 35° F (2° C), may give fairly good germination; otherwise, warm (60° to 80° F; 15° to 27° C) stratification for four to five months, followed by an equal period at 35° to 40° F (2° to 4° C) can be used. Collecting the seed from the tree just as the seed coats turn completely brown but before the seeds drop and the seed coats become hard and dry, followed by immediate planting, has given good germination *(7)*.

Suckers arising around the base of trees cut back to the ground have been successfuully mound layered, and softwood cuttings taken from stump sprouts have been rooted.

LIQUIDAMBAR *(Liquidambar styraciflua* L.)   American sweet gum. Propagation is usually by seeds which are collected in the fall but not allowed to dry out. Stratification for one to three months at about 40° F (4° C) is recommended to overcome seed dormancy. Selected clones are grafted onto *L. styraciflua* seedlings.

Leafy softwood cuttings of partially matured wood can be rooted under mist in midsummer. Treatments with naphthaleneacetic acid have been helpful.

LIRIODENDRON TULIPIFERA   *See* Tulip Tree.

LOCUST, BLACK *(Robinia pseudoacacia* L.)   This is readily propagated by seeds, which are soaked in concentrated sulfuric acid for one hour, followed by thorough rinsing in water, before planting.

LONICERA sp.   *See* Honeysuckle.

MADRONE, PACIFIC *(Arbutus menziesii* Pursh.)   This Pacific Coast evergreen tree is usually propagated by seeds which are stratified for three months at 35° to

40° F (2° to 4° C). Seedlings are started in flats, then transferred to pots. They are difficult to transplant and should be set in their permanent location when not over 18 in. tall. Propagation can also be done by cuttings, layering, and grafting.

MAGNOLIA *(Magnolia sp.) (2, 6, 19, 52)*   Seeds, cuttings, grafting, and layering are utilized in propagating magnolias. Magnolia nursery trees are difficult to transplant, so they should be set out from containers or balled and burlapped, but only in early spring.

Seeds   Magnolia seeds are gathered in the fall as soon as possible after the fruit is ripe, when the red seeds are visible all over the fruit. After cleaning, the seeds should either be sown immediately in the fall or—prior to spring planting—stratified for two to three months at about 40° F (4° C). Allowing the seeds to dry out at any time seems to be harmful. After sowing, the germination medium must not become dry. *M. grandiflora* seeds, and perhaps others, will lose their viability if stored through the winter at room temperature. If prolonged storage is necessary, the seeds should be held in sealed containers at 32° to 40° F (0° to 4° C).

Magnolia seedlings grow rapidly, and generally are large enough to graft by the end of the first season. Transplanting should be kept at a minimum, since this retards the plants.

Cuttings   Some species, such as *M. soulangeana* and *M. stellata,* are successfully propagated commercially by leafy softwood cuttings. These may be taken from late spring to late summer after terminal growth has stopped and the wood has become partly matured.

Excellent rooting can be obtained if cuttings are taken from very young plants, the bases wounded, treated with a root-promoting substance, then rooted in sand in outdoor mist beds. Under such conditions, rooting is rapid and in high percentages, and there is little trouble from diseases.

Leafy cuttings of *M. grandiflora,* taken from late spring to late summer, wounded, and treated with indolebutyric acid at 5000 to 20,000 ppm have rooted well with bottom heat (75° F; 24° C) and under intermittent mist *(35, 109)*.

To obtain survival of the rooted cuttings through the following winter, they should be rooted early enough in the season so that some resumption of growth will occur before fall.

Grafting   *Magnolia kobus* is probably the best rootstock for the oriental magnolias, whereas *M. acuminata* can be used as a stock for either oriental or American species. *Magnolia grandiflora* seedlings or rooted cuttings are used for *M. grandiflora* cultivars.

One-year-old seedlings are potted in early spring and then grafted while the understocks are in active growth during mid- to late-summer. Side or side-veneer grafts are satisfactory, with the union and scion waxed after grafting. Some propagators pot the seedlings in the fall, then bring them into the greenhouse and do the grafting in midwinter. The newly grafted plants may be set on open benches in the greenhouse or placed in closed propagating frames, where they stay for seven to ten days while the union is healing. Air is gradually given until after six weeks they can be removed from the case and the understock cut off above the union.

Layering   Simple or mound layering gives good results. One- or two-year-old shoots arising from the base of stock plants are started in spring, but often two seasons are required to produce well-rooted layers.

MAHONIA *(Mahonia sp.)*   Seeds of most species should not be allowed to dry out; they are stratified through the winter to obtain satisfactory germination. On the other hand, dry seeds of red mahonia *(M. haematocarpa)* will germinate promptly after planting in the spring. Cleaned seeds of fall-planted *M. aquifolium,* over-wintered outdoors, germinate in spring.

Leafy cuttings taken in summer and treated with a high concentration of indolebutyric acid (20,000 ppm for 10 seconds), then rooted under intermittent mist, have given good results with *M. bealei (34).*

MALUS sp.   *See* Crabapple, Flowering.

MANZANITA (Arctostaphylos sp.)   Seed propagation can be done, but with difficulty. Soaking the cleaned fruits in sulfuric acid up to 24 hours before planting will aid germination. Cuttings can be rooted if taken from fall to early spring. Stock material should be sterilized with a weak Clorox solution. Cuttings are rooted under high humidity in polyethylene-covered frames. Vermiculite-perlite (1:1) is a suitable rooting medium, and root-promoting growth regulators are helpful *(55).*

MAPLE (Acer sp.) *(5, 13, 82, 92, 121)*   Various methods of propagation are in use—seeds, grafting, budding, cutting, and layering.

Seeds   Most maples ripen their seed in the fall, but two species—*A. rubrum* and *A. saccharinum*—produce seed in the spring. Such spring-ripening seeds should be gathered promptly when mature and sowed immediately without drying. For other species, stratification, usually for 90 days at 40° F (4° C), followed by spring planting, gives good germination. Fall planting out-of-doors may be done if the seeds are first soaked for a week, changing the water daily. Seeds of the Japanese maple, *A. palmatum,* germinate satisfactorily if they are placed in warm water (about 110° F; 43° C) and allowed to soak for two days, followed by stratification. Soaking seeds of *A. rubrum* and *A. negundo* in cold running water for five days and two weeks, respectively, before planting may increase germination. *Acer* seeds should not be allowed to dry out.

Cultivars of some maples, such as *A. palmatum* 'Atropurpureum,' will reproduce fairly true from seed, especially if the stock plants are isolated. The few off-type plants can be removed from the nursery row.

Cuttings   Leafy Japanese maple cuttings, as well as those of other Asiatic maples *(22),* will root readily in a sand–peat moss medium if they are made from tips of vigorous shoots in late spring and placed under mist. Wounding and relatively strong applications of indolebutyric acid are helpful. IBA at 1000 to 10,000 ppm. applied by the concentrated-dip method, has given good results in rooting leafy cuttings of various *Acer* species under mist in the greenhouse. It is often a problem, however, to successfully overwinter rooted maple cuttings.

Sugar maple *(A. saccharum)* cuttings are best taken in late spring after cessation of shoot elongation, then rooted under mist. Long, thick cuttings are preferred. IBA treatments give variable results. Overwintering is a problem but survival is best with large, vigorous, well-rooted cuttings. In cold climates it may be necessary to pot the cuttings in late summer, harden them off, then store until spring at about 34° F (1° C) *(32).*

Grafting and Budding   *Acer palmatum* seedlings are used as the understock for Japanese maple cultivars, *A. saccharum* for the sugar maples, *A. rubrum* for red maple cultivars, and *A. platanoides* for such Norway maple clones as 'Crimson King,' 'Schwedler,' and the pyramidal forms. There is some evidence of delayed incompatibility in using *A. saccharinum* as a stock for red maple cultivars.

Seedling rootstock plants are grown for one year in a seedbed, then in the fall or early spring are dug and transplanted into small pots in which they grow, plunged in propagating frames, through the second summer. In late winter, the stock plants are brought into the greenhouse preparatory to grafting. As soon as roots show signs of growth, the stock is ready for grafting. Dormant scions are taken from outdoor plants. The side graft or side-veneer graft (see pp. 378, 380) is ordinarily used. While the union is healing, the plants are set in a grafting case with peat moss covering the union. Sometimes grafting wax is used, covering both the scion and the graft union *(46)*.

Maples are also successfully T-budded on one-year seedlings in the nursery row, the buds being inserted from mid- to late summer. The wood is removed from the bud shield, which then consists only of the actual bud and attached bark. The seedling is cut back to the bud the following spring just as growth is starting.

METASEQUOIA (*Metasequoia glyptostroboides* Hu and Cheng)   Dawn redwood *(21)*. Seeds germinate without difficulty, and both softwood and hardwood cuttings root readily. The leafless hardwood cuttings may be lined-out in the nursery row in early spring. Leafy cuttings root easily under mist if taken in summer and treated with indolebutyric acid at 20,000 ppm by the concentrated-dip method *(36)*.

MOCK ORANGE (*Philadelphus* sp.)   The many cultivars of mock orange are best propagated by cuttings, which root easily. Hardwood cuttings planted in early spring or leafy softwood cuttings under glass in early summer can be used. Removing rooted suckers arising from the base of old plants is an easy means of obtaining a few new plants.

MULBERRY, FRUITLESS (*Morus alba* L.)   Some mulberry trees produce only male flowers and hence do not bear fruits. Propagated vegetatively, these are suitable as ornamental shade trees. Some are very rapid growers. They may be propagated by cuttings or by budding or grafting on mulberry seedling rootstocks. Leafy softwood cuttings taken in midsummer will root under mist.

MYRICA (*Myrica californica*)   Pacific wax myrtle. Bayberry *(M. pennsylvanica)*. These are propagated by seed. Remove all wax from the fruits but prevent drying. Seeds will germinate if planted out-of-doors in early fall.

MYRTLE (*Myrtus* sp.)   These are usually propagated in summer by leafy softwood cuttings of partially matured wood rooted under glass. Treatments with indolebutyric acid have been helpful in some instances.

NANDINA (*Nandina domestica* Thun.) *(3)*   Propagation is usually by seed, which should not be allowed to dry out. The embryos in the mature fruits are rudimentary, but will develop in cold storage. Seeds can be collected in late fall, held in dry storage at 40° F (4° C), then planted in late summer, germination starting in about 60 days. Germination tends to take place in autumn regardless of planting date. No low-temperature moist stratification period is necessary. Growth of the seedlings is slow, taking several years to produce salable plants. Suckers arising at the base of old plants may be removed for propagation.

NERIUM OLEANDER   *See* Oleander.

NORFOLK ISLAND PINE (*Araucaria heterophylla* [Sallisb.] Franco.)   This is propagated by seed. Cuttings of side branches will root but produce horizontally growing plants. Cuttings from terminals grow upright.

OAK (*Quercus* sp. *(106)*   Seed propagation is generally practiced. Wide variations exist in the germination requirements of oak seed, particularly between the black oak (acorns maturing the second year) and white oak (acorns maturing the first year) groups. Seeds of the white oak group have little or no dormancy and, with few exceptions, are ready to germinate as soon as they mature in the fall. Seeds of most species of the black oak group have embryo dormancy, requiring either stratification (32° to 35° F; 0° to 2° C) for one to three months of fall planting. Seeds of the following species will germinate without a low-temperature stratification period: *Quercus agrifolia, Q. alba, Q. arizonica, Q. bicolor, Q. chrysolepis,*

Q. douglasii, Q. turbinella, Q. garryana, Q. lobata, Q. macrocarpa, Q. montana, Q. petraea, Q. prinus, Q. robur, Q. stellata, Q. suber, and *Q. virginiana.*

Acorns are often attacked by weevils. Soaking in water held at 120° F (49° C) for 30 minutes will rid the acorns of this damaging pest.

Acorns of many species tend to lose their viability rapidly when stored dry at room temperature. They should be held under cold, moist conditions or stored dry in sealed containers at 32° to 36° F (0° to 2° C).

To obtain lateral root branching—which makes the seedlings more adaptable to transplanting—the acorns can be planted in a box which has a copper wire screen mesh about 6 in. below the seeds. The tip of the tap root, upon contacting this, will be killed, forcing development of many lateral roots *(103).*

Bench grafting of potted seedling stocks in the greenhouse in late winter or early spring is moderately successful. Side grafting or the whip, or tongue, method is ordinarily used, with dormant one-year-old wood for scions. Crown grafting seedlings in place in the nursery row is also practiced occasionally. This is done in the spring just after the stock plants start to leaf out. Scions are taken from wood gathered when dormant and stored under cool, moist conditions until used. Various grafting methods are satisfactory—whip, cleft, or bark. Budding has generally been unsatisfactory. In grafting, only seedlings of the black oak group should be used for scion cultivars of the same group and, in the same manner, only seedlings of the white oak group should be used as stocks for other members of the white oaks. The use of seedlings of the same species is preferable. Although some distantly related species of the oak will unite satisfactorily, incompatibility symptoms usually appear later.

Attempts to propagate oaks by cuttings or layering have usually been unsatisfactory, although some success has been obtained in rooting leafy softwood cuttings of *Quercus robur* 'Fastigata' under outdoor mist in midsummer after treatment with IBA at 20,000 ppm *(38).*

OLEANDER *(Nerium oleander* L.)   Seedlings reproduce fairly true-to-type, although a small percentage of plants with different flower colors will appear. The seeds should be collected in late fall after a frost has caused the seed pods to open. Rubbing the seeds through a coarse mesh wire screen will remove most of the fuzzy coating. The seeds are then planted immediately in the greenhouse in flats without further treatment. Germination occurs in about two weeks. Cuttings root easily under glass if taken from rather mature wood during the summer. Simple layering is also successful. Plant parts are very poisonous.

OLIVE *(Olea europaea* L.)   A fruitless cultivar, 'Swan Hill,' is available for use as a patio or street tree. Cuttings can be rooted under mist if treated with IBA at 2000 ppm or it can be grafted on easily rooted cultivars used as a rootstock. Olives will not survive outdoor winter temperatures below 15° F (−9° C). *(49).*

PAEONIA SUFFRUTICOSA   See Peony, Tree.

PALMS *(63, 77, 128)*   There are several thousand species, in many genera, of ornamental palms. They are propagated by seed, which should be planted as soon as possible after harvesting; they should be prevented from drying. If dry, soaking the seeds in water for 24 hours before planting is helpful. Palm seeds are susceptible to surface molds and should be protected by dusting with a fungicide. Any seeds which float in water should be discarded. A mixture of one-half peat moss and one-half perlite is a good germination medium. Seeds of most species will germinate in one to three months, especially if bottom heat (80° F; 28° C) is maintained, but some may take as long as one to two years.

PARTHENOCISSUS   Virginia Creeper (*P. quinquefolia* Planch.). Boston Ivy (*P. tricuspidata* Planch.). These two ornamental vines can be propagated by seeds planted in the fall or stratified for two months at about 40° F (4° C) before planting in the spring.

Softwood cuttings taken in late summer root easily under glass, as do hardwood cuttings planted outdoors in early spring. Plants may be started by compound layering.

PEAR, 'BRADFORD' (*Pyrus calleryana* Dcne. 'Bradford')   This ornamental pear introduced by the USDA can be propagated by T-budding or root grafting onto *P. calleryana* seedlings. It is not compatible with *P. communis* roots. Leafy cuttings, if treated with IBA, can be rooted under mist, with peat moss, perlite, vermiculite (1:1:1) giving good results as a rooting medium *(1)*. Cuttings are best taken in late summer.

PEONY, TREE   Moutan. (*Paeonia suffruticosa* Andr.) *(54, 126)*. Seed propagation is complicated by "epicotyl dormancy." The seeds should be planted in a moist medium and, after roots have developed, transplanted to pots of soil which are placed in a cold room (40° to 50° F; 4° to 10° C) or outdoors (in winter) for 2½ months. This overcomes dormancy conditions in the shoot tip, which then grows readily upon transfer of the plants to warm (or spring) temperatures. Selected clones do not come true from seed, however. Such cultivars are propagated by grafting in late summer on herbaceous peony *(P. lactiflora)* roots as the understock. The grafts are callused in a sand-peat medium in a greenhouse until fall when they are potted. They should finally be planted deeply so scion roots will form.

PHILADELPHUS sp.   *See* Mock Orange.

PICEA sp.   *See* Spruce.

PINE (*Pinus* sp.) *(85, 93, 112)*   These are ordinarily propagated by seeds. Considerable variability exists among the species in regard to seed dormancy conditions. Seeds of many species have no dormany and will germinate immediately upon collection, whereas others have embryo dormancy. With the latter, stratification at 32° to 40° F (0° to 4° C) for one to three months will increase or hasten germination. Seed coat dormancy also seems to be present in *P. cembra* and *P. monticola*. With these, concentrated sulfuric acid treatment for three to five hours and for 45 minutes, respectively, followed by stratification for three months at 36° F (2° C) will aid germination. Species which have no dormancy conditions and can be planted without treatment include *Pinus aristata, P. banksiana, P. canariensis, P. caribaea, P. clausa, P. contorta, P. coulteri, P. edulis, P. wallichiana, P. halepensis, P. jeffreyi, P. latifolia, P. mugo, P. nigra, P. palutris, P. pinaster, P. ponderosa, P. pungens, P. radiata, P. resinosa, P. roxburghi, P. sylvestris, P. thunbergi,* and *P. virginiana.* However, if seeds of the above species have been stored for any length of time, it would be advisable to give them a cold stratification period before planting as described above. Pine seeds can be stored for considerable periods of time without losing viability if held in sealed containers between 5° and 32° F (−15° and 0° C). They should not be allowed to dry out.

Pinus cuttings are difficult to root, although those of Mugho pine *(Pinus mugho)* root easily if taken in early summer, and *P. radiata* roots well if cuttings are taken from young trees *(14)*. Success is more likely if cutting material is taken from low-growing lateral shoots on young trees in winter. Treatment with indolebutyric acid (see p. 291) is beneficial. Considerable study has been given to the rooting of cuttings of Monterey pine *(Pinus radiata)* because of its importance as a lumber and Christmas tree species. Best rooting was from cuttings taken in early winter.

Wounding, plus a concentrated-dip treatment of IBA at 4000 ppm, was beneficial. A more symmetrical root system could be induced by clipping the ends of the original roots and allowing the root system to develop from the secondary roots. Outdoor rooting under intermittent mist has been successful using IBA at 5 percent, with peat moss, redwood sawdust and *P. radiata* litter (1:1:1) as the rooting medium *(58, 74)*.

Pines can be propagated asexually by rooting needle fascicles (needle leaves held together by the scale leaves, containing a base and a diminutive shoot apex). Rooting is best when the fascicles are taken from trees less than four years old. IBA treatments are helpful *(78)*. By selecting certain seedlings whose cuttings root easily and by using critical timing in taking cuttings, it is possible to set up clones where cutting propagation is commercially feasible. This was shown to be true for the Mugho pine *(45, 100)*.

Side-veneer grafting is used for propagating selected clones; well-established two-year-old seedlings of the same or closely related species should be used as rootstocks. Scions should be of new growth, taken from firm, partly matured wood.

PISTACHE, CHINESE *(Pistacia chinensis* Bunge)   Seeds should be collected from relatively large fruits, blue-green in color, in midfall. Pulp must be removed. Soak fruits in water, then rub over a screen. Seeds that float in water have aborted and should be discarded. Stratification at 40° to 50° F (4° to 10° C) for ten weeks gives good germination. Seedlings exhibit a wide range of variability. T-budding selected clones on seedling *P. chinensis* rootstocks in late summer is used to produce uniform, superior trees. Propagation by cuttings is very difficult. Pistache nursery trees do not transplant well if the roots become exposed, so they are usually handled as container-grown nursery stock *(62)*.

PITTOSPORUM *(Pittosporum* sp.)   These are started by seeds or cuttings. The seeds are not difficult to germinate; dipping a cloth bag containing the seeds for several seconds in boiling water may hasten germination. Leafy semi-hardwood cuttings taken after a flush of growth has partially matured will root readily, particularly under mist. Indolebutyric acid treatments are also beneficial.

PLANE TREE *(Platanus* sp.)   Sycamore. Seeds are ordinarily used in propagation but they should not be allowed to dry out. The best procedure is to allow the seeds to overwinter in the seed balls right on the tree. They may then be collected in late winter or early spring and planted immediately, with prompt germination usually occurring. If the seeds are collected in the fall, then stratification over winter at about 40° F (4° C) should be used.

PLATANUS sp.   *See* Plane Tree.

PLUMBAGO *(Plumbago* sp.)   Seeds sown in late winter usually germinate easily. Leafy cuttings taken from partially matured wood can be rooted without difficulty under glass. Root cuttings can also be used, and old plants can be divided.

PLUMERIA *(Plumeria* sp.)   Leafy cuttings 6 to 8 in. long of this tender shrub widely grown in Hawaii will root readily under mist if treated with 2500 ppm indolebutyric acid.

PODOCARPUS *(Podocarpus* sp.)   These evergreen trees and shrubs, having foliage resembling the related yews *(Taxus)*, make good container plants. They can be propagated by stem cuttings taken in late summer.

POINSETTIA *(Euphorbia pulcherrima* Willd.) *(105)*   Propagation by leafy cuttings under mist is the usual procedure. Low-strength root-promoting chemicals are helpful. Stock plants making moderate growth should be used as a source of cuttings. Disease control is very important, with attention being given to using sanitary procedures throughout the rooting operation. It is best to root cuttings in containers so roots will not be disturbed. Cuttings can be rooted in the greenhouse from spring to fall. Leaves are very poisonous.

POPLAR (*Populus* sp.)   Cottonwood. Aspen. These can be propagated by seeds; they should be collected as soon as the capsules begin to open, and planted at once, because they lose viability rapidly, and should not be allowed to dry out. However, if held in sealed containers near 32° F (0° C), seeds of some species can be stored for as long as three years. There are no dormancy conditions, and seeds germinate within a few days after planting. The seedlings are highly susceptible to damping-off fungi and will not tolerate excessive heat or drying. Poplars are difficult to propagate in quantity by seed.

Hardwood cuttings of *Populus* (except the aspens) planted in the spring root easily. Treatments with indolebutyric acid are likely to improve rooting. Leafy softwood cuttings (of some species at least) taken in midsummer will also root well.

QUAKING ASPEN (*P. tremuloides*)   This can be propagated by removing root pieces, inducing adventitious shoots to form from them in vermiculite, then rooting these shoots as stem cuttings under mist with IBA treatments.

POPULUS sp.   *See* Poplar.

PRIVET (*Ligustrum* sp.)   Seed propagation is easily done. The cleaned seed should be stratified for two to three months at 32° to 50° F (0° to 10° C) before planting. Hardwood cuttings of most species planted in the spring root easily, as do softwood cuttings in summer under glass. Japanese privet (*L. japonicum*) is somewhat difficult to start from cuttings, the best results being obtained with actively growing terminal growth rather than more mature wood.

PRUNUS CAMPANULATA, P. SARGENTII, P. SERRULATA, P. SIEBOLDI, P. SUB-HIRTELLA, P. YEDOENSIS   *See* Cherry, Flowering.

PYRACANTHA (*Pyracantha* sp.)   Firethorn. Propagation is almost always by cuttings. Partially matured, leafy current season's growth taken from late spring to late fall and rooted either in the greenhouse under glass or under mist gives good results. Treatments with root-promoting substances (see p. 291) are beneficial. Large (18-in.) leafy cuttings, taken in early spring, wounded at the base, treated with indolebutyric acid, and rooted under mist produced salable nursery plants—potted in one-gallon containers—in six months (*113*).

PYRUS CALLERYANA   *See* Pear, 'Bradford.'

QUERCUS sp.   *See* Oak.

QUINCE, FLOWERING (*Chaenomeles* sp.)   These are easily started by seeds, which should be fall-planted or stratified for two or three months at 40° F (4° C) before sowing. Clean the seeds from the fruit. Root cuttings can be taken in late fall, cut into 2- to 4-in. lengths, and stored at 35° to 40° F (2° to 4° C) until spring, when they can be lined-out horizontally in the nursery row. Leafy cuttings of partially matured wood may be rooted under mist in late spring. Treatment with indolebutyric acid at about 15 ppm for 24 hours is beneficial. Older plants tend to produce suckers freely at the base; these suckers may be removed and used if they are well rooted.

REDBUD (*Cercis* sp.)   Seed propagation is successful, but seed treatments are necessary, owing to dormancy resulting from an impervious seed coat plus a dormant embryo. Probably the most satisfactory treatment is a 60-minute soaking period in concentrated sulfuric acid, followed by stratification for three months at

35° to 40° F (2° to 4° C). Fall sowing outdoors of untreated seeds may also give good germination.

Softwood cuttings of some *Cercis* species root readily under glass if taken in spring or early summer. Simple layering is also used successfully. T-budding in midsummer on *C. canadensis* seedlings is used commercially for *Cercis* cultivars.

REDWOOD   Coast Redwood. (*Sequoia sempervirens* [D. Don] Endl.). Giant sequoia (*Sequoiadendron giganteum* [Lindl.] Buchh.) or Sierra Redwood. Both species are ordinarily propagated by seed. Seeds of *S. sempervirens* are mature at the end of the first season, but those of *S. giganteum* require two seasons for maturity of the embryos. Cones are collected in the fall and allowed to dry for two to four weeks, after which the seeds can be separated. Seeds may be kept for several years in sealed containers under 40° F (4° C) storage without losing viability. Stratification for ten weeks before planting at about 40° F (4° C) promotes germination of *S. giganteum* seed. Seeds of *S. sempervirens* will germinate without a stratification treatment.

Fall planting may also be done, sowing the seed about ⅛ in. deep in a well-prepared seedbed. A burlap cover on the seedbed, which is removed when germination starts, is helpful. The young seedlings should be given partial shade for the first 60 days.

S. sempervirens may also be propagated by leafy cuttings taken from sprouts arising from the burls around the bases of older trees. Young *S. giganteum* trees are grown for Christmas trees.

REDWOOD, DAWN   *See* Metasequoia

RHAMNUS sp.   *See* Buckthorn.

RHODODENDRON (*Rhododendron* sp.) *(9, 70, 71, 114, 115)*   These can be propagated by seeds, cuttings, grafting, and layering.

Seeds   Seedlings may be used as rootstocks for grafting or for the propagation of the ornamental species. *Rhododendron ponticum* is the principal rootstock for grafting. The seed should be collected just when the capsules are beginning to dehisce, and may be stored dry and planted in late winter or early spring in the greenhouse. Seed to be kept for long periods should be put in sealed bottles and held at about 40° F (4° C). A good germination medium is a layer of shredded sphagnum moss or vermiculite over a mixture of sand and acid peat. The very small seeds are sifted on the surface of the medium and watered with a fine spray. The flats should be covered with glass and always kept shaded. Careful attention must be given to provide adequate moisture and ventilation as well as even heat: 60° to 70° F (15° to 21° C). The plants grow slowly, taking about three months to reach transplanting size. After two or three true leaves form, they are moved to another flat and spaced 1 to 2 in. apart, where they remain through the winter in a cool greenhouse or in cold frames. In the spring the plants are set out in the field in an acid soil, and by fall they are ready to be dug and potted preparatory to grafting in the winter.

Cuttings *(114)*   Rooting cuttings is the chief method of rhododendron propagation. These are best taken, midsummer into fall, from stock plants in full sun grown especially for this purpose. However, stem cuttings—or leaf-bud cuttings—of some hybrids taken in midwinter will root well. Any flower buds should be removed from the cuttings. Treatments with indolebutyric acid at relatively high concentrations are required; IBA in talc at 1 or 2 percent concentration, plus an added fungicide (benomyl) works well. Wounding the base of the cutting on both sides is also beneficial. A rooting medium of two-third sphagnum peat moss and

Figure 18–1   Propagation of rhododendron by leaf-bud cuttings. *Left:* cutting when made. *Center:* root development after several weeks. *Right:* appearance of root ball and new shoot after five months. Courtesy H. T. Skinner.

one-third perlite is suitable. Bottom heat at 75° F (24° C) should be used. Rhododendron cuttings are best rooted under mist in the greenhouse and should be lifted soon after roots are well formed (about three months) or the roots will deteriorate. After rooting and transplanting (into German peat moss, with added fertilizers) the cuttings should be held at 40° F (4° C) for about 20 days, after which the night temperature can be raised to a minimum of 65° F (18° C). Supplementary light at this stage to extend the day length will give good growth response. Plants started from cuttings usually develop rapidly and are free of the disadvantage of suckering from the rootstock, which occurs with grafted plants.

Grafting   A side-veneer graft is most successful. The best scionwood is taken from straight, vigorous current season's growth. After grafting, the plants are kept in closed frames under high humidity and a temperature near 70° F (21° C) until the union has healed. The plants should then be moved to cooler conditions—50° to 60° F (10° to 15° C)—and the top of the stock removed above the graft union. After the plant has hardened, it is transplanted to the nursery row in acid soil and grown for two years, after which it is ready to dig as a salable nursery plant.

Layering   Rhododendrons are easily reproduced by trench and simple layering.

All parts of rhododendron and azalea plants are poisonous, often fatal, if eaten.

RHUS sp.   *See* Sumac.

ROBINIA PSEUDOACACIA   *See* Locust, Black.

ROSE (*Rosa* sp.) *(12, 26, 33)*   All the selected rose cultivars are propagated by asexual methods. T-budding on vigorous rootstocks is most common, although the use of softwood or hardwood cuttings or grafting, layering, or the use of suckers is sometimes practiced. Seed propagation is used in breeding new cultivars, in producing plants in large numbers for conservation projects or mass landscaping, and in growing rootstock plants of certain species, such as *R. canina*.

Seeds (15, 89, 117)   As soon as the rose fruits ("hips") are ripe but before the flesh starts to soften, they should be collected and the seeds extracted. It is best to stratify them immediately at 35° to 40° F (2° to 4° C). Six weeks is sufficient for *Rosa multiflora,* but others—*R. rugosa* and *R. hugonis*—require four to six months, and *R. blanda* ten months *(104). Rosa canina* germinates best if the seeds are held at room temperature for two months in moist vermiculite and then transferred to 32° F (0° C) for an additional two months. Hybrid rose seeds usually respond best to a stratification temperature of 34° to 40° F (1° to 40° C) for 60 to 90 days, although some seeds may germinate with no cold stratification treatment. Germination is probably prevented in rose seeds by inhibitors occurring in the seed coverings, as well as by the mechanical restriction imposed by the massive pericarp (fruit wall) *(61).* The seeds may be planted either in the spring or in the fall in seedbeds or in the nursery row. In areas of severe winters, seedlings are likely to be winter-killed if they are smaller than 4 in. by the onset of cold weather.

Cuttings. Hardwood Cuttings   Hardwood cuttings are widely used commercially in the propagation of rose rootstocks, and to some extent in propagating the strong growing polyanthas, pillars, climbers, and hybrid perpetuals. The hybrid teas and other similar everblooming roses can also be started by cuttings, but more winter-hardy and nematode-resistant plants are produced if they are budded on selected vigorous rootstocks. In mild climates, the cuttings are taken and planted in the nursery in the fall. In areas with severe winters, cuttings may be made in late fall or early winter, tied in bundles, and stored in damp peat moss or sand at about 40° F (4° C) until spring, when they are planted in the nursery row. The rootstocks are ready to bud by the following spring, summer, or fall. The cuttings are made into 6- or 8-in. lengths from previous season's canes of ¼ to ⅜ in. diameter. In commercial practice, large bundles of canes are run through band saws to cut them to the correct length. Disbudding is usually done in rootstock propagation; all buds except the top one or two are removed so as to prevent subsequent sucker growth in the nursery row.

Softwood Cuttings   Softwood cuttings are made from current season's growth, from early spring to late summer, depending upon the time the wood becomes partially mature. Rooting is fairly rapid, occurring in 10 to 14 days. At the end of the season the cuttings may be transplanted to their permanent location, potted and over-wintered in a cold frame, or transferred to the nursery row for another season's growth or to be budded to the desired cultivar.

Cultivars of miniature roses are easily propagated by cuttings of softwood or semi-hardwood under mist. In mild climates, rooting can be done in outdoor beds from early spring to fall. A 1:1 perlite–peatmoss mixture is a good rooting medium for these types of roses *(88).*

Budding   T-budding is the method ordinarily used. The buds are inserted into 3/16- to 3/8-in. diameter rootstock plants. In mild climates budding can be done during a long period, from late winter until fall, but mostly in the spring. Early buds will make some growth during the summer and produce a salable plant by fall. Some propagators break over the top of the rootstock about two weeks after budding to force the bud out. After the bud has reached a length of 4 to 8 in., the top of the stock is entirely removed. In areas with shorter growing seasons, budding is done during the summer. Buds inserted late in the summer either make little growth or remain dormant until the following spring. In this case, the rootstock is cut off just above the bud in late winter or early spring, forcing the inserted bud into growth. Shoots from buds which started in the fall are cut back to ½ in. in the spring. The shoot then grows through the following entire summer, producing a well-developed plant by fall. After the shoot has grown about 6 in., it is generally cut back to 2 or 3 in. to force out side branches.

Budwood may be obtained during the budding season from current season's growth of the desired cultivar, only a day's supply being taken at a time. It is best collected early in the morning, clipping off the leaves immediately and leaving about ¼ in. of the petiole attached to the bud. Lateral buds from the stems producing the flowers are the best to use. Plump, but dormant, buds three or four nodes below the flower are the most desirable. The wood should be at a stage of maturity in which the thorns are easily removed. It is important to use budwood taken only from plants free of *Verticillium*. Buds from diseased wood can infect each budded plant with this fungus *(98)*.

An alternative method for obtaining budwood, which has been widely and successfully used, is to store dormant wood under refrigeration just below freezing (30° to 32° F ; —1° to 0° C) until time for budding. The budwood is collected in late fall after the flowers are shed and the thorns become dark. The leaves are removed by hand, but the thorns are left intact. Sticks 10 to 15 in. long are put up in bundles of 30 or 40 each. The bundles are wrapped tightly as possible in waterproof paper over which a layer of moist wrapping—such as wet newspapers—is placed. Finally, the bundles are covered with another layer of waterproof paper.

Buds are inserted in the wood of the original cuttings rather than into new growth arising from the rootstock.

Rootstocks for Rose Cultivars *(11)*   Most rose rootstock clones have been in use for many years, propagated by cuttings; many of the clones are virus-infected, thus infecting the cultivar top after budding. However, these clonal rootstocks are available with the viruses eliminated by heat treatments. Holding potted rose plants at a dry heat of 98° to 100° F (37° to 38° C) for four to five weeks will rid infected stocks of the virus.

Rosa multiflora   This is a useful rootstock, especially in its thornless forms, for outdoor roses. Several "strains" have been developed, some giving better bud unions and bud development than others *(99)*. Cuttings of this species root readily, develop a vigorous, nematode-resistant root system, and do not sucker excessively. It is adaptable to a wide range of soil and climatic conditions, but does not seem to do well in the southern U.S. Seedlings are used in the eastern part of the U.S. and cuttings on the Pacific Coast. The bark often becomes so thick late in the season that budding is impossible.

Rosa canina (Dog rose)   Although this has not done well under American conditions, it has been the most commonly used stock in Europe. It is usually propagated by seed, since the cuttings do not root easily; however, the seeds are difficult to germinate. The prominent thorns make it difficult to handle. It also tends to sucker. Young plants on this stock grow slowly, but they are long-lived. *Rosa canina* is adaptable to drought and alkaline soil conditions.

Ragged Robin (Rosa chinensis 'Gloire de Rosomanes')   This old French stock is popular in California for outdoor roses, resisting heat and dry conditions well. It is also resistant to nematodes and does not sucker if the lower buds on the cuttings are removed. This stock grows steadily through the summer, permitting budding at any time. The fibrous root system is easy to transplant but requires good soil drainage. In some areas, however, it is difficult to propagate and is injured by leafspot. Owing to its susceptibility to *Verticillium* wilt, it should not be planted on land previously planted to tomatoes or cotton.

Rosa × sp. *Dr. Huey (Shafter)*   This is the principal rootstock in Arizona and the southern San Joaquin Valley California rose districts, replacing 'Ragged Robin' to a large extent. It is useful for late season budding because of its thin bark. It is very vigorous and well adapted to irrigated conditions, and its cuttings root readily. It is very good as a stock for weak-growing cultivars. Defects are its injury from subzero temperatures and susceptibility to blackspot, mildew, and *Verticillium*.

Rosa × *noisettiana* ('Manetti')   This is an old stock, very popular for greenhouse forcing roses. It is also of value for dwarf roses and for planting in sandy soil types. It is easily propagated by cuttings, produces a plant of moderate vigor, and is resistant to some strains of *Verticillium* wilt *(98)*.

Rosa odorata (Odorata 22449)   This is an excellent stock for greenhouse forcing roses. It roots easily from cuttings under suitable conditions, and produces a large symmetrical root system. It is adapted to both excessively dry or wet soil conditions. Since it is not cold-hardy, it should be used only in areas with mild winters. Many clones of this stock are badly diseased and do not root well. The plants are not adaptable to cold-storage handling. It is more susceptible to *Verticillium* than *R. manetti.*

IXL (Tausendschon × *Veilchenblau)*   This stock is used primarily as a trunk for tree roses. It is very vigorous and has no thorns. The canes tend to sunburn and are somewhat susceptible to low-temperature injury.

Multiflore de la Grifferaie   This is useful as a trunk for tree roses, producing desirable straight canes. It is vigorous, extremely hardy, and resistant to borers, but very susceptible to mite injury.

Rosa rugosa   The form which is used as an understock bears purplish-red single flowers. For bush roses it is propagated by cuttings, and for tree roses by seed. The root system is shallow and fibrous and tends to sucker badly, but the plants are very long lived.

Propagation of Tree (Standard) Roses   A satisfactory method of producing this popular form of rose is to use *Rosa multiflora* as the rootstock, which is budded in the first summer to IXL or, preferably, the Grifferaie stock. These are trained to an upright form and kept free of suckers. In the second summer, at a height of about 3 ft, three or four buds of the desired flowering cultivar are inserted into the trunk stock. During the winter, the cane above the inserted buds is removed. The buds develop the following summer, as do buds from the stock, which must be removed. In the fall, the plants may be dug and moved to their permanent location. Tree roses are sometimes dug as balled and burlapped plants, owing to the extensive root system formed during the two years in which the top is being developed. (See Figure 10–30.)

ROSE OF SHARON   *See* Hibiscus syriacus.

RUSSIAN OLIVE   *See* Elaeagnus.

ST. JOHNSWORT (*Hypericum* sp.)   These are easily started by softwood cuttings taken in late summer from the tips of current growth and rooted under high humidity.

SALIX sp.   *See* Willow.

SAMBUCUS sp.   *See* Elder.

SEQUOIA SEMPERVIRENS   *See* Redwood, Coast.

SEQUOIADENDRON GIGANTEUM   *See* Redwood, Giant sequoia.

SERVICE-BERRY (*Amelanchier alnifolia* Nutt.) *(47)*   Seeds show embryo dormancy, which can be overcome by stratification at about 36° F (2° C) for three to six months. Seds should not be allowed to dry out. This is readily propagated by leafy softwood cuttings, taken when the new growth is several inches long, and rooted under mist. Indolebutyric acid at 0.3 percent in talc increases rooting. *Amelanchier* is also easily propagated by root cuttings.

SHRUB ALTHEA   *See* Hibiscus syriacus.

SILK TREE   *See* Albizzia.

SILVER BERRY   *See* Elaeagnus.

SMOKE TREE (*Cotinus coggygria* Scop.)   This can be propagated by leafy softwood cuttings under mist. Tip cuttings taken from new growth in the spring and treated with indolebutyric acid should root in about five weeks. The rooted cuttings are best left in place with the mist discontinued, and transplanted bareroot the following spring. The smoke tree should not be propagated by seeds, since many of the seedlings are staminate plants, thus lacking the showy flowering panicles. Only vegetative methods should be used, with the propagating wood being taken from plants known to produce large quantities of the desirable fruiting clusters.

SORBUS (*Sorbus* sp.) *(68)*   Seeds should be collected as soon as the fruits mature; fleshy parts are removed to eliminate inhibitors. Shallow planting in the fall should give germination the next spring. Cuttings do not root and layering is difficult. Either fall budding or bench grafting (whip graft) is successful. Selected cultivars are best worked on seedlings of their own species, although *S. aucuparia* (European mountain ash) seedlings seem satisfactory as a rootstock for other species.

SPIRAEA (*Spiraea* sp.)   This is usually propagated by cuttings, although some species, such as *S. thunbergi,* are more easily started by seeds, which should not be allowed to dry out. Leafy softwood cuttings taken in midsummer and rooted under high humidity are generally successful. Treatments with one of the root-promoting substances are often of considerable benefit. Some species, such as *S. vanhouttei,* can be started readily by hardwood cuttings planted in early spring.

SPRUCE (*Picea* sp.)   These are ordinarily propagated without difficulty by seeds, either fall-planted or stratified over winter. Most species have embryo dormancy, requiring one to three months' stratification at about 40° F (4° C) for good germination. Seeds of *P. abies, P. engelmannii,* and *P. glauca,* 'Albertaniana,' are among those giving good germination without stratification.

Colorado blue spruce (*Picea pungens* 'Glauca') grown from seed produces trees which usually have a greenish color with a slight bluish cast. Only a small percentage of the seedlings have the very desirable bright blue color. Several exceptionally fine blue seedling specimens have been selected as clones and are perpetuated by grafting. The two best known are the 'Koster' blue spruce (*Picea pungens* 'Glauca Koster'), developed at the Koster Nursery in Holland many years ago, and the compact 'Moerheim' spruce (*Picea pungens* 'Glauca Moerheimii'), which also originated in Europe, about 1930.

Selected clones of spruce are quite difficult to propagate by cuttings, but there are instances in which good percentages of the cuttings have been rooted. Timing is important; cuttings taken in late winter or early spring apparently root best. Treatments with root-inducing chemicals and wounding are also beneficial. High light intensity during rooting is helpful. Medium-fine, slightly acid sand is a good rooting medium. In making cuttings of upright-growing types, terminal shoots should be selected rather than lateral branches, since the latter, if rooted, tend to produce abnormal, sprawling plants rather than the desired upright form.

The 'Koster' blue spruce is propagated commercially by grafting scions on Norway spruce *(Picea abies)* or Sitka spruce *(P. sitchensis)* seedlings *(28).* Norway spruce seeds are sown in the spring; no previous stratification treatment is necessary. Seedlings are grown through two seasons, at which time they should be about 6 in. high. They may be dug and potted preparatory to grafting, either in the fall of the second season or the following spring. After potting, the plants are set in cold-frame beds with peat moss covering the tops of pots. By late summer of the third season, the understocks should be about pencil size or thicker and be ready to graft. This may be done in the fall or in early spring of the fourth season. Scions about 6 in. long are used with needles removed from the basal 2 in. Either the side or the side-veneer graft is used. The completed graft is wrapped with waxed string, then coated with grafting wax or paraffin. The grafted plants are set in benches in a cool greenhouse (50° to 65° F; 10° to 18° C) until the union is completed, which should be in 30 to 45 days.

The top of the understock is gradually cut back to the graft union, although some propagators retain a portion of the understock above the graft union throughout the first season after grafting. By midspring the plants should be set out in the field and staked. Two or three seasons' growth after grafting is required to produce plants of a salable size. The entire grafting procedure is very slow, often with a low percentage of survivals *(75, 91, 122)*.

STAR JASMINE, CHINESE (*Trachelospermum jasminoides* [Lindl.] Lem.)   Leafy cuttings of partially matured wood can be rooted, especially when placed under mist and treated with a root-promoting substance.

SUMAC (*Rhus* sp.)   These are commonly propagated by seeds which are collected in the fall and should not be allowed to dry out. For prompt germination they should be soaked in concentrated sulfuric acid for one to six hours, depending upon the species, then either fall-planted out-of-doors or stratified for two months at about 40° F (4° C) before planting. Not all species have embryo dormancy, however, and the latter treatment may sometimes be omitted (e.g., with *R. ovata* and *R. integrifolia*). Some species, such as *R. aromatica*, need no pre-treatments and will germinate with fall sowing. Some sumac plants bear only female flowers and others only male flowers, whereas still others have both flower types on the same plant. To insure fruiting, plants of the latter type should be propagated asexually. In seed propagation many of the seedlings produced will not bear fruit.

For those species which sucker freely, such as *R. typhina* and *R. copallina*, root cuttings several inches long planted in the nursery row in early spring may be used. Leafy softwood cuttings, at least of some species, as *R. aromatica*, taken in midsummer, root well under plastic if treated with 1 percent IBA mixed with 50 percent captan, 1:1.

SYCAMORE   *See* Plane Tree.

SYRINGA sp.   *See* Lilac.

TAMARISK (*Tamarix* sp.)   These are easily rooted by hardwood cuttings, which are usually made about 12 in. long and planted deeply. Softwood cuttings taken in early summer will also root readily under glass.

TAXUS sp.   *See* Yew.

THUJA sp.   *See* Arborvitae.

TILIA sp.   *See* Linden.

TULIP TREE (*Liriodendron tulipifera* L.)   Yellow poplar. Seed propagation is ordinarily used, the seed being stratified for about two months. Seeds should not be allowed to dry out. A daily varying stratification temperature between freezing and about 50° F (10° C) has given good results, although a constant temperature around 40° F (4° C) would probably be equally satisfactory. Fall planting, with outdoor stratification through the winter, has also given good germination. Seeds

of this species are often devoid of embryos, so cutting tests of each seed lot should be made. Although this is not often done commercially, leafy stem cuttings taken in the summer have been rooted in fairly good percentages. Root cuttings also have been successful.

Young tulip trees are very difficult to transplant, so they should always be propagated into containers or dug, balled, and burlapped for transplanting from the nursery row.

TRACHELOSPERMUM JASMINOIDES *See* Star Jasmine, Chinese.

TRUMPET CREEPER *(Campsis* sp.) This is usually propagated by cuttings, but seeds can also be used. With the latter method, stratification for two months at 40° to 50° F (4° to 10° C) hastens but does not increase germination. Both softwood and hardwood cuttings root readily. *C. radicans* can be started by root cuttings. Layering is also successful.

***TSUGA* sp.** *See* Hemlock.

***ULMUS* sp.** *See* Elm.

VIBURNUM *(Viburnum* sp.) *(8, 44, 83)* This large group of desirable shrubs can be propagated by a number of methods, including seeds, cuttings, grafting, and layering. At least one species *(V. dentatum)* is readily started from root cuttings.

Seeds The viburnums have rather complicated seed dormancy conditions. Seeds of some species, such as *V. sieboldi,* will germinate after a single ordinary low-temperature (40° F; 4° C) stratification period, but for most species a period of two to nine months at high temperatures (68° to 86° F; 20° to 30° C), followed by a two- to four-month period at low temperatures (40° F; 4° C) is required. The initial warm temperatures cause root formation, and the subsequent low temperature causes shoot development. Cold stratification alone will not result in germination. Such rather exacting treatments may best be given by planting the seeds in summer or early fall (at least 60 days before the onset of winter), thus providing the initial high-temperature requirement; the subsequent winter period fulfills the low-temperature requirement. After this, the seeds should germinate readily in the spring. Often, collecting the seeds early, before a hard seed coat has developed, will hasten germination. Viburnum seed can be kept for one or two years if stored dry in sealed containers and held just above freezing. *V. lantana, V. opulus,* and *V. rhytidophyllum* are commonly propagated by seed.

Cuttings Although some viburnum species *(V. opulus, V. dentatum,* and *V. trilobum)* can be propagated by hardwood cuttings, softwood cuttings rooted in sand or perlite under glass or mist are successful for most species. Soft, succulent cuttings taken in late spring root faster than those made from more mature tissue in midsummer, but the latter are more likely to grow on into sturdy plants that will survive through the following winter. Treatments with indolebutyric acid are helpful. One of the chief problems with viburnum cuttings is to keep them growing after rooting. Cuttings made from succulent, rapidly growing material often die in a few weeks after being potted. This trouble may be overcome by not digging the cuttings too soon, allowing a secondary root system to form which will better stand the transplanting shock. It may help, too, to feed the rooted cuttings with a nutrient solution about ten days before the cuttings are to be removed. Cuttings of some species root more easily than others. *Viburnum carlesii* and *V. rhytidophyllum,* for example, are difficult to root, but *V. burkwoodii* and *V. plicatum* var. *tomentosum* root readily.

Grafting Selected types of viburnum are often propagated by grafting on rooted cuttings, layers, or seedlings of *V. dentatum* or *V. lantana.* Often, grafted viburnums will develop into vigorous plants more quickly than those started as cuttings. *V. opulus* 'Roseum' (Snowball) is dwarfed when grafted onto *V. opulus* 'Nanum' cuttings. It is important that all buds be removed from the rootstock so that subsequent suckering from the stock does not occur. The understocks are potted in the fall and brought into the greenhouse, where they are grafted in midwinter by the side graft method, using dormant scionwood. After grafting, the potted plants are placed in a closed, glass-covered frame with the unions buried in damp peat moss.

Sometimes, rather than being plunged into peat moss, the scion and graft union are dipped in melted paraffin, which is satisfactory if only a very thin coating is applied. Close attention must be given the graft during the healing period, avoiding either excessive humidity or high temperatures from direct sunlight.

Grafting can also be done in late summer, using potted understock plants and scion material which has stopped growing and become hardened. The grafted plants are placed in the greenhouse and plunged in slightly damp peat moss in closed frames until the unions heal, after which they are moved to outdoor glass-covered cold frames for hardening-off for the winter.

Layering   Simple layering is widely used, especially in Europe, for propagating most viburnum species. Wood of the previous season's growth will produce roots in 18 to 24 months if layered in the spring. Some species are best layered in midsummer, using current season's wood.

WEIGELA (*Weigela* sp.)   This is easily propagated either by hardwood cuttings planted in early spring or by softwood tip cuttings under glass taken any time from late spring into fall. Treatment with a root-promoting substance promotes rooting of the softwood cuttings.

WILLOW (*Salix* sp.)   Willow seeds must be collected as soon as the capsules mature (when they turn from green to a yellowish color) and planted immediately, since at room temperature they retain their viability for only a few days. Even under the most favorable conditions, maximum storage is four to six weeks. No dormancy occurs, germination taking place 12 to 24 hours after planting if the seeds are kept constantly moist. Willows are difficult to propagate in quantity by seed.

Willows root so readily by either root or stem cuttings that there is little occasion to use other methods. Hardwood cuttings planted in early spring root promptly.

WISTERIA (*Wisteria* sp.)   These may be started by softwood cuttings under glass taken in midsummer. Indolebutyric acid treatments at about 25 ppm for 24 hours will often aid rooting. Some species can be started by hardwood cuttings set in the greenhouse in the spring. Simple layering of the long canes is quite successful. Choice types are often grafted on rooted cuttings of less desirable types. Suckers arising from roots of such grafted plants should be removed promptly. Wisterias do not transplant easily, so young nursery plants are best started in containers. Seeds and pods are quite poisonous if eaten.

WITCH HAZEL (*Hamamelis* sp.)   This is propagated by seed planted outdoors in early fall. Prevent seeds from drying. *H. japonica* seeds should be soaked for a week, changing water daily. Chinese witch hazel (*H. mollis*) is propagated by grafting on potted *H. virginiana* rootstock plants.

XYLOSMA CONGESTUM (Lour.) Merr. *(66)*   This is propagated by rooting leafy cuttings, taken in late summer or early fall, under closed frames, using the first and second subterminal cuttings on the shoot. Cuttings are dipped in IBA at 5000 ppm and rooted in a 1:3 peat moss—perlite mixture with bottom heat (70° F;

21° C). Results under mist have been contradictory. Hardening of the rooted cuttings should be done in a cool, humid greenhouse.

YELLOW POPLAR   *See* Tulip Tree.

YEW (*Taxus* sp.) *(29, 86, 116, 125)*   Most clonal selections of yews are propagated by cuttings, which root without much difficulty. Seedling propagation is little used, owing to the variation appearing in the progeny, the complicated seed dormancy conditions, and the slow growth of the seedlings. Side or side-veneer grafting is practiced for those few cultivars which are especially difficult to start by cuttings, easily rooted cuttings being used as the rootstock.

Seeds   This method is confined in commercial practice almost entirely to the Japanese yew, *Taxus cuspidata,* which comes fairly true from seed if isolated plants can be located as sources of seed. Seed imported from Japan is believed to produce uniform offspring.

To produce good germination, seeds should be given a warm temperature (68° F; 20° C) stratification period in moist peat moss or other medium for three months, followed by four months at a lower temperature (41° F; 5° C). Seedling growth is very slow. Two years in the seedbed, followed by two years in a lining-out bed, then three or four years in the nursery row are required to produce a plant of salable size of the Japanese yew.

Cuttings (90)   *Taxus* cuttings may be rooted outdoors in cold frames or in the greenhouse under mist, the latter giving much faster results.

For the cold frame, fairly large cuttings, 8 to 10 in. long, are made in early fall from new growth with a section of old wood at the base. *Taxus* cuttings seem to respond well to treatments with a root-inducing chemical, indolebutyric acid at relatively high concentrations being particularly effective.

Cuttings may be kept in closed frames through the winter; in climates with severe winters, the frames should be kept covered, especially at times when the ground is frozen but the sun is shining. Rooting takes place slowly during the following spring and summer.

For greenhouse propagation, cuttings should be taken in early winter, after several frosts have occurred, and rooted in sand with bottom heat at about 70° F (21° C) and an air temperature of 50° to 55° F (10° to 13° C). Rooting under mist is suitable for *Taxus* cuttings. Rooting is fairly rapid in the greenhouse, taking about two months, although the cuttings should not be dug too soon. This allows time for secondary roots to develop from the first-formed primary roots. There is evidence that cuttings taken from male plants (at least in *T. cuspidata expansa*) root more readily than cuttings taken from female plants (those that produce fruits) *(29).* Berries and foliage are very poisonous if eaten.

REFERENCES

1   Ackerman, W. L., and G. A. Seaton, Propagating the Bradford pear from cuttings, *Amer. Nurs.,* 130(7):8. 1969.

2   Afanasiev, M., A physiological study of dormancy in seed of *Magnolia acuminata, N. Y. (Cornell) Agr. Exp. Sta. Mem. 208,* 1937.

3   ———, Germinating *Nandina domestica* seeds, *Amer. Nurs.,* 78(9):5–6. 1943.

4   Allen, G. S., and J. N. Owens, *The Life History of Douglas Fir,* Ottawa, Canada: Forestry Service, Environment Canada, 1972.

5   Argles, G. K., Propagating maples (*Acer* species), *Nurs. and Gard. Cent.,* 148(5): 129–33; 148(6):199–201. 1969.

6 ——, The propagation of magnolias, *Nurs. and Gard. Cent.,* 148(10):361–65; 148(11):399–406. 1969.

7 Bailey, C. V., Early collection and immediate sowing increase germination of Basswood seed, *Tree Plant. Notes,* 46:27. June 1961.

8 Barton, L. V., Germination and seedling production in species of *Viburnum, Proc. Plant Prop. Soc.,* 8:126–34. 1958.

9 Bowers, C. G., *Rhododendrons and Azaleas,* 2nd ed. New York: The Macmillan Company, 1960.

10 Brydon, P. H., The propagation of deciduous azaleas from cuttings, *Proc. Inter. Plant Prop. Soc.,* 14:272–76. 1964.

11 Buck, G. J., Varieties of rose understocks, *Amer. Rose Ann.,* 36:101–16. 1951.

12 ——, and E. C. Volz, A handbook for rose growers, *Iowa Agr. Ext. Bul. P117,* 1955.

13 Burton, J. H., The grafting of some maples, *Proc. Plant Prop. Soc.,* 2:71–73. 1952.

14 Cameron, R. J., The propagation of *Pinus radiata* by cuttings, *New Zeal. Jour. Forest.,* 13:78–89. 1968.

15 Carter, A. R., Rose rootstocks—performance and propagation from seed, *Proc. Inter. Plant Prop. Soc.,* 19:172–80. 1969.

16 Carville, L., Propagation of Knaphill azaleas from softwoods, *Proc. Inter. Plant Prop. Soc.,* 17:255–58. 1967.

17 Chandler, C., The propagation of *Larix* from softwood cuttings, *Contrib. Boyce Thomp. Inst.,* 20:231–38. 1959.

18 Chandler, G. P., Rooting daphnes from cuttings, *Proc. Inter. Plant Prop. Soc.,* 19:205–6. 1969.

19 Chase, H. H., Propagation of Oriental magnolias by layering, *Proc. Int. Plant Prop. Soc.,* 14:67–69. 1964.

20 Chester, K. S., Graft-blight; a disease of lilac related to the employment of certain understocks in propagation, *Jour. Arn. Arb.,* 12:79–146. 1931.

21 Chu, K., and W. S. Cooper, An ecological reconnaissance in the native home of *Metasequoia glyptostroboides, Ecology,* 31:260–78. 1950.

22 Coggeshall, R. C., Asiatic maples, their propagation from softwood cuttings, *Arnoldia,* 17:45–55. 1957.

23 ——, Whip and tongue grafts for dogwoods, *Amer. Nurs.,* 111(2):9, 56, 59. 1960.

24 ——, Hybrid lilacs from softwood cuttings, *Amer. Nurs.,* 115(12):7–8, 57. 1962.

25 Corns, W. G., and R. J. Schraa, Dormancy and germination of seeds of silverberry (*Elaeagnus commutata* Bernh.), *Canad. Jour. Bot.,* 40:1051–55. 1962.

26 Crockett, J. U., *Roses.* New York: Time-Life Books, 1971.

27 Cumming, W. A., *Crataegus* rootstock studies, *Proc. Inter. Plant Prop. Soc.,* 14:146–49. 1964.

28 Curtis, W. J., The grafting of Koster spruce, *Cedrus atlantica* 'Glauca,' copper beech, and variegated dogwood, *Proc. Plant Prop. Soc.,* 12:249–53. 1962.

29 Davidson, H., and A. Olney, Clonal and sexual differences in the propagation of *Taxus, Proc. Inter. Plant Prop. Soc.,* 14:156–62. 1964.

30 deRigo, H. T., and W. O. Hawley, Effect of time of rhizomatous propagation of a temperate zone bamboo on shoot growth, *Agron. Jour.,* 58:401–2. 1966.

31 de Wilde, R., Production and breeding of lilacs, *Proc. Inter. Plant Prop. Soc.,* 14:107–13. 1964.

32 Donnelly, J. R., and H. W. Yawney, Some factors associated with vegetatively propagating sugar maple by stem cuttings, *Proc. Inter. Plant Prop. Soc.,* 22:413–31. 1972.

33 Edmunds, F., Commercial production of roses in Oregon, *Proc. Inter. Plant Prop. Soc.,* 13:211–14. 1963.

34 Enright, L. J., Vegetative propagation of *Mahonia bealei, Proc. Plant Prop. Soc.,* 7:69–70. 1957.

35 ———, Responses of *Magnolia grandiflora* and several species of *Berberis* to root-promoting chemical treatments, *Proc. Plant Prop. Soc.,* 8:67–69. 1958.

36 ———, Responses of *Metasequoia* cuttings to growth regulator treatments, *Bot. Gaz.,* 120:53–54. 1958.

37 Fazio, S., Propagating *Eucalyptus* from cuttings, *Proc. Inter. Plant Prop. Soc.,* 14:288–90. 1964.

38 Flemer, W., III, The vegetative propagation of oaks, *Proc. Plant Prop. Soc.,* 12:168–71. 1962.

39 Fleming, R. A., Rootstocks for ornamental trees, *Rpt. Hort. Exp. Sta. and Prod. Lab.,* Vineland, Ontario, pp. 46–49. 1962.

40 Fordham, A. J., *Cedrus deodara* 'Kashmir' and its propagation by cuttings, *Proc. Inter. Plant Prop. Soc.,* 18:319–21. 1968.

41 ———, Vegetative propagation of *Albizia, Amer. Nurs.,* 128(4):7, 63. 1968.

42 Fox, B. S., Propagation of cotoneasters, *Proc. Inter. Plant Prop. Soc.,* 22:213–18. 1972.

43 Frierson, J. L., Burford holly shows value as an understock, *Amer. Nurs.,* 107(9):9, 56. 1958.

44 Giersbach, J., Germination and seedling production of species of viburnum, *Contrib. Boyce Thomp. Inst.,* 9:79–90. 1937.

45 Girouard, R. M., Vegetative propagation of pines by means of needle fascicles —a literature review, *Inform. Rpt., Dept. Environ., Canad. Forest. Serv.,* Quebec. 1971.

46 Goddard, A. N., Grafting Japanese maples, *Plant Prop.,* Inter. Plant Prop. Soc., 16(4):6, 1970.

47 Harris, R. E., The vegetative propagation of *Amelanchier alnifolia, Canad. Jour. Plant Sci.,* 41:728–31. 1961.

48 Hartline, J. B., Holly propagation, in D. Hansell (ed.), *Handbook of Hollies, Amer. Hort. Mag.,* 49(4):213–18. 1970.

49 Hartmann, H. T., 'Swan Hill': a new fruitless ornamental olive, *Calif. Agr.,* 21(1):4–5. 1967.

50 Heit, C. E., Propagation from seed: growing larches, *Amer. Nurs.,* 135(8):14–15, 99–110. 1972.

51   Henny, J., Exbury azaleas, *Proc. Inter. Plant Prop. Soc.*, 13:231–33. 1963.

52   Hess, C., Sr., Magnolias from grafts, *Proc. Plant Prop. Soc.*, 3:113–15. 1953.

53   Hess, C. E., Propagating and overwintering *Cornus florida rubra* cuttings, *Proc. Plant Prop. Soc.*, 5:43–44. 1955.

54   Hicks, H. E., The propagation of tree peonies, *Proc. Plant Prop. Soc.*, 6:31–33. 1956.

55   Hildreth, W. R., The propagation of manzanitas by cuttings, *Jour. Calif. Hort. Soc.*, 30(2):45–47, 53. 1969.

56   Hill, J. B., Juniper grafting—practical and technical aspects, *Proc. Plant Prop. Soc.*, 3:86–93. 1953.

57   ——, The propagation of *Juniperus chinensis* in greenhouse and mist bed, *Proc. Plant Prop. Soc.*, 12:173–78. 1962.

58   Hill, S. R., and W. J. Libby, Outdoor rooting of *Pinus radiata*, *Plant Prop.*, Inter. Plant Prop. Soc., 15(4):13–16. 1969.

59   Hume, E. P., and P. Owens, Rooting of hybrid lilac cuttings in outdoor beds, *Plant Prop.*, Inter. Plant Prop. Soc., 16(2):14–17. 1970.

60   Hume, H. H., *Camellias, Kinds and Culture.* New York: The Macmillan Company, 1951.

61   Jackson, G. A. D., and J. B. Blundell, Germination in *Rosa*, *Jour. Hort. Sci.*, 38:310–20. 1963.

62   Joley, L., Experiences with propagation of the genus, *Pistacia*, *Proc. Plant Prop. Soc.*, 10:287–92. 1960.

63   Kiem, S. C., Propagation of palms, *Amer. Hort. Mag.*, 40:133–37. 1961.

64   Kirkpatrick, H., Propagation of hybrid lilacs from cuttings, *Proc. Plant Prop. Soc.*, 6:81–83. 1956.

65   Kofranek, A. M. and R. A. Larson, eds. Growing azaleas commercially. *Agr. Publ.*, University of California, Richmond, California. 1975.

66   Kubo, E., Propagation of *Xylosma congestum*, *Proc. Inter. Plant Prop. Soc.*, 15:340–41. 1965.

67   Lamb, J. G. D., Trials on propagation of *Chamaecyparis* at Kinsealy. *Proc. Inter. Plant Prop. Soc.*, 20:334–38. 1970.

68   Lawyer, D. A., Propagating *Sorbus*, *Amer. Nurs.*, 127(2):7, 54, 56, 58, 60. 1968.

69   Lanphear, F. O., The seasonal response in rooting of evergreen cuttings, *Proc. Inter. Plant Prop. Soc.*, 13:144–48. 1963.

70   Leach, D. G., *Rhododendrons of the World and How to Grow Them.* New York: Charles Scribner's Sons, 1961.

71   ——, Efficient production of rhododendrons, *Amer. Nurs.*, 122(9):7, 36, 46. 1965.

72   Lee, F. P., *The Azalea Book*, 2nd ed. Princeton, N.J.: D. Van Nostrand Co., Inc., 1965.

73   Li, Hui-Lin, Ginkgo—the Maidenhair tree, *Amer. Hort. Mag.* 40:239–49. 1961.

74   Libby, W. J., and M. T. Conkle, Effects of auxin treatment, tree age, tree vigor, and cold storage on rooting young Monterey pine, *Forest Sci.*, 12:484–502, 1966.

75 Mahlstede, C., A new technique in grafting blue spruce, *Proc. Plant Prop. Soc.,* 12:125–26. 1962.

76 March, S. G., Propagating Ghent and Mollis azaleas, *Amer. Nurs.,* 110(12):98–101. 1959.

77 Martens, O., Palms—propagation, production, and uses, *Proc. Inter. Plant Prop. Soc.,* 21:110–18. 1971.

78 Mergen, F., and B. A. Simpson, Asexual propagation of pines by rooting leaf fascicles, *Silvae Genet,* 13(5):125–64. 1964.

79 McCahon, W., Propagation of hydrangeas, *Proc. Inter. Plant Prop. Soc.,* 14:253–54. 1964.

80 McClure, F. A., *The Bamboos: A Fresh Perspective.* Cambridge, Mass., Harvard University Press, 1966.

81 McDaniel, J. C., A look at some hackberries, *Proc. Inter. Plant Prop. Soc.,* 14:143–46. 1964.

82 McGill, W., The selection of maple understock, budwood, and the timing and placement of buds, *Proc. Plant Prop. Soc.,* 2:64–69. 1952.

83 McMillan-Browse, P. D. A., Notes on the propagation of viburnums, *Proc. Inter. Plant. Prop. Soc.,* 20:378–86. 1970.

84 ———, Propagation of *Aesculus, Plant Prop.,* Inter. Plant Prop. Soc., 17(2):4–6. 1971.

85 Mirov, N. T., *The Genus Pinus.* New York: The Ronald Press Company, 1967. Chap. 5, Morphology and Reproduction.

86 Mitiska, L. J., The propagation of *Taxus* by seeds, *Proc. Plant Prop. Soc.,* 4:69–73. 1954.

87 Moore, J. C., Propagation of chestnuts and camellias by nurse seed grafts, *Proc. Inter. Plant Prop. Soc.,* 13:141–43. 1963.

88 Moore, R. S., Mist propagation of miniature roses, *Proc. Inter. Plant Prop. Soc.,* 13:208–10. 1963.

89 Morey, D., Growing roses from seed, *Pac. Coast Nurs.,* 20(6):29, 59. 20(7):17–18. 1961.

90 Nelson, S. H., Mist propagation of evergreens in the greenhouse during the winter, *Proc. Plant Prop. Soc.,* 9:67–76. 1959.

91 Nienstaedt, H., Fall grafting of spruce and other conifers, *Proc. Plant Prop. Soc.,* 8:98–104. 1958.

92 Nordine, R. M., Collecting, storage, and germination of maple seed, *Proc. Plant Prop. Soc.,* 2:62–64. 1952.

93 O'Rourke, F. L. S., The propagation of pines, *Proc. Plant Prop. Soc.,* 11:16–22. 1961.

94 Orton, E. R., Jr., S. H. Davis, Jr., and L. M. Vasvary, Growing American holly in New Jersey, *N. J. Agr. Ext. Bul. 388,* 1966

95 Penfold, A. R., and I. L. Willis, *The Eucalypts: Botany, Cultivation, Chemistry and Utilization.* New York: Interscience Publishers, Inc., 1961.

96 Pinney, J. J., A simplified process for grafting junipers. *Amer. Nurs.* 131(10):7, 82–84. 1970.

97 Pridham, A. M. S., Propagation of American elm from cuttings, *Proc. Inter. Plant Prop. Soc.,* 14:86–88. 1964.

98 Raabe, R. D., and S. Wilhelm, Budwood as a source of Verticillium wilt in greenhouse roses, *Calif. Agr.,* 20(10): 5–6. 1966.

99 Roberts, A. N., Scion-bud failure in field-grown roses, *Proc. Amer. Soc. Hort. Sci.,* 80:605–14. 1962.

100 Roberts, A. N., and F. W. Moeller, Propagation of Mugo pine successful, *Ore. Orn. & Nurs. Dig.,* 12(1):1–2, 1968.

101 Ryan, G. F., Grafting *Eucalyptus ficifolia, The Plant Prop.,* Inter. Plant Prop. Soc., 12(2). 1966.

102 Salter, C. E., *Clematis armandii* grafting, *Proc. Inter. Plant Prop. Soc.,* 20: 330–32. 1970.

103 Schneider, G., Production of rootstocks for ornamental trees in the container nursery, *Proc. Plant Prop. Soc.,* 10:282–85. 1960.

104 Semeniuk, P., and R. N. Stewart, Low temperature requirements for after-ripening of seed of *Rosa blanda, Proc. Amer. Soc. Hort. Sci.,* 85:639–41. 1964.

105 Shanks, J. B., Poinsettias—greenhouse culture, *Misc. Publ. 696, Md. Agr. Exp. Sta., Dept. Hort.,* 1969.

106 Skinner, H. T., Vegetative propagation of oaks and suggested research techniques, *Proc. Plant Prop. Soc.,* 2:81–85. 1952.

107 ———, Fundamentals of azalea propagation, *Proc. Plant Prop. Soc.,* 4:129–36. 1954.

108 Snyder, W. E., The fundamentals of juniper propagation, *Proc. Plant Prop. Soc.,* 3:67–77. 1953.

109 Stadtherr, R. J., *Magnolia grandiflora* by cuttings, *Proc. Inter Plant Prop. Soc.,* 17:260–62. 1967.

110 Stoutemyer, V. T., The propagation of *Chionanthus retusus* by cuttings, *Nat. Hort. Mag.,* 21:175–78. 1942.

111 Schmidt, C., Propagation of *Ilex aquifolium* from cuttings, *Proc. Plant Prop. Soc.,* 11:318–20. 1961.

112 Ticknor, R. L., Review of the rooting of pines, *Proc. Inter. Plant Prop. Soc.,* 19:132–37. 1969.

113 Tinga, J. H., J. J. McGuire, and R. J. Parvin, The production of pyracantha plants from large cuttings, *Proc. Amer. Soc. Hort. Sci.,* 82:557–61. 1963.

114 Van Veen, T., The propagation and production of rhododendrons, *Amer. Nurs.,* 133(8):15–16, 52–58. 1971.

115 ———, Rhododendrons in America, Portland, Ore.: Sweeney, Krist, Dimm, 1969.

116 Vermeulen, J., The propagation of *Taxus* by cuttings, *Proc. Plant Prop. Soc.,* 4:76–79. 1954.

117 Von Abrams, C. J., and M. E. Hand, Seed dormancy in roses as a function of climate, *Amer. Jour. Bot.,* 43:7–12. 1956.

118 Walter, H., Propagation of Chinese hibiscus, *Proc. Inter. Plant Prop. Soc.,* 17:263–64. 1967.

119   Watkins, J. V., Jacaranda, *Horticulture,* 50(5):22–23. 1972.

120   Weiser, C. J., and L. T. Blaney, The effects of boron on the rooting of English holly cuttings, *Proc. Amer. Soc. Hort. Sci.,* 75:704–10. 1960.

121   Wells, J. S., Rooting *Acer palmatum, Amer. Nurs.,* 98(7):15, 59–64. 1953.

122   ——, Pointers on propagation: propagating Koster spruce, *Amer. Nurs.,* 98(9):13, 48–53. 1953.

123   ——, *Plant Propagation Practices.* New York: The Macmillan Company, 1955, Chap. 31.

124   ——, A propagation program for hollies, *Proc. Plant Prop. Soc.,* 7:92–98. 1957.

125   ——, Propagation of *Taxus*—review, *Amer. Nurs.,* 114(10):11, 12, 91–98. 1961.

126   Wister, J. C., and G. S. Wister (eds.), *The Peonies.* Washington: American Horticultural Society, 1962.

127   Wyman, D., *Crab Apples for America.* Rockford, Ill.: American Association Botanical Gardens and Arboreta, 1955.

128   Yocum, H. G., Factors affecting the germination of palm seeds, *Amer. Hort. Mag.,* 43:104–6. 1964.

SUPPLEMENTARY READING

Crockett, J. U., *Evergreens.* New York: Time-Life Books (1971).

——, *Flowering Shrubs.* New York: Time-Life Books (1972).

——, *Trees.* New York: Time-Life Books (1972).

Dunmire, J. R. (ed.), *Sunset Western Garden Book.* Menlo Park, Ca.: Lane Magazine & Book Company (1967).

Forest Tree Seed Directory (English, French, Spanish). Rome: Food and Agriculture Organization of the United Nations (1961).

Fowells, H. A., *Silvics of Forest Trees of the United States,* USDA Forest Service Handbook No. 271. Washington, D.C.: U.S. Government Printing Office (1965).

Hepting, G. H., *Diseases of Forest and Shade Trees of the United States,* USDA Forest Service, Handbook No. 386. Washington, D.C.: U.S. Gov. Printing Office (1971).

International Plant Propagators' Society, *Proceedings of Annual Meetings.*

Mathias, M. E., and E. McClintock, *A Checklist of Woody Ornamental Plants of California,* California Agricultural Extension Service Manual 32. Berkeley, California: University of California (1963).

Teuscher, H., "Handbook on Conifers," *Plants and Gardens,* Brooklyn Botanic Garden, 25(2):1–105. 1969.

U.S. Department of Agriculture, Forest Service. *Seeds of Woody Plants in the United States.* C. S. Schopmeyer, Ed. Agr. Handbook No. 450. 1974.

Wells, J. S., *Plant Propagation Practices.* New York: The Macmillan Company, 1955.

Wyman, D., *Trees for American Gardens,* 2nd ed. New York: The Macmillan Company, 1965.

——, *Shrubs and Vines for American Gardens,* 2nd ed. New York: The Macmillan Company, 1969.

Propagation of Selected Annuals and Herbaceous Perennials Used as Ornamentals

19

Herbaceous plants are classified as *annuals, biennials,* or *perennials,* although the differences among these types may not be sharp (see p. 6). They may also be classified as *hardy, half-hardy,* or *tender.* In general, the propagation procedures for such plants will depend upon their categories and the locality where they are to be grown.

Seeds of *hardy annuals* can be sown outdoors early in the spring; flowering takes place in summer or fall. In areas of mild winters, seeds of half-hardy annuals can be sown in the fall, and flowering takes place the following spring. In cold-winter areas the seeds of the half-hardy annuals must be sown in spring, but not until danger of freezing is over. In these areas it is best to start the plants indoors several weeks in advance of planting, then transplant them outdoors.

Tender annuals usually require higher temperatures for seed germination and seedling growth, and must be planted only when danger of freezing is over. This group invariably is best handled by starting the plants indoors, then transplanting to the garden.

Seeds of *hardy biennials* are best sown in summer for flowering the following spring. Plants from seed sown in autumn may not be as large as they would have been if started earlier. Seeds may be sown in place or in a nursery area for transplanting to their permanent location the following spring. Seeds of *half-hardy biennials* grown in cold areas must either be sown indoors for later transplanting or dug in fall and held for spring planting, or the plants must be given some protection during winter.

Most *perennials* can be readily grown from seed. The usual procedure is to sow seeds in early summer, shoot growth taking place that same year, with flowering beginning the following season. Seeds of *hardy perennials* are sown in place or seedlings transplanted to their permanent location the following spring. *Half-hardy* and *tender perennial* plants are often started indoors, transplanted outside when danger of freezing is over and treated as annuals, but in mild climates they could be grown out-of-doors as a perennial.

Many cultivars of important perennial plants are clones and must be propagated vegetatively. For these, the most widely used propagation procedure is division (see p. 473); in fact, many herbaceous perennials must be divided every few years or the plants become crowded. Many herbaceous perennials have specialized growth structures such as bulbs, corms, or rhizomes (see Chapter 15). Other propagation methods for herbaceous perennial clones include rooting cuttings or, sometimes, grafting.

In the following lists, seed germination data are given for most plants, including suggested approximate temperatures that should give the most rapid and complete germination, along with the expected germination time *(4, 18)*. If two temperatures are given, separated by a dash, the first is the minimum night temperature, and the second the maximum day temperature. A single figure indicates a constant temperature.

The propagation methods listed will serve as a guide, but some variation from the methods indicated may be necessary with individual cultivars *(3, 4, 5, 10, 18, 19, 22)*.

Seeds of many of the species are available from commercial suppliers attached at proper spacing on long plastic tapes which are laid in shallow trenches and covered with soil for planting. The tape dissolves rapidly upon watering, leaving the seed to germinate.

***ACHILLEA* spp.**   Yarrow. Hardy perennial. Seeds germinate in one to two weeks at 68 F (20° C). Also propagated by dividing clumps.

ACHIMENES   Tender perennial. Seeds germinated in a warm greenhouse can be used for propagating species. Plants grow from small tuber-like rhizomes which can be divided for propagation. Softwood cuttings in spring or leaf cuttings in summer can be rooted. Partially dried leaf scales can be planted.

***ACONITUM* spp.**   Monkshood. Hardy perennial. Seeds sometimes show dormancy and before planting must be moist-chilled below 41° F (5° C) for six weeks. Plants have tuberous roots that can be divided, but once established they should not be transplanted. All parts of the plants are poisonous.

ADIANTUM   *See* Fern.

***AGAVE* spp.**   Many species of succulents, including Century plant. Perennial. Seeds should be sown in sandy soil when ripe. Reproduces vegetatively by offsets (see p. 470) from base of plant; these are removed along with roots and repotted in spring. Some species produce bulbils that can be used for propagation.

AGERATUM HOUSTONIANUM   Ageratum. Half-hardy annual. Seeds germinate in one to two weeks at 68–86° F (20–30° C). May also be propagated by cuttings.

AGROSTEMMA GITHAGO   Corncockle. Hardy annual. Seeds germinate in two to three weeks at 68° F (20° C).

***ALLIUM* spp.**   Ornamental onion; also, onion, chives, and garlic. Propagated by seed. Plants grow from bulbs which produce offsets. Many species produce bulbils.

***ALOE* spp.**   Succulents of the lily family. Propagated by seed in well-drained sandy soil. Germination takes place in three to four weeks at 68 to 75° F (20 to 24° C). Plants produce offshoots that can be detached and rooted. Plants with long stems can be made into cuttings, which should be exposed to air for a few hours to allow cut surfaces to suberize.

ALTHAEA ROSEA   Hollyhock. Half-hardy biennial. Seeds germinate in two to three weeks at 68° F (20° C). Where winters are not too severe sow seeds in summer, transplant in fall for bloom following year, or sow seeds in warm greenhouse in winter and transplant outdoors.

ALYSSUM SAXATILE   Goldentuft. Hardy perennial. Seeds germinate in three to four weeks at 68–86° F (20–30° C). Sow in summer for bloom the following year. Germination may be stimulated by light or exposure of moist seeds at 50° F (15° C) for five days *(2)*. Propagated by division or by softwood cuttings in spring. Double forms must be propagated by cuttings or division.

AMARANTHUS CAUDATUS   Love-Lies-Bleeding. Half-hardy annual. Seeds germinate in two to three weeks at 68–86° F (20–30° C). Light may increase germination *(2)*. Sow in warm greenhouse for later transplanting or out-of-doors when frost danger is past. *A. gangeticus tricolor.* Joseph's Coat. Same as for *A. caudatus.* Sensitive to excess water.

AMARYLLIS BELLADONNA   Perennial. Belladonna lily. Grows from bulbs outdoors in mild areas or in pots in cold climates. Propagate by bulb cuttings or separation of bulbs.

ANCHUSA CAPENSIS   Bugloss. Hardy annual or biennial. Seeds germinate in two to three weeks at 68–86° F (20–30° C). Sow seeds in summer for bloom next year or plant in greenhouse in winter for later transplanting to garden. Seeds may be sensitive to temperatures above 60° F (15° C) *(2)*. *A. azurea.* Perennial. Selected clones best propagated by root cuttings or clump division.

ANEMONE CORONARIA   Poppy anemone. Tender perennials. Seeds germinate in five to six weeks at 68° F (20° C) and may be sensitive to higher temperatures *(2)*. Plants develop clusters of small, clawlike tuberous roots. *A. japonica.* Japanese anemone. Hardy perennial. Since seeds do not come true, cultivars are propagated by division or by root cuttings. Roots are dug in fall and cut into 2-in. pieces which are laid in flats or in a cold frame, then covered with an inch of soil. After shoots appear, plants are potted. *A. pulsatilla.* Pasque flower. Hardy perennial. Seeds germinate in five to six weeks at 68° F (20° C), but may be sensitive to high temperatures *(2)*. Plants can be divided.

***ANTHEMIS* spp.**   Golden Marguerite. Camomile. Hardy perennial. Seeds germinate in one to three weeks at 68° F (20° C). Plants can be divided or propagated by stem cuttings.

ANTHURIUM ANDRAEANUM   Anthurium. Remove offshoots with attached roots from the parent plant or root two- or three-leaved terminal cuttings under mist.

ANTIRRHINUM MAJUS   Snapdragon. Tender perennial, treated as an annual. Seeds germinate in one to two weeks at 55° F (13° C) and may respond to light *(2)*. Some hybrids are best started at 60 to 65° F (16 to 18° C). Seeds germinate well in mist *(4)*. Start indoors for later outdoor planting (in fall in mild climates or in spring in severe winter areas). Softwood cuttings root readily.

***AQUILEGIA* spp.**   Columbine. Hardy perennials. Seeds germinate in three to four weeks at 68–86° F (20–30° C) and may respond to light and three to four weeks moist-chilling at 41° F (5° C) *(2)*.

***ARABIS* spp.**   Rockcress. Hardy perennials. Seeds germinate in three to four weeks at 68° F (20° C) and may respond to light *(2)*. Softwood cuttings taken from new growth immediately after bloom root readily. Plants can be divided in spring or fall.

ARCTOTIS STOECHADIFOLIA African daisy. Half-hardy annual. Seeds germinate in two to three weeks at 68° F (20° C). Sow indoors for later transplanting.

ARMERIA **spp.** Thrift. Hardy evergreen perennials. Seeds germinate in three to four weeks at 68° F (20° C). Best propagated by clump division in spring or fall.

ASCLEPIAS TUBEROSA Butterfly weed. Hardy perennial. Seeds germinate in three to four weeks at 68–86° F (20–30° C). Fresh seed may need chilling. Plants should not be disturbed once established. *A. curassavica.* Bloodflower. Tropical perennial. Propagated by seed or by rooting softwood cuttings. Long taproot makes division difficult.

ASPARAGUS ASPARAGOIDES Smilax. Tender perennial. Propagated by seeds which germinate in three to four weeks at 68–86° F (20–30° C). Sow seeds soon after they ripen, since they are short-lived *(4)*. Cuttings can be made of young side shoots taken from old plants in spring; clumps can be divided. *A. plumosus*, Fern asparagus, and *A. sprengeri*, Sprenger asparagus. Seeds germinate in four to six weeks at 68–86° F (20–30° C). Crack the seed coats with a knife. This can also be propagated vegetatively as described for Smilax.

ASTER **spp.** Hardy perennials. Seeds germinate in two to three weeks at 68° F (20° C). Cultivars are propagated by lifting clumps in fall and dividing into rooted sections, discarding the older parts.

AUBRIETA DELTOIDEA Aubrieta. Hardy perennials, sometimes treated as annuals. Seeds germinate in two to three weeks at 55° F (13° C). Clumps are difficult to divide; cuttings may be taken immediately after blooming.

AUCUBA JAPONICA VARIEGATA Gold dust plant. Propagate by root cuttings or by leafy stem cuttings under mist. Does well in shade.

BAPTISIA **spp.** False Indigo. Hardy perennial. Seed germination at 68° F (20° C) is slow and uneven. Gather seeds when ripe and sow outdoors to overwinter. Clumps are difficult to divide due to long tap root.

BEGONIA **spp.** Begonia. Tropical perennials. Seeds, which are very fine and need light, germinate in two to four weeks at 68° F (20° C). Sow on moist, light medium with little or no covering. Begonia species and Wax begonias are grown by seed, but most other types must be propagated vegetatively. *Tuberous begonias.* Grow from tuberous stems, which can be divided into sections that bear at least one growing point. Leaf, leaf-bud, and short stem cuttings (preferably with piece of tuberous stem attached) will root readily. *Fibrous-rooted begonias* (Wax begonias, Christmas begonias, and others which include cultivars derived from *B. socotrana* and *B. semperflorens*). Propagated by leaf cuttings or softwood cuttings taken from young shoots in spring and summer. *Rhizomatous types* (Various species and cultivars, including Rex Begonia). Plants are divided or rhizomes are cut into sections. Propagation by leaf cuttings is usually used, but stem cuttings will also root. *B. evansiana* produces small tubercles, which are detached and planted.

BELLIS PERENNIS English Daisy. Hardy perennial often treated as annual or biennial. Seeds germinate in one to two weeks at 68° F (20° C) and may respond to light *(2)*. Clumps should be divided every year to prevent crowding.

BOLTONIA **spp.** Boltonia. Hardy perennial. Seeds germinate in two to three weeks at 68° (20° C). Divide plants in spring or fall.

BROWALLIA **spp.** Amethyst Flower. Tender, blue-flowered perennial often treated as annual. Seeds germinate in two to three weeks at 68° F (20° C). Softwood cuttings can be taken in fall or spring. Can be used as flowering pot plant indoors in winter.

CACTUS (10, 11)   Large group of many genera, species, and some cultivars. Tender to semihard perennials. Seed propagation can be used for most species, but seeds often germinate slowly. Sow fungicide-treated seed in well-drained, sterile mixture, and water sparingly, but do not allow medium to dry out. Pieces of stem can be broken off and rooted as cuttings; or small offsets, which root readily, can be removed. Allow sections to dry for a few days to heal (suberize) the cut surface with a corky layer before rooting. High humidity during rooting is unnecessary, but bottom heat is beneficial. Grafting is used to provide a decay-resistant stock for certain kinds and to produce unusual growth forms. For example, the pendulous *Zygocactus truncatus* is sometimes grafted on tall erect stems of *Pereskia aculeata*. Intergeneric grafts are usually successful. A type of cleft graft is used. The stem of the stock is cut off, and a wedge-shaped piece is removed. The scion is prepared by removing a thin slice from each side of the base; this is fitted into the opening made in the stock. The scion is held in place with a pin or thorn. The completed graft is placed under glass and held in a warm greenhouse until healed *(6, 9)*.

CALADIUM BICOLOR   Various cultivars. This tropical perennial, grown for its strikingly colorful foliage, produces tubers. Propagation is by removing the tubers from the parent plant at the end of the four- to five-month dormancy period just before planting. Sometimes the tubers are cut into pieces, each containing at least two buds ("eyes"). Caladiums do best out-of-doors when planted after the minimum night temperature is above 65° F (18° C) or as pot plants maintained with night temperatures of 65–70° F (18–21° C) and day temperatures of 75–85° F (24–29.5° C).

CALCEOLARIA spp.   Tender perennials often grown as annuals. Seeds germinate in two to three weeks at 68° F (20° C). Propagation is also by softwood cuttings.

CALENDULA OFFICINALIS   Pot marigold. Hardy annual; gives winter bloom in mild climates from seed sown in late summer. Seeds germinate in one to two weeks at 68–86° F (20–30° C). Thin plants to 12 in. apart.

CALLA   *See* Zantedeschia spp.

CALLISTEPHUS CHINENSIS   China aster. Half-hardy annual. Seeds germinate in two to three weeks at 68° F (20° C). Plant only wilt-resistant types.

CAMPANULA CARPATICA   Tussock Bellflower. Hardy perennial. Seeds germinate in two to three weeks at 68–86° F (20–30° C) and may respond to light *(2)*. *C. lactiflora*. Bellflower. Hardy perennial. Seeds germinate in two to three weeks at 55–90° F (13–32° C). *C. medium*. Canterbury Bells. Hardy biennial. Seeds, which germinate in two to three weeks at 68–86° F (20–30° C), are sown in late spring or early summer for bloom the following year. *C. persicifolia*. Peach bells. Hardy perennial. Seeds germinate in two to three weeks at 55–90° F (13–32° C) and may respond to light *(2)*. Small offsets can be detached and rooted. *C. pyramidalis*. Chimney bellflower. Hardy perennial, often treated as a biennial. Seeds germinate in two weeks at 68–86° F (20–30° C). Campanulas may also be increased by division and by rooting cuttings.

CANNA spp.   Canna. Tender perennial. Cultivars do not come true from seed. Seeds, which have hard coats and must be scarified before planting, are germinated in a warm greenhouse. Cultivars are propagated by dividing the rhizome, keeping as much stem tissue as possible for each growing point. In mild climates this is done after the shoots die down in the fall or before growth starts in spring. In cold climates the plants are dug in fall, stored over winter, divided in spring, then started in sand or sandy soil for transplanting outdoors when frost danger is over.

CARNATION   See Dianthus.

CATANANCHE CAERULEA   Cupidsdart. Hardy perennial. Seeds germinate in two to four weeks at 68–86° F (20–30° C). Plants may be divided in fall.

CELOSIA ARGENTEA   Cockscomb. Tender annual. Seeds germinate in one to two weeks at 68–86° F (20–30° C) and may respond to light *(2)*.

CENTAUREA CINERARIA and others   Dusty Miller. Tender perennial. Seeds germinate in two to four weeks at 68–86° F (20–30° C). Cuttings can be rooted. *C. cyanus.* Cornflower. Batchelor Button. *C. moschata.* Sweet-Sultan. These are hardy annuals whose seeds germinate in three to four weeks at 68–86° F (20–30° C).

CERASTIUM TOMENTOSUM   Snow-in-summer. Hardy perennial. Seeds germinate in two to four weeks at 68° F (20° C). This is easily propagated by division in the fall or by softwood cuttings in summer.

CHEIRANTHUS CHEIRI   Wallflower. Semihardy perennial often treated as a biennial. Seeds germinate in two to three weeks at 54° F (13° C) and may respond to light *(2)*. Choice plants may be increased by cuttings taken in early summer.

CHRYSANTHEMUM CARINATUM, C. CORONARIUM, and *C. SEGETUM* **hybrids** Many cultivars. Hardy annuals. Seeds germinate in two to four weeks at 68° F (20° C). *C. parthenium.* Feverfew. Hardy perennial usually grown as an annual. Start from seeds, as described above. Plants easily self-seed or can be divided. *C. maximum.* Shasta Daisy. Hardy perennial but often treated as a biennial, since it is short-lived. Plants increase from a series of stolon-like shoots which can be lifted in the fall. Individual pieces have roots and can be transplanted.

   C. morifolium   Garden and greenhouse chrysanthemum and *C. frutescens.* Marguerite. Hardy and semihardy perennials. After flowering, lateral shoots develop from the base of the flowering stems, particularly if the tops are cut back. When the new side shoots are 3½ to 4 in. long and firm but not woody, they are cut off and rooted as softwood cuttings. The best source of new cuttings is a mother block (or increase block) grown in an isolated area away from the producing area. Such plants are grown in programs designed to keep them pathogen- and virus-free and true-to-type *(4)*. Unrooted cuttings can be held for as long as 30 days at 33° F (0.5° C). For outdoor planting, cuttings may be taken in the same way. In areas with mild winters, cuttings can be taken in late winter for rooting and later transplanting to the garden. In cold-winter areas, the plants should be dug in the fall and brought into the greenhouse or cold frame; cuttings should be made in winter. The plants may also be left in place and divided in spring or fall. If cuttings are taken from the ends of stems high above ground they will not be infected with certain soilborne insects and diseases.

CLARKIA **spp.**   Hardy annuals. Seeds germinate in one to two weeks at 54–90° F (13–32° C). Seeds of some strains need light.

CLEOME SPINOSA   Spiderflower. Tender annual. Seeds germinate in one to two weeks at 54–90° F (13–32° C). Seeds may respond to light *(2)*.

CODIAEUM VARIEGATUM   Croton. Tropical perennial. Propagated by leafy cuttings in spring or summer. Tall, "leggy" plants can be propagated by air layering.

COLCHICUM AUTUMNALE   Autumn crocus. Saffron. Hardy perennial that grows as a corm. Seeds are sown as soon as ripe in summer but may require chilling over winter to germinate. Several years are required for plants to reach flowering size.

COLEUS BLUMEI   Tender perennials. Seeds germinate in two to three weeks at 68–86° F (20–30° C), but seedlings will be variable. Selected individuals are propagated by softwood cuttings, which root easily.

CONVALLARIA MAJALIS   Lily-of-the-valley. Hardy perennial that grows as a rhizome, whose end develops a large underground bud, commonly called a "pip." In fall the plants are dug, and the pip, with attached roots, is removed and used as the planting stock. Digging should take place in early autumn, with replanting completed by late autumn. Single pips may be stored in plastic bags in the refrigerator, then planted in late winter for spring bloom.

COREOPSIS spp.   Hardy annuals and perennials. Seeds, which germinate in two to three weeks at 68° F (20° C), may respond to light *(2)*. Perennial clumps can be divided in spring or fall.

CORTADERIA SELLOANA   Pampas grass. Best feathery plumes are found on female plants. Propagated by clump division.

COSMOS BIPINNATUS and *C. SULPHUREUS*   Half-hardy annual. Seeds germinate in one to two weeks at 68–86° F (20–30 C) and may respond to light *(2)*.

CRASSULA ARGENTEA   Jade plant. Can be propagated at any time by leaf-bud or stem cuttings.

CROCUS VERNUS   Dutch crocus. Also other *C.* species. Hardy perennial that grows from a corm. Seeds germinate as soon as ripe in summer; several years are required for plants to flower. When leaves die in fall, plants are dug and corms and cormels are separated and replanted.

CUCURBITA PEPO var. *OVIFERA*   Ornamental gourds. Tender annuals. Seeds germinate in two to three weeks at 68–86° F (20–30° C).

CYCLAMEN spp.   Tender perennials. Plants, which grow from a large tuberous stem are propagated by seeds, which germinate in three to four weeks at 68° F (20° C). Seedlings require one to several years to flower.

CYMBALARIA MURALIS   Kenilworth Ivy. Semi-hardy perennial. Seeds germinate in one to four weeks at 54° F (12° C). Self-seeds readily. Softwood cuttings or clump division may be used.

CYNOGLOSSOM AMABILE   Chinese Forget-me-not. Hardy biennial grown as an annual. Seeds germinate in two to three weeks at 68° F (20° C) and may respond to light *(2)*.

DAHLIA   Tender perennials consisting of hundreds of cultivars. Seeds germinate in two to three weeks at 68–86° F (20–30° C) when planted indoors for later transplanting outdoors. Cultivars must be propagated vegetatively. Plant grows from large tuberous roots. Clumps are dug in the fall before frost and are stored over winter at 30 to 50° F (2 to 10° C), covered with a material such as soil or vermiculite to prevent shriveling. In spring, when new sprouts begin to appear, divide the clumps so that each root section has at least one sprout. Plant outdoors when danger of frost is over. Dahlias can also be propagated by softwood or leaf-bud cuttings.

DELPHINIUM spp.   Hardy perennials, usually propagated by seeds which germinate in three to four weeks at 54° F (12° C). Seeds are short-lived and should be used fresh, or stored in containers at low temperature and reduced moisture. Seeds are usually sown outdoors in spring or summer to produce plants which flower the following year. Delphiniums can be propagated easily by softwood cuttings taken in spring. Clumps can be divided in spring or fall, but such plants tend to be short-lived. *D. ajacis.* Larkspur. Hardy annual. Seeds germinate in three to four weeks at 54° F (12° C). Young plants and seeds can be poisonous if eaten.

DIANTHUS CARYOPHYLLUS   Carnation. Tender to semi-hardy perennial that has many cultivars used in florist's trade. Seeds germinate readily but are used primarily for breeding. Carnations are readily propagated by softwood cuttings *(13)*. With mist, and with growth regulator treatment, rooting can be done almost any time of the year (see p. 291). The best source of cuttings is a mother (or increase) block isolated from the producing area, this block originating from cuttings taken from stock plants maintained under a program designed to keep them pathogen- and virus-free and true-to-type (see p. 197). Such rooted cuttings are produced to a large extent by specialist growers, but commercial benches may be another source, providing careful disease control and selection is practiced in the blocks. Lateral shoots ("breaks") that arise after flowering are removed and used as cuttings. Cuttings root in two to four weeks and may be planted directly to a greenhouse bench or transplanted to peat pots or to a nursery bed. *D. plumarius* and related species. Garden pinks. Hardy perennials, although some kinds are grown as annuals or biennials. Seeds germinate easily in two to three weeks at 68° F (20° C), but may not reproduce the cultivar. Softwood cuttings are taken in early summer and rooted to produce next year's plants. Layering can also be used. *D. barbatus.* Sweet William. Perennial but grown as a biennial. Start by seed planted outdoors in spring. In mild-winter areas transplant to permanent location in fall. In cold-winter areas, overwinter in a cold frame and transplant in spring.

***DICENTRA* spp.**   Bleedingheart. Hardy perennials. Seeds are sown in late summer or fall for overwintering at low temperatures; alternatively, seeds should be stratified for six weeks below 41° F (5° C) before planting. Divide clumps in spring or fall. Stem cuttings can be rooted if taken in spring after flowering. Root cuttings about 3 in. long taken from large roots after flowering can be used.

DICTAMNUS ALBUS   Gasplant. Hardy perennial. Propagation is the same as given for *Dicentra*. Plants should not be disturbed after establishment. Some people are allergic to plants of this species.

***DIEFFENBACHIA* spp.**   Dumbcane. Tropical perennial. Cut stem into 2-in. segments, and place horizontally in sand. New shoots and roots will develop from nodes. If plant gets tall and "leggy," the top may be cut off and rooted as a cutting, or the plant may be air layered. Leaves and stem are poisonous. (Fig. 10–12.)

***DIGITALIS* spp.**   Foxglove. Seeds germinate in two to three weeks at 68–86° F (20–30° C) and may respond to light *(2)*. Sow seeds outdoors in spring, transplant to a nursery row at 9-in. spacing, then transplant to permanent location in fall. Perennial species increased by clump division.

***DIMORPHOTHECA* spp.**   Cape marigold. Half-hardy annual. Seeds germinate in two to three weeks at 68–86° F (20–30° C).

***DORONICUM* spp.**   Leopard bane. Hardy perennial. Seeds germinate in two to three weeks at 68° F (20° C). Divide plants in spring or fall.

***DRACAENA* spp.**   Variable group of tropical perennial foliage plants. Seeds germinate in three to four weeks at 86° F (30° C). Old stems are cut off when dormant and laid horizontally in sand in warm greenhouse. New shoots that develop from nodes are cut off when about 3 in. long to root as cuttings, or they may be left intact to root. Leafy ends of bare shoots can be cut off and rooted as cuttings. Some plants grow stout underground stems, whose half-inch tips ("toes") can be cut off and planted. Can also be propagated by air layering.

ECHEVERIA *See* Succulents.

ECHINOPS EXALTATUS Globe thistle. Hardy perennials. Seeds germinate in one to four weeks at 68–86° F (20–30° C). Plants may be divided in spring. Root cuttings, 2 to 3 in. long, may be made in the fall and planted in sandy soil in a cold frame.

EPIPHYLLUM spp. Leaf-flowering cactus. Tender perennial. Seeds do not germinate well when fresh but will after 6 to 12 months' storage if planted in a warm greenhouse. Propagated readily by leaf cuttings or by grafting to *Opuntia*. See **Cactus.**

ESCHSCHOLZIA CALIFORNICA California poppy. Hardy annual. Sow seeds outdoors in fall in mild climates or in early spring in colder areas. Tends to self-sow.

FATSHEDERA LIZEI Tree ivy. Cross between *Hedera helix* and *Fatsia japonica*. Propagated by stem cuttings or by air layering.

FERNS *(14, 20)* Many genera and species (see p. 64 for life cycle). Dust-like spores are collected from the spore cases on lower sides of fronds. Examine these sporangia with a magnifying glass to be sure they are ripe but not empty. Place fronds with the spores in a manila envelope and dry for a week at 70° (21° C). Screen them to separate spores from the chaff. Transfer to a vacuum-tight bottle and store in a dry, cool place. Sow spores evenly on top of sterilized moist soil mixture, e.g., two-thirds peat moss, one-third perlite in flats, paying particular attention to sanitation. Leave 1 in. space on top and cover with a pane of glass. Use 65 to 75° F (18 to 24° C) air temperature; bottom heat may be helpful. Keep moist, preferably using distilled water to avoid salt injury.

Spores germinate and produce moss-like growth ⅛ in. thick which is composed of many prothallia. Fertilization of the archegonium on the underside of the prothallus occurs in three to six months. In a first transplanting, a small piece of prothallus is removed with tweezers and transplanted to wider spacing in a new flat of soil mixture. The prothallia expand to about ½ in. in diameter and produce tiny sporophyte plants with primary leaves and roots. A second transplanting is made from which the fern plant will grow.

Several vegetative propagation methods are possible. Ferns grow from thick rhizomes which can be divided. Certain species (e.g., *Cystopteris bulbifera*) produce small "bulblets" about the size of a pea on the underside of the leaf. These drop when mature, are planted, and will produce a fern plant by the second year. Other species produce small vegetative buds on the upper surface or edge of the leaves; these detach and form new plants.

FICUS ELASTICA Rubber plant. Tropical perennial. Propagate as cuttings taken from 6- to 12-in. shoots; single buds or "eyes" can be removed and rooted. These are made in spring, inserted in sand or a similar medium, and held in a warm greenhouse. Plants that become too "leggy" can be air layered.

FREESIA spp. Tender perennials. Seeds planted in fall germinate in four to six weeks and will bloom the next spring. Plants grow from corms which are planted in spring and dug in fall. Small cormels are removed at this time and replanted to grow larger for future flowering.

GAILLARDIA spp. Blanketflower. annual and hardy perennials. Seeds germinate in two to three weeks at 68° F (20° C) and may respond to light *(2)*. Perennial kinds are planted in spring to bloom the following year. These may also be started from root cuttings or may be divided in spring or fall but are not long-lived.

GALANTHUS spp.   Snowdrop. Hardy perennial. Bulbs are planted in the fall for bloom the following spring. Offsets are removed when bulbs are dug.

GAZANIA spp.   Tender perennial often grown as an annual. Propagated by seeds sown in spring or by softwood cuttings taken in late summer, rooted in a cold frame, then transplanted in spring. Divide clumps after three or four years.

GENTIANA spp.   Gentian. Many species, mostly hardy perennials, although some are annuals and biennials. Plant fresh seed in the fall to overwinter outdoors. Seeds germinate in one to four weeks at 68° F (20° C), but it is best to hold them at 32° F (0° C) for ten days before planting.

GERANIUM spp.   Cranesbill. True geraniums. Hardy perennial. (*See also* Pelargonium.) Species started by seeds, which germinate in one to six weeks at 54–90° F (12–32° C). Cultivars are best started by stem cuttings.

GERBERA JAMESONII   Transvaal daisy. Tender perennial. Seeds germinate in two to three weeks at 68° F (20° C). It is important to use fresh seed. Remove basal shoots from the rhizome and use as cuttings. Aseptic shoot-tip culture can be used for rapid, large-scale multiplication, starting from a single selected plant.

GEUM spp.   Avens. Hardy perennials. Seeds germinate in three to four weeks at 68–86° F (20–30° C). Propagate also by clump division in spring or fall.

GLADIOLUS   Tender perennial grown from a corm. Seed propagation is used for developing new cultivars. Seeds are planted in spring either indoors for later transplanting or outdoors when danger of frost is over. See p. 491 for details of culture.

GODETIA spp.   Hardy annuals. Sow seeds in early spring; these germinate in two to three weeks at 68° F (20° C).

GYPSOPHILA ELEGANS   Baby's Breath. Annual. Seed germinates in two to three weeks at 68° F (20° C). *G. paniculata.* Hardy perennial. Started by seed as above. Plants can be divided in spring and fall. Double-flowered cultivars are grafted on seedling *G. paniculata* (single-flowering) roots. This can be done in summer and fall, using outdoor-grown plants for rootstocks that are placed in a cold frame for healing of the graft; grafting is also done in winter and early spring, using greenhouse-grown stock plants. In the latter case, scions are taken from 3- to 4-in shoots produced from potted plants about a month after they are put in the greenhouse. For the stock, roots about the size of a pencil are cut off 1 in. below the crown. Cut two edges along base of scion to make a wedge; split upper end of root about 1 in. along one side and almost through, insert scion, then wrap. Cut root to about a 2-in length. Place grafts in closed case to heal; harden-off, then transplant to the field in the spring.

HAWORTHIA   *See* Succulents.

HELENIUM AUTUMNALE   Sneezeweed. Hardy perennial. Seeds germinate in one to two weeks at 68° F (20° C). Cultivars are increased by division. Separate rooted shoots in spring, line-out in nursery, then transplant in fall and winter.

HELIANTHEMUM NUMMULARIUM   Sunrose. Half-hardy perennial. Seeds germinate in two to three weeks at 68–86° F (20–30° C). Cultivars are propagated by softwood cuttings taken from young shoots in spring. Transplant to pots and place in permanent location the following winter or spring. Division of clumps is also possible, but plants tend to be short-lived.

HELIANTHUS ANNUUS   Sunflower. Hardy annual. Seeds germinate in two to three weeks at 68–86° F (20–30° C). *H. decapetalus* and other hardy perennial species are increased by division.

HELIOPSIS SCABRA   Heliopsis. Hardy perennial. Seeds germinate in one to two weeks at 68° F (20° C). Divide clumps in fall.

HELIOTROPIUM spp.   Heliotrope. Tender perennial usually grown as an annual. Seed germinates in three to four weeks at 68–86° F (20–30° C) and may respond to light *(2)*. Start indoors for later spring planting. Take softwood cuttings of side shoots in fall or spring and root at low temperatures (50° F; 10° C) and slightly moist conditions.

HELLEBORUS spp.   Hellebore. Christmas Rose and Lenten Rose. Hardy perennials. Sow seeds as soon as ripe; give six weeks moist-chilling before planting. Plants require several years to produce flowers. Roots of both species are poisonous.

HEMEROCALLIS spp.   Daylily *(8)*. Hardy perennial. Seeds require about six weeks of moist-chilling for germination. Seed propagation is only used to develop new cultivars. Divide clumps in fall or spring, separating into rooted sections, each with about three offshoots.

HESPERIS MATRONALIS   Rocket. Hardy perennial often grown as a biennial. Usually propagated by seeds which germinate in three to four weeks at 68–86° F (20–30° C). Clumps may be divided in fall or spring. Double-flowered cultivars are propagated by softwood cuttings taken in late summer.

HEUCHERA spp.   Alum root. Coral-bells. Hardy perennial. Seed germinates in two to three weeks at 68–86° F (20–30° C). Fungicidal treatment recommended. Light may stimulate germination *(2)*. Divide clumps in fall or spring. Leaf cuttings are made in late fall; entire leaf plus short segment of petiole can be rooted in sand.

HIPPEASTRUM spp.   Amaryllis. Tender bulbous perennial. Remove and pot bulb offsets which will flower the second year. Make bulb cuttings in late summer. Seeds germinate under warm conditions (68–86° F; 20–30° C) and sprouting is slow and uneven. Seedlings take two to four years to produce flowers.

HUNNEMANNIA FUMARIAEFOLIA   Goldencup. Tender perennial often grown as annual. Seeds germinate in two to three weeks at 68° F (20° C). For bloom first year, sow seeds early indoors then transplant outdoors when danger of freezing is over.

HYACINTHUS spp.   Hyacinth. Hardy, spring-flowering perennial; bulbs are planted in the fall. Removal of offset bulbs gives small increase. For commercial propagation, new bulbs are obtained by scoring or scooping mature bulbs. Seeds may be planted outdoors in fall, but up to six years are required to produce bloom.

IBERIS spp.   Candytuft. Hardy annual and perennial species. Seeds germinate in one to two weeks at 68–86° F (20–30° C) but may need light. Root softwood cuttings in summer or divide clumps in fall.

IMPATIENS spp.   Snapweed. Touch-Me-Not. Balsam. Perennials and half-hardy annuals. Seeds germinate in two to four weeks at 68° F (20° C) and may respond to light *(2)*. Perennial species can be started by cuttings.

INCARVILLEA spp.   Incarvillea. Hardy perennial. Seeds germinate in one to two weeks at 68° F (20° C). Divide in fall or, preferably, in spring.

IPOMOEA spp.   Morning Glory. Tender perennial grown as an annual. Seeds germinate in one to three weeks at 68–86° F (20–30°C). Notch seed coats or soak seeds overnight in warm water before planting.

IRESINE spp.   Bloodleaf. Tender perennial. Softwood cuttings root easily. Keep stock plants over winter in greenhouse and take cuttings in late winter or spring.

IRIS spp.   Perennials. There are several different groups of hardy or semi-hardy iris which grow either from rhizomes or from bulbs. Rhizomes are divided after bloom. Discard the older portion and use only the vigorous side shoots. Leaves are trimmed to about 6 in. Bulbous species follow a typical spring-flowering, fall-planting sequence. The old bulb completely disintegrates, leaving a cluster of various-size new bulbs. These are separated and graded, the largest size being used to produce flowers, the smaller for further growth. Seeds, which are used to propagate species and to develop new cultivars, should be planted as soon as ripe after being given a moist-chilling period; germination is often irregular and slow. Removal of embryo from the seed and growing it in artificial culture has given prompt germination in some cases.

IXIA spp.   Corn Lily. Tender, summer- or fall-flowering perennials grown from corms. In cold climates these are dug in fall and stored over winter. Small cormels are removed and planted in the ground or in flats to reach flowering size, as is done with gladiolus.

KALANCHOE spp.   Tropical perennials. See **Succulents.**

KNIPHOFIA UVARIA   Torch lily. Half-hardy perennials. Seeds germinate in three to four weeks at 68–86° F (20–30° C). Divide clumps in spring, putting sections in pots until they are well rooted.

KOCHIA SCOPARIA   Summer cypress. Half-hardy annual. Seeds germinate in one to two weeks at 68–86° F (20–30° C). Transplant outdoors when frost danger is over. Reseeds profusely and may become a weed.

LANTANA SELLOWIANA. L. CAMARA   Lantana. Tender perennials. Seeds germinate in six to seven weeks at 68° F (20° C). Softwood cuttings root easily.

LATHYRUS LATIFOLIUS   Perennial pea vine. Hardy perennial. Seeds germinate in two to three weeks at 68–86° F (20–30° C). Clumps may be divided. *L. odoratus.* Sweet pea. Hardy annual. Seed germinates in two weeks at 68° F (20° C). Notching seed or soaking in warm water may hasten germination. Plant outdoors in fall where winters are mild, in spring where winters are severe. Early flowering Multiflora types will bloom in short days of winter for greenhouse use. Spring- and summer-flowering types bloom only after 15-hour day lengths occur. Cuthbertson types are heat-resistant.

LAVANDULA OFFICINALIS   Lavender. Half-hardy perennial. Seeds, which may be planted in winter, germinate in two to three weeks at 52–90° F (11–32° C). Take cuttings from side shoots in late summer or fall; plant in soil and cover or start in cold frame. Divide clumps in the fall.

LAVATERA TRIMESTRIS   Tree Mallow. Half-hardy annual. Plant seeds outdoors in place as they do not transplant well. Seeds germinate in one to three weeks at 68° F (20° C).

LIATRIS spp.   Gayfeather. Hardy perennial. Seeds germinate in three to four weeks at 68–86° F (20–30° C). Divide tuberous roots in early spring.

LILIUM spp.   Lily. Hardy perennials. These are spring- and summer-flowering plants grown from scaly bulbs; most have a vertical axis, but in some species growth is horizontal with a rhizomatous structure. Lilies include many species, hybrids, and named cultivars. Seed propagation is used for species and for new cultivars. Seeds of different lily species have different germination requirements *(21)*.

Immediate seed germinators include most commercially important species and hybrids (*L. amabile, L. concolor, L. longiflorum, L. regale, L. tigrinum,* Aurelian hybrids, Mid-Century hybrids, and others). Germination is epigeous; a shoot should emerge three to six weeks after planting at moderately high temperatures. Treat seeds with a fungicide to control *Botrytis*. Sow ¾ in. deep in flats during winter or outdoors in a seedbed in early spring. Dig the small bulblets in fall, sort for size, store over winter, and replant with same sizes together. Plants normally grow two years in a seedbed and two years in a nursery row before producing good-size flowering bulbs.

Another group is the *slow seed germinators* of the epigeal type (*L. candidum, L. henryi,* Aurelian hybrids, and others) in which seed germination is slow and erratic; the procedures used are essentially the same as described above.

The most difficult group to propagate are the *slow seed germinators* of the hypogeous type (*L. auratum, L. bolanderi, L. canadense, L. martagon, L. parvum, L. speciosum,* and others). Seeds of this group require three months under warm conditions for the root to grow and produce a small bulblet, then a cold period of about six weeks, followed by another warm period in which the leaves and stem begin to grow. This sequence can be provided by planting the seeds outdoors in summer as soon as they are ripe, or planting seeds in flats and then storing under appropriate conditions to provide the required temperature sequence. Vegetative methods of propagation include natural increase of the bulbs, such as bulblet production on stems (either naturally or artificially), aerial stem bulblets (bulbils), or scaling. These procedures are described in Chapter 15.

LINARIA spp.   Toadflax. Hardy annual and perennials. Seeds germinate in two to three weeks at 54° F (12° C). Perennial species take two years to produce bloom from seed. Clumps can be divided in spring or fall.

LINUM spp.   Flax. Hardy annual and perennial species. Seeds germinate in three to four weeks at 54° F (12° C). Divide clumps of perennial species in fall or spring.

LOBELIA ERINUS   Lobelia. Tender perennial grown as an annual. Seeds germinate in two to three weeks at 68–86° F (20–30° C), but seedling growth is slow. May respond to light *(2)*. Start indoors 10 to 12 weeks before transplanting outdoors after last frost. Mature plants, if potted in the fall and kept in greenhouse over winter, can be used to provide new growth for cuttings to be taken in late winter.

LOBELIA spp.   Hardy perennials. Seeds germinate in three to four weeks at 68–86° F (20–30° C). Divide clumps in fall or spring.

LOBULARIA MARITIMA   Sweet Alyssum. Perennial but grown as hardy annual. Seed germinates in one to two weeks at 68° F (20° C) and blooms appear in six weeks.

LUNARIA ANNUA   Honesty. Biennial, sometimes grown as an annual. Seeds germinate in two to three weeks at 68° F (20° C). *L. rediviva.* Hardy perennial. Propagated by seed as described above. Also increased by division.

LUPINUS HARTWEGII, L. NANUS, and others   Hardy annuals. Seeds germinate in two to three weeks at 68° F (20° C). *L. polyphyllus.* Perennial. Seeds germinate in three to four weeks at 68° F (20° C). Seed coats may be hard and should be scarified. *L. arboreus.* Tree lupine. Hardy perennial. Start from seeds indoors and transplant to permanent location. *L.* 'Russell Hybrid.' Sow seeds in spring or summer, or propagate by cuttings taken in early spring with small piece of root or crown attached.

LYCHNIS spp.   Campion. Mostly hardy perennials but some are grown as annuals or biennials. Seeds germinate in three to four weeks at 68° F (20° C). Clumps can be divided in spring or fall.

LYCORIS spp.   Spider lily. Tender and semi-hardy bulbous perennials. Propagation is by bulb offsets which are removed when the dormant bulbs are dug. These are replanted to grow larger. Bulb cuttings can also be used for increase.

MARIGOLD   *See* Tagetes.

MATTHIOLA INCANA   Common stock. *M. longipetala bicornis.* Evening scented stock. Perennial grown as biennial or annual. Seed germinates in two weeks at 54–90° F (12–32° C) and may respond to light *(2).* Seeds are sown in summer or fall for winter bloom, in late winter indoors for spring bloom, or outdoors in spring for summer bloom.

MESEMBRYANTHEMUM spp.   *See* Succulents.

MIMULUS spp.   Monkey flower. Includes many species of tender to hardy plants. Mostly perennials, but sometimes grown as annuals. Seeds germinate in one to two weeks at 54° F (12° C). Softwood cuttings taken from young shoots can be rooted.

MOLUCELLA LAEVIS   Bells of Ireland. Half-hardy annual. Seed germinates in three to five weeks at 50° F (10° C). Difficult to transplant; sow seeds in place.

MONSTERA DELICIOSA   Often misnamed cut-leaf philodendron. Easily propagated by rooting sections of the main stem, by stem cuttings, or by air layering.

MUSCARI spp.   Grape Hyacinth. Hardy bulbous perennial. Plant blooms in spring; bulbs become dormant in fall when they are lifted and divided. Increase by removing bulb offsets. Seed propagation can also be used.

MUSHROOMS   Cultivars of *Agaricus campestris.* This spore-bearing fungus is propagated commercially by the use of masses of mycelium in blocks of sterile manure, rye grain, or tobacco stems, called "spawn." Spawn is produced by several firms that have developed the detailed techniques required for preparing such propagating material from the spores. Commercial growers generally prefer to purchase spawn rather than produce their own. The growing medium for mushrooms has traditionally been composted horse manure, but "synthetic" compost prepared from a mixture of corncobs and hay, plus a nitrogen-phosphorus-potassium fertilizer, is also satisfactory. Mushroom production is usually done in well-insulated concrete block houses. There are definite temperature and relative humidity requirements, and adequate ventilation is necessary. Light is not required. Successful mushroom growing is not easy and requires considerable specialized knowledge *(16).*

MYOSOTIS SYLVATICA   Forget-me-not. Hardy biennial. Seeds germinate in two to three weeks at 68° F (20° C). Sow in summer and transplant to permanent location the following spring. *M. scorpioides* is a perennial started from seed; division in spring is also used.

NARCISSUS PSEUDO-NARCISSUS and other species   Daffodil. Hardy perennial spring-flowering bulb. Vegetative propagation procedures are described in Chapter 15. Species can be grown from seed, but cultivars do not come true; seedlings require several years to produce flowers. Seeds should be planted in fall so they can be moist-chilled (stratified) over winter.

NASTURTIUM   *See* Tropaeolum.

NEMESIA STRUMOSA   Half-hardy annual. Seeds germinate in two to three weeks at 55° F (13° C). Sow outdoors in spring in cold climates, in fall in mild-winter areas.

NEPETA MUSSINII   Nepeta, Catmint. Hardy perennial. Seeds germinate in two to three weeks at 68° F (20° C). Softwood cuttings of nonflowering side shoots taken in early summer root readily. Cuttings may be inserted directly into soil if protected. Plants may be divided in spring, using newest parts and discarding older portion of clumps.

NICOTIANA spp.   Flowering tobacco. Half-hardy annuals. Seeds germinate in one to two weeks at 68–86° F (20–30° C) and may respond to light *(2)*. Sow indoors four to six weeks before last frost then transplant out-of-doors.

NIEREMBERGIA spp.   Cupflower. Tender perennial, sometimes grown as an annual. Seeds germinate in two to three weeks at 68–86° F (20–30° C). Softwood cuttings removed from new growth in spring root readily. Clumps can be divided.

NYMPHAEA spp.   Hardy water lily. Consists of numerous species and many named cultivars. Plants grow as rhizomes. Clumps are divided in spring. Seeds are used to grow natural species and to develop new cultivars. Tropical water lily hybrids and species grow from tubers. Seeds do not reproduce hybrids. Both kinds of seeds, when used, are planted 1 in. deep in sandy soil, then immersed in water 3 to 4 in. deep. Hardy species should be started at 60° F (16° C), tropical species at 70 to 80° F (21 to 27° C). Vegetative propagation is either from small tubers that can be removed from old tubers in fall or from small plantlets growing from the leaf. A specific method is to remove small tubers around the base of an old one in early fall and store them in slightly moist sand. After two to three weeks they are planted in pots and immersed in water in an indoor tank at 80° F (27° C). When a new shoot has produced two leaves, it is pinched off, potted, and placed in the tank. A continuous succession of new shoots is produced and these shoots can be removed for propagation.

OENOTHERA spp.   Evening-primrose. Hardy perennials, but some kinds are biennial. Seeds germinate in one to three weeks at 68–86° F (20–30° C). Plants can also be increased by division of clumps in the fall.

OPUNTIA   *See* Cactus.

ORCHIDS (12, 17, 23)   Many genera, hybrids, and cultivars are cultivated, and many more are found in nature. Some, such as *Aerides, Arachnis, Phalaenopsis, Renanthera* and *Vanda,* exhibit *monopodial* habit of growth. This means they are erect and grow continuously from the shoot apex. Adventitious roots are produced along the stem and inflorescences are produced laterally from leaf axils. Most others, including *Brassovola, Calanthe, Cattleya, Cymbidium, Laelia, Miltonia, Odontoglossom, Oncidium,* and *Phaius* have a *sympodial* habit of growth, are procumbent, and do not grow continuously from the apex. Their main axis is a rhizome in which new growth arises from offshoots, or "breaks." Pseudobulbs are usually present on plants of this type.

Many orchids are *epiphytes* (i.e., air plants), typically growing on branches of trees. Others *(Cymbidium, Cypripedium, Paphiopedium)* are terrestrial and grow in the ground.

Seed propagation is important in commercial production. Many important cultivars are seedling hybrids, either from species or between genera, resulting from controlled crosses of carefully selected parents. Many such important crosses are between tetraploid and diploid parents to produce triploids. These offspring are sterile and are not in turn usable as parents. Seedling variation may occur and a certain amount of selection is necessary. Five to seven years is required for a seedling plant to bloom. Orchid flowers are hand-pollinated. A seed capsule requires 6 to 12 months to mature. A single capsule will contain many thousands of tiny seeds with relatively undeveloped embryos. Embryo culture is universally used for seed propagation. The procedure is described in Chapter 16. Knudson's C medium is usually used. Arditti *(1)* has summarized the many experiences of testing various nutritional and other factors for orchid seed germination. Orchid seed can be stored for many years if held in sealed containers over calcium chloride at about 36° F (2° C).

Vegetative methods for orchids are generally slow, difficult for many genera, and usually too low-yielding for extensive commercial use. Sympodial species are increased by division of the rhizome while it is dormant or just as new growth begins. Four or five pseudobulbs are included in each section. "Backbulbs" and "greenbulbs" can be used for some genera.

Orchids with long canelike stems, as *Dendrobium* and *Epidendrum,* sometimes produce offshoots ("keiki") that produce roots. Offshoots can also be produced if the stem is cut off and laid horizontally in moist sphagnum or some other medium. Flower stems of *Phaius* and *Phalaenopsis* can be cut off after blooming and handled in the same way. A drastic method of inducing offshoots is to cut out or mutilate the growing point of *Phalaenopsis,* remove the small leaves, and treat the injured portion with a fungicide. Offshoots may then be produced.

Monopodial species are difficult to propagate vegetatively. Cuttings, preferably with a few roots already present, is successful with *Vanda* and *Arachnis.* Air layering is also possible.

Vegetative propagation by proliferation of shoot-tip (meristem) cultures *in vitro* has revolutionized orchid propagation, particularly for *Cymbidium, Cattleya,* and some other genera *(17).* The shoot growing point is dissected from the plant and grown on a special, sterile medium; a proliferated mass of tissue and small protocorms develop which can be periodically divided. Many thousands of separate protocorms can be developed in this way within a matter of months, each of which will eventually differentiate shoots and roots to produce an orchid plant. The procedure is described in Chapter 16.

PAEONIA LACTIFLORA and other hardy perennial species   Herbaceous peony. Numerous hybrid cultivars. Seeds are used for growing species or for breeding new cultivars. Seed propagation is difficult; germination may require one to two years. Sow seeds in fall to give moist-chilling over winter. Roots develop during the first summer, and shoots develop the second spring. Another method is to collect seeds before they become black and completely ripe. Do not allow them to dry out; sow in pots which are buried in the ground for six to seven weeks. Roots will develop; dig up and plant in protected location or under mulch over winter. The best method is to divide clumps in fall; each piece should have at least one bud or "eye," preferably three to five.

PAEONIA SUFFRUTICOSA   Tree peony (see p. 607).

PANSY   *See* Viola.

PAPAVER NUDICAULE   Iceland poppy. Hardy perennial grown as biennial. Seeds germinate in one to two weeks at 54° F (12° C). Sow in permanent location in summer for bloom next year. *P. orientale.* Oriental poppy. Hardy perennial. Very fine seeds, which may respond to light, should be covered very lightly *(2).* Seeds germinate in one to two weeks at 54° F (12° C). Cultivars are propagated by root cuttings. Dig when leaves die down in fall, cut into 3- to 4-in. sections, lay horizontally in sandy soil in a frame or flat covered to about an inch. Rooted cuttings are transplanted in spring. Or dig plants in spring, prepare root cuttings, and plant directly in permanent location. *P. rhoeas.* Corn poppy. Shirley poppy. Hardy annual. Seed germinates in one to two weeks at 55° F (13° C). Sow in late summer for early spring bloom or in early spring for summer bloom. Do not transplant.

PELARGONIUM spp.   Geranium. Many types and cultivars. Tender perennials. Started easily by seed but germination may be erratic. Seeds can be used to propagate true species, some Zonal and Show pelargoniums which come relatively true from seed, and selected F_1 hybrid strains *(4, 7)*. Leafy softwood stem cuttings with three or four nodes root readily. Prune plants back in summer and use new side shoots for cuttings. Pathogen-free stock, identified by culture indexing, should be used and can be supplied by specialists *(7)*.

PENTSTEMON spp.   Beardtongue. Semi-hardy to hardy perennials, sometimes handled as annuals. Seeds germinate in two to three weeks at 68–86° F (20–30° C) but growth is slow and uneven; seeds may respond to light *(2)*. Plants started indoors in early spring and transplanted outdoors later may bloom the first year. Plants are usually short-lived. Softwood cuttings taken from nonflowering side shoots of old plants root readily. Make cuttings in fall to obtain plants for next season. Clumps may be divided.

PEPEROMIA spp.   Tender perennial. Softwood stem, leaf-bud, or leaf cuttings root readily. Plants can be divided.

PETUNIA HYBRIDS   Petunia. Tender perennial often grown as annual. Seeds germinate in one to two weeks at 68° F (20° C). Seeds of double-flowered cultivars and some F_1 hybrids may need light and high temperature (80 to 85° F; 27 to 29° C) for good germination. It is best to start plants indoors for later outdoor planting. Softwood cuttings taken in late summer or fall from side shoots root easily.

PHILODENDRON spp.   Tropical vines. Seeds germinate readily at about 77° F (25° C) if sown as soon as they are ripe and before they become dry. Best propagation methods are use of stem cuttings, rooting sections of main stem, or by air layering.

PHLOX DRUMMONDI   Annual phlox. hardy. Seed germinates in two to three weeks at 68° F (20° C). Start indoors for later outdoor planting, or outdoors after frost.

PHLOX DIVARICATA   Sweet William. Hardy perennial. Expose seeds to cold during winter before planting. Softwood cuttings taken in spring root easily. Divide clumps in spring or fall.

PHLOX PANICULATA   Garden phlox. Hardy perennial. Plants do not come true from seed. Sow seeds as soon as ripe in fall to germinate the next spring. They will germinate in three to four weeks at 68° F (20° C). Grow plants one season and transplant in fall. Softwood cuttings taken from young shoots in spring or summer root easily, but they are subject to damping-off if kept too wet. It is best to propagate from root cuttings. Dig clumps in fall; remove all large roots to within 2 in. of crown (which is replanted). Cut roots into 2-in. lengths and place in flats of sandy soil; cover ½ in. deep. Transplant next spring. Divide clumps in fall or spring.

PHYSALIS ALKEKENGI   Chinese lantern plant. Hardy perennial. Seeds germinate in three to four weeks at 54–90° F (12–32° C) and may respond to light *(2)*. Divide clumps in spring or fall, or propagate by root cuttings.

PHYSOSTEGIA VIRGINIANA   False dragonhead. Hardy perennial. Seeds germinate in two to three weeks at 54–90° F (12–32° C). Divide clumps in fall or spring.

PLATYCODON GRANDIFLORUM   Balloon flower. Hardy perennial. Seeds germinate in two to three weeks at 68–86° F (20–30° C), but plants do not flower for two or three years. Tip cuttings taken in summer and treated with a growth regulator will root under mist.

POLEMONIUM spp.   Jacob's ladder. Hardy perennial. Seeds germinate in three to four weeks at 68–86° F (20–30° C). Divide clumps or root stem cuttings.

POLIANTHES TUBEROSA   Tuberose. Tender bulbous perennial. Propagated by removing offsets at planting time. The small bulbs take more than one year to flower. Divide clumps every four years.

PORTULACA GRANDIFLORA   Rose moss. Half-hardy annual. Seeds germinate in two to three weeks at 68–86° F (20–30° C); they respond to light (2). Reseeds itself.

PRIMULA OBCONICA   Top primrose. Tender perennial grown as an annual. Seeds germinate well in a cool greenhouse or after three to four weeks at 54–90° F (12–32° C). The very tiny seeds respond to light and should not be covered. *Primula malacoides.* Fairy primrose. Seeds will germinate in two to three weeks at 68–86° F (20–30° C). *Primula sinensis.* Chinese primrose. Seeds will germinate in three to four weeks at 68° F (20° C). In these species double-flowering cultivars do not produce seeds but are propagated by cuttings taken in spring, or by division. Other *Primula* spp. Hardy perennials grown outdoors. Seeds germinate in three to six weeks at 68° F (20° C), but some species may require lower temperatures. It is best to collect and sow seeds as soon as they are ripe in fall. Cuttings taken in spring root easily. Clumps can be divided just after flowering.

RANUNCULUS ASIATICUS   Turban or Persian ranunculus. Tender perennial. Seeds germinate in one to four weeks at 68° F (20° C). Grows from tuberous roots which can be divided. *Ranunculus* spp. Buttercup. Hardy perennial. Seeds germinate in one to four weeks at 68° F (20° C). Plant seeds as soon as they are ripe. Divide plants in spring or fall.

RESEDA ODORATA   Mignonette. Hardy annual. Seeds should not be covered; they germinate in two to three weeks at 54° F (12° C) and respond to light (2)

RICINUS COMMUNIS   Castor bean. Soak seeds in water for 24 hours or nick with file before planting. Plant each seed in individual pot and after frost danger is past transplant to outdoor location. Seeds are poisonous.

RUDBECKIA spp.   Coneflower. Hardy annual, biennial, and perennial species. Seeds germinate in two to three weeks at 68–86° F (20–30° C). Perennial kinds are propagated by division. Will reseed naturally.

SAINTPAULIA IONANTHA   African violet (15). Tropical perennial. The very fine seeds, which germinate in three to four weeks at 86° F (30° C), should not be covered. Seedlings are very subject to damping-off. Vegetative methods are necessary to maintain cultivars. Plants can be divided. Leaf cuttings are easily rooted, either in a rooting medium or in water.

SALPIGLOSSIS SINUATA   Painted tongue. Semi-hardy annual. Seeds are difficult to germinate but some will start in one to two weeks at 68–86° F (20–30° C). Start indoors in peat pots.

SALVIA SPLENDENS   Scarlet sage. Tender perennial grown as an annual. Germinate seeds at 68–86° F (20–30° C), then grow at 55° F (13° C) night temperature. Softwood cuttings taken in fall root readily.

SALVIA spp.   Sage. Annual, biennial, and perennial species. Seeds germinate in two to three weeks at 68–86° F (20–30° C) and may respond to light (2). Soak flats thoroughly, and do not rewater until sprouted; they also grow well under mist

(4). Softwood cuttings of young shoots 3 to 4 in. long root readily. Plants can be divided, but such divisions are slow to recover.

SANVITALIA PROCUMBENS   Creeping zinnia. Hardy annual. Seeds germinate in one to two weeks at 68° F (20° C) and may respond to light *(2).* Sow in place in spring, or in fall in mild climates.

SANSEVIERIA TRIFASCIATA and S. 'HAHNII' (dwarf form)   Bowstring hemp. Snakeplant. Tropical perennial. Plants grow from a rhizome, which can be readily divided. Leaves may be cut into sections, several inches long, and inserted into a rooting medium; a new shoot and roots will develop from base of leaf cutting. The variegated form, *S. t.* 'Laurentii,' is a chimera which can be maintained only by division.

SAPONARIA OFFICINALIS   Bouncing Bet. Hardy perennial. Seeds germinate in two to three weeks at 68° F (20° C). The plant spreads rapidly by an underground creeping stem which can be divided. *Saponaria vacaria.* Soapwort. Hardy annual. Seeds germinate in two to three weeks at 68° F (20° C).

SAXIFRAGA spp.   Many interesting and unusual species and hybrids. Mostly hardy perennials. Seeds germinate easily; they are sown preferably when ripe. Some hybrids and cultivars are maintained only by vegetative methods. Most plants grow as small rosettes and are easily propagated by making small cuttings involving single rosettes which are taken after bloom. Plants can be divided in spring or fall. *S. stolonifera,* Strawberry geranium, is a tender perennial that reproduces by runners.

SCABIOSA spp.   Pincushion Flower. Annual and hardy perennial species. Seeds germinate in two to three weeks at 68–86° F (20–30° C). Perennial kinds can be divided.

SCARLET SAGE   *See* Salvia.

SCHIZANTHUS spp.   Butterfly flower. Tender annual. Seeds germinate in one to two weeks at 54° F (12° C) and are sensitive to high temperatures *(2).* Sow seeds in fall for early spring bloom indoors, or sow in early spring to be transplanted outdoors for summer bloom.

SCILLA spp.   Squill. Includes several kinds of bulbous hardy spring flowering perennials. Dig plants when leaves die down in summer and remove the bulblets. *S. autumnalis* is planted in spring and blooms in fall.

SEDUM   *See* Succulents.

SEMPERVIVUM   *See* Succulents.

SENECIO spp.   Cineraria. Tender perennials grown as annuals. Seeds germinate in two to three weeks at 68° F (20° C).

SINNINGIA SPECIOSA   Gloxinia. Tropical perennial. Commonly grown from seeds, which are very fine and require light. Sow in uncovered, well-drained, peat moss medium; they germinate in two to three weeks at 68° F (20° C). Vegetative methods are required to reproduce cultivars. Plant grows from a tuberous root on which a rosette of leaves is produced. This can be divided as described for tuberous begonia. Softwood cuttings or leaf cuttings taken in spring from young shoots starting from the tubers root easily.

SNAPDRAGON   *See* Antirrhinum majus.

STOCK   *See* Matthiola.

STOKESIA LAEVIS   Stokes Aster. Hardy perennial. Seeds germinate in four to six weeks at 68–86° F (20–30° C), and the plants bloom the first year. Make root cuttings or divide clumps in spring.

STRELITZIA REGINAE   Bird-of-paradise. Tropical perennial that grows from a rhizome. This can be divided in the spring, or small offsets can be detached and placed in a potting mixture. Seeds should be sown under warm conditions. Freshly harvested seed should be used to avoid seed coat impermeability.

SUCCULENTS (9, 11)   This is a loosely defined horticultural group that includes many genera such as: Agave, Aloe, Crassula, Echevaria, Euphorbia, Hoya, Kalanchoe, Portulacaria, Sedum, Yucca. These are plants with fleshy stems and leaves that store water, or plants that are highly drought-resistant. Mostly they are half-hardy or tender perennials. Seed propagation is possible, although young plants are often slow to develop and to produce flowers. It is best to germinate seeds indoors at high day temperatures (85 to 95° F; 29 to 35° C). Seedlings are susceptible to damping-off.

Most species root readily by cuttings—either stem, leaf-bud, or leaf. Cuttings should be exposed to the open air or inserted into dry sand for a few days to allow callus to develop over the cut end. A well-drained rooting medium, such as coarse sand, is used, and some protection from drying is needed. Some species can be reproduced by removing offsets. Grafting is possible as described for cacti.

SWEET ALYSSUM   *See* Lobularia maritima.

SWEET PEA   *See* Lathyrus.

TAGETES spp.   Marigold. Tender annuals. Seeds germinate readily in one week at 68–86° F (20–30°C) and sometimes respond to light *(2)*. Sow in place in spring after frost.

THALICTRUM spp.   Meadow rue. Hardy perennial. Seeds germinate in four to six weeks at 68° F (20° C). Sometimes hard seeds are present. Plants can be divided in spring or fall.

THERMOPSIS CAROLINIANA   Tender or hardy perennials. Use fresh seeds, which will germinate in two to three weeks at 68–86° F (20–30° C). Plants can be divided, but it is best to leave them undisturbed.

THUNBERGIA spp.   Clockvine. Tender perennials grown as annuals. Seeds germinate in two to three weeks at 68–86° F (20–30° C), but seedlings grow slowly. Softwood cuttings taken from new shoots root readily.

THYMUS spp.   Thyme. Hardy perennials. Seeds germinate in one to two weeks at 54–90° F (12–32° C). Sometimes germination is promoted by light. May be increased by division or by softwood cuttings taken in sumer.

TIGRIDIA PAVONIA   Tiger flower. Tender bulbous perennials. Plant bulbs in spring and dig in fall when leaves die. Increase by removal of small bulblets just before they are planted. Easily started by seed.

TITHONIA ROTUNDIFOLIA   Mexican sunflower. Tender perennial grown as an annual. Seeds germinate in two to three weeks at 68–86° F (20–30° C).

TOLMIEA MENZIESII   Piggyback plant. New plantlets form on upper surface of leaves. Such leaves are removed and petiole is stuck in rooting medium to depth of new plantlet, which then resumes growth.

TORENIA FOURNIERI   Wishbone flower. Tender perennial. Seeds germinate in two weeks at 68–86° F (20–30° C).

TRACHYMENE CAERULEA   Laceflower. Tender perennial grown as an annual. Seeds germinate in two to three weeks at 68° F (20° C).

TROLLIUS spp.   Globeflower. Hardy perennial. Plant seeds in fall. Increases also by clump division.

TROPAEOLUM MAJUS   Nasturtium. Tender perennial grown as an annual. Plant seeds in place; they germinate in one to two weeks at 68° F (20° C), but are difficult to transplant. Double-flowering kinds must be propagated vegetatively, usually by softwood cuttings.

TULIPA spp. and hybrid cultivars   Tulip. Hardy bulbous perennials. Plant bulbs in fall for spring bloom. Seeds are used to reproduce species and for breeding new cultivars. They germinate readily if planted as soon as ripe. Vegetative methods include removal of offset bulbs in the fall. Different bulb sizes are planted separately, since the time required to produce flowers varies with size. For details of procedure see Chapter 15.

VALERIANA OFFICINALIS   Valerian. Hardy perennial. Seeds germinate in three to four weeks at 54–90° F (12–32° C). New plants can be obtained by clump division.

VENIDIUM FASTUOSUM   Venidium. Half-hardy annual. Seeds germinate in four to six weeks at 68–86° F (20–30° C) *(18);* or sow outdoors at 50 to 55° F (10 to 13° C).

VERBASCUM spp.   Mullein. Hardy perennials and biennials. Seeds are slow to germinate; best temperature is 86° F (30° C). Propagate named cultivars by root cuttings taken in early spring.

VERBENA HYBRIDA   Verbena. Tender perennial grown as an annual. Seeds germinate in three to four weeks at 68–86° F (20–30° C); sometimes promoted by light *(2).* May be propagated by softwood cuttings taken in sumer. *V. canadensis.* Clump verbena. Hardy perennial that blooms first year from seed. Seeds germinate in two to four weeks at 54–90° F (12–32° C). Plants can be propagated by division or by softwood cuttings.

VERONICA spp.   Speedwell. Hardy perennials. Seeds germinate in two weeks at 54–90° F (12–32° C). Plants are increased by division in spring or fall or by softwood cuttings taken in the spring or summer.

VINCA MAJOR   Tender perennial. Propagate by division or by softwood cuttings taken in summer. *V. minor.* Periwinkle. Hardy perennial. Seeds germinate in two to three weeks at 68° F (20° C). Easily propagated by softwood cuttings or by division. *V. rosea.* Madagascar periwinkle. Tender perennial grown as an annual. Propagation is the same as for other species. Started by seed germinated at 68° F (20° C).

VIOLA CORNUTA   Horned Violet. Tufted Pansy. Hardy perennial. Seeds germinate in two to three weeks at 54–90° F (12–32° C). Seeds of some cultivars need light. Vegetative propagation is by cuttings taken from new shoots obtained by heavy cutting back in the fall. Clumps may also be divided. *V. tricolor hortensis.* Pansy. Hardy or semi-hardy, short-lived perennial often grown as an annual. Usually propagated by seeds as described for *V. cornuta* but may also be increased by cuttings or by division. *V. odorata.* Sweet violet. Tender to semi-hardy perennials. Grows by rhizome-like stems, which can be separated from others on the crown and treated as a cutting with some roots present. *Viola* species. Many hardy perennial kinds. These are grown by seeds as above, but germination may be slow and seeds are best exposed to cold before planting. Many species produce seeds in inconspicuous, enclosed (cleistogamous) flowers near the ground, whereas the conspicuous, showy flowers produce few or no seeds. These plants can also be reproduced by cuttings or by division.

YUCCA spp.   Yucca. Tender to semi-hardy perennials. Seeds germinate at 68° F (20° C) but rather slowly and require four to five years to flower. Plants are monocots; some are essentially stemless and grow as a rosette, while others have either long or short stems. Offsets growing from the base of the plant can be removed and handled as cuttings; sometimes entire branches or the top of the plant can be detached a few inches below the place where leaves are borne and replanted in sandy soil. Sections of old stems can be laid on sand or other medium in a warm greenhouse, and new side shoots that develop can be detached and rooted.

ZANTEDESCHIA spp.   Calla. Several species, which have similar propagation requirements. Tropical perennials. Plants grow by thickened rhizomes which produce offsets or rooted side shoots; these are removed and planted.

ZEBRINA PENDULA   Wandering Jew. Easily propagated at any season by stem cuttings.

ZINNIA ELEGANS and other species   Zinnia. Half-hardy hot weather annual. Seeds germinate outdoors in one week at 68–86° F (20–30° C). Sometimes seeds respond to light *(2)*.

REFERENCES

1 Arditti, J., Factors affecting the germination of orchid seeds, *Bot. Rev.* 33:1–97. 1967.

2 Assoc. Off. Seed Anal., Rules for seed testing, *Proc. Assoc. Off. Seed Anal.*, 60(2).1–115. 1970.

3 Bailey, L. H., and E. Z. Bailey, *Hortus Second.* New York: The Macmillan Company, 1941.

4 Ball, V. (ed.), *The Ball Red Book,* 12th ed. Chicago. Geo. J. Ball, Inc., 1972.

5 Booth, C. O., *An Encyclopedia of Annual and Biennial Garden Plants.* London. Faber & Faber, Ltd., 1957.

6 Carter, F. M., Grafting cacti, *Horticulture,* 51(3):34–35. 1973.

7 Dallon, J., Jr., and D. Durkin, Culturing geranium from seed, *Proc. Int. Plant Prop. Soc.,* 21:324–30. 1972.

8 Darrow, G. M., and F. G. Meyer (ed.), *Day Lily Handbook,* American Horticultural Society, Vol. 47, No. 2. 1968.

9 Edinger, P. (ed.), *Succulents and Cactus.* Menlo Park, Ca.: Lane Magazine and Book Company, 1970.

10 Everett, T. H. (ed.), *New Illustrated Encyclopedia of Gardening.* New York: Greystone Press, 1960.

11 Haage, W., *Cacti and Succulents.* New York: E. P. Dutton & Co., Inc., 1963.

12 Hawkes, A. D., *Orchids, Their Botany and Culture.* New York: Harper & Row, Publishers, 1961.

13 Holley, W. D., and R. Baker, *Carnation Production.* Dubuque, Iowa: Wm. C. Brown, 1963.

14   Hull, Helen, Handbook on ferns. *Plants and Gardens* 25(1):1–77. 1969.

15   Kramer, J., *How to Grow African Violets,* 4th ed., Menlo Park, Ca.: Lane Magazine and Book Company, 1971.

16   Lambert, E. B., Mushroom growing in the United States, *U.S.D.A. Farmers' Bull. No. 1875, 1967.*

17   Morel, G. M., Meristem culture: clonal propagation of orchids, *Orchid Digest,* 30(2):45–49. 1966.

18   Northen, H. T., and R. T. Northen, *Complete Book of Greenhouse Gardening.* New York: The Ronald Press Company, 1957.

19   Post, K., *Florist Crop Production and Marketing.* New York: Orange-Judd Publishing Co., Inc., 1949.

20   Roberts, D. J., Modern propagation of ferns, *Proc. Int. Plant Prop. Soc.,* 15:317–21. 1965.

21   Rockwell, F. F., G. C. Grayson, and J. de Graaf. *The Complete Book of Lilies.* Garden City, N. Y.: Doubleday & Company, Inc., 1961.

22   Taylor, N., *Encyclopedia of Gardening,* 4th ed. Boston: Houghton Mifflin Company, 1961.

23   Withner, C. L. (ed.), *The Orchids, A Scientific Survey.* New York: The Ronald Press Company, 1961.

SUPPLEMENTARY READING

Cathey, H. M., "Growing flowering annuals," *United States Department of Agriculture Home and Garden Bulletin* No. 91, pp. 1–16. 1965.

Crockett, J. U., *Annuals.* New York: Time-Life Books, 1971.

———, *Perennials.* New York: Time-Life Books, 1972.

———, *Foliage House Plants.* New York: Time-Life Books, 1972.

Dunmire, J. R., ed., *Western Garden Book.* Menlo Park, Ca.: Lane Magazine and Book Company, 1967.

Graf, A. B., *Exotica 3—Pictorial Cyclopedia of Exotic Plants from Tropical and Near-Tropical Regions,* 6th ed. East Rutherford, N.J.: Roehrs Company, 1973.

Laurie, A., D. C. Kiplinger, and K. S. Nelson, *Commercial Flower Forcing,* 7th ed. New York: McGraw-Hill Book Company, 1968.

Mastalerz, J. W. (ed.), *Bedding Plants,* Pennsylvania Flower Growers. University Park, Pa.: Pennsylvania State University. 1966.

Nehrling, A., and I. Nehrling, *Propagating House Plants,* Great Neck, New York: Hearthside Press, Inc., 1972.

Nelson, K. S., *Flower and Plant Production in the Greenhouse,* Danville, Ill.: Interstate, 1966.

Potter, C. H., *Bedding Plants.* Chicago. Florists' Publishing Company, 1972.

Index